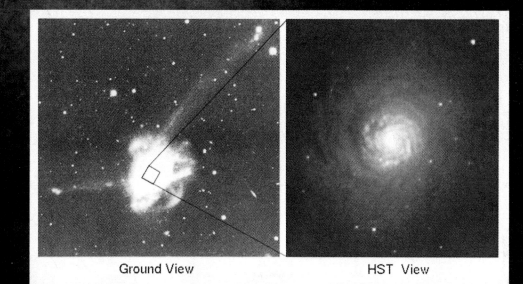

When galaxies collide!
The Hubble Space Telescope
reveals the results.
Courtesy of
B. Witmore (STScI)
and NASA

Ground View   HST View

SEVENTH EDITION

ASTRONOMY

THE EVOLVING UNIVERSE

MICHAEL ZEILIK

The University of New Mexico

JOHN WILEY & SONS, INC.

New York   Chichester   Brisbane   Toronto   Singapore

ACQUISITIONS EDITOR Clifford Mills

DEVELOPMENTAL EDITOR Sean M. Culhane

MARKETING MANAGER Catherine Faduska

SENIOR PRODUCTION EDITOR Nancy Prinz

COVER AND INTERIOR DESIGNER Dawn L. Stanley

MANUFACTURING MANAGER Andrea Price

PHOTO RESEARCH Claudia Smith–Porter/Sean M. Culhane

SENIOR ILLUSTRATION COORDINATOR Sigmund Malinowski

ILLUSTRATOR Boris Starosta

COVER PHOTOS Courtesy of:
    Stock Imagery, Inc. and Masterfile Corporation Photo Library.

COVER SPECIAL EFFECT: Chattum Design Group, Inc.

This book was set in ITC Garamond Light by GTS Graphics and printed and bound by Von Hoffmann Press, Inc. The cover was printed by Phoenix Color Corp.

Library of Congress Cataloging in Publication Data:
Zeilik, Michael.
      Astronomy : the evolving universe / Michael Zeilik. — 7th ed.
         p.   cm.
      Includes index.
      ISBN 0-471-59739-2 (paper)
      1. Astronomy.   I. Title.
   QB45.Z428   1994
   520—dc20   93-39216
                                                                   CIP

Printed in the United States of America

10 9 8 7 6 5 4 3 2

About the Author

Dr. Michael Zeilik works as Professor of Astronomy at the University of New Mexico. In his teaching, he specializes in innovative, introductory courses for the novice, non-science major student. His classes include demonstrations and computer simulations to provide students with concrete experiences of abstract astronomical concepts. He has been supported by grants from the National Science Foundation, NASA, the Exxon Educational Foundation, and the Slipher Fund of the National Academy of Sciences for innovations in astronomy education, delivery of astronomy to the general public, and astronomy workshops for in-service teachers. He is a member of Commission 46, The Teaching of Astronomy, of the International Astronomical Union.

Dr. Zeilik's research activities currently focus on two areas: magnetic activity cycles of sun-like stars in binary systems, and astronomy in the historic and prehistoric Pueblo world. The work on stars involves frequent use UNM's Capilla Peak Observatory, of which he was Director from 1979–1988. He has published over 80 research papers.

Dr. Zeilik earned his A.B. in Physics with honors at Princeton University and his M.A. and Ph.D. in Astronomy at Harvard University. He has been a Woodrow Wilson Fellow, a National Science Foundation Fellow, and a Smithsonian Astrophysical Observatory Predoctoral Fellow. At the University of New Mexico, he has been named a Presidential Lecturer, the highest award for all-around performance by a faculty member.

Dr. Zeilik is listed in *American Men and Women of Science, The Writers Directory, Contemporary Authors, Who's Who in the West,* and *Who's Who of Emerging Leaders in America.* He is a member of the Authors Guild and the Textbook Authors Association.

Preface

When all the stars were ready to be placed in the sky
First woman said, "I will use these to write the laws that
are to govern mankind for all time . . . if they are written
in the stars they can be read and remembered forever."

FRANC J. NEWCOMB, *Navaho Folk Tales*

Astronomy is all about vision, an inner one and an outer one. The outer vision is our perception of the universe. A little of that comes from simple naked-eye observations of the sun, moon, planets, and stars. Such observations by the unaided eye established the foundations of astronomy for thousands of years by detecting the patterns of visible celestial events. Technology transformed that vision, especially in modern times. Yes, most people know that astronomers use telescopes to pull the distant universe into view. But many do not know that current telescopes detect a cosmos in forms of light that were impossible only a few decades ago. We now explore the universe as it shines in radio, infrared, ultraviolet, x-rays, and gamma-rays, as well as the visual. As most astronomers, I am astounded, fascinated, and awed by the richness and variety that the cosmos contains.

But the power of technology would be wasted were it not for our inner vision of the cosmos—the universe in our minds, in our imaginations. In the Navaho creation story, First Woman lays out in the stars the laws of moral guidance for Navahos to establish and maintain harmony in their lives and in the universe. Like other people, astronomers find patterns in nature that reveal a harmony in the cosmos—physical laws rather than moral ones. And like any curious people, we feel compelled to understand these patterns in a rational way. Science is a special way that humans have to comprehend nature, a process based on observations that then turn into mental images—*models*—of how things work. These models must then be testable in a public arena before they are accepted as conditionally OK. But a proper scientific model must at any time be able to be verified or contradicted by observations. The interaction between models and observations deepens our understanding of the cosmos.

When models become powerful enough, they are often elevated to the status of *theories*—concepts that unify diverse marvels of the cosmos. These unifying concepts often come from other areas of science—physics especially, chemistry, geology, and biology. All these help to forge the links that bind our general understanding of the universe, its parts, and its key to the past, present, and future of the cosmos.

If you think that astronomy is simply looking up at the sky to observe celestial objects, you've just got the start of the subject. Astronomy today involves much, much more—from tiny moons to vast galaxies. The ambitious goal is to understand them all. I intend this book to provide you with basic information and conceptual thinking to deepen your vision of unity and connection—to enrich your understanding of the astronomical world.

Goals of this Edition

A *seventh* edition! I really can't believe that an idea that I had some 20 years ago has lived so long and gone so far. Many instructors and students have told me that they've enjoyed using the book, and that feedback has certainly helped me along. I also feel a strong desire to improve this book so that it innovates and evolves as a better learning tool. Finally, astronomy changes rapidly, especially with the advent of new space and ground-based telescopes. Our outer vision grows, and our inner one intensifies. So what's new for this edition?

One, the obvious **updating of the material.** Constant change drives the excitement of astronomy. That is a main reason why astronomy appeals to me, and, I hope, to you. There's always something new and unexpected for us to discover in the universe. This edition includes, for instance, the latest observations of the Big Bang, key discoveries from the *Hubble Space Telescope* (and other space observatories), expanded material on the large-scale structure of the universe and the dark matter issue, the exciting views of Venus from the *Magellan* mission, the start of a new search for extraterrestrial life, a bright supernova in a nearby galaxy visible from the Northern Hemisphere, more about asteroidal collisions with the earth, and the rediscovery of the comet Swift-Tuttle (missing for over 100 years!).

Two, a **rewriting of the material** so as to make the descriptions, concepts, and explanations as clear and concrete as possible. I have taken great care to minimize the use of technical terms and the passive voice. I also gave special attention to the Learning Objectives and Key Concepts to keep them concise and precise.

Three, a **completely new art program.** All the line art has been reconceptualized and redrawn in full color with the goal of understandability by the novice student (while keeping the science accurate). We also have new color photos and images from astronomers and observatories around the world.

Four, an expanded **visible pedagogy.** Most of the features from the sixth edition have been retained. I have included more "At a Glance" marginal tables to which my students responded with great enthusiasm. I have also added "Quick Response" questions at the end of each major section to invite you to think about a major point in that section. At the end of each section is an outline of "The Unifying View." It has two parts: a summary of the key concept for each chapter, as a check that students are getting the "big picture" within the contents of a chapter. The second part restructures the specific concepts around larger, unifying ones. The goal is to show how the subject matter is linked conceptually.

Five, an **enhanced observational orientation.** Chapter 1 contains more and better sky charts. The star charts in Appendix G have been completely redrawn. Fred Schaaf has contributed four seasonal essays (at the end of each part) to guide your observations of the sky. Special star maps have been developed for these essays. Each chapter contains a "See for Yourself!" section that includes other observations or hands-on activities.

Thematic and Topical Structure

This book has two main aspects. One, to describe to your students in readable form the full range of the astronomical universe. Two, to introduce students to how astronomers think about the cosmos so they can gain some understanding about its operation. I hope that your students will become so enticed by the contents that they will become intrigued by the concepts linking and illuminating astronomical phenomena.

This edition is for a one-semester introduction to astronomy. It retains the previous structure of four coherent parts, each focusing on a key subtheme of cosmic evolution. Like the cosmos, each part connects to the others, so you can really approach the parts in any order. The four parts are:

Part 1: Changing Conceptions of the Cosmos

This part accents the evolution of cosmological ideas, from the nonscientific views of the Babylonians to the mind-boggling visions of modern astronomy. It leads off with the simplest observations you can make without a telescope from the earth and ends with the farthest reaches of the visible universe. Part 1 acquaints the reader with the idea of *scientific models,* the conceptual core of modern scientific thought. Scientific models are born from our imagination and experience; they mark the essential creative act of the scientific enterprise. As such models evolve, they shape our changing conceptions of the cosmos. The development of scientific models resounds throughout the book; it is *the* fundamental tool to understand the universe.

Essential to the formation of models are new astronomical observations. In the age of microelectronics and space astronomy, we have greatly expanded our vision of the cosmos. Many space telescopes, even the crippled *Hubble Space Telescope,* mark great changes in the outer vision of 20th century astronomy. In the next century, we can hope that lunar-based telescopes will augment the observational legacy of space telescopes. Supercomputers endow our minds with another way to see the physical processes in the cosmos by simulating astronomical systems. Together, computers and telescopes work as the tools to impart innovative ideas to astronomers.

Part 2: The Planets: Past and Present

Flyby spacecraft and gangly landers have provided new insights to our understanding of the planets. This part focuses on the physical properties of the planets to infer their origin and evolution. It first takes a comparative look at our current knowledge of the planets, especially our earth and moon, as well as Mercury, Venus, and Mars. These planets show different degrees of evolution, with the earth being the most evolved. The other planets—Jupiter, Saturn, Uranus, Neptune, and Pluto—have, in contrast, changed little since their birth.

Space missions have disclosed that the moons of the outer planets are really worlds unto themselves, places of rock and ice scarred by violence in the past. These new worlds provide important clues, along with comets and meteorites, of the early history of the solar system. So planetary evolution traces back to origin—the birth of the solar system from an interstellar cloud of gas and dust. By astronomical standards, that birth was quick, violent, and chaotic—shaping the primitive forms of the planets. This picture implies that many other single stars have planetary systems and that these other worlds resemble the local planets in broad ways.

Part 3: The Universe of Stars

Our sun and its planets swing around in a vast island of stars called the Milky Way Galaxy—our home galaxy. As the nearest star, our sun serves as the close-up model for other stars, especially to understand the physical processes within them. The Galaxy contains some hundreds of billions of stars at various stages of their lives. These stars have been born, like the sun, from clouds of interstellar gas and dust. Modern technology supplies us with views deep into the regions of starbirth, showing us the early lives of stars.

Because stars live long by human standards, we cannot directly observe their evolution. We can build models of stars with computers, and these models provide us with the dimension of time to map out the lives of stars. Ordinary stars, powered by fusion reactions, grow old and blow off material before their demise. Many have violent deaths, leaving bizarre corpses such as neutron stars. The most violent deaths of massive stars are signaled by enormous explosions that build heavy elements and propel them into space to seed the next generation of stars. The span of the lives of stars guide us to an understanding of what will happen to our sun in its old age.

Part 4: Galaxies and Cosmic Evolution

The universe contains galaxies, in which the most visible material of the cosmos resides. This part first examines ordinary galaxies, like our Milky Way, and then hyperactive ones. We are beginning to realize that galaxies with unusual activity may be interacting by gravity with their neighbors, which somehow triggers the fireworks. The show includes long, thin jets of material, confined by magnetic fields, rocketing out of the cores of these galaxies. All these galaxies dwell in clusters with other galaxies, and we have just come to the realization that clusters are laid out in long chains with vast voids in between—the cosmic tapestry. Amid this remarkable layout of visible matter lurks matter that we cannot yet see—the so-called dark matter that may well shape the visible universe.

The cosmic design in the large-scale architecture was imprinted in the awesome explosion in which the universe began. That explosion—the Big Bang—linked the smallest pieces of matter to the universe at large. The Big Bang also left relics that we observe today, evidence of the violence of creation. From the Big Bang some 15 billion years ago, cosmic evolution shaped us at where we are now in the universe—intelligent beings curious about the possible existence of creatures elsewhere. We ponder the future of the human race—the evolution of a technological civilization—and hope that we can learn to manage well ourselves and the earth. Yet, we may be forced to leave our home planet and perhaps travel to the stars.

Your Student's View

This book is for your students, who are probably novices to astronomy. I want them to be able to learn effectively from it. I have designed this four-part structure so that you can investigate each part somewhat independently of the others. Many cross-references, especially to the basic physical and astronomical ideas, should help you to link the parts together. I have made a concerted effort to introduce ideas as concretely as possible. Within parts, I deal with the most familiar first: Chapter 1 (Part 1) with the visible sky; Chapter 8 (Part 2) with the earth; Chapter 13 (Part 3) with the sun; and Chapter 18 (Part 4) with the Milky Way Galaxy. Within chapters, I have tried to present concrete examples before abstract notions.

The *Focus Sections* furnish another linkage throughout the text. I have set these optional Focus Sections off from the main text to enrich ideas by basic mathematics (algebra, trigonometry, and geometry). You will need to decide which of them will be specifically assigned. You will note many more problems at the end of each chapter. Most of them draw on the material in the Focus Sections.

A novice student tends to lose the big picture in the details, so I have placed more weight on the *Part Introductions* and *Epilogues* to carry the larger view. The Parts also have their own objectives to highlight the overall picture. At the end of each part, you will find an outline called "The Unifying View." Here I have tried to help you with the big picture by presenting a *Central Concept* for each chapter in that part and an outline of the crucial ideas organized around *Astronomical Concepts, Information Concepts, Energy Concepts, Motion and Force Concepts,* and *Quantum Concepts.* The impetus for this organization came from the American Association for the Advancement of Science Project 2061 Panel Report on *Physical and Information Sciences and Engineering* by George Bugliarello.

To aid novice science students, I have simplified the language as much as I can by using ordinary English rather than technical jargon. My rule when using technical terms is: *Define every term you use, and use every term you define.* I have worked hard to avoid a one-time use of a technical term just for the sake of completeness. Overall, I have tried to simplify the writing. Through the first six editions, I have achieved the goal of making each more readable; this one improves the readability compared to the sixth edition.

If your students are really baffled by a good strategy to learn astronomy, please tell them to read carefully the special section by Mark Hollabaugh (who teaches at Normandale Community College) on "How To Study Astronomy." It provides a general strategy for studying plus specific guides to each of the learning features of this book.

Note: If you have picked up this book because you are curious about astronomy, you may be interested in taking a college-level course by mail for academic credit. Write to: Independent Study, Continuing Education, The University of New Mexico, 1634 University NE, Albuquerque, NM 87131. The course is called Astronomy 101C; I am the instructor.

Instructor's Resources

This text is but one part of a complete learning environment. You and your classroom make up the major part. To help you, many materials have been developed to support teaching from this text. Your *Wiley* representative should be contacted for more information about them. These resources include:

- **Instructor's Manual** by J. Wayne Wooten of Pensacola Junior College. The manual includes chapter overviews, teaching hints, answers to the end-of-chapter Study Exercises, discussion topics, and class demonstrations.

- **Test Bank** prepared by Professor Wooten and including some questions by myself and Daniel Weeks. These questions have all been classroom tested.

- **MicroTest III Test Generation System** from Chariot Software Group. This software allows you to create quizzes and comprehensive exams quickly and easily. Available in Macintosh and MS-DOS versions.

- **Color Transparency Acetates.** These four-color overhead transparencies are selected from the stunning new art program.

- **Concept Development Slide Set.** This unique 35mm slide set consists of 200 progressive illustrations prepared by Dr. Francis Lestingi of the State University College in Buffalo, NY to further explore basic concepts presented in the text. A second set of slides will present 33 unique images from the text. These are free to adopters.

- **Cosmic Clips: Animations from Astronomers.** A joint project of Wiley and the Astronomical Society of the Pacific, this video contains carefully selected astronomical simulations and animations, many done on supercomputers. The segments have been chosen (by me and Dr. Sally Stephens of the ASP) for visual impact and lucid understanding by the novice student. The video contains a narration aimed at the novice student and also comes with an information booklet for background on each of the clips.

- **Voyager II: the Dynamical Sky Simulator**™ for the Apple Macintosh™. Some illustrations in this book were generated with the Voyager software (version 1.2). The new version is available to adopters at a discount.

- **Dance of the Planets.** This MS-DOS program simulates the motions in the solar system. It is available to adopters at a discount with a coupon.

Acknowledgments

Many people have helped me make this a better book. First and foremost are all my professional colleagues, who have kept me up to date on their work.

Next my special thanks go to the reviewers of this edition. They include the following people:

ROBERT H. ALLEN
University of Wisconsin — La Crosse

HARRY J. AUGENSEN
Widener University

RICHARD BERENDZEN
The American University

JEFFEREY BRAUN
University of Evansville

BRIAN KEITH CLARK
Illinois State University — Normal

STEVE FEDERMAN
University of Toledo

DARREL B. HOFF
Harvard-Smithsonian Center for Astrophysics

MARK HOLLABAUGH
Normandale Community College

DAVID W. HUGHES
University of Sheffield, England

JAMES H. HUNTER
University of Florida — Gainesville

JOHN LAIRD
Bowling Green State University

JOHN L. SCHMITT
University of Missouri — Rolla

ALEX G. SMITH
University of Florida — Gainesville

DAVID THEISON
University of Maryland

JEAN L. TURNER
University of California — Los Angeles

DANIEL WEEKS
University of New Mexico

An extra round of applause goes to Darrel Hoff (whom I have known for over 25 years) for an in-depth review that kept the student firmly in mind. I have admired Fred Schaaf's astronomical writings and am pleased that he contributed distinct sky essays for this edition. Mark Hollabaugh and I share a passion for green chili salsa and for the astronomy of the Native Americans. His student guide to the use of this book is greatly appreciated. John Wayne Wooten, the supplements author, contributed much of his experience as a teacher.

Locally, Claudia Smith-Porter worked tenaciously to acquire the new visual images in the book. Ellen V. Roderick of the Write/Action Company in Santa Fe provided some in-depth editorial work. My son Zachary kept trying to add more material by pounding at the computer keyboard. Luckily, the computer was off most of the time.

Within Wiley, Cliff Mills, my Acquisitions Editor, kept the vision clear. I must give most of my thanks to Sean Culhane, my Developmental Editor, who handled many details of the project and came up with some of the creative ideas that are embodied in this edition. Cathy Faduska, Marketing Manager, developed a strong sales plan and contributed to the look and feel of the book. Joan Kalkut, Associate Editor, handled the ancillary material to complete the learning package. Dawn Stanley created a revised design that kept the strengths of the previous edition while adding more visual impact. The production crew as usual was great! Nancy Prinz, Senior Production Editor, coordinated all their efforts. Ishaya Monokoff of Wiley's Illustration Department deserves my gratitude for finding Boris Starosta, the freelance technical illustrator who shouldered the monumental task of reconceptualizing and redrawing the complete art program. Sigmund Malinowski coordinated my efforts with Boris, who worked to develop figures that were understandable by novice students, astronomically accurate, and many just stunning. Boris has the artistic vision needed for an attractive astronomy book.

Other folks include my correspondence students (who have just the book and their brains) and the students in my regular Astronomy 101 at UNM. I have also received letters from students at other colleges and universities, as well as from instructors. Thank you all!

Any errors in the text are my responsibility. It is a little known fact that minor corrections and changes *can* be made in future printings of this edition. Please keep this fact in mind and send me any errors you may find. When ordering, please request the *latest printing* of the book so your students will have the most correct version.

Your feedback can improve this book! Please send any comments to me at the Department of Physics and Astronomy, The University of New Mexico, 800 Yale Blvd. NE, Albuquerque, New Mexico, 87131-1156, USA. My Internet mail address is: zeilik@chicoma.lanl.gov. I am also on CompuServe as user 72600,3202. I usually log onto the Astronomy Forum.

Michael Zeilik

Abbreviations are often used for the names of major observatories and agencies (particularly in the figure captions). These are:

AAO—Anglo-Australian Observatory
AATB—Anglo-Australian Telescope Board
AUI—Associated Universities, Inc.
AURA—Association of Universities for Research in Astronomy
CTIO—Cerro Tololo Interamerican Observatory
ESA—European Space Agency
ESO—European Southern Observatory
KPNO—Kitt Peak National Observatory
LANL—Los Alamos National Laboratory
NASA—National Aeronautics and Space Administration
NCAR—National Center of Atmospheric Research
NCSA—National Center for Supercomputer Applications (at the University of Illinois, Champaign/Urbana)
NOAO—National Optical Astronomy Observatories
NRAO—National Radio Astronomy Observatory
NSO—National Solar Observatory
SAO—Smithsonian Astrophysical Observatory
STScI—The Space Telescope Science Institute

Note: The Mars Observer (Chapter 10) was last heard from in August, 1993, as it started its automated sequence for entering a Martian orbit. No one yet knows the cause of the mishap or the fate of the spacecraft.

How to Study Astronomy

Mark Hollabaugh
Normandale Community College
Bloomington, Minnesota

Welcome to astronomy! Have you ever wondered why the stars are different colors? Have you visited a planetarium or an observatory? Perhaps you've sat in front of the television and watched live pictures from the space shuttle. Maybe you're thinking of astronomy or physics as a career. Or, like me, maybe you are awestruck by the night sky.

When I was growing up in Michigan, I would go outside on crisp winter nights to watch the northern lights dance across the sky. In August, I counted meteors during the Perseid meteor shower. I could imagine the terror ancient peoples must have felt as the moon turned blood red during a lunar eclipse. I looked through a real telescope for the first time when I was in the seventh grade. A science teacher who spent summers in our town set up his telescope each Friday night in the city park. After a brief peer though the scope, I would run back to the end of the line for another look. I was hooked on astronomy. Later, after studying physics, I did research on comets, asteroids, and cosmic rays.

Throughout these experiences, I have marveled with fascination at the grandeur and intricacies of the universe. Now, I try to pass that fascination on to my students. This is also what Professor Michael Zeilik does in *Astronomy: The Evolving Universe.* My association with Mike began with my interest in Native American astronomy. We also share an interest in helping college students like you to enjoy learning astronomy. Your professor has chosen a superb textbook for you to use in your astronomy course. I'd like to give you some suggestions to assist your learning.

Understanding Your Learning Style

Each of us learns in different ways. Some of us learn best by reading. Others find listening to a lecture to be most helpful. Speaking to others can help us clarify our own understanding. Drawing diagrams or pictures can help us see the relationships between concepts.

When learning something new, it helps to know your own strengths, weaknesses, and learning style. As you begin your study of astronomy, make a list of the kinds of activity that help you learn best. You will find a wide variety of learning resources presented in your astronomy course: lectures, textbook readings, laboratory or other group exercises, videotapes, slides, or videodisks. These are all parts of your "tool kit" for understanding astronomy. Concentrate extra effort on learning activities of the kinds that give you the greatest difficulty. You may, for example, need to learn how to use your textbook as a resource to clarify ideas you don't understand from a lecture.

Memorization may help you correctly answer questions on an examination, but you will not be able to apply the concepts later. Avoid memorizing and seek rather to understand concepts and relate them to one another.

Learning to Learn

Ask yourself some questions about your study habits. Do you set aside a specific block of time each day to study a given subject? Do you work in a quiet place, free from distractions, such as the library? Do you study at a time of the day when you are at your best for learning new material? When you prepare for an examination, have you kept up with your reading of the text so all you need to do is review? Do you attend all class lectures and other activities? Do you concentrate on understanding concepts as opposed to memorizing facts or data? If you have answered yes to these questions, you have developed some excellent study habits! If not, set goals for yourself to change your learning habits. Find out if your school has an office or program that helps students to improve their study skills, and make use of it.

Learning from Lectures

Should I read the book first or go to lecture first? How you personally answer the question depends on your own learning style. It is usually helpful if you have some basic introduction to a topic before you go to your lecture. Look at the syllabus and make a note of the topics for the day. Then scan the *Learning Objectives* for the chapter. Next, read the *Key Concepts* at the end of the appropriate chapter. Then skim the chapter in the textbook. If, for example, the lecture will be on Venus and Mars, page through Chapter 10. Examine the photographs, and make note of any questions you might have about them. Pause to read the *At a Glance* boxes. After class you should, of course, read the chapter more carefully.

Learning by Taking Notes

Taking notes is a two-step process. First, you will make some notes in class as your professor lectures. Try to keep your notes in outline form. Don't worry about making a formal outline. Rather, just indent subtopics under major topics. If your professor has provided a course outline for you, *use it.*

Concentrate on getting down the "big ideas" and explanations of concepts. Your instructor wants you to *understand* astronomy! You can fill in the details later as you read the text.

The second step in taking good notes is editing them. Check for concepts you don't understand and look them up in the textbook. Jot down any definitions you may have missed. In short, edit your notes, making sure you understand the principal ideas. This process will keep you well prepared for taking examinations.

Learning with Other Students

Other students are a most important resource to your learning. You will understand more about astronomy if you explain, elaborate, or defend ideas to others. Forming small, informal study groups out of class is an important strategy. You can use the resources of the textbook to quiz one another. Come to such study groups prepared: edit your notes, review the *Key Concepts,* and make a list of questions for your group. Your instructor may formalize the use of cooperative groups, perhaps by computer networking. Learning together means you learn more. Don't wait to work together until the night before an examination. Make time in your schedule for regular meetings of your study group. Your school may have a computer network available to all students. If so, make use of this facility to stay in touch with students in your class.

Learning from the Textbook

Professor Zeilik has incorporated many useful learning tools into your textbook to help you discover the excitement of astronomy. You should be aware of what your instructor expects of you. He or she may omit some topics. You may be assigned specific learning objectives. It may help you to view your textbook as your primary reference. It supplements the lecture and other activities in the class. The textbook is your teacher away from the classroom. Your professor will expect you to read carefully the assigned sections in *Astronomy: The Evolving Universe.*

Take notes as you read. Simply highlighting text is not particularly effective in placing concepts into your long-term memory. You will learn more, and recall it better later, if your brain is stimulated by the process of jotting down notes as you read. Make notations in the margins of your textbook, or use Post-it™ notes. Later, transcribe these notes in outline form to your notebook. Try to focus on the big ideas and reasons the universe works the way it does.

Let's take a few moments to become familiar with the design of the book by looking at the various sections of a chapter.

We hope you will enjoy your study of astronomy. Approach your class with a sense of awe and curiosity. Try to understand how the universe works. And, as another famous "science officer" has said, "Live long and prosper!"

PART ONE

...NGING CONCEPTIONS OF THE COSMOS

CHAPTER 1
From Chaos to Cosmos

CHAPTER 2
The Birth of Cosmological Models

CHAPTER 3
The New Cosmic Order

CHAPTER 4
...Clockwork Universe

I HAVE OFTEN WONDERED WHAT the first stargazers felt as the heavens reeled above them. What magic did the dance of the heavens weave for them? How did they imagine the cosmos and their place within it?

My work as an astronomer has given me hints to answers to these questions. On my frequent observing trips to Capilla Peak Observatory in New Mexico, I often work anxiously through sun set and dusk to set up my equipment. When all is finally ready, I walk out of the dome to take a curious look at the sky.

As always, I'm stunned by the depth of a dark sky. To the south that night, Scorpius balances on its coiled tail. Faint constellations pattern the sky in an unfamiliar quilt of lights. Mars shines like a red beacon in the sea of stars. I sense then how ancient people watched the sky and wondered.

Part One presents the evolution of conceptions of the cosmos from ancient musings to modern speculations. It starts off with the basic astronomical observations, then investigates how astronomy prompted new ideas about the universe. I confine the tale to Western cultures and highlight crucial episodes in the evolution of astronomical concepts. This perspective will give you some insight into the development of physical ideas that relate to cosmological ones and into the birth of scientific thought—an evolution of human ideas ...eme in the grand scheme of cosmic evolution. ∎

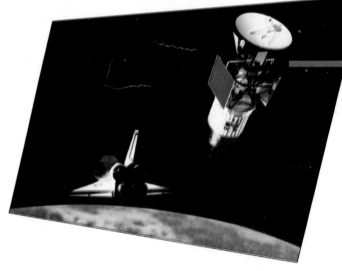

OBJECTIVES. Your instructor may indicate to you the objectives for which you are responsible. Before beginning to read the chapter, study the objectives. *After* you have read the chapter, write out a response to *each* objective in one or two sentences. Although this may seem time-consuming to you, it is an excellent way to summarize and apply what you have learned. At the end of each chapter, you will find review questions keyed directly to the objectives.

CENTRAL QUESTION AND OPENING QUOTATION. Each chapter has a theme. Keeping this theme in mind as you read will help you to understand the unifying idea of each chapter. Try to relate concepts back to the *Central Question.*

LEARNING OBJECTIVES
After studying this chapter, you should be able to:

18-1 Explain at least one astronomical difficulty in trying to figure out the structure of the Galaxy from our location in it.

18-2 Name the important spiral-arm tracers and state how they are used.

18-3 Present the observational evidence for the Galaxy's having a spiral structure, that is, describe what methods astronomers use to work out the positions of spiral arms.

18-4 Sketch the rotation curve of the Galaxy, describe how to find from it the approximate mass of the Galaxy, and argue that a significant amount of the

Galaxy's mass must exist in the halo in an unseen form.

18-5 Describe the sun's orbit around the galactic center and explain the techniques used to find the distance and speed of this orbit.

18-6 Explain how radio astronomers used 21-cm-line observations to trace spiral arms, indicate the limitations of their method, and explain its advantage over optical observations.

18-7 Describe the contents of a typical spiral arm and tell how these evolve.

18-8 Describe the evolution of spiral arms in terms of the density-wave model.

18-9 Outline a model for the evolution of the disk of the Galaxy.

18-10 Outline a model for the evolution of the halo of the Galaxy.

18-11 Make two rough sketches of the entire Galaxy, from a top and a side view, labeling the disk, spiral arms, halo, globular clusters, nucleus, and the sun's position.

18-12 Describe what information radio, infrared, and x-ray observations provide about the nucleus of the Galaxy and make a case for the possible existence there of a supermassive black hole.

Central Question

What evolutionary processes induce the structure of the Galaxy?

AT A GLANCE. Throughout each chapter are boxed lists that give you brief summaries of astronomical topics. These summaries can be helpful to you when you survey a chapter or when reviewing for an examination. Rather than memorize numerical data, use such data to compare differing astronomical objects. For example, examine the boxes on pages 179 and 190, and note that our moon's radius is a little smaller than Mercury's.

18

The Evolution of the Galaxy

A broad and ample road, whose dust is gold, And pavement stars, as stars to thee appear Seen in the galaxy, that milky way . . .

JOHN MILTON, *Paradise Lost*

Like a majestic cosmic pinwheel, our Milky Way Galaxy spins slowly in space. Our sun orbits the center of it along with more than *100 billion* other stars. Because the sun is about 30,000 ly from the nucleus, it completes one revolution roughly every 250 million years.

What is the three-dimensional structure of this enormous system of stars? Optical astronomers probe the structure near the sun, and radio astronomers study regions farther away. The details are not yet in, but astronomers have been able to establish the broad outlines of the Galaxy's structure—a remarkable achievement, considering that we are buried within the Galaxy. These investigations show that the Galaxy does have a spiral structure, a pattern inferred in part from spiral designs observed in other galaxies (Fig. 18.1).

This overall spiral structure of the Galaxy evolves. Theoretical work sees the grand design as a wave pattern; it links many of the galactic entities in the disk by the process of galactic evolution. The Galaxy's structure evolves—but at a rate so slow that we won't see any changes in our lifetimes.

The Galaxy looked very different in the past, especially just after its formation. The evolution of stars, interstellar matter, and the nucleus has driven significant changes. This chapter takes a brief excursion into our Galaxy's history to see how our spiral Galaxy evolved.

What is the three-dimensional structure of this enormous system of stars? Optical astronomers probe the structure near the sun, and radio astronomers study regions farther away. The details are not yet in, but astronomers have been able to establish the broad outlines of the Galaxy's structure—a remarkable achievement, considering that we are buried within the Galaxy. These investigations show that the Galaxy does have a spiral structure, a pattern inferred in part from spiral designs observed in other galaxies (Fig. 18.1).

This overall spiral structure of the Galaxy evolves. Theoretical work sees the grand design as a wave pattern; it links many of the galactic entities in the at a rate so slow that we won't see any changes in our lifetimes.

FIGURE 19.10 The galaxy NGC 3187 in the constellation Leo. This galaxy is a spiral with its arms twisted by the tidal interaction with a nearby galaxy. (Courtesy California Association for Research in Astronomy.)

3

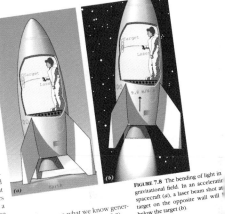

114 ■ CHAPTER ONE FROM CHAOS TO COSMOS

13.5 The Quiet Sun

The term *quiet sun* refers to the seemingly placid day-to-day face of the sun. I use this expression to describe solar phenomena that are characteristic of large regions for long periods of time, such as the flow of energy out of the sun. This energy flow, from core to surface and beyond the atmosphere, controls the environment of the quiet sun.

The Numbers Game

As I have progressed through the astronomical, biological, and sociological factors needed for a rough estimate, the footing has become shakier. I have also ignored some important factors in the analysis, such as the possible stable planetary orbits in a binary star system. I did not intend to give precise results, but rough estimates, because exact answers are not yet possible. The point is to get a feel for reasonable exclusions in the enormous range of values each element might take.

Not evaluating L_{ic}, I come up with

$$N_{ic} = R \cdot P_p P_e N_e P_l P_i L_{ic}$$
$$= 10 \times 0.5 \times 0.1 \times 1 \times 1 \times 1 \times L_{ic}$$
$$\approx 0.5 \, L_{ic}$$

The result depends critically on L_{ic}—how long our candle remains lit. If we long as the sun shines, then L_{ic} is roughly 10^{10} and N_{ic} some 50×10^9. In this case, many a star in the Galaxy has fostered an intelligent civilization!

FIGURE 7.8 The bending of light in gravitational field. In an accelerating spacecraft (a), a laser beam shot at target on the opposite wall will below the target (b).

Let's sum up what we know generally of the interstellar gas (Table 15.2):

1. *H II regions.* Zones of glowing, ionized hydrogen surrounding young, hot stars (spectral types O and B); contain a minor amount of the interstellar gas, perhaps a total of 10 million solar masses in the Galaxy.

2. *H I regions.* Clouds of cool, neutral hydrogen roughly 10 ly in diameter and each containing about 50 solar masses of material; total mass in the Galaxy may be 3 billion solar masses.

The Sun's Structure
At a Glance

Core	Site of fusion reactions; temperature 8×10^6 to 15×10^6 K; energy transport by radiation
Convection zone	Energy transport by convection; temperature below 500,000 K; outer 0.5 of the sun's radius, below photosphere
Photosphere	Origin of continuous and absorption-line spectra; temperature from 6400 to 4200 K; energy transport by radiation
Chromosphere	Energy transport by radiation and magnetic fields; temperature from 4200 to 10^6 K
Corona	Energy transport by radiation and magnetic fields; temperature from 1 to 2×10^6 K; source of the solar wind

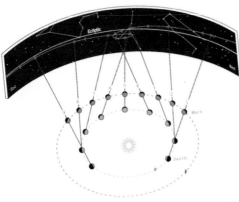

FIGURE 3.4 Retrograde motion of Mars in the Copernican model. This diagram shows the 1995 retrograde. The marks begin on October 15, 1994 and have one-month intervals up to June 15, 1995. As the earth comes around the same side of the sun as Mars, it is moving faster along its path and so passes Mars. During this passing interval Mars appears to move backward (to the west) with respect to the stars. Note that Mars appears in the middle of its retrograde motion just as the earth passes it, so Mars is in opposition to the sun as viewed from the earth. Also, it is closest to the earth, and so appears its brightest in the sky. The same basic diagram applies to Jupiter and Saturn; they undergo retrograde motion when the earth passes them. The illusion of retrograde motion results from the passing situation.

summer solstice port on April 15—just the right date to begin planting key crops, such as corn, beans, and squash.

This example shows that the prehistoric Pueblo people very likely made naked-eye observations that revealed the basic cycles of the heavens.

■ What was Tycho's main contribution to the astronomy of his time?

22.6 Neighboring Solar Systems?

What evidence do we have of other planetary systems? Because a planet shines by reflected light from its parent star, because it is small, and because it lies very close to its local sun, as seen from the earth, a planet's gleam would be lost in the stellar glare. So we cannot *directly* observe other planets outside the solar system with earth-based telescopes. (The Hubble Space Telescope will give us more power for this task.)

The stars are always located in space. Note that the distance from the sun to the planet varies as it moves along its elliptical orbit.

LAW 2. *Law of Equal Areas (1609)* A line drawn from a planet to the sun sweeps out equal areas in equal times. Physically, this law notes that the orbital velocities are nonuniform but vary in a regular way: the farther a planet is from the sun, the more slowly it moves (Fig. 3.15).

LAW 3. *Harmonic Law (1618)* The square of the orbital period of a planet is directly proportional to the cube of its average distance from the sun (Fig. 3.16). This law points out that the planets with larger orbits move more slowly around

FIGURE 1.15 Precession and the change of position of the vernal equinox with respect to the zodiacal stars. (*a*) A sunset scene (similar to Figure 1.13) in 3000 B.C on March 21. Note that the sun lies in Taurus.

INTERACTION QUESTIONS. At the end of each numbered section of text, you will find a question to think about and then answer right away. These questions will give you an *immediate* check on your comprehension of what you have just read. If you are uncertain of your response to the question, reread the material.

The Unifying View

CHAPTER CONCEPTS

CHAPTER 1 From Chaos to Cosmos
The motions of astronomical objects you can see by eye follow distinctive patterns and cycles in the sky over both short and long periods of time. These repeated motions suggest an underlying design to the heavens.

CHAPTER 2 The Birth of Cosmological Models
Scientific models of the cosmos can explain and predict the motions of celestial bodies, especially those of the planets. Early models of the cosmos were centered on the earth.

CHAPTER 3 The New Cosmic Order
A heliocentric model of the cosmos was rediscovered during the 16th century in Europe. But this break with the geocentric tradition required new physical laws and a revolution of the cosmological views of the time.

CHAPTER 4 The Clockwork Universe
Newton's laws of motion and gravitation explain, predict, and unify the motions of the bodies in the solar system. These laws are universal and apply to objects outside of the solar system.

Enrichment Focus

GEOMETRY OF ELLIPSES

The most essential property of an **ellipse** is its eccentricity. To define it exactly, the eccentricity, e, is

$$e = \frac{c}{a}$$

If the sun is at one **focus** (F_1) of a planet's orbit (Fig. 3.14), then R_p, the closest point to the sun, is the **perihelion** point in the orbit. The perihelion distance R_pF_1 is

$$\overline{R_pF_1} = a - c = a - ae = a(1 - E)$$

Note that a is the average distance from one of the foci of a point moving on the ellipse. So for planetary orbits, a is the average distance of a planet from the sun.

As an example, the orbital eccentricity of Mercury is 0.21 and its semimajor axis is 0.39 AU. Then Mercury's perihelion distance is

$$R_p = a(1 - e)$$
$$= 0.39(1 - 0.21)$$
$$= 0.39(0.79)$$

Similarly, the aphelion distance is

$$R_a = a(1 + e)$$
$$= 0.39(1 + 0.21)$$
$$= 0.39(1.21)$$
$$= 0.47 \text{ AU}$$

Kepler discovered law 2 before law 1, when he was grappling with the orbit of Mars. Before finding law 2, Kepler had concluded that the planes of all the orbits pass through the sun. This discovery set the stage for law 1.

Lunar eclipses have a large potential audience: all people on the night side of the earth (because the entire moon is in the earth's shadow). During your lifetime, you can expect to see roughly 50 lunar eclipses, about half of them total. In contrast, since solar eclipses are visible only along a narrow band of the earth, a total eclipse of the sun occurs rarely for any one location on the earth.

Giant H II regions, which surround young, massive stars, are always found near molecular cloud complexes (Fig. 15.11).

Most of the atmosphere's mass is hydrogen (74 percent) and helium (25 percent). All other elements (loosely called *metals* or *heavy elements* by astronomers) make up a mere 1 percent. These abundances are relative to the total *mass*.

If an element's absorption lines don't show up in the visible spectrum, does it mean that this element does not exist in the sun? Not necessarily. Perhaps so little of the element exists that it does not produce detectable absorption lines. Or the strongest absorption lines may be in a region of the sun's spectrum unobservable through the earth's atmosphere. Or the temperature, pressure, and density of the sun's atmosphere may inhibit the formation of the element's spectral lines. If the conditions are.

Key Concepts

1. Celestial bodies participate in cyclic motions; some are short term and others long term in duration (Table 1.3).
2. Relative to the horizon, the stars rise in the east and set in the west every day. The stars make up fixed patterns, called constellations, which do not change over a human lifetime. Different constellations are visible in the night sky at different seasons.
3. Relative to the horizon, the sun rises roughly in the east and sets roughly in the west daily, with the exact position changing with the seasons: its highest p. called the ecliptic that cuts through the constellatio the zodiac; the sun completes one eastward circ the zodiac in a year.
4. Relative to the horizon, the moon rises in the ea sets in the west; relative to the stars, the moon eastward (completing one circuit in about a mo the moon moves around the sky relative to th goes through a cycle of phases, which depenc angle between the sun and the moon.

Electromagnetic energy binding electrons to nuclei of atoms
Radiative energy carried by photons, the greater with higher frequency
Mass is energy ($E = mc^2$)
Transformations occur among the various forms of energy, and energy can be transferred from one location to another
Atoms and molecules absorb and emit light
A planet's amounts of kinetic and potential energy change as it orbits the sun
Mass and energy warp spacetime

TUM CONCEPTS On a very small scale, matter is made of discrete particles
On a very small scale, energy comes in discrete units
Physical phenomena are governed by a few basic interactions of matter and energy, often at the quantum level
Optics is the manipulation of light by its interactions with matter
Electromagnetic energy has both wave and quantum properties

TION AND FORCE Everyday static and moving systems are explained and predicted using Newton's
CONCEPTS laws of motion
Inertia is the basis of natural motion
Net forces result in the acceleration of masses
Accelerations are changes in velocities of masses

ENRICHMENT FOCUS. Astronomers use the laws of physics to describe astronomical phenomena. These laws usually are stated mathematically. You will notice that mathematical topics are discussed in short, compact essays. If your instructor covers these topics, you will want to review the focus boxes carefully. They may even be assigned. In most cases you should actually work through the calculations with pencil and paper and your calculator.

KEY CONCEPTS. When you read a mystery novel, you probably wouldn't read the ending first. However, in reading *Astronomy: The Evolving Universe*, it is well to begin a chapter by studying the *Key Concepts* at the end of it. This will give you a concise overview of the principal ideas in that chapter. The *Key Concepts* also will be *very* useful when reviewing for an examination.

THE UNIFYING VIEW. Sometimes science may seem to you to be a large mass of data and theories. Scientists in any field, however, always attempt to relate the data and theories to fundamental principles. At the simplest or most complex level, everything in astronomy is related to basic principles of physics. At the end of each part, you will see how what you have learned is related to five core concepts: Astronomical, Informational, Energy, Force and Motion, and Quantum. If you have never learned about the inner workings of the atom, you may find the Quantum concepts especially stimulating.

KEY TERMS. Astronomy, and all of science, is rich with specific, technical terms. At the end of each chapter is a list of the key terms that appeared in boldface type in the chapter. Write a brief definition of each term and then check it with the *Expanded Glossary* that follows Appendix G. Simply being able to define a term may not mean you understand it. When you read, you may encounter even nontechnical words you don't fully understand. If you can't determine the meaning from the context, look the word up in a dictionary.

Key Terms

acceleration	centripetal acceleration	inverse-square law	speed
apogee	centripetal force	mass	(universal) law of gravitation
binary star	escape velocity	mechanics	velocity
center of mass	free-fall	Newton's laws of motion	
central force	inertia	perigee	

Study Exercises

1. What was one important astronomical achievement of the Babylonians? (Objective 2-2)
2. Contrast Babylonian and Greek astronomy in terms of their observational achievements. (Objectives 2-1, 2-9, and 2-10)
3. How did Aristotle's model explain (a) the apparent lack of motion by the earth, (b) the daily motion of the stars,

(c) the annual motion of the sun through the zodiac? (Objective 2-4)
4. How did Aristotle argue against a heliocentric model? (Objectives 2-3 and 2-4)
5. In the Ptolemaic model, how did observations set the period of the epicycle? The deferent? (Objective 2-6)

Problems & Activities

1. Suppose you tried to duplicate Eratosthenes' observations and came up with value of 8°. What would you calculate for the earth's circumference and radius? (Objective 2-10)
2. With naked-eye observations, the smallest angle you

can measure is about 1 arcmin. Assume that the stars are fixed on a sphere and consider two stars 1° apart as seen from the sun. How far must the sphere of stars be placed from the sun so that their parallax is 1 arcmin or less? (Objective 2-11)

See for Yourself!

Most binoculars are specified by two numbers, such as "7 × 50", which is read "seven by fifty". The second number is the diameter of the binocular's lenses, in millimeters. In this case, they are 50 mm across. The first number gives the magnification, in this case "seven power", which means that an object appears 7 times larger seen through the binoculars. Use such a pair of binoculars to test this out.

What is the apparent angular diameter of the moon viewed through these binoculars? What is the apparent size-to-distance ratio? Which of these two have

changed, and by how much? What is your resolution limit using these binoculars? What would be the smallest crater you could see on the moon? Examine the moon near first quarter with 7 × 50 binoculars to confirm your estimations.

Rather than buy a small, cheap telescope, I urge you to purchase binoculars for your first investigations of the heavens with an instrument. Don't be put off by the "low" power of binoculars. It's really the light-gathering power that counts. ■

STUDY EXERCISES. At the end of each chapter is a series of study exercises that require you to explain concepts you have encountered in the chapter. The most effective way to use this material is to work on the exercises with other students. You, and your friends, will learn more in the process. If you do not answer a question to your satisfaction, check the objective keyed to the exercise and discuss the material again. Try to explain *why* your answers are correct.

PROBLEMS & ACTIVITIES. These typically require some knowledge of simple geometry or algebra. They test your understanding of chapter objectives with specific questions. Some activities will require you to carry out a short-term project.

SEE FOR YOURSELF. Something in a chapter may catch your imagination and make you want to pursue it further. Your professor may encourage you to do extra projects. These *See for Yourself* activities, at the end of each chapter, present observational and practical opportunities to further your interest in astronomy.

PART TWO *Epilogue*

This part has shown you the new richness in our knowledge of the solar system. But don't let this wealth of information detour you from the main point: What do the present properties of the solar system reveal about its origin and evolution?

By now you realize that the models we have developed to answer this question strike a very tentative chord. But we do have a pretty good idea of the general themes. The formation of the solar system naturally accompanied the birth of the sun from a cloud of interstellar materials. The closeness in age of the sun, moon, earth, and meteorites implies that the process of formation lasted a short time compared to their age of 4.6 billion years—at most a few hundred million years.

To glean the conditions in the solar system's early years, we must look at the fossil objects in it: the interplanetary debris of comets, asteroids, and meteoroids. These objects have evolved little (compared to the planets) and so give us the most direct clues to the past.

That past involves a nebular model for the solar system's formation. The start of the process requires the gravitational collapse of a huge interstellar cloud. What prompts such a collapse? We don't really know yet. But we do have evidence for the existence of enormous clouds between the stars. We can also see very young stars that have been born from such clouds. And that starbirth may well involve the formation of planets. So the fact that stars are common implies that planets are common, too. Stars and planets are connected in starbirth, that is one of the key themes of this book.

Since their birth, the planets have evolved, some more than others. The Jovian ones have changed little since their formation; they have essentially retained their youthful appearances. In contrast, the terrestrial planets have reshaped their interiors and surfaces and transformed their atmospheres. And the earth, a restless planet, has changed the most; its structure now differs greatly from is protoplanet state.

I can't help wondering how many other planets circle other stars in the sky, especially when I observe on a winter night. How many monsters such as Jupiter? How many jeweled Saturns? How many barren places like Mercury? And how many like the earth, where perhaps my counterpart also dreams of inhabited worlds? ■

119

EPILOGUE. "So, what's the point?" You might ask that after reading to the end of a part. The *Epilogues* will give you some insights into the "why" questions in astronomy. If you are philosophically inclined, you may find these sections especially thought-provoking.

CHAOS TO COSMOS

in brightness
y eye from the
u can, with a
l's light varia-
zine *Sky and*
s and times of
ter star (260
(surface tem-
panion, Algol
osities) K star
K). From the
the period,
that of Algol
is 3.7 solar
mass.

ation

s that allow
most cases,
rs can also
e determined. When the luminosities of main-sequence stars are plotted against the stars' masses, the points fall into a definite pattern (Fig. 14.26). For main-sequence stars, the mass determines the luminosity, and the resulting correlation is called the **mass–luminosity relation** (sometimes abbreviated *M–L relation*).

How to view the image formed by a reflector? Newton put a small mirror tilted at 45° to the path of light to direct the focus out the side of the telescope tube (Fig. 6.8a), where an eyepiece is placed. Such a design is called a *Newtonian reflector*. Most research telescopes today use a different optical design called a *Cassegrain reflector* (Fig. 6.8b). All large telescopes built now are reflectors; advances in technology allow them to be smaller, more rigid, and cheaper to construct than the classical reflectors such as the Hale 5-m on Mt. Palomar.

Functions of a Telescope

1. *To gather light.* Whether a reflector or a refractor, a telescope has one primary function: to gather light. A telescope is basically a light funnel, collecting photons (Section 5.3) and concentrating them at a focus. So objects viewed through a telescope appear brighter than they do when seen by the eye alone.

How much light a telescope can collect depends on the *area* of its objective, which is proportional to the square of the diameter. So a mirror with twice the diameter of another can gather four times as much light. That's the reason astronomers want large telescopes—big mirrors that have plenty of light-gathering power (Fig. 6.9).

Light-gathering power has no absolute standard; it's a relative measure. For example, the diameter of your eye in the dark is about 0.5 cm. A telescope with a 50-cm diameter objective (100 times bigger) would have a light-gathering power 10,000 times (100^2) greater than your eye. Similarly, the 5-m Hale telescope at Mt. Palomar outdoes the 4-m Mayall telescope at the Kitt Peak Observatory by more than 50 percent.

2. *To resolve fine detail.* The next most important function of a telescope, is to separate, or *resolve*, objects that are close together in the sky. This ability is called **resolving power,** usually expressed in terms of the minimum angle between same wavelength, the resolving power in angular distance.

Caution. This procedure gives you only the abundance in the *photosphere*. It does not tell you directly the abundance in the *interior*.

FIGURE 10.18 Hubble Space Telescope (HST) image of Mars. This color image was taken when Mars was 85 million km from the earth on December 13, 1990. The resolution is 50 km. (Courtesy P. James. University of Toledo, and JPL/NASA.)

PHOTOS & TABLES. Many new full-color photos have been added to this edition to enhance your understanding of the cosmic wonders. The tables put researched astronomical data at your fingertips.

PROPERTY	SPIRALS	IRREGULARS	DWARF ELLIPTICALS	GIANT ELLIPTICALS
Diameter (ly)	90×10^3	20×10^3	30×10^3	150×10^3
Mass	10^{11}	10^6	Not well known	10^{13}
Luminosity	10^{10}	10^9	10^6	10^{11}
Color	Bluish (disk), reddish (halo and nucleus)	Bluish	Reddish	Reddish
Neutral gas (fraction of mass)	5%	15%	Less than 1%	Less than 1%
Types of star	Young (disk), old (halo and nucleus)	Young	Old	Old

TABLE 19.1 GENERAL PROPERTIES OF GALAXIES OF DIFFERENT TYPES

Note mass and luminosity are given in solar masses and solar luminosities, respectively.

Improve Your Night Vision

SUMMER

Fred Schaaf

A giant triangle of stars shines bright enough and high enough on summer evenings to be seen even in big cities on all but very hazy nights. In addition to being a lovely sight and a guide to other star patterns, this triangle is highly instructive about one important aspect of how stars differ.

The Summer Triangle is composed of Vega, a star in the tiny constellation Lyra the Lyre; Altair, a star in Aquila the Eagle; and Deneb, a star in Cygnus the Swan. In June and July, you'll find the triangle partway up the east sky at nightfall, with Vega on top, Altair to the lower right, and Deneb to the lower left as you face east. In the middle of August and September evenings, first Vega and then Deneb passes near the zenith (the overhead point) as seen from the earth's midnorthern latitudes. Later in the night, the triangle heads down the west sky (because, of course, the earth's rotation carries us east), Vega still leading the way.

The most interesting thing about this huge star pattern is that while its stars appear to be moderately different

in brightness, one of them is actually tremendously d[...]ent—tremendously brighter. The key here, of cours[...] distance. The uninformed observer might assume th[...] the stars—some bright, some faint—were at the sam[...] tance. In reality, although Altair appears a little bri[...] than Deneb, the former is one of our closer stellar n[...] bors, just 16.5 ly away (Vega is 26 ly away, also c[...] Deneb is roughly 100 times farther than Altair and t[...] fore has a spectacularly brighter "absolute magnitude" "luminosity" (measures of the true brightness of star[...] Deneb were as close to us as Altair, it would rival a[...] moon in brightness in our sky! Consider these rea[...] behind the appearances as you gaze upon the Sun[...] Triangle.

Not until the invention of the spectroscope in[...] nineteenth century did astronomers have any clues a[...] how far away various stars were. Other information[...] was not known until quite modern times includes[...] explanations of the shape of our Milky Way Galaxy[...]

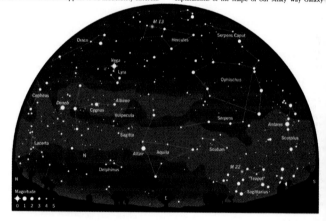

122

IMPROVE YOUR NIGHT VISION. Part of the joy of learning astronomy comes from looking up at the night sky and knowing the names of specific stars and constellations, and locating the planets. The more you "poke around" the night sky, the more you'll understand about astronomy. This element, written by Fred Schaaf, will help you find your way as you journey through the stars. These informative essays, which conclude each part of your textbook, deal with real-life astronomy such as the issue of light pollution.

y we see a softly glowing Milky Way band of light
~tched across the heavens at certain times, especially on

If you want to get a good view of the Milky Way band
urself, pick a summer evening that is fairly haze-free
ok for a deep blue sky before sunset), with the moon
above the horizon (the week between last quarter
on and new moon is best). Then, be sure you are miles
ay from the lights of any sizable city. Under these good
nditions, you'll be amazed to see the dreamy grandeur
the band, extending especially prominently from high
in Cygnus the Swan down to the teapot *asterism* (unof-
al star pattern) in Sagittarius the Archer, low in the
th.

What causes the soft glow? You can answer this for
urself by lifting binoculars to any of the brighter patches
Milky Way: what you see is stars, stars, . . ., and more
s. The Milky Way glow is produced by the combined
iance of distant stars that are mostly too dim to be seen
vidually with the naked eye.

But why is the glow localized to a fairly narrow band,
ch is highest in our sky in summer and winter (though
mer in winter)? A *galaxy* is an immense congregation,
ally consisting of billions of stars, revolving
mon center of gravity. Astrono
ny clues that ours i
ks som
st of the
ugh tow
). The gl
d extendi
within the
look towar
see the bar
ge. And in w
ard dimmer,
d is fainter.

By the way,
a good country
-particularly a
s Great Rift is ac
t. We see similar
es that are positio

There is much
sky. In the hour
, Delta Aquarid m
gust 9 to 14 the mai
wer is even better
er to "See for Yourse
re on meteors, see th

But meteors are be
you do if you have a

in a fairly large city? Binoculars can show you distinct star
colors, double stars, open star clusters, and much more,
which we'll discuss in the other seasonal essays.

Double stars and detail on planets can be seen especially well on many moderately hazy summer nights
because the haze is indicative of a still atmosphere and
therefore a good "seeing"—that is, steadiness of images.
Turn binoculars or telescope on the double star Epsilon
Lyrae near Vega and the double star Albireo in Cygnus. For
the rest of the 1990s, Jupiter will be conveniently placed in
the evening sky for at least part of the summer—no other
planet can show so much detail of cloud structures on its
globe in telescopes.

A giant triangle of stars shines bright enough and
high enough on summer evenings to be seen even in big
cities on all but very hazy nights. In addition to being a
lovely sight and a guide to other star patterns, this triangle
is highly instructive about one important aspec
stars differ.

The Summer Tri
the tiny c

Expanded Glossary

absolute magnitude A measure of the brightness a star would
have if it were to be placed at a standard distance of 10 parsecs
(32.6 light years) from the sun.
absorption (dark) lines Colors missing in a continuous spectrum because of the absorption of those colors by atoms.
absorption-line spectrum Dark lines superimposed on a continuous spectrum.
acceleration The rate of change of velocity with time.
accretion The colliding and sticking together of small particles to
make larger masses.
accretion disk A disk make by infalling material around a mass;
the conservation of angular momentum results in the disk
shape.
active galactic nuclei (AGNs) The nuclei of galaxies that have a
nonthermal continuous spectrum over a wide range of wavelengths and signs of unusual, energetic activity such as radio
jets.
active galaxies Galaxies characterized by a nonthermal spectrum and a large energy output compared to a normal galaxy.
angular speed The rate of change of angular position of a celestial object viewed in the sky.
anorthosite A basaltic mineral composed of calcium and sodium
with aluminum silicate; the predominant mineral of the lunar
highlands.
antapex The direction in the sky from which the sun appears to
be moving relative to local stars; located in the constellation
Columba.
antimatter Elementary particles with the reversed electric charge
or other property compared to that of ordinary matter.
apex The direction in the sky toward which the sun appears to be
moving relative to local stars; located in the constellation
Hercules.
aphelion For a body orbiting the sun, the point on its orbit that is
farthest from the sun.
Aphrodite Terra A large highland region on Venus.
apogee The point in its orbit at which an earth satellite is farthest
from the earth.
apparent magnitude The brightness of a star (or any other celestial object) as seen from the earth; an astronomical measure
of the object's flux.
absolute magnitude A measure of the brightness a star would
have if it were to be placed at a standard distance of 10 parsecs
(32.6 light years) from the sun.
absorption (dark) lines Colors missing in a continuous spectrum because of the absorption of those colors by atoms.
absorption-line spectrum Dark lines superimposed on a continuous spectrum.

acceleration The rate of change of velocity with time.
accretion The colliding and sticking together of small particles to
make larger masses.
accretion disk A disk make by infalling material around a mass;
the conservation of angular momentum results in the disk
shape.
active galactic nuclei (AGNs) The nuclei of galaxies that have a
nonthermal continuous spectrum over a wide range of wavelengths and signs of unusual, energetic activity such as radio
jets.
active galaxies Galaxies characterized by a nonthermal spectrum and a large energy output compared to a normal galaxy.
angular speed The rate of change of angular position of a celestial object viewed in the sky.
anorthosite A basaltic mineral composed of calcium and sodium
with aluminum silicate; the predominant mineral of the lunar
highlands.
antapex The direction in the sky from which the sun appears to
be moving relative to local stars; located in the constellation
Columba.
antimatter Elementary particles with the reversed electric charge
or other property compared to that of ordinary matter.
apex The direction in the sky toward which the sun appears to be
moving relative to local stars; located in the constellation
Hercules.
aphelion For a body orbiting the sun, the point on its orbit that is
farthest from the sun.
Aphrodite Terra A large highland region on Venus.
apogee The point in its orbit at which an earth satellite is farthest
from the earth.
apparent magnitude The brightness of a star (or any other celestial object) as seen from the earth; an astronomical measure
of the object's flux.
absorption (dark) lines Colors missing in a continuous spectrum because of the absorption of those colors by atoms.
absorption-line spectrum Dark lines superimposed on a continuous spectrum.
acceleration The rate of change of velocity with time.
accretion The colliding and sticking together of small particles to
make larger masses.
accretion disk A disk make by infalling material around a mass;
the conservation of angular momentum results in the disk
shape.
active galactic nuclei (AGNs) The nuclei of galaxies that have a
nonthermal continuous spectrum over a wide range of wavelengths and signs of unusual, energetic activity such as radio
(32.6 light years) from the sun.

124

EXPANDED GLOSSARY If you forget the meaning of an astronomical term, look in the *Expanded Glossary* first. The *Expanded Glossary*
contains additional terms that are not explicitly
defined in the text itself.

Brief Contents

Contents

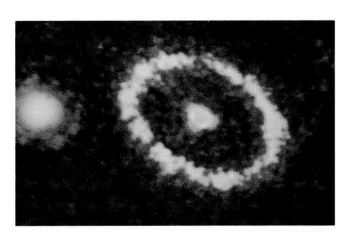

PART ONE

CHANGING CONCEPTIONS OF THE COSMOS

I HAVE OFTEN WONDERED WHAT the first stargazers felt as the heavens reeled above them. What magic did the dance of the planets weave for them? How did they imagine the cosmos and their place in it?

My work as an astronomer hints at answers to these questions. On observing trips to Capilla Peak Observatory in New Mexico, I often work through sunset and dusk to set up my equipment. When all is finally ready, I walk out of the dome to take a curious look at the sky.

The depth of a dark sky is always stunning. To the south on one wonderful night, Scorpius was balanced on its coiled tail. Faint constellations patterned the sky in an unfamiliar quilt of lights. Mars shone like a red beacon in the sea of stars. The sky was still and inspiring. I sensed then how ancient people watched the sky and wondered.

The problem of the design of the cosmos and humanity's place in it has intrigued our species for centuries. The picture of the universe painted by a culture displays its religious and philosophical beliefs. Yet basic astronomical observations set the outlines of the cosmic scheme. The human evolution from fascinated stargazing to fathoming a design of the universe occurs slowly but dramatically.

Part One of this book traces the evolution of conceptions of the cosmos from ancient musings to modern speculations. It starts off with basic astronomical observations, then investigates how astronomy prompted new ideas about the universe. I confine the tale to Western cultures and highlight crucial episodes in the introduction of new astronomical thoughts about the universe in the European tradition. This perspective will give you insight into the development of physical ideas that connect to cosmological ones and into the birth of scientific thought—an event in human history that marks the beginning of the modern era and of today's notions of cosmic evolution. ■

PART OBJECTIVE: To be able to pinpoint how astronomical observations and physical concepts interact to form a scientific model of the cosmos.

LEARNING OBJECTIVES

After studying this chapter, you should be able to:

1-1 Describe the daily motions of the sun, moon, stars, and planets relative to the horizon.

1-2 Describe the seasonal positions of the sun—at sunrise, noon, and sunset—relative to the horizon.

1-3 Describe the motions of the sun and the moon, as seen from the earth, relative to the stars of the zodiac.

1-4 Describe the motions of the planets, as seen from the earth, relative to the stars of the zodiac, with special attention to retrograde motions.

1-5 Tell what astronomical events or cycles set the following time intervals: day, month, and year.

1-6 Describe the astronomical conditions necessary for the occurrence of a total solar eclipse and a total lunar eclipse.

1-7 Describe the phases of the moon in terms of the moon's position in the sky relative to the sun.

1-8 Define the *ecliptic* and tell how to find its approximate position in the sky.

1-9 Argue, from naked-eye observations and simple geometry, an order of the sun, moon, and planets from the earth.

1-10 Identify and describe at least one specific astronomical achievement of a prehistoric culture.

1-11 Make use of angular measure to find positions of celestial objects relative to the horizon and relative to one another.

Central Question
What astronomical objects can you see without a telescope, and how do they change with time?

From Chaos to Cosmos

The stars I know and recognize and even call by name. They are my names, of course; I don't know what others call the stars. Perhaps I should ask the priest. Perhaps the stars are God's to name, not ours to treat like pets. . . .

OLD WOMAN QUOTED BY ROBERT COLES IN

The Old Ones of New Mexico

Do the heavens have an order? Can we make sense of the events in the sky? Pure wonder, which prompted the observations of the early skywatchers, gave way to a desire to find harmony in the seeming chaos of the heavens. This quest drove people to concepts of space and time. The arching heavens, far removed from the earth, displayed cycles of celestial motions that skywatchers read as cosmic clocks. Early astronomers found an order in the heavens, a structure in space, and a sequence in time. A cosmos emerged, shaped by the patterns in the sky.

This chapter deals with naked-eye observations of the sky, like those made by early astronomers—observations you can make without optical equipment. You will be able to sense the regular cycles of motions in the heavens. The chapter does not attempt to explain these motions. (That explanation will come in Chapters 2, 3, and 4.) But long-term, naked-eye observations do establish the periods of celestial cycles with amazing precision. Your recognition of these rhythms will help you to appreciate the first steps taken by ancient astronomers in the development of astronomy and of early concepts of the cosmos.

1.1 The Visible Sky

Have you ever looked carefully and curiously at the night sky far away from a city? If so, at first glance, you may have found no pattern in the stars, and no way to judge their distances, beyond the obvious statement that the stars are far away from the earth.

Constellations

Take the time to study the stars for a while; you'll find that they fall into patterns—designs imposed by your mind (Fig. 1.1). Ancient skywatchers noticed stellar patterns, named them, and passed the names on. Such public patterns are called **constellations.** Early constellations marked groups of stars, with ill-defined boundaries that typically were seen as outlining mythological or realistic figures. Constellations today (88 official ones in all) are established by international agreement of astronomers. (Appendix G contains seasonal star charts of constellations as seen from the midlatitudes of the Northern Hemisphere.)

If you observe the sky nightly, you'll see that the shapes of the constellations don't change. In fact, if you watched them for your whole life, you wouldn't notice any change. The stars appear to hold fixed positions relative to one another.

Angular Measurement

How do you measure how far apart stars appear in the sky? You need a convenient sighting device such as your hand held at arm's length (Fig. 1.2). This simple sextant allows you to measure the **angular separation** or **angular distance** between the stars, that is, the angle between one star and another. Angular measurement is based on counting by 60: a circle is divided into 360 degrees (°), each degree into 60 minutes (arcmin or ') of arc, and each minute into 60 seconds (arcsec or ") of arc. An *arc* is any part of the circumference of a circle.

Angular Measure

At a Glance

One circle contains 360 degrees

One degree equals 60 minutes of arc

One minute equals 60 seconds of arc

Angular diameter is a size-to-distance ratio

Enrichment Focus 1.1

ANGLES IN ASTRONOMY

Angles fashion the basis of simple observational astronomy. If you look carefully at the sky, you observe many angles: between stars, between the moon and stars, between planets and stars, between the sun and the horizon. Some angles are large, such as the circle around the horizon that covers 360° or the 90° from the horizon to a point overhead. But most angles in astronomy are fairly small, such as the 5° between the pointer stars of the Big Dipper or the 0.5° angular diameter of the sun and moon.

Go out on a night with a full moon and hold your hand out at arm's length. Wrap your first three fingers around your thumb, and hold your little finger in front of the moon: you'll see that its width covers the moon! The angular size of your little finger at arm's length is a bit greater than the angular size of the moon. That's how they appear. However, this comparison tells you nothing about the *actual* size of the moon—or the width of your little finger!

Luckily, your finger is close enough to permit you to measure its width; do so with a centimeter rule. Hold your little finger at arm's length again and measure with the ruler the distance from your eye to your little finger, still in centimeters. Divide the width of your finger (its size) by the eye–finger distance; you have then determined the *size-to-distance ratio* of your little finger at arm's length. What value do you get for this ratio?

The size-to-distance ratio is related to the angular size of an object. For small angles (about 10° or less), the size-to-distance ratio is directly proportional to the angular size. Here, *directly proportional* means that the smaller the angular size, the smaller the size-to-distance ratio. Any object with a size-to-distance ratio of 1/57 has an angular size of 1°. What did you find for your little finger at arm's length? You should have found a result between 1/50 and 1/60. So your little finger covers an angle of about 1°, or about twice the angular size of the moon. For convenience, you may use 1/55 as the size-to-distance ratio that corresponds to 1°.

Let's apply this relation. A tennis ball has a diameter (size) of about 6 cm. For a tennis ball to have an angular diameter of 1°, how far away from you must it be placed? Well, since a size-to-distance ratio of 1/55 is required, the ball must be at a distance of 55 × 6 cm = 330 cm. It then covers the same angle as your little finger at arm's length and twice the angle covered by the moon.

You should see that *if* you knew the distance to the moon, you could find out its actual diameter from its angular diameter. How? You know that the moon has an angular diameter of some 0.5°; therefore its size-to-distance ratio is about 1/110. The problem is finding that distance!

We can summarize these concepts in equation form as follows (with $1/55 \approx 0.018$):

$$\text{size} \approx 0.018 \times \text{angular size} \times \text{distance}$$

and if we use s to represent the actual size, A the angular size (in degrees), and D the distance (in units such as meters), then

$$s \approx 0.018\,A\,D$$

The units of s are the same as for D.

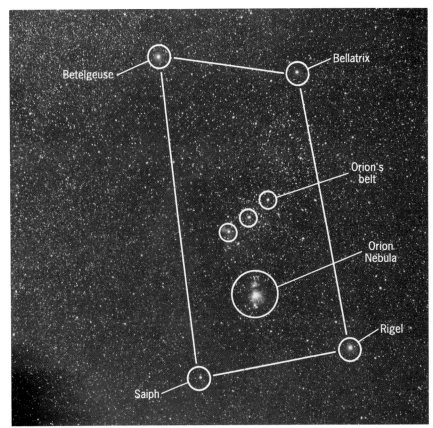

FIGURE 1.1 Time exposure of the stars in the constellation Orion and others nearby. The three bright stars that make a diagonal line pointing to the top right form Orion's belt. Below the center star is the fuzzy patch of the Orion Nebula. The very bright star at bottom right is bluish Rigel; to the upper center is reddish Betelgeuse. An outline is given for Orion, and some of the major features are identified. This photo shows many faint stars invisible to the eye. See the constellation charts for winter in Appendix G to find Orion. (Courtesy of Dennis di Cicco, *Sky and Telescope* magazine.)

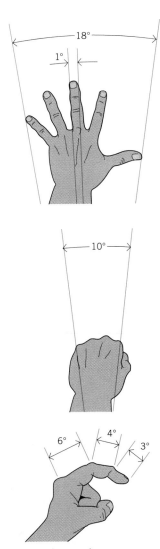

FIGURE 1.2 Angular measurements made with a hand extended at arm's length. The angles are approximate and typical for an average adult. You will need to calibrate your own hand to find its dimensions.

At arm's length, your fist covers about 10° of sky; each fingertip masks roughly 1°. So you can use your hand to measure the angular size of the sun and moon (both about 0.5°, or half a fingertip) and the separation of the pointer stars in the Big Dipper (about 5°, or half a fist: Fig. 1.3).

The sun and the moon each cover a certain angular extent of the sky. The angular measurement of that portion is called the **angular diameter.** The sun and moon appear as disks, and their angular diameter is the angular distance between opposite edges. Angular diameter depends on the distance to the object involved (which you may not know) and its actual diameter (Focus 1.1). For example, imagine that a ball is placed 1 meter (abbreviated m; see Appendix A on units) away from you, and you mea-

sure its angular diameter (Fig. 1.4). Now suppose that the ball is moved three times farther away, to a distance of 3 meters. Then its angular diameter is only one-third of the original measurement. So an angular diameter is a measure of the *size-to-distance ratio* of an object.

The concepts of *angular distance* and angular diameter are related. Suppose you are looking at two objects some distance apart. The farther they are away from you, the smaller the angle be-

FIGURE 1.3 Measuring angles on the sky with a hand at arm's length. The angle between the two pointer stars in the Big Dipper is about 5°, or half a fist at arm's length.

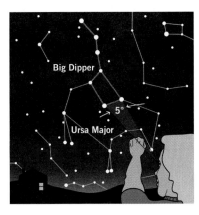

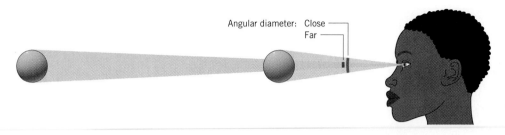

Angular diameter: Close
Far

FIGURE 1.4 The physical relationship between angular diameter and distance. The closer an object is, the larger the angle it covers in the field of view. The farther away, the smaller the angle.

tween them will appear. When you look at the sky, you see the angular distances between celestial objects. That's all you know—angles—unless you can measure physical distances.

Yet, angles are crucial in astronomy. They do provide information about positions of objects in the sky. They also tell a bit more. For example, the angular size of the moon gives us a good idea of the size of the moon compared to its distance (Focus 1.1). So if you know one, you can find out the other from just a measurement of the moon's angular diameter.

Motions of the Stars

Stay out one night and watch the stars from dusk to dawn. They move relative to your horizon—rising in the east, slowly traveling in arcs against the sky, and setting in the west (Fig. 1.5). (The **horizon** is an imaginary line at which the earth and sky would meet if no terrain were in the way. The horizon you'd view at sea comes closest to this ideal.) If you live in the continental United States and face south in December, you'll see Orion east of south early in the evening (Fig. 1.6a). After a few hours, Orion will appear due south at about midnight (Fig.

1.6b). (If you face south, west is to your right, east to your left.)

If you acquire a regular stargazing habit, you'll see that the visible constellations change with the seasons. In winter, consider Orion again in the south at midnight. (Refer to the winter star map in Appendix G.) Look south on the following nights at the *same* time. Orion moves slowly to the west *toward* the sun (Fig. 1.7a). In summer, you can't see Orion at

FIGURE 1.5 The motion of the stars relative to the horizon. The view is to the west. This 2-hour time exposure shows the east-to-west motion of the stars as they set. The most prominent stars here are those of Orion. The streak of a meteor (which spans 10°) starts just below the stars of the belt and cuts across the sword and the Orion Nebula. (Courtesy of Akira Fujii.)

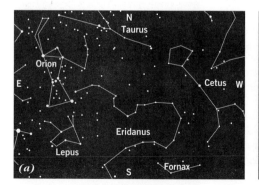

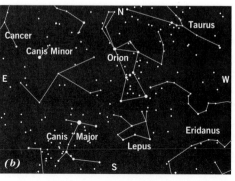

FIGURE 1.6 Motions of the stars on the same night. (*a*) Facing south at a midnorthern latitude on December 21 at about 10 P.M. local time. Note that Orion is still east of due south. (*b*) The same view about 2 hours later, at local midnight. Note that Orion and all the other stars have moved to the west relative to the horizon. The total view is 83° by 57°. (Diagrams generated by *Voyager* software for the Macintosh, Carina Software.)

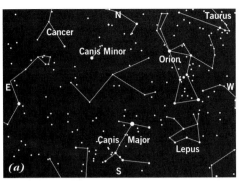

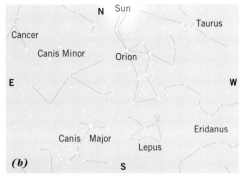

Motions of Stars	
At a Glance	
Daily	East to west, with respect to the horizon
Yearly	East to west, with respect to the sun
Long-term	No visible motions relative to each other

FIGURE 1.7 Seasonal motions of stars. (*a*) View as in Figure 1.6*a* at about local midnight two weeks later. (*b*)View of the daytime sky from the same location on June 21 at local noon. Note a part of the sun's disk at the top, due south. These are the stars you would see if the sun's glare and the blue sky did not obscure them. Compare to Figure 6*b*. (Diagrams generated by *Voyager* software for the Macintosh, Carina Software.)

all because it's next to the sun and is up during the day (Fig. 1.7*b*). (See the summer star map in Appendix G.) In winter, a year later, Orion again lies south at night. All constellations take one year to return to their initial places in the sky relative to the sun.

Caution. Don't confuse this gradual *yearly* change relative to the sun with the much faster *daily* motion relative to your horizon (from east to west)!

If you live in the Northern Hemisphere and often look north, you'll find that some stars never dive below your horizon but trace complete circles above it (Fig. 1.8). These are the **circumpolar stars.** The center of these circles marks the **celestial pole,** the point about which the stars seem to pivot. A modestly bright star called *Polaris* lies close to the north celestial pole. Polaris is now the *north pole star* (Fig. 1.8). (No bright star now falls close to the south celestial pole, so we don't have an obvious south pole star.)

■ What does the concept of angular distance relate to, and how is it used in astronomy?

1.2 The Motions of the Sun

The sun's daily motion—rising along the eastern horizon, tracing an arched path in the sky, and setting along the western horizon—defines the most basic time

FIGURE 1.8 The motion of the stars around the north celestial pole. This long-time exposure shows the circumpolar motions above Star Hill Inn, New Mexico. As the stars move around the north celestial pole, they trace out arcs, which are progressively larger the farther a star lies from the pole. Polaris, the brightest star near the center of the stellar circles, has traced a small arc, which indicates that it is not exactly at the north celestial pole. The trails among the buildings at the bottom were made by flashlights carried by astronomers. (Courtesy of Dennis di Cicco, *Sky and Telescope* magazine.)

Motions of the Sun

At a Glance

Daily	East to west with respect to the horizon
Annually	West to east through the zodiac
Seasonally	Higher (summer) and lower (winter) with respect to the horizon at noon

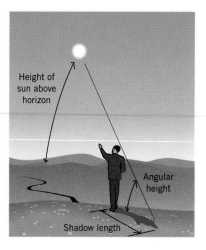

FIGURE 1.9 Shadows and the position of the sun in the sky. The higher the sun relative to the horizon, the shorter the shadow. The shortest shadow of the day occurs at noon. The sun is then due south. The angle between the horizon and the sun is the angular height of the sun, which has its maximum value for the day at noon.

cycle: day and night. Midway between sunrise and sunset, the sun reaches its highest point relative to the horizon, which defines **noon.** The interval from one noon to the next sets the length of the **solar day.**

Motions Relative to the Horizon

Consider your shadow cast on a flat place on the ground. You can use the length of your shadow to study the sun's daily and seasonal motions in the sky relative to the horizon. The tip of the shadow marks the end of an imaginary line that connects the tip itself, the top of your head, and the sun (Fig. 1.9). The shadow points in the direction opposite from the sun's location in the sky. The length of the shadow tells the height of the sun relative to the horizon. When the sun hangs low in the sky, the shadow is longer. At noon the shadow reaches its shortest length for that day. Also at solar noon, in the northern latitudes, the shadow points due north—so the sun lies due south.

Observe your shadow throughout a year. You'll find that the height of the sun in the sky at noon varies with the season. During the summer, the shadow is its shortest at noon on the **summer solstice** (around June 21), the day with the greatest number of daylight hours. At the summer solstice, the noon sun hits its highest point in the sky for the year (Fig. 1.10*a*). In winter at noon, the shadow stretches its longest on the **winter solstice** (around December 21), the day

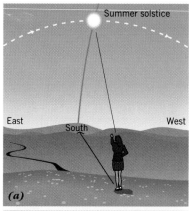

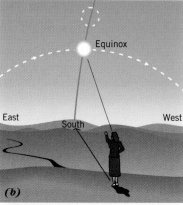

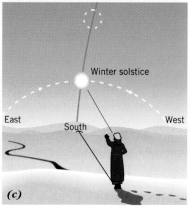

FIGURE 1.10 Seasonal position of the sun at noon, when it hits its highest point in the sky, relative to the horizon, for the day. The angular height varies with the seasons. Measuring with a fist, an observer can find that for a mid-northern latitude, the sun is highest at the summer solstice (*a*), lower at the fall equinox (*b*), and lowest at the winter solstice (*c*).

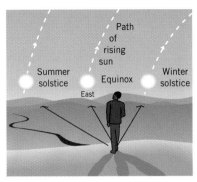

FIGURE 1.11 The seasonal positions of the sunrise points along the horizon. The sunrise positions from the summer solstice to the winter solstice for an observer at midnorthern latitudes. The view would be symmetrical facing west at sunset.

with the fewest daylight hours. The noon sun has dropped to its lowest point in the sky for the year (Fig. 1.10c). On the first day of spring and fall, you cast a shadow with a length between its summer minimum and winter maximum—these days

(a)

(b)

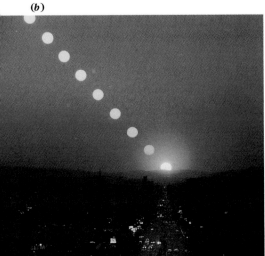

FIGURE 1.12 Changing positions of sunset points near the fall equinox. (a) Sunset sequence on 20 September over a street in Salt Lake City. Exposures were made at 5-minute intervals. (b) Similar sequence of exposures made on 22 September, the day of the fall equinox. Note how far south the sunset point has moved in two days. (Courtesy of Keith Finlayson.)

are the **spring** and **fall equinoxes:** around March 21 for spring (vernal) and September 21 for fall (autumnal) (Fig. 1.10b). The cycle registered by your shadow defines the second basic unit of time: the year of seasons. (*Note:* The dates given here are for seasons in the Northern Hemisphere; the seasons are reversed in the Southern Hemisphere.)

The seasonal change in the sun's noon position is related to a cyclical change in the sunrise and sunset points. Consider sunrise. At the summer solstice, for midnorthern latitudes, the sun rises at a point the farthest north of east for the year (Fig. 1.11). For about four days, the sun seems to rise at the same place on the horizon—it appears to stand still. After the summer solstice, the sunrise position moves southward. At the winter solstice, the rising point reaches as far south of east as it will get for the year. The sun again stands still for a few days, and then moves northward. Between the solstice points lies the point for sunrise of the equinoxes. Only at the equinoxes does the sun rise due east and set due west. At the equinoxes the sunrise and sunset points show the greatest angular speeds for the year. The span between two days is dramatic (Fig. 1.12). A symmetrical pattern occurs for the sunset points around due west.

Motions Relative to the Stars

The sun also moves with respect to the stars—a motion somewhat hard to observe, for you can't see the stars during the day. Try this: pick out a bright constellation, such as Gemini, visible just above your western horizon right after sunset (Fig. 1.13a). (We are using the horizon here to fix the sun's position and to block out the sun's direct light.) Look again at the *same time* about two weeks

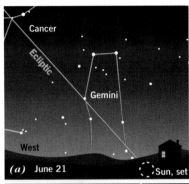

(a) June 21 · · Sun, set

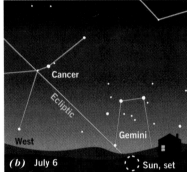

(b) July 6 · Sun, set

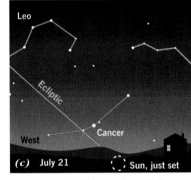

(c) July 21 · Sun, just set

FIGURE 1.13 The motion of the sun relative to the stars. Find a constellation (such as Gemini, shown here) near the western horizon at sunset (a). About a week later, look for the same stars just after sunset (b). They will have moved down the western horizon. Eventually, they will set before the sun goes down (c). So the stars seem to move to the west with respect to the sun. Or you can think of the sun as moving east with respect to the stars. (Diagrams generated by *Voyager* software for the Macintosh, Carina Software.)

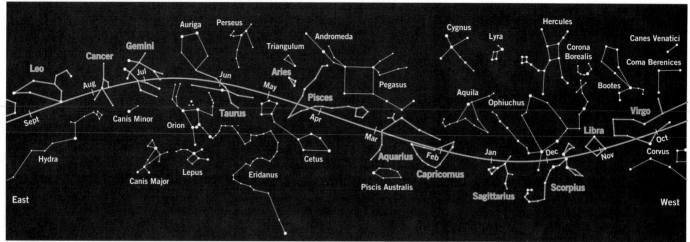

FIGURE 1.14 The ecliptic and zodiacal stars. If there were no atmosphere, you could see the sun's changing position with respect to background stars. The sun's path is called the ecliptic. This diagram indicates the sun's position by date along the ecliptic and shows the major constellations. The zodiacal constellations are indicated. Note that the sun moves west (right) to east (left) among the stars and returns to the same position on the ecliptic in one year.

later. Because the sun has moved *east* relative to the stars, the constellation will appear to have shifted to the west (Fig. 1.13*b*). And about two weeks later, again at the same time, the sun will have moved out of Gemini and into Cancer (Fig. 1.13*c*). Relative to the stars, the sun appears to move to the east. In one year the sun returns to the same position relative to the stars, so it circuits 360° in a year, or about 1° a day.

The speed with which an object covers a certain angular distance is called its **angular speed.** Relative to the stars, the sun moves west to east at an angular speed of roughly 1° per day. (In contrast, the sun moves 360° per day from east to west relative to the horizon.) Imagine that you recorded the sun's po-

sition among the stars for a year. If you drew an imaginary line through these points, you'd trace out a complete circle around the sky: this is called the **ecliptic** (Fig. 1.14). The traditional 12 constellations through which the sun moves define the **zodiac.** (Note that the sun also travels through a part of the constellation Ophiuchus, which is *not* a zodiacal constellation.)

Caution. Don't confuse the eastward motion of the sun with respect to the stars in a year with the much faster westward motion of the sun relative to the horizon in a day!

The sun's position along the ecliptic is labeled with reference to the constellations of the zodiac (Table 1.1). For

TABLE 1.1 THE ZODIACAL CONSTELLATIONS

CONSTELLATION	TRADITIONAL PATTERN	APPROXIMATE DATE FOR SUN'S LOCATION WITHIN CONSTELLATION
Aries	Ram	April 30
Taurus	Bull	May 30
Gemini	Twins	July 5
Cancer	Crab	July 30
Leo	Lion	September 1
Virgo	Virgin	October 11
Libra	Balance	November 9
Scorpius	Scorpion	December 3
Sagittarius	Archer	January 7
Capricornus	Goat	February 8
Aquarius	Water Bearer	February 25
Pisces	Fishes	March 27

instance, to say that the sun is "in Taurus" specifies a place along the ecliptic. The zodiac probably arose from a desire to mark the sun's position with respect to the stars. (You can't see the sun in a constellation during the day, but you can get a rough idea near sunrise and sunset.)

The sun's location in the zodiac also roughly indicates the time of year. Twice yearly, in spring and fall, the equinoxes occur. The summer and winter solstices mark two other key times. Nowadays the sun lies in Aries in the spring, Gemini in the summer, Virgo in the fall, and Sagittarius in the winter (Fig. 1.14). These seasonal locations change slowly with time; for the Babylonians 5000 years ago, the sun was in Taurus in the spring.

Precession of the Equinoxes

In 3000 B.C. the sun appeared in Taurus at the vernal equinox, the first day of spring (Fig.1.15a). Today you see the sun in Pisces at the start of spring. In the passage of 5000 years, the position of the vernal equinox has moved to the west out of Taurus, through Aries (Fig. 1.15b), and into Pisces (Fig. 1.15c). That is, in 5000 years the position of the vernal equinox has moved through two constellations. So to circuit the whole zodiac will take six times as long, or about 6 × 5000 = 30,000 years. (A more precise calculation gives 25,780 years.) This slow westward drift of the equinoxes with respect to the stars is called the **precession of the equinoxes.** (Chapter 4 will provide an explanation of this motion.)

The most dramatic effect of the precession of the equinoxes is to change the zodiacal location of the sun at the equinoxes and solstices. Precession has another less obvious effect: the celestial poles move in the sky, with the result that the north pole star changes (Fig. 1.16). The north celestial pole now lies near the star Polaris. About 5000 years ago, the north celestial pole was near the star Thuban (in Draco). About 12,000 years from now, precession will have carried the north celestial pole near to the bright star Vega in Lyra.

Precession is hard to observe without a telescope, because it takes place so slowly. But if a culture kept astronomical records for a few centuries, its astrono-

mers could notice the shift in the equinoxes and solstices with respect to rising stars.

Note. It is standard practice to say that the sun is "in" a constellation. But the stars that form the constellation are far from the sun and are not necessarily close together in space; they appear close together to us because they're located roughly in the same direction along our line of sight. The statement "the sun is in Pisces" means that the sun is in the same direction in the sky as the pattern of stars we call Pisces.

■ The sun takes one year to circle the zodiac. In which direction does it travel relative to the stars?

1.3 The Motions of the Moon

If you carefully watch the moon for a few hours on a clear night, you can spot two of its celestial motions. First, like the sun and the stars, the moon rises in the east and sets in the west. Second, the moon also journeys eastward against the backdrop of the zodiacal stars. Here's how to observe this eastward motion (Fig. 1.17). Wait until the moon appears close to a bright planet or star (Fig 1.17a). On the same night, observe the moon and planet again an hour later (Fig. 1.17b). Repeat the observation after another hour has passed (Fig. 1.17c). The moon

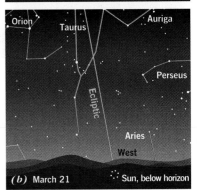

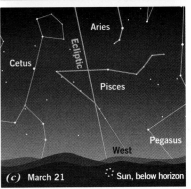

(a) March 21 — Sun, below horizon
(b) March 21 — Sun, below horizon
(c) March 21 — Sun, below horizon

FIGURE 1.15 Precession and the change of position of the vernal equinox with respect to the zodiacal stars. (a) Sunset scene (similar to Fig. 1.13) in 3000 B.C. on March 21. Note that the sun lies in Taurus. (b) Sunset view for 500 B.C. on the same date. The sun now appears in Aries. (c) The sunset scene in A.D. 2000. Then the sun will lie in Pisces. Note that the equinox point moves to the *west* along the ecliptic. (Diagrams generated by *Voyager* software for the Macintosh, Carina Software.)

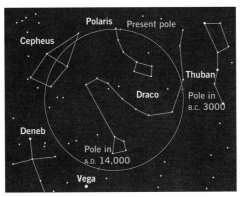

FIGURE 1.16 Precession and the change of position of the north celestial pole with respect to the stars. The pole's motion completes a circle counterclockwise in the sky in about 26,000 years. About 12,000 years from now, the pole will be near the bright star Vega.

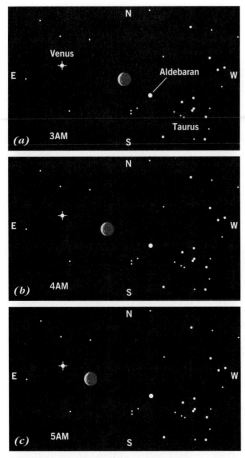

FIGURE 1.17 The motion of the crescent moon relative to Venus and the bright star Aldebaran (in Taurus). Note that east is to the left. The field of view is 9° by 7°; the moon covers an angle of about 0.5°. (*a*) The scene at 3 A.M. (*b*) An hour later. (*c*) And another hour later. (Diagrams generated by *Voyager* software for the Macintosh, Carina Software.)

will have moved to the *east* with respect to the planet (both will have moved westward in the sky with respect to the horizon). If you measure the moon's rate of motion, you'll find an angular speed of about 0.5° per hour (Fig. 1.18). At this rate, the moon circuits the zodiac in about 27 days. (Although the moon's path does not fall right on the ecliptic, it lies close to it; so the moon stays within the zodiac.) This eastward motion of the moon relative to the stars causes it to rise *later* every day (by about 50 min).

Watch the moon for a few nights; you'll note that the amount of its surface that is illuminated—its **phase**—follows

FIGURE 1.18 The motion of the moon relative to Venus as both set over Tulsa, Oklahoma, in 1988. The beginning of this sequence starts at the top left of the photo. Study the position of the moon with respect to Venus from the image at the top down to the one at the lower right. Note how the moon moves to the east (upward and to the left) with respect to Venus; so it lies close to the ecliptic. (Courtesy of Bill Sterne, Jr., Sterne Photography.)

a regular sequence. When the moon rises at sunset, and its face is completely illuminated—it's a *full moon*. About 14.5 days later, the moon is *new* and is not visible in the sky. A few days later, the moon reappears in the west at sunset, partially illuminated—a *crescent moon*. About two weeks later, the moon rises in the east as a *full moon* at just about the time the sun sets. A complete cycle of phases—say, from one full moon to the next—takes about 29.5 days. Thus we have a third basic unit of time: the month of phases.

The different phases of the moon are related to specific alignments of the sun and the moon in the sky. At new moon, the angular distance of the sun and moon is small—less than a few degrees. At first quarter, the moon lies 90° east of the sun (Fig. 1.19*a*). So if you were to point to the setting sun with one arm and to the moon with the other, the angle between your arms would be 90°. At full, the moon is 180° from the sun (Fig. 1.19*b*); at last quarter, it is 90° west of the sun (Fig. 1.19*c*). First and last quarters refer to the *position of the moon in the sky*—one-quarter of a full circle away from the sun—*not* to the amount of illumination; the moon at quarter phase looks half full.

■ What was the impact of the moon's phases on the calendar?

1.4 The Motions of the Planets

If you observe the sky often, you can quickly pick out heavenly bodies that don't belong to the constellations. Five such objects wander in regular ways through the stars of the zodiac: the **planets** Mercury, Venus, Mars, Jupiter, and Saturn. (Uranus, Neptune, and Pluto cannot be seen without a telescope.) Viewed by naked eye, the planets look pretty much like stars, though they twinkle less than stars do. At times, some planets are brighter than the brightest stars; all planets vary in brightness. It is their motions, however, that really separate the planets from the stars.

Retrograde Motion

The planets display a peculiar motion that sets them apart from all other objects in the sky. For example, suppose you observe Mars every night for a few months near the time when Mars appears brightest in the sky. (This occurs about every two years.) At first, Mars moves slowly eastward (with respect to the stars) through the zodiac close to the ecliptic (Fig. 1.20). In this regard, it moves like the sun and the moon. But later, Mars falters in its eastward motion

Motions of the Moon

At a Glance

Daily	East to west with respect to the horizon
Monthly	Eastward with respect to the stars
Rate	Covers 360° in about 27 days around zodiac

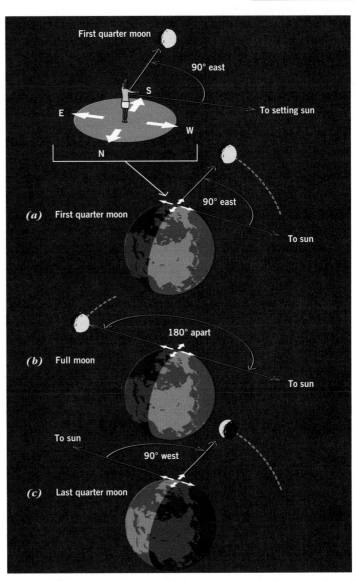

FIGURE 1.19 The orientation of the sun and moon in the sky as viewed by an observer on the earth for different phases of the moon. At first quarter (*a*) the moon is 90° east of the sun (due south as the sun sets). At full (*b*) it is 180° away (in opposition, so the full moon rises as the sun sets). At last quarter (*c*) the moon is 90° west of the sun (due south as the sun rises in the east). The view here is facing south from a mid-northern latitude.

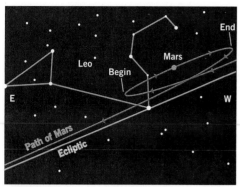

FIGURE 1.20 The retrograde motion of Mars, relative to the stars, during its 1995 opposition. The line across the center marks the ecliptic; shown is the outline of the zodiacal constellation Leo. "Begin" and "End" mark the points on Mars' path in the sky when the retrograde motion begins (January 1995), and ends (March). Note that during its retrograde motion Mars moves from *east to west;* otherwise it moves from west to east. (Diagram generated by *Voyager* software for the Macintosh, Carina Software.)

with respect to the stars and stops for a short time.

Then for about three months, Mars moves *westward*—which is the opposite of its normal direction relative to the stars. After that, its westward motion slows down and stops. Finally, Mars resumes its normal eastward course. The planet's backward motion to the *west* is called **retrograde motion.** In the middle of its retrograde motion, Mars shines its brightest.

All planets loop along or near the ecliptic with periodic retrograde motion—but generally they are not all in retrograde motion at the same time or for the same duration. For instance, Mars takes about 83 days to undergo its complete retrograde cycle; Saturn takes 139 days (Table 1.2). The sun and the moon *never* move retrograde, and, in general, the planets travel eastward along or near the ecliptic.

Elongations, Conjunctions, and Oppositions

The alignment of the sun and a planet in the sky at the time of retrograde motion divides the visible planets into two groups. Mercury and Venus make up one; they never stray very far in angular distance along the ecliptic from the sun. Because they stick close to the sun, they are not visible except as morning and evening "stars" (Fig. 1.21). That means that Mercury and Venus gleam over the western horizon after sunset and pop above the eastern horizon before sunrise. (But keep in mind that they are *not* stars.)

You can use your fist to measure the maximum angular separation of Mercury or Venus from the sun. For Mercury, the average maximum separation is 23° (about $2\frac{1}{2}$ fists); for Venus, about 46° (about $4\frac{1}{2}$ fists). When either planet is at its greatest angular separation from the sun, it has reached **maximum elongation** (Fig. 1.22). (Mercury's maximum elongation varies substantially; it can be as large as 28°. The figure given here is its *average* value.)

Mercury and Venus begin their retrograde motions after they hit maximum elongation east of the sun as evening stars. They then move westward, pass

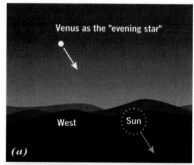

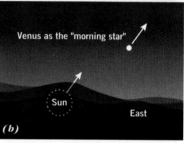

FIGURE 1.21 Retrograde motion of Venus. This diagram shows how Venus moves east to west with respect to sun and ecliptic during retrograde motion. Venus goes through retrograde motion when it switches from being an evening "star" (*a*) to a morning "star" (*b*): that is, from being on the east side of the sun to being on the west side. So it moves retrograde after its greatest eastern elongation.

TABLE 1.2 MOTIONS OF THE VISIBLE PLANETS			
PLANET	**TYPICAL DURATION (days)**	**TIME BETWEEN OCCURRENCES (days)**	**PERIOD AROUND ECLIPTIC, Eastward (years)**
Mercury	34	116	≈ 1
Venus	43	584	≈ 1
Mars	83	780	2
Jupiter	118	399	12
Saturn	139	378	30

NOTE: The duration of retrograde varies somewhat from the values above from retrograde to retrograde. This last column gives time interval for the planet to circuit the ecliptic once.

the sun, and reappear as morning stars west of the sun. When Mercury or Venus lies close to the sun in the sky, the planet and the sun are aligned in **conjunction.** (Whenever two celestial objects appear to come close together in the sky, they are said to be in conjunction.)

The second group of planets visible to the naked eye comprises Mars, Jupiter, and Saturn. In contrast to Mercury and Venus, these planets move freely along the ecliptic with respect to the sun. And they retrograde when they stand in **opposition** to the sun: opposite the sun in the sky. At opposition, the sun and the planet are separated by 180° (similar to when the moon is full). Then the planet rises as the sun sets. If you were to point one arm to Mars at opposition and the other at the sun, your arms would be 180° apart (they'd make a straight line). A planet at opposition crosses the middle of its retrograde loop and shines its brightest since it is closest to the earth.

In summary, Mercury and Venus can *never* be in opposition to the sun; they retrograde after passing their greatest eastern elongation and continue until reaching greatest western elongation. Mars, Jupiter, and Saturn can be in opposition or conjunction, but they retrograde *only* at opposition.

Relative Distances of the Planets

The time each planet spends in its retrograde motion and the time each one takes to circle the ecliptic provide clues to the relative distances of the planets from the earth (Table 1.2). Assume for a

moment that the planets move at the same speed. The one that appears to move at a faster angular rate is closer than another that moves at a slower angular rate. So the slowest-moving planet is the most distant from the earth; the swiftest is the nearest.

Here's an analogy (Fig. 1.23). Suppose you are watching the lights of two airplanes at night and want to estimate the planes' relative distances from you. Assume that both planes are flying at the same speed. The one that appears to move faster—at a greater angular speed—must be the closer of the two. Apply the same argument to the moon, sun, and planets: the fastest (the moon) is closest to the earth; the slowest (Saturn) is the farthest. (The assumption that the speeds are the same turns out not to be correct, but the basic conclusion is right because the differences in speeds are not very great.) The motions appear centered on the earth—**geocentric.**

But what about retrograde motion of the planets? The evolution of our ideas about the motions of the planets—including retrograde—marks a central question of the next three chapters: How do the planets move?

■ **When does Saturn exhibit its retrograde motion?**

1.5 Eclipses of the Sun and Moon

An eclipse of the sun (a **solar eclipse**) occurs when the moon passes in front of the sun (Fig. 1.24). Although the moon is smaller in diameter than the sun, it's closer to the earth by an amount that makes the angular diameter of the sun and moon almost the same—about 0.5°. So the moon may just cover the sun's disk when it passes directly between the sun and the earth, as it may do at new moon.

Why don't eclipses happen each and every month? Mainly because the moon's path in the sky relative to the stars does *not* usually coincide exactly with the ecliptic; it is tilted at an angle of about 5°. The ecliptic and the moon's path cross at two points. Only at or near these points will the sun and the moon

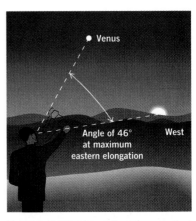

FIGURE 1.22 Measuring the greatest eastern elongation of Venus at sunset. At this time the angle between Venus and the sun is about 46°. The same observation for Mercury gives a maximum angle of about 23°.

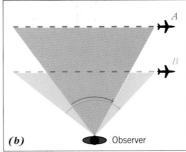

FIGURE 1.23 Judging relative distances from angular speeds. (*a*) The scene is at night, so all you can see are the lights from the planes. In the same time, plane *A* covers a smaller angle than plane *B*. (*b*) So plane *B* is the closer, if both planes fly at the same speed.

Motions of the Planets

At a Glance

Daily	East to west with respect to the horizon.
General	Eastward along the ecliptic
Occasional	Westward in retrograde loops

Total Solar Eclipses

At a Glance

DATE	DURATION (min)	LOCATIONS
Nov 3 1994	4.6	Central South America
Oct 24 1995	2.4	Southern Asia
Mar 9 1997	2.8	Siberia, Arctic
Feb 26 1998	4.4	Central America
Aug 11 1999	2.6	Europe, Asia

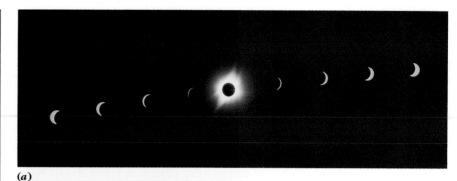

(a)

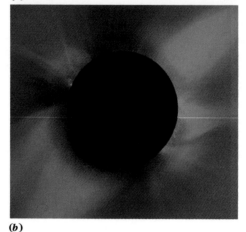

(b)

FIGURE 1.24 A total eclipse of the sun on July 11, 1991. As viewed from the earth, the moon covers the sun's visible disk, so the sun's outer atmosphere (its corona) is visible. (*a*) The progress into and out of the eclipse in 8-minute intervals. (Courtesy of Bill Sterne, Jr., Sterne Photography.) (*b*) View of the eclipse at totality. Note the structure in the corona. (Courtesy of NSO/Sacramento Peak.)

be so close to overlap that an eclipse occurs. If the moon is more than 0.5° above or below the sun, it will pass by without blocking out the sun's disk. No eclipse will occur, not even a partial one, when the moon cuts out only a part of the sun.

When the sun and moon are exactly lined up, the new moon completely covers the sun, and we have a *total solar eclipse* (Fig. 1.25). During a total solar eclipse, the moon's shadow is about 300 kilometers (abbreviated *km:* see Appendix A on units) wide. Only people in this narrow band on the earth see a total eclipse as the shadow sweeps by. Those just outside the central band see a partial solar eclipse, in which the moon does not completely cover the sun. (Because the distance of the moon from the earth varies, its angular size during an eclipse can be smaller than the sun's. In this case, we have an *annular* eclipse.)

An eclipse of the moon (a **lunar eclipse**) occurs when the moon passes directly through the shadow cast by the earth (Fig. 1.26). Then the sun's illumination is cut off from the moon (Fig. 1.27). A total eclipse of the moon takes

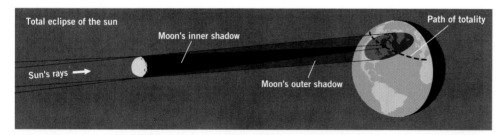

FIGURE 1.25 Alignment of moon, sun, and earth for a total solar eclipse. The moon must be new and on or very close to the ecliptic. The moon's central shadow is usually long enough to hit the earth; a total eclipse is seen in the path made as the central shadow moves along the earth.

FIGURE 1.26 Alignment of sun, moon, and earth for a total lunar eclipse. The moon must be full and on or close to the ecliptic and pass through the earth's inner shadow. The total lunar eclipse is visible to everyone on the night side of the earth.

place only when the moon is *full*—when the earth lies directly between the sun and the moon. Also, the moon must be close to the ecliptic; otherwise it will miss the earth's shadow.

Lunar eclipses have a large potential audience: all people on the night side of the earth (because the entire moon is in the earth's shadow). During your lifetime, you can expect to see roughly 50 lunar eclipses, about half of them total. In contrast, since solar eclipses are visible only along a narrow band of the earth, a total eclipse of the sun occurs rarely for any one location on the earth.

Eclipses take place less frequently than other celestial events described here. Their spectacular nature motivated people to study them, however, and we have reliable records dating as far back as 750 B.C. The Greeks, for example, found that eclipses were predictable from the motions of the sun and the moon. They also noted that solar eclipses prove that the moon is closer than the sun because the moon blots out the sun, but the reverse never happens. In addition, they decided that the sun must be larger than the moon (which it is!).

Now you can see how the ecliptic got its name: only when the moon lies on or close to the *ecliptic* can eclipses occur.

■ **Why don't we have a solar eclipse every month?**

1.6 Prehistoric Astronomy

We have few records of the astronomy of ancient peoples. Astronomical knowledge was probably passed on by word of mouth in symbolic stories, and records of astronomy appear in such oral traditions. The Polynesian islanders, for example, knew how to navigate from Tahiti to Hawaii by starwatching. In the pueblos of the Southwest, skywatchers today still keep their seasonal watch. Ancient peoples likely had much more astronomical knowledge than we usually concede, and some of this learning was incorporated in their art and architecture. I will treat just one here: the Pueblo Indians of the U.S. Southwest.

FIGURE 1.27 A total lunar eclipse. The full moon lies almost in the center of the earth's shadow in this time exposure taken on December 9, 1992. The reddish color comes from dust in the earth's atmosphere, which scatters blue light but allows red light to pass. (Courtesy of Dennis di Cicco, *Sky and Telescope* magazine.)

Sunwatching of the Southwestern Pueblos

After their conquest of Mexico, the Spanish turned north in the sixteenth century on a fruitless, obsessive search for the fantastic Seven Cities of Cibola, said to be made of gold. They did not find golden cities. But they did encounter the adobe villages (Fig. 1.28), which they called *pueblos,* of the native peoples who had lived in them at least a thousand years before the arrival of the Europeans. Many pueblos disappeared in historic times (from A.D. 1540 onward); those that survived are the cultural connection

Total Lunar Eclipses

At a Glance

DATE	LOCATIONS
Apr 4 1994	Eastern North America, South America, Europe
Sep 27 1996	North and South America, Europe, Africa
Sep 16 1997	Europe, Africa, Asia, Australia
Jan 21 2000	North and South America, Europe, Africa

to the people called the *Anasazi*, who occupied a vast area in the Southwest. The modern Pueblos provide insights into the past because they are descendants of the Anasazi (although we don't know from which specific Anasazi villages).

Sunwatching plays a central role in the agricultural and ceremonial life of the pueblos. The seasonal cycle of the sun sets the calendar for rituals and for specific crop plantings—essential activities for survival in the high desert environment. A religious officer, called the *sun priest,* watches daily from a special spot within the pueblo or not far outside it. The sun priest carefully observes sunrise (or sunset) relative to features at the horizon. He knows from experience what horizon points mark the summer and winter solstice and the times to plant crops (Fig. 1.29). These he announces within the pueblo—ahead of time, so that ritual preparations can be carried out.

Along with horizon calendars, some Pueblo sunwatchers make use of buildings so located that sunlight passing through windows in certain rooms hits special markings on the walls at vital times of the year. As the sun moves along the horizon, sunlight falls on the wall in different positions. So light and shadow casting, along with horizon features,

FIGURE 1.28 1896: Preparation for ceremonial activity as part of the Zuni celebration of Shalako, which is scheduled by observation of the phases of the moon. The Zuni clowns, called Mudheads, gather in a plaza at the pueblo. (Photo by Ben Wittick, from the collection of the Museum of New Mexico, Negative No. 16446.)

form the basis of the sunwatching for calendrical purposes—practices still carried out today.

One critical aspect of historic Pueblo astronomy is that the sun priest must *anticipate* ceremonials. The Pueblo culture requires preparation time for a ceremony to permit participants to achieve the proper frame of mind. Anticipation entails an astronomical problem: forecasting of the solstices. Because the sun does not move noticeably relative to

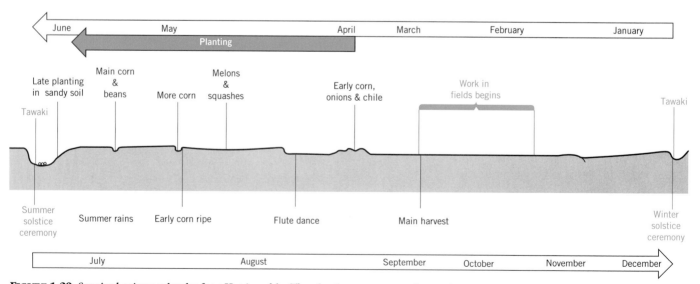

FIGURE 1.29 Sunrise horizon calendar for a Hopi pueblo. The planting season runs from winter to summer solstice (right to left along top). Ceremonial and harvest events, such as the August Flute Dance, are scheduled between the summer and winter solstices (left to right across bottom). The Hopi word "tawaki" indicates the sun's position on the horizon at the solstices.

the horizon around June 21 and December 21, unaided naked-eye observation cannot tell which day is, in fact, the solstice.

One way to forecast the solstice is to observe the sunrise point while it is still moving perceptibly along the horizon—at least a week before the solstice. Watching for two weeks (or more) before is even more reliable. The sun priest can then do a day tally to forecast the day on which the solstice will occur. Of course, this requires a few years of sunwatching to establish the countdown intervals for specific horizon markers. But once established, the horizon calendar remains fixed, and the knowledge can be handed down to others. (Precession does *not* change the seasonal sunrise and sunset points.)

Anasazi Sunwatching: Ancient Pueblo Calendars

Around A.D. 1000, the Anasazi prospered and built community houses of up to five stories, containing hundreds of rooms. An Anasazi building at Hovenweep National Monument may have housed a solar calendar. Now bridging Colorado and Utah, the Hovenweep ruins date from about A.D. 1100. The dwellings contain many towerlike structures located on rocky outcrops (Fig. 1.30a). Most of the buildings have small portals built into them. The portals, typically 10 centimeters (abbreviated cm) in diameter, are made with such attention to their view that many are angled in the walls. In three ruins there are rectangular rooms that are late additions to each structure. Astronomer Ray Williamson has proposed that some portals in these add-on rooms were used for solar observations, with alignments for the solstices and planting and harvesting times. From an astronomical view, the portals define narrow angular views on the horizon and permit narrow shafts of light to enter the rooms.

In the structure called Hovenweep Castle, the possible observing room lies at the south end of a D-shaped tower. Two portals are in the west side; one opens to the northwest for the summer solstice sunset, the other to the southwest for the winter solstice sunset (Fig.

(a)

Hovenweep Castle

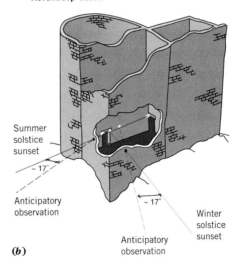

Summer solstice sunset

~ 17°

Anticipatory observation

~ 17°

Winter solstice sunset

Anticipatory observation

(b)

FIGURE 1.30 Ruins of Hovenweep Castle viewed from the west. (a) Arrow indicates the summer solstice sunset porthole. We can tell that the room was added later to the structure because its stonework differs from that of the adjacent tower. (Photo by M. Zeilik.) (b) Hovenweep Castle: interior and exterior views of the anticipatory and solstitial observations for winter and summer through the two westside port holes. Note that the first light through the portholes occurs when the sun lies 17° before the solstitial sunset positions, about two months prior to the solstices.

1.30b). The summer solstice sunset falls on the north side of the lintel of the east door. About 67 days before the summer solstice, the sunlight first enters the room, the beam striking the west side of the lintel of the north door. As the sun moves north, the sunset point tracks 2.4 meters from one lintel to the other, a linear distance that corresponds to an angle of 17° on the horizon.

A similar pattern works for the winter solstice portal. At the winter solstice sunset, the light strikes the east side of the north doorway's lintel. The beam first appears 70 cm to the right of this position, or at an angle of about 17°. So the sunset beam enters the room some 70 days before the winter solstice. For both solstices, the western horizon lacks features for an effective horizon calendar.

If markers were placed on the inside north wall, the tracks of the sunlight would make it quite easy to anticipate both solstices. The same technique could also set the planting cycle, which runs from the middle of April to the summer solstice. In fact, sunlight first enters the summer solstice port on April 15—just the right date to begin planting key crops, such as corn, beans, and squash.

This one example of many shows that the Anasazi people very likely made naked-eye observations that revealed the basic cycles of the heavens.

■ **What are the observing duties of the sun priest in Pueblo culture?**

Key Concepts

1. Celestial bodies participate in cyclic motions; some are short term (hours, days) and others long term (months, years: see Table 1.3). Most of these motions are periodic; that is, they repeat after a given period of time has elapsed. These repetitions give the impression that the sky has an order.

2. Relative to the horizon, the stars rise in the east and set in the west every day. The stars make up fixed patterns, called constellations, which do not change over a human lifetime. Different constellations are visible in the night sky during different seasons—except for the circumpolar groupings, which are always visible. The stars act as the background for the motions of other celestial objects.

3. Relative to the horizon, the sun rises roughly in the east and sets roughly in the west daily, with the exact position changing with the seasons: its highest point in the sky defines noon. The height of the noon sun varies with season—it is highest at the summer solstice, lowest at the winter solstice, and midway at the equinoxes. Relative to the stars, the sun moves eastward along a path called the ecliptic, which cuts through the constellations of the zodiac; the sun completes one eastward circuit of the zodiac in a year.

4. Relative to the horizon, the moon rises in the east and sets in the west; relative to the stars, the moon moves eastward (completing one circuit in about a month). As the moon moves around the sky relative to the

TABLE 1.3 SUMMARY: CELESTIAL MOTIONS VISIBLE WITHOUT OPTICAL AID

OBJECT	DAILY MOTION	LONG-TERM MOTION
Sun	E to W in about 12 hours from sunrise to sunset. Length of day varies from season to season.	W to E along ecliptic 1° per day. Height of sun in sky at noon is maximum in summer, minimum in winter.
Moon	E to W in about 12 hours 25 minutes from moonrise to moonset. Moonrise is about 50 minutes later each day.	W to E within 5° of ecliptic. It takes 27.3 days to travel 360° relative to the stars. Phases repeat in cycles of 29.5 days.
Planets	E to W in about 12 hours from rising to setting; interval varies depending on the rate of a planet's motion with respect to the stars.	W to E within 7° of ecliptic. Average speed along ecliptic varies, fastest for Mercury and slowest for Saturn. Retrograde motion from E to W at a time specific to each planet.
Stars	E to W in about 12 hours from star rise to star set. Star rise is about 4 minutes earlier each day. Circumpolar stars never set; their motions center on the celestial pole.	In fixed positions with respect to one another. Relative to the sun, a constellation returns to the same position in 1 year. Position of the celestial pole changes slowly, returning to its initial position in about 26,000 years.

sun, it goes through a cycle of phases, each of which depends on the angle on the sky between the sun and the moon.

5. Relative to the horizon, the planets rise in the east and set in the west daily: generally, the planets move on or close to the ecliptic eastward with respect to the stars; occasionally, the planets loop westward relative to the stars in retrograde motion. Of the five planets visible without a telescope, Mars, Jupiter, and Saturn retrograde only when in opposition to the sun, Mercury and Venus when in conjunction with the sun and moving from evening "star" to morning "star" (which also means that Mercury and Venus never move far from the sun).

6. A lunar eclipse occurs when the full moon passes through the earth's shadow; a solar eclipse occurs when the new moon passes between the sun and the earth. Solar eclipses show that the moon is closer to the earth than the sun is.

7. The relative angular speeds of the planets with respect to the stars allow us to estimate their relative distances from the earth: the slower planets have the greater distances.

8. Historic and ancient Pueblo people in the U. S. Southwest paid close attention to the motions in the sky, especially the seasonal motion of the sun (necessary for agriculture), and left behind records of their knowledge in their art and architecture. Other cultures throughout the world did the same; they were well attuned to the rhythms of the sky.

9. The measurement of angles is fundamental to astronomy. Most angles in astronomy are small, and the size of the angle is related to the size-to-distance ratio of an object (or the separation of two objects and their distance from us). We cannot find the actual size of an object unless we know its distance.

Key Terms

angular diameter
angular distance
angular separation
angular speed
celestial pole
circumpolar stars

conjunction
constellation
eclipse (lunar/solar)
ecliptic
equinox (spring/fall)
geocentric

horizon
maximum elongation
noon
opposition
phases (moon)
planets

precession of the equinoxes
retrograde motion
solar day
solstice (summer/winter)
zodiac

Study Exercises

1. Tell how you can find the ecliptic in the sky. Tell how you can find the zodiac. (Objectives 1-4 and 1-8)
2. Draw a schematic diagram of the retrograde motion of a planet relative to the stars. Be sure to indicate the directions east and west clearly. (Objective 1-4)
3. What celestial bodies *never* show retrograde motion? (Objective 1-3)
4. Into what two groups can the planets be divided on the basis of retrograde motion? (Objective 1-4)
5. When Mars is at opposition, at about what approximate time will it rise? Set? (Objective 1-4)
6. For what two reasons did ancient astronomers believe (correctly) the moon to be closer to the earth than the sun is? (Objectives 1-6 and 1-9)
7. (a) You go outside at about 9 P.M. and face south.

The moon is up and off to your right, near the horizon. Is it rising or setting? What is its phase? (b) The next night you go out again at 9 P.M. Where is the moon? Is it higher, lower, or not up at all? Did it move east or west with respect to the stars? Has its phase changed? If so, how? (Objectives 1-3, 1-5, and 1-7)

8. Describe the changing position of the rising sun on the eastern horizon throughout a year, with special emphasis on the solstices and equinoxes. (Objective 1-2)
9. What phase must the moon be in for a solar eclipse to occur? A lunar eclipse? (Objective 1-6)
10. On what observational basis can you argue that Mars must be closer to the earth than Saturn? (Objectives 1-9 and 1-11)

11. You are a novice sun priest and must establish your own day count to anticipate the winter solstice. How will you do it? (Objectives 1-2 and 1-10)

12. At what time of year would the daily angular speed of the sun along the horizon (at rising or setting) be the greatest? The lowest? (Objective 1-2)

13. Consider observing the sun at sunset from a midnorthern latitude. How would the setting point change from winter to summer solstice? (Objective 1-2)

14. Is the sun always overhead at noon, when observed from a midnorthern latitude? If not, when is it overhead? (Objective 1-2)

15. Imagine that from a midnorthern latitude, you measured the length of a shadow cast by a telephone pole at noon on the day of the spring equinox. A week later, you measure the shadow again. Is it shorter or longer? (Objectives 1-1 and 1-2)

Problems & Activities

1. The moon circuits the zodiac once in 27.3 days. What is its average angular speed per day? Per hour?

2. What is the size-to-distance ratio of the sun? How many diameters of the sun would fit in the earth–sun distance?

3. How does the size-to-distance ratio of the sun compare to the size-to-distance ratio of the moon? Which object is at the greater distance? Can you make any confident statement about the actual size of the moon relative to the sun?

4. A stick 10 cm high casts a shadow 3.5 cm long at noon. What is the sun's altitude? *Hint:* Draw a diagram to scale. Or you can divide the height of the stick l by the shadow length s. This ratio is the tangent of the angle of the sun's height h above the horizon

$$\tan h = \frac{l}{s}$$

Calculate the ratio and use a trigonometric table or a pocket calculator with trigonometric functions to find h.

5. The sun circuits the zodiac in about a year. What is its average angular speed per month? Per day? Per hour? How does it compare to the moon's average angular speed?

6. Use a ruler to measure the size of the moon in Figure 1.18. Because the moon has an angular size of 0.5°, divide this value by the linear size you measure to find the scale of the photo. Then use this scale to estimate how far the moon had moved in angle relative to Venus between the exposures at the top left and the tenth one down.

See for Yourself!

To calibrate the angular size of your fist, find a spot that has a clear horizon. Then use your fist to measure the angular distance from the horizon to a point overhead. That angle is 90°, so divide that by the number of fists to find how many degrees are covered by your fist. What is the size-to-distance ratio of your fist at arm's length?

You can make your own horizon calendar by tracking the sun's different seasonal positions at sunset along the horizon. Find a place that has a clear view of your horizon as well as a few prominent landmarks. Make a sketch of the general horizon features. You can use your fist to get an approximate angular scale. Standing in the same spot each day, make a mark on the drawing of the sun's setting position; also write down the date and time. Do this at least weekly for at least three months centered around the solstice or the equinox. (As an alternative, you can take a series of sunset pictures just before the sun dips below the horizon.) Be persistent! You need careful long-term observations to come to firm conclusions. ■

2-1 Describe the essential aspects of a scientific model and evaluate cosmological models in the context of scientific model making.

2-2 Cite at least one important astronomical achievement in the records of the Babylonians.

2-3 Describe the physical basis of Aristotle's cosmological model and tell how it influenced his picture of the cosmos.

2-4 Describe the geometric devices and physical ideas that Aristotle used to explain basic astronomical observations in the context of his model.

2-5 State the assumptions and physical basis for Ptolemy's cosmological model.

2-6 Sketch the basic Ptolemaic model for the motions of Venus, the sun, and Mars, and show—using Venus and Mars as examples—how the epicycle and the deferent were used to explain retrograde motion.

2-7 Indicate the geometric devices and physical ideas that Ptolemy used to explain the main celestial motions and any variations in the major cycles.

2-8 Evaluate the essential assets of the Ptolemaic model that led to its wide, long-term acceptance; as part of this appraisal, be able to construct a simplified version of the model.

2-9 Compare and contrast a mythical and a scientific cosmological model with specific examples from Babylonian and Greek astronomy.

2-10 Use at least one specific case to show how geometric and aesthetic concepts influenced Greek ideas about the cosmos.

2-11 Describe the difference between a sun-centered model and an earth-centered one with respect to an annual stellar parallax.

Central Question
What is a scientific model, and how did early cosmological models explain and predict astronomical observations?

The Birth of Cosmological Models

Marduk bade the moon come forth;
entrusted night to her,
made her creature of the dark, to measure time;
and every month, unfailingly, adorned her
with a crown.

FROM *Enuma Elish*, TRANSLATED BY THORKILD JACOBSON

How have people pictured the **cosmos**? Older cosmologies generally paid little attention to the details of the celestial motions even if the cycles were carefully observed. Most ancient cultures viewed the cosmos as finite and geocentric, closed off by a shell of stars. And order in the cosmos tended to be explained in terms of religious myths.

In the fourth century B.C., the Greeks first attempted to take cosmological ideas beyond myths. Grappling with the vexing problem of planetary motion, Greek thinkers refined their cosmological picture with geometric devices intended to account for the celestial cycles. These designs marked the first earnest models: mental images with features like those observed in nature. Early cosmological thought culminated in the geocentric model of Claudius Ptolemy. So well did Ptolemy succeed that no one challenged his model for more than 1400 years.

This chapter examines early cosmologies and highlights the close correspondence of these images to actual observations. You'll learn how some early cultures tried to make sense of what they saw in the sky. You'll see the contrast between cosmological ideas grounded in myths and ideas developed as scientific models. You'll follow the evolution of thought that resulted in the first comprehensive model of the cosmos, one that eventually seeded the birth of models used today.

2.1 Scientific Models

Chapter 1 described naked-eye astronomical observations. As you read, you probably felt the urge to place these observations into a grand design, a model for the operation of the heavens. People in general want to make sense of what they see in the world. And this natural drive gave birth to **scientific models**—conceptual plans to explain the workings of nature. Scientific models lie at the heart of the workings of modern science. This mental framework did not, however, always underpin people's ideas about the world. Ancient models tended instead to be mythical.

A scientific model evokes a mental picture that is based on geometric ideas, physical concepts, aesthetic notions, and basic assumptions. It draws on analogy and metaphor of what is seen in nature. In this process, a model strives to come to grips with a seemingly chaotic world by casting it in familiar terms.

Certain elements always enter into a scientific model (Fig. 2.1). First, our sense impressions of the world provide raw information and stimulate our curiosity. Then we try to interpret these impressions by geometry, physics, and aesthetics in astronomical models. Our choices here are filtered by whatever assumptions, both explicit and implicit, we apply to the situation. In astronomy, geometry outlines a visual framework for the model and the shapes of objects within it. Physics deals with the motions and interactions of various components within the model. And aesthetic ideas—gut judgments of what seems beautiful—compel us to select the simplest, most pleasing models from those we can imagine. Simplicity also forces us to include only the essential aspects of explanation. Scientific models are mental, essential, simple, and beautiful.

Finally comes the crucial test: How well do the features of the model correspond to observations? If the accord turns out well (within the recognized errors of observations), the model is confirmed as workable. If not, various aspects of the model are modified to get a better match.

So a scientific model has two key features: *it explains what is seen, and it predicts observations accurately*. A model's predictions of future events or descriptions of past events must relate directly to observations and do so with ample accuracy to be convincing. Although all good scientific models contain both explanation and prediction,

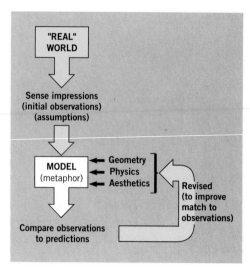

FIGURE 2.1 Schematic flowchart of the process of scientific model making: the feedback loop allows us to revise the model by comparing predictions to observations. This revision cycle is a crucial part of the process.

one feature may overshadow the other in a particular model.

A mature model's predictions must be numerical, to permit direct comparison with measurements. But keep in mind that every observation has an error attached to it. For instance, at best the human eye can see an angle as small as 1 arcminute. So all unaided observations of the sky include an error of at least this amount, which is about one-thirtieth the angular size of the moon. Instruments such as telescopes can improve on this natural limitation; but even telescopes today spawn errors in every observation. A model is deemed sufficient if its numerical predictions and explanations fall within the recognized range of intrinsic errors in observations—as judged by the standards of the day.

To confirm a model, we look at how well its predictions fit our best observations. The search for confirmation can make or break a model. All scientific models are necessarily tentative, containing the seeds of their own destruction. For scientific models *must* change when new or more accurate observations become available. As models evolve, we acquire a deeper understanding of the cosmos, the order in the physical world. The conception of models is the creative act of science.

To understand as completely as possible the scientific models in this book, you should be able to:

1. State the aesthetic, geometric, and physical bases of the model.
2. State clearly the assumptions behind the model and evaluate how well they are supported.
3. State the key observations the model attempts to explain and evaluate how well it succeeds.
4. Describe the relative importance of various aspects of the model, making clear the connections between parts.
5. Indicate how the model deals with new observations and suggest predictions it can make for new situations.

Some models prove to have a great deal of power by tying together disparate observations. Others direct scientists into new areas never imagined. When a model proves to be very robust and potent, it may be elevated to the status of *scientific theory*.

■ What is *the* critical test of any scientific model?

2.2 Babylonian Skywatching

About 1600 B.C. the Babylonians compiled the first star catalogs and began long-term records of planetary motions. By 800 B.C. Babylonian astronomers had fixed planetary locations with respect to the stars of the zodiac and were keeping records of these positions on clay tablets. The objects of their early observations included Venus, Jupiter, and Mars. These records often spanned several centuries.

That the Babylonians became such careful observers was in part a result of state support for the calendar and for astrology. These political needs induced techniques for predicting planetary positions and for recording long-term observations. Meanwhile, the records the Babylonians kept enabled them to find the basic cycles of periodic celestial motions. They even made detailed observations of solar and lunar eclipses, including one that occurred in 240 B.C. Such records enabled the Babylonian astronomers to predict eclipses, at least in a rough way.

The Babylonians went beyond recording these basic cycles. Their records

helped them discover the *variations* in them. For instance, they recognized that the angular size of a planet's retrograde loop and the duration of this motion change from one retrograde loop to the next (Fig. 2.2). The Babylonian astronomers had tables of the major cycles and variations on clay tablets. They used these tables of periods to predict future planetary positions and retrograde motions. This procedure required no explanation of the cycles, only a knowledge of their existence over a long period.

The Babylonian astronomers also served as priests who fostered the continuity of astronomical knowledge. But religion divorced Babylonian cosmology, the grand picture of the universe, from astronomy. In the cosmic picture, the gods created, ordered, and controlled the world. These divine functions were explained in religious myths. The Babylonian tale of genesis, the *Enuma Elish,* deals with the forming of the world from initial chaos by the god Marduk (later identified with Jupiter), who fashioned the stars from an unordered swirl of primeval waters.

Because tables of astronomical cycles functioned strictly for predictions and lacked any geometric or physical framework, Babylonian astronomers

Scientific Models
At a Glance

Explain	Current and past observations
Predict	Future observations
Verifiable	By a variety of observations
Changeable	To match observations better

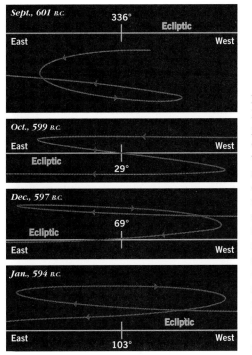

FIGURE 2.2 Retrograde motions of Mars in Babylonian times: Mars appears at different positions along the ecliptic (indicated in degrees from the vernal equinox) for different times of retrograde motion indicated by the dates. The dates are for the *middle* of the retrograde motion. The positions along the ecliptic are measured from 0° at the position of the vernal equinox on the ecliptic. Note how the shape and size (and so the duration) of the retrograde loop vary. The Babylonian astronomers knew of these variations. The motions shown here are relative to the stars; east is to the left, west to the right. (Courtesy of Owen Gingerich.)

could predict but not explain celestial motions in terms of physical causes. They never developed actual scientific models of the heavens, just records of their patterns.

■ What was one of the reasons for the accuracy of Babylonian observations?

2.3 Greek Models of the Cosmos

The early Greek philosophers did what the Babylonian astronomers never dreamed of: they devised geometrical, physical models of the cosmos. The Babylonians made sense of the world through myths. The Greeks had myths, too; but their curiosity drove them to develop models as a means of grasping reality. A look here at Greek astronomy and cosmology will show how scientific model making arose in Western culture.

Harmony and Geometry

The ideas of Pythagoras (who lived in the sixth century B.C.) forged the aesthetic basis of much of Greek cosmology. The Pythagorean cosmos first incorporated some aspects of a scientific model. In it, the earth has a spherical shape for reasons of symmetry and perfection—a sphere is the most perfect of all solids. Here we see how geometry and beauty intertwine and add power to a mental picture.

One spherical shell enclosed the cosmos and held the stars. Smaller spheres within it carried the planets around. The stellar sphere, driving all other spheres, rotated daily east to west—an explanation of the daily motion of the sky. At the same time, the planetary spheres rotated slowly at different rates from west to east. Their rotational periods matched the time for each planet to circuit the zodiac. These rotations explained the eastward motion of the planets relative to the stars as seen from the earth. The earth lies motionless in the center—this model is *geocentric*.

The Pythagorean model relied strongly on the notions of harmony and symmetry. Such aesthetic ideas play an essential role in scientific models.

Though this early model contained forceful aesthetic and geometric elements, it lacked physical ideas. And its correspondence to observations was crude. For example, the basic model failed to account for retrograde motions of the planets.

Roughly a century later, the philosopher Plato (427–347 B.C.) posed ideas derived from Pythagorean concepts. Plato, too, saw the perfection of the cosmos in the form of a sphere. So he assumed—demanded!—that all heavenly bodies move at a uniform rate around circles. (A uniform rate is what you'd expect from the rotation of a sphere.) Plato also succinctly stated the problem of the planets: the goal of an astronomical model was "to save the appearances"— to devise a model that explained the observed motions. This goal preoccupied astronomers for centuries. It also burdened them with an aesthetic ideal that turned out to be wrong.

A Physical Geocentric Model

Aristotle (384–322 B.C.), the most famous of Plato's pupils, also based his geometric model of the cosmos on the idea of uniform, circular motions. He even tried to incorporate the retrograde motions of the planets. But the results were a mess—a model with 56 spheres centered on the earth (Fig. 2.3). Even with this complexity, the model did not match observations very well. Geometry by itself was not enough.

Yet Aristotle's model was more scientific than earlier efforts because it did incorporate—for the first time—physical ideas of motion. What were these physical concepts? Aristotle viewed the cosmos as divided into two distinct realms: a region of change near the earth and an eternal region in the heavens. The realm of change contained bodies made of four basic elements—earth, air, fire, and water. Each element had its own natural motion toward its place of rest in the cosmos: earth to the center, fire to the greatest heights reaching toward the moon, air below fire, and water between the earth and the air. In contrast, the heavens were made of an immutable, crystalline material, which did not obey terrestrial laws.

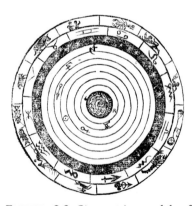

FIGURE 2.3 Geocentric model of the cosmos in the tradition of Aristotle: a medieval picture of the Greek philosopher's system of geocentric spheres. The earth lies in the center; above it are the natural realms of water, air, and fire below the sphere of the moon. Beyond the moon are the heavenly spheres to which the planets and the stars are attached. Only one sphere for each planet is shown; the actual scheme used a number of spheres to account for all the motions of each planet. (Courtesy of the Yerkes Observatory.)

The ancient Greeks believed that each realm had different versions of **natural motion:** motion without forces. In the heavens, the celestial spheres rotated naturally. So *no* forces were needed to move the planets around the earth. In the terrestrial realm, earth, air, fire, and water each had its natural motion. For example, the natural motion of earthy material was toward the center of the cosmos. But here **forced motions** also could occur. For instance, to keep a cart moving, a person must continue to push it. When this effort is no longer exerted, the cart rolls a bit but will soon stop. Such motions, Aristotle reasoned, required a **force,** a push or a pull, to keep them going. So here we meet a crucial physical distinction between *forced* and *natural* motion.

These physical ideas shaped Aristotle's image of the cosmos. He argued that the earth must be stationary and in the center of the universe. How so? First, the natural motion of earthy material to seek the center of the cosmos was taken to explain the location of the earth there. Second, Aristotle reasoned that if the earth rotated, bodies thrown upward would not drop back to their point of departure. Yet, he noted, heavy objects thrown upward do indeed return to their starting place. Conclusion: the earth does not rotate. (Note how physical ideas of natural and forced motion influenced Aristotle's incorrect model of the cosmos. You'll see in this part how concepts of motion profoundly affect cosmological models.)

A model gains credibility when it explains observations in a natural way. In this respect, Aristotle's model was successful, for it explained two important observations: the lack of a stellar parallax and the spherical shape of the earth. Aristotle noted that if the earth moved around the sun, the stars must display an annual shift in their positions, called *stellar parallax* (described below: see Fig. 2.4). No one had observed this change, so Aristotle concluded that the earth did not move around the sun, a conclusion that reinforced his geocentric view. (Stellar parallax does occur, but it is too small to detect with the unaided eye.)

Aristotle followed Pythagorean ideas and added observations of his own

(such as the curve of the earth's shadow on the moon during a lunar eclipse) to conclude that the earth was a sphere. This idea led to a calculation of the earth's size by Aristotle, who cited a diameter of 5100 km, and by a later Greek, Eratosthenes, who found a diameter of 13,400 km (Focus 2.1).

A Contrary View: A Sun-Centered Model

After the time of Alexander the Great (356–323 B.C.), the Greek scientific tradition centered on the great library at Alexandria. Here worked astronomers such as Eratosthenes, who calculated the earth's size, and Aristarchus, who proposed a sun-centered, or **heliocentric,** model of the cosmos. Aristarchus lived in the third century B.C. His heliocentric model had the earth rotating on its axis once a day to explain the daily motion of the sky. The earth also moved around the sun in one year, which explained the annual motion of the sun through the zodiac.

Unfortunately, when the library at Alexandria burned, all the major writings of Aristarchus were destroyed. We know of his ideas from comments by others and from one fragment of a work. In it, Aristarchus figured out the earth–sun distance relative to the earth–moon distance and inferred that the sun was a body much larger than the earth. But we have no evidence that Aristarchus worked out planetary motions in any detail using his heliocentric scheme.

This early heliocentric model was attacked on two fronts: it contradicted Aristotle's physics by stating that the earth moved, and it required a stellar parallax, which was not observed. For these reasons, and also because of the dominant influence of Aristotle's ideas, Aristarchus' model languished.

Stellar Parallax in a Finite Cosmos
Parallax is an apparent shift in the positions of a body (or bodies) because of the motion of the observer. When the body in question is a star, we use the term **stellar parallax.** In a heliocentric model, stellar parallax arises from the earth's motion around the sun; it is called

Enrichment Focus 2.1

SURVEYING THE EARTH

Around 200 B.C. the Greek astronomer Eratosthenes, believing the earth to be round, calculated its circumference. Here's how Eratosthenes reasoned. While on vacation from his studies at the library in Alexandria, he visited Syene (now Aswan), Egypt. He noted that at the summer solstice, sunlight fell directly down a well at noon (Fig. F.1), indicating that the sun was directly overhead. At noon on the same date the following year, back in Alexandria (located directly north of Syene) Eratosthenes observed that a sundial shadow indicated the sun to be about 7° south of directly overhead. Because the earth's circumference totals 360°, the astronomer concluded that the distance from Syene to Alexandria must be 7/360 of the circumference (Fig. F.2).

To find the length of the circumference, Eratosthenes needed the distance from Alexandria to Syene. The trip by camel between these two cities took about 50 days. The average camel then traveled about 100 stadia per day; so the trip covered 5000 stadia. [The *stadium* (plural *stadia*) was a unit of length measuring about 1/6 km; its exact value varied in the ancient world.] Eratosthenes calculated that the earth's circumference was 360/7 × 5000 = 250,000 stadia. Take the length of a stadium as 1/6 km; then the earth's circumference comes out to 42,000 km—surprisingly close to the modern value of 40,030 km. (This coincidence may be chance. Although Eratosthenes measured the angular separation of Syene and Alexandria very accurately, we do not know how well he knew the distance between the cities, nor do we know the exact length of a stadium in kilometers. In any case, he got the ratio right.) Divide the circumference by 2π and you have the radius: about 6700 km. Double the radius to get the diameter—13,400 km.

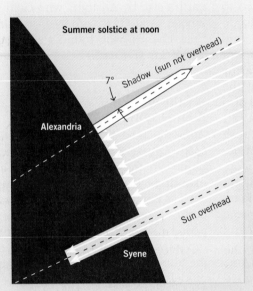

FIGURE F.1 Eratosthenes' solar observations at noon at Alexandria and Syene (Aswan). He noted that at Syene on the summer solstice, the noon sun's light came down from directly overhead. In another year he saw that in Alexandria on the summer solstice, the noon sun's light came down at an angle of 7° from the vertical.

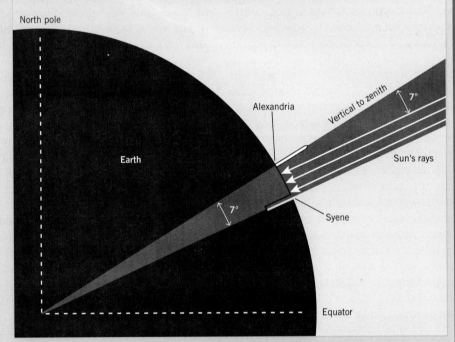

FIGURE F.2 The geometrical basis for calculating the earth's circumference. If the sun is so far away from the earth that its incoming rays are parallel, only a curved earth could simply account for the difference in the sun's noon angle at the two locations: in this case, the 7° between Alexandria and Syene at noon.

heliocentric stellar parallax. The details of heliocentric parallax differ, depending on whether the stars in space are believed to be confined to a thin shell (as in the Greek picture) or spread throughout space (as in modern concepts; see Chapter 14).

Parallax occurs when observations take place from two separate locations. Consider, for example, a merry-go-round. As it turns, a rider on it sees nearby objects as shifting in their positions relative to more distant ones. That shift is parallax. The greatest angular shift of any single nearby object occurs from the two positions separated by the diameter of the merry-go-round. And the shift recurs in a periodic cycle.

Now imagine that the stars are stuck in a thin shell that closes off the cosmos, as in the model of Aristarchus (Fig. 2.4). Pick out two stars close together on the celestial sphere (*A* and *B* in Fig. 2.4). Observe them when they are in the southern sky at midnight (position *2* in Fig. 2.4); they will appear some angular distance apart (angle *A2B*). Just after sunset three months later, observe the stars again (position *3* in Fig. 2.4); they will appear closer together (angle *A3B* is less than angle *A1B*), partially because you're now seeing them at an angle rather than face-on. Observe them again six months later (position *1* in Fig. 2.4); their angular separation is *A1B,* which is the same size as angle *A3B*.

Imagine viewing these stars over a six-month cycle: from positions *3, 1,* and *2.* You'd see the stars close together (*3*), farther apart (*1*), and then closer together again (*2*). This yearly cyclical shift in angular position is heliocentric stellar parallax. Note that the size of the earth's path compared to the size of the shell of stars determines the size of the shift: the smaller the ratio, the smaller the shift. Of course, if the earth were in the center of the stellar shell and did not move, no shift would occur.

Greek astronomers did *not* observe this parallax. So they viewed the heliocentric model as inconsistent with observations and in part rejected it on this basis. A geocentric model predicts *no* such parallax. (We now know that heliocentric parallax is too small to detect without a telescope, because the stars are very far away: Section 14.2).

Expanding the Geocentric Model

The distinguished astronomer Hipparchus lived and worked at Rhodes from 160 to 127 B.C. To the basic geocentric model, Hipparchus added the geometric devices of **eccentrics, epicycles,** and **deferents** to explain aspects of planetary motions that had been shrugged aside by earlier observers. Each of these devices accounted for observed features of the motions: the epicycle and deferent together explained the retrograde motions; the eccentric accounted for the variable motion of the planets and sun through the zodiac. So Hipparchus used geometry "to save the appearances" and stuck to the assumption of uniform, circular motions. Let's see how he did it.

Hipparchus assumed that the planets move at a uniform speed on circular paths. Yet he knew from observations that the planets' motions are not uniform but vary: the average angular speed of a planet is faster in one region of the zodiac and slower in the opposite region. Hipparchus explained this variation with an *eccentric* (Fig. 2.5*a*). The earth was displaced from the center of the planet's circular motion. Then the planet under consideration appeared to go faster through the zodiac when it was closer to the earth and slower when farther away (Fig. 2.5*b*). But its actual motion along the circle remained at a constant angular rate as seen from its center rather than from the earth.

But the eccentric did *not* explain retrograde motion. For that, Hipparchus used the combination of an *epicycle* and *deferent* (Fig. 2.6). The deferent was a large circle, sometimes centered on the earth, sometimes offset (eccentric). The center of the epicycle, always a smaller circle, moved along the circumference of the deferent. The planet was fixed to the epicycle, so its motion comprised its circuit around the epicycle *plus* the epicycle's circuit around the deferent. As the

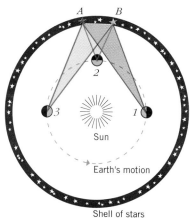

FIGURE 2.4 Stellar parallax in a finite, heliocentric model. *A* and *B* are two stars fixed on the nearby stellar sphere. The earth moves around the sun in a year, causing a change in the observed angle between the stars. The largest angle, *A2B,* occurs at position *2;* there are intermediate angles, *A1B* and *A3B,* at positions *1* and *3,* respectively.

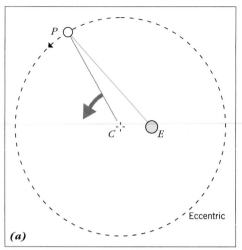

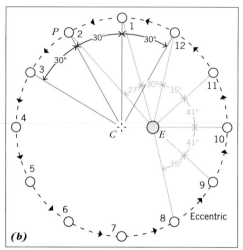

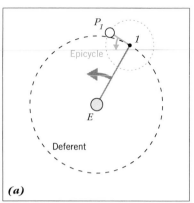

FIGURE 2.5 The eccentric (*a*). A planet (*P*) revolves with uniform circular motion about the center (*C*) of its path. As observed from *C*, the planet moves at a constant angular speed. The planet's motion as seen from the earth is not uniform because the earth (*E*) is displaced from the circle's center. (*b*) The effect of the eccentric for modeling a planet's motion. As seen from the circle's center, the planet moves through equal angles in equal amounts of time—a constant angular speed. But as seen from the earth, the planet covers different angles in the same times. In particular, the planet has the fastest angular speed when closest to the earth (at point *10*) and the slowest when farthest away (at point *4*).

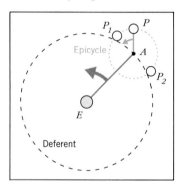

FIGURE 2.6 The epicycle and the deferent. A planet (*P*) is attached to a small circle (epicycle) whose center rides on a larger circle (deferent). The earth (*E*) lies in the center of the deferent. The radius of the epicycle turns in the same direction as the radius of the deferent. So when the planet moves on the inside of the deferent (from *P₁* to *P₂*), it moves opposite its normal motion with respect to the stars; this is its *retrograde motion.*

planet moved on the part of the epicycle interior to the deferent, it moved in the opposite direction from the deferent motion and so imitated the backward swing of retrograde motion (Fig. 2.7). As a bonus, while the planet is in the middle of its retrograde motion, it is closest to the earth and shines the brightest. Note that all motions were uniform as viewed from the center of the epicycle or deferent.

The motion of the deferent represented the planet's general motion from west to east through the zodiac. The model matched up with observations because the periods of the epicycle and the deferent were set from observations of the time each planet took to circuit the zodiac once (the deferent's period) and

the time between retrograde motions (the epicycle's period).

■ According to Plato, and as reinforced by Aristotle and Hipparchus, what is the one goal of an astronomical model?

2.4 Claudius Ptolemy: A Total Geocentric Model

Two-and-a-half centuries after Hipparchus, Claudius Ptolemy (Fig. 2.8) worked in Alexandria and so had access to records in the great library. There, he molded the astronomical tradition into a comprehensive model that would endure for centuries.

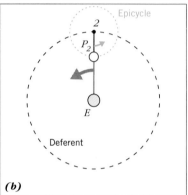

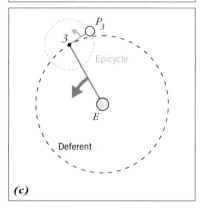

FIGURE 2.7 The retrograde motion of a planet, such as Mars, as seen from the earth, produced by an epicycle and deferent. At *P₁*, the planet begins its retrograde motion. At *P₂*, it is in the middle of its retrograde motion. At *P₃*, its retrograde motion ends. Since the planet is closest to the earth in the middle of its retrograde loop, it appears at its brightest then. The period of the deferent is the average time Mars takes to circuit the ecliptic; the period of the epicycle is the time between retrograde motions.

We know little of Ptolemy's life, not even the dates of his birth and death. His observations indicate that he worked around A.D. 125. Ptolemy's most influential astronomical work—the *Almagest*—was the first professional astronomy textbook. Many of the geometric devices that Ptolemy used in the *Almagest* did not originate with him, but he was the first to design a complete system that accurately predicted planetary motions, with errors of usually not more than 5° and often less. His careful application of geometry nicely described the celestial motions.

A Geocentric Model Refined

At the start of the *Almagest,* Ptolemy pays his respects to Aristotle and then defines the problem: use geometry to describe astronomical observations, especially the major cycles and their variations, known since Babylonian times. Note that Ptolemy, like Hipparchus, follows Plato's dictum "to save the appearances."

Ptolemy assumes that the earth is spherical and in the center of the cosmos. The earth has no motions. It is *much* smaller than the outer sphere of stars. In addition, he assumes that uniform motion around the centers of circles is the aesthetically pleasing motion for the celestial spheres. So Ptolemy's model has a physical basis: that of Aristotle.

In practice—and this is a key point—Ptolemy ends up *violating* this uniform motion precept. Recall that there are two main variations in the motions of the planets: retrograde motion and the variable motion through the zodiac. Also, retrograde motions show variations in size, shape, and duration (Fig. 2.2). How can uniform circular motion account for these frequent variations?

The Equant and Nonuniform Motion

Think of the problem of the motion of the planets as a model-building puzzle that needs a different device for each observed aspect. The two main aspects are (1) eastward motion through the zodiac—with variations, and (2) retrograde motion—also with variations! Like Hipparchus, Ptolemy modeled the first with a planet moving around the earth at uniform speed on a circle. But he offset the earth from the center of the circle—the eccentric—to account for variations. For the second, again like Hipparchus, Ptolemy used smaller circles (epicycles) moving on the larger ones (deferents). Then, for variations in the retrograde motions, Ptolemy invented a new geometrical device: the **equant.**

To make an equant, Ptolemy started with an eccentric. Here the earth lies away from the center of the circle of a planet's deferent. Ptolemy then imagined another point, *not at the circle's center,* from which the motion would appear uniform. This imaginary point, the *equant* point (Fig. 2.9*a*), lay opposite the center of the deferent from the earth (which is at an eccentric point). If you stood there, you would see the planet move around the sky at a uniform angular speed relative to the stars. From the earth and from the circle's center, though, the motion is *not* uniform (Fig. 2.9*b*).

Imagine a planet's epicycle as lying on the outside of the face of a clock with one hand, which drives the epicycle around. The ideal is to have the hand at

FIGURE 2.8 Claudius Ptolemy: ". . . we shall only report what was rigorously proved by the ancients." (Courtesy of the Granger Collection.)

the center of the clock. The equant, though, positions the hand off-center. The hand turns uniformly and so moves the epicycle around as viewed from the point of attachment. (See Fig. 2.10.) But as seen from the center, the motion is nonuniform.

You do not need to understand the nitty-gritty details of the equant to appreciate Ptolemy's contribution to the evolution of cosmological models. The main point is that with the equant, celestial motions *no longer had to be uniform around the centers of circles*. The equant was a nonphysical, totally geometric device that broke the fundamental assumption that planetary motion had to be uniform along circles.

Why did Ptolemy violate this aesthetic ideal? Probably because he demanded that his model fit observations reasonably well. With the equant, he could better match the model's predictions to the observations, especially the annoying variations of the planets in their overall retrograde cycles. This accord apparently struck Ptolemy as more important than the traditional precept of uniform, circular motion—a crucial step to a robust scientific model.

Ptolemy's Complete Model

When completed, Ptolemy's model looked fairly simple (Fig. 2.11), but it treated Mercury and Venus differently from the rest of the planets. Recall (Section 1.4) that Mercury stays within about 23° of the sun, and Venus stays within 46°. To explain these facts, Ptolemy made Mercury's epicycle smaller than Venus' and demanded that the centers of these epicycles always lie on the line connecting the earth and the sun. So Mercury and Venus were constrained to stay near the sun (Fig. 2.11*a*).

In contrast, Mars, Jupiter, and Saturn can be anywhere along the zodiac relative to the sun. So the centers of their epicycles could be anywhere on the perimeter of the deferents. To ensure that these planets retrograde only at opposition, Ptolemy set the radii of their epicycles parallel to the earth–sun radius (Fig. 2.11*b*).

Note that in Ptolemy's model the

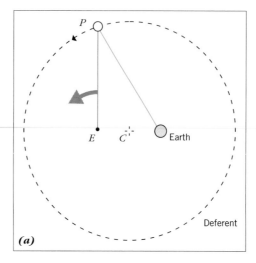

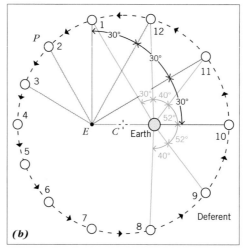

FIGURE 2.9 Ptolemy's equant: For clarity, the epicycle is *not* shown. (*a*) Here the earth is at the eccentric point. The equant point is an imaginary point (*E*) on the opposite side of the center from the earth. (*b*) A planet moves on the deferent circle so that as viewed from the equant (*E*), it would appear to cover equal angles in equal times and so move at a uniform angular speed. But as seen from *C* or the earth, the planet moves through different angles in equal times, which means that its angular speed varies. The motion illustrated here ignores the motion around the epicycle, which generates retrograde motion.

heavenly spheres are *actual, physical spheres*. The motion of these spheres—rotation—was assumed to be natural and to drive all the heavenly motions. No force was required.

Although Ptolemy's model lost acceptance centuries ago, you should not condemn it. This model remained in use for some 1400 years mainly because it worked—it predicted planetary positions to the accuracy needed (a few degrees) by astronomers who had no telescopes. And it agreed with early Greek physics and aesthetics. It survived because no other satisfactory model was advanced to compete with it. And more than any other astronomical book, the *Almagest* showed that the motions in the heavens, though complex in appearance, could be described in terms of simpler regularities by geometry.

The Size of the Cosmos

Ptolemy's cosmos was finite. How large was it? Ptolemy gives the distances to the sun and the moon in terms of earth radii, as worked out by Aristotle and Hipparchus. Going farther out, he assumes that no space is wasted between the heavenly spheres. Then, with the earth–moon distance set at 59 earth radii, the distances to Saturn can be laid out with the celestial spheres nesting tightly together. The sphere of the stars then lay 20,000 earth radii distant. It was a small cosmos—only about the distance we know today of the earth from the sun! But it did mark a reasonable attempt to establish the distance scale of the known universe.

In fact, this model had *no fixed distance scale* because of its focus on angular speeds. An angle gives a size-to-distance ratio; an angular speed gives a *rate-to-distance ratio* (Focus 2.2). But we can't know actual sizes or speeds unless we know distances! For any of the planets, we can adjust its speed around its circles to fit the observations for any reasonable range of distances. The whole system can be made larger or smaller in an arbitrary way.

■ Ptolemy's model was accepted in scientific circles for more than 1,400 years. Why?

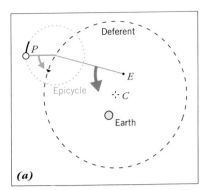

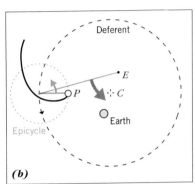

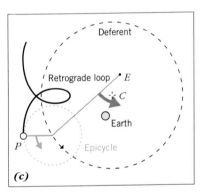

FIGURE 2.10 The equant combined with the epicycle and the deferent to model the actual retrograde motions of a planet. Note that if the planet undergoes retrograde at a different position along the deferent, the shape and size of its retrograde loop will differ from that shown here. This modeling explains observations such as those shown in Figure 2.2. (Adapted from a diagram by Owen Gingerich.)

FIGURE 2.11 The overall Ptolemaic model for the planets. (*a*) The Ptolemaic model for the moon, Mercury, and Venus. The centers of the epicycles of Venus and Mercury are fixed along a line with the sun; this explains why they are never very far from the sun. The size of each epicycle accounts for the measured size of the maximum elongation angles. (*b*) The Ptolemaic model for Mars, Jupiter, and Saturn. To account for the observed sizes of the retrograde loops, the epicycles of these planets decrease in size, so that Mars has the largest and Saturn the smallest. For the retrogrades to occur at opposition, the radii of these epicycles must align with the earth–sun radius.

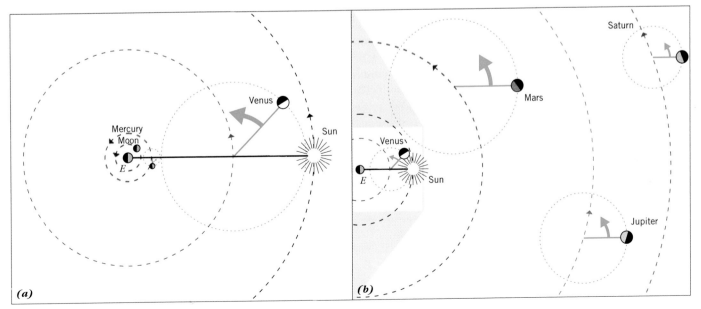

Enrichment Focus 2.2

ANGULAR SPEED AND DISTANCES

Focus 1.1 dealt with angular sizes and distance. Recall that the angular size (for small angles) gives a size-to-distance ratio. The moon's angular size of 0.5° means it has a size-to-distance ratio of about 1:110, so that 110 lunar diameters fit within the earth–moon distance.

Angular speeds are analogous. For instance, the moon moves eastward relative to the stars at about 0.5° per hour. So its speed-to-distance ratio is about 1:110 per hour. Since the

moon's angular size is 0.5°, we then know that the moon has an angular speed of one diameter per hour. But we don't know the moon's diameter—and so its speed is unknown—unless we can measure or estimate its distance. That's the key issue!

Again, by analogy to Focus 1.1, we can summarize these concepts in equation form for small angles:

actual speed ≈ 0.018 × angular speed × distance

and if we use v to represent the actual speed, a the angular speed (in degrees per time, such as hours), and D the distance (in units such as meters), then

$$v \approx 0.018\, aD$$

Note that the distance units of v are the same as the distance units of D, and the time units of v are the same as those of a. So if a is in degrees per hour and D in meters, then v comes out in meters per hour.

TABLE 2.1 A SUMMARY OF PTOLEMY'S MODEL

OBSERVATION	EXPLANATION
Sky	
Motion of entire sky E to W in about 24 hours	Daily motion E to W of the sphere of stars, carrying all other spheres with it.
Sun	
Motion yearly W to E along ecliptic	Motion of sun's sphere W to E in a year.
Nonuniform rate along ecliptic	Eccentric (earth displaced from center of sun's circle).
Moon	
Monthly motion W to E compared to stars	W to E motion of the moon's sphere in a month.
Planets	
General motion W to E through zodiac	Motion of deferent W to E. Period is set by observation of period of planet to go around ecliptic.
Retrograde motion	Motion of epicycle in same direction as deferent. Period is the time between retrograde motions.
Variations in speed through zodiac, in retrograde motions	Eccentrics, equants.
Mercury, Venus	
Average greatest elongation of 23° and 46°	Size of epicycles set by those angles.
Appear near to sun	Centers of epicycles on earth–sun line.
Mars, Jupiter, Saturn	
Retrograde at opposition, when brightest	Radii of epicycles aligned with earth–sun radius.

NOTE: Error of predictions was usually 5° or less, occasionally larger.

Key Concepts

1. Scientific models are the core of creativity and understanding in the scientific process. Models are designed to explain and predict what is observed; astronomical models contain geometric, physical, and aesthetic elements. Scientific models are never complete or final—they evolve continuously.

2. The retrograde motions of the planets vary from planet to planet and also for the same planet at different times. Constructing an explanation of retrograde motion was a most vexing puzzle for the construction of early cosmological models.

3. Babylonian astronomers were excellent observers of celestial cycles but never developed scientific models for the cycles whose periods they observed and used for predictions. Instead, they used mythical models to explain what happened in the skies.

4. The Greeks made the first scientific models, which were based on aesthetic notions and on a desire to explain observations. These early models included the central concepts of harmony, symmetry, and simplicity, as well as the use of geometry "to save the appearances." The celestial motions were locked into the concept that required them to occur at a uniform rate along circles.

5. Aristotle developed a complete cosmological model based on physical ideas. Different concepts of motion applied to the terrestrial and celestial realms (no forces were believed to be needed to turn the celestial spheres). The model was geocentric. It did not predict the celestial motions very accurately but did account for the major cycles.

6. Aristarchus proposed a heliocentric model that never gained headway because it violated Aristotle's physics and predicted an annual stellar parallax, which was not observed. In this sun-centered model, the earth moved around the sun, as did the other planets, while the moon moved around the earth.

7. Ptolemy built up a comprehensive cosmological model (Table 2.1) based on the physics of Aristotle and on the geometric ideas of the Greeks. For its completeness, he invented the equant, which violated the precept that celestial motions had to be uniform and circular. His model predicted planetary positions to acceptable accuracy (5° or less in general, usually about 1° or better).

8. The assumed or observed nature of natural and forced motions strongly influences the development of cosmological models. For instance, geocentric models place the earth at the center in part because the "natural" motion of earthy material is to fall toward the center of the cosmos. And the celestial motions were viewed as natural, from the rotation of spheres.

9. The basic aspects of a geocentric model explain in a reasonably simple way most of the major and minor motions of the sky by various geometric devices. The model was strong enough to survive for a long time—more than a thousand years. However, because such a model focuses on angular speeds of celestial objects, it does not have a fixed distance scale. And observation of an annual stellar parallax would disprove the model.

Key Terms

cosmos	epicycle	forced motion	heliocentric stellar parallax	scientific model
deferent	equant	geocentric	natural motion	stellar parallax
eccentric	force	heliocentric	parallax	

Study Exercises

1. What was one important astronomical achievement of the Babylonians? (Objective 2-2)

2. Contrast Babylonian and Greek astronomy in terms of their observational achievements. (Objectives 2-2, 2-9, and 2-10)

3. How did Aristotle's model explain (a) the apparent lack of motion by the earth, (b) the daily motion of the stars, and (c) the annual motion of the sun through the zodiac? (Objective 2-4)

4. How did Aristotle argue against a heliocentric model? (Objectives 2-3 and 2-4)

5. In the Ptolemaic model, how did observations set the period of the epicycle? The deferent? (Objective 2-6)

6. How did Ptolemy treat Mercury and Venus differently from the other planets? (Objective 2-6)

7. How did Ptolemy violate his own precept of uniform, circular motion? (Objectives 2-5, 2-6, and 2-7)

8. What observation was Ptolemy trying to explain with each of the following geometric devices: (a) epicycle, (b) eccentric, and (c) equant? (Objectives 2-6 and 2-7)

9. State one strength of Ptolemy's model. (Objectives 2-1 and 2-8)

10. How does a geocentric model differ from a heliocentric one in its prediction of an annual stellar parallax? (Objective 2-11)

11. What is the *crucial* observational test for a heliocentric model? (Objectives 2-1 and 2-11)

12. What is the function of geometry in building scientific models? (Objective 2-1)

Problems & Activities

1. Suppose you tried to duplicate Eratosthenes' observations (Focus 2.1) and came up with value of 8°. What would you calculate for the earth's circumference and radius?

2. With naked-eye observations, the smallest angle you can measure is about 1 arcminute. Assume that the stars are fixed on a sphere and consider two stars 1° apart as seen from the sun in a heliocentric model. How far must the sphere of stars be placed from the sun so that their parallax is 1 arcmin or less? *Hint:* What is the size-to-distance ratio of 1 arcmin?

3. The sun does not move at a uniform speed along the ecliptic. Over the half-year centered on the summer, it covers 180° in 186 days; over the winter half, 180° in 179 days. Using the eccentric to model this motion in a geocentric picture, calculate how far away from the center of the circle of the sun's motion the earth must be located. *Hint:* Draw a circle with a radius of, say, 10 cm. Use half the circle to show the summer interval, and half the winter. Assume that the sun's motion along the circle is uniform. Then how far would the earth have to be offset from the center of the circle to explain the sun's observed motion?

4. Suppose that in a heliocentric model, such as that of Aristarchus, the shell of stars had a radius of 10 earth–sun distances. If two stars, located in the plane of the earth's orbit, are observed to be 6° apart when the earth is closest to them, how far apart will they be in angle when the earth is on the side of its orbit farthest from them? Do you think this difference would be observable with the naked eye?

5. What is the size-to-distance ratio of the moon? How could you make an estimate of the moon's distance from the earth if you knew, say from Eratosthenes (Focus 2.1), the diameter of the earth in units like stadia or kilometers? What, then, is the moon's speed along a circular path?

6. Aristarchus is reported to have concluded that the moon has a diameter one-fourth that of the earth. Using Eratosthenes' diameter for the earth, at what distance does that place the moon?

7. What is the size-to-distance ratio for the sun? How many solar diameters fit in the earth–sun distance? If you assume that the moon is the same size as the earth, what size does that make the sun? What, then, is the sun's speed along a circular path?

See for Yourself!

You can carry out Eratosthenes' technique by finding a friend who lives far north or south of you. Use a stick in level ground to measure the angular height of the sun at local noon (see Activity 4 in Chapter 1) on the same day. Call your friend and compare results; the angles should differ. A good atlas will give you the distance between your locations. Find the circumference and radius of the earth.

To do so, recognize that the ratio of the difference in the angle of the sun at noon at the two locations divided by 360° is equal to ratio of the distance between the locations (say in kilometers) divided by the earth's circumference:

$$\frac{\text{difference in angle}}{360°} = \frac{\text{difference in locations}}{\text{earth's circumference}}$$

Then solve for the earth's circumference:

$$\frac{\text{earth's}}{\text{circumference}} = \binom{\text{difference}}{\text{in locations}} \times \frac{360°}{\text{difference in angle}}$$

Then divide the circumference by 2π to get the radius. How does it compare to the value given in Appendix C? ■

Central Question
How did Copernicus' sun-centered model explain the motion of the planets and ignite a revolution in cosmological thought?

The New Cosmic Order

"He was rather weird," was Hans Castorp's view. "Some of the things he said were very queer: it sounded as if he meant to say that the sun revolves around the earth."

THOMAS MANN, *The Magic Mountain*

Astronomy languished with the decline of Greek civilization in the first few centuries after Ptolemy. Most of the Greek astronomical works were unknown in Europe until the twelfth century, when Arabic manuscripts of Greek thought were translated into Latin. In the thirteenth century, Alfonso the Great of Spain sponsored the publication of the lists that became known as Alfonsine Tables, handy compilations of planetary positions based on the Ptolemaic model. The essential Ptolemaic model still worked well enough; most practicing astronomers had no reason to discard it.

One astronomer among many was unhappy: Nicolaus Copernicus, whose great work *De revolutionibus orbium coelestium* (On the revolutions of the heavenly spheres) was published in the year of his death. In *De revolution ibus,* the earth was shaken from a static place at the center of the cosmos into a path around the sun—just another planet with the rest. The model of Copernicus was *heliocentric*.

Copernicus' view did not immediately wrench the minds of astronomers from their geocentric notions. Some, like the Danish observer Tycho Brahe, considered Copernican claims but rejected them. Others, like Johannes Kepler, were struck by essential harmonies in Copernicus' model. The revolution in astronomy after Copernicus marked the first major shift in our concept of the earth's place in the cosmos—from geocentric to heliocentric. This revolution also injected a new concept into the models—that of a *physical* force working in the heavens. Ultimately, these shifts shaped modern views of the solar system and the cosmos.

FIGURE 3.1 Nicolaus Copernicus: "In the center rests the sun." (Courtesy of The Granger Collection, New York.)

3.1 Copernicus the Conservative

Nicolaus Copernicus (1473–1543) initiated an intellectual revolution (Fig. 3.1) by developing a new heliocentric model of the cosmos when the old one seemed adequate to most people, especially the astronomers of the time. This idea sparked a firestorm during the Reformation and Renaissance years and eventually transformed the earth's place in the cosmos.

The Heliocentric Concept

While studying medicine and law in Italy, Copernicus became familiar with the works of Aristotle, Pythagoras, and Plato. An offshoot of Plato's philosophy asserted that the sun is the source of the godhead and of all knowledge. These ideas not only singled out the sun as a body quite different from the planets but also encouraged Copernicus to ponder a heliocentric model.

In about 1514, Copernicus wrote a summary of his new model and circulated it to some friends. In this general outline of the heliocentric model (Fig. 3.2), the sun replaced the earth at the center of the cosmos, the earth revolved around the sun yearly, and also rotated on its axis daily. (These ideas had been outlined earlier by Aristarchus: Section 2.3.)

Copernicus still assumed that all celestial motions must be uniform, circular motions. He viewed Ptolemy's equant as a violation of this ideal and so attacked the Ptolemaic model as "not sufficiently pleasing to the mind." Recall (Section 2.4) that the equant required that a planet move *nonuniformly* as seen from the center of its deferent. The new model would reinstate the uniform circular motion that Ptolemy, in practice, gave up.

Just what compelled Copernicus to offer a new model is still somewhat of a puzzle. The basic Ptolemaic model worked well enough, by the demands of the day, in its planetary predictions. Contrary to some tales, the Ptolemaic model did *not* require the continual addition of circles to match observations and so had *not* become a monstrosity by medieval times. Copernicus took 20 years to develop his model. Yet his predictions came out no better than those based on the Ptolemaic model.

However, Copernicus asserted that his model surpassed the Ptolemaic one on aesthetic grounds—the notions of harmony and order. What were the aesthetically pleasing aspects of his model?

The Plan of *De Revolutionibus*

Long after Copernicus had privately distributed his ideas, his work became known to two astronomers at the University of Wittenberg: Georg Rheticus and Erasmus Reinhold. Fascinated by the new model, Rheticus visited the aging Copernicus, who showed him a final manuscript copy of *De revolutionibus*. Rheticus begged Copernicus to allow its publication. (Not until Copernicus' lifetime had printed texts, rather than manuscript copies, become common.) Copernicus finally agreed. The book was published in April 1543, after much trouble with the printer. Meanwhile Copernicus had suffered a stroke and was confined to his bed. Although the book was delivered to him before he died in June, he probably did not read it.

How fortunate that he didn't, for the final version of his life's work had been renamed! The original one was simply *De revolutionibus;* the published title sported two more words: *De revolutionibus orbium coelestium.* An unsigned preface, added without the astronomer's knowledge, stated that the work contained a new hypothesis about a heliocentric cosmos, for use in computing planetary positions. But it was not to be taken as *reality.*

For some years the preface was attributed to Copernicus and was taken to imply that he had not believed in the reality of his model. But Johannes Kepler (whose contributions are discussed in Section 3.4) discovered that the author of the preface was Andreas Osiander, a Lutheran clergyman who oversaw the completion of the book's publication. Osiander probably felt that a disclaimer was necessary to protect the book from criticism by those who believed that the Bible taught that the earth was motionless. He may have also changed the title to emphasize the motions of the heavens rather than those of the earth.

De revolutionibus took after the *Almagest* in outline and basic intention. It explained planetary motions using only *uniform circular motions*—the tradi-

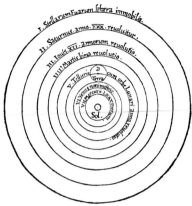

FIGURE 3.2 Simplified diagram of the Copernican model emphasizing one main departure from the Ptolemaic model: the sun is placed in the *center* of the cosmos, with the planets, including the earth, revolving around it. Like Ptolemy, Copernicus placed the sphere of stars beyond the sphere of Saturn. (Courtesy of Owen Gingerich.)

tional precept of the Greeks. Copernicus thought that although Ptolemy's model matched observations, it clashed with the Greek ideal because it used equants. Copernicus was greatly offended by the equant; it constituted his *major* objection to the Ptolemaic model. In a basically conservative mood, he wished to devise a model that was faithful to uniform circular motion and to eliminate the equant. So Copernicus decided to discover a natural explanation for retrograde motion and a new harmony for the celestial spheres that would relate the planets' distances to their periods around the sun.

■ In what essential way were *De revolutionibus* and the *Almagest* similar?

3.2 The Heliocentric Model of Copernicus

In the introduction to *De revolutionibus,* Copernicus lays out his assumptions. First, he requires that the planets move in circular paths around the sun at uniform speeds. Second, he assumes that the closer a planet is to the sun, the greater is the speed of its revolution. For instance, because Mercury is closer to the sun than the earth is, Mercury travels around the sun faster than the earth does. Except for the position of the sun, these ideas are *identical* to Ptolemy's assumptions. Then Copernicus treads on different ground; here's a short summary of his ideas.

1. All the heavenly spheres revolve around the sun, and the sun is at the center of the cosmos.
2. The distance from the earth to the sphere of stars is much greater than the distance from the earth to the sun.
3. The daily motion of the heavens relative to the horizon results from the earth's motion on its axis.
4. The apparent motion of the sun relative to the stars results from the annual revolution of the earth around the sun.
5. The planets' retrograde motions occur from the motion of the earth relative to the other planets.

Let's expand on these points to emphasize key differences between the Copernican and Ptolemaic models.

In point 1, Copernicus asserts, for philosophical and aesthetic reasons, that the cosmos is heliocentric rather than geocentric. But as the Greeks had realized from Aristarchus' model, a heliocentric scheme demands a phenomenon that had never been observed, namely, a stellar parallax (Section 2.3). Point 2 addresses this issue: if the stars are very far away, the parallax would be too small to be detectable by naked-eye observations. (Later, telescopes would show that Copernicus' intuition about the vast distances to stars was correct.)

In point 3, Copernicus declares that the daily westward motion of the skies results from the earth's rotation (from west to east) rather than from the rotations (from east to west) of celestial spheres. This idea has aesthetic appeal, for it replaces the rotations of *many* celestial spheres with the rotation of just *one* sphere, the earth. (You may wonder how the earth rotates without objects flying off its surface. This physical objection was voiced against the sixteenth-century Copernican model, and Copernicus had no good answer for it. Now we explain it by gravity and inertia: Chapter 4.) Copernicus suspected that the cosmos had more than one center of motion but could not prove it.

Point 4 explains how, in a heliocentric model, the sun appears to move around the earth. Here's an analogy. Imagine you're walking slowly counterclockwise around a lamppost. Look in the direction of the lamppost and note the background behind it. You'll see the background slowly change: the lamppost appears to move counterclockwise (the same direction as you're going) with respect to background objects. Now imagine that the lamppost is the sun, you are the earth revolving around the sun, and the background consists of the stars of the zodiac (Fig. 3.3). From the earth, you see the sun in a constellation, say Leo. As the earth revolves counterclockwise, the stars behind the sun change. After one month the sun appears in Virgo, one constellation to the east. The sun seems to have moved, relative to the

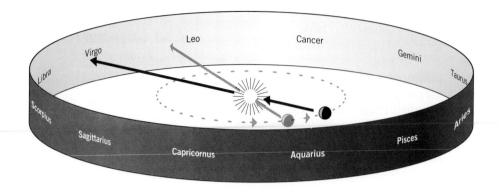

FIGURE 3.3 The sun's apparent motion through the zodiac in a heliocentric model. As the earth travels around the sun (counterclockwise; west to east), the line of sight to the sun and toward the background stars moves in the same direction (eastward). For example, start with the sun in Leo. A month later the earth will have moved eastward far enough to put Virgo behind the sun. The sun seems to move *eastward* through the zodiacal stars at the rate of about one constellation a month.

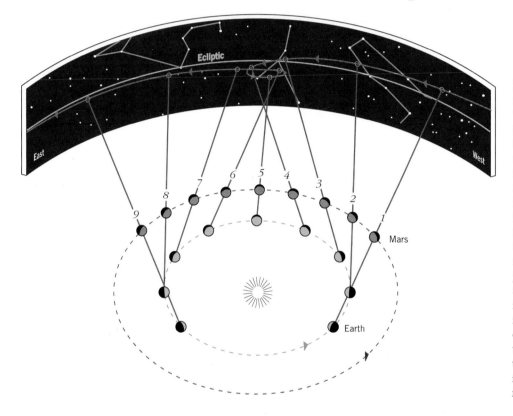

FIGURE 3.4 The 1995 retrograde motion of Mars in the Copernican model. The numbered positions begin on October 15, 1994, and have one-month intervals up to June 15, 1995. As the earth comes around the same side of the sun as Mars, it is moving faster along its path and so passes Mars (at 4, 5, and 6). During this passing interval, Mars appears to move backward (to the west) with respect to the stars. Since Mars appears in the middle of its retrograde motion just as the earth passes it, Mars is in opposition to the sun as viewed from the earth. Also, it is closest to the earth, and so appears its brightest in the sky. The same basic diagram applies to Jupiter and Saturn; these planets undergo retrograde motion when the earth passes them. The illusion of retrograde motion results from the passing situation.

stars, counterclockwise. Actually, the earth has moved; it's the line of sight from the earth to the sun that has changed.

Point 5 makes retrograde motion a natural result of the planets' revolutions. Retrograde motion arises from the planets chasing one another around, the faster (inner) planets regularly passing the slower (outer) ones. An analogy: when you pass a car on a highway, the slower car appears to move backward with respect to background objects. Similarly, as the earth speeds past another planet, that planet seems to move westward

(backward) against the backdrop of stars.

Retrograde Motion Explained Naturally

In the Ptolemaic model, epicycles moving on deferents generated the retrograde motions. Since retrograde motion arose naturally in his model, Copernicus eliminated five Ptolemaic epicycles and made a simpler model in which the relative motions of the earth and planets produce retrograde motions. When the earth passes any of the outer planets or when the earth is passed by the inner ones, retrograde motion occurs. The *passing* is the key to retrograde motion in this model.

Take Mars as an example (Fig. 3.4). Its retrograde motion occurs at opposition. In a heliocentric model, opposition happens when the earth is in line between the planet and the sun. As the earth approaches, the line of sight from the earth to Mars moves eastward. But as the earth passes Mars, the line of sight swings westward relative to the stars. As the earth moves on, the line of sight eventually moves eastward again. Mars undergoes the illusion of retrograde motion as the earth passes it. Note (Fig. 3.4) that the earth and Mars are closest together in the middle of the retrograde motion. So Mars should be brightest in the sky then—and it is! The same explanation applies to the motions of Jupiter and Saturn.

The identical chase-and-pass scenario results in the retrograde motions of Venus and Mercury. These inner planets pass the earth because they move faster around the sun. Consider Venus. When Venus moves around the back of the sun (from greatest western elongation), it appears to move eastward. But as Venus catches up to the earth (from the east side of the sun) and passes it, you observe Venus moving *westward* with respect to the stars (toward the west side of the sun). So for the two inner planets, retrograde also arises from relative motion, but the role of the earth is reversed. Note that the passing of *any* two planets produces the illusion of retrograde motion of one as seen from the other. If you stood on Venus and viewed the earth, it would appear to undergo retrograde motion (Fig. 3.5)!

Copernicus aimed for *simplicity* in his model. This aesthetic goal is common to scientific models. It is used as an essential standard by which to judge the best among competing ideas.

Planetary Distances

Using the planets' observed *synodic periods,* Copernicus achieved another basic goal: the establishment of the order and distances of the planets from the sun. (The **synodic period** is the time required for a planet in some alignment with the sun as seen from the earth—such as opposition—to move around the sky and return to that same alignment.) From the synodic periods, Copernicus calculated the **sidereal periods:** the periods of revolution of the planets with respect to the stars as seen from an observer at the sun (Focus 3.1). Examples of synodic periods are (1) Venus from one greatest western elongation to the next (584 days) and (2) Mars from one opposition to the next (780 days). Their sidereal periods are 225 days and 687 days.

Copernicus found that the planetary order from the sun falls into a natural sequence when based on sidereal periods: Mercury, with the shortest period, is closest to the sun; Saturn, with the longest, is the farthest away. He savored a harmony in this sequence, which later scientists—such as Kepler and Galileo—considered to be an essential elegance of the heliocentric model.

Copernicus also calculated the distances of the planets from the sun relative to the earth–sun distance. [Our term **astronomical unit,** abbreviated AU, is simply one earth–sun distance; the value was not well determined until the nineteenth century.] The figures Copernicus

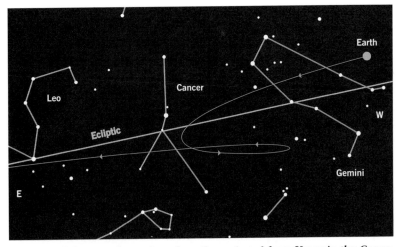

FIGURE 3.5 Retrograde motion of *earth as viewed from Venus* in the Copernican model. The geometry is the same as in Figure 3.4, with Venus the planet interior to the earth and the imaginary observing platform. The retrograde motion occurs as Venus passes the earth on the same side of the sun, so that the earth is in opposition as seen from Venus. Shown here is the ecliptic and three zodiacal constellations (Leo, Cancer, and Gemini, from east to west). (Imaginary view was produced by *Voyager* software for the Macintosh.)

TABLE 3.1 RELATIVE DISTANCES OF THE PLANETS		
PLANET	**COPERNICUS' VALUE (AU)**	**MODERN VALUE (AU)**
Mercury	0.38	0.387
Venus	0.72	0.723
Earth	1.00	1.00
Mars	1.52	1.52
Jupiter	5.22	5.20
Saturn	9.17	9.54

used for the distances came from direct observations and simple geometric arguments (Table 3.1); he was very pleased to be able to relate the motions of the planets to their distances from the sun. Let's see in detail how this is done, for it forged a crucial difference from the Ptolemaic model.

Relative Distances of the Planets

The planets' distances from the sun, relative to the earth–sun distance (the AU), can be found directly from observations in the Copernican model. The method, which assumes circular orbits, differs for planets interior and exterior to the earth.

For Mercury and Venus, the inner planets, the method rests on the observed maximum elongation angle of the planet from the sun (see Section 1.4). Remember that this angle averages about 23° for Mercury and 46° for Venus.

Here's the procedure for Venus. In a heliocentric model, maximum elongation occurs when the line of sight from the earth to Venus hits tangent to Venus' orbit (Fig. 3.6). So the line from the sun to the line of sight makes a right angle. Draw a triangle with one side the sun–Venus distance *(SV)*, another side the sun–earth distance (*SE* equals 1 AU), and the third side the earth–Venus line *(EV)*. You've drawn a right triangle.

Now observe the elongation angle, *SEV.* Let *SE* be one unit long. Then draw a triangle to scale with *SEV,* and Venus' distance from the sun *(SV)* will be 0.72 the earth's *(SE).* Since *SE* is 1 AU, Venus is 0.72 AU from the sun. Similarly, use 23° as the maximum elongation angle for

Mercury to find its distance as 0.39 AU. Note that just one observation establishes the distances of the inner planets in a heliocentric model.

A different method applies to the outer planets. Take Mars as an example. Start with some alignment of the earth and Mars (*A* in Fig. 3.7*a*). Wait one sidereal period of Mars. By the time the earth has returned to the same position in its orbit relative to the stars as seen from the sun (Focus 3.1), our planet—moving faster—will have gone around more than once (*B* in Fig. 3.7*b*). Find the position of Mars from these two observations from the earth by extending the lines of sight for both. The point at which the lines intersect is the position of Mars in its orbit (*M* in Fig. 3.7*c*).

Suppose you have made an accurate scale drawing of these observations. Calling the earth–sun distance 1 AU, you can also measure the sun–Mars distance in AU. Or you can solve the triangle *MSB* for side *SM* using trigonometry, because you know all the angles. A similar technique works for the other outer planets. These distances (Table 3.1) are fixed by the planets' sidereal periods and the geometry of the heliocentric model.

Problems with the Heliocentric Model

Copernicus eliminated epicycles to explain retrograde motion and at the same time expelled the equant. Yet the heliocentric model did not predict planetary positions any better than the Ptolemaic model. In fact, the Copernican model was not even simpler, judged on the basis of a count of circles. Copernicus eliminated five, but he was forced to add many smaller ones to accommodate the *variations* in planetary motions. (Ptolemy had accounted for these with the equant.) So by a simple count of the total of circles involved, the complete Copernican model is somewhat more complicated than Ptolemy's original model.

Copernicus also sidestepped the fact that his model violated Aristotelian physics. And he did *not* offer new physical ideas to support his model; those still needed development. Copernicus had created a new geometric model, not a physical one.

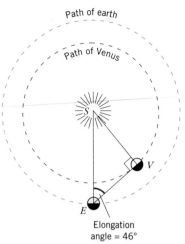

FIGURE 3.6 The distance of Venus from the sun. As seen from the earth at its time of greatest elongation, Venus makes an angle (*SEV*) of about 46° from the sun. The line of sight from the earth then is tangent to the orbit of Venus. So the radius of Venus' orbit drawn to this point makes a right triangle with one side the sun–Venus distance (*SV*) and the other the earth–sun distance (*ES*).

Caution. Don't fall into the trap of thinking that the Copernican model included a force between the sun and the planets. It did not! Planetary motions were believed to arise from the natural motions of the celestial spheres—uniform rotation. As in the Ptolemaic model, this was understood as natural motion without forces.

The Impact of the Heliocentric Model

Although Copernicus' claims clashed with tradition, a few astronomers in the middle of the sixteenth century tested the new setup; they knew that the venerable Alfonsine Tables (based on the Ptolemaic model) contained some inaccurate positions.

Foremost among those willing to try the new model was Erasmus Reinhold, who had encouraged Rheticus, his fellow professor at Wittenberg, to visit Copernicus. Reinhold made up planetary position tables based on the new model; these were adopted by almanac makers as the Prutenic Tables. Yet Reinhold did not swing completely to the Copernican model. He saw it as more geometrical rather than physical. Because Copernicus held to uniform circular motions, his model was geometrically similar to Ptolemy's (except for the different placement of earth and sun). What astronomers would switch if the new model offered no advantages over the old?

And why did Copernicus bother to develop his model at all? Certainly, he was motivated by aesthetic reasons: he saw a new harmony in the "design of the universe and fixed symmetry in its parts" that was "pleasing to the mind." The "fixed symmetry" referred to the spacing of the planets as determined by observations. In the Ptolemaic model, each planet's circle is independent of the others. Overall, Copernicus' model had a unity lacking in the Ptolemaic one, so it was simpler in grand design.

Owen Gingerich, a contemporary historian of science at Harvard University, has scoured the world for more than a decade for original editions of *De revolutionibus*. He has examined more than 260 of them, paying special attention to any marginal notes added by sixteenth-century owners and readers. Gingerich's survey found that Copernicus' book was widely read with a high degree of technical comprehension, though sometimes selectively.

In 1616 the Catholic Church placed *De revolutionibus* on the Index of Prohibited Books because of specific passages, which were supposed to be changed by the censors. Very few of the copies that Gingerich saw had been changed, however, with most of the censored books being found in Italy. Hence, the printing of Copernicus' ideas resulted in their rapid spread in Europe, intellectual sparks that kindled new ideas.

■ How did Copernicus' model deal with the unobserved stellar parallax?

3.3 Tycho Brahe: First Master of Astronomical Measurement

Tycho Brahe (1546–1601) was born in Denmark into a family of noble standing. While he was a child, his uncle—who had no children—stole him from his parents, supported the boy, and decided that his adopted son should become a lawyer. Young Tycho was sent off to study law, but secretly he worked on astronomy, spent his allowance on astronomy books, and sneaked out at night to make observations.

Later in his career, Tycho (Fig. 3.8) rejected the heliocentric model on both physical and observational grounds. He proposed an alternative model that mixed the classic geocentric model with the Copernican one. His work is infused with the attitudes typical of a professional astronomer just after the publication of *De revolutionibus*. It revealed that the reality of the Copernican model had escaped most astronomers, who saw the heliocentric model as merely a geometric device for calculating planetary positions. Copernicus invited this stance because his model violated Aristotelian physics and offered no alternative physical basis.

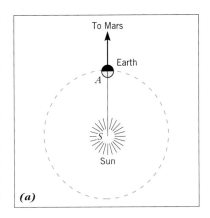

(a)

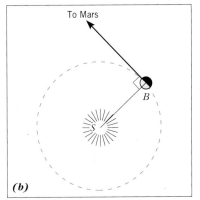

(b)

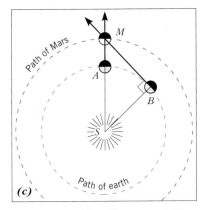

(c)

FIGURE 3.7 The distance of Mars from the sun determined in a heliocentric model. (*a*) Start with the earth at *A* and Mars at opposition. We know the angle of Mars relative to the sun, but we don't know the distance. (*b*) Wait one sidereal period of Mars. Then Mars returns to the same place in its orbit (as seen from the sun), but the earth has gone around almost twice (to *B*). (*c*) The line of sight from the earth (at *B*) to Mars crosses the Martian orbit at the same spot (*M*) that the earlier one did (from *A*). Call the earth–sun distance 1 unit. Then the distance from the sun to Mars is known to this scale because all the angles are known.

Enrichment Focus 3.1

SIDEREAL AND SYNODIC PERIODS IN THE COPERNICAN MODEL

From the earth you can see a planet in a particular alignment with the sun—opposition, for example. The time interval for that planet to return to the same alignment with respect to the sun in the sky is called the *synodic period*. In contrast, consider the time it takes for a planet to return to a specific location in the zodiac as seen from the sun. That's the planet's *sidereal period*.

The synodic period is a geocentric property, while the sidereal period is a heliocentric one. How can you transform synodic periods into sidereal ones? Assume that the planets move in circles and that the closer a planet's path is to the sun, the faster the planet moves. So its sidereal period is shorter. You then have a chase situation, in which an inner planet always catches up to an outer one.

Imagine that two adjacent planets start out lined up (Fig. F.3), with the outer planet in opposition as seen

TABLE F.1 SYNODIC AND SIDEREAL PERIODS COMPARED

PLANET	SYNODIC PERIOD: View from Earth (days)	SIDEREAL PERIOD: View from Sun (days)
Mercury	116	88
Venus	584	225
Earth	—	365
Mars	780	687
Jupiter	399	4,330
Saturn	378	10,800

from the interior one. How long until they are aligned again? Note that the inner planet will make one circuit around the sun and then catch up to the outer one to reform the alignment. In other words, the inner planet laps the outer one. So as seen from the inner planet, the synodic period will be longer than its sidereal one.

The inner planet moves at $360/P_i$ degrees per day, where P_i is the sidereal period of the inner planet. The outer planet, moving at the slower rate of $360/P_o$ degrees per day, lags behind the inner planet more and more each day, where P_o is the sidereal period of the outer planet. At the end of one day, the inner planet has gained an angle of $(360/P_i - 360/P_o)$ degrees on the outer one. This day's gain is equal to $360/S$ degrees, where S is the synodic period of either planet seen from the other. (These periods are the same.) In algebraic terms,

$$\frac{360°}{S} = \frac{360°}{P_i} - \frac{360°}{P_o}$$

or, by dividing by 360°

$$\frac{1}{S} = \frac{1}{P_i} - \frac{1}{P_o}$$

Take the earth as the inner planet and any planet exterior to it as the outer one. Since the earth's sidereal period is one year, $P_i = 1$; then

$$\frac{1}{S} = 1 - \frac{1}{P_o}$$

where P_o is the sidereal period of the outer planet, and S, its synodic period *as seen from the earth,* is in years.

Now consider the earth to be the outer planet and choose any inner one. Then $P_o = 1$ is the earth's sidereal period, and

$$\frac{1}{S} = \frac{1}{P_i} - 1$$

where all the quantities are in years. Use Venus as an example. The synodic period of Venus is 585 days, or $585/365 = 1.60$ years. So

$$\frac{1}{P_{Venus}} = 1 + \frac{1}{1.60}$$
$$= 1.62$$
$$P_{Venus} = 0.62$$

This transformation from synodic to sidereal periods played a key role in the Copernican model, because it places the planets in a *natural* order (Table F.1) from the sun, from the shortest (Mercury) to the longest (Saturn) periods.

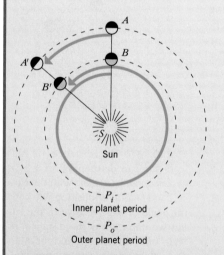

FIGURE F.3 Synodic and sidereal periods in the Copernican model. As well as traveling a shorter distance, the inner planet moves faster along its path, and so must have a shorter sidereal period (P_i) than the outer planet (P_o).

The New Star of 1572

In November 1572 a new star, called then a **nova,** burst into view in the constellation of Cassiopeia. At its brightest, it could be seen during the day. (Today we know that a nova is a star normally too faint to be seen that suddenly increases many times in brightness. Such an apparently new star is called a nova from the Latin *stella nova,* "new star.")

Tycho observed this new star for more than two years and collected observations from all over Europe of the star's position in the sky. Using others' data as well as his own, Tycho showed that the nova was in the same place in Cassiopeia regardless of where it was observed. That is, the nova did *not* have an observable angular shift from different places on the earth. So it was at a great distance from the earth. Tycho concluded that the nova had to lie in the sphere of stars, thus rebutting the Aristotelian tenet that the heavens were immutable.

Tycho's work on the new star of 1572 catapulted him to fame as an astronomer, and King Frederick II of Denmark offered Tycho an observatory on the island of Hven. With royal funds—estimated by Tycho to be more than a ton of gold—Tycho built on Hven the first modern astronomical observatory (Fig. 3.9), Uraniborg, Castle of the Heavens. Here he worked in grand style with the finest observing equipment (designed by himself) and a crew of assistants. His instruments had an accuracy of 1 arcmin (as good an angular precision as can be had with the naked eye).

At Uraniborg, Tycho began the first *comprehensive* observational program. The planets were observed not only at times of importance, such as during retrogrades, but also in between notable events. No one had tried for this continuity before. Tycho eventually amassed records of precise planetary positions covering the years 1576 to 1591.

Tycho's Hybrid Model

Tycho concluded that the nova of 1572 made the classic geocentric model untenable, so he devised his own. He had read about the Copernican model, but he could not accept the idea of the earth revolving around the sun. He thought that "the earth, that hulking, lazy body" was "unfit for a motion as quick as that of the ethereal torches." In Tycho's model, the moon and sun revolve around the earth, but all the other planets revolve around the sun (Fig. 3.10). This was geometrically the same as the Copernican model—only the one center of motion was changed.

Also, Tycho had tried to measure a stellar parallax (Section 2.3) but he failed to detect it. The lack of a detectable parallax fortified Tycho's belief that the

FIGURE 3.8 Tycho Brahe: part of the astronomer's nose was cut off in a fencing duel and replaced with a silver piece, visible in this portrait. Around the figure are arrayed the coats of arms of the royal families to which Brahe was related. A playwright in England may have seen this portrait, for he used the names Rosenkranz and Guildenstern in his play *Hamlet.* (Courtesy of The Granger Collection, New York.)

FIGURE 3.9 View of Uraniborg, Tycho's research observatory, at the observatory grounds on the island of Hven. (Courtesy of Owen Gingerich.)

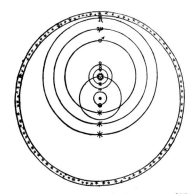

FIGURE 3.10 Simplified version of Tycho's model for the cosmos. It is a geocentric modification of the Copernican one. The sun moves around the earth, and all the planets circle around the sun. (Courtesy of Owen Gingerich.)

Copernican model was invalid. (We now know that the largest heliocentric parallaxes are almost 100 times smaller than Tycho could have detected with his best instruments.)

When Frederick II died, Tycho fell from royal favor because of his autocratic ways. So he moved his equipment to Prague, where he worked for Emperor Rudolf of Bohemia and took on the young Johannes Kepler as an assistant. When Tycho died in 1601, an era in astronomy ended. Another was born when Kepler created a new astronomy based on Tycho's observations.

■ What was Tycho's main contribution to the astronomy of his time?

3.4 Johannes Kepler and the Cosmic Harmonies

Ptolemy's model, because it kept the earth in the center of the cosmos, upheld traditional physical ideas. In contrast, the Copernican model violated concepts of Aristotelian physics. So the heliocentric model lacked a physical basis, judged by the standards of the time.

Johannes Kepler (1571–1630) forged new ideas about planetary motion that cast the foundation of modern cosmological concepts. Kepler (Fig. 3.11) reshaped the Copernican model into a truly *physical* one in which the sun determines the planetary orbits by a force between the sun and the planets. Kepler also simplified the Copernican model, perfected its usefulness for the prediction of planetary positions, and moved it to the realm of modern ideas about the solar system.

The Harmonies of the Spheres

Born on December 27, 1571, in the small town of Weil der Stadt in southwestern Germany, Kepler in 1589 entered the University of Tübingen. He was an outstanding student. In his last year at Tübingen, he was selected to replace a teacher of mathematics at a Protestant high school in Graz, Austria, where he then made his first key astronomical discoveries.

One day while teaching, Kepler was struck by the most compelling insight of his life. As he was telling his class about conjunctions of Jupiter and Saturn, which occur at regular intervals, he saw that the ratio of the radii of two circles— one drawn around a triangle and one drawn inside it (Fig. 3.12)—is about 2 to 1. That is almost the same as the ratio of the distances of Saturn and Jupiter from the sun in the Copernican model, 1.83 to 1. Kepler found this result exciting, for Saturn was the first planet in from the sphere of stars, and the triangle was the simplest plane figure. He felt he had found a geometrical design for the solar system.

While still at Tübingen, Kepler had been delighted by the harmonies he found in the Copernican model, which he adopted. One such harmony appears in the spacing of the planets and their sidereal periods around the sun. Copernicus, however, had not explained this relationship. Kepler thought he could: perhaps a geometric layout determined the planets' spacings.

Kepler approached this idea by first considering plane geometric figures. Then he realized that the situation was actually three-dimensional and required solid figures. From classical geometry Kepler knew that only five regular solids existed, those with faces having the same kinds of regular plane figure. The known planets numbered six, separated by five spaces. So Kepler felt that these five regular solids might establish the spacing of the planetary spheres. He worked out a nesting scheme of solids and spheres (Fig. 3.13) that gives distances from the sun within a few percent of those given by Copernicus. Kepler's intuition later proved to be wrong—the discovery of any new planet shoots it down—but it energized his later study of planetary motions. It led to radical results.

With great excitement, Kepler wrote to his former astronomy teacher, Michael Maestlin, about not only this idea but also another: the sun must have some power or force to propel the planets in their orbits. This conviction seemed reasonable because Mercury, for example, moves faster in its orbit than Venus. A force from the sun would push

FIGURE 3.11 Johannes Kepler, as a young man: "Astronomy has two ends, to save the appearances and to contemplate the true form of the edifice of the world." (Courtesy of Owen Gingerich.)

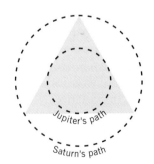

FIGURE 3.12 Kepler's schematic picture of the paths of Jupiter and Saturn drawn within (Jupiter) and around (Saturn) a triangle based on the cycle of conjunctions of Jupiter and Saturn. The ratio of radii of the two circles is almost the same as that of the distances of Jupiter and Saturn from the sun in AU in the Copernican model.

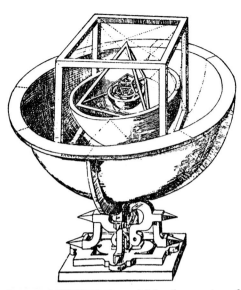

FIGURE 3.13 Kepler's model for the spacing of the planets based on the five regular solid figures. During Kepler's lifetime six planets were known, so it made sense to suggest that five spaces existed between them. Kepler knew that only five regular solids exist. By arranging these solids between the spheres holding the planets, Kepler thought he had explained the "edifice of the world." (Courtesy of Owen Gingerich.)

Mercury harder than Venus, because Mercury is closer to the sun. Kepler failed to work this idea out at the time, but it was firmly implanted in his mind.

In 1594 Kepler published the results of his work in the *Mysterium cosmographicum* (The cosmic mystery), which announced the details of his nesting scheme and declared his adherence to the Copernican model. It was the first such treatise by a professional astronomer. His intuition that the sun physically directs planetary motions set the stage for the next advance of astronomy in terms of physical laws.

■ What was one of Copernicus' harmonies adopted by Kepler?

3.5 Kepler's New Astronomy

Kepler was not yet satisfied with his geometric model. He needed better observational data to calculate the thicknesses of the nesting shells. To obtain them, Kepler turned to Tycho Brahe, who had read the *Mysterium cosmographicum* but did not like its mystical approach to astronomy. When Tycho received Kepler's request for observations, he invited Kepler to come to Prague to discuss the matter personally. Tycho and Kepler met on February 4, 1600, an encounter fateful for them both and for the progress of astronomy. Their personalities clashed so strongly that they could not work on astronomy together.

Later, Tycho was at a party with a baron and drank too much. Because it was impolite to leave during the affair, Tycho did not relieve himself and as a result suffered a urinary infection that led to his death. Just before he died however, Tycho relented in his hostility to Kepler. He consented to Kepler's use of his observations but urged Kepler to support his cosmological model with these data. After Tycho's death, Kepler was promoted to Tycho's position and allowed access to his unpublished records. With them, Kepler first attacked the problem of the orbit of Mars.

The Battle with Mars

Kepler first tried to match Tycho's observations with a heliocentric scheme that had a circular orbit, an eccentric, and an equant. With this model, he could fit Tycho's data to within 2 arcmin along the ecliptic. But his predictions above and below the ecliptic were wide from the mark. This discrepancy struck Kepler as important—he realized that he must treat the orbit in three dimensions. He then moved the center of the earth's orbit and fitted positions above and below the ecliptic correctly, but was 8 arcmin off along the ecliptic. Kepler rejected this result, for he knew that Tycho's observations were accurate to 1 arcmin. This demand—that his model's predictions be as good as the observations—marked the first time anyone seriously set this tough standard in astronomy.

Frustrated, Kepler pondered the notion that the sun drove the planetary motions. But by what force? Influenced by William Gilbert's book *De magnete,* Kepler imagined the sun showering out a magnetic force that propelled the motion of the planets. This idea required that a planet move faster in its orbit when it was closer to the sun.

Kepler found that a magnetic force idea worked out *if* the orbit of Mars was *elliptical* rather than circular. He wrote, "With reasoning derived from physical principles agreeing with experience, there is no figure left for the orbit of the planet except a perfect ellipse." Kepler had found one of his three famous laws of planetary motion, which rested on a *physical* explanation of Tycho's precise observations.

Properties of Ellipses

To understand Kepler's laws, you need to have in mind the basics of the geometrical properties of ellipses. (Focus 3.2 provides the details.) Ellipses are a lot like circles, so their basic properties are fairly easy to grasp. You can make an ellipse by taking a loop of string and holding it down on a board with two tacks. Keeping the string taut with a pencil held against it, draw a curve around the tacks (Fig. 3.14). That curve is an ellipse. The two tacks mark the two *foci* of the ellipse. Each point on the ellipse has the property that the sum of its distances from the two foci is the same as the sum of those distances for every other point. The line through the foci to both sides of the ellipse is called the **major axis.**

The amount by which an ellipse differs from a circle is defined as an el-

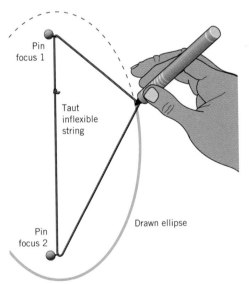

FIGURE 3.14 Drawing an ellipse with a loop of string around two tacks on a board. The pencil must hold the loop taut at all times to draw the proper curve.

lipse's **eccentricity.** Imagine that you moved the two tacks at the foci closer together. The ellipse would become more circular. In fact, when the two tacks coincided, the ellipse would become a circle, with an eccentricity of zero. Moving the two tacks farther apart would increase the eccentricity until the tacks reached the limits of the string. The eccentricity would then be 1. Note that the eccentricity increases as the distance between the foci does. The most essential property of an ellipse is its eccentricity.

Kepler's Laws of Planetary Motion

Today, Kepler's fame rightly rests on his three laws, which broke the tradition of uniform motion on circular orbits. Yet **Kepler's laws of planetary motion** were small fragments of his wider vision for harmonies in the physics of celestial motions. Let's look briefly at each law in modern terminology. (The dates given here are those of publication not of discovery.)

LAW 1. *Law of Ellipses (1609)* The orbit of each planet is an ellipse, with the sun at one focus (Fig. 3.15). The other focus is located in space. The distance from the sun to the planet *varies* as the planet moves along its elliptical orbit.

LAW 2. *Law of Equal Areas (1609)* A line drawn from a planet to the sun sweeps out equal areas in equal times. Physically, this law holds that the orbital speeds are nonuniform but vary in a regular way: the farther a planet is from the sun, the more slowly it moves (Fig. 3.16).

LAW 3. *Harmonic Law (1618)* The square of the orbital period of a planet is directly proportional to the cube of its average distance from the sun (Fig. 3.17). This law reveals that the planets with larger orbits move more slowly around the sun and so implies that the sun–planet force decreases with distance from the sun.

What do these laws mean?

Well, for the first time we can talk about actual *orbits* of the planets and the patterns related to them. Law 1 points

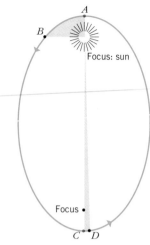

FIGURE 3.15 Kepler's first law. The shape of the planetary orbits is elliptical (greatly exaggerated here), with the sun located at one focus of the ellipse. The closest point along the orbit to the sun is called the *perihelion;* the farthest, the *aphelion.* The sum of the distances of any point on the ellipse from the two foci is a constant.

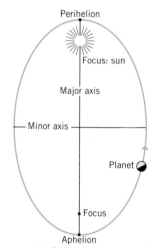

FIGURE 3.16 Kepler's second law. Consider two equal periods of time: in one the planet is closest to the sun (*AB*), and in one it is farther away (*CD*). At *AB*, the planet–sun distance is shortest, and the planet moves fastest in its orbit. At *CD,* the planet moves more slowly because it is farther from the sun. In both cases, the areas (shaded areas *AB* to the sun and *CD* to the sun) covered by the line drawn from the planet to the sun are equal.

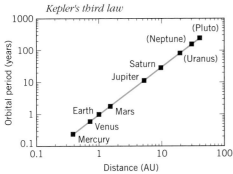

Kepler's third law

FIGURE 3.17 Kepler's third law. A planet's orbital period (once around the sun) is related to its average distance from the sun. For each (now known) planet, this diagram plots the orbital period (in years) squared to its distance (in AU) cubed. The line indicates the values expected from the modern formulation of Kepler's third law. Note how well the points for the planets fall along the same straight line, even Uranus, Neptune, and Pluto, which were not known in Kepler's time.

out that the shape of the orbits is elliptical, so a planet's distance from the sun varies. (Note that the sun does *not* lie in the center of the ellipse but at one focus.) It also shows that the planetary orbits follow a special and regular geometrical shape.

Law 2 notes that as a planet's distance from the sun varies, so does its orbital speed: the closer a planet is to the sun, the faster it goes. You get the sense that the sun pulls a planet toward it; then the planet whips around the sun and slows down as it moves away—a force from the sun acts upon it. This law also allows us to find the planet in another part of the orbit once it has been observed at some position.

Law 3 states that the more distant a planet's orbit is from the sun, the slower

Enrichment Focus 3.2

GEOMETRY OF ELLIPSES

Ellipses are related to circles. If you draw a curve as shown in Figure 3.14, the two tacks mark the two *foci* (F_1 and F_2) of the ellipse. Each point on the ellipse has the property that the sum of its distances from the two foci (F_1 to P and F_2 to P in Fig. F.4) is the same. The line through the foci to both sides of the ellipse (R_a to R_p) is called the **major axis.** Half this length is the **semimajor axis,** usually designated by a; the major axis has a length of $2a$. The distance from the center of the ellipse to a focus is designated c, so the distance between the two foci is $2c$.

The most essential property of an **ellipse** is its eccentricity. To define it exactly, the eccentricity, e, is

$$e = \frac{c}{a}$$

If the sun is at one **focus** (F_1) of a planet's orbit (Fig. F.4), then R_p, the closest point to the sun, is the **perihelion** point in the orbit. The perihelion distance R_pF_1 is

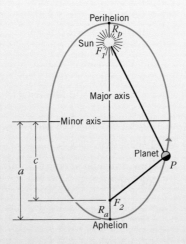

FIGURE F.4 The geometrical properties of an ellipse.

$$\overline{R_pF_1} = a - c = a - ae = a(1 - e)$$

In comparison, R_a, the farthest point from the sun, is the **aphelion** point.

The aphelion distance R_aF_1 is

$$\overline{R_aF_1} = a + c = a + ae = a(1 + e)$$

Note that a is the average distance from one of the foci of a point moving on the ellipse. So for planetary orbits, a is the average distance of a planet from the sun.

As an example, the orbital eccentricity of Mercury is 0.21, and its semimajor axis is 0.39 AU. Then Mercury's perihelion distance is

$$\begin{aligned} R_p &= a(1 - e) \\ &= 0.39(1 - 0.21) \\ &= 0.39(0.79) \\ &= 0.31 \text{ AU} \end{aligned}$$

Similarly, the aphelion distance is

$$\begin{aligned} R_a &= a(1 + e) \\ &= 0.39(1 + 0.21) \\ &= 0.39(1.21) \\ &= 0.47 \text{ AU} \end{aligned}$$

its average orbital motion will be. Look at Table 3.2. You can see that a greater distance from the sun means that a planet takes much longer to orbit the sun once. You can also note that if you square the orbital period (in years) and divide by the cube of the distance (in AU), the ratio comes out the same for all the planets—unity, in these units. The fact that this ratio is the same for such different orbits—a common pattern—suggests a similar physical cause.

Both laws 2 and 3 imply that a force between the sun and the planets weakens with increasing distance. Hidden in law 3 is an exact description of how the sun–planet force decreases with greater distance, but Kepler was not able to figure it out. You can catch a glimpse of this by examining Figure 3.17, which plots the data from Table 3.2. Note that all the points fall along the same line. (Later, Isaac Newton did figure out the force: Section 4.4.)

Law 3 can be written algebraically as

$$\frac{P^2}{a^3} = k$$

where P is the orbital period, a the average distance from the sun, and k a constant. If P is in years and a in AU, then k equals 1. For *any* body orbiting the sun (even spacecraft!), law 3 can be used to find the average distance from the orbital period or the orbital period from the average distance. For example, if $a = 4$ AU, then $P = 8$ years. This result depends only on a; no amount of eccentricity of the orbit can change the orbital period.

And, if the orbiting mass is much smaller than the mass it orbits, the result is independent of the mass of the orbiting object.

The New Astronomy

As he announced in *Astronomia nova* in 1609, Kepler had developed an astronomy based on geometry *and* on physics (a force between the sun and the planets). Elliptical orbits satisfied both his intention and the observations of Tycho Brahe. Using ellipses, Kepler calculated the Rudolphine Tables (1627), which supplanted both the Prutenic Tables based on the Copernican model and the Alfonsine Tables based on the Ptolemaic one.

Kepler's Rudolphine Tables completely revised the heretofore mediocre science of astronomy. Earlier, a few degrees of error between predictions and observations was considered to be "good enough." Kepler's calculations were some 10 times more accurate than those based on either the Copernican or the Ptolemaic model (Fig. 3.18). The reason was that the older models assumed uniform, circular motions for the planets, whereas in fact, the motions are neither circular nor uniform.

Kepler broke the ancient spell of perfect circles and uniform motion that had mesmerized astronomers for centuries. He rewove the fabric of the heavens into a pattern that proclaimed the end of an ancient era and the birth of astronomy as a physical science.

■ **What physical activity of all the planets does each of Kepler's laws deal with?**

TABLE 3.2 MOTIONS OF THE PLANETS

PLANET	PERIOD, P (years)	P^2	DISTANCE, a (AU)	a^3	P^2/a^3
Mercury	0.24	0.058	0.39	0.059	0.97
Venus	0.62	0.38	0.72	0.37	1.0
Earth	1.0	1.0	1.0	1.0	1.0
Mars	1.9	3.6	1.5	3.4	1.1
Jupiter	12	140	5.2	140	1.0
Saturn	29	84	9.5	86	0.98

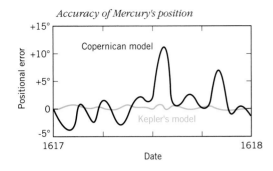

FIGURE 3.18 The accuracy of predictions of planetary positions: difference between predicted and actual positions of Mercury in 1617 and 1618 in Kepler's model (heliocentric, elliptical orbits, nonuniform speeds) and in the Copernican one (heliocentric, circular orbits, uniform speeds). In the Copernican model, the error is as large as 10°. For Kepler's scheme, the maximum error is 10 times smaller. (Courtesy of Owen Gingerich and Barbara Welther.)

Key Concepts

1. Copernicus proposed his heliocentric model mainly for aesthetic reasons; he especially disliked the equant in the Ptolemaic model, which violated the traditional precept of uniform heavenly motions about circles. Copernicus believed that, overall, his model was more pleasing than a heliocentric one.

2. The Copernican model (Table 3.3) views the daily motion of the sky as arising from the earth's rotation, the sun's eastward motion along the ecliptic as a reflection of the earth's revolution around the sun, and retrograde motion as an illusion when one planet passes another. The order of the planets is set by their periods of revolution about the sun; the planets' distances are set by observations and simple geometrical layouts.

3. Copernicus' model did not predict planetary positions any better than the Ptolemaic model did, nor was it simpler in terms of the total of circles used. It also violated the accepted physics of the day (that of Aristotle), and it required a stellar parallax, which was

TABLE 3.3 COMPARISON OF THE PTOLEMAIC (P), COPERNICAN (C), AND KEPLERIAN (K) SYSTEMS

OBSERVATION	EXPLANATION
Motion of entire heavens daily from E to W	P: Motion of all heavenly spheres E to W C: Reflection of rotation of earth from W to E K: Same as C
Annual motion of sun W to E through zodiac	P: Rotation of sun's sphere W to E in a year C: Reflection of annual revolution of the earth about the sun K: Same as C
Nonuniform motion of sun through zodiac	P: Circular path of sun eccentric, with uniform speed C: Same as P K: Elliptical orbit of the earth, with nonuniform speed
Retrograde motions of the planets	P: Epicycles and deferents C: Relative motions of planets, including earth, around sun K: Same as C
Variations in retrograde motions	P: Equant, eccentrics C: Small epicycles, eccentrics K: Nonuniform orbital motion and elliptical shape of orbits
Distances of planets	P: Arbitrary as long as angular relationships are correct C: Relative distances set by observations K: Distances related to periods
"Cause" of planetary motions	P: Natural motion of celestial spheres; no force C: Same K: Magnetic force from sun
Accuracy of prediction	P: Error typically 5° or less, sometimes 10° C: Same as P K: Error generally about 10', sometimes as large as 1°

not observed. All these points made the Copernican model quite unattractive in principle to contemporary astronomers.

4. Tycho Brahe typified the reaction of practicing astronomers toward the Copernican model. He thought it an interesting geometric idea but could not accept a moving earth. Tycho tried to observe stellar parallax but didn't detect it. So he eventually developed a hybrid model with the earth at the center of the cosmos. Tycho did, however, compile a years-long series of accurate and continuous observations that formed the basis of Kepler's incisive work.

5. Kepler used Tycho's observations and tried to match them with the Copernican model. His failure to do so within the errors of Tycho's fine observations drove him to decide that planetary orbits are elliptical (first law); that the planets move around the sun nonuniformly but predictably (second law); and that the farther a planet is from the sun, the longer it takes to complete one orbit (third law). Kepler held that the sun exerts a force that keeps the planets in their orbits.

6. Kepler's model predicted planetary positions much more accurately than either the Ptolemaic or the Copernican model, mainly because Kepler broke with the long tradition of uniform, circular motions for celestial objects. He was driven to this radical approach by his faith in the reliability and accuracy of Tycho's observations. Here we see how good observations can work to change a scientific model.

7. Kepler's notion that astronomy should be based on physics as well as on geometry produced the crucial concept that planetary motions and the layout of the solar system arose from a force between the sun and the planets. The planets actually moved on orbital paths, and the sun compelled and controlled their motions.

Key Terms

aphelion

astronomical unit (AU)

eccentricity

ellipse

focus

Kepler's laws of planetary motion

major axis

nova

perihelion

semimajor axis

sidereal period

synodic period

Study Exercises

1. What was Copernicus' primary objection to the Ptolemaic model? (Objectives 3-1 and 3-2)

2. How did Copernicus account for retrograde motion in his model? (Objective 3-3)

3. If you were on Mars, when would you see Jupiter retrograde? The earth? (Objective 3-3)

4. What is the difference between a planet's synodic and sidereal periods? How can sidereal periods be found from synodic ones? (Objective 3-4)

5. In your opinion, what was the major advantage, if any, of the Copernican model over the Ptolemaic one? (Objectives 3-1 and 3-5)

6. What force keeps the planets moving around the sun in the Copernican cosmos? (Objectives 3-6 and 3-8)

7. Give at least two differences between the models of Copernicus and Kepler. (Objective 3-9)

8. What two advantages did Kepler's model have over that of Copernicus? (Objective 3-9)

9. According to Kepler's second law, a planet travels slowest at what point in its orbit? (Objective 3-10)

10. In what sense did Kepler develop a *new* astronomy? (Objective 3-7)

11. Keeping in mind that the orbit of Mercury is an ellipse with a large eccentricity, explain the fact that the maximum elongation angle of Mercury can vary over a few degrees. (Objectives 3-8 and 3-10)

12. Imagine a new planet is discovered between Mars and Jupiter. Using Table 3.2, what would you predict that this planet would have for the ratio of P^2/a^3?

Problems & Activities

1. The maximum elongation angle of Mercury averages 23°. Calculate the average Mercury–sun distance in AU. *Hint:* sin *SEM* = *SM/SE*.

2. As an alternative procedure for Problem 1, draw on graph paper a circle with a radius of 10 cm. Mark the earth at some position on the circumference and the sun at the circle's center. Now follow Figure 3.6. Draw a line between the sun *(S)* and the earth *(E)*. With a protractor, find an angle of 23° outward from the earth–sun line and draw a line marking it. Find on this line the position at which another line from the sun strikes it at 90°. Measure the distance from the sun to this point. What is your result for the sun–Mercury distance? It should be close to 4 cm, or 0.4 AU, on this scale.

3. Take the sun–Mercury distance you calculated in Problem 1 or 2 and use one of Kepler's laws to find Mercury's orbital period.

4. Mercury's synodic period is 116 days. What is its sidereal period? Compare your answer here with your answer in Problem 2.

5. The semimajor axis of the orbit of Mars has a size of 1.52 AU and an eccentricity of 0.09. Calculate the perihelion and aphelion distances of Mars. Assume that the earth's orbital eccentricity is zero; then what are the closest and farthest possible distances of Mars at opposition?

6. Jupiter's average distance from the sun is 5.2 AU. What is its sidereal period?

7. Suppose you lived on Mars. What would be the maximum elongation of the earth from the sun?

8. The synodic period of Saturn as seen from the earth is about 378 days. (a) What is the synodic period of the earth as seen from Saturn? (b) What is the earth's maximum elongation angle as seen from Saturn? (c) What is Saturn's sidereal period around the sun?

9. Compare Mars, Jupiter, and Saturn on the basis of their respective speeds around the sun. Consider the orbits to be circular. Calculate for each an orbital speed in units of AU per year. What is the trend that you see?

See for Yourself!

Make up a scale model of the distances in a heliocentric system as it would be seen at the time of Copernicus. Use the earth–sun distance (the AU) as the standard unit. Position Mercury, Venus, the earth, Mars, Jupiter, and Saturn in the correct relative distances from the sun. Don't try to scale the sizes of the planets, just their distances.

For this model to work out, measure or estimate the size of the room that you will use. Use one-tenth of its size as the length of the astronomical unit. Then your model will be almost as large as the room for the distance of Saturn from the sun (Table 3.1). ■

Central Question
How did Newton's laws of motion and gravitation furnish a unified, physical model of the cosmos?

4

The Clockwork Universe

I am much occupied with the investigation of physical causes. My aim in this is to show that the celestial machine is to be likened not to a divine organism, but rather a clockwork....

JOHANNES KEPLER

What makes the world go 'round? For the planets, according to Isaac Newton, it's gravitation. By linking the cosmos with gravity, Newton achieved the goal that had eluded Kepler: a model of the cosmos run by a single force.

Newton arrived at this goal after crucial groundwork had been laid by Copernicus, Kepler, and Galileo. Copernicus violated the physics of the day and offered no substitute; this gap was an embarrassment for the Copernican model. Kepler sought to establish a clockwork cosmos driven by a single force. His magnetic force finally failed; but his concept of a force directly influenced Newton's vision. Galileo's concern with terrestrial motions also shaped Newton's ideas. And Galileo's use of the telescope provided support for the heliocentric model.

Newton erased the old physical separation between the earth and the heavens. He linked gravity (terrestrial physics) to the orbital motion of the planets (celestial physics). The links he forged were his laws of motion and gravitation, which sprang from his concept of natural motion. With Newton's ideas, the universe took on a new aspect. No more would there be the closed, finite universe of Ptolemy, Copernicus, and tradition. In Newton's grand vision, the universe grew to an infinite expanse, driven by a single force: gravitation.

FIGURE 4.1 Galileo Galilei. "By denying scientific principles, one may maintain any paradox." (Mezzotint by S. Sartain, 1852, courtesy of The Granger Collection, New York.)

4.1 Galileo: Advocate of the Heliocentric Model

The Italian scientist Galileo Galilei (1564–1642) tried to establish celestial physics on a firm experimental and mathematical basis. In contrast to Kepler, Galileo (Fig. 4.1) concerned himself almost exclusively with terrestrial motions, especially those of falling bodies. He felt that to establish the laws of terrestrial motions would do more to cement the structure of the Copernican cosmos than any observations, including those made with a telescope. He did make use of telescopic observations, but he held the physics at a higher level. He failed to achieve his goal, but his discoveries guided the work of Newton.

The Magical Telescope

Galileo used his telescope to bolster the Copernican model. He did not invent optical lenses, nor was he the first to use them in a telescope. Rather, he promoted his astronomical ideas by exploiting the novelty and shock value of telescopic observations. Galileo's report of his observations in his book *Siderius nuncius* (The starry messenger) was widely circulated and brought him fame.

By Galileo's time, glass lenses had been in use for about 300 years. Their origins are not clear, but with them eyeglass makers could correct defects of vision. In 1609, Galileo heard that a Dutchman had constructed a spyglass that made distant objects appear to be nearby. Although Galileo had little experience with optics, he set to work immediately to duplicate the instrument (Fig. 4.2) and succeeded in constructing an optical device that made objects appear thirty times closer than when viewed with the naked eye (Chapter 6). He put the device to astronomical use at once. Within a few weeks in 1609 and 1610, he made a series of astronomical discoveries that marked a new era in astronomy. Although Galileo was not the first person to build a telescope, he first recognized that the telescope increased our power to perceive reality.

When Galileo turned his telescope to the moon, he saw that the moon's surface does not conform to the Aristotelian ideal for a perfect heavenly body. Rather than being smooth and spherical, it is rough, with chains of mountains and valleys and many craters (Fig. 4.3). Galileo then determined the height of a lunar mountain from the length of its shadow: approximately 6 km was the astonishing (and correct) result.

Galileo next peered at the stars. His instrument fragmented the faint band of the Milky Way into innumerable stars, more than can be distinguished with the unaided eye. This observation refuted the Aristotelian idea that the sky contained a set number of stars that cannot change.

Then Galileo hit on what he considered to be his most important discovery—four new "planets." Actually they were not new planets but rather were the four brightest moons of Jupiter. (At least seventeen moons are now known: Chapter 11.) To his amazement, Galileo found that these four bodies revolved around Jupiter (Fig. 4.4). From continuous nightly observations, he estimated the orbital periods of the Jovian satellites. Here he found another argument against tradition, for Jupiter and its satellites resembled a miniature solar system. This fact required a second center of revolution in the cosmos and contradicted the Aristotelian doctrine that only the center of the universe—the earth—could be the center of revolution.

Historically, Galileo may not have been the first person to report on moons orbiting Jupiter. An old Chinese record notes that Kan Te, an astronomer in China in the fourth century B.C., made many observations of Jupiter. In one of his books, Kan Te states that Jupiter looked as if it had "a small reddish star attached to it." This may be a record of the brightest of Jupiter's moons, which can, under ideal circumstances, be visible to the unaided eye. If so, Kan Te recorded this moon some 2000 years before Galileo observed it with a telescope!

The Starry Messenger

Galileo's *Siderius nuncius* was published on March 12, 1610. Written in Ital-

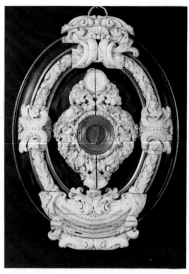

FIGURE 4.2 The lens from Galileo's largest telescope. It was accidentally broken by him and mounted in this ivory frame in 1677. (Courtesy of the Istituto e Museo di Storia della Scienza; photo by Franca Principe.)

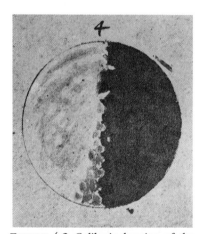

FIGURE 4.3 Galileo's drawing of the moon as seen through his telescope. Note the craters and mountains, especially along the boundary between the illuminated and dark regions. (Courtesy of the Yerkes Observatory.)

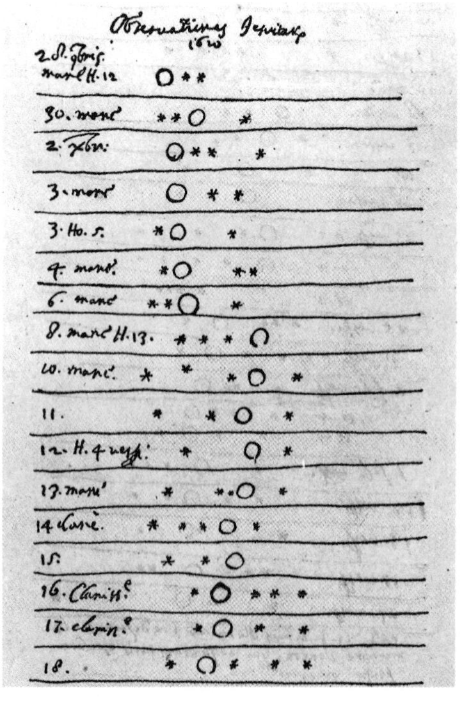

FIGURE 4.4 Some of Galileo's drawings of early telescopic observations of the moons of Jupiter. The large circle represents Jupiter; the "stars" are the moons. Their changing positions next to the planet indicated that they were revolving around Jupiter, rather than being just background stars. These observations appear to be copies of Jesuit observations made in 1620. (Courtesy of the Yerkes Observatory.)

ian rather than Latin (except for the title), the book was accessible to the general public, and it sold as fast as it could be printed. These telescopic discoveries, especially the news of Jupiter's moons, not only brought fame to Galileo but also created a demand for the telescopes produced in his workshop. He was still cautious in print: although he used his observations to demolish the Aristotelian cosmos, he did not openly advocate the Copernican scheme or support it with his observations.

Critics quickly spurned Galileo's work by invoking atmospheric phenomena or flaws in the lenses. Such arguments seem strange to us today, but they appeared reasonable enough to seventeenth-century people. The art of lens making was not far advanced, and many lenses produced ghost images. (Galileo's telescopes produced images that were far inferior to those in a cheap modern telescope.) Opponents conceded that Galileo might have honestly reported what he had seen but did not believe that his observations had a direct connection with reality.

Yet, Galileo was convinced by the regularity of the telescopic phenomena that he was viewing reality and not illusion. In 1611 Kepler, who was well regarded in scientific circles, backed up Galileo with observations of his own and also with his theory of optics, which supported the validity of telescopic observations. He also showed that the moons of Jupiter obeyed his third law and so demonstrated that their orbits were like those of the planets around the sun.

Galileo's Discoveries and the Copernican Model

Goaded by the opposition, Galileo continued to scan the skies for new marvels. He observed sunspots. Defects on the supposedly perfect sun, the sunspots dealt another blow to the tenets of Aristotelian cosmology. In 1613 Galileo published his results in *The Letters on Sunspots* and made his first direct, printed declaration of his belief in the Copernican model.

Note that none of these observations *directly* confirm the Copernican model. Rather, they work as evidence *against* the old Aristotelian–Ptolemaic model. Galileo's observations of the phases of Venus (Fig. 4.5) could support the Copernican model. Half of Venus, which has a spherical shape, is illuminated by the sun. In the Copernican model, we observe Venus from outside its orbit (Fig. 4.5), and Venus moves around the sun faster than the earth does. So we view Venus from different angles relative to the sun, and—like our moon—Venus shows different phases, from crescent to full. (The full phase is very hard to observe because it occurs when Venus lies on the other side of the sun when viewed from the earth.) Galileo observed part of this phase cycle of Venus.

Now in the Ptolemaic model, Venus does show limited phases. Because Venus must stay along an epicycle whose center must lie fixed along the earth–sun line (Section 2.4), the range of phases is less than in the Copernican model. For instance, Venus can never be full because it cannot lie opposite the sun. And the phases near full also do not take place.

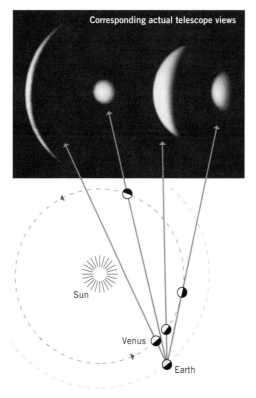

FIGURE 4.5 The phases of Venus in the Copernican model. The range of phases is restricted from our view because Venus' orbit is interior to the earth's. In particular, we can never observe a "full" Venus. The actual view through a telescope at each relative position of the earth and Venus is shown above each position of Venus. (Photo courtesy of the Lowell Observatory.)

Caution. The observations of the phases of Venus do favor the Copernican model over the Ptolemaic one, but they do *not* in general show that a heliocentric model is better than a geocentric one. For example, Tycho Brahe's geocentric model permits the same cycle of phases as the model of Copernicus.

The Crime of Galileo

Galileo publicly declared himself a Copernican in 1613. His intellectual vehemence and enthusiasm irritated many of his opponents, some of whom enjoyed substantial power in the Church. In 1616 an official of the Inquisition apparently warned Galileo to stop teaching the Copernican model as truth rather than as a hypothesis. Galileo's position was held to be contrary to Holy Scripture. At the same time, *De revolutionibus* was placed on the Index of Forbidden Books until "corrections" had been made. (As indicated in Chapter 3, only a few percent of the copies were actually changed.) These two events initiated a complex chain: Galileo, an obedient Catholic, avoided teaching the Copernican model as true; he also entered into arguments about scientific truth versus revealed truth. His argumentative spirit promoted his downfall.

In 1623 Cardinal Barberini, an erstwhile friend of Galileo and a patron of the sciences, was elected pope and took the name Urban VIII. Galileo thought he saw his chance and had a long audience with the new pontiff, during which Urban discussed the decree forbidding the heliocentric teachings. Feeling that he had the support of the pope and also of the Jesuit astronomers, Galileo wrote the *Dialogue on the Two Chief World Systems.* With the approval of the Roman Catholic censors in Florence, the book was published in 1632, after a few minor corrections had been made. The book supposedly relates an objective debate about the relative merits of the Ptolemaic and Copernican systems, with the judgment being rendered in favor of the traditional view. In reality, the text is a thinly veiled polemic favoring the Copernican cosmos.

Immediately Galileo found himself in trouble, for his opponents swiftly countered his claims with theological arguments. The pope himself was incensed because his favorite arguments had been put in the mouth of Simplicio, the supporter of the geocentric viewpoint in the *Dialogue.* (Galileo had not drawn Simplicio kindly, and although there had been a sixth-century commentator on Aristotle named Simplicio, Galileo's readers knew the word also meant "simpleton.") Some copies of the book were seized before they left the printers, and the Inquisition summoned Galileo to Rome and forced him to recant his scientific beliefs. His friends were afraid to come to his support. As punishment he was placed under perpetual house arrest. The *Dialogue* remained forbidden to Roman Catholics until 1835, when the works of Galileo, Copernicus, and Kepler were finally removed from the Index.

Galileo's trial by the Inquisition, although motivated as much by internal Church politics and Galileo's abrasive personality as by conviction that the Copernican system was wrong, frightened intellectuals in regions where the Church exerted power. The confrontation with the Church surprised Galileo, who thought he had divorced theology from science.

In October 1992, the Church decided to clear Galileo of his official condemnation of 1633. The Vatican finally admitted to error in its action over 300 years ago.

■ **What one telescopic observation of Galileo most strongly supported the Copernican model?**

4.2 Galileo and a New Physics of Motion

Despite his observational triumphs, Galileo still felt the Copernican system lacked the anchor of a physical understanding of motion like the one Aristotle had provided for his model. Kepler had taken an important step to a new physics by his mathematical description of *celestial* motions. Galileo took the next step by devising a mathematical description of *terrestrial* motions, particularly motions of falling bodies. Galileo aimed to justify the Copernican scheme by his systematic, experimental search for physical ideas of motion. In unraveling the details of motion produced by applied forces, he was inventing the careful study of **mechanics.**

To set up the new science of mechanics, Galileo needed an important tool: a description of the motion of falling bodies, that is, the motion of masses under the influence of gravity. To understand Galileo's contribution, you need a precise physical description about such motion.

Acceleration, Velocity, and Speed

When you step on a car's accelerator, the vehicle does just that: it *accelerates*. If you start out at rest, the car goes from zero speed to faster speeds, as you can tell from the speedometer. **Speed** is the rate of your motion. If you are cruising on the highway, you can pass a slower car by stepping on the accelerator to reach a faster speed (Fig. 4.6a). In both instances your speed changes; that's one aspect of the meaning of *acceleration*.

But when you travel in a car you don't simply race along at some speed alone. You must also go in specific directions; otherwise, you'd never make it to your destination. Suppose, for example, that you travel from Albuquerque to Santa Fe, New Mexico; you go from one place to another at some *speed* in some *direction*. *Velocity* involves the direction as well as speed of motion. If you take one hour to drive from Albuquerque to Santa Fe, a distance of about 100 km, your speed is about 100 km per hour (km/h). When you drive back at the same speed, your velocity is different because you're heading in a *different direction*. In the first case, your velocity is 100 km/h *north*; in the second, 100 km/h *south*. So **velocity** is the rate of change of position in a certain direction. It's not simply speed, which tells only how fast you're going, but not where you're heading!

Imagine now that you're driving around a circular race track at a constant speed. Are you accelerating, even though your speed is constant? Yes! Why? Because your car is constantly changing *direction* as you negotiate the

circle of the track (Fig. 4.6b). Your speed does not change, but your direction does. So your velocity changes, and you accelerate. In fact, since your direction is constantly changing, you are undergoing a constant acceleration (Focus 4.1).

To sum up. You are accelerating if your speed changes while your direction remains the same. And you accelerate if *either* or *both* your speed and your direction change.

Note that speed and velocity have different *physical* meanings. *Speed* is the average rate of travel and is measured in distance per unit of time—such as miles or kilometers per hour (mph or km/h) or even just meters per second (m/s). Velocity is speed with something added: direction. **Acceleration** is the rate at which *velocity* (not speed!) changes. It is measured in distance per unit of time per unit of time, such as meters per second per second (m/s/s). What happens when you hit the brakes of a car? You accelerate, *opposite* the direction of motion. We commonly call this *deceleration*.

A car with a fairly powerful engine can accelerate from zero to 100 km/h in about 10 seconds. What is its acceleration? It averages 10 km/h/s (kilometers per hour per second) in the *forward* direction. If it has excellent brakes, the car can stop in about 4 seconds from 100 km/h. What's its deceleration? It averages about 25 km/h/s in the *backward* direction.

Natural Motion Revisited

Aristotle divided motions into two categories: forced and natural (Section 2.3). An example of Aristotelian natural motion is the fall of a rock dropped to the earth. The rock follows the supposedly natural tendency of earthy material to seek the central point of the cosmos; its motion requires *no* force. What about forced motion? Throw a rock at a right angle to the natural downward motion! Aristotle would require a continuously acting force to keep up this unnatural, forced motion horizontal to the ground. As you know, however, after you have thrown an object horizontally, its motion continues for a while after the object leaves your hand. This tendency for the

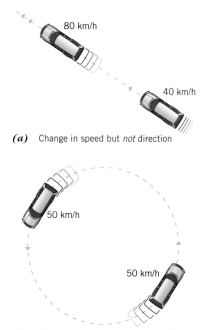

(a) Change in speed but *not* direction

(b) Change in direction but *not* speed

FIGURE 4.6 Acceleration. (*a*) A car accelerates if its speed changes while its direction remains the same. (*b*) A car also accelerates if it moves at a constant speed around a circular track, because the direction of its motion is always changing.

motion to continue, even when the force has stopped, is called **inertia.** It presented a nagging problem to Aristotelian physics, because inertia contradicts the basic premise about forced motion.

Galileo inverted Aristotle's ideas of natural and forced motion. He concluded that the downward motion of objects results from an attractive force: gravity. In addition, he viewed the horizontal motion of objects through the air as that from their inertia. This inertial motion, he argued, was a natural one and would continue if no forces, such as air resistance, countered it. So for Galileo, the downward motion of dropped objects was a forced motion, not a natural one.

To cast this revision, Galileo first had to arrive at a new version of inertia. He explained his idea this way (Fig. 4.7): if a perfectly smooth ball were placed on a frictionless flat surface that sloped, the ball would roll down the slope forever (assuming the surface were infinitely long). The greater the slope, the faster the ball would roll. By the same argument, a ball traveling up a slope would eventually stop. If, however, the surface had no slope and the ball were placed on the level with no horizontal velocity, the ball would remain at rest. A ball on a level surface, if pushed, would move straight ahead at a constant velocity forever, since it is not slowed down by an ascent or speeded up by a descent. So inertia is a natural tendency for a body in motion to remain in motion.

Forced Motion: Gravity

Using this concept of inertia, Galileo dealt with the problem of falling bodies. He viewed falling *not* as natural motion (as had Aristotle) but rather as *motion due to a force:* gravity. Motion under the influence of gravity only is called **free-fall.** Drawing on both his experiments and intuition, Galileo concluded that free-falling motion took place at a *constant acceleration,* with the object's velocity changing at a constant rate downward as it fell. (The acceleration of

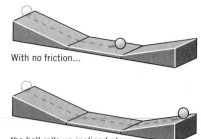

With no friction...

the ball rolls up inclined plane until it reaches the height at which it started;...

Or if it rolls onto a flat surface, it keeps going.

FIGURE 4.7 Galileo's concept of inertia (natural motion). The trick is to ignore friction in this thought experiment of rolling a ball down an inclined plane. It will roll up an equal distance onto a plane with a similar incline. If the ball rolls down the plane onto a flat surface, it will keep moving along with no change in speed.

Enrichment Focus 4.1

SPEED, VELOCITY, AND ACCELERATION

You must understand the differences among these terms, because they are physically real, not just semantics. You may perhaps see the distinctions more clearly from simple equations.

If you travel over some distance in a certain amount of time, then your speed is

$$\text{speed} = \frac{\text{distance}}{\text{time}}$$

where we really mean the *average* speed over the time interval. For now, let's assume a constant speed. Then you are probably more familiar with this relationship in the form

$$\text{distance} = \text{speed} \times \text{time}$$

Now specify a *direction* to the motion. Then we are talking about velocity

rather than speed, but the relations are the same for a constant velocity:

$$\text{velocity} = \frac{\text{distance}}{\text{time}}$$

and

$$\text{distance} = \text{velocity} \times \text{time}$$

We can change velocity by changing speed, direction, or both. The rate of change in the velocity is called *acceleration,* which is defined by

$$\text{acceleration} = \frac{\text{change in velocity}}{\text{time interval}}$$

In some special cases, like bodies falling near the earth's surface, the acceleration is *constant.* Then we have

$$\text{velocity} = \text{acceleration} \times \text{time}$$

Now imagine an object, say a mass at the end of a string, moving in a circle at a constant speed. The direction of the motion is constantly changing, so the mass is accelerating. The magnitude of the centripetal acceleration is

$$\text{centripetal acceleration} = \frac{(\text{circular speed})^2}{\text{radius}}$$

and its direction is toward the center of the circle. This is not at all an intuitive result! But it was correctly derived by Newton.

gravity at the earth's surface has a value of 9.8 m/s/s and is usually denoted by *g*. This means that for every second of fall, an object gains 9.8 m/s of speed. You can recall this number by its approximate value, 10 m/s/s.)

Galileo also concluded that in the absence of air resistance, *all* falling masses have the *same* acceleration. When dropped, they reach the same velocity after the same elapsed time. They also fall the same distance in the same time. His conclusion—that all masses fall with the *same acceleration* near the earth—directly contradicted the traditional teaching that a massive body falls faster than a less massive one and so moves a greater distance in a given time.

What do you think from your own experience? Try dropping a lightweight ball and a piece of paper from the same height. The ball hits the ground first as the paper drifts down through the air. Now crumble the paper so that it has about the same shape and size and as the ball. Drop both again. What's the result? Both reach the ground in about the same time!

In the case of *constant* acceleration, such as that of falling bodies, the velocity increases as time passes. That is, a dropped object falls with greater and greater speed, the closer it gets to the ground. When you throw a ball near the earth's surface, its path describes an arc (Fig. 4.8). This pattern results from a combination of the ball's falling motion (which occurs with a constant acceleration) and its horizontal motion (which occurs at pretty much a constant speed, because air friction has little effect on the ball during its time in the air). The curved path results from a force in one direction only: that pulling the ball to the ground, toward the center of the earth. So the ball's path results from a combination of two types of motion: a natural, inertial one in the horizontal direction, and a forced one vertically.

Legend has it that Galileo climbed to the top of the Leaning Tower of Pisa and dropped two different objects that hit the ground at the same time, to the astonishment of the skeptical crowd gathered around the tower's base. As far as historians can determine, Galileo did *not* actually attempt this experiment. The demonstration would have been a risk, for air friction probably would have kept the different masses from striking the ground together. The rumored Pisa experiment may not be the only one that Galileo did not do, even though he reported the results in his writings. Some of his experiments were only mental exercises, but Galileo intuitively reached the correct results more often than not. And he did perform many experiments to contradict Aristotle's concepts about motion. Eventually, Galileo's views on motion became a pivot on which modern physics turned.

Galileo's Cosmology

In his *Dialogue on the Two Chief World Systems,* published in 1632, Galileo compared the traditional model to the Copernican one. In fact, he displayed the planetary system as a pure Copernican scheme with no evidence at all of the ideas of Kepler (such as elliptical orbits). Nor did Galileo attempt to apply his own terrestrial mechanics to the motions of the planets. The physics of the heavens and earth remained separate.

His ignorance of physics and of ellipses provides evidence that Galileo paid little attention to the use of the Copernican system in predicting the planetary positions. What struck Galileo was the Copernican order of the universe, rather than the detailed astronomy of planetary motions. He apparently felt, with the same conviction as Copernicus, the harmony of the planetary order established from observation. Of course, Galileo relied heavily on his telescopic observations along with the general arguments of Copernicus (Section 3.1).

Galileo placed the stars in a thick shell far beyond the planets. This stellar shell, a holdover from Greek ideas, was included in the original Copernican model. However, the 1632 *Dialogue* keeps open the possibility that the universe is not spherical and finite but is open and infinite, with the stars sprinkled "through the immense abyss of the universe." (In fact, as early as 1563, some accounts of the Copernican model already had stars scattered through unlimited space.) The idea of the cosmos as an

Motion
At a Glance

Speed	Rate of change of position
Velocity	Speed with direction
Acceleration	Rate of change of velocity

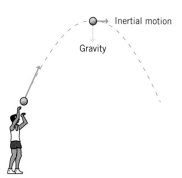

FIGURE 4.8 Path of an ball thrown in the air. Gravity pulls it straight down, and its inertia keeps it moving horizontally. Both motions take place simultaneously. The result is a curved path (called a *parabola*), if air friction can be ignored.

infinite universe rather than a closed space would be supported by Newton.

■ **Standing on the observation deck of a tall building, you drop two perfectly round and smooth spheres of equal size; one is aluminum and the other is lead. Assuming no air resistance, which one will hit the ground first?**

4.3 Newton: A Physical Model of the Cosmos

Despite his interest in—and insight into—terrestrial motions, Galileo did not worry about the details of celestial motions. Clearly, he believed that the model of Copernicus was on the right track. But until the general concepts of motion were understood, Galileo thought it premature to consider physical explanations of planetary motion that would provide a physical foundation for the Copernican model.

Sir Isaac Newton (1642–1727) emerged as the genius destined to fuse the terrestrial and celestial realms and so to end the long-standing separation first assumed by the Greeks (Fig. 4.9). The publication, in 1687, of Newton's *Principia,* containing his new physics of motion and the concept of gravitation, provided a unified physical view of the universe. At last, Newton's discovery of gravitation solved the old puzzle of the motions of the planets.

The Prodigious Young Newton
In the small English village of Woolsthorpe, on December 25, 1642, Newton's widowed mother gave birth to a sickly child. The fragile baby was so small at birth that it is said he could have fit into a quart mug!

At age 18, Newton enrolled in Trinity College at Cambridge University. He first intended to study mathematics as applied to astrology, but Professor Isaac Barrow, who sensed Newton's abilities, encouraged him to study physics. Then in 1665 the bubonic plague overwhelmed England, the university shut down, and Newton returned to Woolsthorpe. There, in quiet isolation, he made discoveries in mathematics, optics,

and the science of mechanics. As he wrote, "In those days I was in the prime of my invention, and minded mathematics and philosophy more than at any other time since."

This fertile period in Woolsthorpe generated the legend of the falling apple. As is common in a creative flash of genius, Newton—whether he in fact saw an apple fall or not—linked two seemingly unrelated phenomena: the fall of an apple and the orbit of the moon. He was puzzled by the fall of objects, such as apples, and wondered about the force that attracts masses, such as the moon, to the earth's center.

On returning to Trinity, Newton showed his work to Barrow, who soon actually resigned his position so that Newton could have it. Newton, pursuing his interest in optics, invented one version of the *reflecting telescope* (Chapter 6), which uses a mirror as the primary light gatherer. His design was communicated to the Royal Society of London, for which he constructed a small reflector (Fig. 4.10). This application of his ideas resulted in Newton's election as a Fellow of the society. The association was not completely happy, however, for when his work on light was also communicated to the society, such bitter controversy arose that Newton resolved never to publish again.

FIGURE 4.9 Sir Isaac Newton. "I have laid down the principles of philosophy; principles not philosophical but mathematical. . . ." (Courtesy of the Bettmann Archive.)

FIGURE 4.10 Newton's reflecting telescope. A curved mirror at the back brings the light to a focus. The lens on the side near the top is used to view the image, which is brought out to the side by a small, flat mirror. (By permission of the President and Council of the Royal Society, London.)

The Magnificent *Principia*

Newton broke his vow about 10 years later, when Edmond Halley (1656–1742) requested his advice on the problem of elliptical orbits described by Kepler's laws. Halley (Fig. 4.11) queried Newton on the nature of the force between the sun and the planets that produced such orbits and was surprised to hear that Newton had solved the problem in exact detail. Newton, however, had misplaced the solution and could not find it at the moment; he promised Halley that he would send it along later.

Recognizing the importance of Newton's discovery, Halley cajoled his introverted friend to make the studies public, promising to oversee and finance their publication. Stimulated intellectually by Halley's support, Newton labored for nearly two years to complete his *Philosophiae naturalis principia mathematica* (The mathematical principles of natural philosophy). Published in 1687, the *Principia* solved the vexing problem of planetary motions.

Forces and Motions

Before dealing with Newton's analysis of orbits, let's reflect on our own experiences with forces. Imagine that you have a small cart that moves along a flat surface with little friction. Consider the cart at rest. Then push the cart quickly, applying a force for a brief time. What happens? The cart accelerates from zero velocity to some velocity. And the direction of the acceleration (and the resulting velocity) is the same as the direction of the initial force. This example shows that a force can produce accelerations (a rate of change in *velocity,* remember!).

Now imagine that you push the cart, again at rest, with a force greater than the one you applied the first time. What happens? Right—the cart achieves a greater acceleration and reaches a greater velocity than before. In fact, if you apply twice as much force as the first time, the cart will accelerate twice as much. (The acceleration is still in the same direction as before.) If you use three times the force, you get three times the acceleration, and so on. Hence, the acceleration of any object is *directly proportional to the force* exerted on it. "Directly proportional" means here that the acceleration increases in the same way the force increases. (What would happen if you applied four times the force?)

Let's add bricks to the cart until it has twice its original mass and reapply our initial degree of force. What occurs when we apply one unit of force? The acceleration is still in the same direction as the applied force, but it is *less,* because there is more mass to be moved. In fact, if you apply the *same* amount of force as you did the first time, the cart will have *half* the acceleration for *twice* the mass. With three times the mass and the same force, the resulting acceleration would be one-third as much, and so on. The acceleration of any object is *inversely proportional to its mass*—the greater the mass, the less the acceleration (for the same amount of force). "Inversely proportional" means here that the acceleration *decreases* as the mass *increases.*

To sum up. The acceleration of an object depends directly on the amount of force applied to it and inversely on the mass of the object.

"Amount of force" means the *net* force on the object. Why net force? Because several forces may act on an object, and they may balance and cancel out if they act in opposite directions. Consider a tug-of-war between matched teams. Both exert the same amounts of force on the rope, but in opposite directions. The net force is zero because the opposite, equal forces balance. They have achieved an equilibrium. This example emphasizes that forces have *direction* and can work either together or against each other.

Newton's Laws of Motion

In the *Principia,* Newton defines mass, velocity, and acceleration. He then expounds his three axioms, or **laws of motion.** In modern terms, these famous laws are as follows.

LAW 1. *The Inertial Law* A body at rest or in motion at a constant velocity along a straight line remains in that state of rest or motion unless acted on by a net outside force (Fig. 4.12).

FIGURE 4.11 Edmond Halley. After he inspired Newton to put together the *Principia,* Halley used Newton's laws to compute the orbit of the comet that bears his name. (Courtesy of the Bettmann Archive.)

LAW 2. *The Force Law* The rate of change in a body's velocity due to an applied net force is in the same direction as the force and proportional to it, but is inversely proportional to the body's mass (Fig. 4.13).

LAW 3. *The Reaction Law* For every applied force, a force of equal size but opposite direction arises (Fig. 4.14).

Newton's first law takes a logical step beyond Galileo's concept of inertia by postulating that constant, uniform motion is the *natural* state of moving mass anywhere in the universe. The first law tells you whether a net force is acting on an object: look for a change either in an object's speed or in the direction of its motion, or in both speed and direction.

The second law extends the recognition of a force to its results. The direction of the change in motion is the same as the direction of the applied force. Also, the amount of acceleration depends directly on the size of the force. So exerting a force (simply as a push or a pull) accelerates an object; that is, it slows it down, speeds it up, or changes the direction of its motion.

For example, suppose you are drifting in space next to two different objects. You push one. It accelerates momentarily and moves away, traveling in the direction of your push. You measure its acceleration, its change in velocity.

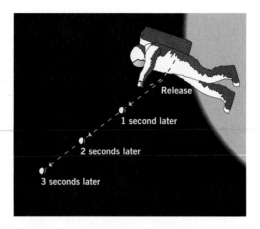

FIGURE 4.12 Newton's first law. An object thrown by an astronaut moves at a constant speed along a straight line (no friction slows it down). So it covers *equal distances* in *equal intervals* of time.

FIGURE 4.13 Newton's second law. A net force applied at an object changes its speed, direction of motion, or both in the direction of the applied force. Here, gravity pulls on an apple and accelerates it to the ground. As this apple accelerates, it falls faster so that it covers a greater distance in the same amount of time. Here are shown intervals of 0.1 second.

The Newtonian Model

At a Glance

Heliocentric

Gravitational forces for heavenly motions

Infinite in size, nearby stars farther from earth that the sun is

Variations in eastward motions: Earth's orbit around sun elliptical

Retrograde motions: Illusion from the passing of planets

Variations in retrograde motions: Elliptical orbits

Now push the other mass with the *same* force. You measure and find its acceleration is half the acceleration of the first. (Its resulting velocity is half as much.) You've applied the same force to both objects, but the accelerations differ. So the second must have twice the mass of the first. Newton's second law provides a means to measure mass.

In algebraic form, Newton's second law is

$$a = \frac{F}{m}$$

where F is the net applied force, m the object's mass, and a the acceleration resulting from the force. Note that, like velocities, forces have directions, and so do accelerations. An object will accelerate in the *same* direction as an applied force. The unit of mass in the International System of Units (Appendix A) is the **kilogram.** If we apply enough force to ac-

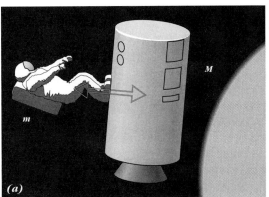
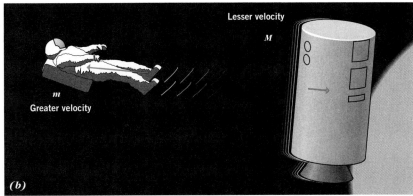

FIGURE 4.14 Newton's third law. Imagine you are in space next to a spacecraft that has a larger mass than you (*a*). Push off it momentarily; it pushes back on you with a force equal to the force you apply to it, but in the opposite direction. So you and the spacecraft move apart (*b*). If you push off from a spacecraft more massive than you, you end up moving with a greater velocity than the spacecraft because your lesser mass is being accelerated by the same force applied to the more massive object.

celerate one kilogram to 1 m/s/s, we've applied a force of 1 **newton**—named in honor of Isaac Newton!

A word about equations in this book: you will find very few equations in the main body of this text. *Their main function is to serve as a mathematical shorthand of selected, crucial physical and astronomical concepts.* In this context, I expect you to grasp them conceptually, with an understanding at least in words of their essential meaning. The Enrichment Focus sections show additional equations and manipulate them, which your instructor may require.

Newton's third law recognizes that forces interact in simultaneous pairs. If you were at rest in space and pushed against a massive spacecraft, the ship would react to your applied force with an equal but oppositely directed force, pushing you away (Fig. 4.14). Now the forces applied to you and to the spacecraft were the same, but the resulting accelerations were different. According to Newton's second law, the acceleration is greater for you than for the spacecraft because your mass is less. As a result, you would move quickly away from the spacecraft, which would hardly budge. Also, you and the spacecraft would move in *opposite* directions.

This third law marks Newton's most original contribution to the understanding of motion in general and of

gravitation in particular. Here Newton reasoned that gravitation is an *interaction* among the sun and the planets, and among all the planets. In fact, of all masses in the cosmos!

■ **What one aspect do Newton's laws have in common?**

4.4 Newton and Gravitation

From the platform of his ideas about force and motion, Newton attacked the problem of the planets by devising the law of gravitation. Two questions needed to be answered: In what *direction* does the force of gravity act? and What is the *amount* of the force? The first question involves a recognition of the general nature of the force, and the second involves a recognition of the physical properties that determine the force's strength. To answer these two questions, Newton combined his laws of motion with Kepler's planetary laws to arrive at a law of universal gravitation.

Newton demonstrated that the type of force that causes the elliptical orbits of Kepler's first law is a **central force,** one directed to the center of the motion. Also, he showed that planets moving in orbits under the influence of a central force follow Kepler's second law: the law of areas. Finally, Newton showed from the geometric properties of ellipses that the force is described by a specific type of force law and then he rederived Kepler's third law with this force law. Thus Newton ensured that his procedure fell in line with planetary laws known in his time.

Centripetal Acceleration

How do the moon and the famous apple enter into this scheme? Newton recognized that gravity caused the apple's fall. Might it not be that the earth's gravity, pulling on the moon, also keeps the moon in its orbit? For simplicity, assume that the moon's orbit is circular. The direction of the moon's orbital motion changes constantly. The moon stays on a curved path around—rather than moving along a straight line away from—the earth (Fig. 4.15). Newton's first law tells us that a force acts on the moon; moreover, according to the second law, this force results in an acceleration toward the earth's center. Such a centrally directed force is called a **centripetal force.** (*Centripetal* means "directed toward the center.") The resulting acceleration toward the center is called **centripetal acceleration.**

Let's try to get a feel for centripetal force and acceleration. You experience centripetal force when you turn a corner in a car. The faster you go around the corner, the greater the acceleration. But the radius of the turn also affects the acceleration. For instance, if you make a long, gentle curve—even at 90 km/h—you turn the steering wheel gently to change the car's direction. Here the turn has a large radius. But if you take a sharp turn at 90 km/h, you must turn the wheel sharply. Here the turn has a smaller radius. So centripetal force depends on both the speed of an object around a circle and the radius of the circle (Focus 4.1).

Now apply the concept of centripetal force to the moon. Consider first a billiard ball hitting the middle of each of the side cushions of an imaginary, square billiard table. At each collision, the direction of the ball's velocity changes, so it must accelerate from an applied force. The direction of that force is toward the center of the table for each collision. Consider next a pentagonal

table (Fig. 4.16*a*); the force of collision at each side still points to the center, but the angle of strike and rebound is smaller than for a square.

Finally, imagine a circular table with its infinite number of sides (Fig. 4.16*b*). The ball always touches the cushion (the angle between the cushion and the path is zero), and at every point, the force on the ball points to the center of the circle, and so its acceleration is also in that direction. Now, imagine the lunar orbit as circular. By analogy to the circular table, at every point in this orbit, a centripetal force acts on the moon (Fig. 4.16*c*). That force keeps the moon bound to the earth.

What causes the centripetal force that holds the moon in its orbit? Newton generalized from the apple to the moon: "I began to think of gravity extending to the orb of the moon." In this statement he expressed the new insight that *every body in the universe attracts every other body with a gravitational force.* This statement became the first *universal physical law.*

Newton's Law of Gravitation

What, in Newtonian terms, does **gravitation** mean? First, it means that *all* masses in the universe attract all other masses. (The gravitational force can only attract; it does not repel, despite some claims to the contrary in science fiction stories.) Second, Newton's law of gravitation says that—if you consider just two masses as an example—the amount of the gravitational force depends directly on the amount of material *each* mass has. So if you doubled the mass of one and kept the distance between the two the same, the force would also double. If you doubled the other mass, the force

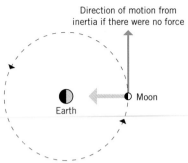

FIGURE 4.15 Central force and the moon's orbit. Since the moon moves neither at a constant speed nor along a straight line, according to Newton's first law, a force must be acting on it. This force pulls the moon toward the earth in a direction toward the center of the moon's orbit.

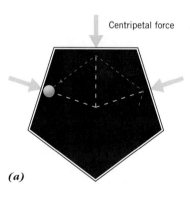

(a)

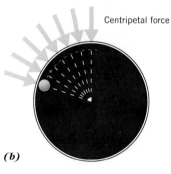

(b)

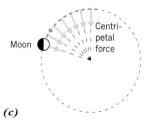

(c)

FIGURE 4.16 Centripetal force and circular motion. (*a*) A ball moving on a table with five sides strikes the center of each side. The force at each impact is centrally directed, that is, centripetal. (*b*) On a circular table, a centripetal force is exerted on the ball at every point of contact in its motion around the circumference. (*c*) As in the case of the ball, the moon orbiting the earth requires a centripetal force acting on it at each and every point along the orbit.

would double again. So doubling both masses results in four times the initial force. (The kind of material that makes up the masses does not matter.)

Third, masses at greater separations have *less* gravitational force than those closer together, and the drop-off of force with distance happens in a special way—as the *inverse square of the distance*. Consider, for example, a mass at different distances from the earth. The mass at 1 unit of distance from the earth experiences one unit of force (Fig. 4.17). Now move the mass away by two units of distance. The force is *less*. How much less? By $\frac{1}{2}$ squared, which is $1/(2 \times 2)$, or $\frac{1}{4}$ as much as when the mass and the earth were 1 unit apart (Fig. 4.17). Now move the mass to half the unit distance. The force is *greater*, by 1 over $\frac{1}{2}$ squared, or 4 times as much (Fig. 4.17).

Now consider any two masses 1 meter apart. A certain amount of gravitational force attracts them. Move the masses until they are 2 meters apart. The force is less by $\frac{1}{2}$ squared, or $1/(2 \times 2)$, $\frac{1}{4}$ as much as when the masses were 1 meter apart (Fig. 4.18). At 3 units apart, the force is $\frac{1}{9}$ as much; at 4 units, $\frac{1}{16}$ as much. Note in all these examples that the masses remain the same. (What if the distance is 5 meters?)

You'll run across a number of relations in this book. Some are *direct*, such as that between the gravitational force and the masses of the objects involved. If you double one mass or the other, then the force doubles. For this kind of direct relation, I'll use the phrase "directly proportional to." (Can you think of another kind of direct relation?) In contrast, the relationship between the force and the distance between the two objects is an *inverse* one. As the distance increases (or decreases), the force decreases (or increases). For such inverse relations, I'll use the phrase "inversely proportional to."

Caution. Gravity cannot be understood in terms of a simple inverse relation. Rather, it's expressed by an *inverse-square* relation. Here, the force is inversely proportional to the *square* of the distance, not just the distance by itself! So the gravitational force between two

masses decreases much faster than you'd expect intuitively, probably on the basis of a simple inverse relation.

In modern algebraic form, Newton's law of gravitation is

$$F = \frac{Gm_1 m_2}{R^2}$$

where F is the gravitational attraction between two spherical bodies m_1 and m_2, whose centers are separated by a distance R. The symbol G is a constant, a number whose value is assumed not to vary with time and location in the universe; its value relates the size of the force to the sizes of the other quantities. The value for G in the International System of units, (SI units, see Appendix A), is 6.67×10^{-11}. (Here I am expressing the constant in powers-of-ten notation, also in Appendix A.) In this system, a unit of gravitational force is the *newton (N)*—the same as for forces in general. Recall that one newton is the amount of force needed to accelerate a mass of one kilogram (kg) at a rate of one meter per second per second (m/s/s). This force is *small*—between two 1-kg masses 1 meter apart, the amount of gravitational force is a mere 6.67×10^{-11} N. That's 0.0000000000667 newton!

But gravitation has a far reach. Newton used the moon and the apple to test the validity of this law. He knew that the earth's gravity at its surface (1 earth radius from the center) caused the apple to fall. The earth–moon distance is about 60 earth radii, so if an **inverse-**

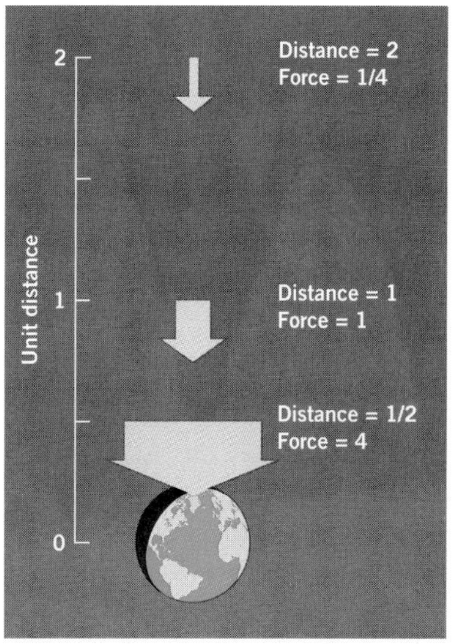

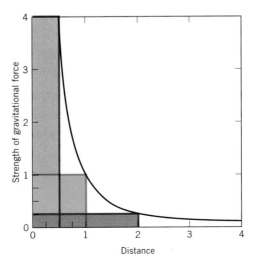

FIGURE 4.17 Moving the same mass at three different relative distances from the earth. For each distance, the thickness of the arrow indicates the relative amount of the gravitational force between the mass and the earth.

FIGURE 4.18 An inverse-square law in graphical form. This graph shows how the strength of the gravitational force changes with the distance between masses: at a separation of 1 unit of distance, the force has 1 unit of strength. When the masses are twice as far apart, a distance of 2, the force has only 1/4 the original strength.

Enrichment Focus 4.2

NEWTON, THE APPLE, AND THE MOON

How did Newton connect the apple's fall to the moon? He knew that the distance to the moon was approximately 60 earth radii. He also surmised that if the earth's influence extended to the moon, it grew weaker as the distance increased. How much did the force weaken? Newton guessed that it weakened as the inverse square of the distance (Fig. F.5). An apple dropping from a tree is 1 earth radius from the center of the earth. So, comparing the accelerations of the moon and the apple, Newton predicted that the moon's acceleration must be $1/60^2$ less, or 1/3600 as large. The acceleration due to gravity at the earth's surface is 9.8 m/s/s. For the moon, then, the predicted acceleration is 9.8 m/s/s divided by 3600, or about 2.7×10^{-3} m/s/s. This acceleration of the moon toward the earth is a centripetal acceleration, for the necessary force, gravity, pulls the moon toward the earth's center.

Note that Newton did not need to know the mass of the apple or the mass of the moon, because the accelerations of these bodies do not depend on their masses. And remember that Galileo (Section 4.1) had reached the same conclusion from his own experiments. Newton made crucial use of this conclusion in the moon–apple

test. How did Newton's *predicted* value compare with the *actual* rate of the moon's fall? Newton had found that for a circular orbit, the centripetal acceleration has a value of

$$a = \frac{V^2}{R}$$

where a is the centripetal acceleration, R the radius of the orbit, and V the orbital velocity. For the moon, R equals 3.84×10^8 m. Its velocity is the distance it travels in one orbit divided by the period for one orbit, or

$$V = \frac{2\pi R}{P}$$

One day contains $24 \times 60 \times 60 = 8.64 \times 10^4$ s, so the moon's orbital velocity in a sidereal month of 27.3 days is

$$V = \frac{2\pi(3.84 \times 10^8 \text{ m})}{(27.3 \text{ days})(8.64 \times 10^4 \text{ s/day})}$$

$$= 1.02 \times 10^3 \text{ m/s}$$

Then the moon's orbital acceleration is

$$a = \frac{V^2}{R} = \frac{(1.02 \times 10^3 \text{ m/s})^2}{3.84 \times 10^8 \text{ m}}$$

$$= 2.7 \times 10^{-3} \text{ m/s}^2$$

This result comes close to the prediction made from an inverse-square law.

Newton did not have the modern values used in these calculations for the period and size of the moon's orbit and the acceleration by gravity. But he chose a value of the moon's distance that made his results compare closely. Newton felt assured that his approach was correct even though he fudged the figures slightly.

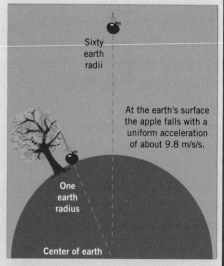

FIGURE F.5 The moon and the apple. The apple is 1 earth radius from the earth's center and it falls with an acceleration of 9.8 m/s/s. If an inverse-square law for gravity is correct, then the moon accelerates at a rate of 1/3600 less than the apple.

square law correctly describes gravitational forces, the acceleration of the moon toward the earth must be $1/60^2$ or 1/3600 as much as the acceleration of the apple (Focus 4.2).

Newton compared his predicted centripetal acceleration with the centripetal acceleration derived from observations of the moon's orbit. As he put it, the predicted and the observed accelerations were "pretty nearly" the same. Newton concluded that the cause of the moon's centripetal acceleration is the same as that of the apple's: the earth's

gravity. This force, extended out to the distance of the moon, keeps the moon in its orbit. In other words, the moon's motion results from two independent actions—its natural, inertial motion (along a straight line) and its centripetal acceleration from the earth's gravity. In fact, the moon moves about one kilometer in a second along its orbit while it falls 0.14 cm toward the earth in the same time (Fig. 4.19). That 0.14 cm is about the height of the lowercase letters in this book! In the same second, an apple near the earth's surface falls about 5 meters.

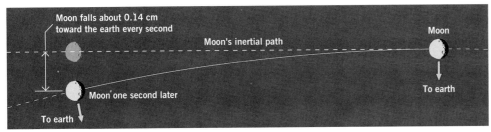

FIGURE 4.19 The moon's orbit. In one second, the moon moves about 1 km. If it followed just its inertial path, this single kilometer would be a straight line. During the second, the earth's gravity causes the moon to fall about 0.14 cm toward the earth. The combination of the inertial motion and the accelerated one keeps the moon on an elliptical orbit around the earth.

Caution. An apple orbiting the earth at the same distance as the moon would have the *same centripetal acceleration* as the moon (even though the *forces* would differ)! So how far would the apple fall in one second?

Consider Galileo's law of free-fall from Newton's standpoint. Imagine dropping a cannonball and a tennis ball. The earth exerts a much greater gravitational force on the cannonball than on the tennis ball because of the cannonball's greater mass. However, when the two are dropped, they fall side by side and land at the same time (in a vacuum). Gravity's effect is the same on both objects; more precisely, the *acceleration* of each is the same because the two spheres speed up at the same rate. Although the forces are different, the accelerations turn out to be the same. Why? Because the extra force on the cannonball is exactly offset by its greater inertia to changes in motion. (This result influenced Einstein in the development of his theory of gravitation: Chapter 7.)

This concept applies to other bodies. On the moon, for example, the acceleration of gravity is about one-sixth less than at the surface of the earth. If you drop a hammer and a feather from the same height, they will hit the lunar ground at the same time. Not only do both objects fall with the same acceleration but, without any atmosphere, air friction does not complicate the result. Apollo astronauts did such an experiment and proved Galileo's prediction.

Finally, Newton's third law requires that the gravitational force of the earth on the moon be exactly equal to that of the moon on the earth (even though the masses differ). They are *interacting* bodies. (Question: How does the gravitational force of the earth on the apple compare to that of the apple on the earth?)

■ Cite one commonly experienced example of centripetal force.

4.5 Cosmic Consequences of Universal Laws

The gravitational force of the earth causes the centripetal acceleration of the moon. This was Newton's central discovery: the earth's gravity keeps the moon swinging around. Newton's vision resulted in a new understanding of the planetary orbits. The most obvious: the sun's gravity locks the planets in their elliptical orbits. Newton had found the physical interaction between the sun and the planets first sought by Kepler. He derived a new form of Kepler's third law that included the gravitational constant and the masses of the interacting bodies (Focus 4.3). Related to basic physical laws, the revised third law became a potent tool in determining the mass of the sun, as well as the masses of the planets that have at least one satellite—quantities never known before! In fact, mass turns out to be a key physical property of astronomical objects. And Newton's ideas provide us with the masses of stars—and even galaxies!

Newton answered in detail the ancient question of how the planets moved. And he answered it precisely. His predictions of planetary positions were far more accurate than those of his

Gravitation

At a Glance

Force between all masses of any material

Acts over large distances

Proportional to: Product of masses

Inversely proportional to: Distance between masses squared

predecessors. Newton's ideas provided the physical support sorely needed for the Copernican model.

The Earth's Rotation

One objection to the Copernican model: objects not tied down to the earth should fly off because of its rotation. Newton noted that gravity holds things down. Another objection was that dropped objects should not land at their starting position; rather, they should be left behind by the turning earth. Newtonian physics explained that objects on the earth have inertia, which they retain when falling. That is, unattached objects do *not* lose their forward inertial motion but continue to move with the ground beneath them. When you throw an object straight up, it is moving in the direction of the earth's rotation while it is in the air; moreover, it retains the inertia it had while in your hand. So the object aloft keeps up with the turning earth, because no force is acting on it to change its eastward velocity. When it drops, therefore, it falls right back into your hand.

The Precession of the Earth's Axis

Section 1.2 described the precession of the equinoxes; here I'll explain it briefly using Newton's law of gravitation. You can understand this effect if you picture the earth as spinning like a top. If you place a spinning top on a table, its axis of rotation moves in a circle—an effect called *precession*. The precessional motion occurs in response to gravity trying to pull the top down so it falls over.

The earth, because it is spinning, is not a sphere but oblate—it has an equatorial bulge. (Newton actually predicted roughly the amount of the oblateness.) Now, the moon's orbit does not lie in the same plane as the earth's equator, but near the ecliptic. In general, the moon lies above or below the earth's equator. So the moon's gravitational force attracts the earth unevenly—it tends to pull the bulge closest to it more than the part of the bulge on the opposite side of the earth. The net force tries to tip the earth's axis of rotation from its orientation in space. The spinning earth responds by a precession of its axis. This motion makes the earth's poles describe a circle in the sky, so what stars are closest to the poles changes over 26,000 years. The earth's equator projected into space also moves, which we see as the precession of the equinoxes. (The sun's gravity also takes part, but its effect is about $\frac{3}{8}$ that of the moon.)

Enrichment Focus 4.3

THE MASS OF THE SUN

By using his laws of motion and gravitation, Newton reworked Kepler's third law so that it had the form

$$P^2 = \frac{4\pi^2}{G(M_{sun} + m_{planet})}a^3$$

where m_{planet} is the mass of the planet, and M_{sun} the mass of the sun. Compare this to Kepler's third law in the form

$$P^2 = ka^3$$

with $k = 4\pi^2/G(M_{sun} + m_{planet})$. Note that the constant relates to the masses of the two bodies involved.

You can use the earth's orbit to find the sun's mass. The earth–sun distance, a, is 1.50×10^{11} m. The earth's period, P, is 365.25 days, or 3.16×10^7 s. Because the mass of the sun is much larger than that of the earth, we can approximate $m_{earth} + M_{sun}$ by M_{sun}. That is, $m_{earth} + M_{sun} \approx M_{sun}$. Then Newton's revision of Kepler's third law gives

$$M_{sun} = \frac{4\pi^2}{G}\frac{a^3}{P^2}$$

and substituting in the values, we have

$$M_{sun} = \frac{4\pi^2}{6.673 \times 10^{11}}\frac{(1.496 \times 10^{11})^3}{(3.156 \times 10^7)^2}$$

$$= 1.99 \times 10^{10} \text{ kg}$$

So the mass of the sun is about 3.3×10^5 times that of the earth.

The Earth's Revolution and the Sun's Mass

You can use the earth's orbit to find the sun's **mass.** The earth–sun distance, a, is 1.50×10^{11} m. The earth's period P is 365.25 days. With this information, you can use Newton's second law and the law of gravitation to work out the mass of the sun (Focus 4.3).

The knowledge of the sun's mass bolstered the Copernican model in the framework of Newtonian physics. Newton's laws show that the sun has roughly 3.3×10^5 times the mass of the earth. And Newton's third law of motion requires equal gravitational forces between the sun and the earth—the force of the sun on the earth equals that of the earth on the sun. However, the second law demands that the earth's acceleration be much greater than the sun's (in fact, 3.3×10^5 times greater!). This law is fulfilled only because the sun has a small acceleration in one year. So the earth orbits the sun rather than the other way around. That is, sun and earth interact gravitationally, and because of their mutual attraction, both orbit around a common point called the *center of gravity* or the *center of mass.* As seen from this point, the earth's centripetal acceleration is 3.3×10^5 times greater than the sun's; so the earth has greater change of velocity than the sun. But the sun also moves!

The **center of mass** is the *balance point of two objects connected together.* Consider two unequal masses at the ends of a rod (Fig. 4.20). The balance point is closer to the more massive of the two. If you threw the spinning rod in the air, the two masses would spin around the center of mass. Gravity works to bind two masses together, as the rod does here, and two masses linked by gravity also have a balance point, a center of mass. For the earth and sun, the center of mass is very close to the sun because the sun is much more massive than the earth.

Gravity and Orbits

Newton's laws allow us to picture planetary orbits (and Kepler's laws) in simple physical concepts. Each planet's path is an ellipse, with the sun at one focus, and each planet has a unique semimajor axis,

eccentricity, and orbit period. Imagine a planet at aphelion, when it moves most slowly in its orbit (Fig. 4.21). As the planet moves toward the sun, its speed increases in response to the sun's gravitational force. The planet whips around the sun at perihelion at its fastest speed. As it moves away from the sun, the gravitational pull of the sun retards its speed. So it slows down, until it turns the corner at aphelion and falls in again. This natural rhythm is a consequence of Kepler's second law. A good analogy here is the swing of a mass at the end of a string that makes up a pendulum. The earth's pulls on the pendulum's bob. The back-and-forth motion makes up a period. The longer the string, the longer the period. And the bob moves fastest at the bottom of its swing (when closes to the earth) and slowest at the top of its arc.

Newton's physics also correctly describes the orbits of comets around the sun. Through the Middle Ages, astronomers believed that comets were objects confined to the earth's atmosphere. Later Tycho showed by parallax observations that comets are not atmospheric phenomena. Later still, Newton and Halley decisively demonstrated that comets orbit the sun in accord with the law of gravitation. In fact, Halley correctly predicted the return of the comet that bears his name, but he did not live to see the comet return in accordance with his prediction. (More on comets in Chapter 12.)

Newton's ideas also led to the discovery of Neptune in 1846, long after the first publication of the *Principia.* Neptune was the first planet to be found because of its gravitational effects on another. Newton's laws predicted the existence of Neptune *before* the planet was observed.

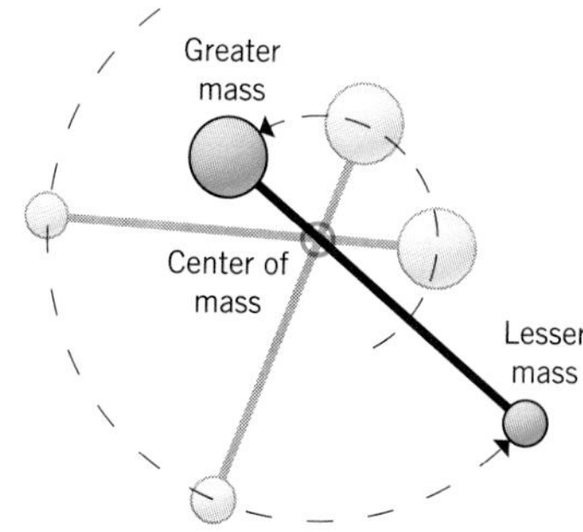

FIGURE 4.20 Motions of two masses at the end of a rod as the rod is thrown in the air. Both spheres revolve around the center of mass in the same period of time.

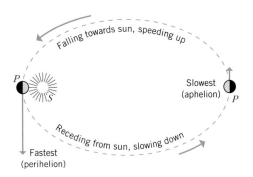

FIGURE 4.21 The elliptical path of a planet around the sun and the variation of its orbital speed as it completes one orbit.

The discovery of Neptune rested on observed irregularities in the orbit of Uranus (a planet discovered with a telescope in 1781). Astronomers had noted small discrepancies between the observed positions of Uranus and those predicted from Newton's laws. Such irregularities occur because all planets attract one another. Jupiter's tug, for instance, causes the orbit of Uranus to differ from what we would expect if Uranus and the sun were the only attracting bodies. And the existence of Neptune was revealed before the planet was discovered when Uranus' motion was seen to deviate from the pertubations from the effects of the known planets.

Applying Newton's laws to explain the deviations of Uranus as attributable to a planet beyond Uranus, the Englishman John C. Adams (1819–1892) in 1845 estimated where the unknown planet should be in the sky. He communicated his result to the Astronomer Royal, Sir George Airy (1801–1892), hoping that a search for the new planet would begin immediately. But Airy did not believe the result, partly because he thought the gravitational force deviated from an inverse-square law at large distances.

At almost the same time in Paris, Urbain J. J. Leverrier (1811–1877) made similar calculations and transmitted them to Johann Galle (1812–1910) in Berlin. Galle found the planet on September 23, 1846—the very night he received the predicted position! Newton's laws triumphed. In the twentieth century, discrepancies in Neptune's orbit stimulated the search for other planets. The search resulted in the discovery of Pluto (Chapter 11), even though we now know that the mass of Pluto is too small to affect Neptune.

An equally dramatic discovery, which extended the validity of Newton's laws beyond the bounds of the solar system, was the observation of **binary star** systems in the late eighteenth century. (A binary star system consists of two stars, held together by their mutual gravity, orbiting around their center of mass.) William Herschel (1738–1822) and his sister, Caroline Herschel (1750–1848), observed many pairs of stars over a long time, looking for heliocentric parallax.

Unexpectedly, the Herschels observed the orbital motions of some pairs. These observations showed that some pairs of stars lie close together in the sky not simply because they both happened to be in the same direction but because they are actually linked by gravitational forces.

Calculations of the orbital periods of binary star systems and the separation between the stars confirmed that the motions of these stellar pairs follow Kepler's laws. These laws can be derived from Newton's laws of motion and gravitation. Since binary star systems obey Kepler's laws, they indirectly confirm Newton's laws. (Chapter 15 has more details on binary star systems. In the same way that we found the mass of the sun, we can use binary systems to find the masses of stars—a powerful extension of Newton's laws.)

Orbits and Escape Speed

From Newton's laws of motion and gravitation arises the concept of **escape speed,** the minimum speed an object needs to attain to be free of the gravitational bonds of another. Let's see what that means.

Consider a ball at the edge of a table (Fig. 4.22*a*). Let it drop, and it falls straight to the ground. Now give the ball a push (Fig. 4.22*b*). As it passes the end of the table, it falls down, but the push makes it travel some distance horizontally before it hits the ground. Push it harder and it goes farther horizontally (Fig. 4.22*c*).

Now imagine, as Newton did, a giant cannon placed on the top of a very high mountain and aimed parallel to the ground (Fig. 4.23). Fire a cannonball. It travels some distance, then falls to the ground. If you use more powder in the cannon, the ball travels farther along the earth's curve before it hits the ground. If you put a large enough charge in the cannon, the ball goes completely around the earth in a *circular* orbit, returning to the cannon (orbit *C* in Fig. 4.23). The inertial motion of the ball *exactly* compensates for the falling due to gravity.

The ball's circular orbit defines a crucial path around the earth, for the ball (or any mass) requires a specific speed—no more, no less—for the cir-

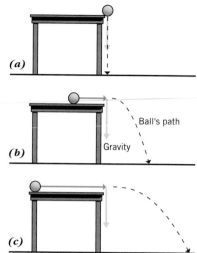

FIGURE 4.22 Motion of a ball near the surface of the earth. (*a*) Dropped from the edge of a table, the ball falls straight down. (*b*) When pushed along the table before it reaches the edge, the ball moves horizontally (along a parabola) before it hits the ground. (*c*) When pushed harder, it moves a greater distance before it strikes the ground.

cular orbit. A somewhat lower speed results in an *elliptical* orbit. The position of the top of the mountain then marks the farthest point from the earth in such an orbit—the **apogee** point.

Now dump an even larger charge in the cannon. With a starting speed greater than that needed for a circular orbit, the cannonball also travels around the earth in an elliptical orbit. But increasing the starting speed makes the orbit more eccentric. The point at the top of the mountain will now be the point in the orbit closest to the earth—the **perigee** point (orbit *D* in Fig. 4.23). Eventually the orbit becomes so eccentric that the semimajor axis is infinitely long, and it would take the ball an infinite time to return (orbit *E* in Fig. 4.23). So it never does. The speed that is necessary to achieve this result is called the escape speed. (Note from Kepler's second law that when the ball is an infinite distance away, its speed relative to the object it left is zero.) Any speed *larger* than the escape speed produces the same effect: the ball leaves the gravitational grip of the earth, never to return.

What determines the escape speed from the surface of an object? Consider

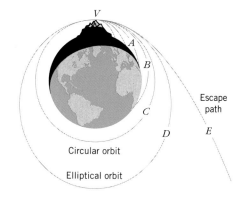

FIGURE 4.23 Launching earth satellites, according to Newton. From the top of a high mountain (*V*), cannonballs fired at a low speed hit the earth after traveling some distance (*A* and *B*). A certain minimum speed places them in a circular orbit (*C*); a somewhat greater speed an elliptical orbit (*D*); and above the escape speed, an escape orbit (*E*).

the earth. Its escape speed is 11 km/s. Suppose that you kept the earth at its present radius but increased its mass. Its escape speed would increase. Now suppose that you kept its mass the same but decreased its radius. The escape speed from the surface would again increase. Basically, *an object's mass and radius determine its escape speed from its surface.* A greater mass or a smaller radius (for the same mass) results in a larger escape speed. A smaller mass or a greater radius (for the same mass) results in a smaller one (Focus 4.4).

Today we use rockets rather than Newton's imaginary cannon to place satellites in orbit around the earth. Or to

Enrichment Focus 4.4

ESCAPE SPEED

From Newton's laws of motion and gravitation arises the concept of *escape speed,* the minimum speed an object needs to be free of the gravitational bonds of another. Algebraically, the formula is

$$V_{esc} = \sqrt{\frac{2\,G\,M}{R}}$$

where V_{esc} is the escape speed, M the mass of the object you want to escape, and R its radius. (Assume that the mass is spherical.) Note the mass of the escaping object does *not* enter into the calculation of escape speed. (Why not?)

Let's calculate the earth's escape speed. The mass of the earth is 6.0×10^{24} kg, its radius is 6.4×10^6 m, and $G = 6.7 \times 10^{-11}$ in SI units (Appendix A). So

$$V_{esc} = \sqrt{\frac{2\,G\,M}{R}}$$

$$= \sqrt{\frac{2(6.7 \times 10^{-11})(6.0 \times 10^{24})}{6.4 \times 10^6}}$$

$$= 1.1 \times 10^4 \text{ m/s} = 11 \text{ km/s}$$

For objects in the solar system, the radius and mass of the earth are convenient units to use. With these units, the escape speed in kilometers per second is

$$V_{esc} = 11\sqrt{\frac{M}{R}}$$

This form of the equation emphasizes that escape speed *increases* as *mass increases* or as *radius decreases*. For example, increasing the mass by 4 with the same radius doubles the escape speed. So will decreasing the radius by 4, if the mass stays the same. What if you did both? Then the escape speed increases to four times the original amount. So the escape speed is directly proportional to the square root of the mass and inversely proportional to the square root of the radius.

send them off with greater than escape speed to journey to other planets—and even out of the solar system!

In terms of a scientific model, the predictive success of Newton's laws was astounding. Time after time, the numbers and not just the general ideas came out right. Planetary positions could be predicted precisely, as could the motions of planetary satellites. And Newton's ideas were *universal*. No longer were separate laws needed for falling bodies, the moon, and binary stars. One simple force binds them all: gravity!

Newton's Cosmology

For the Greeks, Ptolemy, and Copernicus, the cosmos—enclosed by the sphere of fixed stars—appeared to be finite and bounded. Galileo had left open the possibility that the universe might be infinite but did not strongly push his case. Newton held that his laws required that the universe be infinite in extent. His argument went like this: if the universe were finite, gravitation would eventually pull all matter to the center. As a result, only one large mass would exist. However, we see other masses—stars, for example. In an infinite universe, the matter would be pulled into an infinite number of small condensations. Newton concluded, therefore, that the universe is infinite.

Not all things in the heavens and on earth sat happily in Newton's infinite universe. The innermost planet, Mercury, posed an annoying problem: the axis of its orbit rotates in space, and this motion cannot be explained completely by the attraction of existing planets. Some astronomers thought that the excess rotation was caused by a planet between Mercury and the sun. Although observations of this hypothetical body—called Vulcan—have been reported, they have so far proved to be mistaken. The supposed Vulcan has never been seen, even when modern observation techniques have attempted to catch its swift flight. (The problem of Mercury's orbit was later solved by Einstein: Chapter 7).

Newton was sorely disturbed by the mutual forces among the planets, which he thought must eventually lead to the disintegration of the solar system. To avoid this awful event, Newton envisioned the hand of God occasionally descending to reset the clockwork mechanisms of planetary motions, like a conscientious artisan making adjustments. The order of the mechanical universe—ordained by the Divine Being—was maintained by His intervention, and the expanses of Newtonian space were benevolently watched by the distant God.

■ **What discovery about the orbits of comets was impelled by Newtonian physics?**

Key Concepts

1. Galileo used his telescopic observations to promote the Copernican model (which he believed in for aesthetic reasons), even though none of his data gave direct proof of its validity. Galileo recognized that the Copernican model lacked a physical basis and sought one by analyzing the motions of free-falling objects near the earth. To this end, he developed a new concept of inertia. He concluded that falling objects have accelerated motion and so are subject to a force: gravity.

2. With his laws of motion and gravitation, Newton abolished the ancient physical distinction of motions in the terrestrial and celestial realms. The result was a physically unified cosmos (Table 4.1), linked by gravity and powerful laws of explanation and prediction of motions.

3. Newton's laws of motion described natural motion and forced motion in a new way. He viewed natural motion as that of a body at rest or moving at a constant speed along a straight line. Forces result in accelerations (changes in velocities); forces act in mutual pairs; and a net force results in acceleration—forced motion.

4. From these ideas about motion, Newton concluded that planetary orbits result from a centrally directed or centripetal force; that force is the gravitational

TABLE 4.1 EVOLUTION OF COSMOLOGICAL MODELS

ASPECT OF MODEL	PTOLEMY	COPERNICUS	KEPLER	NEWTON
Planetary motions	Uniform, circular	Uniform, circular	Nonuniform, elliptical	Nonuniform, elliptical
Force on planets?	No, natural motion	No, natural motion	Yes, magnetic	Yes, gravitation
Kind of cosmos	Geocentric, finite	Heliocentric, finite	Heliocentric, finite	Centerless, infinite

force between the sun and the planetary masses. Newton used the moon's orbit to confirm that the force of gravity varies with the inverse square of the distance.

5. Newton's law of gravitation describes the strength of the gravitational force as (a) depending directly on the product of the masses of the two objects and (b) inversely as the square of the distance between their centers. The direction of the force pulls the masses together. This description of gravitation resulted in elliptical orbits for the planets.

6. Newton's physics justified the Copernican model as revised by Kepler, with the correct force, gravity; the earth's rotation and revolution were naturally explained. Newton's work also resulted in very accurate predictions of planetary motions along elliptical orbits, as well as a revised version of Kepler's third law. The final result was a pleasing, and accurate physical foundation for a heliocentric system.

7. The discovery of Neptune confirmed the predictive validity of Newton's law of gravitation within the solar system. The discovery of binary stars whose motions follow Kepler's laws verified the validity of Newton's law of gravitation beyond the solar system.

8. Newton's laws provide the mathematical tools for placing satellites in orbit around the earth, for setting the trajectories of spacecraft in the solar system, and for defining the escape speed of an object from a planet or even the solar system itself.

9. Newton's cosmology demanded that the universe be infinite in extent; if it weren't, gravity would bring all the material together, which had not been observed and did not appear to be remotely likely.

Key Terms

acceleration	centripetal acceleration	inertia	newton
apogee	centripetal force	inverse-square law	Newton's laws of motion
binary star	escape speed velocity	kilogram	perigee
center of mass	free-fall	mass	speed
central force	gravitation	mechanics	velocity

Study Exercises

1. Use one of Galileo's major telescopic discoveries to support the Copernican model and refute the traditional (Aristotelian–Ptolemaic) one. (Objective 4-1)

2. What important discovery of Kepler's about planetary orbits was ignored by Galileo? (Objective 4-3).

3. Describe Galileo's concept of inertia and contrast it to the Aristotelian one. (Objective 4-2)

4. You have two balls of the same size and shape. One is lead, the other wood. You drop them together. What happens? Explain. (Objectives 4-2, 4-5, and 4-6)

5. Imagine that you're out in space and push away from you an object having a mass identical to your own. What happens? Explain. (Objective 4-5)

6. Describe two ways in which Newton's model of the cosmos differs from that of Copernicus. (Objective 4-9)

7. Answering the main physical objections to the Copernican model, use Newtonian physics to argue

that the earth rotates on its axis and revolves around the sun. (Objective 4-10)

8. Describe one way in which Newton and Kepler had similar ideas about the cosmos and one way in which their ideas differed. (Objective 4-9)

9. Give a simple example, different from those in the text, of each of Newton's laws of motion. (Objective 4-5)

10. Imagine you hold a ball in your hand with your arm out to your side. You walk along quickly toward a target on the floor. You release the ball just as it reaches a point above the target. Use Newton's concept of inertia to predict where the ball will strike: behind, on, or in front of the target. (Objectives 4-5 and 4-6)

11. Consider an object shot upward from the earth with *less* than escape speed. Describe its motion. (Objective 4-12)

12. Consider a special satellite sent into space to orbit the earth at the same distance as the moon. How would the centripetal acceleration of such a satellite compare to the centripetal acceleration of the moon? (Objectives 4-7 and 4-8)

13. When you see Mars at opposition to the sun in the sky, the earth and Mars lie closest together on the same line from the sun. Consider the gravitational attraction of Mars on the earth at that time. How does it compare with the gravitational attraction of the earth on Mars at the same time? (Objectives 4-5 and 4-7)

14. In words, explain the physical meaning of Kepler's third law of planetary motion. (Objective 4-11)

Problems & Activities

1. Refer to Appendix B to find out the distances of the planets from the sun (in AU) and the masses of the planets (relative to the mass of the earth, when the earth's mass = 1.0). Place these numbers in a table from Mercury outward, and then calculate the force of gravity on each planet from the sun.

2. From Newton's second law, $F = ma$, you can see that the acceleration of a mass is $a = F/m$. Hence, to find the centripetal acceleration of the force of gravity, divide both sides of Newton's law of gravitation by m, the mass of the accelerated object (say the earth moving around the sun). As in Problem 1, refer to Appendix B to find the relative distances of the planets and calculate the relative centripetal acceleration for each.

3. Use information from Appendix B to calculate the escape speed from the moon.

4. Imagine you have a force of 10 newtons applied to a mass of 1 kg. What happens when you apply it to a mass of 5 kg? 10 kg?

5. What is the gravitational force between you and another person standing 10 meters apart? 20 meters? 100 meters? *Hint:* Estimate your mass!

6. A satellite near the surface of the earth completes its orbit once in 90 minutes. What is its centripetal acceleration? Compare your result to the acceleration of gravity near the earth's surface.

7. Use the information about the satellite in Problem 6 to find the mass of the earth. *Hint:* Apply Kepler's third law.

8. What is the approximate distance between the earth and Jupiter at opposition? Estimate this distance in AU and then convert to kilometers using 1 AU $\approx$ 1.5 $\times$ 10^8 km. Jupiter's brightest moon is called Ganymede. Its distance from the center of Jupiter is about 1.1 $\times$ 10^6 km. Given the smallest angle observable by the eye (about 1 arcmin), could Ganymede be seen as an object separate from Jupiter during Ganymede's maximum elongation from Jupiter?

See for Yourself!

The four largest and brightest moons of Jupiter are those seen by Galileo with his telescope. Their order outward from Jupiter is Io, Europa, Ganymede, and Callisto (see Appendix B.7). Their orbital periods range from 1.8 to 17 days. A small telescope or powerful binoculars will provide you with a way to observe these moons for yourself (Fig. A.1). You can find their configurations relative to Jupiter in the *Astronomical Almanac* or the *Observer's Handbook* for the current year. *Astronomy* and *Sky and Telescope* magazines provide this information each month.

You can easily compare the motions of the inner two moons during an observing session of a few hours on one night. Sometimes you will see a shadow of a moon on Jupiter as the moon moves between Jupiter and the sun. Or you may even see a moon vanish behind Jupiter's disk! If you have a choice of observing seasons, look while Jupiter is at opposition, and so the closest to the earth. ■

FIGURE A.1 Jupiter and three of its brightest moons as seen through a small telescope. (Photo by M. Zeilik.)

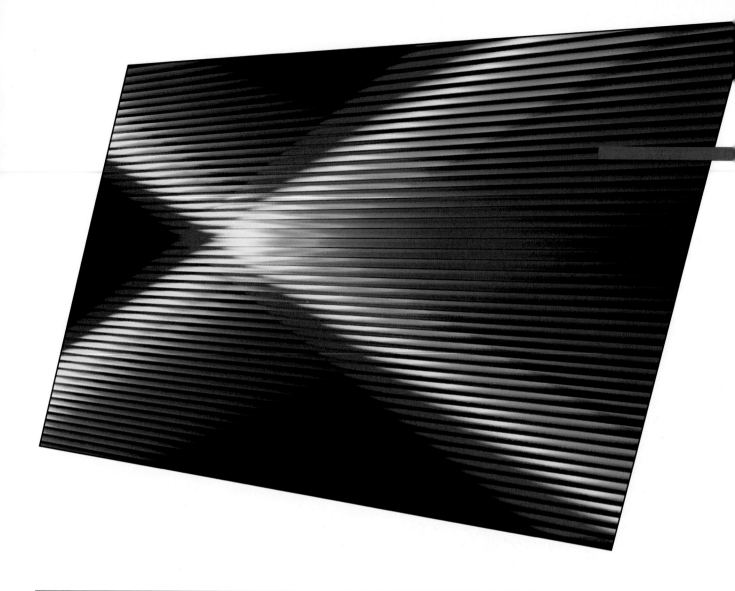

LEARNING OBJECTIVES
After studying this chapter, you should be able to:

5-1 Describe the differences in the appearance of continuous, absorption, and emission spectra as seen through a spectroscope.

5-2 Explain how an understanding of spectra made it possible for astronomers to determine the chemical compositions of stars and the physical conditions inside their atmospheres.

5-3 Use Kirchhoff's rules to relate the three basic spectral types to the physical conditions of their production.

5-4 Describe how wavelength, frequency, and speed of light are related.

5-5 Describe the relationship between energy and frequency or wavelength for light and apply it to the spectrum.

5-6 Briefly describe the electromagnetic spectrum with examples from each major region.

5-7 Use an energy-level diagram to explain in general how atoms absorb and emit light.

5-8 Use the energy-level diagram of a hydrogen atom to explain how the Balmer series is produced, both as emission and absorption lines.

5-9 Explain in simple physical terms how absorption lines occur in spectra.

5-10 Describe the concept of the conservation of energy and apply it to ordinary and astrophysical situations.

5-11 Describe the concept of energy and the various types of energy, and apply these to light and astrophysical situations.

Central Question
How do atoms produce light, and what does light reveal about the sun and stars?

The Birth of Astrophysics

We understand the possibility of determining their [celestial bodies']
shapes, their distances, their sizes and motion, whereas never, by any
means, will we be able to study their chemical composition.

AUGUST COMTE, *Cours de Philosophie Positive* (1835)

Astronomers had for centuries charted the positions of the stars and planets. Their attention focused on the question of how the planets moved, a problem solved by Isaac Newton (Chapter 4). In contrast, the stars played a passive role, forming a mere backdrop to the complex activity of the planets.

But how to find the physical makeup of the stars and planets? Auguste Comte saw no hope of accomplishing this. The nineteenth-century French philosopher had arrived at his pessimistic view for hard practical reasons: astronomers had no means to bring a piece of a star or planet into the lab for examination; all they had at hand was the light funneled into their telescopes.

The birth of astrophysics at the end of the nineteenth century gave astronomers physical insight about light. Experimental attempts to understand light unexpectedly led to a study of atoms. The atoms in stars (and the sun) emit starlight (and sunlight), so an analysis of light reveals information about the physical conditions and chemical compositions of stars. Astronomers could finally penetrate the environments of the distant stars and the sun, and so prove Comte wrong. Astrophysics transformed our conception of the cosmos. No more were the stars regarded as simple points of light. Now they appeared as other suns made of the same stuff found in our sun and the earth.

5.1 Sunlight and Spectroscopy

Matter makes up the sun and stars—matter in the form of atoms. Atoms can give off and absorb light; the sunlight we see originates from atoms. The structure of atoms can be investigated by analyzing the light they emit and absorb. We can also investigate the physical environment of atoms by analyzing light. To understand more of the sun (and stars), you need to understand how atoms and light interact.

Atoms and Matter

In the eighteenth and nineteenth centuries, chemists discovered that substances can be divided into two classes: chemical **elements** and chemical **compounds.** *Elements* cannot be broken by chemical reactions into simpler substances. (See Appendix F for a periodic table of the elements.) They are the most basic substances, such as hydrogen (H), helium (He), carbon (C), and oxygen (O). Although 92 elements occur in nature (and more have been created in the laboratory), many are rare; only a few dozen are common. Most substances you encounter are not elements but *compounds,* made of two or more elements. For example, water (H_2O) is a chemical compound composed of the elements hydrogen and oxygen (Fig. 5.1*a*).

A **molecule** is the smallest unit of a compound that still has the chemical properties of the compound. However, a compound can be broken down into elements, so a unit of matter smaller than a molecule must exist. An **atom** is the smallest unit of an element that displays the chemical properties of the element. A compound is created when atoms join together to form molecules of the compound.

A Model of the Atom

Physicists developed a useful model of the atom in the twentieth century. The modern concept of an atom (Fig. 5.1*b*) pictures a tiny, dense **nucleus** surrounded by rapidly moving *electrons.* The study of electricity revealed that matter has two opposite charges, positive and negative. Like charges repel each other and opposite charges attract. **Electrons,** which carry a negative charge, are low-mass particles (a mere 10^{-30} kg; see Appendix C). **Protons** and **neutrons,** which are about 2000 times as massive as electrons, make up an atom's nucleus. The protons are positively charged, and the neutrons have no electrical charge. The nucleus of an atom, because of the protons it contains, is positively charged and attracts the negatively charged electrons.

This attractive electrical force binds the electrons to the nucleus and holds the atom together. (A strong *nuclear*

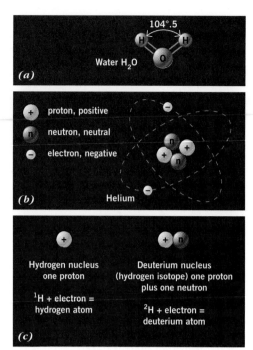

force binds the protons and neutrons together in the nucleus, despite the mutual repulsion of the positively charged protons.) The electrons whiz around the nucleus in orbits, but their distance is very great in terms of the size of the nucleus. Most of an atom is empty space! If the proton of a hydrogen atom were the size of a marble, the electron would be about a kilometer away from the nucleus.

Elements differ from one another because their atoms differ. The nucleus makes the key difference; the atoms of different elements contain different numbers of protons. The nucleus of a hydrogen atom, for example, contains one proton; helium, two (Fig. 5.1*b*); and carbon, six. (The table of the elements in Appendix F summarizes the elemental properties.) The number of electrons orbiting the nucleus of a normal atom is the same as the number of protons in the nucleus, so the atom carries no net electric charge. The electrons determine the chemical properties of atoms.

Nuclei that contain the same number of protons but a *different* number of neutrons are called **isotopes** of the element. For example, heavy hydrogen, sometimes called *deuterium,* has one proton and one neutron, whereas ordinary hydrogen has one proton and no neutrons (Fig. 5.1*c*). They are the same

FIGURE 5.1 Molecules, atoms, and isotopes. (*a*) The chemical structure of a molecule of water, H_2O. (*b*) The structure of a helium atom. (*c*) The difference between a hydrogen atom and deuterium, an isotope of hydrogen, lies in the neutron.

Atoms

At a Glance

Nucleus	Protons and neutrons, positive charge
Elements	Determined by number of protons
Isotopes	Determined by number of neutrons
Electrons	Orbiting nucleus, equal to number of protons for neutral atom

element, but each of the atoms in a deuterium molecule has more mass than the atoms of ordinary hydrogen, because of the extra neutron. (The table in Appendix F gives only the most common isotope of each element.) Most elements contain approximately the same number of protons and neutrons, or a few more neutrons. For example, atoms of the most common isotope of carbon contain six protons and six neutrons.

Although an atom is neutral electrically, this condition can be changed by, for instance, taking away or adding electrons. If the number of electrons is less than or greater than the number of protons (so the net charge is negative or positive), the atom becomes an **ion.** In most astronomical situations, we find *positive* ions, which have resulted from the *loss* of one or more electrons.

Simple Spectroscopy

Breaking up light into its component colors to detect how atoms and light interact is called **spectroscopy.** You do a little spectroscopy automatically when you look at a rainbow. Next time you see one, look carefully at the colors, which run from red to violet. What's happening? Raindrops are dispersing sunlight into a continuous band of colors.

You can use a prism to do the same at home or in a lab (Fig. 5.2). Use a slotted piece of cardboard to pass a beam of sunlight through a prism. Let the light coming out of the prism fall on a white sheet. You'll find the sunlight spread into an array of colors, with red light bent the least and violet the most. This sequence of colors is called a **spectrum.** A spectrum with no breaks in it is called a **continuous spectrum;** it has a continuum of unmixed shades of color. White light contains all these colors. You can think of white light as a mixture of all colors at equal intensity.

An inexpensive device to disperse white light into its colors is called a *diffraction grating.* Ask your instructor to show you how to use one in class. If you can borrow one for a while, you can examine various light sources with it around campus or town. An ordinary incandescent light bulb shows a continuous spectrum (Fig. 5.3a). A fluorescent

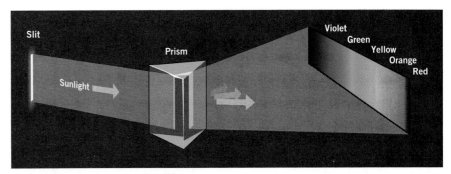

FIGURE 5.2 Experiment with a prism and sunlight to produce a spectrum: sunlight passes through a slit in front of a prism, which disperses the light into its colors, from violet to red.

(a)

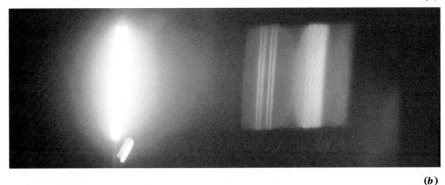

(b)

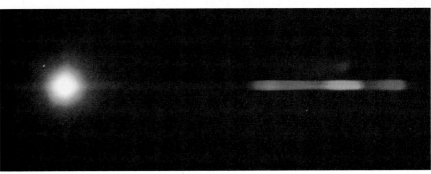

(c)

FIGURE 5.3 Views of various spectra (object is at the left, its spectrum on the right). (a) Photograph of a spectrum from an incandescent light made through a diffraction grating. (b) Another diffraction grating spectrum; this one is of a fluorescent light. (c) The moon and its spectrum as viewed through a diffraction grating. (Spectra photos courtesy of Dean Zollman and the American Association of Physics Teachers.)

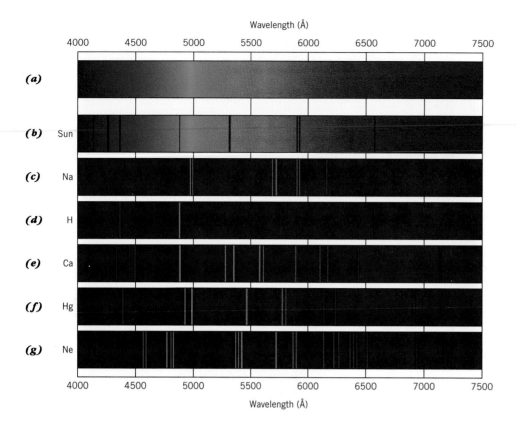

FIGURE 5.4 Major spectral types: the shorter wavelengths lie to the left, with wavelength increasing to the right. At top (*a*) is a continuous spectra such as that from an incandescent light bulb. Below it (*b*) is the absorption-line spectrum of the sun; only the most prominent dark lines are shown from the following elements: H, hydrogen (the Balmer series); Ca, calcium; Fe, iron; and Na, sodium. Below the sun's spectrum are the emission-line spectra of selected elements: sodium (*c*), hydrogen (*d*), calcium (*e*), mercury (*f*), and neon (*g*). Note how some bright lines of sodium, calcium, and hydrogen line up with the dark lines in the sun's spectrum, which indicates the presence of these elements in the sun's atmosphere.

light shows mostly distinct, bright lines (Fig. 5.3*b*). If you look at the full moon, you'll see mostly a continuous spectrum with some well-defined dark lines (Fig. 5.3*c*).

Suppose you put another slotted piece of cardboard (or slit) behind your prism or grating to let just a *single color* pass through. You'd then see a bright *line* of that color, not a complete, continuous spectrum. This line is called a **spectral line.** If you let sunlight pass through a prism and slit, and then magnify the spectrum with a telescope, you'll see a continuous spectrum filled with dark lines (Fig. 5.4*b*). Light is missing from some of the colors. The dark lines, from which light of these colors has been removed, or absorbed, are called **absorption lines,** and a spectrum exhibiting them is called an **absorption-line spectrum** (Fig. 5.4*b*). When light is emitted only at certain colors, as in the case of a neon sign, we speak of **emission lines** or **emission-line spectra** (Fig. 5.4*c*–5.4*g*).

Straight lines appear in a spectrum because the slit admitting the light to the prism is a straight-line source of light. Each line is an image of the slit in the light of one color. You would also see a line spectrum if you looked at a straight hot wire. But if the slit, or other source of light, were curved in an S shape, the lines in the spectrum would also be S-shaped.

■ What is a spectrum?

5.2 Analyzing Sunlight

To unravel the message of sunlight, let's consider a few experiments with a **spectroscope,** an instrument for observing fine details in a spectrum.

For the first experiment, point a spectroscope at a pure neon advertising sign. You'll see an emission-line spectrum with the brightest lines in the red region (Fig. 5.4*g*). That's why such a sign appears red to our eyes. A mercury street light (which uses mercury gas) has its strongest bright lines in the yellow and green regions (Fig. 5.4*f*). These lines combined result in the greenish color of mercury street lamps. In contrast, sodium street lamps cast a ghastly yellow because the brightest lines from hot sodium gas fall in the yellow region of the spectrum (Fig. 5.4c).

For the second experiment, put sodium (in the compound form of table salt) in a hot, colorless flame. Visually, you'll see the flame turn yellow. Through the spectroscope, you'd see a series of bright lines, the brightest, a pair, in the yellow region (Fig. 5.4c). With the spectroscope you can prove that different chemical elements (and molecules) give off different patterns of bright lines. *Each chemical element displays a unique arrangement of bright lines when excited in the gaseous state.* A particular spectral pattern demonstrates the presence of a particular element. Each pattern is unique to an element, as fingerprints are to people.

The atoms in the gas produce the observed bright lines. Different gases composed of different atoms produce different patterns of spectral lines. So the bright-line spectrum of a hot gas is characteristic of the atoms in the gas. The pattern of the lines reveals which elements the gas contains.

For the third experiment, look at sunlight with a spectroscope. You'll see dark lines against a bright, continuous background of color (Fig. 5.4b). A pair of dark lines in the yellow region of the spectrum particularly stands out. Astronomers call this pair the **D lines.** Could these dark lines be related to the bright lines of sodium observed when salt is put in a flame? Could the dark lines in the sun's spectrum also come from sodium in the sun?

To test this idea, pass light from a glowing solid (which gives off a continuous spectrum) through the sodium flame. A pair of dark lines appears in the spectrum at exactly the same position in the yellow regions as the pair in the sun's spectrum. So it must be sodium making these dark lines, *removing* yellow light from the solar spectrum. Now pass sunlight through a sodium flame into the spectroscope. What do you expect? That the added light of the sodium flame will make the lines less dark? Wrong! The pair of lines becomes even darker! The sodium in the flame absorbs even more of

the sun's yellow light. That result proves that sodium made the lines. But why did the lines get *darker*?

Kirchhoff's Rules

From lab experiments such as those described above, the German physicist Gustav Kirchhoff (1824–1887) formulated empirical rules of *spectroscopic analysis,* the determination of the physical state and the composition of an unknown mixture of elements. **Kirchhoff's rules** are briefly stated as follows:

RULE 1. A hot and opaque solid, liquid, or highly compressed gas emits a continuous spectrum (Fig. 5.5a). Example: filament of a light bulb.

RULE 2. A hot, transparent gas produces a spectrum of bright lines (emission lines). The number and colors of these lines depend on which elements are present in the gas (Fig. 5.5b). Example: a neon sign.

RULE 3. If a continuous spectrum (from a hot, opaque solid, liquid, or gas) passes through a transparent gas at a lower temperature, the cooler gas will cause the appearance of dark lines (absorption lines), whose colors and number will depend on the elements in the cool gas (Fig. 5.5c). Example: light from the sun.

Essentially, the first rule says that an opaque, hot material produces a continuous spectrum. Recall that atoms, instead, usually emit and absorb at discrete colors. However, when the atoms are jammed together so densely that the material becomes opaque to light, a continuous spectrum appears. For instance, the sun is so hot that it is gaseous, but it is still opaque enough to produce a continuous spectrum.

A solar spectrum shows dark lines against a continuous background. What causes these dark lines? By Kirchhoff's third rule you expect a cooler and less dense gas to lie between the visible surface of the sun and us. Such a gas could be in the atmosphere of the earth or in the atmosphere of the sun. Although a few lines in the solar spectrum are produced by gases in the earth's atmosphere (water vapor in particular), most dark lines are produced in the solar atmosphere. Light from the continuous spectrum passes through this atmospheric layer, which absorbs it to produce the dark lines. The positions of the lines in the spectrum, their colors, tell which elements are present. The solar composition determined from the dark lines relates only to the region of the sun that produces the absorption spectrum (its atmosphere).

Caution. Dark lines do not mean the *complete* absence of light at those colors; rather, the lines appear dark in contrast to the continuous spectrum.

With these ideas in mind, you should now understand what happens when sunlight passes through the sodium vapor to make darker D lines. The sodium in the flame is *cooler* than the sun's visible surface, so the sodium *absorbs* the light in the D lines. Though hot, the sodium gas removes some specific wavelengths from the beam of sunlight. That's why the sodium D lines got darker when sunlight passed through the sodium flame. This experiment also proves that the colors that are missing from the continuous solar spectrum were absorbed by the atoms in the cool gas of the solar atmosphere.

Whether the atoms in the gas emit or absorb depends on the physical conditions in the gas. Emission requires high temperatures in a transparent gas; absorption occurs when a continuous spectrum passes through a cooler transparent gas. In either case, the pattern of emission lines is the same as the pattern of absorption lines for a gas with the same chemical composition.

Here's the fundamental point concerning emission and absorption lines: *the lines absorbed by a gas from a continuous spectrum are the same lines emitted by the gas when energy is put into it.*

The statement above is one way to describe the conservation of energy, which we now discuss. **Energy** allows events to happen in the universe. The concept of energy has fundamental importance because, although energy

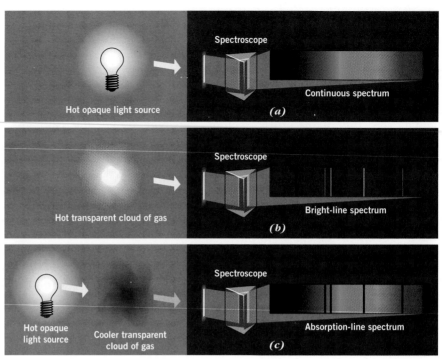

comes in many forms, the total amount of energy that an object has can be described and assigned a number. If you have a bunch of objects, the sum of their individual energies is the total energy of the group. Moreover, the law of **conservation of energy** says that the total energy of a group of objects, isolated from the rest of the world, has a constant value.

The Conservation of Energy

Energy exists in many different forms. Three common ones are *kinetic energy,* which is due to motion; *potential energy,* which is stored up by position under an applied force; and *radiative energy,* which is carried by light. This chapter deals mostly with radiative energy. All forms of energy are measured in the same unit: the **joule** (J). You do a joule of work when you exert a force of one newton over a distance of one meter. (Table 5.1 gives the energy outputs of some familiar objects and events.) If you lift an apple from the ground to over your head, you've done about one joule's worth of work. You have also changed the apple's energy; if you release it, it will fall to the ground. The apple has acquired potential energy

FIGURE 5.5 Basic spectral types as viewed through a spectroscope. (*a*) Continuous spectrum with no lines from a hot opaque solid (the light filament). (*b*) Emission-line spectrum with a few lines from a hot transparent gas. (*c*) Absorption-line spectrum with the same spectral lines as in (*b*); in this case, the continuous spectrum emitted by the light has passed through a transparent cooler gas with the same chemical composition as in (*b*).

TABLE 5.1 SOME ENERGY OUTPUTS	
ENERGY SOURCE	**TOTAL ENERGY (J)**
Sun's radiation (1 year)	10^{34}
Volcanic explosion	10^{19}
H-bomb	10^{17}
Thunderstorm	10^{15}
Lightning flash	10^{10}
Barrel of oil	10^{9}
Battery (D cell)	10^{4}
Baseball pitch	10^{2}
Typing (per key)	10^{-2}
Flea hop	10^{-7}

(from your work on it), which is transformed to kinetic energy when the apple falls. Though the forms of the energy have changed, the total is still the same—it has been conserved.

Kinetic Energy

Kinetic energy is that associated with motion. You can understand it in terms of work. To stop a moving object requires an obvious effort, and the more massive the object or the faster its motion, the greater the effort (work!) needed to stop it. For instance, it is easier to grab and stop a bicycle traveling at 5 km/h than one moving at 10 km/h. But you'd not want to try that with a car, even at 5 km/h!

The kinetic energy of a moving object depends on its mass and its speed in a special and precise way: it is directly proportional to the mass and to the square of the speed (Focus 5.1). So a bicycle moving at 10 km/h has *four* times the kinetic energy than one that travels at 5 km/h; and at 15 km/h, it has *nine* times the kinetic energy of one that goes at 5 km/h. Investigators use this concept when unraveling an automobile accident; with the brakes locked, a car that has been clocked at 100 km/h will skid four times as far as one moving at 50 km/h, if the cars have equal mass. The faster car has four times the kinetic energy, and so stopping it takes four times the work.

You must distinguish between the kinetic energy of a large object, such as a baseball, and the average kinetic energy of the particles that make it up. A baseball's kinetic energy allows it to go somewhere, but that property doesn't affect the microscopic motions of its atoms, which move small distances at random within the ball. The motion of the atoms makes up the *thermal energy* of the ball—a measure of the total kinetic energy of a large collection of particles. When you add energy to a gas, the average value of the random speeds of its constituent molecules increases. What have you done? You have used a hot flame, say, to raise the thermal energy of the gas. **Heat** is the thermal energy that is transferred from one body to another. Heat always flows from a hotter body to a colder one.

Temperature measures the average kinetic energy of particles in, say, a gas. That, in turn, depends on the average speed of the particles. The temperature scale that indicates this average speed most directly is called the **kelvin** (K) scale; at 0 K, all microscopic motions have ceased. (See Appendix A for temperature units.) On this scale, your body temperature is about 310 K.

To sum up. *Heat* is the flow of thermal energy from a hotter body to a cooler one; *temperature* measures the average kinetic energy from the random motions of these particles. These are all manifestations of *energy*, or the ability to do work.

You should now be able to tell the difference—physically—between a fast baseball and a hot baseball (and a fast, hot baseball!).

Potential Energy

Potential energy is the stored energy of position as in a stretched rubber band. The possibility of motion characterizes potential energy. Fuels have potential energy, which is released when they are burned. On the chemical scale, changes within and between the electrical charges of molecules result in the release of the energy. These, too, are related to positions and are a form of potential energy. Why? Basically because electrical charges create forces between themselves. Likewise in an atom, the electrons

Enrichment Focus 5.1

KINETIC ENERGY

The kinetic energy of an object depends on both its mass and its speed. In algebraic form, the expression for kinetic energy is

$$KE = \frac{1}{2} mv^2$$

where *KE* is the kinetic energy (in joules), *m* the mass of the moving object (in kilograms), and *v* its speed, which is squared. Note that moving at a higher speed results in a large increase in kinetic energy. Many highways in the United States have speed limits of 55 mph. If a car travels at 75 mph instead, its kinetic energy is roughly doubled: $(75/55)^2 \approx 2$.

Let's do an example. In its orbit, the earth moves at about 30 km/s. What is its kinetic energy? The earth's mass (Appendix C) is about 6×10^{24} kg, so

$$KE = \frac{1}{2}(6 \times 10^{24} \text{ kg})(30 \times 10^3 \text{ m/s})^2$$
$$= \frac{1}{2}(6 \times 10^{24} \times 9 \times 10^8)$$
$$= 2.7 \times 10^{33} \text{ J}$$

That amount of kinetic energy is equivalent to the amount of radiative energy that the sun emits in a month!

---

orbiting the nucleus have specific kinetic energies that are due to their motion. They also have specific potential energies that correspond to their distances from the nucleus and their electrical attraction to the protons there.

One important example of potential energy relates to gravitational attraction. On the earth, all masses feel the earth's pull, although some, such as an apple held in your hand, do not immediately respond to the force. When you drop the apple, however, it gains kinetic energy, and the higher the point from which you drop it, the greater the kinetic energy it has attained by the end of its fall. The position above the earth's surface is related to the total amount of potential energy the object has. Energy is transformed from one form to another (potential to kinetic), yet the total energy of the apple remains constant, for as it loses potential energy, it gains an equal amount of kinetic energy.

Experiments have shown that energy is naturally transformed from one form to another, but that as long as all the energy is carefully tallied, the total amount remains the same—it is, again, conserved. One way to paraphrase the concept of the conservation of energy, as stated above, is: *energy cannot be created or destroyed; it may be transformed, but the total does not change.* The absorption and emission of light by atoms obey this fundamental conservation law. You will see the conservation of energy applied throughout this book. (This principle is modified somewhat because of Einstein's discovery that matter is actually a form of energy: Section 7.2.)

■ **Which of Kirchhoff's rules is exhibited by an incandescent light bulb?**

5.3 Spectra and Atoms

How can the physical conditions in stars be discovered from their spectra? The spectral code of the patterns of lines was cracked in a great revolution of physics in the twentieth century: the quantum theory and its explanation of the nature of the atom and of light. But to understand the power of spectroscopy, you must first know something about light.

Light and Electromagnetic Radiation

Today, we view light physically as having both particle and wave properties. Light sometimes behaves as waves and sometimes as particles, depending on how it is observed. Both models are needed to describe the properties of light completely. In either case, light carries energy—radiant energy. I will use both models; let me emphasize first some of the wavelike features of light.

Energy

At a Glance

Energy	Changes the condition of matter
Examples	Kinetic, potential, radiative
Heat	Microscopic kinetic energy
Conservation	Energy mutates, but the total remains the same

Waves

You probably have some experience with waves. Imagine that you're at a beach with waves arriving at regular intervals. You time them with your watch. The number of waves that arrive in a given period is the **frequency.** For instance, one wave hitting the shore every minute is a frequency of one per minute. The distance between the crests (peaks) of these waves is their **wavelength.** For example, if the crests are 10 meters apart, the wavelength is 10 meters. The *wave velocity* indicates how fast and in what direction the waves are traveling. Waves that move 10 meters in a minute have a velocity of 10 m/min.

Note that waves do *not* involve the mass motion of material over long distances. If you are in a boat on a lake with no currents, waves from passing motorboats cause your boat to bob up and down. But it stays in the same location. A wave carries energy of motion from one place to another, but it does not transport material.

To sum up. Waves have three fundamental properties (Fig. 5.6): wavelength, frequency, and velocity. The *wavelength* is the distance between two successive crests of a wave, the *frequency* is the number of waves that pass you each second, and the *velocity* is the distance covered in one second by a crest traveling in a certain direction.

The units of velocity are length per time. When you multiply frequency (number per time) and wavelength (length), the result equals wave velocity (length per time): frequency times wavelength equals velocity. Or, as an equation

$$f \times \lambda = v$$

where f is the wave's frequency, λ the wavelength (in meters, say), and v the velocity (in m/s). This fundamental rule applies to waves of all kinds. It tells you that for waves traveling through some material at constant velocity, a change in frequency requires a change in the wavelength.

For light in a vacuum, the velocity is the *same for all wavelengths:* 299,79 km/s. This value is usually designated b a lowercase c and is more easily recalle by its approximation, 300,000 km/s (3 $\times$ 10^8 m/s). So for light waves only,

$$f \times \lambda = c$$

where again λ is the wavelength and f the frequency. Because c is constant for all kinds of light, different wavelengths must have different frequencies.

The Electromagnetic Spectrum

Visible light is one type of wave produced whenever electric charges are accelerated. Radio waves constitute another type: a transmitter uses electricity to move electrons rapidly back and forth in an antenna; when the electrons are accelerated in a periodic fashion, they produce radio waves at a certain wavelength. Such radiated energy, generated by accelerated electrical charges, is termed *electromagnetic radiation.* The range of all different wavelengths of electromagnetic radiation makes up the **electromagnetic spectrum** (Fig. 5.7); visible light covers only a small part of the electromagnetic spectrum. All forms of electromagnetic radiation have the *same* physical nature.

Although the wavelength of electromagnetic radiation is sometimes measured in meters or centimeters, scientists often use special wavelength units for different regions of the electromagnetic spectrum. For example, visible light is usually measured by astronomers in *angstroms* (abbreviated Å), where 1 Å is 10^{-10} m. Physics people tend to use *nanometers* (abbreviated nm), where 1 nm is 10^{-9} m, so 1 Å = 0.1 nm. In the infrared region the unit of length commonly used is the *micrometer* (abbreviated μm), with 1 μm equal to 10^{-6} m.

Sometimes frequencies are used instead of wavelengths. In the radio region, astronomers commonly talk in terms of frequency rather than wavelength. The unit of frequency is the *hertz* (abbreviated Hz), which is one cycle (or vibration) per second. When you see a wave go by, from peak to peak, it has gone through one cycle. Even in the

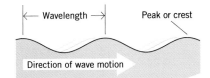

FIGURE 5.6 Some properties of waves. Every wave has a peak and a trough: the distance from one peak of a wave to the next is its wavelength. The number of waves that pass by each second (per second is one *hertz*) is the frequency.

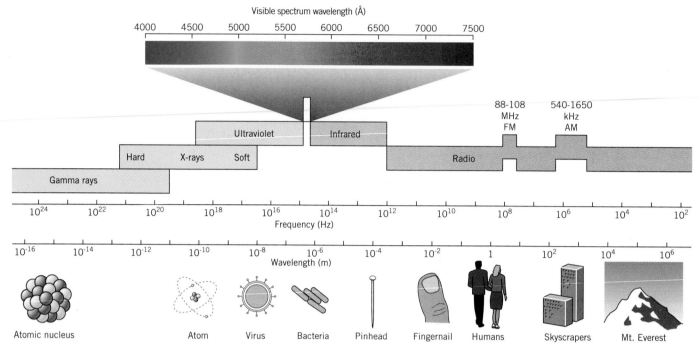

FIGURE 5.7 The electromagnetic spectrum from the shortest (gamma rays) to the longest (radio) wavelengths. Note that visible light makes up just a tiny piece of the complete spectrum. (Adapted from a diagram by E. J. Chaisson.)

radio region, many cycles go by in a second. For example, the AM band (your car radio) covers the range from 540 to 1650 kHz (kHz is the abbreviation for *kilohertz*, or 1000 Hz). The FM band ranges from 88 to 108 megahertz; 1 *megahertz* (abbreviated MHz) is 1 million hertz. Radio astronomers work at such high frequencies that they use the unit called the *gigahertz* (abbreviated GHz), 1 billion hertz. Police radar used in the United States to trap speeders typically operates at a frequency of about 10 GHz.

In a sense, a spectroscope works like a radio tuner. Both devices arrange a part of the electromagnetic spectrum by spreading it out from shorter to longer wavelengths. A certain wavelength can then be selected for examination. You make a selection on a radio by tuning it to a specific station that broadcasts at a certain wavelength. A spectroscope also can be "tuned in" to look at a particular line of some region of wavelengths.

Atoms, Light, and Radiation

Although a wave model for light was championed during the nineteenth century, it did not explain how atoms produced light. Astronomers were puzzled by spectral lines from stars. Why should stars emit light at certain wavelengths and not at others? The pattern of lines of hydrogen appeared in a simple series in the visible region of the electromagnetic spectrum. This series of hydrogen lines was named the **Balmer series** (Fig. 5.8), after the Swiss mathematician Johannes J. Balmer (1825–1898), who in 1885 devised an empirical formula describing the regular sequence of lines. Because the Balmer series followed a patterned sequence, the inference was that emission and absorption were related to a

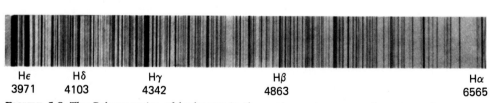

Hε	Hδ	Hγ	Hβ	Hα
3971	4103	4342	4863	6565

FIGURE 5.8 The Balmer series of hydrogen in the sun's spectrum over the range of wavelengths from about 4000 Å (blue end) to 6600 Å (red end). The Balmer series lines from the first (alpha) to the fifth (epsilon) one are marked from right to left (longer to shorter wavelength). (Courtesy of the Observatories of the Carnegie Institution of Washington.)

simple structure of atoms. But what was that structure?

In 1911 the New Zealand-born physicist Ernest Rutherford (1871–1937) proposed a model for the atom in which all the positive charge is concentrated in a tiny nucleus, with a surrounding negatively charged cloud of electrons. Rutherford imagined that the mutual attraction of unlike charges holds the atom together. If an atom loses one or more of the electrons, it becomes an ion. Physicists had noticed that an element's spectrum changes when it is ionized. For example, when hydrogen is ionized, it no longer produces the Balmer series of lines. The physicists inferred that the arrangement of electrons somehow determines its light-emitting properties.

In 1900 the German scientist Max Planck (1858–1947) announced a revolutionary idea about atoms and light. Perplexed by experimental work on spectra, Planck developed a startling new model for light: radiating matter emits light in discrete chunks of energy that he called quanta. From this basis Planck explained some of the observed properties of radiation.

Taking up this quantum idea, Albert Einstein (1879–1955) showed that each **quantum,** usually called a **photon** when talking about light, carries an amount of energy that is directly related to its frequency (Focus 5.2). Light of higher frequency (shorter wavelength) transports more energy than light of lower frequency. For example, an x-ray photon carries much more energy than a photon of visible light. This relationship between energy and frequency is a direct one: a photon with *twice* the frequency carries *double* the **radiative energy.** Similarly, a photon with half the frequency transports half as much energy. For example, a photon with a wavelength of 5000 Å carries 4×10^{-19} J; light at 2500 Å has 8×10^{-19} J.

With this idea in mind, you should now realize that the sun's continuous

Enrichment Focus 5.2

ENERGY AND LIGHT

The energy-level diagram of an atom describes a situation analogous to that of objects pulled by gravity. Instead of gravitational force, however, we have the force of electrical attraction between the positively charged nucleus and the negatively charged electrons. When an electron falls from one energy level to a lower one, it loses potential energy, which must manifest itself in some form, in this case a photon (radiative energy). To reverse the process requires the electron to gain energy, again produced from the absorption of a photon having exactly the energy required to raise the electron to the next level.

The amount of energy carried by light depends on its frequency (or wavelength). The fundamental relation is

$$E_{photon} = hf$$

where E is the energy in joules, h a

constant (called *Planck's constant*) with a value of 6.63×10^{-34} J · s, and f the frequency in hertz. Note that light with twice the frequency carries twice the radiative energy; three times the frequency, three times the energy; and so on.

Since wavelength and frequency are related by the wave equation

$$f = \frac{c}{\lambda}$$

we can rewrite the energy equation as

$$E = \frac{hc}{\lambda}$$

with the wavelength λ in meters.

A few examples. Consider police radar operating at a microwave frequency of 10 GHz. What is the energy of a single photon? Since 1 GHz = 10^9 Hz, then

$E = hf$
$= (6.63 \times 10^{-34}$ J · s$)(10 \times 10^9$ Hz$)$
$= 6.63 \times 10^{-24}$ J

Contrast this to light at 5000 Å in the visible part of the spectrum. Since 1 Å = 10^{-10} m, then

$E_{photon} = \frac{hc}{\lambda}$

$= \frac{(6.6 \times 10^{-34} \text{ J} \cdot \text{s})(3 \times 10^8 \text{ m/s})}{(5000 \text{ Å})(10^{-10} \text{ m/Å})}$

$= 4.0 \times 10^{-19}$ J

so that the visible photon has about 10^5 times more energy than the microwave one. However, an intense beam of microwaves (for example, in a microwave oven) can carry more energy than a beam of visible light, say from a flashlight. Each microwave photon carries less energy than each visible light photon, but the oven generates many photons.

spectrum is, in fact, an energy spectrum, from lower (infrared) to higher (ultraviolet) energies. You have seen that atoms emit and absorb certain colors at specific wavelengths. For each wavelength you can imagine a photon carrying off a certain amount of energy. Atoms must absorb and emit energy in the form of whole photons. This quantum model emphasizes that in the emission or absorption of light, an atom loses or gains energy in *discrete* amounts, as the energies of individual photons.

Solving the Puzzle of Atomic Spectra

The Danish scientist Niels Bohr (1885–1962) meshed Planck's quantum and Rutherford's atomic pictures. The resulting scheme, known as the **Bohr model of the atom,** explained and predicted the absorption and emission of photons.

Bohr pictured atoms as having many possible arrangements of electrons. He imagined that the emission or absorption of light arises from a **transition** between electron arrangements. In line with Planck's theory and observed spectra, Bohr realized that only certain electron arrangements are permitted. Similarly, electron transitions can occur only between these special arrangements. You can visualize this atomic model by using concepts of energy. For a given nuclear charge (which is positive and varies from element to element), electrons have available a large number of possible states with specific energy values or **energy levels.** Each level corresponds to a certain orbit the electrons can have—a certain total energy, which is the sum of its potential and kinetic energies.

Consider the hydrogen atom: it has one proton for a nucleus and one electron attached by the electric force between itself and the proton. (Other atoms have more protons and electrons but are held together in the same way.) The electron has a certain total energy. The essence of quantum theory is that electrons remain in stable states of specific energies, and each has a particular orbit (Fig. 5.9*a*). For example, the electron in the lowest energy level (called the **ground state**) securely orbits the

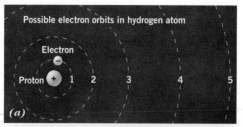

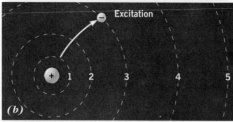

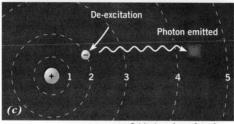

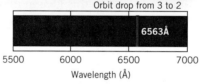

FIGURE 5.9 Energy orbits for hydrogen. (*a*) The first five orbits for an electron; energy increases outward from the nucleus. (*b*) The addition of energy can excite an electron to a higher orbit, such as from 1 to 3. (*c*) When an electron drops an orbit, it loses energy, usually as a photon; here in the drop from 3 to 2, a photon at 6563 Å is emitted.

nucleus. The electron must gain energy to move out to larger orbits. But not just any amount of energy. The orbits, and so energy levels, follow strict spacing rules, and only an exact amount of energy can be gained by an electron as it moves from one orbit to the next.

Energy can be added to the atom either by collision with another particle or by absorption of a photon with sufficient energy. The electron jumps up one or more energy levels (Fig. 5.9*b*). The atom is then *excited* and the process of reaching this state is called **excitation.** The excited condition does not last long; the electron drops to a lower level in about 10^{-8} s. To descend to a lower energy level, however, the electron must lose some energy (Fig. 5.9*c*). The electron achieves such a loss by emitting a photon with an amount of energy *equal to* the amount it needs to lose. This energy is directly related to the photon's frequency.

If an atom gains enough energy, the electron flies away from the nucleus.

Atoms and Light

At a Glance

Photons	Produced and absorbed by electron transitions
Transitions	Electrons move between discrete energy levels
Energy levels	Stable values of energies for electrons
Excitation	By collisions, absorption of photons

The atom is then *ionized,* and the process is called **ionization.** The loss of an electron changes the energy arrangements available and also changes the atom's spectrum. For a given atom and electron in an atom, a certain minimum energy is necessary to break the electron loose from the electrical grip of the nucleus—the **ionization energy.** For hydrogen, the ionization energy is a mere 2.2×10^{-18} J.

Energy Levels

So far, I've been describing energy levels in terms of orbits of an electron. This description is generally valid, but it may mislead you into imagining that the electron orbits are sharp circles. They are *not.* The electrons actually orbit in fuzzy clouds, many of which are not even spherical. A better way to talk about how atoms emit and absorb light is to conceive of *energy levels* themselves without images of orbits of any kind. This picture is especially useful for atoms more complex than hydrogen.

As an analogy, imagine moving a bowling ball up and down the stairs. Each step is an incremental change in the ball's energy, and only full-step changes are permitted. If we try to add less than one step's worth of energy, the ball remains at its initial level. If we add exactly one step of energy, the ball moves up one step. When the ball loses energy, it descends in steps until it hits the floor at the bottom of the staircase; this is the ground level, the state of lowest energy. The natural tendency of an atom is to attain this ground level. The key point is this: since electron transitions can occur only between certain energy levels (stairs), an atom must produce a particular pattern of lines (colors of light).

Let's apply the energy-level concept to hydrogen. Every upward step requires the absorption of energy (Fig. 5.10*a*) and every downward one the emission of energy (Fig 5.10*b*). The greater the energy difference needed for a transition, the higher the frequency (hence the shorter the wavelength) of the photon produced from that transition. All transitions to and from the lowest energy level (level 1) involve large energy changes (short-wavelength photons), so they correspond to wavelengths in the ultraviolet range of the spectrum. This set of lines, called the **Lyman series,** cannot be seen by the eye because it lies in the ultraviolet region; it can, however, be detected by photography. The set of transitions down to and up from the second energy level (level 2) is the *Balmer series* (Fig. 5.10*c*); it lies mainly in the visible region of the spectrum. The hydrogen absorption lines in the sun's spectrum are those in the Balmer series.

Keep in mind that the Balmer series involves the set of spectral lines arising from transitions between the *second* energy level of hydrogen and the levels *above* the second. The transitions are *upward* from level 2 for absorption, and *downward* to level 2 for emission. Note again that the *larger* the transition, the *higher* the energy of photon emitted, and the shorter its wavelength. So the H_β photon, for example, which comes from a downward transition from level 4 to level 2, has a wavelength lower than that of H_α (4861 versus 6563 Å: Fig. 5.10*c*).

Also, even though one atom can emit (or absorb) only one line at a time, many atoms can undergo different transitions at the same time, with the result that many lines are visible at once. Note that in all these cases the atoms are conserving energy by transforming it among various forms: radiative, potential, and kinetic.

Other Atoms

I have concentrated on the hydrogen atom because it has the simplest characteristics of all the elements and plays a dominant role in the cosmos. Also, the Bohr model correctly describes the emission and absorption of light by hydrogen atoms.

The simple Bohr quantum model needs drastic modification to work with other elements. Despite complications, one essential point remains: each atom and each ion (even of the same atom, for more than one electron can be lost from bound states) has its own unique set of electron energy levels. So each has its unique energy-level diagram. Because

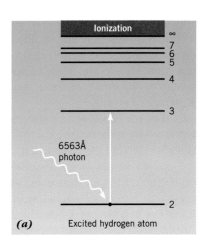

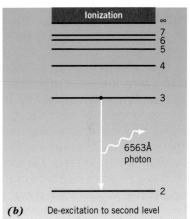

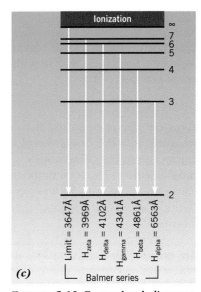

FIGURE 5.10 Energy-level diagrams for the Balmer series of hydrogen. The energy-level differences are drawn to scale. (*a*) Absorption of a 6563-Å photon by an excited atom boosts an electron upward to level 3 from level 2. (*b*) When the electron drops from level 3 to 2, the atom emits a 6563-Å photon. (*c*) The Balmer series from 3647 to 6563 Å involves transitions to and from the *second* energy level.

spectral lines are produced by electronic transitions between energy levels, each element or ion has a unique set of spectral lines. Hence, the study of spectra reveals key information about the internal structure of atoms.

Let me give schematic examples with two hypothetical elements, A and B (Fig. 5.11). Note the different spacings of the energy levels of each element. Also note the correspondence between energy differences of each transition and color (or wavelength). Specifically, transitions with the smaller amounts of energy (smaller differences between energy levels) result in longer wavelengths of light (in the red region of the spectrum).

In a sense, atoms are tuned. They respond to energy input by resonating at their natural frequencies. In a sense, the energy levels are like the keys on a piano, which when struck cause their respective wires to resonate. You can go up or down the keyboard in whole steps or in half steps or multiples of these, but no interval smaller than a half step can be played. The natural resonators of atoms are their energy levels, so that the light emitted and absorbed by atoms provides information about their natural resonances, much as the playing of the keys of a piano tells you how the instrument is tuned.

Caution. Do not confuse the *energy* of photons with the *brightness* of emission lines. The energy determines the wavelength (color). The total number of photons emitted at a given wavelength sets the brightness or intensity of the line.

To recap. The electrons in atoms are stable only at certain energies. Add sufficient energy, and an electron moves to a higher energy level. Shortly it drops to a lower energy level and emits a photon. The bigger the downward jump, the more energy the photon has. Any jump, however, produces or absorbs a photon with a specific energy and wavelength. Atoms absorb and emit photons, producing absorption and emission lines, by means of this discrete jumping of electrons.

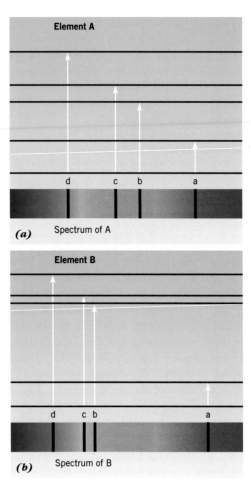

FIGURE 5.11 Different spectra from hypothetical element A (*a*) and element B (*b*), which have different sets of energy levels. Note how energy *differences* and wavelength correlate; the *larger* the difference, the *shorter* the wavelength of the absorbed photons.

■ Does a short wavelength of light indicate a small or large frequency? A small or large energy?

5.4 Spectra from Atoms and Molecules

So far I have represented spectra by a picture or diagram with the spread-out colors so that lines are visible (as in Fig. 5.3). Astronomers also show spectra by using graphs that plot brightness or intensity against wavelength (Fig. 5.12). You will see pictorial and graphic representations in this book, so you should know how to use both. A continuous spectrum (Fig. 5.12*a*) has no sharp dips or peaks, even though the brightness or intensity may vary. Emission lines appear as sharp peaks (Fig. 5.12*b*); absorption lines as sharp dips (Fig. 5.12*c*).

Atoms have another important process of producing or absorbing radiation. Imagine an ionized gas, say of hydrogen.

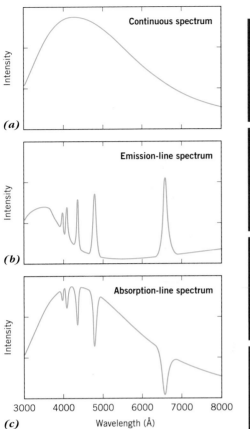

FIGURE 5.12 Graphical representations of the intensity or brightness profiles of the main spectral types: (*a*) continuous, (*b*) emission line, and (*c*) absorption line. The intensity is plotted versus wavelength, so that emission lines rise above the level of the continuous spectrum and absorption lines fall below it. Sharp peaks correspond to emission lines; sharp dips to absorption lines. The emission lines in (*b*) are those of the Balmer series of hydrogen. The same lines are shown in absorption in (*c*) against the background of the continuous spectrum of (*a*).

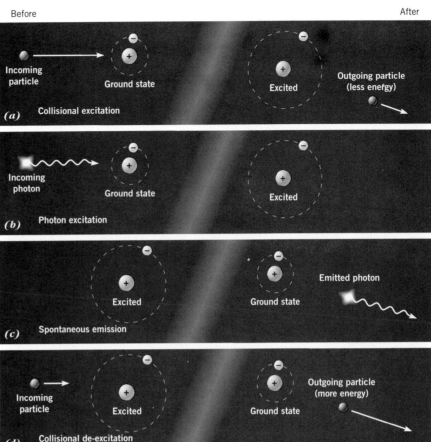

FIGURE 5.13 Atomic processes that transform energy. (*a*) An atom is excited from the ground state when struck by a particle in collisional excitation. The outgoing particle has less kinetic energy. Its loss has raised the electron one energy level. (*b*) An atom is excited from the ground state when it absorbs a photon. The energy of the photon goes into exciting the atom. (*c*) Once an atom has been excited, it usually emits a photon in *any* direction when the electron drops from the excited to the ground state. (*d*) A particle strikes an excited atom in collisional deexcitation. The electron loses energy and drops to the ground state. The particle gains an equal amount of kinetic energy, and no photon is emitted.

The electrons have been ripped away from the protons, leaving hydrogen ions. The free electrons and protons exert mutual attraction, and, because the electrons have much less mass than the protons, their paths bend as they speed by the protons. If a photon encounters an electron near a proton, the electron may jump suddenly to a path of higher energy, absorbing the photon. During this absorption, the electron stays free from the proton. Similarly, an electron skirting around a proton may lose energy and emit a photon. Because an electron has a continuous range of energies over which it can change, such transitions result in *continuous* (rather than discrete) emission and absorption.

What can excite atoms for emission or absorption? Imagine a box filled with a gas. The atoms of the gas collide with one another (and with the sides of the box). Heat up the gas. The atoms then move faster and knock together harder. Collisions can bump electrons into higher energy levels; the harder the collision, the higher the level. This process is called **collisional excitation** (Fig. 5.13*a*). Some of an atom's kinetic energy is transferred to an electron of another atom. Photons can also excite atoms when absorbed, but only certain photons, namely those with energies corresponding to the *difference* in energies of two energy levels of the atom. This process is called **photon excitation** (Fig. 5.13*b*). Whether excited by collisions or by photons, the excited atom usually radiates quickly, sending off a photon in any direction (Fig. 5.13*c*). In some cases, a particle strikes an excited atom and carries off its energy before the excited atom has emitted a photon (Fig. 5.13*d*).

Absorption-Line Spectra

Now return to the sodium vapor experiment with the quantum model in mind. When sodium chloride is placed in a flame, individual atoms of sodium are released from the salt. Some atoms collide with others and are excited. As the electrons return to lower levels, they emit photons of yellow light at wavelengths of 5890 and 5896 Å—the sodium D lines (Fig. 5.4c).

Let's follow a sodium D-line photon (5890 Å) on its way from the sun to the spectroscope through the sodium vapor cloud. If a photon of just this energy encounters an unexcited sodium atom, the sodium atom will *absorb* the photon. The gas contains enough sodium atoms to absorb almost all the photons that try to pass through. The sodium atoms don't absorb any other wavelength in the visual region, so other photons with different wavelengths will pass right through the gas, preserving the continuous spectrum. A dark line appears at 5890 Å because these photons

were absorbed by the sodium gas.

What happens to the sodium atom that has been excited by the photon absorption? An excited atom quickly emits a new 5890-Å photon. The original 5890-Å photon was headed directly for the spectroscope slit before it was absorbed by a sodium atom. The brand-new 5890-Å photon can be reemitted in any direction (Fig. 5.14). Very rarely will it be emitted in the same direction in which the original photon was traveling; only very few of these new photons enter the spectroscope slit. That's why the sodium line in the sun's spectrum becomes darker when passed through the sodium vapor. If you observe the sodium vapor from the side without the sun behind it, you would see only emission lines from the reemitted 5890-Å photons.

This basic idea of how atoms in transparent gases produce line spectra applies to the sun (Fig. 5.15) and also to stars, because their spectra resemble the spectrum of the sun. The absorption lines are made as follows. Low down in

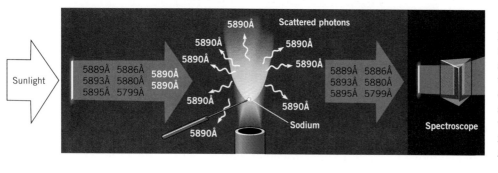

FIGURE 5.14 Paths of photons from the sun through a hot sodium flame and into a spectroscope. Photons with wavelengths of 5890 Å are scattered out of the beam, with the result that an absorption line is observed at this wavelength. If you observed the flame from the side, you would see the 5890 Å line in emission.

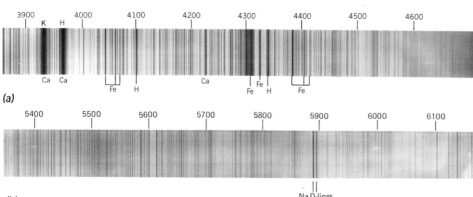

FIGURE 5.15 The sun's spectrum. (*a*) The short-wavelength end of the spectrum, showing absorption lines of calcium, iron, and hydrogen. The wavelength scale is at the top in angstroms (Å). Note the very strong lines of calcium between 3900 and 4000 Å; these are the K and H lines of singly ionized calcium. (*b*) The green-to-red region of the spectrum, showing the D lines of sodium. Kirchhoff showed that these two especially strong lines in the yellow region are made by sodium. (Courtesy of the Observatories of the Carnegie Institution of Washington.)

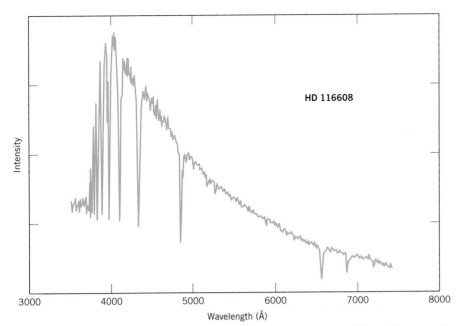

FIGURE 5.16 Spectrum of a star hotter than the sun (called HD116608). The general shape of this intensity graph shows the continuous part of the spectrum; the sharp dips are absorption lines. The strongest of these are from the Balmer series. (From *A Display Atlas of Stellar Spectra* by B. Margon, Department of Astronomy, University of Washington.)

the sun's atmosphere, the solar gas is compressed and opaque; here the continuous spectrum is produced. Above this region, the gas is transparent and cooler; these layers of the sun's atmosphere are the last barrier through which the continuous radiation must pass before it escapes. Atoms and ions at this level in the atmosphere absorb light at their characteristic wavelengths from the continuous spectrum and create the absorption lines in the sun's spectrum. The atmospheres of stars produce absorption-line spectra by similar atomic processes.

Most stars have dark-line spectra (Fig. 5.16). Here is one plotted as intensity against wavelength for a star that has a higher temperature than the sun at its surface. The overall trend is the star's continuous spectrum; note that it peaks around 4000 Å. The sharp dips in the spectrum are absorption lines; the strongest ones here are those from the Balmer series of hydrogen. Compared to the sun's spectrum, this star has much more intense Balmer lines. (Chapter 14 will explain this difference.)

By matching the spectral lines of stars to those made in labs from known elements (Fig. 5.17), astronomers inferred that stars contained almost all elements found on the earth. But one important difference emerged: in contrast to the earth, stars and the sun contain mostly hydrogen. (Details will be found in Chapter 13 for the sun and Chapter 14 for other stars.) The similarity of spectra and chemical composition gave astronomers a new insight about the nature of the stars: stars are other suns.

Spectra of Molecules

A molecule contains two or more atoms bound together by electric charges. Consider a very simple model of a molecule, one with two atoms connected by a fairly rigid spring (Fig. 5.18). Now, if energy is added to the molecule (but not enough to break the bond), how can the molecule react? First, it can spin around the middle of the spring—one form of kinetic energy—at a large variety of possible energies. As for atoms, the molecule has a restricted number of discrete energy states. If it absorbs energy (by photons or collisions), it can achieve a higher energy level. If it loses energy, say by emitting a photon, it will drop to a lower energy state.

Rotation is not the only response

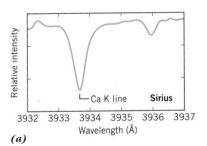

(a)

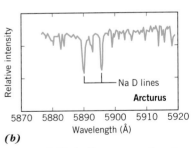

(b)

FIGURE 5.17 Stellar spectra showing familiar absorption lines. (*a*) Intensity profile of the K line of singly ionized calcium absorption in the spectrum of the bright star Sirius, a line also prominent in the sun's spectrum. (Adapted from a diagram in *Sample Spectral Atlas for Sirius* by R. Kurucz and I. Furenld, Smithsonian Special Report #387, 1979.) (*b*) Section of the spectrum of Arcturus, showing the sodium D lines. (Courtesy of J.E. Gaustad, Swarthmore College Observatory.)

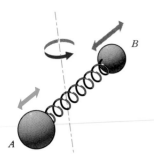

FIGURE 5.18 Schematic representation of a molecule made of two atoms. The bond between them can act like a spring, permitting the atoms to vibrate when the molecule is excited. Or, if enough energy is added, the two atoms can rotate about their center of mass.

the molecule can have. The atoms can vibrate at the ends of the springlike bond. Again, the possible energy states are discrete. The addition of energy moves the molecule up in energy levels; the release moves it down. So a molecule can absorb and emit photons from changes in vibrational energy, which is another form of kinetic energy.

Molecules can absorb and emit photons and create absorption and emission spectra by changing energy levels in vibrational or rotational states (Fig. 5.19). In general, these energy changes are much smaller than those for the electrons in atoms, so the wavelengths are much longer than those for visible light. In fact, most molecules observed by astronomers have their spectral lines in the infrared or radio range of the electromagnetic spectrum. Some very cool stars, though, show lines from molecules in the visible region of the spectrum.

So molecules interact with light by changing discrete energy states, conserving energy just as atoms do, while transforming kinetic energy into electromagnetic energy.

■ **How does the sun produce its absorption-line spectrum?**

FIGURE 5.19 Spectrum of a hydrogen molecule. Note the series of lines in the red end of the spectrum. (Courtesy of Bausch & Lomb.)

Key Concepts

1. A spectroscope spreads light into its component colors, and this display is called a spectrum. Three general types of spectrum appear through a spectroscope (Table 5.2): continuous, with a smooth spread of colors; emission, which shows discrete, bright lines; and absorption, in which dark lines appear in a continuous spectrum.

2. The sun and most other stars have absorption-line spectra. By matching the dark lines in a star's spectrum with those produced by elements in the lab, the

TABLE 5.2 BASIC SPECTRAL TYPES

SPECTRUM TYPE	APPEARANCE	PHYSICAL CONDITIONS
Emission	Distinct bright lines against a dark background	Hot, transparent gas
Continuous	Smooth blend of all colors	Hot, opaque gas, liquid, or solid
Absorption	Distinct dark lines against a bright background of colors	Light from a hot, opaque gas, liquid, or solid passing through a cooler, transparent material

chemical composition of a far-away object can be inferred; this technique works because each element has a unique set of spectral lines. These lines are determined by the arrangement of the electrons bound to the nucleus of the atom.

3. Kirchhoff's rules of spectral analysis describe the physical conditions under which each type of spectrum is produced (Table 5.2): (a) continuous spectrum from a hot opaque gas, liquid, or solid; (b) emission-line spectrum from a hot transparent gas; (c) absorption-line spectrum from a continuous spectrum passing through a cooler, transparent material (usually a gas).

4. Light has wave properties; red light has longer wavelengths than blue light. Light also carries radiative energy: the shorter the wavelength, the greater the energy. So blue light has more energy per photon than red light. The total range of wavelengths of light is called the electromagnetic spectrum, which goes from very long radio waves to very short gamma rays. All forms of electromagnetic radiation travel at the same speed, about 300,000 km/s.

5. In an energy-level model for the hydrogen atom, Balmer lines arise from an electron starting in the second level and jumping to any higher level (for absorption lines) or dropping from a higher level to the second level (for emission lines).

6. The quantum model for the emission and absorption of light by atoms views the electrons as playing a critical role; electrons can inhabit only stable regions called energy levels. Electrons can move to higher levels only if they gain energy; when they drop to lower levels, they give off energy (usually as photons); the greater the difference between levels, the greater the energy in the emitted photons. Electron transitions between energy levels produce and absorb light.

7. Spectroscopic analysis reveals the properties of stellar atmospheres, whose physical environments and chemical compositions can be inferred like those of the sun.

8. A key rule in the operation of the universe is that energy is conserved, although it can mutate into many forms. Atoms work to transform potential and kinetic energy into radiative energy, and also do the reverse. Molecules can do the same.

9. Atoms are the basic units of matter. The number of positively charged protons and electrically neutral neutrons in an atom's nucleus determines the element. The number of negatively charged electrons attached to the nucleus equals the number of the protons. The electrons and their energy levels give atoms their ability to manage light, and so reveal the universe to us.

Key Terms

absorption line	element	ionization energy	potential energy
absorption-line spectrum	emission line	isotopes	proton
atom	emission-line spectrum	joule	quantum
Balmer series	energy	kelvin	radiative energy
Bohr model of the atom	energy level	kinetic energy	spectral line
collisional excitation	excitation	Kirchhoff's rules	spectroscope
compound	flux	Lyman series	spectroscopy
conservation of energy	frequency	molecule	spectrum
continuous spectrum	ground state	neutron	temperature
D lines (of sodium)	heat	nucleus	transition
electromagnetic spectrum	ion	photon	wavelength
electron	ionization	photon excitation	

Study Exercises

1. You throw ordinary table salt, which contains sodium, into the flame of a Bunsen burner. What kind of spectrum do you see when you look at the vaporized salt with a spectroscope? (Objective 5-1)
2. How do we know that sodium exists in the sun? (Objectives 5-1, 5-2, and 5-3)
3. The spectra of most stars look like the spectrum of the sun: the absorption type. How can particular elements be identified in these spectra? (Objective 5-2)
4. Suppose you had a box of positive hydrogen ions. What kind of *line* spectrum would they produce? (Objective 5-9)
5. The Balmer lines in the sun's spectrum are absorption lines. Do electrons jump up or fall down energy levels to produce them? (Objective 5-8)
6. Arrange the following kinds of electromagnetic radiation in order from the least to the most energetic: x-rays, radio, ultraviolet, infrared. (Objective 5-5)
7. Use Kirchhoff's rules to explain why the spectrum of the moon resembles that of the sun. (Objective 5-3)

8. What happens to a hydrogen atom when it absorbs light with enough energy to knock off its electron? (Objective 5-9)
9. Use Kirchhoff's rules to predict what would happen if the emission-line spectrum from a hot, transparent gas passed through another transparent gas that was hotter and had a different chemical composition. (Objective 5-3)
10. Consider a planet orbiting the sun. At what point is its kinetic energy the greatest? The least? What is the value of the total energy of the planet at these two points? (Objectives 5-10 and 5-11)
11. Does light of different wavelengths travel at the same speed or at different speeds? (Objective 5-4)
12. What region of the electromagnetic spectrum contains electromagnetic radiation of the longest wavelengths? (Objective 5-6)
13. In the energy-level model, when an atom absorbs a photon, what generally happens to one of the electrons? (Objective 5-7)

Problems & Activities

1. Compare the wavelengths of the calcium H and K lines (Fig. 5.15a) to the sodium D-lines (Fig. 5.15b). Which lines have the shorter wavelength? Do these photons have a lower or higher energy? Roughly, by what percentage is their energy greater or less?
2. Calculate the wavelengths of electromagnetic radiation at the following frequencies: (a) 100 MHz (about the middle of the FM band), (b) 1000 kHz (AM band), and (c) 10 GHz (a common frequency for a radio telescope).
3. Calculate the ratio of the energies of each frequency in Problem 2: 100 MHz to 1000 kHz, 1000 kHz to 10 GHz, and 100 MHz to 10 GHz.
4. Compare the kinetic energies of cars moving at 10, 30, and 60 km/h.
5. What is the energy of the first line in the Balmer series of hydrogen?
6. Assume that a 100-watt light bulb converts all its electrical energy into light at a wavelength of 5500 Å. One watt equals the output of one joule per second. How many photons does the light bulb give off every second?

See for Yourself!

You can examine spectra yourself by building a simple spectroscope and using it to observe light produced under different physical conditions. Actually, new gratings are available that allow you to see most major spectral lines without the need for building a spectroscope. Ask your instructor about a grating. The grating has parallel lines inscribed on it at very close spacings. These lines use the wave nature of light to produce spectra.

You will need a shoebox, two single-edged razor blades, and a diffraction grating to build the spectroscope. First paint the inside of the box a flat black, or cover the interior with a dark, dull fabric. Then cut a hole 1 inch (2.5 cm) square in the center of each end of the shoebox (Fig. A.2). Both openings should be the same distance from the upper edge. Inside the box over one of the holes, tape a grating positioned so that the lines run up and down. Next, cut out a 2-inch (5-cm) square of cardboard. With a razor blade, make a vertical slot that is one-quarter inch (10 mm) wide and 1 inch (2.5 cm) long. Tape a razor blade on one side of the slit so that its edge is parallel to the slit. Place a small piece of cardboard tight against the taped blade, and tape another blade on the other side as close to the cardboard spacer as possible. The blades should be parallel and the opening between them narrow.

Remove the cardboard spacer from between the blades and mount them on the inside of the box opposite the diffraction grating. Test the spectroscope by aiming the end with the slit at a light bulb. You should see a continuous spectrum. To see the spectrum completely, you'll have to look through the grating at an angle.

Now you can put your spectroscope to work. Go out one night and observe different street lights (such as mercury and sodium ones; you might also find xenon) and display signs that look like neon. (You'll find that not all illuminated tubes contain neon!) Compare what you see to the color plate of spectra in the book.

Observation of the sun's spectrum involves some care. NEVER LOOK AT THE SUN DIRECTLY! Instead, look at the sun's reflection off a piece of glass or any shiny surface. What do you see? Look for dark, vertical lines. Compare them to the photos of the solar spectra in the book. Can you match the lines of any elements? You should be able to see the lines for sodium and calcium. ■

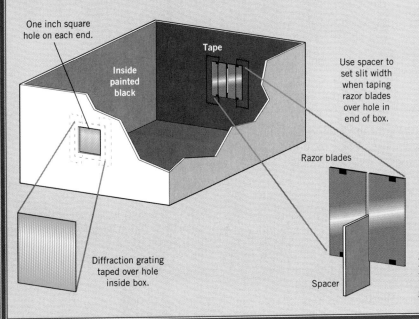

One inch square hole on each end.

Inside painted black

Tape

Use spacer to set slit width when taping razor blades over hole in end of box.

Razor blades

Diffraction grating taped over hole inside box.

Spacer

FIGURE A.2 Constructing a simple box spectroscope with a diffraction grating.

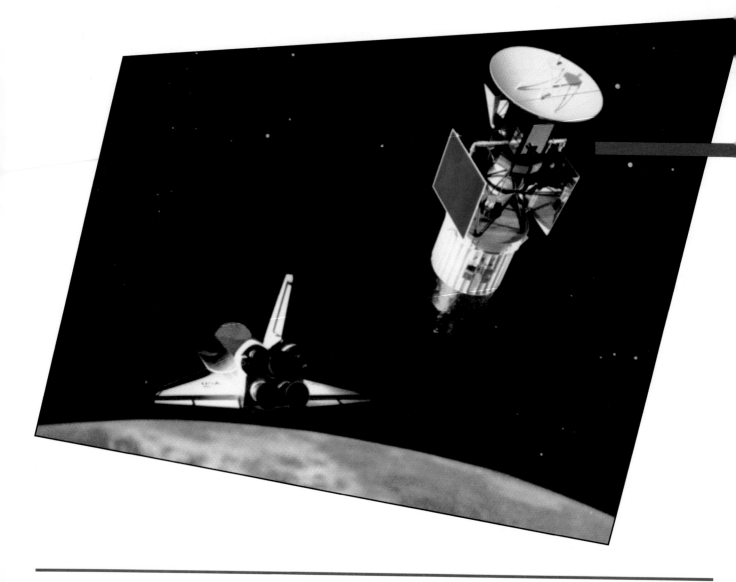

Central Question
How do astronomers collect the light from celestial objects?

Telescopes and Our Insight into the Cosmos

All astronomical research must in the end be reducible to a visual observation. . . .

AUGUST COMTE, *Cours de Philosophie Positive*

This chapter turns from astronomical concepts to astronomical tools. Telescopes (Fig. 6.1) and other equipment, such as spectroscopes, spring from advances in technology. The technological development of astronomical tools affects not only *what* we observe but also *how* we observe. It expands, deepens, and sharpens our perceptions of the cosmos. What and how we observe act as a prelude to and confirmation of our models of astronomical objects.

Technical advances in this century have been so great that now astronomers rarely labor alone. Usually, they work in concert with colleagues whom they may never see during the course of a research project. The technology has grown too complex for any one observer to master, and the astronomer now often works with engineers and computer experts. This chapter looks at some of the developments that have provided new ways of observing the universe and considers how they have influenced our astronomical models.

Observations, by which the effectiveness of models are judged, accelerate the evolution of astronomical ideas. From such judgments some models are discarded, new ones are proposed, and a few are finally adopted. Not all new observations have dramatic effects. In many cases, the change of view brought on by the change of vision is slow and subtle. Slowly or swiftly, new observations compel new conceptions of the cosmos.

6.1 Observations and Models

How do observations and models interact? The process usually goes like this (Section 2.1): models spring from observations, whether straightforward or subtle. Basic astronomical observations naturally drove people to create models of the cosmos. The Ptolemaic model marked the first detailed attempt at a scientific explanation of these observations (Section 2.4). It used geometric, physical, and aesthetic ideas to *explain* what was seen, and it also *predicted* planetary positions. To be useful, a prediction must have definite numbers attached to it. So

FIGURE 6.1 Capilla Peak Observatory's 61-cm telescope with a CCD camera at the back end. (Photo by M. Zeilik.)

how well a model corresponds to actual observations becomes crucial to its acceptability.

The issue of "how well" revolves around the techniques of observation, as well as gut judgment of how good is good enough. For example, astronomers knew for centuries that Ptolemaic predictions often missed the actual planetary positions by several degrees. This discrepancy did not bother astronomers in those earlier times, for a few degrees away was considered to be acceptable. Their observations did not aim at any greater accuracy. In particular, the Ptolemaic model provided sufficient accuracy for planetary conjunctions.

Not until Tycho Brahe (Section 3.3) did nontelescopic observations reach the limits imposed by the human eye. Tycho's technical achievement compelled Kepler (Section 3.5) to take a disparity of only 8 arcmin seriously. The failure of the Copernican model to fit the data with *circular* orbits resulted in Kepler's devising a model with *elliptical* orbits. The success of Kepler's modification formed a critical link in the evolutionary chain to Newton's idea of gravitation (Section 4.4).

Prior to Tycho's observations and the astronomical use of the telescope by Galileo, the Ptolemaic and the Copernican models explained the motions of the planets equally well. And both made predictions just about as badly. Because these models were equally confirmed observationally, astronomers had to choose between them on the basis of aesthetic and philosophical beliefs.

In fact, the crucial observation needed to distinguish between the simple heliocentric and geocentric models, that of *heliocentric parallax* (Section 2.3), was not made until the 1830s, more than two centuries after the introduction of the Copernican model! Only by then had the techniques of measuring stellar positions become accurate enough to detect heliocentric parallax, which amounts to less than 1 arcsec for even the nearest star. (Recall that 1 arcsec is only 1/3600 of a circle: that's the angular diameter of a U.S. penny placed at a distance of about 4 km.) Copernicus guessed correctly that the stars were *very* far from the earth compared with the earth–sun distance.

To sum up. Observations form the building blocks for model making, and they can also act as the driving force that causes models to be discarded. This destructive effect often is seen when accepted models fail to account plausibly for new observations.

One more point. Astronomy, in comparison to physics and chemistry, works more as an *observational* science than an *experimental* one. In a laboratory experiment investigators can isolate certain traits and keep some fixed while changing others to see how the variables affect the outcome. Not so in astronomy. We cannot change or control our subjects to study their physical characteristics. We can only work with what is given—for the most part, the light from celestial objects.

You should not, however, get the impression that astronomy totally lacks experimentation. Astronomers do experiment in basic ways. First, we can observe the same objects in different ways. Usually this means using different telescopes that operate at different wavelengths. That's the importance of technological innovation, for it provides new tools for new experiments. Second, laboratory experiments such as those of spectroscopy can help us to analyze observations. And of course we make use of experiments in physics and chemistry that provide basic data on properties of matter and radiation. Third, within the solar system, we can use space probes to measure the actual conditions in interplanetary space, in the neighborhoods of planets, and on some planetary surfaces.

Finally, we can play with theoretical models backed up by whatever observations we have. Theoretical model making may also be tied to technology; for example, electronic computers can manipulate quickly very detailed models of astronomical objects. With the advent of supercomputers and computer graphics, we have begun to explore visually many realms of theoretical ideas. Such numerical simulations promote insights to the complex physics of astronomical systems—in essence, a new way to conceive them. Computer graphics also provide the power to manipulate data so that we can understand them in new ways.

■ **How have observations over the centuries affected development of cosmological models?**

6.2 Visible Astronomy: Optical Telescopes

As extensions of the human eye, telescopes amplified the power of detection without at first extending the spectral range of our vision. Today we can sense much more than the visual part of the electromagnetic spectrum (Section 5.3). This section treats only optical telescopes, those that manipulate light detectable by the eye. Before I deal with telescopes, let me discuss a little about **optics:** how the direction of light is controlled. Some knowledge of basic optics is necessary to understand how telescopes work.

The Basis of Optics

When traveling through empty space or a uniform medium, light moves in a straight line. To change its direction, we act on it in specific ways. Although light has characteristics of both waves and particles, geometric optics pictures it as particles moving along straight-line paths, which are called *light rays*. Using lenses, mirrors, and prisms, we can change the direction of light rays. We can even break white light into its component colors (light of different wavelengths). How light rays are affected by bouncing off or passing through materials is the essence of optics. Light must be

manipulated to change its path from its natural way, which is a straight line.

When light crosses the boundary from one transparent material to another (from air to glass, for example), its direction changes (Fig. 6.2*a*). This bending of light rays is termed **refraction.** Refraction occurs because light travels at different speeds in different media. Consider what happens when a beam of light—a bunch of rays—passes through the air to hit glass at some angle (Fig. 6.2*a*). The first part of the beam to strike the glass enters it and slows down. The rest of the beam continues to move, but at a faster speed, and so gains on the light already in the glass. This catch-up results in the front of the beam turning toward the glass. Here's an analogy: imagine a row of musicians in a marching band turning a corner. To maintain a straight line, the people on the inside march more slowly than those on the outside. The entire row turns around some angle and remains straight.

Key point. The amount of refraction depends on the wavelength of light, with blue light bent more than yellow, and yellow more than red for most kinds of glass. The shorter the wavelength, the greater the amount of refraction.

You have all looked in a mirror, so you know that flat surfaces return light by bouncing back the rays. This process is **reflection** (Fig. 6.2*b*). A light ray bounces off a polished surface the same way a ball bounces off a smooth wall: the ball rebounds at the same angle at which it hit. Similarly when light is reflected: the incoming and outgoing angles are the same. Reflection does *not* depend on the wavelength of the light; red and blue light are reflected the same way at the same angle.

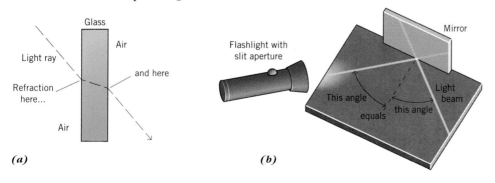

FIGURE 6.2 The refraction and reflection of light. (*a*) When light rays cross the boundary between two different materials, their paths are bent. The ray here bends when it enters the glass from the air and also when it exits into air. (*b*) The reflection of light. Note that the incident angle and the reflected angle are the same. This holds true even if the reflecting surface is curved.

(a)

(b)

Note that reflection and refraction are common to *all* waves. You can even see these effects with water waves.

Optics and Images

The goal of optics is to make images by refraction and reflection. An **image** forms when light rays are gathered together in the same relative alignment they had when they left an object. You can recognize an image of an object as a visual representation of the object itself.

How can refraction form an image? Suppose that light travels through a glass prism (Fig. 6.3a). Because the sides of a prism are not parallel, a light ray does not come out along its original path but is bent toward the prism's base. Now place two prisms base to base (Fig. 6.3b), such that two light rays from a point source converge to a point. However, rays entering the two prisms at different heights and angles converge at different points. To get all rays to come to the same point requires a smoothly curved surface. Such a piece of glass is a **lens** (Fig. 6.4), which refracts light to form an image.

A lens brings rays from a very distant point source to a point image at its focus. From an object of finite size, the lens makes an image of the object by focusing parallel rays from each point of the source onto separate points in the image. This image is generally (but not always) smaller than the object and upside down (Fig. 6.5). For objects at large distances, the distance from the lens to the image is roughly the same for all objects. This distance, from the lens to the image, is termed the **focal length.**

How can a mirror make an image? You know that an image from a flat mirror is undistorted. An irregularly curved mirror, such as one in a fun house, creates a distorted image. A smoothly curved mirror—whose surface, for instance, follows the curve of a parabola—brings all the light to a focus (Fig. 6.6). This curved surface directs the light rays to a focus according to the basic law of reflection: the incoming angle equals the outgoing angle.

One of the basic properties of a lens or mirror is the ratio of its focal

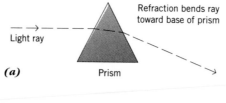
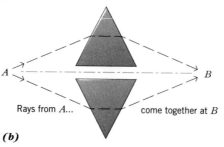

FIGURE 6.3 Prisms as crude lenses. (*a*) Because its sides are not parallel, a prism refracts a light ray toward its base. (*b*) Placed base-to-base, two prisms refract rays from *A* to meet at *B* in a crude focus to make an image.

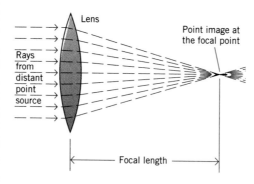

FIGURE 6.4 A lens forming a point image from a point source. A lens acts like base-to-base prisms (Fig. 6.3). Because its surface is smoothly curved, it forms a sharp focus at the focal point. The distance from the lens to the focal point, for a distant source, is the focal length.

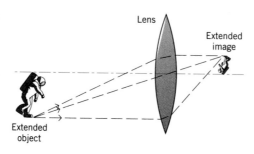

FIGURE 6.5 A lens forms at the focal point an extended image, upside down, of an extended object. If the object were a point source of light, such as a star, the image would also be a point.

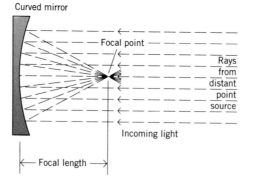

FIGURE 6.6 A smoothly curved mirror reflects incoming light to a focus. Astronomical mirrors have their reflecting material on their front surfaces and usually follow a *parabolic* curve to make a clean focus.

length to its diameter, called the **f-ratio** (Focus 6.1). You find the f-ratio by dividing the focal length by the diameter of the lens or mirror (taking care that both have the same units, such as centimeters or inches). For example, at Capilla Peak Observatory of the University of New Mexico, we have a 60-cm-diameter mirror with a focal length of 900 cm; so the f-ratio is 900/60 or 15.

The smaller the f-ratios, the brighter the image at the focus, just as for a camera lens. (If you have a camera with an adjustable lens, you can see that the f-ratios are in a sequence: 2.8, 4, 5.6, 8, 11, 16, and 22, for example. Stepping down from one f-ratio to the next larger one, say from 16 to 22, decreases the brightness of the image by a factor of 2.) So, other factors being equal, a telescope with a smaller f-ratio produces a brighter image for viewing or for recording. However, the total light gathered still depends on the area of the objective.

Telescopes

Basically, a telescope gathers up light and allows you to examine an image at a focus. To make a telescope you need a lens or mirror, called the **objective,** to

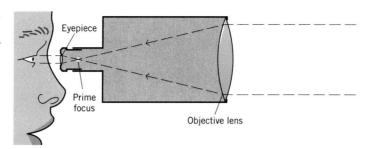

FIGURE 6.7 The design of a simple refracting telescope. An objective lens gathers the incoming light and brings it to a focus called the *prime focus.* An eyepiece allows viewing of a magnified image. Galileo's telescopes worked on the same basic design.

bring light to a focus. A lens called an **eyepiece** placed just beyond the focus allows visual examination of the image, or a camera can be placed at the focus to photograph the image, or the light can be diverted into a spectroscope.

There are two basic types of telescope, distinguished by the objective: **refracting telescopes** (or *refractors*) use a lens (Fig. 6.7), and **reflecting telescopes** (or *reflectors*) use a mirror. Galileo's telescope (Section 4.1) was a refractor, Newton's (Section 4.3) a reflector. Early large telescopes of good quality

Enrichment Focus 6.1

PROPERTIES OF TELESCOPES

The main optical properties of a telescope revolve around its objective, its main lens or (more commonly) mirror. The objective will have a certain diameter, D, and a certain focal length, f, over which distance the objective brings light to a focus. The f-ratio is defined by the focal length divided by the objective's diameter:

$$\text{f-ratio} = \frac{f}{D}$$

The objective's diameter determines the light-gathering power (LGP) by its area $(=\pi R^2)$, which is proportional to $\pi D^2/4$ for a circular aperture. Since LGP has no absolute standard, you just

need to compare the ratio of the areas (diameters squared) to find the relative performance. For example, to compare 5- and 4-meter telescopes:

$$\text{LGP} = \left(\frac{5 \text{ m}}{4 \text{ m}}\right)^2$$

$$\text{LGP} = \frac{25}{16} = 1.6$$

The objective's diameter again plays a critical role for the theoretical resolution (TR), the smallest angle discernible by the telescope. Using units of meters for diameter and the wavelength of the light, we have:

$$\text{TR (arcsec)} = 2.52 \times 10^5 \frac{\lambda}{D}$$

So, for example, an optical telescope with a 0.1-m (10-cm) objective at a wavelength of about 5500 Å (5.5 × 10^{-7} m) has a theoretical resolution of

$$\text{TR (arcsec)} = (2.52 \times 10^5)\frac{5.5 \times 10^{-7}}{0.1}$$

$$= 1.4 \text{ arcsec}$$

The atmosphere usually allows this resolution to be achieved in practice. At the best sites, the seeing limit of the atmosphere can be as fine as $\frac{1}{3}$ arcsec.

FIGURE 6.8 The Great Refractor of Harvard College Observatory. Built in the middle of the nineteenth century, this refracting telescope has a 15-inch objective lens. The objective is at the front end, and the eyepiece is at the rear. In its day, it was one of the largest telescopes in the world. (Courtesy of Owen Gingerich.)

tended to be refractors (Fig. 6.8), but images viewed through a refractor are flawed by color halos: that is, when one color is in focus, the other colors are not. This is because simple lenses act essentially as prisms, and different colors have different focuses. Reflected light, however, does not break up into colors, so the image formed by a reflector does not have color halos. Later lenses have eliminated most of the refractor's color problems, but Newton's solution was to design the reflector.

How to view the image formed by a reflector? Newton put a small mirror tilted at 45° to the path of light to direct the focus out the side of the telescope tube (Fig. 6.9a), where an eyepiece is placed. Such a design is called a *Newtonian reflector.* Most research telescopes today use a different optical design called a *Cassegrain reflector* (Fig. 6.9b). All large telescopes built now are reflectors. Advances in technology allow them to be smaller, more rigid, and cheaper to construct than the classical reflectors such as the Hale 5-meter on Mt. Palomar.

Functions of a Telescope

1. *To gather light.* Whether a reflector or a refractor, a telescope has one primary function: to gather light. A telescope is

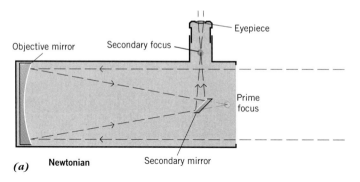

(a) **Newtonian**

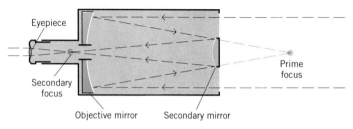

(b) **Cassegrain**

FIGURE 6.9 Designs of reflecting telescopes. The main problem is viewing the image made by the objective mirror. Newton placed a small, flat mirror at an angle to reflect the light out to one side, where an eyepiece is positioned (*a*). This is called a *Newtonian reflector.* A common modern design uses a small convex mirror to reflect the light back down through a hole in the objective mirror for viewing (*b*). This is called a *Cassegrain reflector.*

basically a light funnel, collecting photons (Section 5.3) and concentrating them at a focus. So objects viewed through a telescope appear brighter than they do when seen by the eye alone. This difference indicates a telescope's **light-gathering power**.

How much light a telescope can collect depends on the *area* of its objective, which is proportional to the square of the diameter. So a mirror with twice the diameter of another can gather four times as much light. That's the reason astronomers want large telescopes—big mirrors that have plenty of light-gathering power (Fig. 6.10).

Light-gathering power has no absolute standard; it's a relative measure. For example, the diameter of your eye in the dark is about 0.5 cm. A telescope with a 50-cm-diameter objective (100 times bigger) would have a light-gathering power 10,000 times (100^2) greater than your eye. Similarly, the 5-meter Hale telescope at Mt. Palomar outdoes the 4-meter Mayall telescope at the Kitt Peak Observatory by more than 50 percent.

2. *To resolve fine detail.* The next most important function of a telescope, is to separate, or *resolve,* objects that are close together in the sky. This ability is called a telescope's angular **resolution,** usually expressed in terms of the minimum angle between two points that can be clearly separated. For example, a 10-cm telescope has a resolution of 1.4 arcsec at visual wavelengths. If the telescope is aimed at two stars that are more than 1.4 arcsec apart, you will see two separate star images. If the angular separation of the stars is less than 1.4 arcsec, you will see a single, elongated image. For the same wavelength of light, the resolution depends inversely on the objective's diameter. So a mirror *twice* the size of another has *twice* the resolution; that is, it can resolve objects half as far apart in angular distance. A 5-cm telescope would have a resolution of 2.8 arcsec at visual wavelengths.

You can think of resolution in terms of size-to-distance ratios (Focus 1.1). At best, the unaided human eye can see an angular separation of about 1 arcmin,

which corresponds to a size-to-distance ratio of about 1/3600. So we can see that the moon (angular size about 30 arcmin) has a disk with features. But Mars, even when closest to the earth, has an angular size of only 18 *arcsec*, so we cannot see Mars as a disk without a telescope. From the resolution just given for a 10-cm telescope, you'll note that with such an instrument we could see Mars as a disk and some of the larger features on the surface. The better the resolution, the more *information* we can obtain about a celestial object.

Even the best optics in a telescope have limited resolution because of the wave nature of light. Water waves bend when they strike the edge of a barrier; this process is called *diffraction.* A similar bending occurs with light waves. The objective of a telescope acts as a hole in a barrier to the light waves. As the light waves travel through this opening, they bend around the edges (just as water waves do). This bending spreads out the waves and ultimately determines the **theoretical resolution** of any telescope. As discussed in Focus 6.1, the limit to a telescope's resolving power depends directly on the wavelength of light (longer waves are bent more than shorter ones), and it varies inversely as the size of the aperture (the smaller the size, the greater the bending for the same wavelength of light).

The theoretical resolution of a 5-meter telescope is a fine 0.02 arcsec. But ground-based telescopes never attain such performance. The resolution of a big telescope is limited not by the telescope's optics but by the earth's atmosphere. You have probably noticed that stars twinkle. The twinkling comes from turbulence in the air that makes the atmosphere act like a huge, distorted lens. The motion of blobs of air, like the shimmering above a hot road, distorts and blurs images seen through a telescope.

Even on the best of nights, the 5-meter Hale telescope does not resolve better than a 10-cm telescope. At Kitt Peak (Fig. 6.11), for example, it is a rare night when star images are smaller than 1 arcsec. The limit that the earth's atmosphere sets on the resolving power makes a strong case for a large telescope

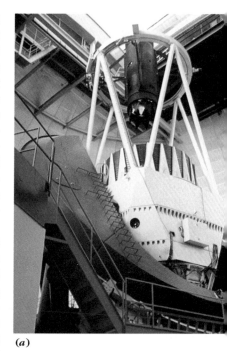

(*a*)

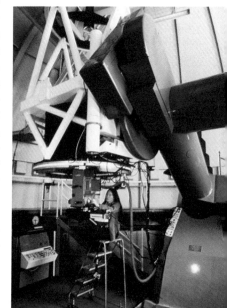

(*b*)

FIGURE 6.10 Reflecting telescopes. (*a*) The 4-meter telescope at Cerro Tololo Interamerican Observatory in Chile. (Courtesy of NOAO/CTIO.) (*b*) The 1.3-meter telescope at Kitt Peak. It is optimized for use in the infrared. (Courtesy of NOAO/KPNO.)

in space. There a telescope's resolution would be limited by the optics of the device, not by the atmosphere.

3. *To magnify the image.* The *least* important of a telescope's functions is its **magnifying power,** as represented by the apparent increase in the size of an object compared to visual observation. Magnifying power depends on the focal length of the objective and the focal length of the eyepiece. Changing the eyepiece on a telescope changes its magnifying power as follows: the *shorter* the focal length of the eyepiece, the *greater* the magnifying power. For example, if you put in an eyepiece with *half* the focal length of an earlier one, you *double* the magnifying power (Focus 6.1).

An eyepiece with a shorter focal length would give a higher magnifying power, as high as you might like. But there is no point in using a magnifying power any greater than necessary to see clearly the smallest detail in the image, which is set by the resolution. Extremely high magnification merely makes the fuzziness worse.

Caution. Don't develop the impression that astronomers observe the skies with their eyes. They do not. A light-sensing device, called a **detector,** is usually placed at the focus. The detector may be a photographic plate (light-sensitive materials on glass rather than film). That's how many of the photos in this book were made. Or it may be an electronic detector similar, for example, to a television camera. These days astronomers rarely, if ever, use their eyes directly for observations.

To sum up. The three principal functions of a telescope are to gather light, to resolve fine detail, and to magnify the image. Of these, I cannot overemphasize the importance of *light-gathering power.* Most astronomical objects are extremely faint. Without a telescope you can see about 6000 stars in a dark sky, but even a small 15-cm telescope allows you to see some *half million* stars. That is the real power of a telescope—to enable us to see objects that we otherwise would not know existed.

FIGURE 6.11 View of the top of Kitt Peak mountain. The dome of the Mayall 4-meter telescope is in the foreground. (Courtesy NOAO/KPNO)

Next Generation of Telescopes

Advances in design technology, computers, and material have propelled ground-based astronomers into a new era of telescopes. One variation of the old reflector uses a lightweight mirror with a short focal length made with new casting techniques. Such a 3.5-meter mirror is the heart of the Astronomical Research Consortium (ARC), which built the Apache Point Observatory in New Mexico (Fig. 6.12). Its mirror is five times lighter but just as stiff as conventional mirrors of the same size. Hence, the support structure is smaller and lighter, and the whole telescope cost less to build. This new telescope—as well as other new ones coming on line—is also designed for remote observing. The astronomer can "stay

Telescopes

At a Glance

Primary function	To gather light
Primary designs	Reflectors and refractors
Primary ground-based types	Optical, radio, infrared
Primary space-based types	Ultraviolet, x-ray, gamma-ray, infrared
Primary choice for amateur	Small (10- or 20-cm) reflector

home" while communicating by phone and computer while an on-site technician oversees the operation.

Telescopes with very large mirrors—up to 8 and 10 meters—are also planned or under construction. In many of these the mirror will be made in pieces, rather than as a single unit. The Keck Telescope is now the largest ground-based optical telescope. Located on Mauna Kea, Hawaii, at an elevation of 4145 meters (Fig. 6.13*a*), the Keck has innovative designs—especially its 10-meter mirror, which consists of 36 hexagonal segments (Fig. 6.13*b*) with a total weight of 14.4 tons. Each segment has a diameter of 1.8 meters and a thickness of 75 mm. The segments are individually controlled by an active system to maintain the mirror's shape under a variety of stresses. The mirror control system can move each segment as little as one-thousandth the thickness of a human hair! The dome and the telescope's structure minimize local **seeing,** or blurring due to atmospheric unsteadiness, to take advantage of the site's natural potential—occasionally as good as 0.3 arcsec. A clone of the Keck will be built next to the current telescope and designed so that both can be used simultaneously.

■ Of the three main purposes for all telescopes, which is the most critical for astronomers?

6.3 Invisible Astronomy

Your eye senses only a tiny sliver of the electromagnetic spectrum (Section 5.3). When you have your teeth x-rayed, the dentist's x-ray machine does not glow brightly when it's on. But the film placed in your mouth senses the x-rays and gives an internal picture of your teeth. When you stand next to an almost-dead fire, the coals look black. But your skin senses heat—infrared radiation—from the coals.

Why work on invisible astronomy? Electromagnetic radiation outside the visible region of the spectrum is often produced by processes different from those that generate visible light. Such processes take place under particular astrophysical environments. Invisible as-

(a)

(b)

FIGURE 6.12 One example of the new generation of telescopes. The 3.5-meter Astronomical Research Consortium telescope at Apache Point, New Mexico. (*a*) View from the exterior. The design is based on the latest in telescope technology and features a short telescope inside a rotating building. The main mirror had not yet been installed when this photo was taken. (Courtesy of New Mexico State University.) (*b*) Interior view of the telescope. Note the lightweight design of the support structure. Instruments were positioned at special locations on the sides at the bottom of the telescope. (Photo by M. Zeilik.)

(a)

(b)

FIGURE 6.13 The Keck Telescope, the largest optical telescope in the world. (*a*) View of the dome on Mauna Kea, Hawaii. (Courtesy of the California Research Association for Research in Astronomy.) (*b*) View looking down the telescope at the main mirror, which is made of 36 hexagonal segments. (Photo by Roger Ressmeyer-Starlight for the California Association for Research in Astronomy.)

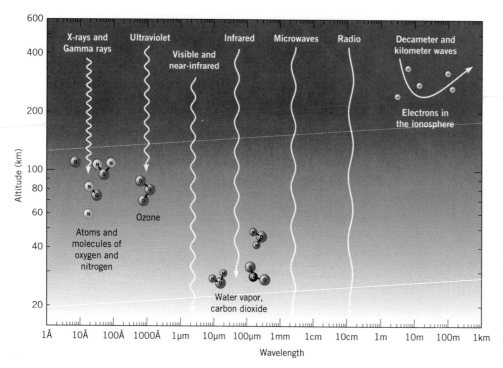

FIGURE 6.14 The transparency of the earth's atmosphere from short-wavelength gamma rays to long-wavelength radio waves. The scale at the left indicates the altitude down to which the radiation at various wavelengths penetrates. Water vapor does most of the absorption in the infrared.

tronomy captures these photons and probes regions not seen in visible light, revealing different information about astronomical objects and even different astronomical objects!

Invisible astronomy has a less obvious aspect: the radiation from space may or may not get to the earth's surface. Our atmosphere effectively absorbs large blocks of the electromagnetic spectrum, especially ultraviolet, x-rays, gamma rays, some infrared, and short-wavelength (millimeter) radio waves (Fig. 6.14). What produces **atmo-** **spheric absorption**? In the case of infrared radiation, it is primarily absorbed by the water vapor in the atmosphere, which is found concentrated in the lower portions, below 20 km. The ultraviolet and x-ray radiation is primarily absorbed in the ionosphere, at an altitude of 100 km, well above the levels that can be reached by balloons and airplanes. The radiation that is absorbed may have journeyed through space for millions or billions of years, only to be snuffed out in the last 0.001 second of its trip, never making it to the earth's surface.

One way to avoid atmospheric absorption is to go above it. This is space astronomy. (In this book, space astronomy includes rockets, balloons, and airplanes, as well as satellites and spacecraft.) So invisible astronomy has two natural divisions: work that can be done from the ground (such as projects using radio and infrared) and that which must be accomplished in space.

Ground-Based Radio

Radio astronomy was born in 1930 when Karl Jansky (1905–1950) undertook a study for the Bell Telephone Company of sources of static affecting transoceanic radiotelephone communications (Fig. 6.15). Jansky identified one source of

Spectral Bands

At a Glance

PORTION OF ELECTROMAGNETIC SPECTRUM	WAVELENGTH (cm)
Radio	10^8 to 10^{-1}
Infrared	10^{-1} to 10^{-4}
Visible	8×10^{-5} to 4×10^{-5}
Ultraviolet	4×10^{-5} to 10^{-6}
X-ray	10^{-6} to 3×10^{-10}
Gamma ray	10^{-10} to 10^{-13}

noise as a celestial object: the Milky Way in Sagittarius. Jansky's discovery was published in 1932 but had little impact on the astronomers of the day.

However, an American radio engineer, Grote Reber, read Jansky's work and decided to search for cosmic radio static in his spare time. By the 1940s, Reber had made detailed maps of the radio sky. These maps of the radio intensity in different parts of the sky are one kind of *contour map*. Sensing that a new astronomy was in the making, Reber took an astrophysics course at the University of Chicago to learn more about astronomy and to discuss his discoveries with astronomers. Only a few were impressed.

World War II forced technical developments in radio and radar work. Accidentally, John S. Hey in Britain discovered that the sun emitted strong radio waves. After the war, Hey continued his astronomical pursuits at radio wavelengths. So did other groups in Britain, the Netherlands, and Australia. Radio astronomy was reborn as part of the technological fallout from research by scientists forced to deal with the practical problems of war.

A common type of **radio telescope,** a radio dish (Fig. 6.16), functions like an optical reflecting telescope. Essentially, it's a radio-wave reflector with a detector (a radio receiver) at the focus of the dish, which reflects and focuses radio waves, acting much like a mirror in a reflecting telescope. Unlike optical astronomers, radio astronomers cannot see sources at the focus. However, the radio receiver that detects the incoming radio waves translates the signal into a voltage that can be measured and recorded. A computer then generates a map of the radio intensity over a region of the sky.

Our atmosphere allows some millimeter, centimeter, and longer wavelengths to reach the ground. These can be observed both day and night. Radio telescopes can even observe on cloudy days at the longer wavelengths. Radio telescopes are typically much larger than optical telescopes, and so they can catch more radiation. Thus radio telescopes are usually more sensitive than optical instruments. These are a few of the ad-

FIGURE 6.15 Early radio astronomy. Karl Jansky with the radio antenna in Holmdel, New Jersey, that he used to discover radio waves from space. (Courtesy of AT&T Bell Labs.)

(a)

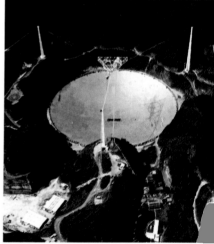

(b)

FIGURE 6.16 Radio dishes. (*a*) One antenna of the Very Large Array (VLA). It has a diameter of 25 meters. The antennas move along tracks to attain positions in different places along the array. (Photo by M. Zeilik.) (*b*) The largest radio dish in the world, the Arecibo antenna in Puerto Rico. Located in a valley and fixed in orientation to the sky, this antenna spans a diameter of 300 meters. (Courtesy of the National Astronomy and Ionospheric Center, Cornell University.)

vantages of radio compared to optical astronomy.

Resolving Power and Radio Interferometers

Radio telescopes have one *major* drawback: poor resolution. How so? Resolution depends on both the diameter of the objective (the light-gathering surface) and the wavelength of the gathered light. Radio waves are much longer than visible light, typically 100,000 times as long. So if an optical and a radio telescope had the same diameter, the radio telescope would have 100,000 times *less* resolution and much less ability to show details in radio sources. For example, a radio telescope with the same resolution as the 5-meter Hale telescope would have to have a diameter 100,000 times as great, about 500 km! Clearly, radio telescopes need to be large to have decent resolving power, but it is also obvious that a single dish of this size cannot be built on earth!

There's a method for making a small radio telescope mimic a large one. Imagine two radio telescopes placed, say, 10 km apart. By synchronizing the signals received by both, the pair can be made to act like a single dish with a diameter of 10 km, but only for a thin strip across the sky. (That's because the two instruments act like two small pieces at the opposite ends of a strip of a larger radio dish.) To get good resolution in a small, more or less circular region of the sky requires an array of coordinated radio telescopes. One such telescope, called the *Very Large Array (VLA),* is in New Mexico (Fig. 6.17). Completed late in 1980, the VLA (which can have its antennas spread out over a few tens of kilometers) has a resolution, at centimeter wavelengths, equivalent to that of a moderate-sized optical telescope. Such complex devices are called **radio interferometers.**

How much resolution does a radio interferometer have? Basically, that depends on the separation of the antennas and the wavelength observed (Focus 6.1). If the antennas are 1 km apart and operate at a wavelength of 1 cm, the resolving power is about 2 arcsec, almost as good as optical telescopes and much better than a single dish could achieve.

How good can the resolution get? In the technique known as *very-long-baseline interferometry (VLBI),* the signals received by very distant antennas (even devices located on different continents) are recorded on magnetic tape and combined later in a computer. The maximum baseline extends to the diameter of the earth, so a resolution of 2×10^{-4} arcsec is possible. In the future, radio telescopes in space, separated by larger baselines, will give even better resolution. Some proposals envision one element of an interferometer on the moon—a separation some thirty times greater than is possible on the earth. Res-

FIGURE 6.17 Aerial view of the central section of the "Y" of the VLA. The antennas are spread out and electronically linked together to achieve high resolving power. (Courtesy of NRAO/AUI.)

olutions could approach 10^{-6} arcsec and we would be able to see (at radio wavelengths) the details on the disks of nearby solar-type stars!

Recently, the United States has begun use of the *Very Large Baseline Array (VLBA),* centered in New Mexico. Resembling the VLA in function, the VLBA will incorporate VLBI techniques in a coordinated array that will span the United States from the Virgin Islands to Hawaii and include an antenna in Iowa (Fig. 6.18). The completed array includes ten 25-meter antennas (similar to those of the VLA) with separations as great as a few thousand kilometers. The receivers will be synchronized by atomic clocks, and the data, recorded on magnetic tapes, will be shipped to a central computer in New Mexico for processing.

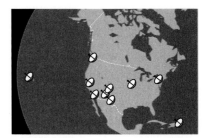

FIGURE 6.18 The Very Long Baseline Array (VLBA). This map shows the geographic layout of the VLBA antennas.

Ground-Based Infrared

Water and carbon dioxide in the earth's atmosphere absorb much of incoming infrared radiation. The infrared astronomer can observe at only a few restricted wavelength ranges: 2–25, 30–40, and 350–450 μm (recall that 1 μm equals 10^{-6} m). Such observations are best made from high sites in dry climates (such as New Mexico), where the atmosphere above the telescope contains little water vapor.

How does an *infrared telescope* differ from an optical one? Mainly in the detector at the telescope's focus. Because our eyes and photographic film sense infrared radiation poorly, special infrared detectors are required. Sensitive ones suitable for astronomical work have been around only since the 1960s. A common infrared detector is a *bolometer,* a tiny chip of germanium (about the size of the head of a very small nail) cooled to very low temperatures, about 2 K. When infrared radiation strikes a bolometer, it heats up, and its resistance to an electric current changes. Such changes are measured, and the amount of variation indicates how much infrared energy the bolometer is absorbing. We now have small arrays of infrared detectors that can produce infrared images.

Infrared observing has at least two distinct advantages over optical observing. First, infrared radiation is less hindered than visible light by interstellar dust (Section 15.2). Second, cool celestial objects (3000 K and cooler) give off most of their radiation in the infrared. Typically, such cool objects cannot be seen in visible light but can be detected in the infrared (Fig. 6.19). So infrared astronomy brings the cold universe into view.

Space Astronomy

What about the parts of the infrared spectrum that do not penetrate to the ground? And ultraviolet light and x-rays? Such radiation can be detected only above the earth's atmosphere, from airplanes, balloons, rockets, satellites, spacecraft, or lunar observations. To these telescopes, the sky looks very different—often dramatically!—from that which we perceive by eye.

For example, most of the far infrared (wavelengths longer than about 40 μm) does not make it to the ground. But at altitudes of 15 to 20 km or so, very little of the earth's atmosphere remains. Far-infrared observations can be made at these altitudes from airplanes or balloon-borne telescopes (Fig. 6.20).

Low-energy infrared astronomy had a space champion: the *Infrared Astronomical Satellite* or *IRAS.* An international collaboration of the United States, the United Kingdom, and the Netherlands, *IRAS* mapped the sky at far infra-

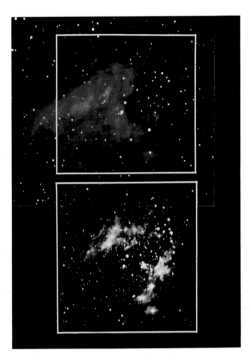

FIGURE 6.19 Optical and infrared views of star-forming regions. The upper image is the region of ionized hydrogen gas surrounding a cluster of hot, young stars. The region of emission is called Messier 17. The dusty region just to the right marks the location of a dense, molecular cloud. The lower image of the same region (inside white box) reveals a cluster of newly born stars, deep within a molecular cloud, detected by an infrared array camera. The false colors here are blue to represent emission at 1.2 μm, green at 1.6 μm, and red at 2.2 μm. (Observations by I. Gatley and C. Lada; courtesy of NOAO/KPNO.)

red wavelengths, from 12 to 100 μm, with a 57-cm telescope. The satellite functioned for about nine months until it ran out of the liquid helium that kept the infrared detectors cooled to 2.5 K. (Direct sunlight in space heats a spacecraft, and it must be cooled to work well in the infrared; essentially *IRAS* was an infrared telescope in a thermos bottle orbiting the earth.) A stunning success, *IRAS* scanned almost all the sky close to three times. It detected more than 200,000 infrared sources, which are kept in a computer archive. Many of these related to the process of starbirth in the Milky Way (more in Chapter 15).

For ultraviolet astronomy, methods of light gathering and detection remain similar to optical astronomy. Special television cameras respond well to ultraviolet light, so detection presents no serious problem. Because glass absorbs ultraviolet light, refracting telescopes cannot be used, but reflectors work perfectly well. All ultraviolet astronomy must be done from space, for the absorbing layer of the atmosphere is higher than balloons or aircraft can reach. Probably the most successful *ultraviolet telescope* is the *International Ultraviolet Explorer (IUE),* launched in 1978 and still operational— far beyond its expected life. IUE has a 0.45-meter telescope and detectors that

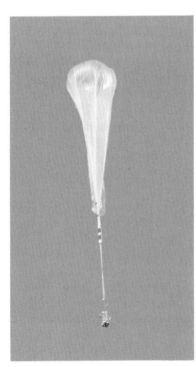

FIGURE 6.20 Balloon astronomy. The 1-meter, far-infrared telescope of the Harvard–Smithsonian Center for Astrophysics and the University of Arizona at launch, carried aloft by a helium balloon to a height of about 25 km. (Photo by M. Zeilik.)

work in the wavelength range from 1150 to 3200 Å. The data are accessible to all astronomers in a computer data bank.

For the high-energy realm of x-ray and gamma-ray astronomy, gathering and focusing the light require special techniques. The most productive of x-ray telescopes was the *Einstein Observatory,* launched in 1978 and used up to 1981. This 58-cm telescope produced high-resolution x-ray images in the wavelength range from 3 to 50 Å. Although the telescope is now defunct, its observations are accessible in an archive of magnetic tapes. They still produce new discoveries when processed by computer and later examined by astronomers.

The current functional x-ray telescope is called *ROSAT* (the *Roentgen Satellite*), launched in 1990; it is a joint project of the United States, Germany, and the United Kingdom. It can provide high-resolution, x-ray images covering areas about 5 arcmin square. *ROSAT's* first priority is a complete survey of x-ray sources in the sky. In April 1991, the shuttle *Atlantis* deployed the *Gamma Ray Observatory (GRO)* of NASA in earth orbit at an altitude of 450 km. *GRO* is designed to detect, with great sensitivity, gamma rays over a wide range of energies (and even some of the most energetic x-rays). It has four separate instruments, each designed for a specific type of gamma-ray observation.

Finally, optical astronomy in the future will involve large space telescopes. The main advantage of this development (besides the lack of weather in space) is that a space telescope is not limited by the atmosphere; rather, it can be used at its theoretical limit of resolving power. The current U.S. space telescope, named the *Hubble Space Telescope (HST),* has a 2.4-meter mirror. Its primary goal was observations of faint objects with high resolution, especially at ultraviolet wavelengths. The best resolution with *HST* was designed to be about one-tenth of an arcsecond. In addition, *HST* was intended to operate over the wavelength range from 1200 Å to 1 μm, depending on the instruments used for sensing the light. The range of the electromagnetic spectrum accessible to *HST's* view

FIGURE 6.21 The *Hubble Space Telescope (HST).* This telescope has a 2.4-meter mirror and was placed 400 km above the earth by the U.S. space shuttle *Discovery.* (Courtesy of NASA.)

amounts to about a thousand times more than that for a ground-based telescope.

Those were the hopes and dreams of astronomers. But they did not work out as planned. On April 24, 1990, the space shuttle *Discovery* transported *Hubble Space Telescope* into space. The astronauts deployed it the next day (Fig. 6.21). At a developmental cost of $1.5 billion, the *HST* is the most expensive astronomical project ever. (For comparison, the VLA cost a bit less than $80 million when it was completed in 1981.) Astronomers were shocked to find out in summer 1990 that the main mirror was flawed.

Simply put, the overall shape of the 2.4-meter primary mirror is too shallow, from edge to center, by about 2 μm. (If the mirror were the size of the United States, the error would amount to a meter or two.) As small as this error seems, it is large compared to the wavelengths of ultraviolet and visible light. So the images of stars (and other objects) cannot be brought into good focus. Although as mentioned above, the original specifications called for starlight to fall into an image with a radius of 0.1 arcsec, the actual performance is some seven times worse—about that of a ground-based telescope at a good site.

What happens now? In the short term, the ability to make high-resolution images is hampered. But astronomers are clever people; they have devised computer image reconstruction techniques that have made it possible to sharpen the raw data. The analogy is the VLA; new computer programs have made its radio images much better than even its original designers had imagined.

The long term may present solutions to the imaging problems. Keep in mind that *HST* was designed for a 15-year mission, with new and upgraded instruments ferried up by future shuttles. New internal optics have been designed to compensate for the mirror's flaws. Some may be installed by the end of 1993. Although working at somewhat less than anticipated limits, *HST* still will be able to discover the unexpected, especially for close by, bright objects. You'll see a fair number of *HST* images in this book, greatly improved by image processing.

Eventually (probably in the next century), **space astronomy** will include observatories on the moon (Fig. 6.22). Just those aspects of the lunar environment that make it tough on people result in a great place for astronomy. One, the lack of atmosphere not only eliminates weather problems but permits access to all wavelengths at a resolution only limited by the telescope. Two, because moonquakes are weak and infrequent, the moon has a firm, seismically stable surface on which to construct telescopes. (A problem of a telescope in low earth orbit is that it is difficult to point it in the right direction in space and keep it on target.) Three, the background electromagnetic noise is very low (especially for radio emissions on the moon's far side, which is shielded from terrestrial radio noise). In contrast, even low earth orbit is a dirty, noisy environment for astronomy.

What kind of telescope would take best advantage of the moon? A grand observatory would operate as an interferometer (like the VLA) at ultraviolet, optical, and infrared wavelengths. If it had telescopes spread over a diameter of

about 10 km, a lunar array would achieve resolutions about 100,000 times better than ground-based telescopes. That amount of resolving power would pick out a U.S. dime on the earth's surface from the moon! Such an instrument would be able to see earthlike planets around nearby sunlike stars.

Unique astronomy can be done with modest instruments on the moon. Even a simple 1-meter telescope would have ten times the resolving power of its earth-based equivalent. And it could easily monitor changes in light (ultraviolet to infrared) from astronomical objects continuously over the 14-day lunar night. A network of small, automated telescopes could record a number of objects on a regular, long-term basis to provide new insights into their physical natures.

New Views of the Sky

The technology of new telescopes, especially those in space, offers the ability to paint dramatic views of a sky not visible to the eye. Let me contrast some of these.

First, the optical sky (Fig. 6.23*a*)—that visible to the eye and optical telescopes. Here we see the great span of the Milky Way, an insider's view of the galaxy in which we reside. The dark lanes

FIGURE 6.22 Artist's conception of a small array of optical telescopes on the moon. These instruments would work as an optical interferometer, and their data would be sent to a central processing unit (building in center) for image production. (Courtesy of JPL and the California Institute of Technology.)

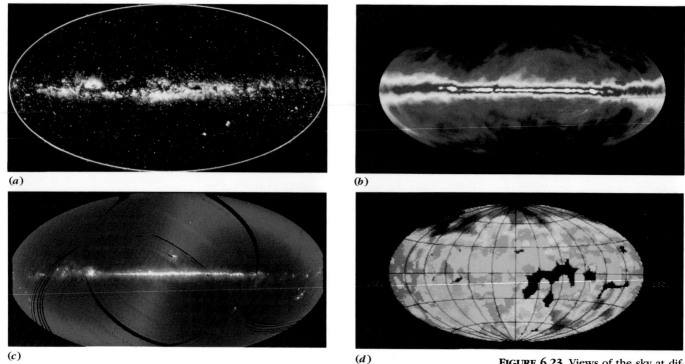

(a)

(b)

(c)

(d)

through the central plane are areas blocked out by dust. The center of the Milky Way lies in the center of the image but is not visible because of the dust. The Magellanic Clouds, two companion galaxies to our own, are visible below the plane to the right of center.

Second, the radio sky at a wavelength of 21 cm (Fig. 6.23b). Hydrogen atoms in interstellar space emit at this wavelength. The colors represent the relative amounts of hydrogen—black and dark blue the least, red and white the most. The filaments of emission above and below the plane of the Milky Way may be made by explosions of massive stars or by winds blowing off of them.

IRAS revealed the infrared sky (Fig. 6.23c). Combined here are the emissions at 12 (blue), 60 (green), and 100 μm (red). Distinct regions are visible amid the overall infrared glow. The bright area at the lower right marks the star-forming region in Orion. The bright patch below the plane at the middle right is the Large Magellanic Cloud (also visible in the optical view). Most of the emission in the galactic plane comes from a combination of hot and cool dust particles. Finally, the x-ray sky provides a high-energy view (Fig. 6.23d). This image is a composite of many rocket flights with x-ray telescopes; the wavelength range is 40–80 Å. Blue and black show the region of least flux (see scale at the bottom); the overall trend gives an idea of the background glow of the x-ray sky. Note the general increase in intensity as you move away from the galactic plane upward.

■ What is the critical limitation of a single-dish radio telescope?

6.4 Image Collection and Processing

You may have the romantic image of an astronomer working with eyes glued to the telescope to take spectacular photographs. Indeed, photography is still done. But today the astronomer usually sits in a warm room and "observes" on a TV monitor that shows an enhanced view through the telescope. Most optical telescopes do very little direct photography. Images can be made by detectors of other kinds. In most cases, including that of photography, the data are later manipulated by computer in digital form. This procedure, called **image processing,** plays a central role in astronomy today, especially with invisible astronomy (look back at Fig. 6.23 for examples).

FIGURE 6.23 Views of the sky at different wavelengths: the images are oriented as a view of the Milky Way with the central plane that of the galactic plane. (*a*) The visible sky, a composite of many astronomical photographs. Note the dark dust lanes through the central plane. (Courtesy of the Lund Observatory, Sweden.) (*b*) The radio sky at 21 cm. This emission comes from neutral hydrogen in clouds in the Milky Way. (Images assembled by C. Jones and W. Foreman from observations by A. A. Stark and colleagues; and M. N. Cleary, C. Heiles, and F. Kerr and colleagues.) (*c*) The infrared sky, combined *IRAS* observations at 12, 60, and 100 μm (blue, green, and red). The blue S-shaped band is emission from dust particles within the solar system. Other emission is from the Milky Way, except for the Large Magellanic Cloud below and at right from the plane. (Courtesy of NASA/JPL Infrared Processing Center.) (*d*) The x-ray sky at wavelengths from 40 to 80 Å. The false-color scale for intensity is given at the bottom. Note how the emission increases away from the plane of the Milky Way, in contrast to the other wavelengths, where the emission tends to be concentrated in the plane. (Courtesy of D. McCannon, D. N. Burrows, W. Sanders, and W. Kraushaar, University of Wisconsin, Madison.)

The computer can display an image of information otherwise not detectable by eye, and the brain can process this information rapidly by visual cues. This technique is called *computer visualization.* Computer-generated images often use *false colors* to show information in a visually informative fashion. These colors are false in the sense that they are not the actual colors of the object in the visual range of the spectrum. Rather, they are codes to a specific property, such as different intensities of a signal.

Understanding Intensity Maps

Many of the figures in this book are *intensity maps,* pictures of how the intensity of some kind of radiation (radio, visible, infrared) varies over some region of the sky. Such maps may show a lot of wavy, connected lines labeled by numbers; these are called *contour maps.* Or they may show regions of different colors as false-color maps. What do they mean? Here's a familiar analogy: a record of atmospheric pressure across the United States, with high and low pressure systems indicated (Fig. 6.24). What do the contours in this weather map tell you?

The weather stations around the United States report their local pressures. Each is put on the map. Then a contour line is drawn that connects all stations giving the *same* reading, provided it does not cross a station of higher or lower reading. Then another contour (of higher or lower pressure) is drawn in, and so on. Contour lines *cannot* cross. Why? Because if they did, it would mean that the *same* place has two *different* pressures, and that's impossible!

At certain places the pressures hit a maximum (high) or a minimum (low). At the center of each is a last contour surrounding the region of highest or lowest pressure. You can imagine the center of a high, for example, as the peak of a pressure mountain. The contours around the peak tell you how the pressure falls from the peak. If the contour lines are close together, the pressure drops quickly over a short distance (the fall-off is steep). If the contour lines are spread out, the pressure diminishes slowly. You have a kind of ridge.

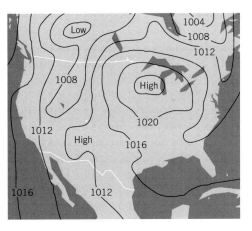

FIGURE 6.24 Weather map showing surface pressure over the United States. The contour lines connect regions with the same pressure readings: "High" marks a peak in the surface pressure, "Low" a local minimum. Units are in millibars.

Image processing can display astronomical data in a variety of formats, depending on what aspects the astronomer wants to emphasize. The simplest one is a contour map like the weather map, which shows high and low regions clearly (Fig. 6.25*a*). A variation of this is a gray-scale map, in which the density of the filled areas gives an indication of signal's intensity (Fig. 6.25*b*). Finally, the most elaborate is a *false-color map,* in which the computer generates different colors for different levels of intensity (Fig. 6.25*c*). But why use false colors, which must be recognized as processing artifacts rather than as properties of the object observed? False colors are used because they convey more visual information than black and white or gray scales. Your eye can detect about 400 shades of red, for instance, but only some 30 shades of gray.

Note that the original image may be collected in many different ways. These days, that information is put into digital form to enable processing and visualization by computer.

Photography

Photography is still the old standby for gathering a large amount of storable information in a short time. Astronomers use special light-sensitive coatings on glass plates, whose size depends on the type of telescope employed. Special wide-field telescopes commonly require very large plates. Glass plates do not bend, can be measured accurately, and can survive a long time. Long time exposures add up the light striking the plate, with the result that very faint objects eventually show up clearly. However, even the most sensitive plates convert only a small percentage of the photons striking them into images. So photography does not make very efficient use of limited time on a telescope.

Despite low efficiency, photography still has one great advantage: the plates are big and cover a large area at a telescope's focus. Hence, a single photograph's information content can be enormous, especially from a wide-angle telescope. Also, plates preserve information with little loss over a long time, providing a record that an astronomer a century later can examine—probably for a purpose completely different from the original intention! Today, specialized optical devices can convert the information on a photograph into digital form for computer enhancement.

Charge-Coupled Devices (CCDs)

The development of solid-state electronics has resulted in small computers with large memories and detectors that match the digital nature of computer processing. The natural match that would make astronomers most happy involves a two-dimensional detector that captures most of the photons striking it. Microelectronics may have produced the dream detector: *charge-coupled devices* (*CCDs* for short), which are used in video cameras.

A CCD is a thin silicon wafer a few millimeters on a side. It contains a large number of small regions, each of which makes up a picture element, called a *pixel*. A typical CCD chip contains a million pixels arranged in a grid. Each pixel converts photons to electrons and builds up charge over a long time. A small computer then moves the charge out of the chip in a regimented way, keeping track of the number of charges accumulated in each pixel. That set of charges is then displayed (again by computer) on a video monitor, and the image is the brightest for the pixels with the greatest amount of the charge, and darkest for the pixels with the least charge. Computer processing after data acquisition can bring out certain aspects in which the astronomer has the most interest (Fig. 6.26).

Because they are flat, two-dimensional detectors, CCDs can gather a large amount of information during one exposure. They share this advantage with photographic plates (though they are a *lot* smaller than such plates right now). CCDs have two distinct advantages over plates. First, they collect electrons at a rate that is directly proportional to the number of photons. Second, they are

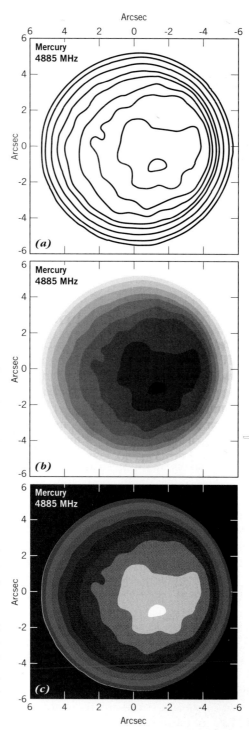

FIGURE 6.25 Variations on processing intensity maps. (*a*) Contour map produced by the VLA of the radio emission from the planet Mercury at a wavelength of 6 cm. Note the peak ("high") lower right of center. (*b*) The same observations as in (*a*) displayed as a gray-scale map. The darker the region, the greater the radio emission. They show the same trend. (*c*) The same observations as in (*a*) and (*b*) displayed as a false-color map. Here blue is the weakest emission, and dark red is the strongest. Note the peak of emission to the right of center. Again, because the image processing was somewhat different, the contours are not exactly the same as in Figure 6.25. (Courtesy of NRAO/AUI; VLA observations by J. Burns, D. Baker, J. Borovsky, G. Gisler, and M. Zeilik; image processing by J.-H. Zhao and M. Ledlow.)

very, very efficient—in some regions of the spectrum (notably the red), they come close to detecting 100 percent of the photons and converting them into electrons.

Along with the small computers that control their operation, CCDs are the best bet (in my opinion) for expanding the capabilities of "small" telescopes for doing frontier astronomy. One telescope can see only a small part of the sky at a time, and the number of telescopes on the earth is extremely limited. Most of them are small, with apertures less than 100 cm. The great efficiency of CCDs enhances the power of these small telescopes and will revive the status of small observatories. For example, in a one-hour exposure, the Capilla Peak Observatory 60-cm telescope with a CCD camera (Fig. 6.1) can display objects that once were visible only on a one-hour ex-

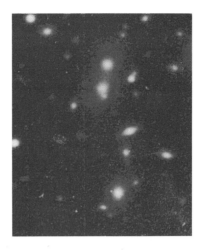

FIGURE 6.26 Example of image processing. This CCD image of the Corona Borealis cluster of galaxies (a nearby cluster) was digitally processed by computer to enhance important details. Blue is the lowest intensity; white, the highest. Almost every visible object is a galaxy. Note that many have red halos around a white core, an indication that material surrounds the galaxies. (Courtesy of the Institute for Astrophysics, University of New Mexico; image taken by J. Moody.)

posure with a photographic plate on a telescope 10 times larger (such as the Hale 5-meter)!

■ In what way are CCDs far superior to photographs for astronomical imagery?

Key Concepts

1. Astronomy is primarily an observational science. Telescopes amplify astronomers' observing power by revealing faint objects, expanding the range of the electromagnetic spectrum that we can perceive, and increasing our ability to see fine detail. These instrumental abilities drive the generation and evolution of astronomical models.

2. Optical telescopes use lenses (refraction) or mirrors (reflection) to gather light and bring it to a focus; usually a photographic plate or light sensor is placed at the focus to detect the light. Telescopes that use only lenses are called refracting telescopes, and those that use a mirror as the objective are called reflecting telescopes.

3. Regardless of its type, a telescope has three basic functions: to gather light (its most important function), to resolve details, and to magnify the image. The light-gathering power of a telescope depends directly on the area of its objective; resolution depends inversely on the diameter of the objective and directly on the wavelength of light being observed (as well as on the turbulence in the earth's atmosphere); magnifying power depends on the ratio of the focal length of the objective and of the eyepiece.

4. Invisible astronomy uses light outside the visible region of the spectrum, such as radio, infrared, ultraviolet, x-rays, and gamma rays. Many of these wavelengths are blocked by the earth's atmosphere, so telescopes must go above it (by balloon, airplane, or rocket) to reveal these ranges of the electromagnetic spectrum.

5. Radio telescopes are limited in resolution because radio wavelengths are very long (compared to visible light). Special electronic techniques allow widely separated radio telescopes to function as a single one; this technique has been used in the design of the VLA in New Mexico and the new VLBA, which are both radio interferometers.

6. Space telescopes have opened up our view of the infrared, ultraviolet, and x-ray heavens. They are not hindered by bad weather, the atmosphere, or the atmosphere's lack of transparency to certain kinds of electromagnetic radiation. Of locations near the earth, the moon promises to be the best all-around site for telescopes of the future.

7. New detectors (such as CCDs) and techniques for the computer processing of images have effectively transformed "small" telescopes into large ones and so extended their usefulness to probe deep into the cosmos. They have made large telescopes even more powerful in some ways that rival the current generation of space telescopes. Computers that control telescopes, collect data, and then process the data into information now play an integral role in new telescopes and detectors of all types.

Key Terms

atmospheric absorption
detector
eyepiece
f-ratio
focal length
image

image processing
lens
light-gathering power
magnifying power
objective
optics

radio interferometer
radio telescope
refracting telescope
refraction
reflecting telescope
reflection

resolution
seeing
space astronomy
theoretical resolution

Study Exercises

1. How did Galileo's observations support the Copernican model and refute the traditional one? (Objective 6-1)

2. What telescopic observations were critical in the confirmation of the Copernican model? (Objective 6-1)

3. What is the *most* important function of a telescope? (Objective 6-2)

4. What are the advantages and disadvantages of radio telescopes? How do they differ from optical ones? How are they similar? (Objectives 6-4, 6-5, and 6-6)

5. Suppose you're on a TV quiz show. They show you optical, infrared, and radio telescopes with the same size objective and ask you to list the devices in order of *increasing* resolving power. What is the correct order? (Objectives 6-3, 6-4, and 6-5)

6. Imagine that you are going before a congressional hearing to justify the expense of putting a large telescope in space. What arguments would you use to persuade the skeptical committee? (Objective 6-10)

7. Describe one advantage of an infrared telescope over an optical one. (Objective 6-9)

8. Why do x-ray telescopes have to be put above the earth's atmosphere? (Objectives 6-8 and 6-10)

9. What kinds of astronomical object can be best studied with a space telescope? With a large (5- to 10-m) ground-based optical telescope? (Objective 6-10)

10. Imagine that you want to buy a telescope on a limited budget. What one property of the telescope should be most important in making your decision? (Objectives 6-2 and 6-3)

Problems & Activities

1. Suppose you have a reflecting telescope with a 900-cm focal length and an f-ratio of 15. What is its theoretical resolving power at visual wavelengths? (*Hint:* Compare it to a telescope with a 10-cm mirror.) What is its light-gathering power compared to that of your eye?

2. How much greater is the light-gathering power of the 5-meter Hale telescope than that of the 2.4-meter *Hubble Space Telescope*? Note that the ground-based Hale telescope can never quite achieve its complete light-gathering power because the earth's atmosphere absorbs some of the light.

3. A typical telescope for amateur astronomy has a 30-cm-diameter objective and an f-ratio of 10. What is the telescope's focal length?

4. A 10-cm telescope has a theoretical resolution of 1.4 arcsec at visible wavelengths. What is the theoretical resolution of the *Hubble Space Telescope* at the same wavelengths? What is the size-to-distance ratio that it can resolve? How far away could it see the disk of a planet like the earth?

5. The maximum possible telescope separation of the VLA is about 40 km. What is its resolution when operating at a wavelength of 2 cm? How does this compare to a 10-cm optical telescope?

6. Imagine that you have a small telescope equipped with interchangeable eyepieces. An eyepiece with a 20-mm focal length gives you a magnifying power of 150. What focal length eyepiece do you need to have a magnifying power of 75? Of 250?

7. What is the smallest feature that you can observe on the moon with the unaided eye? With a 20-cm telescope operating at its theoretical resolution? With any telescope that is limited by the atmospheric seeing, which under good conditions can be 1 arcsec? *Hint:* Use size-to-distance ratios; assume that the moon is about 400,000 km from the earth.

8. Imagine two lights separated by one kilometer. With your unaided eye, how far away from you could these lights be placed and still be visible to you as separate lights? If you used a 10-cm telescope at its theoretical observing power, what would be their distance from you?

9. Consider an optical interferometer placed on the moon with a maximum separation of 10 km. What would be its theoretical resolution? How far away could it see the disk of a planet like the earth?

See for Yourself!

Most binoculars are specified by two numbers, such as "7 × 50," which is read "seven by fifty." The second number is the diameter of the binocular's lenses, in millimeters. In this case, they are 50 mm across. The first number gives the magnification, in this case "seven power," which means that an object seen through the binoculars appears seven times larger than it is. Use such a pair of binoculars to test this out.

What is the apparent angular diameter of the moon viewed through these binoculars? What is the apparent size-to-distance ratio? Which of these two have changed, and by how much? What is your resolution limit using these binoculars? What would be the smallest crater you could see on the moon? Examine the moon near first quarter with 7 × 50 binoculars to confirm your estimations.

Rather than buy a small, cheap telescope, I urge you to purchase binoculars for your first investigations of the heavens with an instrument. Don't be put off by the "low" magnifying power of binoculars. It's really the light-gathering power that counts. ■

LEARNING OBJECTIVES
After studying this chapter, you should be able to:

7-1 State the principle of equivalence and illustrate it with a concrete example.

7-2 Show how the principle of equivalence leads to the local cancellation of gravitational forces.

7-3 Compare and contrast Aristotle's, Newton's, and Einstein's concepts of natural motion for bodies falling near the earth and of the motions of heavenly bodies.

7-4 Describe what is meant by the term *spacetime* and give some common examples.

7-5 Argue that concepts of natural motion must be coupled to a notion of the geometry of spacetime.

7-6 Sketch Hubble's law in graphical form and indicate what observations are needed to make it.

7-7 Describe Hubble's law for the line-of-sight (radial) velocities and distances of galaxies and, from it, find a value of Hubble's constant, and apply Hubble's law in algebraic form.

7-8 Interpret Hubble's law as a consequence of the uniform expansion of the universe.

7-9 Use Hubble's law to estimate the age of the universe, stating clearly your assumptions.

7-10 Use the concept of escape speed to relate the future of the universe to its geometry: hyperbolic, spherical (closed), or flat.

7-11 State Einstein's relation between matter and energy, apply it to astrophysical situations, and use it in algebraic form.

Central Question
How did Einstein's ideas about gravitation lead to an evolving model of the cosmos?

Einstein and the Evolving Universe

What is inconceivable about the universe is that it is at all conceivable.

ALBERT EINSTEIN

The preceding chapters have highlighted important changes in people's ideas of the universe. Two influential schemes, Aristotle's (Chapter 2) and Newton's (Chapter 4), rested on physical laws but differed dramatically in their concepts of natural motion. These different foundations resulted in distinct pictures of the nature of the universe. Newton's cosmos was infinite, yet tied together by a universal physical law; Aristotle's cosmos was finite, with different laws for the terrestrial and celestial realms. Both models had one common aspect: the universe was static.

Newton's grand model gained the authority of success after success (Section 4.5). His infinite space, established firmly by his laws and intertwined with gravitation, seemed invincible. Albert Einstein (Fig. 7.1) greatly admired Newton's synthesis. But Newton had failed to discuss the nature of gravitational force, although he did describe its effects. In addition to pondering gravity, Einstein was puzzled by the nature of light. From these two seemingly separate realms—gravity and light—Einstein synthesized a new view of the universe in his theory of relativity. One surprising result: in Einstein's vision, *gravity is no longer a force.*

In Einstein's theory, space, time, mass, and natural motion take on new relationships from which a new conception of the cosmos emerges—one in which the universe itself *evolves*. This chapter looks again at the concepts underlying Newtonian physics to find the themes of relativity and the consequences of relativistic ideas for the cosmos.

7.1 Natural Motion Reexamined

You live in a Newtonian world. By this statement I mean that you see the world the way Newton did. If an object falls or moves, you assume that a force is acting on it. You are unconsciously applying Newton's laws to what you observe. You infer that a force is acting because you see its effect, a change in motion:

acceleration. You deem the force to be real because you can see the acceleration that results from it.

Newton's Assumptions

Take gravity as an example. How do you know it's a force? Pick up a stone; you feel its weight. Drop it. As it falls, its velocity continually increases. The stone exhibits a constant acceleration. What has changed? The stone's *natural motion,* which is, according to Newton, to be at either rest or moving at a *constant speed along a straight line* (Section 4.3). In contrast, Aristotle (Section 2.3) believed that a falling stone displayed the natural motion of earthy material, so *no force* was acting.

The key point. Newton's definition of natural motion leads directly to his concept of force, particularly gravitational force. Newton saw natural motion as ordinary, inconsequential, and so needing no explanation. To him, accelerated motion demanded explanation. After all, without accelerated motion, apples would not fall from trees, the moon would not orbit the earth, and the earth would not revolve around the sun.

To cope with accelerations, Newton had to deal with forces. Forces result in accelerations, changes in velocity. This emphasis on *rate of change* of velocity rather than velocity itself is critical. All velocities are relative, so you can always make any given velocity disappear. For example, when you have stopped just before entering a highway, other cars go by you at 90 km/h (55 mph). You accelerate to 90 km/h, too. Then you've matched the velocities of the other cars, with the result that compared to yours, they have zero velocity. The original 90-km/h difference has vanished.

In any situation, an observer can match velocities with any other observer. Relative velocities can always be made to disappear. If not, astronauts wouldn't be able to dock their spacecraft! Newton believed that forces were real, that they could not be made to disappear arbitrarily. By connecting forces to accelerations (and *not* to velocities), Newton seemed to have nailed down their absolute reality.

FIGURE 7.1 In 1931, Albert Einstein and wife visited the Hopi House at the Grand Canyon, Arizona, where they posed with Hopi Indians. (Courtesy of the Museum of New Mexico, Negative No. 38193; photo by El Tovar Studio.)

Motion and Geometry

However, Newton had made an assumption that you may not have noticed. In his definition of natural motion, he talks of velocity along a *straight line.* What's a straight line? The shortest distance between two points. You probably have an intuitive picture of a straight line drawn on some flat surface (like a blackboard). You then have in mind the same kind of straight line that Newton envisaged. That's the geometric assumption behind Newton's definition of natural motion: *straight line* means a line on a *flat* surface. Newton really didn't have any choice, for in his time a flat geometry was thought to be the only possible geometry.

Can you imagine straight lines that are on a curved surface rather than a flat

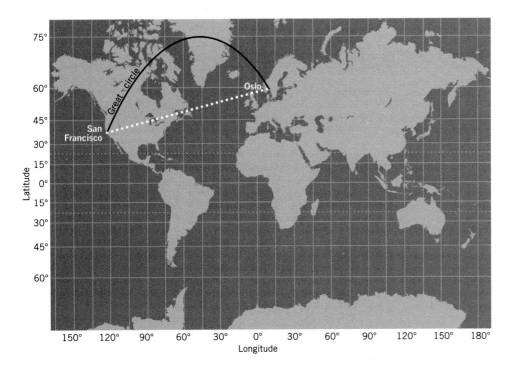

FIGURE 7.2 Straight lines and geometry on the earth. This Mercator map artificially makes the round earth appear flat. So straight lines on such a map are not the shortest distances between two points on the earth. If you use a ruler on this map to draw a straight line between San Francisco and Oslo, you have not shown the path that is the shortest distance between these cities. That path is a great circle line, which appears curved on this flat map.

one? Consider the earth's surface. A straight line on the earth's surface differs from that on a flat surface. Look at a flat (Mercator) map of the world (Fig. 7.2). You might think that traveling on a straight line of the same latitude is the shortest distance between two points of the same latitude. But it's not. Airplanes, for instance, do *not* travel along latitude lines to go the shortest distance. Instead they travel part of a *great circle,* a circle on the earth's surface whose center is the earth's center.

You can make your own straight lines on the earth with a string. Stretch the string tightly against a globe with ends at the two places you want to connect (Fig. 7.3). Note that this path curves relative to latitude lines! On a flat, Mercator map, such a line does not look straight. But it is—on a *curved* surface.

The point here is that the geometry of a curved surface is *not* the same as that of a flat surface. In his definition of natural motion, Newton assumed that our universe had the same geometric properties as a flat surface. Einstein challenged this assumption.

■ **According to Newton, when you are driving on the highway, at the same speed as the other cars, what is their velocity relative to you?**

7.2 The Rise of Relativity

At the start of this century, Albert Einstein created a new vision of the cosmos. This picture emerged in two stages: the **special theory of relativity,** then the general theory. The special theory dealt with the laws of physics as seen by *unaccelerated* observers: those who experience uniform motion. The general theory goes beyond the special theory to cope with the nature of gravitation and so accommodates the case of *accelerated* observers.

Mass and Energy, Space and Time

Einstein showed in his special theory that mass and energy are two aspects of the same essential stuff. He was able to derive the famous formula relating mass and energy to the speed of light:

$$E = mc^2$$

where E is the energy in joules, m is the mass in kilograms, and c is the speed of light in meters per second. This relationship means that you can think of all matter as a form of energy and all kinds of energy as possessing mass. This mass produces gravity and also shows up as inertia. Small masses convert to large

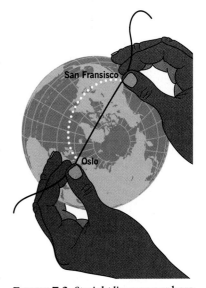

FIGURE 7.3 Straight lines on a sphere. You can check that great circle distances on a globe are the shortest distances (and so straight lines in a spherical geometry) by stretching a string along the globe's surface to connect two points.

amounts of energy. For instance, the complete transformation of 1 kg of matter to energy results in 9×10^{16} J, which is enough energy to keep a 100-watt light bulb lit for about 30 *million years!*

The special theory also resulted in the unification of the concepts of space and time. Scientists had looked at these as separate aspects of the universe (similar to the ancient astronomical separation of earth and sky). Newton, for instance, dealt with space and time as totally unrelated. In Einstein's view, all events in the universe involve space and time together: **spacetime,** which has four dimensions (three of space and one of time).

Einstein made special note of what seems so obvious: when anything happens in the world, it takes place in *both* space and time. He called such happenings *events*. You mark an event by noting where it took place (space) and when it took place (time). Events make up points in spacetime.

Light takes a finite time to reach us. Therefore we do not see in real time events that are spread out in space. Instead, we see distant objects as they were at some instant in the past. Look up at the moon. You see it as it was about a second ago. When you look at the sun, your view is 8 minutes old. The stars at night give you a deeper look into the past. Sirius, as you see it now, shines as it did some 9 years ago. As you look out into space, you look back in time. You now receive light emitted from different objects at many different times. In this sense, a telescope acts like a one-way time machine—it peers only into the past.

Einstein used these connections between mass and energy and space and time to underpin his concept of gravity, the general theory of relativity.

The General Theory of Relativity

From 1905 to 1915 Einstein tackled the case of *accelerated* observers in spacetime. This investigation led to the **general theory of relativity,** which is Einstein's theory about the nature of gravitation.

Einstein rethought gravitation by questioning Newton's concepts of mass and inertia. He saw that Newton defined mass by two different operations: in the second law (Section 4.3) and in the law of gravitation (Section 4.4). Suppose you apply a known force to an object and measure its acceleration. Using the force and acceleration, Newton's second law gives you the object's mass. Mass determined this way is called **inertial mass.** Now take the same object and weigh it. Weight is a force, the amount of gravitational force with which the earth (in this case) attracts the object. Mass measured this way is called **gravitational mass.**

Newton believed (but did not make the distinction) that an object's inertial mass and gravitational mass were the same. He knew this from Galileo's experiments with falling bodies (Section 4.2) as well as from his own careful experiments. These results demonstrated that near the earth, *all masses fall with the same acceleration.* Experimentally, this equality of gravitational and inertial mass holds true to a very high degree of accuracy: no difference in the limits of sensitivity with which tests can be done has ever been detected.

The Principle of Equivalence

Einstein felt that the equality of gravitational and inertial mass was no accident. He saw it as a fundamental fact about the universe and gave it a special place in the general theory as the **principle of equivalence:** *you cannot distinguish accelerations due to gravitation from accelerations due to forces of other kinds.* Or, stated differently, the acceleration of gravity is *independent* of the mass of an object.

Here is Einstein's own example of the principle of equivalence (Fig. 7.4). Imagine yourself on the earth in a spacecraft with *no* windows. If you were to drop objects in the spacecraft and measure their accelerations, you would find that they all fell with the same acceleration, 9.8 m/s/s. Now suppose that without your knowledge, you and the spacecraft were placed out in space and constantly accelerated at 9.8 m/s/s. Repeat your experiments dropping other objects. They will accelerate at 9.8 m/s/s.

Where are you? You'd probably conclude that you were still on the earth—unless you could look out of the spacecraft. You cannot, by your experiments, distinguish between the effects due to a gravitational force and those due to the force of a rocket engine.

The principle of equivalence has profound consequences: you now have a way to cancel out gravity locally. Put yourself in an elevator in a tall building. Let the elevator free-fall (Fig. 7.5). You find yourself weightless; gravity has vanished! Instantly transport yourself in space, far away from any large masses. Your condition is the same—you are weightless without gravity.

Now, you may think that the falling elevator is some kind of cheat: What happens when the elevator hits the ground? Well, abruptly everything in the elevator has weight! But imagine that the elevator were to continue to fall straight into an airless tunnel through the earth. What would happen? It would fly (with weightless conditions inside) through the earth's center and come out the other side to the same height above the ground that it had started from at the other end. It would then reverse direction and fall back, and continue to swing back and forth, taking about 84 minutes for each complete swing. In a sense, it would be orbiting through the earth. During these fly-throughs, the conditions for weightlessness would exist inside the elevator.

Weightlessness and Natural Motion

By free-falling, an observer can make gravity disappear. How strange! Newton saw gravity as mysterious, though he was astute enough to describe its effects. To Einstein, falling objects were not mysterious at all. The objects are simply following their *natural motion in spacetime*. In Einstein's view, gravity is not a force. Here's a rule of thumb in relativity: *when you are weightless, you are following your natural motion in spacetime.* Having weight is unnatural.

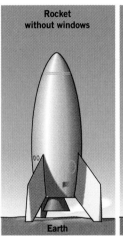
Rocket without windows
Earth

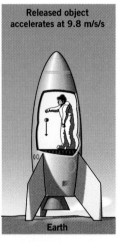
Released object accelerates at 9.8 m/s/s
Earth

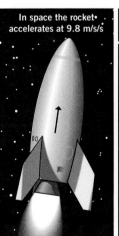
In space the rocket accelerates at 9.8 m/s/s

Released object accelerates at 9.8 m/s/s

FIGURE 7.4 The principle of equivalence. Imagine that you are in a small spacecraft with no windows. Drop a mass and measure its acceleration. If the spacecraft is on the earth, the acceleration will be 9.8 m/s/s. If the ship is in space far away from the earth and its engine is accelerating it at 9.8 m/s/s, the same experiment will yield the same results. Under the conditions of this experiment, you could not determine whether you were on the earth or in an accelerating ship.

(a) Earth *(b)* Space

FIGURE 7.5 The local elimination of gravity, using the principle of equivalence. An observer in an elevator sees no effects of gravity (*a*) if the elevator free-falls at the earth's surface; a dropped object has no measurable acceleration. Out in space, far away from all masses (*b*), the same experiment in a spaceship moving at a constant velocity produces the same result.

Caution. Don't confuse mass and weight! You know that if you are far enough away from any large mass, you will be *weightless*. What does that mean? Simply that if you placed a scale beneath you, it would read zero. Now, when you stand on a scale on the earth, what are you measuring? In Newton's terms you are reading the amount of *gravitational force* exerted on you by the earth's mass. So *weight is a force*.

Mass is related to the inertial properties of matter. Forces can tell you how much mass you have. For example, to move around weightless in a spaceship, you still have to put out an effort. Suppose you want to go from one side of the ship to the other. The easiest way is to push yourself off a wall with a small amount of force. You'll then drift to the other side, and to stop you'll need to push against the opposite wall. If you measure the amount of force and your acceleration, you can use Newton's second law to calculate your mass—your *inertial* mass. No matter where you are—on the earth, the moon, or in space—your inertial mass is *always* the same.

But, you might point out, the moon is also in free-fall, moving under the influence of gravity. If the moon's motion around the earth is its natural motion, why is the orbit curved rather than straight? To answer this question requires a look at geometry and its relationship to physics.

■ According to Einstein's ideas, what does weightlessness indicate about your motion?

7.3 *The Geometry of Spacetime*

How are geometry and physics related? Newton assumed that the geometry of the universe was flat. Einstein made no such assumption. He put this question up to *experimental* confirmation, for observations to tell us the geometry of spacetime.

Euclidean Geometry

Geometry derived from the practical surveying techniques of the Egyptians. So geometry, as we understand it, was developed from experience. For years people held that Euclidean geometry—and *only* Euclidean geometry—applied throughout space to physical measurements. Newton believed this assumption; he had no choice, for no other geometry had been yet devised.

Because of Euclid's parallel-line postulate (essentially, that two parallel lines when extended to infinity remain the same distance apart and will never meet), Euclidean geometry is flat. Thus we have the Pythagorean theorem for right triangles: the statement that the sum of the angles of any triangle *equals* 180° (Fig. 7.6*a*) is a key property of a **flat geometry.**

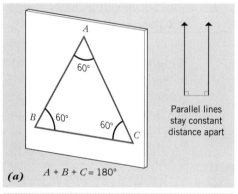

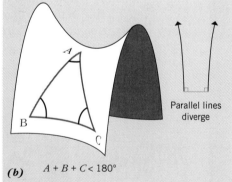

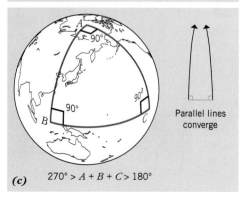

FIGURE 7.6 Properties of different geometries. (*a*) In a flat geometry the sum of the angles of a triangle always equals 180°. (*b*) In a hyperbolic (open) geometry the sum of the angles is always less than a 180°. (*c*) In a spherical geometry the sum of the angles is always greater than 180°. For example, the triangle drawn from two points on the earth's equator to the north pole contains 270 degrees.

Non-Euclidean Geometry

In 1829 the Russian mathematician Nikolai I. Lobachevski (1792–1856) pointed out that Euclid's parallel-line postulate was not unique. Instead, Lobachevski proposed a new postulate that allowed parallel lines to diverge but still resulted in a self-consistent geometry. In two dimensions his geometry has properties similar to the surface of a saddle (Fig. 7.6*b*), on which all parallel lines diverge when extended, affording a curved geometry sometimes termed **hyperbolic geometry.** When a triangle is drawn on a hyperbolic surface, the sum of its angles is *less than* 180°. Hyperbolic geometry is infinite, because extended parallel lines never meet. Both hyperbolic and flat geometries are for this reason called **open geometries.**

In 1854 Georg F. B. Riemann (1826–1866) devised yet another self-consistent geometry having a different parallel postulate. Riemann allowed parallel lines to converge when extended. The surface of a sphere has this **spherical geometry** or **closed geometry** (Fig. 7.6*c*), in which the sum of a triangle's angles is *greater than* 180°. Consider, as an analogy, the earth's surface with its imaginary lines of longitude and latitude. Two lines of longitude, both perpendicular to the equator (and so parallel to each other at the equator), intersect at the poles.

Both these non-Euclidean geometries are characterized by their curvature, which does not typify a flat, Euclidean geometry. The hyperbolic geometry has a negative curvature because it bends away from itself. It extends infinitely far. Spherical geometry, in contrast, curves in on itself. Because of its positive curvature, spherical geometry is finite but unbounded: it has a definite size, but no edge.

Consider again the earth's surface. You can travel around the earth's surface as many times as you like and in any direction you want without ever discovering a boundary. Yet the surface of the earth has a definite area. (Careful—I'm using two-dimensional examples here because they are easy to visualize, but the physical world exists in the four dimensions of spacetime. That's harder to picture, but the idea is the same.)

Local Geometry and Gravity

Newton had only one geometry at his disposal. By Einstein's time three general categories of geometry—hyperbolic, spherical, and flat—were available. Which one was the right choice to apply to the physical world? And how does this choice relate to gravity? Let's look at geometry and physics in a local region and then for the entire universe.

Imagine that you live in two dimensions on the earth's surface. You have no concept of a third dimension and no experience of it. You cannot conceive of an "up" that is off the surface. You and a friend do an experiment. You both stand on the equator some distance apart (Fig. 7.7). You both walk away from the equator on paths that are at right angles to the equator. And you both believe that the geometry of the world you live in is flat.

As you walk—being very careful to keep on lines at right angles to the equator—something strange happens. You are moving closer together! Yet by all the precepts of the Euclidean geometry you learned in school, you should be traveling on parallel lines and staying the same distance apart. What's happening?

You might still stick with the belief that your world is flat. Then you could explain your convergence by saying that there is a strange force of attraction bringing you and your friend together. You might even call this force "gravity" and see it as mysterious. Then you would be thinking like Newton.

Or you could say to your friend, "Perhaps our assumption about this world's being flat is wrong. Maybe it's actually curved. We are moving on straight lines at right angles to another line. These lines should be parallel. Yet we move closer together. That's not what should happen on a flat world. So the experiment is telling us that the world is not flat but curved. Then our approach has a natural explanation: it's our world informing us that it's curved."

No need for a mysterious force! An experiment has revealed the local geometry of the world in the region in which the experiment was performed.

The Curvature of Spacetime

How is spacetime curved? In Einstein's general theory of relativity, the local

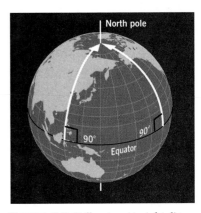

FIGURE 7.7 Following straight lines in two dimensions on a spherical geometry. Lines at right angles to the equator converge at the pole.

distribution of mass (and energy) determines the geometry of spacetime. All objects produce a curvature of nearby spacetime. And that curvature shows itself by accelerated motion, which Newton would say was caused by gravitational forces.

Einstein's theory also predicts that light rays will be bent by the curvature of spacetime near massive objects. To see this effect, let's go back into our Einstein elevator (Fig. 7.8*a*), which is accelerating by its rocket engine at 9.8 m/s/s. Put a laser in the front of the elevator to fire across its width at a target level on the other side. Will the light beam hit the target? No, it will strike below it (Fig. 7.8*b*).

Can you see the reason? As the light travels from one side of the elevator to the other, the elevator continues to accelerate "upward." The light path appears curved to an observer inside the elevator. Now, by the principle of equivalence, if you moved the elevator to the earth's surface, you shouldn't be able to tell the difference by an experiment inside the elevator. The light path will also bend, and by the same amount as before. So light paths should be deflected in all regions of curved spacetime, near all massive objects.

You should notice that the amount of bending in the elevator example above will be extremely small. The reason: light travels fast and crosses the width of the elevator in an extremely short time. Suppose, for instance, that the elevator were 3 meters wide. Then the light transit time would be a mere 10^{-8}s! During this time, the elevator would have moved only 5×10^{-6} m. So testing this prediction of general relativity on the earth would be a difficult job.

Spacetime Curvature in the Solar System

Where is the most warped region of spacetime in the solar system? Where the most mass is located: at the sun. Relativity predicts that the sun's mass deflects light rays away from Euclidean, straightline paths. How to test this? Take a picture of the stars near the sun during a total solar eclipse. Later, when the sun is in a different part of the sky, take a picture of the stars. Compare the angular

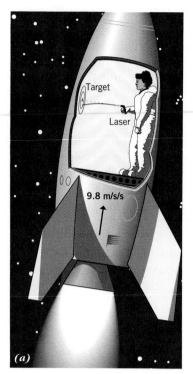
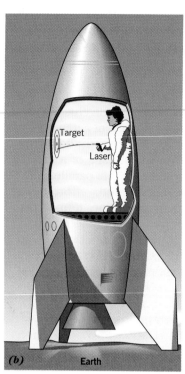

FIGURE 7.8 The bending of light in a gravitational field. In an accelerating spacecraft (*a*), a laser beam shot at a target on the opposite wall will hit below the target (*b*). By the principle of equivalence, the same will happen if a spacecraft is at rest on the earth.

separation of two stars close to but on opposite sides of the sun. With the sun present, the angular separation of the stars is *greater* than when the sun is not there (Fig. 7.9). Is this shift visible? Yes! Such observations have been made by many people over many years. They come out very close to Einstein's prediction from general relativity of 1.75 arcsec. (That's about 0.001 the angular diameter of the sun.)

It may strike you as a bit odd that the stars appear farther apart in angle when the sun lies between them. After all, aren't the light rays bent inward? Yes, after they pass the sun. On earth, we think that the light has traveled on a straight line. Our line of sight goes straight out as if the light path had not been bent. The light has been refracted (Section 6.2). The sun acts as a gravitational lens.

What's happening? In Einstein's view, light pursues a straight line in spacetime. But the geometry in which it travels is curved by the sun's mass. You can picture the sun's mass as creating a warp in the geometry of spacetime (Fig.

7.10). Light crossing the warp appears to us to take a curved path, although it is actually traveling on the shortest distance between two points through the nonflat region of spacetime. Thus we refer to the **curvature of spacetime.**

You can imagine the situation as analogous to that of a golf ball moving along a poor putting green filled with depressions. Between depressions, the surface is flat and the ball moves along a straight line. When it crosses a depression, the ball follows a path that bends, compared to the path on the flat surface. The amount of bending depends on the depth of the depression and the speed of the ball. In Einstein's general theory of relativity, mass and energy create local warps in spacetime. The amount and density of mass and energy determine how steep the depression is. As an object moves through a warped region of spacetime, it follows a straight-line, natural-motion path. In warped spacetime, this path appears curved to us. But no force acts on the object. This is essentially how Einstein understood the nature of the gravitational force locally.

Einstein considered the orbits of the planets in the same way. All the planets move on straight-line paths in spacetime. We view these paths as elliptical orbits in three dimensions. The paths appear curved to us because spacetime is warped, not flat. The amount of warping decreases as the distance from the sun increases. That's why the earth, for example, follows an orbit more strongly curved than that of Jupiter.

To sum up the difference in Newton's and Einstein's concepts of gravity: Newton concentrates on *forces,* Einstein on *courses.* Newton sees gravity as a force acting instantaneously between all matter, a force whose strength depends on the mass of the attracted object and the distance. Einstein focuses on the paths of objects in spacetime; such paths, for free-falling objects, don't depend on the mass of the object.

Experimental Tests of General Relativity

Einstein's model for gravitation would be no more than a fascinating idea if it did not make numerical predictions that

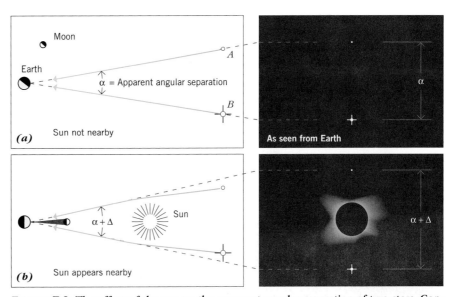

FIGURE 7.9 The effect of the sun on the apparent angular separation of two stars. Consider two stars (*A* and *B*) that have a measured angular separation (α) when the sun is not nearby. During an eclipse, when the sun lies between the two stars, they appear to be farther apart (angle α plus Δ) because of the bending of the light paths by the curvature of spacetime near the sun.

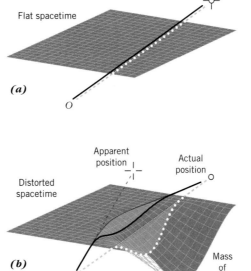

FIGURE 7.10 Einstein's explanation for the change in the stars' angular separation during an eclipse. (*a*) When the sun is not near the line of sight to the star, the region of spacetime through which the light rays travel to an observer (at *O*) is flat. (*b*) When the sun is near the line of sight, the region of spacetime near the sun is distorted. The light rays travel the shortest distance through this curved region. This path refracts that light so that an observer at *O* sees the star shifted to an apparent position away from the sun.

could be tested experimentally, predictions of effects unknown or unexplained by Newton's gravitation. Let's look at two essential solar system tests: the deflection of light in curved spacetime and the advance of Mercury's orbit. Both these phenomena occur because, in the solar system, spacetime is strongly distorted by the sun's mass.

Results for the deflection of starlight from measurements during solar eclipses have been mentioned already. Radio waves and light are essentially the same, so these deflections should also occur for radio signals from very distant celestial objects. The sun eclipses a few of these as it moves along the ecliptic. Radio astronomers, using interferometry (Section 6.3), can accurately (to about 0.1 arcsec) measure the angular deflection of radio waves. This experiment has demonstrated that the observed deflection is within 10 percent of the value predicted by general relativity.

Astronomers have known for a long time that the major axis of Mercury's orbit does not remain fixed in space with respect to the stars (Fig. 7.11). The major axis rotates around in the plane of the orbit. Part of this shifting arises from the gravitational attraction of the other planets on Mercury. But when this effect and others are taken into account, there remains a residual shift of 41 arcsec per century (which means that the orbit turns through an extra 360° in about 3 million years).

What causes the perihelion advance of Mercury's orbit? Perhaps an undiscovered planet, sometimes called Vulcan, orbiting within the orbit of Mercury. But no such planet has ever been definitively observed. General relativity predicts a motion because of the strong curvature of spacetime close to the sun. (Motion of the major axis due to curvature also happens for Venus, the earth, Mars, and the other planets, but it is much smaller for these planets than for Mercury.) The predicted value for Mercury is 43 arcsec per century. So the observed and predicted results agree to within a few percent.

■ What is the chief difference between Euclidean and non-Euclidean geometry?

mind that three basic geometries are possible: hyperbolic, spherical, and flat. Remember also that Einstein does not state which geometry must apply; he leaves this open to experimental confirmation. How to find out?

Cosmic Geometry

Imagine again living in two dimensions on the surface of a sphere. Start at any point and walk in a straight line away from it. Eventually you'll return to your starting point because a sphere has so much curvature that it comes back on itself. Suppose the universe has the same geometric properties (in spacetime) as a sphere's surface. If we sent out light signals, they would eventually return to us. Why? Just as for a two-dimensional surface, our four-dimensional spacetime, if curved enough, will close back on itself. (If the universe is closed, it does *not* have an edge, because it has no boundary. Consider again a sphere's surface: you can go around it many times and never find an edge.) We don't know for sure that our universe *is* closed—that depends on whether it contains enough matter and energy to curve it sufficiently. But if it does contain enough, then it will be finite and unbounded, just as the surface of a sphere is finite and unbounded.

Now turn this idea around. The density of mass (and energy) determines the curvature of spacetime. If this density has a certain critical value (or greater), then the universe curves back on itself and is closed. Einstein's general theory gives this critical density: it is roughly 5×10^{-27} kg/m^3. (That's about one hydrogen atom for every cubic meter of space.)

In 1917 Einstein constructed a model of the universe with a closed ge-

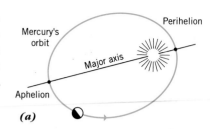

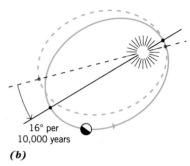

FIGURE 7.11 Perihelion advance of Mercury's orbit. The major axis of the orbit (*a*) rotates in space with respect to the stars (*b*), covering an angle of about 16° every 10,000 years.

7.4 Geometry and the Universe

So far I've described Einstein's picture of gravity in a small region of spacetime, such as near the sun. But his general theory applies also to the universe. With it, the geometry of the cosmos relates to the dynamics of the whole cosmos. Keep in

General Relativity

At a Glance

Deals with	Accelerated observers
Natural motion	Weightlessness
Gravitation	Result of curved spacetime
Spacetime curved by	Matter and energy

ometry. How different from Newton's model, which had to be infinite (Section 4.5)! But Einstein's model was also static because of special conditions he imposed on it. General relativity actually has two natural models of the universe, one expanding and one contracting. These are dynamic models; in a static model, the universe does *not* expand or contract. Nothing was wrong with general relativity, just Einstein's specific application of it this first time. A few years after Einstein proposed his model, astronomers discovered it was *wrong.* They found that the universe is not static but *expanding!*

A Quick Tour of the Universe

Before investigating the impact of general relativity's on cosmological ideas, let's look at the content and scale of the universe. (The rest of this book describes the contents of the astronomical universe in detail.) Because the universe is a big place, we need a long measuring stick. I'll use the most natural one: light travel time. You're probably familiar with the term **light year** (abbreviated *ly*), the distance light travels in a year. That amounts to about 9.5×10^{15} m.

The nearest star, Alpha Centauri, lies some 4 light years away. The stars near the sun and all those you can see in the night sky make up a disk of stars called the *Milky Way Galaxy.* The gravity of its contents—mostly stars, some 10^{12} of them—holds the Galaxy gracefully together. The Galaxy has a diameter of roughly 100,000 ly. The nearest galaxy that resembles our Milky Way Galaxy is the Andromeda Galaxy (Fig. 7.12)—so named because it appears in the constellation Andromeda.

The Andromeda Galaxy and the Milky Way Galaxy are parts of a local group of galaxies, bound by gravity. This neighborhood set of galaxies has a diameter of some 3 million light years (Mly). In this century, astronomers have found that the universe contains many clusters of galaxies (Fig. 7.13). The largest clusters are tens of millions of light years in diameter. Clusters of galaxies are spaced out, on the average, by hundreds of millions of light years. We see those galaxies in all directions, and with our

Einstein's Model

At a Glance

Space and time	Integrated into four dimensions
Geometry	Non-Euclidean (curved) possible
Universe	Finite or infinite; expanding
Natural motions	Straight lines in curved spacetime

FIGURE 7.12 The Andromeda Galaxy, the nearest galaxy to ours with the same overall structure, is about 2 million light years from us. Most of the visible matter is in the form of stars. (Courtesy of Jim Riffle, Astro Works Corporation; photo taken with an Astromak™ telescope.)

biggest telescopes we can detect them at distances of several *billion* light years.

Light travels at a fast but finite speed. So when we look out into space, we are looking back in time. The farther out we peer, the deeper into the past we see. The light we receive now from clusters of galaxies left them many millions to billions of years ago. Astronomers have found that these clusters of galaxies are all moving away from one another. This motion tells us that the universe is expanding.

The Expanding Universe

To understand how astronomers discovered the expanding universe, you'll need to know just a little about galaxies (more in Chapter 19). Here's the main point: galaxies are made (mostly) of stars, bound together by gravity. Galaxies are spread throughout the visible universe, each one a marker of a distant region of space and time. In this century, astronomers found that these cosmic markers are moving apart, which implies that the universe is expanding.

How so? Observations with large telescopes provide data about two properties of galaxies: their distances and their velocities relative to us. The velocities measured, called *radial velocities,* are those along our line of sight only. Radial velocities can be either approaching or receding. For all distant galaxies, the radial velocities are those of *recession.* Remarkably, the distances and radial velocities of galaxies are tied together—the surprising evidence that the universe is expanding. (Chapter 19 deals in detail with how distances and radial velocities of galaxies are measured.)

Hubble's Law

Beginning in 1912, Vesto M. Slipher (1875–1969) began a project of measuring the radial velocities of galaxies. By 1928 he had observations of more than forty galaxies, and a trend emerged: most galaxies appeared to be moving away from the Milky Way Galaxy.

At about the same time, Edwin P. Hubble (1889–1953) had determined the distances to some galaxies and had noted an unexpected direct relationship

FIGURE 7.13 The central region of the nearby Coma cluster of galaxies (distance about 270 Mly). This image has been processed to show the true colors of the galaxies. Note the two supergiant galaxies near the center with their distinct reddish color. About 300 galaxies are visible here. The bluish star to the upper right is in our Milky Way. (Courtesy of L. A. Thompson, Institute for Astronomy, University of Hawaii.)

between radial velocity and distance: the farther out a galaxy, the greater its radial velocity. That is, the radial velocity is directly proportional to the distance. In later collaborative work Hubble and Milton L. Humason (1891–1972), using the 100-inch (2.5-m) telescope, added more data to support the trend. This relationship, now known as **Hubble's law,** states that the distance to a galaxy and its recessional velocity are directly proportional (Fig. 7.14). The number connecting the distance and velocity is called **Hubble's constant** and is usually indicated as H. It's the slope of the line on a plot of radial velocity versus distance (such as Fig. 7.14). The steeper the slope, the larger the value of H.

Today, astronomers believe that H lies between 15 and 27 km/s/Mly. The exact value is disputed; Section 19.4 deals with the controversy. This book uses 20 km/s/Mly. This value means that for every million light years we look away from our Galaxy, the radial velocity

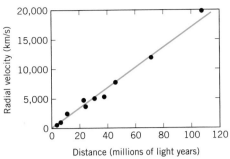

FIGURE 7.14 Hubble and Humanson's original plot of the distance–velocity relationship, now known as Hubble's law: solid circles indicate the measured recessional velocities and distances for some nearby galaxies. The straight line indicates the trend of the data points; its slope is too steep compared to modern data.

of other galaxies is 20 km/s higher. For example, a galaxy at 5 Mly travels at 5 × 20, or 100 km/s radial velocity, whereas one at 10 Mly moves at 10 × 20, or 200 km/s.

In equation form, Hubble's law is

$$H = \frac{v}{d}$$

where v is the recessional velocity (in km/s), d the distance (in Mly), and H is Hubble's constant. Conceptually, this relation simply says that H is the ratio of v to d. And if you know one of them and H, you can find the other.

The discovery of this trend caused a radical revision of Einstein's static model for the universe, which did not allow for the change in distances between galaxies at all. Einstein later admitted that his 1917 static model was the "biggest mistake of my life."

The Meaning of Hubble's Law

At face value, Hubble's law leaves us fixed in the center of the universe. Having been thrust away from the center by Copernicus, astronomers felt somewhat uncomfortable at being repositioned there. Was our Milky Way Galaxy now enthroned again, and the universe centered on it?

A simple argument demonstrates that our Galaxy does not really have a privileged status. From our viewpoint, the rest of the universe appears to be retreating from our Galaxy. But if the expansion is *uniform,* then the view from any other galaxy will be the same (Fig. 7.15). Transported to another galaxy with the usual bundle of tools, an astronomer would plot the same Hubble's law as from our Galaxy. Another galaxy appears, to those in it, as the "center" of the expansion, so no privileged position actually exists. If the galaxies suddenly turned around and moved back, they'd all arrive together at the same time.

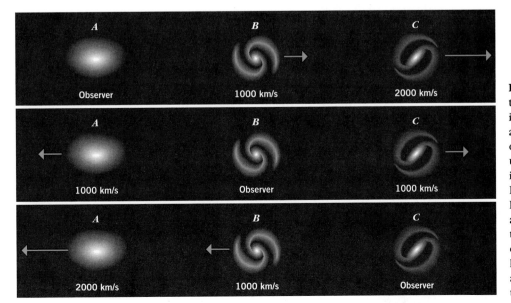

FIGURE 7.15 Uniform expansion in two dimensions. Imagine three galaxies (*A, B,* and *C*) equal distances apart and arranged in a line. Locate yourself on galaxy *A*. If these galaxies expand uniformly, then *B* appears to be receding at 1000 km/s and *C* at 2000 km/s. Now imagine yourself on galaxy *B*. From there, *A* appears to move away at 1000 km/s and so does *C*. In contrast, the view from galaxy *C* has *B* receding at 1000 km/s and *A* at 2000 km/s. An observer on any galaxy sees a Hubble-law relation between distance and recessional velocity.

Enrichment Focus 7.1

HUBBLE'S CONSTANT AND THE AGE OF THE UNIVERSE

You can infer from the measured value of H the time since the **expansion of the universe** began. To do this simply, you need to make the assumption that the expansion rate is uniform. (This assumption does not apply to the real universe. For any realistic model, velocities must decrease with time. However, this assumption allows a simple calculation of the oldest possible age of the universe from the present value of H.)

At a constant velocity, the distance traveled d is equal to the velocity v multiplied by the time t. In algebraic form (Focus 4.1),

$$d = vt$$

Conversely, the travel time is

$$t = \frac{d}{v}$$

For example, if you travel in a car from Boston to New York, a distance of 400 km, at a constant speed of 80 km/h, the trip takes 5 hours (400/80 = 5).

Hubble's law written as an equation is

$$v = Hd$$

Here v is the radial velocity of a galaxy, d the distance to the galaxy, and H Hubble's constant. If H has the units kilometers per second per million light years (km/s/Mly) and d is given in millions of light years (Mly), then v comes out in kilometers per second (km/s). Now compare Hubble's law written as

$$t = \frac{d}{Hd}$$

$$= \frac{1}{H}$$

to the trip formula

$$t = \frac{d}{v}$$

The travel time equals $1/H$. If H is 20 km/s/Mly, then $1/H$ equals 1.5×10^{10} years. To get this result, we must put in the conversion factors to put everything in the same units:

$$t = \frac{1}{H}$$

$$= \frac{1}{20 \text{ km/s/Mly}}$$

$$\times 10^6 \text{ ly/Mly} \times 10^{13} \text{ km/ly}$$

$$= 4.5 \times 10^{17} \text{ s}$$

$$= \frac{4.5 \times 10^{17} \text{ s}}{3 \times 10^7 \text{ s/y}}$$

$$= 1.5 \times 10^{10} \text{ y}$$

So the expansion of the universe started about 1.5×10^{10} years ago!

This conclusion is valid *only* if our assumption of uniform expansion is correct. However, even if the expansion were a bit faster in the past than it is now, the value of 1.5×10^{10} years is still in the vicinity of the actual time since the expansion began.

This analysis implies that the expansion of the universe that we see now began at a finite time in the past and at the same time for all parts of the cosmos. Having a value of Hubble's constant of roughly 20 km/s/Mly, we can estimate when the expansion began (Focus 7.1): 15 billion years ago. Astronomers call this beginning of the universe's expansion the Big Bang.

Caution. When first confronted by the expansion of the universe, people commonly have the misconception that the universe expands "into empty space." That's wrong! It is *space itself* that is expanding. And it does not expand into "something" or even "nothing;" the expression "expand into" is meaningless in this context.

■ **What observational proof exists for the expansion of the universe?**

7.5 Relativity and the Cosmos

With the discovery of the expanding universe, cosmologists began devising models that would account for the expansion in a manner consistent with Einstein's general theory of relativity and Hubble's law. In these models, Hubble's constant and the geometry of spacetime (whether flat, hyperbolic, or spherical) are physically interrelated. Let me show you how, using the concept of *escape speed* (Section 4.5 and Focus 4.4).

Escape Speed and the Critical Density

Consider throwing a ball off the earth's surface. If the ball's speed is less than the escape speed, it will slow down and fall back, regardless of how large or small its mass is. A ball thrown with a speed greater than the escape speed will slow down a bit while it is still influenced strongly by the earth's gravity and then coast outward as the earth's gravity affects it less and less. If it doesn't run into anything, the ball will travel out to infinity and never return to the earth.

Now consider the universe and the galaxies within it. Any one galaxy in the universe is analogous to the ball. The galaxies were once all "thrown away" from one another, for we now see an expansion. Consider a galaxy at some distance from us. Newton showed the net effect of all matter spread within some space: it acts as if this total mass were concentrated at the center (as long as the matter is distributed uniformly). If there is enough mass within that space, the escape speed will be faster than the expansion speed. Note that "enough mass within that space" means a high enough density. (The meaning of *density* here is related to how closely galaxies are packed together; the closer they are, the higher the density.)

If the density is large enough, the galaxies will not have escape speed; the expansion velocities will decrease with time and eventually reverse as gravity herds all the galaxies together. If the average density is too low, the galaxies will have more than escape speed; gravity will never bring the galaxies all together, and the expansion will continue indefinitely.

Hyperbolic, flat, and spherical geometries (Fig. 7.16) correspond physi-cally to the cases of greater than, equal to, and less than escape speed. From Einstein's general relativity and the observed value of H, we can calculate the density required for the cosmos to be flat, a density known as the *critical density*. This *calculated* density can then be compared to the *observed* cosmological density (which must include energy as well, since $E = mc^2$). *If the observed cosmological density is greater than the critical density, the universe is spherical and closed. If less, the universe is hyperbolic and open. If they are exactly equal, the universe is flat.* If H equals 20 km/s/Mly, the critical density is 5×10^{-27} kg/m^3.

Whether the universe is open or closed depends on the density of energy and matter within it. A large enough density curves spacetime so that it wraps around itself to close the cosmos. So if we can observe the average density of matter and energy in the universe, we will have an *experimental* basis for finding out the appropriate geometry of spacetime! Observations of Hubble's constant combined with general relativity provide a way to find out whether the universe is open or closed.

The Future of the Universe

What's the future for the universe? In each geometry, some special moment marks the start of the expansion. In a closed universe, the rate of expansion slows and eventually stops; then the universe contracts. In a hyperbolic universe, the rate of expansion doesn't decrease as rapidly, and it never stops. Even after in-

Expanding Cosmos	
At a Glance	
Evidence	Redshifts of galaxies
Pattern	Hubble's law
Rate (now)	Hubble's constant
Future	Expand forever, or contract

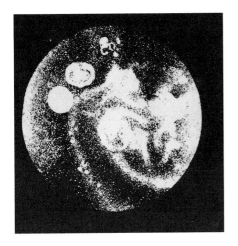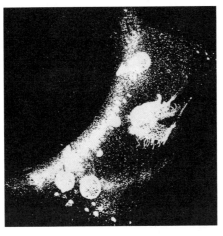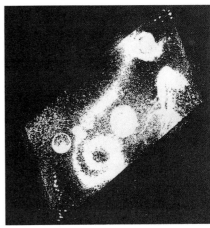

FIGURE 7.16 Views of different geometries for the cosmos. These computer-generated images show closed (left), hyperbolic (middle), and flat (right) geometries. (Courtesy of the Lawrence Berkeley Laboratory, University of California.)

finite time, the galaxies are still moving apart at a finite velocity. In the borderline case of a flat universe, the expansion slows down just enough to permit the universe to come to a stop after infinite time of infinite expansion.

We seem to have two possible cosmic destinies. In one, the expansion grinds on forever, and the universe gradually thins out. In the other, the expansion slows down, stops, and reverses, and the universe collapses. During the time of diminishing size, the galaxies rush together into a dense conglomeration with (theoretically) zero radius. The universe eventually would crush every-thing into high-energy light—and many weird forms of mass (Chapter 21). This final collapse is sometimes called the Big Crunch.

Results to date have been a mixed bag, some supporting an open, others a closed universe. The case is not yet set-tled; the issue will arise again in Chapters 19 and 21. The crucial problem is that of nonluminous matter, which may fill space enough to make the actual density greater than the critical density.

■ **What role does density play on the escape speed of galaxies with respect to the expansion of the universe?**

Key Concepts

1. Newton's concept of natural motion (and so of forces) rested on the assumption that the geometry of the universe was Euclidean (flat). Einstein, in developing his theory of gravitation—the general theory of relativity—challenged this assumption by allowing any geometry to describe the cosmos.

2. Before developing the general theory, Einstein worked out the special theory of relativity, which relates to the laws of physics as seen by observers moving at constant velocities (no acceleration). Consequences of the special theory are the unification of space and time (in four-dimensional spacetime) and the intimate relationship between matter and energy and their transformation.

3. The general theory deals with gravitation; it is based on the principle of equivalence, which states that accelerations due to gravity cannot be distinguished from accelerations from other forces. The principle of equivalence implies that gravity disappears in free-fall, when one is weightless. Free-falling is then viewed by Einstein as natural motion; gravity appears when the spacetime in which free-fall takes place is curved rather than flat.

4. Einstein's concept of gravity as the curvature of spacetime has been tested and found to predict accurately such effects in the solar system as the bending of starlight near the sun. Mass and energy cause the warping of spacetime.

5. The general theory allows the cosmos to have any one of three general geometries: hyperbolic, spherical (closed), or flat (Table 7.1). Which one actually applies can be determined by observations.

6. Astronomers have found that the galaxies are moving apart in a systematic way: the farther away a galaxy is, the faster it recedes. This relationship, expressed in Hubble's law, implies that the cosmos is expanding. The rate of expansion is described by Hubble's constant, which has a value between 15

TABLE 7.1 GEOMETRY OF THE UNIVERSE AND OBSERVATIONS

ASPECT	HYPERBOLIC	SPHERICAL	FLAT
Extent in space	Infinite	Finite	Infinite
Extent in time	Infinite	Finite	Infinite
Topology	Unbounded	Bounded	Unbounded
Average density[a]	$<5 \times 10^{-27} \text{ k/gm}^3$	$>5 \times 10^{-27} \text{ kg/m}^3$	$5 \times 10^{-27} \text{ kg/m}^3$
Age now	>15 billion years	<15 billion years	15 billion years
Future	Expansion forever	Expansion stops; collapse	Expansion forever

[a]Based on a value of $H = 20$ km/s/Mly.

and 27 km/s/Mly; the value is the same measured in different directions in space, which implies that the expansion is uniform. The beginning of this expansion was a cosmic explosion called the Big Bang.

7. From the value of the Hubble constant, we can estimate the time since the beginning of the expansion and so the "age" of the universe. If it is assumed that the Hubble constant has been constant and has a value of 15 km/s/Mly, then the expansion started at most 15 billion years ago.

8. If the geometry of the cosmos is hyperbolic or flat, the universe will expand forever; if closed, the expansion will stop at some time in the future, whereupon the universe will contract. To find out which is the case, compare the *critical density* (derived from Hubble's constant and general relativity) to the actual average density of the cosmos. If

the actual is greater than the critical, the universe is closed; if less than the critical, it is open.

9. The geometry of the universe determines its future; the actual density of the universe relative to the critical density (determined from Hubble's constant and general relativity) indicates the overall geometry. If open or flat, the universe will expand forever, whereas a closed universe will eventually stop expanding and then contract.

10. The concept of escape speed is one way to understand the physical meaning of the geometrical curvature of the cosmos. If the Big Bang had enough energy to accelerate matter to great enough speeds, the universe will expand forever (open). If not—that is, if the energy was too small, hence the speeds too low—matter will eventually collapse (closed). If the energy were just right, the speeds would just equal the escape speed, and the expansion could go on forever (flat).

Key Terms

closed geometry
curvature of spacetime
expansion of the universe
flat geometry

general theory of relativity
gravitational mass
Hubble's constant
Hubble's law

hyperbolic geometry
inertial mass
light year
open geometry

principle of equivalence
spacetime
special theory of relativity
spherical geometry

Study Exercises

1. Describe standing on the earth in Newton's terms and in Einstein's. (Objectives 7-1, 7-2, and 7-3)
2. Describe orbiting the earth in Newton's terms of gravity and contrast this to Einstein's description. (Objectives 7-1, 7-2, and 7-3)
3. Describe the motions of the planets around the sun in Newton's terms and in Einstein's. (Objectives 7-1, 7-2, and 7-3)
4. In Einstein's model, how is the geometry of the universe related to its average density? (Objective 7-9)
5. Suppose you were an astronomer in the Andromeda Galaxy with the same tools as an earthbound astronomer. You measure the recessional velocities from galaxies and your distance to them. What result would you expect? (Objectives 7-6 and 7-7)

6. Compare physically the case of the universe expanding and then contracting (a closed geometry) to the situation of firing a rocket off the earth with less than escape speed. (Objective 7-10)
7. In what sense is a telescope a time machine? (Objective 7-4)
8. Consider the following situations and state whether the condition of weightlessness characterizes each one: (a) a spacecraft orbiting the earth, (b) the moon orbiting the earth, (c) the earth orbiting the sun, (d) Jupiter orbiting the sun, and (e) a spacecraft orbiting the sun. For each of these cases, what can you conclude with respect to whether the object is following a straight-line path in spacetime? (Objectives 7-2, 7-3, 7-4, and 7-5)

9. How can the value of Hubble's constant be used to infer an age for the universe? Describe the procedure; calculations not needed! State what assumptions are made. (Objectives 7-8 and 7-9)
10. In what sense is free-falling motion natural motion in Einstein's view? (Objective 7-3)
11. State verbally Einstein's relation between matter and energy. Does it violate the conservation of energy? (Objective 7-11)

Problems & Activities

1. Suppose you hear on the radio that astronomers have discovered a galaxy 500 Mly away. Use Hubble's law to estimate how fast the galaxy is receding.
2. For $H = 20$ km/s/Mly, at what distance would a galaxy's recessional speed be 150,000 km/s, or half the speed of light?
3. What would be the inferred age of the universe if the Hubble constant were 30 km/s/Mly?
4. Imagine that the mass of your body were completely converted to energy. About how much energy would be released? Assume this conversion takes place in one second. How many 100-watt light bulbs could be lit up?
5. The Andromeda Galaxy has an angular diameter of about 3°. What is its size-to-distance ratio? How many of its diameters fit into its distance from us. Our Galaxy has a diameter of some 100,000 ly. What do you estimate as the distance to the Andromeda Galaxy?
6. Suppose you placed the Andromeda Galaxy at a distance of 200 Mly. What would be its angular size? What would be its recessional speed?

See for Yourself!

The accompanying table gives the recessional speeds and distances for a small sample of bright galaxies — those that are fairly close to us. Plot these points in graph paper, using distance on the horizontal (x) axis, and recessional speeds on the vertical; (y) axis. When you've plotted all the points, draw a "best fit" straight line through them. Find the slope (rise over run) of this line. Then answer the following questions:

1. What is the Hubble constant from these data?
2. What is the recessional speed of a galaxy at 50 Mly?
3. The galaxy Messier 65 has a recessional speed of 700 km/s. What would be its distance?

DATA ON SELECTED GALAXIES

GALAXY NAME (NGC NUMBER)*	DISTANCE (Mly)	RECESSIONAL SPEED (km/s)
55	10	120
628	55	520
2903	31	490
3521	42	650
4472	72	850
1313	17	270
4486	72	1190
4945	23	280
5055	35	570
4631	39	630

*New General Catalog

PART ONE *Epilogue*

I have guided you on a long trip through space and time. I hope that you have found it as fascinating as I do. I am amazed at how our cosmological ideas grow from simple observations of the sky to working models of the universe.

What distinguishes current concepts of the cosmos from those of the past? Basically, the demand that conceptual models both explain and predict what is observed. Such was not the case in some ancient cosmologies, such as the Babylonian. Ptolemy's geocentric picture represents the first try at a complete scheme intended to explain and predict planetary motions. But Ptolemy did not believe that his model corresponded to the real world, that the planets really moved on epicycles and deferents.

Starting with the work of Copernicus, astronomers began to devise models that corresponded closely to reality, as determined by observations. Kepler believed that a force from the sun actually pushed the planets in their orbits; he made the first try at a physical unification of the cosmos. Newton achieved that goal with gravitation and his laws of motion. But Newton would talk only about the effects of gravitation, not its cause. Einstein took the deeper step in the general theory of relativity. He envisioned gravitation as a manifestation of the curved geometry of spacetime, rather than a force. In so doing, he rediscovered the importance of geometry in the physical world—an attitude first taken by the Greeks. So Einstein's ideas were truly radical and transformed markedly our model of the cosmos.

That model relies on the work of modern astronomers with large telescopes. We have discovered that the universe contains innumerable galaxies, each holding billions of stars. These galaxies are moving away from us and from one another: The universe is expanding. By applying Einstein's theory of relativity to the expanding universe, we can consider its past and ponder its future.

Scientific models form the core of contemporary scientific thought and will be used throughout this book. When you run across such a model, be sure that you can state the aesthetic, geometric, and physical bases of the model, the assumptions behind it and their reasonableness, and the key observations the model tries to explain and how well it does so. You should also be able to make predictions from the model.

Finally, remember that *no scientific model is ever final*. It is always subject to revision or complete replacement. This fact underlies the evolution of models, as you've seen in this part with our conceptions of the cosmos. ■

The Unifying View

ASTRONOMICAL CONCEPTS

All we see directly are angular relationships among objects on an apparently two-dimensional sky.

The measurement of distances is crucial to all aspects of astronomy:
   Angular sizes are size-to-distance ratios.
   Angular speeds are speed-to-distance ratios.
   Light conveys to us information about astronomical objects.
   The astronomical cosmos has an order, unified by physical laws.

INFORMATION CONCEPTS

We perceived the world through our senses, sometimes aided by instruments.
   Telescopes are the main instruments used by astronomers to deepen our view of the cosmos.

Data have no meaning until transformed into information.
   Observations are data that have been interpreted.
   All observations involve some error or uncertainty; the smaller the uncertainty in an observation, the more information the observation contains.

Science is a knowledge-based system for finding meaning in the patterns of nature inferred from data.

Model building is the essential process by which science discovers knowledge and reveals information.
Models explain and predict observations.
Models are metaphors, not reality.
Models are never complete.
Models are best when they are beautiful.

ENERGY CONCEPTS

Energy is the ability to do work and transform matter.

The total energy of a system is conserved; it has a constant value.

Energy comes in many forms, such as potential, kinetic, thermal, radiative:
Gravitational potential energy.
Kinetic energy of moving masses, such as planets.
Electromagnetic energy binding electrons to nuclei of atoms.
Radiative energy carried by photons, the greater with higher frequency.
Mass is energy ($E = mc^2$).

Transformations occur among the various forms of energy, and energy can be transferred from one location to another.
Atoms and molecules absorb and emit light, usually at special energies.
A planet's amounts of kinetic and potential energy change as it orbits the sun.
Mass and energy warp spacetime.

MOTION AND FORCE CONCEPTS

Everyday static and moving systems are explained and predicted using Newton's laws of motion:
Inertia is the basis of natural motion.
Net forces result in the acceleration of masses.
Accelerations are changes in velocities of masses.
Forces act in pairs.

Newton's laws of motion plus his law of gravitation explain and predict the motions of the masses in the solar system in a unified way.
Gravitation acts as the inverse square of the distance between masses.
Knowledge of these laws permits us to send spacecraft to other solar system bodies in a predictable way.
Kepler's laws are an alternative expression of forces and gravitation.

Einstein unified space and time in the cosmos:
By relating the geometry of spacetime to natural motions, he created a new theory of gravity.
The general theory of relativity predicted an expanding cosmos.

QUANTUM CONCEPTS

On a very small scale, matter is made of discrete particles.

On a very small scale, energy comes in discrete units.

Physical phenomena are governed by a few basic interactions of matter and energy, often at the quantum level.
Optics is the manipulation of light by its interactions with matter.

Electromagnetic energy has both wave and quantum properties.

Improve Your Night Vision

SUMMER

Fred Schaaf

A giant triangle of stars shines bright enough and high enough on summer evenings to be seen even in big cities on all but very hazy nights. In addition to being a lovely sight and a guide to other star patterns, this triangle is highly instructive about one important aspect of how stars differ.

The Summer Triangle is composed of Vega, a star in the tiny constellation Lyra the Lyre; Altair, a star in Aquila the Eagle; and Deneb, a star in Cygnus the Swan. In June and July, you'll find the triangle partway up the east sky at nightfall, with Vega on top, Altair to the lower right, and Deneb to the lower left as you face east. In the middle of August and September evenings, first Vega and then Deneb passes near the zenith (the overhead point) as seen from the earth's midnorthern latitudes. Later in the night, the triangle heads down the west sky (because, of course, the earth's rotation carries us east), Vega still leading the way.

The most interesting thing about this huge star pattern is that while its stars appear to be moderately different in brightness, one of them is actually tremendously different—tremendously brighter. The key here, of course, is distance. The uniformed observer might assume that all the stars—some bright, some faint—were at the same distance. In reality, although Altair appears a little brighter than Deneb, the former is one of our closer stellar neighbors, just 16 ly away (Vega is 25 ly away, also close). Deneb is roughly 100 times farther than Altair and therefore has a spectacularly brighter "luminosity" (measure of the true energy output of stars). If Deneb were as close to us as Altair, it would rival a half-moon in brightness in our sky! Consider these realities behind the appearances as you gaze upon the Summer Triangle.

Not until the invention of the spectroscope in the nineteenth century did astronomers have any clues about how far away various stars were. Other information that was not known until quite modern times includes the

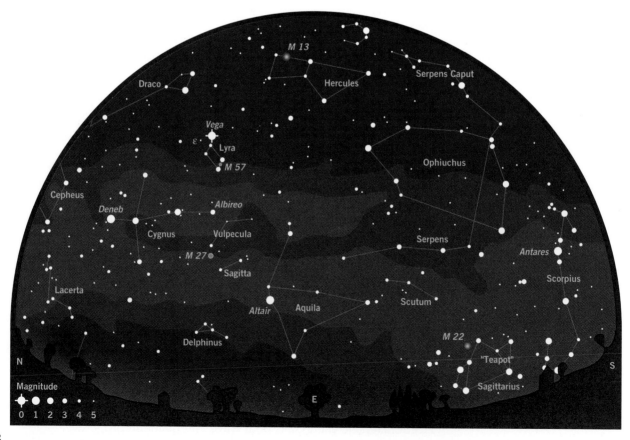

explanations of the shape of our Milky Way Galaxy and why we see a softly glowing Milky Way band of light stretched across the heavens at certain times, especially on summer evenings.

If you want to get a good view of the Milky Way band for yourself, pick a summer evening that is fairly haze-free (look for a deep blue sky before sunset), with the moon not above the horizon (the week between last quarter moon and new moon is best). Then, be sure you are miles away from the lights of any sizable city. Under these good conditions, you'll be amazed to see the dreamy grandeur of the band, extending especially prominently from high up in Cygnus the Swan down to the teapot *asterism* (unofficial star pattern) in Sagittarius the Archer, low in the south.

What causes the soft glow? You can answer this for yourself by lifting binoculars to any of the brighter patches of Milky Way: what you see is stars, stars, . . . , and more stars. The Milky Way glow is produced by the combined radiance of distant stars that are mostly too dim to be seen individually with the naked eye.

But why is the glow localized to a fairly narrow band, which is highest in our sky in summer and winter (though dimmer in winter)? A *galaxy* is an immense congregation, usually consisting of billions of stars, revolving around a common center of gravity. Astronomers have learned from many clues that ours is a spiral galaxy—a structure that looks somewhat lens shaped from a side view because most of the stars are concentrated in an equatorial plane, though toward the middle there is a central bulge (Fig. 18.1 on page 414). The glow of distant stars in our galaxy looks like a band extending around the heavens because we ourselves lie within the equatorial plane of the galaxy. Only

when we look toward the middle of the galaxy, in Sagittarius, do we see the band widen greatly, as a result of the central bulge. And in winter, we are looking away from the center, toward dimmer, outlying parts of the Milky Way—so the band is fainter.

By the way, when you look at the summer Milky Way on a good country night you will notice much structure in it—particularly a huge dark split that begins up in Cygnus. This Great Rift is actually caused by clouds of unlit gas and dust. We see similar lanes in our views of other spiral galaxies that are positioned edge-on with respect to us.

There is more to see and learn from in the summer sky. In the hours after midnight in the last week of July, Delta Aquarid meteors are impressive. From about August 9 to 14 the main part of the famous Perseid meteor shower is even better (for information on the Perseids, refer to "See for Yourself" at the end of Chapter 12; for more on meteors, see the Autumn essay at the end of Part Two).

But meteors are best seen far from city lights. What can you do if you have a hazy summer night and/or live in a fairly large city? Binoculars can show you distinct star colors, double stars, open star clusters, and much more, which we'll discuss in the other seasonal essays.

Double stars and detail on planets can be seen especially well on many moderately hazy summer nights because the haze is indicative of a still atmosphere and therefore of good "seeing"—that is, steadiness of images. Turn binoculars or telescope on the double star Epsilon Lyrae near Vega and the double star Albireo in Cygnus. For the rest of the 1990s, Jupiter will be conveniently placed in the evening sky for at least part of the summer—no other planet can show so much detail of cloud structures on its globe in telescopes.

PART TWO

THE PLANETS: PAST AND PRESENT

PART TWO

THE PLANETS: PAST AND PRES

T HE SPACE PROGRAMS OF the United States and the then Soviet Union energized my life with their amazing results. I watched in awe as the TV showed the pitted moon from a spacecraft plunging into it. And I recall myself and fellow graduate students, beers in hand, unable to look away from the fuzzy images of Neil Armstrong stepping gently onto the lunar surface. Flyby and robot lander missions, though, easily outstripped these first moments of men in space. Their fine photos provided a new vision of our planetary neighbors.

Part Two focuses on the members of the solar system. I start with the earth, the planet we know best. Our world provides the model and method to examine the other earthlike planets, including our companion, the moon. Then the focus shifts to the alien worlds, the planets like Jupiter. These bodies are compared in an evolutionary context, aided by the recent information from space missions. Then we take a brief look at the debris that floats among and around the planets: rings, comets, asteroids, and meteoroids. This part closes with current models of the formation of the solar system—as a natural aftermath of the birth of the sun.

Everything we see now has evolved since the birth of the solar system. By delving into the present physical properties of all heavenly bodies, we can infer what they might have been like in the past and what their futures might bring. Our familiarity with the earth colors our picture of the evolution of the other planets, a scenario outlined by basic physical and chemical ideas. The goal is to fit the facts together to infer how the whole solar system has evolved. The modern model for the solar system's origin and evolution implies that planets may form as a natural result of starbirth. So many other solar systems may exist in the Milky Way Galaxy—worlds on which other creatures may also wonder about the puzzle of the cosmos around them. ∎

PART OBJECTIVE: Outline an evolutionary scenario for each of the major objects in the solar system and for the solar system as a whole.

8-1 Describe and apply one method for determining the earth's mass and density; apply the concept of escape speed to the earth.

8-2 Sketch the interior structure of the earth, indicating the composition of each general region, and argue that the earth's interior structure implies that it all must have been molten at one time.

8-3 Argue from at least two observations that the earth's core probably has a metallic composition.

8-4 State the estimated age of the earth, with a range of uncertainty, and explain the method by which this age is inferred using radioactive decay and half-life.

8-5 Describe the properties of magnetic fields, apply these to the overall structure of the earth's magnetic field, and present a possible model for the field's source.

8-6 Describe at least two ways in which the earth's atmosphere affects astronomical observations and two ways in which it affects the earth's surface environment.

8-7 Explain how the earth's atmosphere acts like an insulating blanket that keeps the earth's surface relatively warm.

8-8 Outline a possible model for the evolution of the earth's crust and interior, with an emphasis on heat flow.

8-9 Outline a possible model for the evolution of the earth's oceans that ties in with a broader view of the earth's history.

8-10 Outline a possible model for the evolution of the earth's atmosphere; in-dicate how humankind affects the atmosphere now.

8-11 Summarize the physical processes that affect the evolution of the earth's atmosphere, crust, and interior (especially the outward flow of heat) and serve as a model for the evolution of earthlike planets.

8-12 Describe the interactions of charged particles and magnetic fields, and apply these ideas to the earth's magnetosphere, emphasizing its interaction with the solar wind and to other astrophysical situations.

Central Question
What are the basic physical features of the earth, and how have they changed since our planet's formation?

The Earth:
An Evolving Planet

We travel together, passengers on a frail spaceship, dependent on its vulnerable reserve of air and soil; all committed for our safety to its security and peace; preserved from annihilation only by the work and love we give our fragile craft.

ADLAI E. STEVENSON

The earth is a tiny planet, whirling around one ordinary star. And no longer the center of the universe, as people believed until fairly recently. But that change in cosmic position does not mean we should value the earth any less. For the earth is our delicate ship, protecting us on our long passage through space.

As our home, the earth is a perfect planet because we evolved on it. It has just the right range of temperatures, just the necessary atmospheric composition, just the ideal amount of water to foster living creatures. Yet this planet did not always make such a comfortable abode. Our earth has changed tremendously since its birth, and we are a part of that change. A primary scientific goal of examining the earth in fine detail is to infer the history of the earth and to mull over its future and our impact on its future.

This chapter looks at the physical makeup of the earth. We live here, so we know this planet in more detail than any other. Our present understanding of the earth indicates that it is a highly evolved planet. It has altered dramatically in its physical structure since its formation, mainly from the flow of energy from its interior to its surface. We will use our home planet as the basis of comparison for understanding the makeup and evolution of the other earthlike planets.

FIGURE 8.1 The whole earth from space, showing Africa and Saudi Arabia. Note the heavy cloud cover, which is typical for the earth. The spiral cloud patterns arise from the global circulation of the atmosphere. (Courtesy of NASA.)

8.1 The Mass and Density of the Solid Earth

Since the time of Pythagoras, astronomers have known that the earth's general shape is round (Fig. 8.1). In his *Almagest* (Section 2.4), Ptolemy compiled an extensive list of observations indicating that the earth was spherical. Christopher Columbus (1451–1506) knew that the earth was round and thought he could find a shortcut to the Indies by sailing to the west. His mistake was to use

Ptolemy's value for the earth's circumference, which was smaller than the actual value. So the voyage was much longer than Columbus had expected, and he didn't get where he wanted to go (nor did he ever realize where he ended up)!

The earth is not exactly spherical, because it rotates around its axis. The diameter through the equator is about 12,756 km, while the diameter through the poles is about 12,714 km. The earth's equatorial radius, which is a standard unit of size in the solar system, is 6378 km. Recall it by rounding off to 6400 km.

Note. Most figures given in the text are rounded off to one or two significant figures; those in tables and the "At a Glance" features are more accurate.

How to find out the earth's mass? Recall (Section 4.2) that Galileo found that all bodies at the earth's surface have the same acceleration due to gravity, g. Newton's law of gravitation (Section 4.4) relates this acceleration to the earth's mass and radius and to the gravitational constant, G. So if we know G (which can be measured in the lab), the earth's radius, and g (which can be measured at the earth's surface), we can figure out the earth's mass from Newton's law. It comes out to some 6×10^{24} kg—or about the mass of 10^{21} cars!

Another way to find the mass is to apply Newton's form of Kepler's third law (Focus 4.3) to the orbital properties of an artificial satellite in earth orbit.

Knowing both the earth's mass and its volume (from its radius), we can find out a key physical property: its *average density*. **Density** is a measure of how well matter is packed into a given volume, or mass per unit volume. Consider two suitcases, both the same size. Imagine filling one with small rocks and the other with peanuts, both packed as tightly as possible. Close and lift the suitcases. Which one has more mass? Right, the one filled with rocks—it has more mass in the same amount volume. So the suitcase filled with rocks is denser than the one jammed with peanuts.

If you divide the earth's mass by its volume, you obtain a density of about 5500 kg/m³, or 5.5 times the density of water (which is 1000 kg/m³ and makes a good density reference). This average density indicates that the earth, in bulk, consists of a combination of rocky and metallic materials. (Most rocks have a density between 2000 and 4000 kg/m³; pure iron has a density of 7800 kg/m³.) Note that different materials in general have different densities. Iron is more dense than water; wood, less so. If you place both materials in water, the wood floats and iron sinks.

Rocks near the earth's surface average 2400 kg/m³, about half the average density of the earth. This difference implies that the core of the earth is denser than the surface average. Present estimates indicate that the earth's core has a density of some 12,000 kg/m³. This high density implies that the core contains dense materials, such as iron. The weight of the overlying layers compresses the core, causing the material there to have a density higher than normal (higher, that is, than it would have if it were uncompressed).

You will see in these chapters that a solid object's density gives information about that object's composition and internal structure. These data, in turn, provide hints about the object's origin and so the origin of the solar system.

■ **What key physical property can be determined from the mass and volume of a celestial body?**

8.2 The Earth's Interior and Age

From the earth's general properties (such as mass, size, and density), we can develop models of its interior. It has

The Earth

At a Glance

Equatorial radius	6378 km
Mass	5.97×10^{24} kg
Bulk density	5520 kg/m³
Escape speed	11.2 km/s
Atmosphere	Nitrogen, oxygen

three distinct layers: the *core,* the *mantle,* and the *crust* (Fig. 8.2). The **core** makes up the central zone and extends more than halfway to the surface; its radius is about 3500 km. It probably combines mostly iron and nickel in an alloy form. Because of the melting properties of this alloy, the inner core is solid but the outer core is molten. The temperature may exceed 6000 K.

Above the core extends the **mantle,** roughly 2900 km thick. The mantle material is rock made of iron and magnesium combined with silicon and oxygen (a silicate mineral called *olivine;* rocks are generally made of **silicates**, minerals containing compounds of silicon and oxygen). The temperature within the mantle varies, from about 3800 K at the base to 1300 K at the top (Fig. 8.3). Within this range of temperatures, the mantle behaves like a plastic, even though it is solid rock. Under slow, steady pressure, such material flows like a liquid.

Encasing the mantle is the **crust,** the solid surface layer, which varies in depth from 8 km (under oceans) to 70 km (under continents). Most of the crustal material consists of rocks that have solidified from molten lava and are called **igneous rocks.** These rocks are *basalt:* silicates of aluminum, magnesium, and iron. They comprise the ocean basins and the subcontinental sections of the crust. The continental masses are mostly *granite:* silicates of aluminum, sodium, and potassium. Because the granite has a lower density than the basalt, the continental plates float on the basalt. Also, because the mantle is denser than

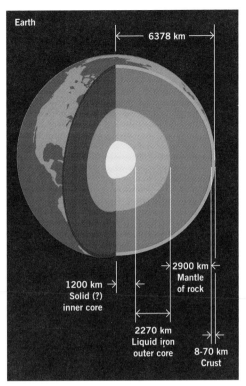

FIGURE 8.2 Model of the earth's interior, showing the sharply-defined core, mantle, and crust, each with its characteristic composition and density. The core and mantle have roughly constant thickness. The crust differs in thickness by almost a factor of 10. It is thinnest under the ocean basins (about 8 km) and thickest under the continents (up to 70 km). Note how large the radius of the core is compared to the overall radius of the earth.

the basalt and granite, the entire crust floats on the mantle.

You may wonder how we have any knowledge of the earth's interior, since we can't see into it. Geologists infer the structure by studying the shock waves from earthquakes. These waves, which travel through the interior and are affected by it, convey clues to the interior's physical state. We can get a firm idea of the interiors of other earthlike planets only if we also have information from shock waves going through them.

To sum up. The earth's interior is **differentiated** (Fig. 8.2). This means that it

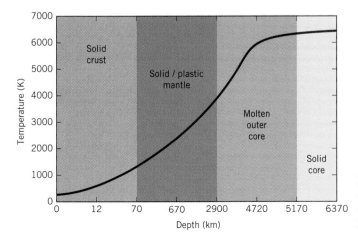

FIGURE 8.3 Model of the temperature profile of the earth's interior. The temperature at first rises quickly through the crust and mantle, then levels off in the core.

has distinct layers, with the least dense materials at the surface and the most dense at the center. Basically, the interior consists of two zones: one metal-rich (the iron core), the other silicate-rich (the mantle and crust).

This separation naturally occurs within a mixture of materials of different densities that is fluid, at least in part. That is the process of differentiation: the lower-density materials float on the higher-density ones. How did the earth get this way? According to present models, the earth was formed such that its material was initially well mixed. Imagine the interior then heating up until it was mostly molten: the denser materials (iron) would have settled at the core, with the less dense materials (silicates) forming a froth on top.

What heated the interior? One source is stored internal energy from formation (Section 8.7). Another is the heat generated by radioactive decay (Focus 8.1). As radioactive isotopes decay, they produce fast-moving particles that collide with the atoms in the rocks and so transfer their energy to it. The transfer heats the rocks and the temperature of the rocks increases. In the past, the earth had more radioactive material than now—about six times as much. The heating from the decay of so much radioactive material could melt the interior enough for it to flow and separate into regions of different density.

Geologists estimate the earth's age at 4.6 billion years from **radioactive dating** (Focus 8.1), which uses the natural decay rate of radioactive isotopes. The oldest rocks on the earth's surface are *not* this old. The oldest known whole rocks, from Western Greenland and Canada's Northwest Territories, are very close to 4.0 billion years old. (The most ancient rocks at other locations are all about 3.8 billion years old.) These rocks are igneous rocks, which means that they solidified from once-molten material. Geologists estimate that about a half-billion years had to elapse for the crust to melt and then cool to solidify the first rocks. So the earth's age is that of the oldest rocks plus the time to form them.

This estimate falls close to that for meteorite material (4.55 billion years) and lunar material (4.6 billion years), determined by the same radioactive dating techniques. The near-coincidence of these ages implies that the solar system formed along with the sun about 4.6 billion years ago (Chapter 12). The exact times of these cosmic events are still being debated, but most geologists agree that the earth is 4.5 to 5 billion years old, with 4.6 billion as the best estimate (with an error range of about 0.1 billion years). So the decay properties of radioactive isotopes result in an age of billions of years. Similar techniques give like results for the moon (Section 9.4) and meteorites (Section 12.3).

■ To use an edible analogy: Which condiment is the more differentiated, oil and vinegar dressing or mayonnaise?

8.3 The Earth's Magnetic Field

You can visualize the earth's magnetic properties by imagining a giant bar magnet located in the core. The magnetic field protrudes from the *south magnetic pole* in the Southern Hemisphere and returns to the *north magnetic pole* in the Northern Hemisphere. The magnetic axis, which connects the magnetic poles, tilts about 12° from the spin axis and does not pass through the earth's center (Fig. 8.4). The part of this magnetic field that is parallel to the earth's surface orients a compass needle so that it points to the north and south magnetic poles.

Magnetic Fields and Forces

A field is another scientific model. It is a way of describing space that is somehow modified by the presence of matter. For example, we can talk about the gravitational field of the earth and how it affects falling objects. *Magnetic fields* are regions of space modified by electrical charges: like magnetic poles (such as two north poles) repel, and unlike poles (a north and a south) attract. Such fields play key roles in many astronomical situations.

Magnetic fields come from electrical charges in motion. The circulation of

The Earth's Magnetic Field

At a Glance

Type	Two poles, north and south
Strength	About 0.4×10^{-4} T at surface
Source	Dynamo in conducting, fluid core
Effects	Magnetosphere, auroras

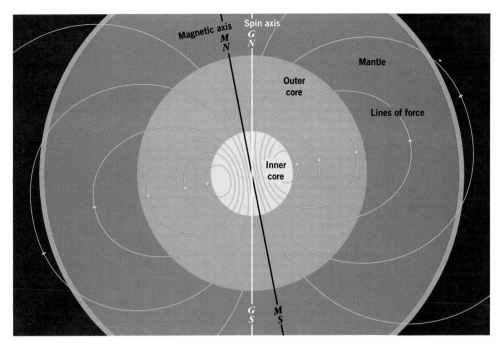

FIGURE 8.4 Model of the earth's magnetic field. The magnetic axis (*MN-MS*) is *not* aligned with the spin axis (*GN-GS*), but is tilted about 12°. Note the continuity of the magnetic field lines between the two magnetic poles. The convention in this book is that the direction of the field lines is out from the south magnetic pole and into the north magnetic pole.

electrons around iron nuclei inside a bar magnet (Fig. 8.5*a*) sets up the bar's magnetic field. In an electromagnet, an electric current, a flow of charged particles (electrons), passes through a loop of wire to create a magnetic field. (A doorbell rings when the button is pressed because a current is sent to create briefly a magnetic field that moves a metal striker to hit the bell.)

The magnetic fields around the earth, the sun, and some planets have two poles and so are called *dipole* fields. This characteristic, dipolarity, allows us to think of these fields as arising from a giant bar magnet buried in the sun or planet (Fig. 8.5*b*), even though they really come from circulating electric currents.

To visualize a magnetic field, take a very small compass and move it around a magnet. The changing direction of the compass needle shows the direction of the magnetic lines of force, called **magnetic field lines.** We usually draw the lines of force so that their spacing indicates the relative strength of the magnetic field: the closer the spacing, the stronger the field. The farther away you are from the magnet, the weaker the field, as indicated by the spreading out of the spacings of the magnetic field lines (Figs. 8.4 and 8.5).

Astronomers generally use the

gauss (abbreviated *G*) as the unit to measure magnetic field strength. The strength of the earth's magnetic field at its surface is about 0.4 G. The SI unit called the *tesla* (abbreviated *T*), with $1 \text{ T} = 10^4 \text{ G}$. The SI unit is the one used in this book (Appendix A).

Variations and Origin

The earth's magnetic field changes with time both in direction and in intensity. The magnetic poles have reversed polarity at least nine times in the past 3.5 million years and many more times in earlier ages (Section 8.5). We have not yet found any particular pattern in these reversals.

The source of the earth's magnetic field is its metallic core, which is both liquid (in part) and a good electrical conductor. It acts like a giant electrical generator—a virtual dynamo—and a large electromagnet. The liquid core generates electricity and creates a magnetic field. (In an electrical dynamo in a generating plant, the reverse process occurs—a spinning magnet moves electrons to make an electrical current.) In general, electric currents create magnetic fields.

Basically, *organized* fluid motions in a conductor will generate a magnetic field. The hot, liquid part of the earth's core contains convective flows, where

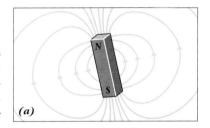

(*a*)

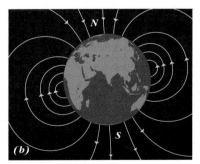

(*b*)

FIGURE 8.5 Magnetic field configurations for dipoles. The spacing of the field lines indicates the strength of the field: the closer the lines, the stronger the field. (*a*) The magnetic field of a bar magnet is a dipole with a magnetic north (*N*) and south (*S*) pole. (*b*) The earth's magnetic field has a dipolelike form with north and south poles.

Enrichment Focus 8.1

RADIOACTIVITY AND THE DATING OF ROCKS

The radioactive dating technique works because of the breakdown of the nuclei of radioactive elements, such as uranium. When these nuclei decay, they break apart into simpler nuclei. As they do, they release energy in the form of high-speed particles. In *alpha decay,* the isotope emits an *alpha particle,* which has two protons and two neutrons. (An alpha particle is the nucleus of a helium atom.) In *beta decay,* the nucleus expels a high-speed electron, which is called a *beta particle* for historic reasons. The alpha and beta particles carry kinetic energy; by collisions, they transfer this energy to electrons and atomic nuclei in the surrounding rock. And so they heat it. The more radioactive material there is in a rock and the faster it decays, the more a rock will be heated.

How fast is that decay? You cannot estimate when any given atom will decay because the process is random, but for a large number of atoms, you can determine a gross rate of disintegration. (An analogous random process is the popping of popcorn. It is impossible for you to predict which kernel will pop next, but you can estimate when the entire batch will be finished.)

Half a piece of uranium-238 (^{238}U) decays to lead in 4.5 billion years, half again in the next 4.5 billion years, and so on. The length of time required for half the material to disintegrate is called the **half-life** of the element (Fig. F.6). So you can calculate the amount of original uranium left at

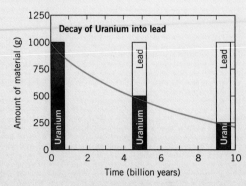

FIGURE F.6 The radioactive decay of uranium into lead, which illustrates the concept of half-life.

any time, even though the decay time for any one uranium atom cannot be specified.

By reversing this idea, you can estimate the age of a rock. Given a rock sample containing ^{238}U and lead and knowing the half-life of the uranium, we can calculate the age of the sample by noting how long it would have taken to form the observed amount of lead by radioactive decay. (To use this technique, you must be able to estimate the amounts of various isotopes of lead originally in the rock at formation. It also involves a sequence of complicated and precise chemical measurements.)

Uranium is not the only element that can be used in radioactive dating. Others that also can serve as radioactive clocks include rubidium (^{87}Rb), which decays to strontium (^{87}Sr), with a half-life of 47 billion years, and potassium (^{40}K), which decays to the inert gas argon (^{40}Ar), with a half-life of 1.3 billion years. Other isotopes are measured as well in order to find out the relative composition of the original

rocks. Whatever elements are used, the derived age is the time elapsed since the rocks last solidified.

For example, suppose a rock sample now contains equal numbers of potassium-40 and argon-40 atoms. If there were no argon atoms in the rock originally, they must all have come from decay of potassium-40. Exactly half have decayed (and half remain), so the rock must be one half-life old, 1.3 billion years. How old is a rock that contains seven times as many argon-40 atoms as potassium-40 atoms? If all the argon came from decay of potassium, the remaining potassium-40 is one-eighth of the original. So three half-lives must have elapsed ($\frac{1}{8} = \frac{1}{2} \times \frac{1}{2} \times \frac{1}{2}$), and the rock must be $3 \times 1.3 = 3.9$ billion years old.

Note that the age inferred in this way is the time since the rock last *solidified*. If a rock once formed has melted and resolidified, the age given by radioactive dating will err on the low side, since it will not have "counted" the portion of the argon gas that escaped during melting.

---

hotter material rises and cooler materials fall. Because it is liquid, the outer core also rotates nonuniformly—the inner regions are carried around faster than the outer one. This kind of motion causes the convective regions to curl around rather than flowing up and down (Fig. 8.6). These motions—millimeters per second!—generate magnetic fields eas-

ily. The earth's rotation helps to stir flows in the core that whirl around to generate electrical currents. This **dynamo model** for the earth's magnetic field, if correct, implies that any planet that exhibits a strong magnetic field must have a substantial fluid interior that is capable of conducting current, and it must rotate rapidly.

Caution. The dynamo model has yet to be worked out in detail. For instance, there is little agreement on how the fluid core flows, what drives its motions, and how these flows generate the complex field at the surface. For now, take it as a basic working model with many details lacking.

The Magnetosphere

Earth-orbiting satellites have detected two doughnut-shaped belts of protons and electrons trapped by the terrestrial magnetic field. These are called the *Van Allen radiation belts* (Fig. 8.7), after their discoverer, James A. Van Allen of the University of Iowa. The sun supplies the charged particles that are trapped in the Van Allen belts. (The flow is called the *solar wind,* Section 13.5.)

In fact, the earth's magnetic field affects the flow of charged particles from the sun for many hundreds of earth radii out into space. The region so affected is called the earth's **magnetosphere** (Fig. 8.8). The magnetosphere acts like a buffer between the earth and the solar wind, which flows over and around it. In turn, the solar wind compresses the earth's field on the day side and stretches it on the night side in a long magnetic tail.

To understand the formation of the magnetosphere for the earth (and other

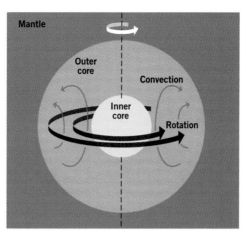

FIGURE 8.6 Fluid motions in the earth's core. The fluid currents in the core are driven by the overall nonuniform rotation, where the inner parts rotate at a higher angular speed than the outer parts, combined with the convection in small regions. These organized currents, in turn, are believed to generate the earth's magnetic field.

FIGURE 8.7 A simplified model of the Van Allen radiation belts, which encircle the earth. These belts contain charged particles, mostly protons and electrons from the sun. The numbers on the contour lines indicate the average density of particles per cubic meter. (Adapted from a NASA diagram.)

FIGURE 8.8 The earth's magnetosphere, created by the interaction of the earth's magnetic field with the flow of charged particles in the solar wind. A long magnetic tail forms downstream, with a thin *plasma sheet* in its core and a *plasma mantle* around the earth, both trapped by the magnetic fields. Note that the magnetic fields have one polarity above the plasma sheet and the opposite below it. The magnetic field lines cross polarity at the *magnetic neutral point.* Note how the impact of the solar wind compresses the Van Allen belts on the sunward side.

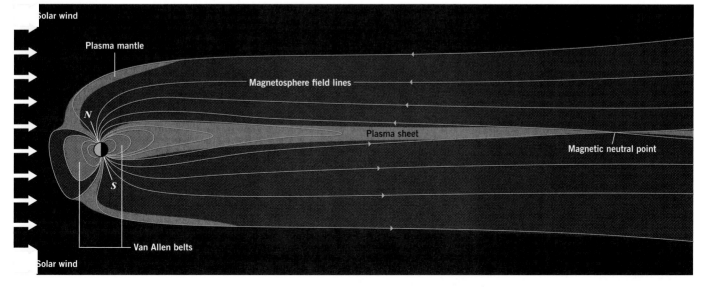

planets), you need a basic idea of how electrically charged particles and magnetic fields interact. They behave such that the particles find it difficult to cross the field lines. If charged particles move parallel to the field lines, they feel no force. But if they move perpendicular to the field lines, they experience a force at right angles to both the field lines and their motion. This force results in the particles having spiraling paths along the field lines. The direction of the spiral twist depends on whether the particle is positively or negatively charged (Fig. 8.9).

Interaction of Magnetic Fields and Plasmas

This linking of magnetic fields and charged particles allows us to grasp what happens to a magnetic field immersed in an ionized gas, called a **plasma.** All, or almost all, of the particles in a plasma are ionized (Section 5.3), so a plasma contains positively charged particles (ions) and electrons in *equal* proportions. As a plasma moves through a magnetic field, the charged particles are trapped in spiral paths along the magnetic field lines. The particles can move along the field lines but cannot cross them. The plasma becomes bound to the magnetic field.

At the same time, the plasma captures the magnetic field and holds it within the plasma. Once an organized flow of charged particles has been set up, it generates its own magnetic field (as does an electromagnet). This field from the moving particles maintains the magnetic field that first caused the current flow. So the original field is reinforced. Then if a plasma moves in bulk, it carries the magnetic field lines with it. How? Those particles are generating a magnetic field, and when the plasma as a whole moves, the particles carry that field with them.

To sum up. Magnetic fields can control the flow of plasmas. In turn, plasmas can carry along magnetic fields. This interaction forms the magnetospheres around planets.

Interaction with the Solar Wind

The magnetosphere has a dynamic interaction with the solar wind. The plasma seeps into the magnetosphere along its boundary and accumulates in its tail. Eventually this buildup of plasma breaks off and flows away with the wind. The magnetic field lines in the magnetosphere come together, break, and reconnect in new patterns.

A drippy faucet provides an analogy to the interaction of the solar wind and the magnetosphere (Fig. 8.10), with the surface tension of the water acting like magnetic field lines. As water drips out of a faucet (Fig. 8.10*a*), a drop builds up, elongates, and then breaks off. The water remaining attached to the faucet springs back to its original shape. Similarly, as the solar wind injects plasma into the magnetosphere (Fig. 8.10*b*), it builds up in mass and energy until a critical amount is reached. Then the field lines in the central part of the tail pinch off, close up, and reconnect—a process called **magnetic reconnection.**

A bundle of plasma containing its own magnetic field lines shoots out with the solar wind into interplanetary space. Other plasma and field lines inside the breakage point snap back toward the earth, injecting into the atmosphere

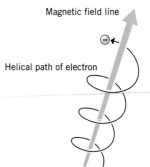

FIGURE 8.9 Effect of a magnetic field on the motion of a charged particle. In the presence of a magnetic field, an electron revolves clockwise around a magnetic field line because of the force of the field on the particle. So an electron in a uniform field moves along a spiral path (a helix). A positively charged particle (such as an ion or proton) turns in the opposite direction.

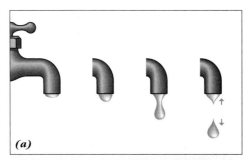

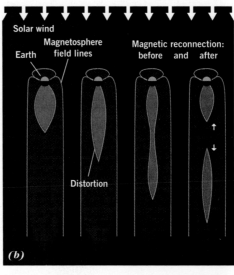

FIGURE 8.10 Magnetic reconnection. (*a*) In analogy to the magnetic process, water dripping from a faucet breaks into drops when the weight overcomes the surface tension of the water. (*b*) As the magnetosphere fills with plasma, the tension of the magnetic field lines breaks at the neutral point, releasing a blob of plasma downstream of the solar wind flow. Another blob snaps back to the earth to create an aurora.

high-energy particles, mostly electrons, to create an **aurora.** These electrons collide with molecules and atoms in the air and excite them so that they emit light. Auroras at the north and south magnetic pole regions are shimmering crowns that make visible to us the effect of the earth's magnetic field on charged particles from the sun (Fig. 8.11). These two luminous ovals surround each geomagnetic pole and attest to the magnetic dynamo deep in the core. Spectacular displays of auroras are visible from the ground and from low earth orbit (Fig. 8.12).

■ How is plasma formed?

8.4 The Blanket of the Atmosphere

Our atmosphere is a blessing. It provides oxygen for breathing, shields us from cancer-causing ultraviolet radiation of the sun, furnishes a thermal blanket to keep the surface warm, and spreads the heat around the earth. Understanding the earth's atmosphere provides information useful for the study of other planetary atmospheres in the solar system.

Relative to the total number of atoms and molecules available, our atmosphere contains approximately 78 percent molecular nitrogen (N_2), 21 percent molecular oxygen (O_2), 1 percent argon (Ar), 0.03 percent carbon dioxide (CO_2), and traces of other elements. Water vapor (H_2O) is present in variable amounts, sometimes as much as 4 percent near the surface.

The weight of the upper atmospheric layers makes the lower portion denser than the upper, just as a sandwich at the bottom of a pile is squashed by the weight of the sandwiches above it. The gas pressure at sea level on the earth's surface, where the entire atmosphere is piled above it, is called *one atmosphere* (1 atm) of pressure. (The **pressure** of a gas is the force it exerts on an area of surface.) The atmospheric pressure and density decrease with height, rapidly at first, then more slowly. At sea level, the pressure is about 10^5 newtons per square meter (or about 15 pounds per square inch)—a car tire is inflated to about twice this pressure. The SI unit for pres-

sure is the *pascal,* which equals one newton per square meter. So the atmospheric pressure at the surface is about 10^5 Pa (Appendix A). The tires on my car state that their maximum rated pressure is 240 kPa.

High in the atmosphere, molecules and atoms can escape into space. Particles of low mass are more likely to leave because, at a given temperature, they have on the average, the highest speeds. If a particle heads outward with at least the earth's escape speed (Section 4.5), and if it doesn't bump into another particle, it can voyage out of the earth's gravitational grip. Such escapes occur in a region about 1000 km up.

Seeing

Even on the clearest nights, atmospheric turbulence makes telescopic images flicker and so imposes a limitation on a telescope's effective resolving power (Section 6.2). The extent to which the atmospheric turbulence affects the image is noted in terms of **seeing.** When the seeing is very good, stellar images are sharp, steady pinpoints, about 1 arcsec or smaller in diameter. At times of bad seeing, the images (a few arcsec in diameter) waver like candle flames in a gentle breeze.

The turbulence that creates seeing occurs from differences in ground heating. If you look over a black surface on a hot, sunny day, you will see the air shimmer, distorting the view behind it. A similar process in the lower atmosphere distorts the view of astronomical objects.

FIGURE 8.11 Overall view of a quiet auroral oval in November 1981. This satellite view is from a distance of 23,000 km at a longitude of 5° west and 78° north. The crownlike structure is shaped by the earth's magnetic field. (Courtesy of L. A. Frank, University of Iowa.)

FIGURE 8.12 The aurora viewed from the ground in Alaska (Courtesy of Syun-Ichi Akasofu, Geophysical Institute, University of Alaska).

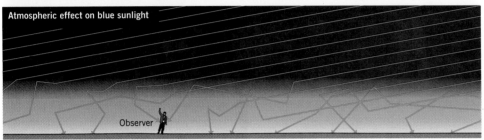

FIGURE 8.13 The blue sky. Air molecules let red light from the sun pass through relatively unhindered, but blue light is scattered in all directions. Looking toward the sun, you see red light directly. Because blue light scatters all around the air, in every direction you look you see the scattered blue light and so a blue sky.

Extinction and Reddening

The atmosphere absorbs and scatters some of the light that penetrates it. This reduction of light is called **atmospheric extinction.** The closer an object appears to be to the horizon, the greater the atmospheric thickness through which the object's light must pass, and so the dimmer the object becomes. Because of atmospheric extinction, the rising full moon has about half the brightness of the same moon when it is overhead.

Scattering of sunlight makes the sky blue. Air molecules scatter blue light (mostly from the sun) more than red. The atmosphere depletes a beam of light of its shorter (bluer) wavelengths, which are scattered around the sky. In any direction you look, you see blue light, and so the entire sky appears blue (Fig. 8.13). Light of longer wavelengths reaches you directly along the line of sight. The sinking sun appears reddish because sunlight passes through a lot of atmosphere before reaching you. Along this path, most of the blue light scatters out, leaving mostly red light to strike your eye (Fig. 8.14).This process is called **atmospheric reddening.**

Albedo

All opaque bodies shine in reflected sunlight, and so does the earth when viewed from space. Different materials reflect light in different amounts. For instance,

FIGURE 8.14 At sunrise, the scattering of blue light occurs to the largest extent, and the sun appears distinctly reddish. The effect is enhanced when the atmosphere is dusty or smoggy. The exposures in this sequence showing the sunrise over Tulsa, Oklahoma were taken 4 minutes apart. (Courtesy of Bill Sterne, Jr., Sterne Photography).

look back at Figure 8.1. You'll note that the oceans appear the darkest, land-masses not as dark, and the clouds the brightest. Clouds (and snow and ice) are good reflectors of sunlight; water absorbs much more than it reflects.

A celestial body's overall reflecting ability is called its **albedo,** the ratio of the light reflected to the incoming light. If an object reflected all the light that struck its surface, its albedo would be 1; if it absorbed it all, the albedo would be 0. The clouds in the earth's atmosphere greatly reflect visible light, and about 35 percent of the incident light is reflected back into space; the earth's albedo equals 0.35. The atmosphere and surface absorb the other 65 percent. (These are average figures; the earth's albedo varies as a result of changes in snow and cloud cover.)

The Greenhouse Effect

About 50 percent of the unreflected sunlight strikes the ground, which absorbs the radiative energy. The ground heats up, and the air in contact with it also becomes warmer. However, if this direct solar radiation were the only source of heat, the temperature at the ground would reach a frigid 255 K (or $-18\,°C$). Water would always be frozen, and life wouldn't exist! The average temperature at the earth's surface is actually higher, about 288 K ($+15\,°C$).

How does this heating happen? Visible light from the sun gets through to the surface and heats the earth. The earth emits in turn infrared radiation. If the infrared radiation simply escaped into space, the earth would be too cold for life. But this radiation doesn't escape completely. Instead, the infrared is absorbed by the earth's atmosphere (mostly by water vapor and carbon dioxide). The atmosphere heats up by absorbing this radiation. Some energy goes off into space (about 10 percent); the rest radiates back to the ground and further heats it.

Both direct sunlight and infrared radiation from the atmosphere heat the earth's surface. The atmosphere acts like a blanket, insulating the ground from space and so helping to warm the earth. Water vapor plays the major role in the

earth's atmosphere, with carbon dioxide playing a minor role.

This warming of the ground by the atmospheric trapping of infrared radiation is often called the **greenhouse effect** by analogy to one process that keeps a greenhouse warm: glass is transparent to visible light but opaque to infrared. Sunlight enters the greenhouse and warms the interior, which in turn emits infrared. This heat can't radiate through the glass, so it stays to help warm up the interior.

Key point. Any planetary atmosphere that is more or less transparent to visible sunlight but opaque to infrared will act to keep the planet's surface warmer than it would be if the planet had no atmosphere.

Caution. The greenhouse effect is misnamed. The absorption of infrared radiation is *not* the main process that keeps the greenhouse warm. The air inside, heated by contact with the hot inside of the greenhouse, cannot escape. This inhibition of convective cooling occurs with any roof. However, the expression "greenhouse effect" is so ingrained in the astronomical vocabulary that I'll use it in this book.

Ozone Layer

About 30 km up in our atmosphere, the oxygen in the air strongly absorbs ultraviolet light from the sun. The absorbed energy promotes the binding of three oxygen atoms to make the ozone molecule, O_3, and an **ozone layer** forms. Some ultraviolet light does not make it to the earth's surface because of this ozone-generating absorption process.

Most life on the earth has developed sheltered from ultraviolet light by the formation of the ozone layer. Even mild exposure to ultraviolet results in a painful sunburn for many people. Small doses over a long time can promote skin cancer.

Human activity has resulted in a sharp drop in the ozone content of the atmosphere. As measured from Antarctica from the 1950s to the 1980s, the amount of total ozone decreased by about half, resulting in the so-called ozone hole above that region (Fig. 8.15). Compounds called *chlorofluorocarbons* (compounds of carbon, chlorine, and fluorine: abbreviated CFCs) released in aerosol propellants, solvents, refrigerants, and other industrial applications are the culprits in the depletion of atmospheric ozone.

The final thieves are chlorine atoms. In a process that takes a year or two from first release of the gases, each chlorine atom destroys some 100,000 molecules of ozone. Each year about a million tons of CFCs are emitted as a result of human actions; these remain in the atmosphere for roughly 100 years. The emissions so far could result in concentrations about ten times that right now, and so decrease the ozone amount even more. Many nations have agreed to severely limit the use of CFCs by 1996. Yet, because of the longevity of the CFCs, much damaging material is already in the air.

Atmospheric Circulation

Sunlight heats the ground and drives the winds; and global wind and cloud patterns result from the earth's rotation. These processes work to drive the atmospheric circulation on any planet. The overall pattern is then sculpted by land and water masses.

Differences in heating of air result in convection, which carries heat from hotter to colder regions (Fig. 8.16). The process of **convection** is a way to carry energy in fluids (gases and liquids). Sunlight heats the ground and the air in contact with it. The heated air expands and its density decreases; so it rises. Denser, cooler air descends to replace it, and a convective cell is established (Fig. 8.16*a*). Local convective flows produce cumulus clouds. You have felt the convective up- and downdrafts if you have flown on a plane that has pierced through a cumulus cloud layer.

Such convective currents work on a global as well as a local level. Generally, the air over the poles is cold and dense; that over the equator is hot and less dense. Why? Because the tropics receive more solar energy than the poles. The temperature differences result in a north–south circulation. Convective cells are set up in these regions (Fig. 8.16*b*). The pattern of earth's atmospheric convective cells depends somewhat on that of land and ocean masses.

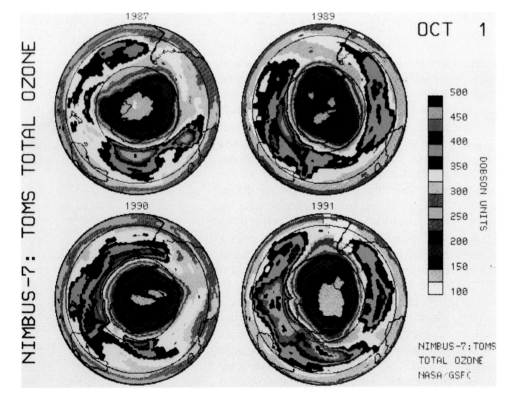

FIGURE 8.15 The ozone hole in the earth's atmosphere above the Antarctic. These Nimbus 7 images taken in 1987, 1989, 1990, and 1991 show the relative concentrations of ozone. The scale at the right gives the false colors relating to different amounts of ozone (measured in Dobson units). The pink shows the ozone hole; purple indicates very low ozone values. The area of the holes for all four years is about the same. "TOMS" means the Total Ozone Mapping Spectrometer. (Courtesy of NASA.)

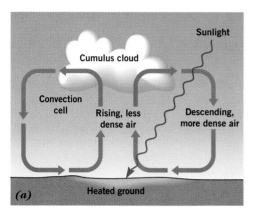

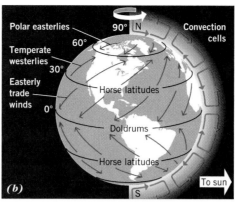

FIGURE 8.16 Atmospheric circulation. (*a*) Convection currents cause the development of cumulus clouds in the earth's atmosphere as heated (and so less dense) air rises and displaces cooler air, which has a higher density and so sinks. (*b*) Solar heating and the earth's rotation result in a global air circulation pattern. The large convection cells have their airflow twisted by the earth's rotation to create an overall flow of winds near the earth's surface. Note the symmetry about the equator.

The earth's rotation then produces spiral motions of these flows: counterclockwise in the Northern Hemisphere and clockwise in the Southern Hemisphere. These spiral flows are called *cyclones*. In general, the faster a planet rotates, the stronger the cyclonic effects. The earth rotates rapidly, so this effect, combined with convective flows, fixes the atmospheric circulation. Similar processes dictate the atmospheric flow of other planets.

■ Given the composition and functions of our atmosphere, why is the "greenhouse effect" such a serious concern?

8.5 The Evolution of the Crust

The earth's surface divides roughly into two levels: the continents and the ocean basins, with an average height difference of about 5 km. Erosion and water transport of materials in a relatively short time (a few million years) should generally erase the difference in height between the oceanic and continental plains. Given the great age of the earth (Section 8.2), the fact that these levels remain distinct implies that somehow the mountain heights and ocean depths are regularly replenished. How?

Planetary Evolution and Energy

Heat flow from the earth to space drives planetary evolution. So, to understand the evolution of the earth, you need to know a bit about energy and heat. Energy and its transformation mark a central idea of modern science. The universe and all its parts are built of energy and matter. It's the energy that motivates the matter—and us, when we say we are energized. **Energy** is ability to do work; it comes in many forms (Section 5.2), such as *kinetic energy*—that of motion.

A mass such as a planet contains a large number of individual particles, which move small distances at random within the planet. Their motions make up the **thermal energy** of the planet— a measure of the total kinetic energy of a large collection of particles, the internal thermal energy a body contains. **Heat** is the thermal energy that is transferred from one location to another. *Heat always flows from a hotter body to a colder one.*

Each major part of the earth—interior, crust, atmosphere, and oceans— has a different temperature. So heat flows between them (from hotter to cooler) and activates their changes. What are the sources of the thermal energy? Some, in the earth's interior, was deposited there at the planet's formation; some is generated by radioactive decay. All the internal energy gradually flows to the surface. There, some energy is added by sunlight to the air, ground, and oceans. Finally, this energy flows into space, which is much colder than the earth.

Today, the total outward heat flow averages a mere 0.06 joule per square meter at the surface. If you could capture all of it over a square meter, it would take about two weeks to raise a cup of water to boiling temperature, 100 °C! Yet, because the earth has a large surface area, the amount flowing out per second is

---

The Solid Earth's Evolution

At a Glance

Cause	Heat flow to surface
Process	Convection in mantle
Effect	Slow motions of crustal plates
Duration	Billions of years

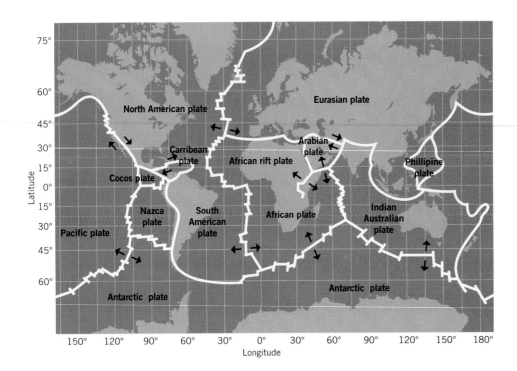

FIGURE 8.17 The major plates in the earth's crust. The ocean plates grow along the midoceanic ridges and result in seafloor spreading across the ridges (arrows).

large (about 10^{13} joules per second). The earth stores a vast amount of thermal energy to keep evolving as it does. And its internal temperature decreases from center to crust so that the heat naturally flows outward.

Continental Drift

Zones of active volcanoes and frequent earthquakes cluster along the chains of young mountain ranges and submarine ridges. Earthquakes and volcanic activity are associated with mountain and island building. This activity shows up in the ocean basins, especially in the Atlantic. Modern sonar measurements have revealed a *midoceanic ridge,* an almost continuous submarine mountain chain that twists through the ocean basins (Fig. 8.17). The midoceanic ridge indicates that crustal evolution takes place in the ocean basins.

These facts fall together as a coherent picture in the model of **continental drift,** the idea that the present continents were at one time a larger landmass that fragmented into pieces that drifted apart. Today's evidence points to two primordial landmasses: one in the Southern Hemisphere, the other in the Northern (Fig. 8.18). These may have broken from

a single, supercontinent. Note that the face of the earth looked quite different only 120 million years ago, when the

240 million years ago

120 million years ago

Present

FIGURE 8.18 Continental drift. These computer-generated models show the trend of continental drift from 240 million years ago, through 120 million years ago to now. The position of the continents is based on magnetic evidence. The present configuration of the continents will change just as dramatically during the next few hundred million years. (Based on maps prepared by A. M. Ziegler and C. S. Scotese of the University of Chicago's Paleographic Atlas project.)

present continental masses were crowded together.

Evidence from the magnetic characteristics of the ocean floors near ridges supports the continental drift model. If the continents do move apart, then the seafloor spreads. Oceanographic cruises across the world have found that the seafloor material contains remnants of ancient magnetism. When lava solidifies to form igneous rock, the iron minerals in the rock align with the earth's magnetic field. The directions and reversals of the direction of the earth's magnetic field in the past are preserved in the rock. On both sides of the mid-Atlantic ridge, the reversal patterns are identical—each side is a mirror reflection of the other. So the seafloor rocks act as a magnetic tape (a very slow-moving one!) that preserves the record of the past changes in the earth's magnetic field.

The alignment of magnetic field reversals indicates that new material emerges from a rift in the center of the ridge and gives the rate of expansion of the seafloor. The movement is about 3 cm per year at its fastest speed across the mid-Atlantic ridge. (That's about the amount your fingernails grow in a year!) This rate amounts to enough movement to push apart the Old World and the New World in a few hundred million years. Current measurements, made with special radio astronomy techniques (called VLBI: Chapter 6) show that along a line between Massachusetts and Sweden, the North American and Eurasian plates move apart at 1.1 cm per year.

Volcanism and Plate Tectonics

New material oozing out from the earth's interior along oceanic ridges accounts for the renewal of the ocean plains. This process marks one example of **volcanism,** by which molten material (produced by internal heating) rises through a planet's mantle and crust to the surface. We tend to think of volcanism as violent and sporadic, but it also acts slowly as a key source of modifying the earth.

The continental plates float like large rafts on the basaltic basin material. Where one continental plate crashes into another, the impact raises mountains (Fig. 8.19). This process builds up moun-

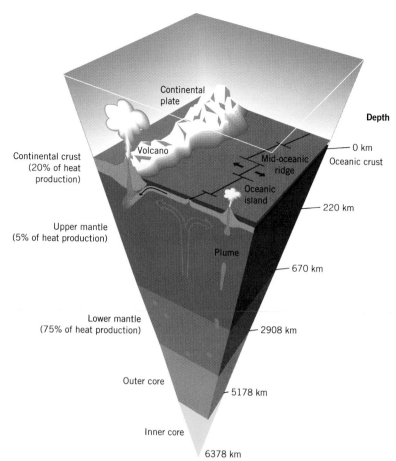

FIGURE 8.19 Interactions of oceanic and continental plates. The oceanic plates gain material from the outflow at oceanic ridges. As these plates expand, they crash into continental plates. Here mountains are built and earthquakes occur. Most of the heat is generated in the lower mantle, which also releases plume material that flows upward into the upper mantle and crust.

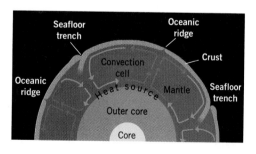

FIGURE 8.20 Simple model of convection in the mantle. Hotter, less dense material rises from the lower mantle, flows horizontally, and then descends to create large convection cells in the mantle. The upflow adds material to the crust at the ocean ridges and pulls down material at the seafloor trenches. The horizontal flow moves the crustal plates.

tains while erosion wears them down. In some regions, one plate may force another to fold under and descend into the mantle in *seafloor trenches* (Fig. 8.20). The plate's descent eliminates surface material essentially at the same rate it is created. Because the plates' creation and destruction zones make natural fault areas, earthquakes and volcanoes predominate along the lines of collisions. The earth's ocean basins, because they have been recently formed, are the youngest parts of the earth's crust, while the center of continents are usually the oldest.

This model of the earth's crustal activity and evolution—seafloor spreading and the creation and destruction of the crust—is called **plate tectonics.** Much of the geologic action takes place where plates collide, such as the arc where the Pacific plate encounters the North American one. Here earthquakes happen along fault lines (such as the San Andreas fault) and new, volcanic mountains (such as Mt. St. Helens) thrust above the ground. The stretching, thinning, and splitting of the earth's crust produce *rifts* in the continents.

What moves these plates? Heat flow! One model (Fig. 8.20) pictures the upper part of the mantle as divided into large convective whirls of flowing rock in the upper 700 km. The mantle's plasticity allows a slow flow upward, horizontally, and downward. At the region of horizontal flow, friction between the plate and the mantle drags the plate along with the mantle's flow. The upwelling magma supplies new materials to the plate. The energy source for such convection comes from heating by radioactive decay or from internal currents left from the time when the earth's core formed. The flows here are very slow, averaging only about a few centimeters per year. The resulting motion of the plates is jerky, as friction holds plates fixed until the stresses exceed the strength of the crustal rocks.

Whatever the exact details, the main point is that heat convection in the mantle drives plate tectonics on the surface. All the major tectonic features are expressions of this flow, which occurs over tens of millions of years or so.

Key point. The earth's crust has evolved since the planet's formation and is changing right now. The two main processes are volcanism and tectonics. When investigating other earthlike planets, you should look for evidence of their crustal evolution.

■ In what way do seafloor rocks act like a magnetic tape?

8.6 Evolution of the Atmosphere and Oceans

No other planet in the solar system has the earth's combination of an extensive atmosphere plus oceans of water. The earth's favorable environment for life results from the transfer of solar heat around the globe by the atmosphere and the oceans. This fine thermostat has worked for hundreds of millions of years. However, the atmosphere and the oceans have changed over time, influenced by and influencing the earth's biological evolution. The future of life on the earth depends critically on the fate of its fluid system.

Origin and Development of the Oceans

By the current model of the formation of the planets of the solar system (Chapter 12), the earth had no oceans when it formed 4.6 billion years ago. The primeval surface may have been about a few thousand kelvins—certainly too hot for water to be liquid! When the surface had cooled to 373 K (100 °C), so that water could condense, perhaps one or two continents existed on the surface, and the rest was the initial ocean basin. This large tub contained very little water then, only a mere few percent of the present volume.

The rest of the oceanic water probably came from the earth's interior. Volcanism bears to the surface a variety of gases, such as carbon dioxide and water vapor. The steam issues from water that was trapped in the solid earth when the planet formed. The present rate of water gassing out from the interior (about 10^{11} kg per year) accounts for the amount of water in the oceans today, if it has been constant for 4.5 billion years. It may have been faster in the past, when the earth was hotter. The earth's oceans now cover 71 percent of its surface with an average depth of 4 km. But the array of the oceans was very different in the past because of plate tectonics.

Evolution of the Atmosphere

Outgassing from the earth's interior also drove the development of the earth's atmosphere. In fact, the earth's atmosphere evolved from the chemical interplay of the solid and fluid earth.

Here is one reasonable model of atmospheric evolution. If our ideas of planetary formation are correct (Section 12.6), the earth's first atmosphere contained mostly hydrogen and helium. But these gases, which are extremely low in mass, escaped because their typical speeds were greater than the earth's escape speed. Our second atmosphere arose mostly from volcanism, which spews out carbon dioxide, sulfur dioxide, hydrogen, nitrogen, water, methane (CH_4), and ammonia (NH_3). (In addition, some inert gases, such as helium and argon, come from the decay of radioactive materials.)

The outgassed materials (which still enter the atmosphere today) interact with the oceans, surface materials, and biomass in complex ways. For example, carbon dioxide is now added to the atmosphere by volcanoes, organic decay, and the combustion of fossil fuels. Carbon dioxide is taken out by plants and is dissolved in the oceans (much as carbon dioxide is added to water to make soda water). The balance, however, has changed with time, so the atmospheric composition has evolved.

The earth's second atmosphere may well have started out with a large amount of carbon dioxide. In about a billion years, the carbon dioxide ended up in the oceans and rocks. So 3 billion years ago the atmosphere consisted mostly of methane and other hydrogen–carbon compounds.

At this time the atmosphere contained little free oxygen, so the earth had no ozone layer. Ultraviolet light readily penetrated and broke up methane, ammonia, and water. The hydrogen from these molecules fled into space. Some of the oxygen freed from the water combined with some of the carbon from the methane and gradually eliminated the elemental carbon. Some of the oxygen atoms eventually formed the ozone layer. Nitrogen from the ammonia became the main constituent of the atmosphere.

The high abundance of atmospheric oxygen was produced (and is now maintained) by biological activity.

Geologic evidence indicates that the transformation to an oxygen-rich atmosphere began roughly 2 billion years ago, when plant activity and photosynthesis bloomed (Chapter 22). About 1 billion years ago the atmosphere may have contained only 10 percent of the present amount of free oxygen. About 600 million years ago, the oxygen content suddenly increased to present levels, with a proliferation of life.

Evolution of the Earth's Surface Temperature

How hot it gets at the earth's surface depends on how much energy is received from the sun and how effectively the greenhouse effect operates. Less solar energy results in lower temperatures. A larger greenhouse effect (more carbon dioxide and water vapor in the atmosphere) delivers higher temperatures. The balance of the input and outgo of energy fixes the temperature.

People have disturbed the natural carbon dioxide balance by extracting fossil fuels from the earth and burning them for energy, which adds to the carbon dioxide in the atmosphere. In addition, our destruction of forests has eliminated a substantial portion of the green plants, which take in atmospheric carbon dioxide, and has added some of their carbon to the atmosphere. The ocean can absorb only a part of the excess.

Our activities have a net result of increasing the percentage of carbon dioxide in the earth's atmosphere (Fig. 8.21). Observations indicate an increase of 5 percent over the past 30 years. Although we are not sure what the impact of this increase will be, it may result in a global temperature increase of about

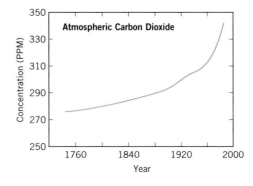

FIGURE 8.21 The growth of atmospheric carbon dioxide with time since the Industrial Revolution. This graph shows the yearly average without the seasonal variations. The carbon dioxide concentration is measured in parts per million (ppm).

2 °C by A.D. 2020 if the overall water vapor and cloudiness do not change. This heating could have serious effects on climate over the long term.

■ **What accounts for the water in our oceans today?**

8.7 The Earth's Evolution: An Overview

The earth is the most evolved of the earthlike planets in our solar system. It's an active world now (powered by internal heat) and shows no sign of quitting. Geologic evidence implies that it's been active since birth—a birth that formed a hot earth. How? Let's again use the concept of the *conservation of energy.*

Consider an apple held in your hand. When you drop the apple, it gains kinetic energy, and the higher the point from which you drop it, the greater the kinetic energy it has attained by the end of its fall. The position above the earth's surface relates to the total amount of potential energy the object has. Energy transforms from one form to another (potential to kinetic), yet the total energy of the apple remains constant, for as it loses potential energy, it gains kinetic energy.

Now imagine a collection of solid debris in space, with a range in sizes. Consider what happens when gravity brings them together. Their potential energy transforms to kinetic energy. They collide and stick together—a process called **accretion.** The resulting mass gains internal energy, and so its temperature increases. If the earth (or any planet) formed by accretion, it was originally hot. Note that in the later stages of accretion, once a surface has formed, late-arriving debris punches out impact craters.

Let's step back to scan the earth's overall evolution (Fig. 8.22 and Table 8.1). It falls into four large-scale stages.

One, the accretion to the earth of smaller bodies in the cloud of gas and dust that eventually formed the sun and planets (Fig. 8.22*a*). The buildup probably took place quickly, in only a few million years. An intense bombardment of large extraterrestrial objects played a main role then, leaving a pockmarked planet of more or less uniform composition (since each body was made of about the same combination of materials). The atmosphere at the time of accretion was rich in hydrogen and inert gases.

Two, some tens of millions of years later, radioactive heating melted the interior, which differentiated (Fig. 8.22*b*).

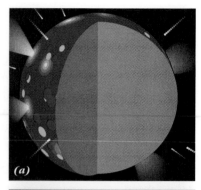
(a)

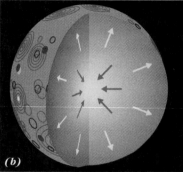
(b)

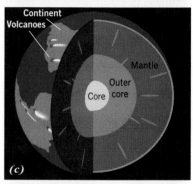
(c)

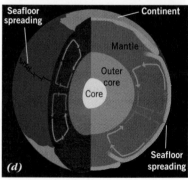
(d)

FIGURE 8.22 One model for the overall evolution of the earth. (*a*) The earth forms by the accretion of planetesimals; late-arriving objects crater the surface. (*b*) The interior, heated by radioactive decay, differentiates. (*c*) Water condenses, the first continents form, and intense volcanic activity occurs. (*d*) The crust thickens and crustal plates roam across the surface.

TABLE 8.1 A POSSIBLE SEQUENCE FOR THE EARTH'S EVOLUTION

STAGE	EVENTS
I. 4.6 billion years ago	1. Formation of earth by accretion from the solar nebula 2. Rapid internal heating from accretion and radioactive decay 3. Blowing off of primeval atmosphere
II. 4.5 billion years ago	1. First differentiation of interior; formation of crust and core 2. Outgassing (of carbon dioxide, water, methane, ammonia, etc.) to make a second atmosphere 3. Infall of large objects to fracture crust in places 4. Ocean basins form and begin to fill with water
III. 3.7 billion years ago	1. First tectonic movements, seafloor spreading, formation of volcanoes and mountains 2. Cooling and thickening of crust 3. Ocean basins mostly filled
IV. 600 million years ago	1. Continuation of processes of stage III at slower rate 2. Enlargement of ocean basins 3. Formation and subsequent breakup of original continental mass

Denser materials sank to the core, lighter ones rose to the crust. The original atmosphere of hydrogen and helium was lost. Perhaps an intense solar wind blew off the primeval accretion atmosphere. Or the low-mass gases may have simply escaped into space. This atmosphere was replaced by one containing much water, methane, ammonia, sulfur dioxide, and carbon dioxide. How? By volcanic activity caused by interior heating. The earth's surface cooled enough for rain to fall and the oceans to form.

Three, about a billion years later, the first continents appeared (Fig. 8.22c). Plate tectonics began; mountains grew, only to yield to the weathering of wind and rain. Slowly even the atmosphere evolved. Four, roughly 2.2 billion years ago, crustal cooling thickened the crust enough to allow plate activity as we see it today. At the end of this stage, which began 600 million years ago (Fig. 8.22d), the earth looked much as it does today. During this time, plate tectonics reordered the continents and ocean basins.

To sum up. The earth's atmosphere, crust, and interior have changed greatly since the formation of the planet. The interior's evolution is driven by internal heat: some from radioactive decay, some stored from its formation by accretion. The heat caused internal layering. As heat flows outward (and finally out to space), it drives tectonics and volcanism. In the distant past, the crust was also shaped by impacts of solid bodies from space; few such impacts occur now. But if the incoming object is large enough, it would create a catastrophic explosion (Chapter 22). The atmosphere also transforms the crust by weathering: wind and water erosion. Some material is also slowly lost to space.

The earth's structure and evolution, which we know in more detail and in greater depth than those of any other planet, serve as a model for the investigation of similar bodies within the solar system, those that are called the **terrestrial planets,** and include the earth's moon.

■ **What did the earth's first atmosphere consist of?**

Evolutionary Processes	
At a Glance	
External	Impacts of bodies from space
Internal	Volcanism and tectonic
Surface	Wind and water erosion
Driven by	Internal heat, sunlight

Key Concepts

1. We use the physical properties of the earth, such as mass and density, to develop a model of its interior, which can be probed by analyzing earthquakes. The earth is divided into zones that are denser the deeper they lie below the surface—the interior is differentiated. The core is the densest part and probably consists of metals such as nickel and iron. The mantle and the crust are made of rocky materials.

2. From radioactive dating of the oldest rocks using isotopes of known half-lives, we estimate that the earth's age is 4.6 billion years, with an uncertainty of about 0.5 billion years. The earth formed with the rest of the solar system at about this time.

3. The earth's atmosphere consists mainly of nitrogen molecules (78 percent) and oxygen molecules (21 percent). The pressure at the surface at sea level (1 atmosphere) comes from the weight of all the air above it. In its outermost region, the atmosphere slowly leaks into space where particles move faster than the earth's escape speed and the density is low so that collisions are infrequent.

4. The atmosphere affects the light that enters it from space, especially sunlight: ultraviolet is absorbed by oxygen molecules, forming the ozone layer; blue light is scattered (making the sky blue), and most of the visible light arrives at the surface (heating the ground). Celestial objects close to the horizon are viewed through more atmosphere and so appear redder and dimmer.

5. Carbon dioxide and water vapor trap the infrared radiation emitted by the ground and so keep the earth warmer than it would be if the infrared escaped directly into space. The water vapor has a greater effect than the carbon dioxide. This warming is called the greenhouse effect and makes our planet habitable. Carbon dioxide added to the atmosphere by people may result in global warming.

6. The earth's crust evolves by seafloor spreading and the motions of crustal plates, driven by convective flows below the surface that result from the transport of internal thermal energy. The ocean basins are the youngest parts of the earth's crust; the continents are older in comparison. Tectonics and volcanism, the main processes driving the evolution of the crust, are powered by the heat flow from the interior.

7. The water in the oceans, and the gases in the atmosphere, probably came from the release of materials from the earth's crust and interior. Such water and gases interact and have also changed because of the development of life on our planet.

8. The evolution of the atmosphere is influenced by outgassing from the crust by volcanism, escape of gases into space (if the particles of the gas move faster than escape speed), biological activity, and electromagnetic radiation from the sun. The atmosphere also interacts with the oceans.

9. The interior and crust of the earth have been greatly modified by the generation and outflow of heat since the planet's formation some 4.6 billion years ago by the gravitational accretion of small masses. The heating from the transformation of gravitation energy caused the interior to differentiate. Some of that stored heat now flows out to the surface from the hotter core.

10. The earth's magnetic field is generated by a dynamo set up by regular fluid motions in the fluid part of its metallic core. The overall structure is that of a dipole, similar to a bar magnet. The field interacts with the plasma of the interplanetary solar wind to create a magnetosphere. The field has varied dramatically over tens of millions of years by switching polarities.

11. The earth is the detailed model for the other planets like it in the solar system; these comprise the terrestrial planets, which we believe formed by accretion. The heat released in gravitational accretion, plus that from radioactive decay, drive volcanism as the heat flows to the surface. Impacts by interplanetary debris and erosion, along with volcanism, have modified the surfaces. Atmospheres are renewed by outgassing from the interior; some gases escape into space. The most important factor in the vigor and length of a planet's evolution is its mass.

Key Terms

accretion	crust	igneous rocks	pressure
albedo	density	magnetic field lines	radioactive dating
atmospheric extinction	differentiated	magnetic reconnection	seeing
atmospheric reddening	dynamo mode	magnetosphere	silicates
auroras	energy	mantle	terrestrial planets
continental drift	greenhouse effect	ozone layer	thermal energy
convection	half-life	plasma	volcanism
core	heat	plate tectonics	

Study Exercises

1. Explain how you could determine the earth's mass by jumping off a building. (Explain it; don't do it!) (Objective 8-1)

2. Contrast the composition of the earth's core to that of its crust. (Objectives 8-2 and 8-3)

3. Discuss uncertainties in the statement "The earth's age is 4.6 billion years." (Objective 8-4)

4. Make a simple argument to demonstrate that the earth's core must be denser than its crust. (Objective 8-3)

5. Describe two effects of the earth's atmosphere on sunlight passing through it. (Objectives 8-6 and 8-7)

6. Suppose the amount of water vapor in the atmosphere suddenly increased by a large amount. What would happen to the earth's surface temperature? (Objective 8-7)

7. How can volcanoes affect the evolution of the earth's oceans and atmosphere? (Objectives 8-8, 8-10, and 8-11)

8. What was the composition of the earth's first atmo-

sphere? What happened to it? What was the composition of the earth's second atmosphere? Where did it come from? What happened to it? (Objective 8-10)

9. Give two ways in which the oceans affect the atmosphere. (Objective 8-9)

10. Where did the oceans' water come from? (Objective 8-9)

11. In what way is the earth's magnetic field able to trap the particles of the solar wind? (Objectives 8-5 and 8-12)

12. What part of the earth's crust is the youngest? The oldest? (Objective 8-8)

Problems & Activities

1. Consider uranium-238, which has a half-life of 4.5 billion years. If this isotope of uranium arrived on the earth's surface when the earth formed, how much of the original amount remains today? How much will remain 4.5 billion years from now?

2. The acceleration due to gravity at the earth's surface is 9.8 m/s/s. The earth's radius is about 6400 km. Use Newton's second law and law of gravitation to calculate the earth's mass. *Hint: $F = ma = mg$*, where g is the acceleration due to gravity at the earth's surface.

3. Imagine that the earth's rotation is very gradually slowing (Section 9.5), so that billions of years from now, the earth will rotate at only one-thirtieth its present rate. If the dynamo model is correct, what will happen to the earth's magnetic field when this substantially slower rotation has been reached?

4. Imagine that the mass of the earth were increased four times but the radius remained the same. What would happen to the earth's density? Its escape speed?

5. Find the radius of the earth's core relative to the total radius if the core density is 10,000 kg/m^3, the mantle density is 2500 kg/m^3, and the average density is 5500 kg/m^3.

See for Yourself!

The half-life of a radioactive isotope of an element is a key property, but probably not one that you have had the opportunity to measure. At the University of New Mexico, we have some fast-decaying isotopes with short half-lives. We can measure the relative amount of the isotope by tracking its activity with a Geiger counter and measuring the counts per minute (cpm) with it as time passes.

Here's a table of typical results:

TIME (MIN)	COUNTS PER MINUTE
Start (0)	1380
5	1150
10	1000
15	900
20	720
25	600
30	520

Draw a graph of these results. Put "Time (min)" along the horizontal axis, starting from zero; place tic marks at intervals of 5 minutes. Label the vertical axis "Radioactivity (cpm)"; set it to zero at the bottom and 1400 at the top. Place tic marks at intervals of 100 (with small divisions of 10). Plot a dot for each entry in the table and draw a smooth, freehand curve between them. From the curve, estimate the half-life of this isotope. ■

LEARNING OBJECTIVES

After studying this chapter, you should be able to:

9-1 Compare the moon and Mercury in terms of their general surface and physical properties.

9-2 Describe the moon's major surface features and indicate a possible formation process for each.

9-3 Describe Mercury's major surface features and indicate a possible formation process for each.

9-4 Compare and contrast the surface environments (such as temperature, atmosphere, surface features, escape speed) of the moon and Mercury to each other and to the earth.

9-5 Sketch a model for the structure of the lunar interior and compare it to the earth's.

9-6 Sketch a model of Mercury's interior and contrast it to the interiors of the earth and the moon.

9-7 Compare and contrast the magnetic fields (or lack thereof) of the moon, Mercury, and the earth.

9-8 Outline a possible history for the moon's evolution in light of Apollo results, present evidence for each of the major stages, and apply the general processes for the evolution of terrestrial planets.

9-9 Compare and contrast the evolution of the moon and Mercury to each other and to the earth, applying the general processes for the evolution of the terrestrial planets.

9-10 Compare and contrast two models of the moon's origin, using Apollo results to support or refute the models.

9-11 Describe the process of cratering of planetary surfaces and tell how craters can be used to infer the relative ages of surfaces.

9-12 Use Newton's law of gravitation to explain the nature of *tidal forces,* describe generally how the moon generates the earth's tides, and apply tidal forces to other astrophysical situations.

Central Question
What processes have driven the short evolution of the small, airless worlds of the moon and Mercury?

The Moon and Mercury: Dead Worlds

Everyone is a moon and has a dark side which he never shows to anybody.

MARK TWAIN, *Pudd'nhead Wilson's New Calendar*

Through a small telescope, the moon strikes you as a stark world, tantalizingly close. Fascination with this neighbor bred tales of traveling to the moon. The stories ranged from bird-borne excursions to the cannon-powered voyage described by Jules Verne, in which the astronauts, after their launch from Florida, circled the moon and returned home by plunging into the sea. Verne had the right idea! NASA carried out his vision, with the addition of a lunar landing, in July 1969. Many earlier dreams were fulfilled by the event: the first visit to an alien world.

The solar system has another body with a similar face: Mercury. Like the moon, Mercury is a small, airless world pockmarked with many craters and scoured by intense sunlight. Both the moon and Mercury are dead worlds. Their interiors are now cooler than the earth; no heat drives surface activity. Without atmospheres, these planets have no wind or water to wear down their landscapes. Their heyday of evolution has passed.

This chapter compares these tiny worlds to each other and to the earth to provide an insight into their evolution. The emphasis falls on our moon because of the Apollo missions, which give solid clues to allow us to reconstruct the moon's history and to infer that of Mercury by comparison.

9.1 The Moon's Orbit, Rotation, Size, and Mass

The moon revolves around the earth in an elliptical orbit at an average distance of 30 earth diameters (Fig. 9.1), or about 384,400 km. Because the moon's orbit is fairly eccentric, the shortest distance from the moon to the earth, called *perigee,* is as little as 356,400 km. The greatest distance, called *apogee,* reaches as far as 406,700 km.

The moon's distance can be measured very precisely. In 1969 the Apollo 11 astronauts deposited special reflectors on the moon's surface. Later Apollo missions put down other such reflectors. These devices reflect back to the earth laser light bounced off the moon from the earth. Timing the round trip accurately measures the earth–moon distance to within 3 centimeters! (To put this into a terrestrial perspective, it is equivalent to measuring the distance between Los Angeles and New York to 0.4 mm.)

The One-Faced Moon and Lunar Day

If you have looked at the moon, even occasionally, you have grown used to the sight of the same lunar face (Fig. 9.2). The moon always shows us the same face because it rotates on its axis (Fig. 9.3) with a period of 27.3 days—the same as its period of revolution about the earth with respect to the stars. This matching of rotation and revolution rates is called **synchronous rotation.** Until satellites orbited the moon, we had no direct view of the far side, the side turned away from the earth. Lunar orbiters photographed the far side (Fig. 9.4), which looks quite different from the near side (Fig. 9.2): it is almost completely cratered, with little of the dark-colored, smoother areas that cover so much of the near side. We now even have views of the moon's polar regions (Fig. 9.5), which are also not visible from the earth.

You can prove to yourself how the moon must rotate once per revolution to keep the same face to the earth. Place a globe on a table. Start to circle around the table, imagining your head to be the moon and keeping your eyes on the globe. You will soon see that your head (and body) need to turn around once for one revolution around the table. (Synchronous rotation is a feature of almost all the moons in the solar system.)

If you stood on the moon's near side, you would see the earth suspended, never rising or setting, against the stars. With the earth always in sight, astronauts on the near side can communicate directly here by radio. By contrast, any astronaut on the far side would never see the earth and would have to rely on a lunar orbiter to relay radio signals to earth.

The moon keeps the same face to the earth, but *not* to the sun. It rotates

once with respect to the sun in 29.5 days, so the lunar day is 29.5 earth days long, which also equals the time between corresponding phases of the moon. Suppose you were standing at the equator of the moon at full moon. Where would the sun be? Directly overhead! When will the sun be directly overhead again? At the next full moon!

The Moon's Physical Properties

Knowing the distance from the earth to the moon, you can find its physical diameter from its angular size. The result is 3476 km, about one-fourth the earth's diameter. If the earth were the size of your head, the moon would be about the size of a tennis ball. On the same scale, the diameter of its orbit would be about 12 meters.

As you'd expect, the moon's mass is much less than the earth's: it is 1/81 times the mass of the earth, or 7.4×10^{22} kg. From the mass and radius, the bulk density comes out to be a little more than 3300 kg/m^3, about the same density as the rocks in the earth's mantle. This low density implies that the moon's interior contains only a small percentage of metallic materials. The moon is mostly rock, similar to those of the earth's mantle.

Tides and the Moon

You may be familiar with the periodic rise and fall of ocean tides, which are controlled by the moon, so that typically two high tides and two low tides occur each day. I'll give you a general explanation of tidal dynamics using Newton's law of gravitation (Section 4.4), but be warned that it does not provide a complete, detailed explanation about tides.

Imagine that the earth's surface is level and covered completely with a layer of water. Consider for a moment the moon's gravitational attraction at three points lined up with the moon (Fig. 9.6a). Recall that the force of gravity decreases as the *inverse square of the distance* between masses. So the moon's gravitational force must be greater at A than at B and greater at B than at C. The greater the force acting on the same mass, the greater its acceleration. So a mass at A has a greater acceleration than

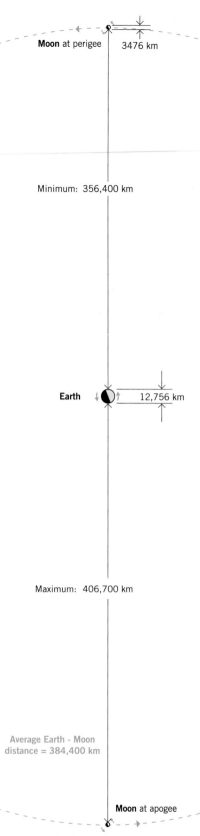

FIGURE 9.1 The moon's orbit relative to the earth: looking down on the orbital plane. The *minimum* perigee and *maximum* apogee distances are shown for this fairly eccentric orbit. (The actual values of apogee and perigee vary monthly.) The earth and moon are drawn to the same scale as the distances.

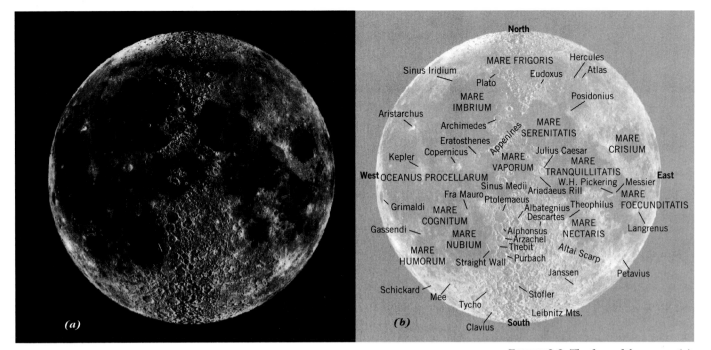

FIGURE 9.2 The face of the moon. (*a*) The moon as seen from the earth through a telescope at full phase. The darker regions are lowlands; the lighter areas, highlands. Note that most of the craters appear in the highlands. Some of the larger craters have rays emanating from them. This special composite photograph shows relief that is invisible under the flat lighting at full moon (when its noon on the moon!). (Courtesy of Lick Observatory.) (*b*) Map of the major features on the moon's near side at full phase.

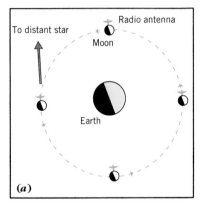 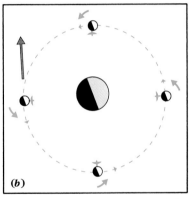

FIGURE 9.3 The moon's rotation. The moon keeps the same face toward the earth because its rotation period is the same as its period of revolution with respect to the stars. Imagine a radio antenna on the moon. (*a*) If the moon did not rotate with respect to the stars, the antenna would not always point at the earth. (*b*) But because it does rotate at a rate equal to its revolution, our imaginary antenna always points to the earth. The moon exhibits synchronous rotation.

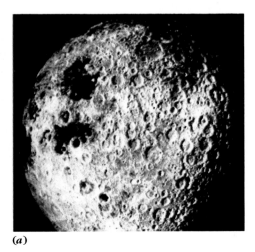 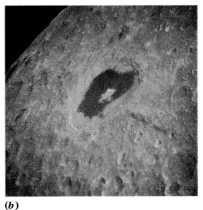

FIGURE 9.4 Views of regions of the moon not visible from the earth. (*a*) Apollo 16 photo of the easternmost part of the moon visible from the earth. The smooth dark area to the left is Mare Crisium. The right half of the picture, rich with craters, is typical of the ancient terrain that covers 85 percent of the lunar surface. (Courtesy of NASA.) (*b*) Close-up of the moon's far side. The smallest craters are about 1 km across. The large dark region near the center is a crater with a dark floor about 200 km across; it is called Tsiolkovsky. (Courtesy of NASA.)

a mass at *B*, and a mass at *B* has a greater acceleration than a mass at *C*.

The *difference* between these accelerations is crucial. Because of the difference, the water at *A* bulges ahead of the earth (point *B*), and the water at *C* lags behind the earth and forms a bulge on the side of the earth opposite the moon. So we have two high tides: one on the side of the earth toward the moon, and one on the opposite side. At any point in the earth's oceans, these two high tides take place about a half-day apart.

Tides result in **tidal friction.** As water moves about in the ocean basins, both high tide peaks almost line up with the moon (one on the side of the earth under the moon, and one on the side opposite). However, the tidal bulges are not right on a line to the moon (Fig. 9.6*b*); the closer bulge (*A* in Fig. 9.6*b*) lies ahead of the moon, the farther bulge (*B* in Fig. 9.6*b*) lags behind. This displacement occurs because of the earth's rotation and the friction of the water flowing in the ocean basins. Because the earth rotates faster than the moon revolves around it, the friction pulls the line between the tidal bulges ahead of the line between the center of the earth and moon. In return, that friction slows down

the rotation rate of the earth. Also, the gravitational attraction of bulge *A* on the moon causes the moon to accelerate in its orbit. This acceleration results in the moon moving away from the earth.

Note that it is the *difference* in gravitational forces that accounts for the tides; this difference is what is referred to as a **tidal force.** It pulls bodies apart. The curious aspect of tidal forces is that they vary as the *inverse cube of the distances* between the masses. Tidal forces grow very strong as two bodies come closer together. You will come across tidal forces in many other situations in the astronomical universe.

Note that tidal forces are interactions, so the earth also exerts a tidal force on the body of the moon. This interaction keeps the same side of the moon facing the earth.

History of the Moon's Orbit

The orbit of the moon has changed because gravity ties the earth and moon together. Tides are one consequence of this coupling. (See Focus 9.1 for details.) Tidal interactions of the earth and moon slow down the earth's rotation rate at a fraction of a second (about 1.4×10^{-3} s) a day per century. This decrease results in an increase of the earth–moon distance of about 4 cm per year (Fig. 9.7)—about the length of your little finger. In the future the moon will move so far from the earth that the length of the month (longer than now) will equal the length of the day (also longer). That future day will be 55 present (24-h) days long.

The moon once orbited much closer to the earth. Its smallest orbit occurred about a billion years ago. At that time, the month was about 6.5 hours long, the day 5 hours long, and the moon only 18,000 km from the earth. The moon would have covered 11°, which is 22 times its present angular diameter! Even as recently as 375 million years ago, the moon orbited at about half its current distance from the earth.

■ How does the tidal friction created by the moon affect the earth's rotation?

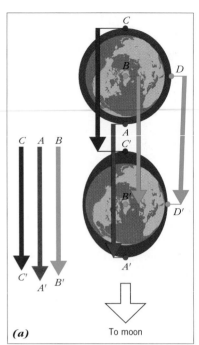

(a) To moon

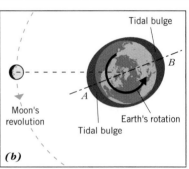

(b)

FIGURE 9.6 Tides and the moon. (*a*) Imagine a smooth earth covered with water. Consider the water at *A*, *C*, and *D* pulled by the moon's gravity. How far the water falls in some time depends on its distance from the moon; water at *A* falls farther than that at *C*. Water at *D* and water at the earth's center (*B*) fall the same distance (because the acceleration on them is the same). So as the earth falls toward the moon (from *B* to *B'*), water on one side falls a bit more (from *A* to *A'*) and on the other side a bit less (from *C* to *C'*). Water flows from *D* to supply the bulges; the depth at *D'* is less than at *A'* or *C'*. (*b*) The tidal bulges on the earth and the moon's motion. The bulges spin ahead of the moon because the earth rotates faster than the moon revolves and friction carries the bulges with the earth's surface. This friction slows down the earth's rotation. Meanwhile, the bulge at *A* acts gravitationally to accelerate the moon in its orbit.

FIGURE 9.5 The north polar regions of the moon, imaged by the Galileo spacecraft. The maria visible from the earth encircle the bottom of the view. (Courtesy of NASA.)

9.2 The Moon's Surface Environment

Because the moon has smaller mass and radius than the earth, its surface gravity is less. How much less? A planet's surface gravity is simply the acceleration of any object at its surface from the planet's gravitational pull. The moon has a mass 1/81 that of the earth and a radius 0.27 that of the earth, so the moon's surface gravity is one-sixth that of the earth. Objects weigh one-sixth as much on the moon as they do on the earth.

The moon has a tenuous no atmosphere. Why? The answer involves **atmospheric escape.** Gravity holds down any atmosphere of a celestial body. An atmosphere consists of a gas of molecules and atoms, moving at various speeds. The temperature of the gas is a measure of the average speed of the particles in it. Typically, gas particles collide frequently. But in the thin upper layers of an atmosphere, far fewer collisions occur. If a gas particle has escape speed here, it speeds off into outer space. For the moon the escape speed is only 2.4 km/s. At typical lunar temperatures, even fairly massive gas particles at the surface have escape speed. The main natural constituent of the lunar atmosphere is neon, which has a high mass. Yet, all the gases totaled up give a surface density a mere 10^{-15} that of the earth's. A basketball stadium on the earth contains as much total mass of air as does the entire lunar atmosphere!

Most gases have escaped from the moon's gravitational grasp since its formation. Some material from the solar wind travels near the moon and stays briefly, but it does not amount to much. The exhaust from *one* Apollo landing dumped more gases into the atmosphere than had existed there! But these gases will not stay around very long either: for example, oxygen dumped at the moon's surface escapes in about 100 years.

We can estimate an airless planet's surface temperature from its energy budget. Suppose that only sunlight heats the planet's surface (Fig. 9.8). The surface absorbs some of the incoming sunlight and reflects some back into space. The moon reflects 7 percent of the light that hits it and absorbs the remaining 93 percent. (Even though a full moon seems to shine brightly in the sky, the moon's surface is really quite dark.) The absorbed sunlight heats the surface; it radiates mostly infrared. The balance between incoming sunlight and outgoing infrared sets the surface temperature of the sunlit side, which hits 375 K (about 100° C).

At night, solar heating stops, and the infrared—because no atmosphere traps it—radiates away into space. Since the lunar night is so long (about 15 earth days), the surface has time to cool, and the temperature plummets to 100 K (about −175° C). The large noon-to-midnight temperature difference occurs because the moon rotates slowly and has no atmosphere. We expect the same kind of large temperature variation from any airless, slowly rotating planet.

Without a notable atmosphere, the moon has no shield from the lethal x-rays and ultraviolet radiation from the sun or from the small, solid particles coming in from space. The moon's cratered surface presents a fierce, unfriendly place for people. The Apollo astronauts, imprisoned in their bulky life-support systems, found the moon bleakly beautiful but uninviting.

■ **If the moon had its current mass but were the same size as the earth, would your weight be more or less than it is on the earth?**

9.3 The Moon's Surface: Pre-Apollo

Until the Apollo astronauts walked on the lunar surface and sampled it, the study of the moon was confined to views

FIGURE 9.7 Tides and the earth's orbit. As the earth's spin slows down (Fig. 9.6b), the moon's *apogee* distance from the earth increases at a current rate of 4 centimeters per year. Effect (and also the eccentricity of the orbit) exaggerated here for clarity.

The Moon

At a Glance

Equatorial radius	1738 km
Mass	7.35×10^{22} kg
Bulk density	3340 kg/m³
Escape speed	2.4 km/s
Atmosphere	Neon, helium (thin!)

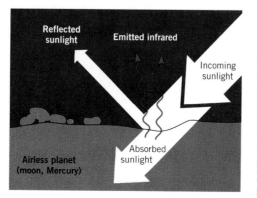

FIGURE 9.8 The energy budget for an airless planet. The ground reflects and absorbs incoming sunlight. The absorbed light heats the ground, which emits mostly infrared to space. The balance of loss and gain fixes the local surface temperature. The relative amounts of absorbed and emitted sunlight are shown by the widths of the respective arrows.

MOMENTUM AND ANGULAR MOMENTUM

The concept of momentum rests on that of the inertia of matter. Consider the following. A bicycle and a truck are coming at you at the same speed. Which would be easier for you to stop? The bicycle, because it has less mass. Now imagine two bicycles coming at you, one with twice the speed of the other. Which is easier to stop? The one moving more slowly. These examples show you that momentum depends on both the mass and the velocity of the object involved. In fact, momentum is defined as the product of an object's mass and velocity:

$$\text{momentum} = \text{mass} \times \text{velocity}$$

If we let p be the momentum, m the mass, and V the velocity, then

$$p = m \times V$$

Note that velocity has a direction, so momentum does too, the same as that of the velocity. You can think about momentum like this: once you get a mass moving, you have to put out an effort to stop it.

Consider a spinning object, such as the earth rotating about its axis. What keeps it spinning? Its inertia about its spin axis. The faster it spins and the more mass is spinning, the harder the object is to stop. This spinning momentum is called **angular momentum.**

You can think of angular momentum as the tendency for bodies, because of inertia, to keep spinning (rotating) or orbiting (revolving). Angular momentum of a body is determined by its mass (but now the distribution of the mass, around the center of motion, complicates the picture), the velocity (around the center of motion), and the radius (the distance of the mass from the center of motion). For a single particle moving in a circle, we can write

$$\text{angular momentum} = \text{mass} \times \text{circular velocity} \times \text{radius}$$

Let L be the angular momentum, m the

mass, V the circular velocity, and r the radius. Then

$$L = m \times V \times r$$

For a body such as the earth, which is made up of many particles moving at different velocities at different distances from the axis of rotation, we must add up the angular momenta of all the particles.

The key aspect of angular momentum is that it is *conserved*. If no twisting forces, called *torques,* act on an object, its angular momentum remains the same. (A torque is a force applied through a lever, such as a torque wrench used to tighten spark plugs in an engine. The owner's manual for my truck specifies a torque of 20 newton-meters.)

You can test this idea by performing a simple experiment (Fig. F.7). Tie a ball to the end of a string and whirl it around your head at a constant speed. Now, with your free hand, grab the loose end of the string. Very slowly pull the end through your hand, shortening the string until it is at half the distance you started with. The ball will move with double its circular speed.

Note that as the distance from the center of spin decreases, the rate of spin increases. But no torques have been applied, so the angular momentum is the same with the string at the different lengths. This is an example of the conservation of angular momentum.

Tidal forces affect the earth–moon system. The total angular momentum of this system consists of two parts: the spin angular momentum (of the earth and the moon rotating about their axes) and the orbital angular momentum (of the moon revolving about the earth). The **conservation of angular momentum** says that the sum of these must remain constant. Because the moon rotates slowly and its mass is small, the moon's spin adds very little to the total and can be ignored. So we need to consider only the spin of the earth and the revolution of the moon. Because of tidal friction, the earth's rotation is slowing down, so its spin angular momentum decreases. For the system's total angular momentum to remain constant, the moon's orbital angular momentum must increase. This can happen only if the moon moves away from the earth.

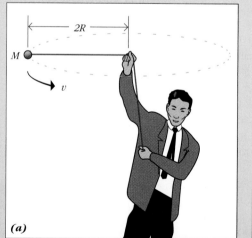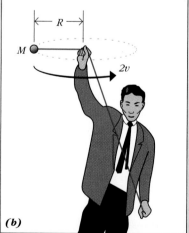

FIGURE F.7 Illustration of the concept of the conservation of angular momentum. (*a*) A mass at the end of a string is made to swing in a circle. (*b*) When the length of the string is decreased, the circular speed of the mass increases. Here at half the radius, the mass moves at twice its original circular speed.

from the earth, from orbiting satellites, or from lunar landers. Galileo first studied the moon carefully with a telescope in 1609 (Section 4.1). He named the dark areas **maria** (Latin for "seas"; the singular form is **mare.**)

Viewed through a small telescope or binoculars, the moon reveals a fascinating terrain. Lunar craters give the moon a pockmarked face. Mountains and craters irregularly rim the moon's edge. Some mountains stand alone in the maria, while others are linked in long ranges. Bright rays flower from some craters. The moon impresses you as a rough, old world that has indeed suffered the incredible violence that forged its blemished surface.

Maria and Basins

Most of the maria lie in the northern half of the moon's hemisphere on the side toward the earth; few are on the far side. The maria appear dark compared with the rest of the surface. Their dark irregular stretches form the face of the "man in the moon."

The maria have other general features that give clues to their formation. Most maria look circular, and some are interconnected. They have smooth surfaces compared with the brighter, cratered regions. Also, the maria have lower elevations, by about 3 km, than the rest of the surface. So the maria are called the lunar **lowlands;** the other areas, the **highlands.** The lowlands cover some 20 percent of the moon's total surface; the highlands, more than 70 percent.

The vast flat extents of the maria are solidified lava flows. If you look carefully at the regions around the maria, you can find craters flooded in by the dark material from the maria (Fig. 9.9). This feature indicates that the lava flow that formed the maria occurred *after* the formation of the lunar crust and certain craters. The lunar maria fill up large, shallow **basins** on the moon's surface, which range in size from 700 to 3000 km.

For example, Mare Imbrium Basin (Fig. 9.10*a*), on the moon's near side, is roughly circular in outline and has a diameter of about 1200 km. The moon's most striking basin is Mare Orientale (Fig. 9.10*b*) with its concentric rings of

FIGURE 9.9 Close-up of the surface of Mare Imbrium. Note the smoother, darker, and lower terrain here compared to the background highlands. This oblique view shows the large crater Copernicus at the top horizon. The largest crater near the center is called Pytheas, almost 17 km in diameter. This viewing angle accentuates the relief within the mare. (Courtesy of NASA.)

mountains. The outer rim of mountains (called the Cordillera Mountains) has a diameter of about 970 km and rises to 7 km. Surrounding this basin for about 1000 km outward lies a blanket of lighter material covering the older lunar surface. If most of its filling material were removed, Mare Imbrium would look like Mare Orientale.

Unusual concentrations of mass lie beneath most maria. These are called **mascons.** One mascon, for example, under Mare Imbrium, has a mass of 1.6×10^{19} kg and an average density of 3700 kg/m^3. (Note that this density is *greater* than the average density of the moon.) If this mascon were spread out to cover California, it would be 12 km thick! Some common process has produced large amounts of dense material under the maria.

Craters

Craters (from the Greek word meaning "cup" or "bowl") are round depressions that litter the moon almost everywhere.

(a)

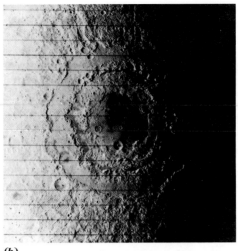

(b)

FIGURE 9.10 Lunar basins. (*a*) Mare Imbrium, in the moon's northern hemisphere on the near side (Fig. 9.2*b*). Note the few craters on the mare's surface and the flooded, semicircular region at the top. (Courtesy of Yerkes Observatory.) (*b*) The Orientale Basin on the moon's far side. The Cordillera Mountains ring the basin like a bulls-eye, an indication of the impact process of the basin's formation. (Courtesy of NASA.)

Their bowls range widely in size: many are smaller than a pinhead; five have diameters greater than 200 km (about the size of Connecticut). Craters are generally round. The heights of the rims of lunar craters are small compared with the diameters, and the floors are depressed compared with the surrounding landscape.

Most lunar craters (perhaps all!) were formed by **impact cratering.** Solid objects from space slammed into the moon's surface and punched out the craters. The explosion initially makes the rock deform like a fluid. An impact pro-

FIGURE 9.11 The crater Aristarchus, about 60 km wide, with its ray system. Note the ripples in the surface just outside the crater; these were formed from the shock of impact, as was the central peak. (Courtesy of NASA.)

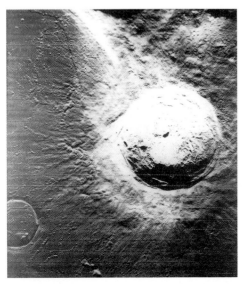

duces a crater with undulating slopes, usually covered by debris, with a rippled terrain around it (Fig. 9.11). If the debris and crater wall were put back into an impact crater, they would fill it up. Material blasted out by an impact falls in streaks, leaving a raylike pattern (Fig. 9.12). Large chunks of ejected (thrown-out) material can create small secondary craters around the main impact crater.

What projectiles made these impacts? Small, solid bodies orbit the sun; they are called *meteorites* when they strike the earth. If they pass close enough to the moon, they will collide with it. But we have seen no new, large craters form recently. (One of the youngest is some 2 million years old.) So the era of heavy-impact cratering is long past. The Apollo missions aimed to find out *when* it occurred.

Cratering

Craters scar every terrestrial planet and nearly every satellite in the solar system. That cratering has happened throughout the solar system provides clues to processes that formed the terrestrial planets. And cratering serves as a tool by which to infer the relative ages of planetary surfaces.

The infall of massive objects onto a planet's surface creates impact craters. The great abundance of craters on the planets and satellites implies that a storm of ancient impacts blasted them long ago. Since that time, erosion and crustal evolution have modified the original pattern and changed the planets' surfaces.

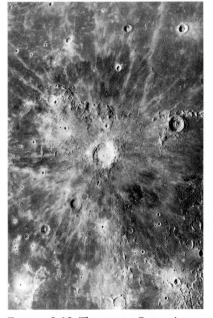

FIGURE 9.12 The crater Copernicus (center; see Fig. 9.2*b* for its location), with its extensive ray system formed by the impact throwing material out of the crater. Copernicus is about 93 km in diameter and is believed to have formed about 900 million years ago. (Courtesy of the Observatories of the Carnegie Institution of Washington.)

In addition, if we can estimate the rate at which cratering occurred, the number of craters will give a guide to the age of the surface on which they are found.

A few major points about cratering. On any planet's surface, small craters greatly outnumber the larger ones. The size of a crater is related to the *kinetic energy* of the projectile that formed it. The largest of these must have been rocks greater than 100 km in diameter. These objects strike planetary surfaces with speeds of 10 km/s or more. They bang into the surface with enormous amounts of energy: some 10^{23} J for the mass that formed the crater Copernicus on the moon. For comparison, the eruption of Mt. St. Helens in May 1980 blew up with an energy of 10^{17} J, an explosive yield equivalent to 35 megatons of TNT. So large craters involve explosions a *million* times greater!

Upon impact, the projectile's kinetic energy converts into an explosive force below the ground at a depth only a few times the diameter of the infalling mass (Fig. 9.13). A shock wave from the impact spreads through the rock, deforming it and throwing it outward. The volume blown out is much larger than that of the projectile, so small objects result in big holes. Because the explosion occurs below the ground, even projectiles hitting at oblique angles leave round craters. (Examine the photos and notice how lunar craters are round.)

Even on an airless world, craters, once formed, can be wiped out by a number of processes: later impacts, materials thrown up from younger, nearby craters, and lava flows. Large craters are less likely to be obliterated by these processes than small ones, so generally the largest craters on a surface are the oldest. Overall, the oldest part of the surface will have the highest density of craters on the surface. We can rank the planets and moons according to the relative amounts of modifications of their surfaces.

Our moon stands in stark evolutionary contrast to the earth. The oldest regions of the lunar highlands have not changed much since their formation. The analysis of cratering here combined with radioactive dating of Apollo samples tells us that a little more than 4 billion years ago the cratering rate was 1000

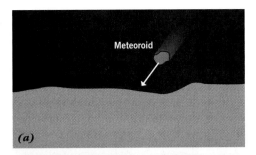

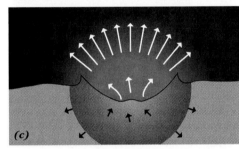

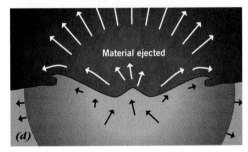

FIGURE 9.13 Formation of a crater by impact of a solid body. (*a*) A meteoroid strikes the surface at a speed of some tens of kilometers per second. (*b*) The initial explosion ejects surface material at a high speed. (*c*) Shock waves through the rock compress and fracture it. The rebound of the rock throws material out of the crater. (*d*) The rim of the crater is folded back by the rebound. Ejected material can form rays and smaller craters surrounding the main impact.

times greater than now. It tapered off, reaching close to the present rate about 3 billion years ago. The moon preserves the record of the *intense bombardment* that, as you will see, occurred throughout the solar system during its first billion years.

■ Were the maria formed before or after the lunar crust was formed?

9.4 Apollo Mission Results

The Apollo program obtained about 400 kg of lunar material from six different sites. Scientists have examined only about 10 percent of it in detail. Already these samples, and other experiments in the Apollo program, have provided the first deep understanding of a planet other than the earth. We can now sketch out many physical details of the moon and outline its history.

The Lunar Surface

The Apollo samples reveal the physical and chemical nature of the lunar surface. The very top layer (1 to 20 m deep) is a porous, somewhat adhesive layer of debris (Fig. 9.14). It consists of fine particles (called **lunar soil**) and larger rock fragments. The soil samples contain a large amount of mostly round pieces of glass (Fig. 9.15), which makes the surface slippery, as the Apollo astronauts found by slipping around!

The moon rocks mostly fall into three categories: dark, fine-grained rocks (Fig. 9.16*a*) similar to terrestrial basalts (magnesium–iron silicates), called **mare basalts;** light-colored igneous rocks (Fig. 9.16*b*) with visible grains, which are called **anorthosites** (aluminum–calcium silicates)—by far the most common rock on the surface; and **breccias,** rock and mineral fragments bound together (Fig. 9.16*c*).

What do the observed characteristics imply? Moon rocks are igneous rocks, so they formed from the solidification of lava. The rate at which lava cools determines the grain sizes of igneous rocks: fast cooling results in small grains, slow cooling in large ones. So the dark rocks (found in the lowlands) cooled faster than the light-colored ones (found in the highlands). In addition, the light-colored rocks are less dense than the dark ones. These two density classes arise from different compositions. It indicates that the anorthosites formed from partial melting and slow cooling of low-density silicates at a relatively low temperature (about 1300 K) inside the moon. In contrast, the mare basalts were hotter, they cooled more rapidly, and they were formed from more dense materials.

We can infer how the breccias formed. Imagine new igneous rocks on the moon's surface. Bodies from space pound into these rocks, fragmenting and heating them. The heat binds some fragments together to make breccias. The

loose material left over makes up the lunar soil.

In a few key ways, moon rocks differ strongly from the earth's igneous rocks. For example, they contain more titanium, uranium, iron, and magnesium. Compared with earth rocks or meteorites, moreover, they are depleted of elements that would condense at relatively low temperatures (less than 1300 K). These elements, called **volatiles,** include sodium, potassium, copper, argon, and chlorine. Moreover, the moon rocks turned out to be bone dry, whereas earth's surface rocks always have some water locked up in their minerals.

In one critical way, on a nuclear level, the moon resembles the earth. The isotopic composition (Section 5.1) of oxygen (relative abundances of oxygen-16, -17, and -18) in the lunar samples is the *same* as that for the earth. Studies of the oxygen isotopic compositions of meteorites, on the other hand, show a distinct variation among samples. Some meteorites are thought to be primitive materials from the formation of the solar system. Indirect evidence suggests that the variation in oxygen isotopes corresponds to condensation in different parts of the

FIGURE 9.14 The moon's surface consists of fine particles, called lunar soil, with larger rock fragments mixed in. Note the footprints left by the astronauts. (Courtesy of NASA.)

FIGURE 9.15 Close-up of lunar soil. Note the glass spheres that form when the surface melts upon an impact and droplets of silicates cool quickly. The coarse rock fragments are only 1 to 2 mm in diameter. (Courtesy of NASA.)

solar system. The similarity for earth and moon shows that these bodies formed in the same general region of the solar system.

Finally, compare the bulk densities of the moon and earth: 3300 kg/m³ versus 5500 kg/m³. This difference implies that the moon overall contains less metals and probably does *not* have a large metallic core, as the earth does (Section 8.2). We don't even know for sure that the moon has a well-defined core!

Moon rocks are dated by the same radioactive decay techniques used to date earth rocks (Focus 8.1). One caution: these methods give the time that has elapsed since the rock last solidified. If a rock has been heated enough to melt it after its original formation, radioactive dating begins with that reheating.

Only a few rocks and some fragments from the lunar soil have ages as great as 4.6 billion years. The anorthosites from the highlands generally are the oldest: 4 billion years. The mare basalts are the youngest: some only 3.2 billion years old and a few as old as 3.8 billion years. These ages imply that the moon formed a little more than 4.6 billion years ago. After formation, the present highlands solidified, about 4 billion years ago. Then the lava flows that made the maria took place, some 3 billion years ago.

The Moon's Interior

The Apollo missions probed the moon's interior with the same shock-wave method used to look into the earth. They found the moon to be an inactive world (compared with the earth). Few moonquakes occur, and those few take place about 800 km below the surface. Such shocks release at most about 100,000 J, barely a tremble. If you stood directly over the strongest moonquake so far recorded, you would not even feel the ground shake. In contrast, the great San Francisco earthquake of 1906 released about 10^{17} J! Geologically the moon is a quiet world.

Such low activity indicates that the moon is cold and solid to a depth of about 1100 km (Fig. 9.17). This region makes up the moon's crust and its mantle, which very likely consists of silicates

(a)

(b)

(c)

FIGURE 9.16 The main types of lunar rock. (*a*) Typical mare basalt. The holes were air bubbles in the lava. The pit in the center resulted from an impact by a small meteorite. (*b*) Rock containing anorthosite (white area). This sample from Apollo 15 is known as the Genesis Rock. (*c*) Lunar breccia. The dark areas are melted rock. Large pieces of other rock are visible embedded in it. (All courtesy of NASA.)

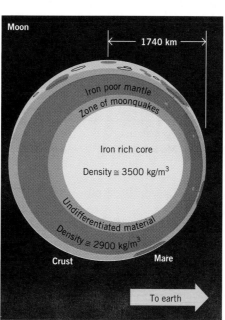

FIGURE 9.17 Interior structure of the moon inferred from Apollo seismic and heat flow measurements. Note the relatively large core of iron-rich silicates, the rocky mantle, and the relatively thick crust. Metals make up a small percentage of the total composition.

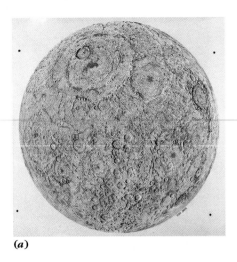

(a)

(b)

(c)

FIGURE 9.18 Model for the moon's evolution. (*a*) 3.9 to 4 billion years ago: the surface has solidified and is cratered by the infall of planetesimals; most of the surface is saturated with craters. (*b*) 3.0 to 3.2 billion years ago: fractures in the surface allow magma from the interior to flow out and fill the lowland basins with material darker than the highlands. (*c*) Now. Many of the large rayed craters blossomed on the surface after the formation of the maria. Parts of the maria have grown lighter in color from the material blown out by crater formation. (Paintings by D. Davis and D. Wilhelm, U.S. Geological Survey.)

only a little more dense than the surface. The mantle is too cool to allow plate tectonics. Encasing the mantle is the crust, which is layered. On the surface lie the rocks and soil sampled in the Apollo missions.

A well-defined lunar core, if such exists, probably makes up the inner 500 km of the moon. It may be hot, about 1500 K, and molten in whole or in part. Evidence for a hot core comes from the Apollo measurement of heat flowing up through the lunar surface—about one-third that of the earth's outflow. So some part of the moon's interior is relatively hot. Since we know that the mantle is solid and cool, we can identify the core as the source of the heat outflow.

Is the core liquid like the earth's? Probably not. The moon's magnetic field is very weak, only 10^{-4} times as strong as that of the earth. If the dynamo model for a planet's magnetic field is correct (Section 8.3), the moon's core cannot be completely molten or composed mainly of metals. (Also, the moon's density is too low to permit a substantial metallic core.) On the other hand, some surface rock samples are magnetized much more than you would expect from such a weak magnetic field. Recall (Section 8.3) that iron minerals in an igneous rock preserve the magnetic field present at the time of solidification. So the moon's magnetic field was stronger in the past than it is now.

Lunar History and Evolution

From the Apollo results we can concoct a scenario of the moon's history since its

formation. The inferred sequence of events (Fig. 9.18 and Table 9.1) relies heavily on the dating of the lunar rocks.

About 4.6 billion years ago, the moon formed by the accretion of chunks of material. These pieces continued to plunge into the moon after most of its mass had gathered. During the first 200 million years after formation, these projectiles from space bombarded the surface and heated it enough to melt it. Less dense materials floated to the surface of the melted shell; volatile materials were lost to space. The crust began to solidify from this melted shell about 4.4 billion years ago. From 4.1 to 4.4 billion years ago, the crust slowly cooled as the bombardment from space tapered off. The debris from this later bombardment made many of the craters now found in the highland areas.

EVENT	TIME (billions of years ago)	PROCESSES
Formation	4.6	Accretion of small chunks of material
Melted shell	4.6–4.4	Melting of outer layer by heat from infall of material and/or radioactive decay; volatile elements lost
Cratered highlands	4.4–4.1	Solidification of crust while debris still falls in to crater it
Large basins	4.1–3.9	Reduced infall, but formation of basins by impact of a few large pieces; outflow of more basalts from lava below solid crust
Maria flooding	3.9–3.0	Flooding of basins by lava produced by radioactive decay
Quiet crust	3.0–now	Bombardment by small particles to pulverize and erode surface

TABLE 9.1 EVOLUTION OF THE MOON

Below the surface, the moon's material remained molten. About 4 billion years ago, a few huge chunks smashed the crust to produce basins that later became maria. (Most of these large impacts happened on one side of the moon, which now faces us today because of tidal forces.) For example, the Mare Orientale Basin formed some 4 billion years ago when an object about 25 km across smashed into the moon. Only later did the basins fill in. As the crust lost its original heat, short-lived (rapidly decaying) radioactive elements reheated sections of it. From 3.0 to 3.9 billion years ago, lava from the radioactive reheating punctured the thin crust beneath the basins, flowing into them to make the maria.

For the past 3 billion years, the lunar crust has been inactive. However, small particles from space have incessantly plowed into the surface since its solidification. These sand-sized grains scoured the surface, smoothed it down, and pulverized it. Continued bombardment by larger bodies churned the fragmented surface. Impacts melted the soil, which swiftly cooled to form breccias and glass spheres. The moon's surface today resembles a heavily bombarded battlefield—constantly blasted, fragmented, stirred up, and melted.

The moon is a low-density, rocky world that cooled quickly. Why? Because it has a large heat-radiating surface compared to its heat-producing (or storing) volume and mass. The basic rule is: *the smaller the planet, the faster it loses thermal energy*. Compared to the earth, the moon had only a brief episode of volcanism, and it never developed tectonic activity.

■ In what general way does the moon's interior composition *not* resemble the earth's?

9.5 The Origin of the Moon

The implications of the Apollo cargo so far have clarified few aspects of the moon's origin. None of the rival explanatory ideas has completely claimed supremacy. I will present just two: binary accretion and the giant impact model, which is the strongest contender to date.

The Binary Accretion Model

The **binary accretion model** views the moon as created out of the same cloud of material as the earth, rather than born directly from our planet. Dust particles grew from a gradual condensation of gas. These eventually accreted into the young moon and earth. The moon formed so close to the earth that earth's gravity held the moon in a close orbit. The leftover bits and pieces from the formation of the two planets fell into the moon and heated its primitive surface.

The composition of lunar minerals discredits this idea. The moon and the earth are somewhat different in composition. And the moon has no water. In addition, the moon has no dense iron core. If the embryonic environment and materials of the two bodies were the same, how did the earth accumulate iron and the moon not? On the other hand, the oxygen isotopic composition of the earth and the moon are the same; this fact supports some binary accretion models.

The Giant Impact Model

Finally, we come to the **giant impact model,** a recent hybrid that relies heavily on binary accretion ideas. This model envisions a glancing impact of a Mars-sized body into the young earth after the core had formed (Fig. 9.19). The tremendous impact vaporized the earth's mantle and spewed some of the earth's mass (and a large fraction the impacting body's mass) into orbit around the earth (Fig. 9.20). The metal-rich core of the impacting body remained intact and joined up with the earth; what was its rocky mantle remained in orbit. The orbiting material condensed, creating a ring of debris. The moon accreted from this material (Fig. 9.21). The essential events were over in about a day.

Note that the chemical differences and similarities of the earth and moon are cleverly taken care of: the similarities arise from the ejected terrestrial material, the differences from those of the impacting body. The vaporization releases vol-

The Moon's Surface	
At a Glance	
Main features	Highlands (cratered), lowlands (filled-in basins)
Relative ages	Highlands older than lowlands
Formation	Of craters and basins by impacts
Oldest rocks	About 4.6 billion years

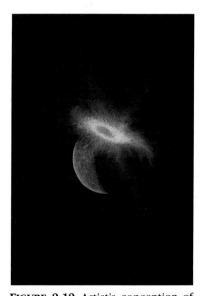

FIGURE 9.19 Artist's conception of the giant impact model of the moon's origin. A Mars-sized body hits the protoearth at grazing angle. Some material from both bodies remains in orbit in a disk around the earth and accretes to form the moon. (Painting by W. Hartmann.)

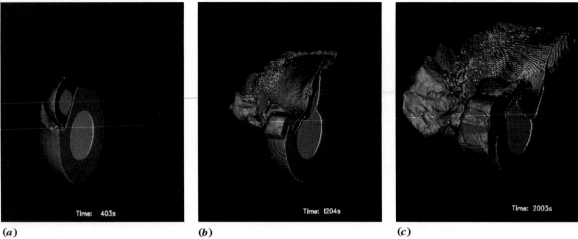

(a) (b) (c)

FIGURE 9.20 Supercomputer simulation of the impact of two bodies in a giant impact like the one that may have formed the moon. The target body represents the young earth. Both bodies have metal cores (red and pink) and rocky mantles (brown and green). After the collision (a), a plume of projectile and target material spreads out (b and c). Eventually the plume condenses into dust particles that accrete to form the moon at a initial distance of about 10 earth radii. Gravitational attraction brings together the clumps of material. This sequence provides a close-up view of the collision. (Courtesy of M. E. Kipp, Sandia National Laboratories, and H. J. Melosh, University of Arizona.)

atiles (such as water) from the material that forms the moon, so explaining their absence. The material ejected from the earth's mantle ensures that the oxygen isotopic abundances match. And the angular momentum is accounted for properly if it is assumed that the body struck the earth on a glancing blow at a speed of about 10 km/s.

Note that the formation of the moon must take place far enough from the earth so that the earth's tidal forces do not inhibit the accretion of masses. Recall (Section 9.1) that tidal forces pull objects apart. So two small masses will be separated by tidal forces if they are too close to the earth. In fact, if a single mass were placed too close to the earth, tidal forces could rip it apart (remember that the tidal force increases as the inverse *cube* of the distance).

You'd expect, then, that a planet has a minimum distance from its center within which a mass would be disrupted

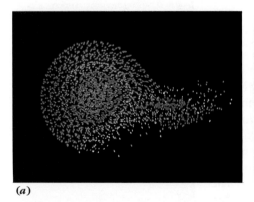
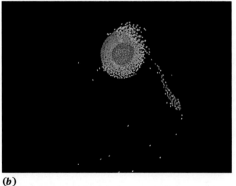
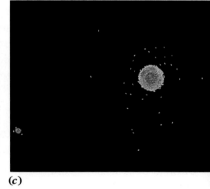

(a) (b) (c)

FIGURE 9.21 Computer model of events after the impact of two bodies to simulate forming the moon. The incoming object has a mass of 0.15 earth mass (slightly more than that of Mars). Elapsed time, in hours after the impact, is displayed in the upper right corners. Both bodies have an iron core (purple) and a silicate mantle (red). Following the collision (a), the impacting body spreads out. Gravitational attraction brings the clumps together. The iron core separates from the rocky mantle of the impacting body (b), and some 4 hours after the collision, it is swallowed up by the earth. (c) About 20 hours later, a silicate lump with a mass about like that of the moon orbits the earth. It is visible at the lower left. Note that in each frame we are viewing a larger area of space. (Courtesy of A. G. W. Cameron and W. Benz.)

by tidal forces. This distance is called the **Roche limit,** after the French mathematician Edouard Roche (1820–1883), who first worked it out. Now, the Roche limit depends on the densities of the bodies involved and the strength of the materials of which they are made. As a good rule of thumb, the Roche limit is roughly 2.5 times the radius of the parent planet. Our moon is about ten times farther than this limit. Any accretion of materials that formed the moon took place beyond the Roche limit.

Overall, the giant impact model has the fewest serious problems and seems so far to be the best bet for a reasonable explanation of the moon's origin.

■ **What key physical property must be explained by any lunar model?**

9.6 Mercury: General Characteristics

Mercury's surface resembles that of the moon in many ways, so we can apply what we know about the moon to Mercury. From such a comparison of these inactive, desolate bodies, we can glean some ideas about the processes that drive the evolution of planets without atmospheres or active interiors.

Orbit and Rotation

Mercury speeds around the sun once every 88 days in a very eccentric orbit. When closest to the sun (at *perihelion*), its distance from the sun is 46 million km; at its greatest distance (at *aphelion*), it lies 70 million km out. This difference means that sunlight at perihelion falls on the planet 2.3 times more intensely than it does at aphelion!

Study of the surface of Mercury is difficult because of its small angular size and poor visibility from the earth (remember, it always stays close to the sun: Section 1.4). But persistent observation reveals faint, dark, and apparently permanent markings (Fig. 9.22). In 1877 Giovanni Schiaparelli (1835–1910) constructed the first map of the Mercurian surface. Because the observed surface features did not seem to change (the

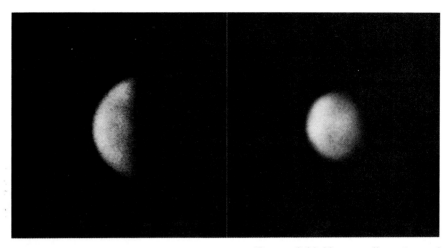

same face was observed), he concluded that the rotation rate of Mercury equaled its revolution rate (as for our moon). Schiaparelli thought Mercury always presented one side to the sun as it rotated once in 88 days—it had synchronous rotation.

This long-accepted view changed radically in 1965, when radio astronomers discovered that Mercury rotates in about 59 days. This value means that Mercury's rotation period is two-thirds its orbital period. This ratio results from tidal interactions between the sun and Mercury (much as the tidal interactions of the earth and moon have resulted in the moon keeping one side toward the earth). How long, then, is the solar day on Mercury—the time from noon to noon? It's about 176 terrestrial days, just twice the length of Mercury's year (Fig. 9.23).

Size, Mass, and Density

Mercury turns out to be a tiny world, only 4900 km in diameter. That's only about 40 percent larger than our moon. Mercury has no known natural satellite. Mariner 10 sped past Mercury three times in 1974 and 1975. The spacecraft's acceleration from Mercury's gravity was accurately measured so we could calculate Mercury's mass: about 0.06 that of the earth, or about 3.3×10^{23} kg.

For such a small planet, Mercury has a fairly large mass. That means its bulk density is high: 5400 kg/m³, essentially the same as the earth's.

Comparing Mercury's interior with that of the earth, we expect a large iron

FIGURE 9.22 Two excellent views of Mercury taken by earth-based telescopes. Note the vague, dusky features. (Photos by C. Knuckles, New Mexico State University Observatory.)

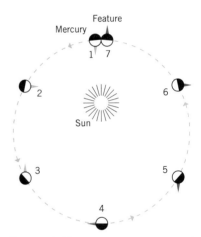

FIGURE 9.23 Mercury's rotation and solar day. Radar measurements in 1965 showed that the rotation period is 59 days with respect to the stars. So relative to the sun, a feature on Mercury at noon (1) ends up at midnight (7) after one revolution. After the next revolution, it is at noon again (1). So the solar day on Mercury is twice the orbital period or 176 days.

alloy core (Fig. 9.24) and a rocky mantle. Although we haven't yet measured seismic waves to confirm this interior model, it's probably correct by analogy to the earth.

Surface Environment

Suppose that you stood on Mercury's equator at perihelion at noon. You would need to be made of sturdy stuff, because the surface temperature would be about 700 K! The surface temperature drops to 425 K at sunset and reaches about 100 K at midnight. This range of temperatures is the widest known in the solar system. (On earth, the day/night variation rarely exceeds 20 K.)

The long, hot solar day and low escape speed (4.3 km/s) make it unlikely that Mercury has any atmosphere. Gas molecules, even the most massive ones, would easily be heated to escape speed. So no atmosphere could be expected to last long. The Mariner 10 space probe detected an atmosphere (of sorts) of helium and hydrogen (probably picked up from the solar wind). But the surface atmospheric pressure was very small, a little less than 10^{-15} that of the earth.

Later ground-based observations indicate that sodium and potassium vapors exist in the atmosphere on the day side. The sodium and potassium are probably released from the rocks, which absorb ultraviolet light from the sun. Overall, the sodium vapor makes up the major part of Mercury's very thin atmosphere. No substantial atmosphere means no insulation from space, so the range of noon-to-midnight temperatures on Mercury is severe.

Mercury's Surface

Mercury's surface is like that of our moon (Fig. 9.25)! There are some differences, though: fewer large craters (20 to 50 km in size), no mountains, and many shallow, scalloped cliffs, called **scarps** (Fig. 9.26), reaching lengths of hundreds of kilometers and rising to 1-km heights. In addition, there are fewer basins and large lava flows, and more relatively uncratered plains amid the heavily cratered regions.

Mercury's highlands are riddled with craters like the moon's bleak highlands (Fig. 9.27). Light-colored rays spring from some of the craters, an indication that these formations resulted from violent impacts during Mercury's stormy past. Some craters are larger than 200 km in diameter, comparable to the biggest lunar craters.

And what of large maria basins, such as found on the moon's near side? Mariner 10 found a large one: the Caloris Basin (Fig. 9.28). Since the Caloris Basin sat on the sunrise line, only about half of it was photographed. It probably has an overall diameter of some 1300 km. The basin is bounded by rings of mountains about 2 km high. The Caloris Basin has a crinkled floor, perhaps representing fractures from rapid cooling of lava. Also visible are older craters flooded by the lava outpouring from the Caloris impact. In size and structure, the Caloris Basin resembles the moon's Mare Orientale. Mercury's surface has more than 20 large, multiringed basins—many very old and covered with later impact craters. Violent impacts marred the surface in the past.

These similarities do *not* mean that the moon and Mercury are identical. Their surfaces differ in at least three ways: Mercury's surface has scarps hundreds of kilometers long; even the most heavily cratered regions are not completely saturated with craters; and Mercury has fewer small craters than would

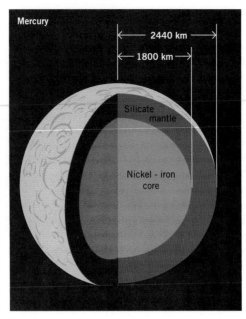

FIGURE 9.24 Model for Mercury's interior. Note the large metallic core (relative to the size of the planet) and the small, rocky mantle. Compare to the earth (Fig. 8.2).

FIGURE 9.25 The cratered surface of Mercury photographed by Mariner 10. Note the heavy cratering, which resembles that of the light-colored regions on the moon. The horizontal lines and small black dots are artifacts of the imaging process. (Courtesy of NASA.)

Mercury

At a Glance

Equatorial radius	2430 km
Mass	3.30×10^{23} kg
Bulk density	5430 kg/m³
Escape speed	4.3 km/s
Atmosphere	Sodium, potassium (thin!)

be expected if the craters had formed at the same rate as on the moon. From the last two points, we can infer that the cratering objects had different ranges of sizes for Mercury and for the moon. That is likely because the gravitational effects of the sun on nearby Mercury would influence the distribution in the sizes of orbiting bodies.

Mercury's scarps vary in length from 20 km to more than 500 km and have heights from a few hundred meters to 1 km. Individual scarps often travel over different types of terrain. If the regions photographed by Mariner 10 represent Mercury's overall surface, the characteristics of the scarps imply that Mercury's radius has shrunk some 1 to 3 km (at which time the scarps formed). How? Probably from cooling of the planet's core, its crust, or both, much as the skin of a baked apple wrinkles as it cools.

Magnetic Field

Mariner 10 detected a weak planetary magnetic field, about 0.01 times the strength of the earth's magnetic field. Small as this sounds, it's sufficient to carve out a magnetosphere in the solar wind (Fig. 9.29). Mercury's field is a di-

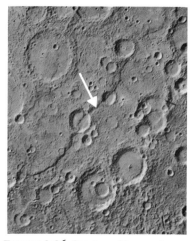

FIGURE 9.26 Scarp on Mercury's surface (arrow) more than 300 km long. It extends diagonally from upper right to lower left. (Courtesy of NASA.)

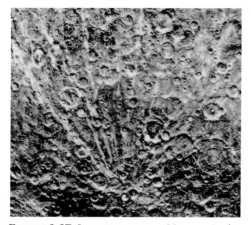

FIGURE 9.27 Impact craters on Mercury in the south polar region. The craters with the brightest rays are the youngest. The largest craters here are about 200 km across. (Courtesy of NASA.)

FIGURE 9.28 Mercury's Caloris Basin (arrow). Only a part shows in the left half of this photo; the rest is in shadow. The basin is about 1300 km across and is rimmed by mountains 2 km high. (Courtesy of NASA.)

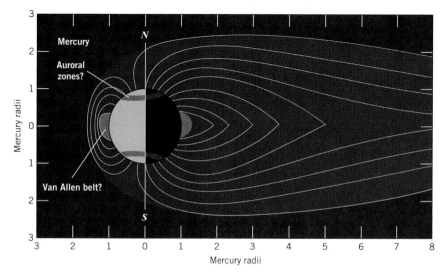

FIGURE 9.29 As inferred from Mariner 10 observations, Mercury's magnetosphere may trap solar wind particles to form Van Allen-type radiation belts. Reconnection events may cause storms that form auroral zones on the surface of the planet because there is no atmosphere to stop the particles. (Adapted from a diagram by D. Baker, J. Borovsky, G. Gisler, J. Burns, and M. Zeilik.)

pole, more or less aligned with its spin axis. In general, Mercury's magnetic field is similar to the earth's, only weaker. And that's the problem. Mercury, like the earth, has a metallic core. But it's presumed to be relatively cold and solid now because a small planet loses heat quickly. Recall (Section 8.3) that the earth's magnetic field arises from swirling motions in its hot, liquid metallic core. The churning is thought to be driven by the earth's spinning. Because Mercury rotates much more slowly than the earth and is thought to have a cool core, no one really expected it to have a planetwide magnetic field.

What's the explanation? No one knows for sure. Perhaps part of the core is fluid. Or the dynamo model may be incorrect; after all, Venus, which has a much larger and presumably hotter core than Mercury, has no detectable magnetic field. (But Venus rotates even more slowly.) Or, perhaps the field is left over from an older time. Maybe it originated from the planet's interaction with the solar wind. Or perhaps small planets with cool metallic cores produce magnetic fields by a mechanism not yet imagined.

■ For Mercury, a solar day lasts two years! How and why?

9.7 The Evolution of the Moon and Mercury Compared

What forces have shaped the surfaces of Mercury and the moon? Impact cratering! Both the moon and Mercury lack substantial atmospheres, so weathering does not erode the surfaces. Both are tiny worlds with relatively cool interiors compared to the earth's interior. So neither has much (if any) volcanic activity now, and neither has undergone continual surface evolution, as the earth experiences from tectonics (Section 8.5).

The lack of an atmosphere and the short period of crustal evolution are related to the small masses of Mercury and the moon. Their surface gravities are so low that most gases have escape speeds,

and an atmosphere is lost. The small masses also imply that internal heating from radioactive decay would be less than that for the earth, and the flow of heat outward would be so fast that both bodies would cool off quickly. The earth is hot at its interior, and the outward flow of heat sets up currents in the plastic mantle. Both Mercury and the moon lack this combination of factors to drive the evolution of their crusts.

Without surface erosion and crustal evolution, the moon and Mercury retain evidence of their early years. The similarities of surface features suggest like histories. For example, the Caloris Basin on Mercury resembles maria on the moon. The lunar maria were formed by the impacts of large bodies, which created large basins. The inflow of lava from the interior filled them to make the maria. Such processes likely built the Caloris Basin.

Using lunar analogies, we can set up a working model for the evolution of Mercury. Once formed by accretion, Mercury probably went through the following general stages: (1) heating of the surface (by impacts and/or radioactive decay) and formation of a solid crust, (2) heavy cratering, (3) formation of impact basins, (4) filling in of basins by volcanism, and (5) low-intensity cratering. We can't date this sequence as we can for the moon because we do not have rock samples from Mercury's surface. But a lunar comparison suggests that the intense sculpting of Mercury's surface took place about 4 billion years ago, not long after the planet formed.

The moon and Mercury may have an affinity in their origins. Mercury has a *much* higher percentage of metals than is predicted by models for the origin of the solar system (Chapter 12). As a differentiated terrestrial planet, Mercury has too small a mantle. A giant impact early in Mercury's history could have stripped off a rocky mantle, leaving mostly a metallic core behind. Hence, the formation of the moon and Mercury may have involved violent impacts in the early days of the solar system.

The moon and Mercury are dead worlds, their dramatic evolution ended early. Compared with the earth, they are

Small Planet Evolution

At a Glance

Formation	By accretion
Volcanism	Short episode early on
Internal evolution	Brief because of rapid cooling
Main process	Impact cratering

fossil planets rather than living ones: they have passed through only the first stage of earthlike planetary evolution; they are the least evolved of the terrestrial planets.

■ What is similar about the early days of the moon and Mercury? How can we know so much about their history today?

Key Concepts

1. The moon and Mercury are the two smallest terrestrial planets. Their bulk densities indicate that they are made of mostly rocks (moon) and metals (Mercury). Their small sizes mean that they cooled off quickly.

2. The most prominent surface features of the moon are craters, formed by the impacts of solid bodies from space some 4 billion years ago. Large basins were also created by the impact of large masses.

3. The moon has very thin atmosphere, because the escape speed is so low. The surface temperature ranges widely, from 375 K at noon to 100 K at midnight, over the lunar day.

4. The tidal interactions of the moon and the earth keep the same side of the moon toward the earth, raise the ocean tides, and cause the moon to move away from the earth at the rate of a few centimeters a year. Tidal forces result from the differences in gravitational forces; they pull bodies apart.

5. The main regions on the moon are the highlands (covered with low-density, light-colored rocks) and the lowlands (covered with darker, higher-density rocks). The Apollo samples, dated by using radioactive decay, show that the highlands are older than the lowlands. The maria are lava flows that filled in huge impact basins.

6. Moon rocks differ from earth rocks in that (a) they contain more titanium, uranium, iron, and magnesium; (b) they have fewer volatiles; and (c) they contain no water. Moon rocks and earth rocks are similar in their density and isotopic abundances of oxygen.

7. The moon may or may not have a well-defined core. If such a core exists, it consists simply of dense rocks with few metals. The core is hot enough to produce a fair heat outflow at the moon's surface. The moon's mantle is now cool and rigid.

8. Two models contend for the explanation of the origin of the moon: binary accretion and giant impact. The giant impact model presents the fewest problems. It depicts a large body striking the young earth with a glancing blow. Material from the earth's mantle and the impacting body create a ring of debris around the earth. This material accretes to make the moon.

9. Mercury has a bulk density almost the same as that of the earth, so it probably has a large metallic core, which is cool. The mantle is likely made of rocky materials. A giant impact early in its history may have knocked off most of the mantle.

10. Mercury has a very thin atmosphere of helium, sodium, and potassium. These gases continually escape to space. The lack of an insulating atmosphere results in temperature extremes that range from 425 K at noon to 100 K at midnight.

11. Mercury rotates three times for every two revolutions around the sun, so that the same side faces the sun at alternate perihelia. This two-thirds ratio of rotation to revolution results from tidal interactions with the sun.

12. The most prominent features on Mercury's surface are craters (formed by impacts), basins (formed by larger impacts), and scarps (formed by the shrinkage of the planet as it cooled). The differentiation of Mercury's interior indicates that in the past it was hot enough to be liquid.

13. Mercury has a much stronger magnetic field than we would expect from a dynamo model, given the planet's slow rotation and cool core. This field produces a small magnetosphere that interacts with the solar wind.

14. The surface of Mercury resembles that of the moon, so the processes that shaped it and the timing of the sequence of important events probably follow those of the moon. Impact cratering played the major role. Volcanism lasted for a brief period as these masses cooled rapidly. No plate tectonics have occurred.

Key Terms

angular momentum	conservation of angular momentum	lunar soil	scarps
anorthosites	craters	mare (pl. maria)	synchronous rotation
atmospheric escape	giant impact model	mare basalts	tidal forces
basins	highlands	mascons	tidal friction
binary accretion model	impact cratering	Roche limit	volatiles
breccias	lowlands		

Study Exercises

1. Suppose you stepped out of Transplanetary Airlines flight 101 onto the moon's surface. You look slowly around, at both the ground and the sky. How does what you see differ from what you would see on the earth? (Objectives 9-2 and 9-4)

2. Suppose you stepped out of the same carrier's flight 102 onto the surface of Mercury. How would the scene differ from that on the earth? (Objectives 9-3 and 9-4)

3. Argue from a comparison of average densities that the moon cannot have a metallic core like the earth's; that Mercury should have a metallic core. (Objectives 9-1 and 9-6)

4. How were most of the craters on the moon formed? On Mercury? Back up your statement with specific evidence. (Objectives 9-2, 9-3, and 9-11)

5. What specific evidence do we have that the moon's lowland regions (maria) formed *after* the highlands? (Objective 9-8)

6. You are writing a grant proposal to NASA to do research on the origin of the moon. Describe the theory you plan to support in the best light possible. (Objective 9-10)

7. Compare the characteristics of the Orientale Basin on the moon to the Caloris Basin on Mercury. (Objectives 9-2, 9-3, and 9-4)

8. In *one* sentence, describe how the surfaces of the moon and Mercury *differ*. (Objectives 9-2, 9-3, and 9-4)

9. Neither the moon nor Mercury has a substantial atmosphere. Why not? (Objective 9-4)

10. In *one* sentence, describe the difference between the interiors of the moon and Mercury. (Objectives 9-5, 9-6, and 9-7)

11. In *one* sentence, compare the evolution of the surface of the moon and the surface of Mercury. (Objective 9-9)

12. Mariner 10 photographed only about half of Mercury's surface. Imagine that in the future a new flyby mission photographs the rest. It finds a region almost devoid of craters. What statements could you make about the age and history of this region? (Objective 9-11)

13. How does it happen that the earth has *two* tidal bulges, one on the side facing the moon and another on the opposite side of the earth? (Objective 9-12)

14. Imagine you are in space near the earth. You have with you two spheres of equal mass. You place them in a line, one closer to the earth than the other. How would the tidal force of the earth on these spheres affect the distance between them? (Objective 9-12)

Problems & Activities

1. Calculate the surface gravity of the moon. Then calculate the surface gravity of Mercury compared to that of the moon.

2. Calculate the escape speed from the moon. Then compare the escape speed from the moon to that of Mercury.

3. When a mass falls into a place from a great distance, it hits the surface with the escape speed from that planet. Imagine a 100-kg mass falling onto the moon and Mercury. Describe the kinetic energy at impact on each planet. Which body do you expect to have larger craters?

4. At Mercury's evening (eastern) elongation on March 13, 1992, the planet had an angular diameter of about 8 arcsec. What was its size-to-distance ratio then? Use Mercury's physical diameter given in this chapter to calculate Mercury's distance on this date.

5. Imagine that the moon were twice its current distance from the earth. By how much would the moon's tidal force on the earth change? What implication does this have for tides on the earth as the moon moves away from the earth?

See for Yourself!

Using binoculars or a small telescope, examine the surface of the moon around first quarter phase. With a good telescope and high power (about 300 magnification), you can do your own visual moonwalk over the varied features of the moon.

Note the darker regions of the maria and the lighter highland regions. Try to identify some of them using Figure 9.2. See if you can spot some of the larger craters, such as Tycho and Copernicus. You should have no trouble scanning the major maria. Look especially for Mare Imbrium, Mare Serenitatis, and Mare Tranquillitatis (where Apollo 13 landed). Make a count of the total individual maria that you can distinguish.

A few of the lunar mountain ranges should also be visible, especially the Appennines next to Mare Imbrium and close to Copernicus. The Leibnitz mountains near the large crater Clavius are also fairly easy to see. ■

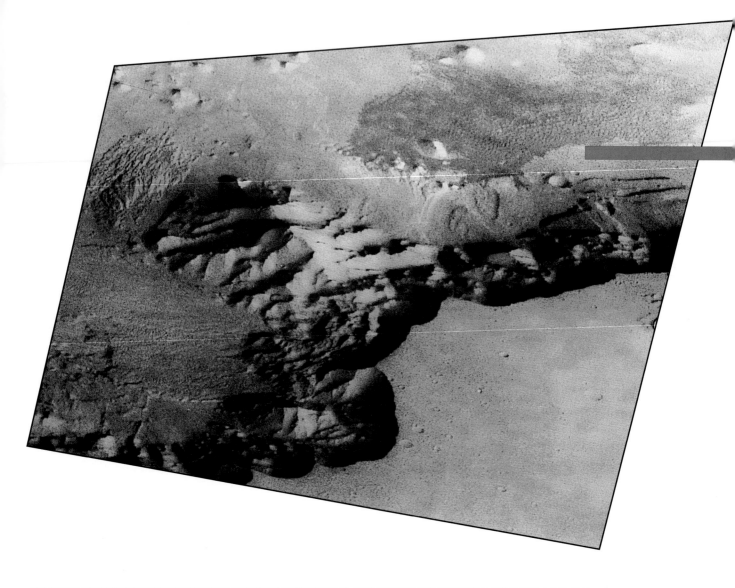

Central Question
What forces have shaped the evolution of Venus and Mars, planets with hot interiors and substantial atmospheres?

Venus and Mars: Evolved Worlds

Venus invited me in the evening whispers
Unto a fragrant field with roses crowned. . . .

JOHN CLEVELAND

Venus, the brilliant light of love; Mars, the red sign of war. Apart from the moon, these two worlds come the closest to the earth in space. And these planets resemble the earth more than any others—they are truly terrestrial planets.

Yet in many ways the similarities are superficial; the differences go much deeper. Volcanism abounds on Venus, reshaping the surface. And it's blistering hot there—hotter even than Mercury at noon! A thick, dense atmosphere of carbon dioxide presses heavily on the barren ground. On Mars, the atmosphere also consists mainly of carbon dioxide. But it's thin, offering no protection from the incoming solar ultraviolet and no hindrance to the outgoing infrared. Mars is mostly a cold desert, where the water ceased to flow tens of millions of years ago. And volcanism ceased before then.

How did Venus and Mars end up so different from the earth? Why does neither show the vast sweep of plate tectonics? What forces shaped their evolution? This chapter deals with the comparative evolution of these earthlike worlds.

FIGURE 10.1 Venus, photographed in blue light at crescent phase. Note the smooth, unbroken clouds. (Courtesy of Palomar Observatory, California Institute of Technology.)

10.1 Venus: Orbital and Physical Characteristics

Viewed from the earth, Venus outshines every celestial body except the sun and the moon. If you know where to look, you can even spot Venus during the day! Venus' brilliance comes from its unbroken swirl of clouds, which reflects 76 percent of the incoming sunlight back into space (Fig. 10.1). This cloud cover completely frustrates any attempt to view the planet's surface features through optical telescopes.

We have pierced Venus' cloudy veil to uncover the surface environment. Radar beams, bouncing off the surface, paint its terrain. Spacecraft have sped by the planet, orbited it, and plunged onto its forbidding surface. Surface temperatures reach highs of 750 K. The atmosphere, which presses at an unrelenting 90 atm onto the ground, contains mostly carbon dioxide. (Recall that the pressure at sea level at the earth's surface is 1 atm.)

Venus has gained the reputation of being the earth's twin sister. In terms of size (the diameter of Venus is 12,100 km; earth, 12,800 km), average density (Venus, 5200 kg/m^3; earth, 5500 kg/m^3), and mass (Venus, 0.82; earth, 1.00), the sisterhood is appropriate. In most other ways, however, Venus has an environment hugely different from the earth's— indeed, a veritable volcanic hell!

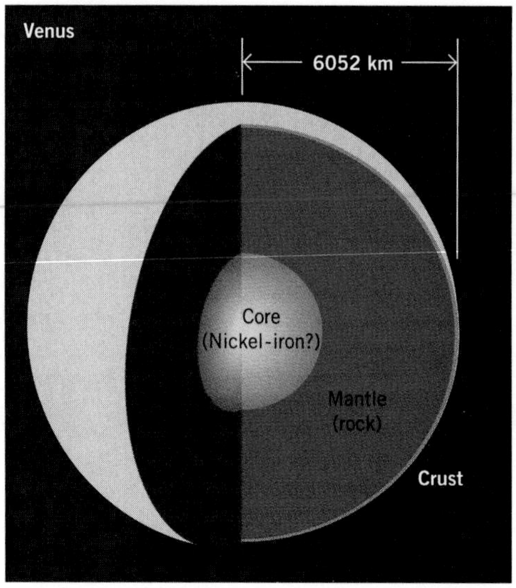

FIGURE 10.2 Model for the interior of Venus; note how closely it resembles that of the earth (Fig. 8.2) with its large, rocky mantle and metallic core.

Revolution and Rotation

Second planet from the sun at an average distance of 0.72 AU, Venus completes one orbit in 225 days. The orbit of Venus is the most circular of all the planets, so Venus' distance from the sun varies by only 1.4 percent.

Thwarted by the lack of a surface view, astronomers before 1961 had no real idea of Venus's rotation rate. Using the *Doppler shift* (see Section 10.2), radar astronomers have found that Venus rotates once every 243 days, *retrograde.* **Retrograde rotation** means that the planet spins from east to west, rather than west to east as does the earth. So on Venus, the sun (though invisible through the clouds) rises in the west and sets in the east. This long retrograde rotation results in a solar day on Venus of 117 terrestrial days.

Size, Mass, and Density

We can measure the earth–Venus distance directly by radar. Then, from the planet's angular diameter, we can calculate its physical diameter of some 12,100 km, only 5 percent smaller than the earth.

Like Mercury, Venus has no known natural moon, so we can accurately find the mass of Venus only when a spacecraft passes or orbits it. In an orbit, we apply Kepler's third law (Section 3.5). During a flyby, we measure the acceleration of the spacecraft (from the Doppler shift of its radio signals) and use Newton's law of gravitation. Venus' mass

comes out to be about 0.82 times that of the earth, or 4.9×10^{24} kg.

With the mass and size known, we can calculate Venus' bulk density: 5200 kg/m^3, almost the same as that of the earth. We guess that the interior of Venus closely resembles the earth's interior (Fig. 10.2): a rocky crust, a large rocky mantle, and a metallic core. Because Venus has a lower bulk density than the earth, we infer that it has a somewhat smaller core.

Magnetic Field

This interior model has a severe problem. A metallic core, liquid in part, implies that Venus should have a planetary magnetic field. Because Venus rotates 243 times more slowly than the earth, we expect its internal dynamo to be weaker; hence the magnetic field should be less intense than the earth's but still there. But, no probe to date has detected *any* magnetic field. If one exists, it must be at least 10,000 times weaker than the earth's magnetic field! That's much weaker than we would expect from a simple dynamo model.

What a magnetic mess—tiny, high-density Mercury has a planetary field, and Venus does not! As you shall see, Mars has a barely detectable magnetic field (Section 10.5). Perhaps a simple dynamo model does not apply to all planets. One other possibility: we know that the earth's magnetic field periodically reverses its polarity (Section 8.3). In the middle of a reversal, the magnetic field is

---

Venus

At a Glance

Equatorial radius	6052 km
Mass	4.87×10^{24} kg
Bulk density	5240 kg/m^3
Escape speed	10.4 km/s
Atmosphere	Carbon dioxide, nitrogen

essentially zero. Venus may have a very weak field right now because it is midway through a reversal.

A very weak magnetic field means that the interaction of Venus with the solar wind differs from that of the earth. Without the buffer of a magnetic field, the solar wind runs right into the upper atmosphere of Venus. As the wind flows around and past the planet, it carries off some of the atmosphere.

■ **If we had two moons, one like our present moon and the other like Venus (but the same size as our moon), which would be brighter in the sky?**

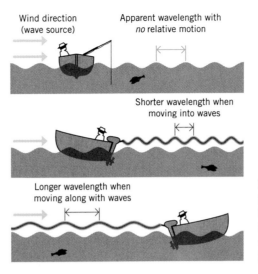

FIGURE 10.3 Wavelength and relative motion for water waves. Movement into the waves makes their wavelength appear shorter; if you move with them, their wavelength seems longer.

10.2 Interlude: The Doppler Shift

Radar observations of the Doppler shift from Venus let us measure the planet's rotation rate. (The same method is used to find out the rotation rate of Mercury: Section 9.6.) Essentially, the Doppler shift (see Focus 10.1 for details) provides us with a cosmic speedometer that allows the measurement of speeds of objects moving toward or away from us. Let's see how it works.

Concept

The Doppler shift occurs with waves of all kinds—light, sound, and even water. Here's an example with water waves. Imagine you are out fishing in a small motorboat (Fig. 10.3). You have been sitting in one spot for a while with little luck, and the rhythm of the waves has lulled you almost to sleep. You decide to move for better fishing. First you go into the wind (the wave source). You notice that you bob up and down *more* frequently than when you were at rest; the wavelength appears to have gotten *shorter.* Just for fun, you drive the boat away from the wind. You discover that your bobbing is less frequent; the waves seem to you to have a *longer* wavelength. The explanation is simple: when you moved in the direction of the waves, they had to catch up with you; when you went into them, you went to meet them.

That's the essence of the **Doppler shift:** when you are moving *toward* a wave source, the waves appear to be more frequent and shorter in wavelength; in contrast, when you move *away* from a wave source, the waves appear to be less frequent and the wavelength longer. It's only the *relative velocity along the line of sight*—called the **radial velocity**—that causes the Doppler shift (for speeds much slower than that of light). Since velocities are relative, it makes no difference whether you're moving, the source is moving, or both. The Doppler shift is an apparent shift in the received wavelength when the source and receiver move with relative radial velocities.

You have seen how the Doppler shift works for water waves (Fig. 10.4*a*). You have probably heard the Doppler shift with sound waves (Fig. 10.4*b*). As a truck approaches you with its horn blasting, its pitch is higher than when the car is at rest. Just as the truck passes you, its pitch is unshifted. Then as the truck moves away, its horn seems to put out a lower pitch. The same effect happens with light waves (Fig. 10.4*c*), where we can determine the Doppler shift by measuring the *change in the wavelength* of lines in a spectrum. The emission or absorption lines are shifted toward the blue end of the spectrum for an object approaching you, and toward the red for one that is receding. These changes are called the **blueshift** and the **redshift.**

Application to Planets

Now apply this idea to radar observations of a planet (Fig. 10.5). We send

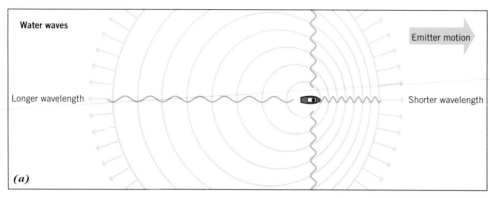

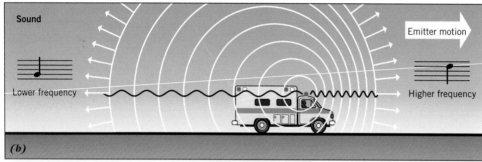

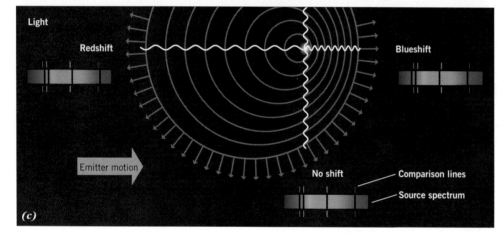

FIGURE 10.4 The Doppler shift for waves of different types. (*a*) In water waves, a moving source compresses the waves in the direction of motion and spreads them out in the opposite direction. (*b*) Sound waves from a moving truck's horn undergo a similar effect, increasing and decreasing in pitch as the truck approaches and then passes. (*c*) For light waves, the Doppler shift causes a change in the position of lines in the spectrum: a blueshift for approach, and a redshift for recession.

radar signals of an exactly known frequency from the earth. They strike a rotating planet all over its surface, and some waves bounce off and return to the earth for reception. Imagine viewing the planet along its equator. Then some of the reflected waves come from the edge of the planet that is moving toward us from the planet's rotation; these are shifted to higher frequencies and shorter wavelengths (*blueshift*). The other edge moves away from us; waves reflected here are lower frequencies and longer wavelengths (*redshift*).

When the reflected waves are received at the earth, the radio telescope needs to be tuned to slightly different

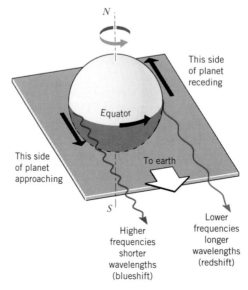

FIGURE 10.5 Doppler shift from a rotating planet. Viewed along the equator, the left side of the planet is approaching the observer, and the right side receding. Viewed from above the north pole, radar waves striking the left side are shifted to shorter wavelengths (blueshifted); those hitting the right move to longer wavelengths (redshifted).

Enrichment Focus 10.1

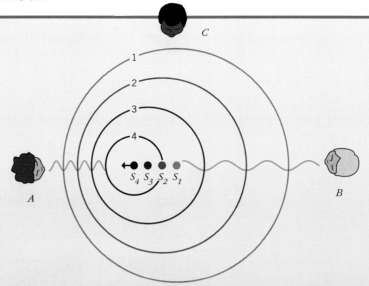

THE DOPPLER SHIFT

The Doppler shift is an apparent change in wavelength from the relative motion of a source and receiver. It allows astronomers to find the line-of-sight velocities of luminous objects without knowing their distance. Let's look at the Doppler shift for light waves, which you can't see directly. You can, however, see colors of light, which are related directly to wavelength. Keep in mind that red light has a longer wavelength than blue.

Imagine a stationary light source giving off just one particular wavelength every second (Fig. F.8). Each wave travels outward with velocity c, the speed of light. When the source is not moving, all the waves are concentric and are separated by the wavelength of emission.

Now imagine that the source moves from point S_1 to point S_2 in one second, and so on. At each point (S_1, S_2, S_3, S_4) it emits a wave (1, 2, 3, 4) that travels out at c. In the direction of its motion, the source catches up a bit with the wave it has just emitted, so for an observer at A, the wavelength appears shorter. This observer sees the distance between the waves as compressed and so observes a Doppler shift to the short-wavelength end of the spectrum. In contrast, an observer at B sees the waves as spread apart, a Doppler shift to the long-wavelength end of the spectrum. An observer at C, at right angles to the direction of the source's motion, measures no change in the wavelength and consequently no Doppler shift. Note that only the velocity along the line of sight contributes to the Doppler shift. It is also important to remember that the Doppler shift does *not* depend on the *distance* between the observer and the source, only on their *relative radial velocities*.

The astronomer takes advantage of the Doppler shift to determine the radial velocities of celestial objects (such as stars) relative to the earth by using spectral lines to measure wavelength shifts. The procedure is to take a spectrogram of the object and at the

FIGURE F.8 The Doppler shift for light waves. A light source travels at a constant speed from right to left (starting at S_1 *and moving to* S_4). It emits a flash of light four times (these are the waves numbered 1 through 4) at each position (S_1 through S_4). The waves are viewed by observers at A, B, and C. The person at A sees a blueshift (the waves appear bunched together), the person at B a redshift (the waves appear spread apart), and the one at C no shift, compared to that emitted by the source at rest.

same time superimpose a comparison spectrum from a local laboratory source (Fig. F.9). The comparison source is at rest with respect to the telescope, and the normal (zero relative velocity) placement of lines to which the shift can be measured. With the measured shift and the value of c, the astronomer calculates the relative velocity between the source and earth by the expression

$$\frac{\Delta\lambda}{\lambda_0} = \frac{V_r}{c}$$

where V_r is the relative radial velocity, λ_0 the rest wavelength of the observed line, $\Delta\lambda$ the observed shift in the line ($\Delta\lambda = \lambda_{observed} - \lambda_0$), and c the speed of light.

Here's an example. A strong dark line from absorption by calcium has a rest wavelength of 3933 Å. Suppose you measure this line in the spectrum of a moving object and find that it has shifted to 3937 Å. Is the object moving away from or toward you? Away, since the wavelength is longer. How fast?

$$\Delta\lambda = 3937 - 3933 = 4 \text{ Å}$$

$$V_r = \frac{\Delta\lambda}{\lambda_0} \times c$$

$$V_r = \frac{4}{3933}(3.0 \times 10^8 \text{ m/s})$$

$$= 3.1 \times 10^5 \text{ m/s} = 310 \text{ km/s}$$

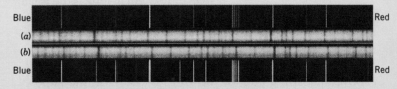

FIGURE F.9 Doppler shift in the spectrum of a star. The spectra are those from the star Arcturus taken (*a*) in July and (*b*) in January. Above and below the spectra of the star, which contain dark lines, are the emission-line spectra of a source at rest made at the telescope; these serve as the comparison spectra for the wavelengths. The shift in position of the observed lines relative to the comparison's lines establishes the amount of the Doppler shift. (Courtesy of Palomar Observatory, California Institute of Technology.)

frequencies to pick up the waves reflected from each side of the planet. The *difference* in frequency between the original signals and the received ones tell us how fast the planet is rotating.

Note that radar waves are used for Venus because we cannot see the surface with visible light. For planets whose surfaces we can see, regular optical telescopes can be used with spectroscopes to measure the Doppler shifts (Focus 10.1).

The Doppler shift is named after Christian J. Doppler (1805–1853), who first noticed the effect in sound waves. Later the French physicist Armand Hippolyte Fizeau (1819–1896) applied the Doppler shift to light waves and recognized its importance in astronomical applications. In honor of these two men, some scientists like to call the phenomenon the *Doppler–Fizeau shift*. Most people call it simply the *Doppler shift*, and that is the name used here. You will come across astronomical applications of the Doppler shift throughout this book.

■ In what daily situation have you heard the Doppler shift for sound?

10.3 The Atmosphere of Venus

The atmosphere of Venus differs remarkably from ours—a key clue to understanding the planet's evolution. Substantial atmospheres directly affect the surface of a terrestrial planet. And an atmosphere's composition allows us to infer some aspects of a planet's history.

Clouds

The yellowish-white clouds of Venus conceal the surface. The cloud tops, which you see in a telescope, reach about 65 km above the surface. (The highest clouds above the earth go up to about 16 km.) The Venutian cloud tops flow with the upper atmosphere, in patterns similar to jet streams of the earth. Ringing the equator, the clouds whiz around at roughly 100 m/s—fast enough to orbit the planet in only 4 days (Fig. 10.6). In addition to this planetwide cir-

(a)

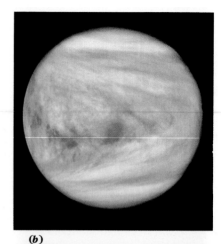

(b)

(c)

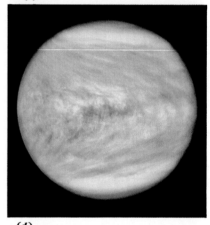

(d)

culation, winds blow from the equator to the poles in large cyclones 100 to 500 km in diameter. They culminate in two giant vortices that cap the polar regions. What drives such fierce winds in the upper atmosphere of Venus? It's believed that solar heating does the trick, as on the earth.

The NASA Pioneer Venus probes found that clouds float in two broad layers (Fig. 10.7). The upper cloud deck has a thickness of some 5 km. Below the upper deck sits a thin haze layer. Below that, at a 50-km height, is the lower cloud deck, by far the denser layer. The clouds here contain liquid droplets and perhaps solid crystals. Below 50 km, the clouds gradually thin out; below 30 km, and down to the surface, the atmosphere is clear of any particles. In the densest part of the clouds, the visibility is less than 1 km, dangerous to fly through but still not a pea-soup fog. (Such a low visibility would close most airports on the earth.) The clouds of Venus aren't particularly murky, just very thick.

FIGURE 10.6 Motions in the upper cloud layers of Venus. The clouds circle the planet once in about 4 days, as shown in this sequence imaged by the Pioneer Venus orbiter. (*a*) May 2, 1980. Note the Y-shaped marking in the clouds; it opens to the west of Venus. The stem of the Y lies as a broad band above the equator. (*b*) Nine hours later, the vertex of the Y is near the west side of the planet. (*c*) Four hours later, the vertex is out of view. (*d*) On the following day, the arms of the Y have changed form. (Courtesy of NASA.)

What are the clouds made of? The best idea sees the upper-level clouds as concentrated solutions of sulfuric acid. (Sulfuric acid, H_2SO_4, is used in car batteries.) Models of the clouds designed to match their observed infrared spectrum imply that they contain a solution of 90 percent sulfuric acid mixed with water. This basic idea has been confirmed by the Pioneer Venus missions, so that both liquid droplets and solid crystals in the clouds are known to consist basically of sulfur and sulfuric acid with other, unidentified materials (probably sulfur compounds).

Although the atmosphere and clouds of Venus do contain some water vapor, it doesn't amount to very much compared with the total amount of water on the earth. If all the earth's water (in both the atmosphere and oceans) were spread in a uniform layer over the earth's surface, that sheet would be 3 km thick. All the water in the atmosphere of Venus (none exists on the surface now because it's so hot) would amount to a layer only 30 cm thick. So our earth's surface and atmosphere have about 10,000 times as much water as Venus does.

Atmosphere and Surface Temperature

Since 1932 we have known that Venus' atmosphere contains carbon dioxide, but we did not know how much until recently. Interplanetary probes by both the United States and the former Soviet Union indicate that the atmosphere contains about 96 percent carbon dioxide, 3 percent nitrogen, some argon, and traces of water vapor (varying from 0.1 to 0.4 percent), oxygen, hydrogen chloride, hydrogen fluoride, hydrogen sulfide, sulfur dioxide, helium, and carbon monoxide.

The Soviet Venera descents found the surface pressure to be 90 atm and the sunlit surface temperature about 740 K. This high temperature probably results from the effective trapping of surface heat (Section 8.4), because carbon dioxide and water vapor absorb infrared radiation well. An extreme greenhouse effect keeps Venus very hot: only about 3 percent of the sunlight hitting the upper atmosphere of Venus makes it to the sur-

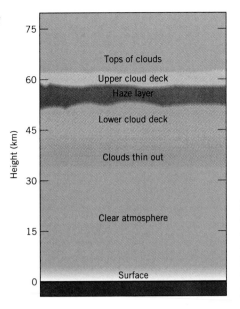

FIGURE 10.7 The structure of the clouds in the atmosphere of Venus. The clouds, consisting sulfuric acid and water, make up two decks (extending from 45 to 65 km) with a layer of haze between them. The air is clear from the surface up to the lower cloud deck. (Based on NASA data.)

face. The clouds reflect most of it (80 percent) and absorb most of the rest. Given that so little sunlight heats the surface, the atmosphere must be a very good insulator to result in the high temperatures recorded. Note that in contrast to the earth, carbon dioxide (because of its high abundance) plays the main role in the greenhouse effect on Venus.

Atmospheric winds on Venus blow from the day side to the night side and from equatorial to polar regions. The wind flow carries heat. Along with the very effective atmospheric insulation, this action helps to keep the temperatures fairly constant over Venus' surface. They vary about 10 K or less from day to night side, so it doesn't cool off much at night. In fact, the overall average surface temperature is a blistering 740 K (about 470 °C)! That's about the temperature of a home electric oven during its self-cleaning cycle.

■ **Why is the surface temperature of Venus so high, even on its night side?**

10.4 The Active Surface of Venus

We've investigated Venus' surface by examining material brought back by landing probes and by analyzing the reflections obtained by bouncing radar (which penetrates the clouds) off the surface.

Atmosphere of Venus

At a Glance

Surface pressure	90 atm
Surface temperature	740 K
Upper cloud deck	65 km
Clouds	Made of sulfuric acid

Radar mapping has taken place both from the earth and from Venus orbiters. The goal has been to find evidence for processes like volcanism and plate tectonics that modify the surface. (*Note:* Many features on Venus bear the names of famous women and female mythological characters.)

The former USSR landed four spacecraft that sent back close-up photos. The pictures showed slabby rocks about tens of centimeters in size (Fig. 10.8). A few rocks have small holes that were once filled with gas, implying a volcanic origin. The rocks rest on loose, coarse-grained dirt. Direct measurements showed that most of the rocks are volcanic basalt, like those lining the earth's ocean basins or lying on the earth's ocean floor near midoceanic ridges. These specimens provided early evidence of local volcanism.

Highlands and Lowlands

What about the general lay of the land? We have a good idea from radar mapping, which reveals a varied terrain: mountains, high plateaus, canyons, volcanoes, ridges, and impact craters. Overall, Venus looks fairly flat. Elevation differences are small, only 2 to 3 km, except for a few highland regions. The continents there reach up to only some 10 km, compared with a 25-km difference on Mars, and 20-km one on the earth. The surface of Venus is remarkably flat compared to the earth. Only some 10 percent of the mapped surface extends above 10 km. In contrast, about 30 percent of the earth's surface reaches above 10 km (from the bottom of the ocean basins).

The southern and northern halves of the mapped face of Venus differ remarkably. The northern region is mountainous, with uncratered *upland plateaus;* these resemble continents on earth. The southern part has relatively flat *rolling terrain,* which appears to consist of vast lava plains.

The great northern plateau of Venus, called *Ishtar Terra,* measures some 1000 km by 1500 km (Fig. 10.9). (That's larger than the biggest highland plateau on the earth, the Himalayan plateau.) This great plateau may have been built up from thin lava flows over an uplifted section of older crust. Mountain

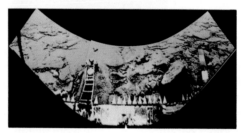

FIGURE 10.8 The surface of Venus photographed by Venera landers. The horizon is visible at the top right and left, and the lander at bottom center. (*a*) Close-up of the surface of Venus, taken by Venera 13, which touched down in the lowlands southeast of Beta Regio. The color here is real; the atmosphere removes blue and yellow from sunlight, leaving red and orange. Note the slabs of rock, patches of dark soil, and pebbles close to the edge of the lander. (*b*) Computer-processed image of the scene in (*a*), giving a view as if illuminated by white light. The gray rocks may contain iron. (*c*) Venera 14 view in 1982; parts of the lander are visible in the foreground. The flat expanses of rock are unbroken by any soil patches, and fewer pebbles (about a centimeter in size) are visible. This image has been computer processed to make the horizon appear flat. (Courtesy of NASA and J. B. Garvin; original images from the former USSR Academy of Sciences.)

ranges border Ishtar on the east, northwest, and north. The eastern range, called *Maxwell Montes,* contains the highest elevations on Venus: up to 11 km. A volcanic cone lies near Maxwell's center—evidence of volcanism here.

Most of the surface consists of flat, volcanic plains that are marked by tens of thousands of volcanic domes and shields. The southern half of Venus' rolling lava plains is punctuated by craters, both large (up to 100 km in diameter) and smaller ones (less than 10 km). About a thousand have been identified. They are probably impact craters created by solid bodies from space crashing into the surface. (Small craters—less than 10 km—do not form on Venus because the dense atmosphere completely vaporizes small pieces of incoming debris before they can reach the ground.) In general, the craters are shallow from erosion by the thick atmosphere. A few have been filled in with lava.

Volcanoes and Tectonic Activity

Beta Regio, which contains at least two separate volcanoes, is an enormous volcanic complex that formed along a great north–south rift zone (Fig. 10.10). The

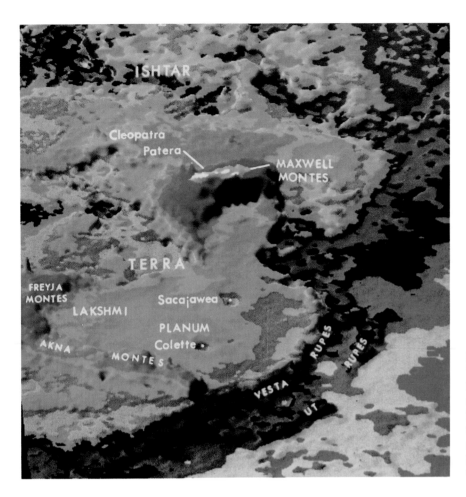

volcanoes here bear gentle slopes; they are called **shield volcanoes.** (Instead of exhibiting a sharply uplifted cone, shield volcanoes are relatively flat, like an armor shield.) Often shield volcanoes have a collapsed central crater at the summit; these two have them. The two volcanoes are called *Theia Mons* (the southern one) and *Rhea Mons* (the northern one). Theia Mons has a diameter of 820 km, a height of 5 km, and a summit crater 60 km by 90 km. In contrast, the island of Hawaii on the earth (which is a shield volcano island) is 200 km across and 9 km high.

Whether any volcanoes are now active is unclear. Volcanoes discharge gases from inside a planet. On earth, this **outgassing** contains mostly water vapor, carbon dioxide, nitrogen, and sulfur gases—all found in the atmosphere of Venus. Yet, the Magellan orbiter (see below) did not find firm evidence of *current* activity. Overall, Venus' surface has large volcanoes and other evidence of volcanism; much of the surface is covered by lava plains.

Now, the large difference on the earth between ocean basins and conti-

nental masses is continually maintained by plate tectonics. The lack of such overall differences on Venus suggests an absence of planetwide crustal plates. However, tectonic work on Venus seems to be localized in zones spread around the planet, especially near highland regions.

Magellan at Venus

In August, 1990, the Magellan spacecraft settled in an orbit around Venus, and one month later began its mission of high-resolution radar mapping—100 meters at best, or about 10 times better than before! This mission lasted almost a year, and it will take many more years to ana-

205

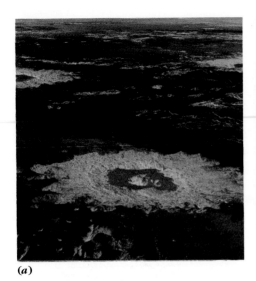

(a)

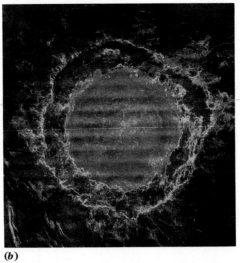

(b)

FIGURE 10.11 Magellan images of impact craters on Venus. (*a*) Three large impact craters visible in the Lavinia region; they range in diameter from 30 to 50 km. Note the bright material, ejected upon impact, surrounding each crater and the central peak in each. Howe crater appears centered in the lower part of the image. It is about 37 km in diameter. Danilova, diameter 48 km, is at the upper left; Aglaonice, diameter 63 km, is at the upper right. Colors are simulated to represent the visual range on Venus. (*b*) The double-ring Mead crater has a diameter of 275 km. Its original interior has been filled in by lava. (Both courtesy of JPL and NASA.)

lyze the data. We have images of the surface of Venus with a clarity never before achieved—a surface molded by extensive and recent volcanism. Yet, no plate tectonics! That's a surprise, because Venus and earth should have lost heat at the same rates. It's the heat flow from the mantle that pushes the plates in the crust.

Amid the volcanic land forms lie stunning impact craters (Fig. 10.11*a*). They generally show central peaks, terraced walls, shocked surfaces, and flooded floors. But because of the dense atmosphere, they lack extensive ray systems of exploded debris; for the same reason, no crater smaller than 3 km in diameter has been seen. The largest known crater, called Mead (after Margaret Mead), spans 275 km in a double-ring structure (Fig. 10.11*b*). Its interior lacks a central impact peak; instead, it is filled to the brim with lava. It resembles the filled impact basins of the moon and Mercury.

Overall, Venus appears to be a fresher, more violent world than could have been guessed from earlier images—a planet with a largely young surface, no more than a few hundred million years. (That's about twice the average age of the earth's surface.) The degree of volcanism seems to be more widespread than on the earth (Fig. 10.12*a*). The entire surface may have been reshaped over the past several hundred million years by extremely runny lava flows. Rift valleys, which form at the separations of crustal plates, show exten-

sive surface fracturing (Fig. 10.12*b*). Also visible are folded and faulted regions that resemble some mountain-building regions on the earth. Such features indicate local tectonic activity. The high degree of activity that seems to characterize Venus now (Fig. 10.12*c*) should continue for eons to come.

But where are the crustal plates? The earth's crust has many rigid plates bounded by narrow zones where lava oozes up from the mantle. Venus has no such plates. Perhaps it is a "one-plate planet," with a crust that never became rigid. Like dough, Venus' crust is not stiff enough to slide around in pieces. Instead, it wrinkles and puckers locally from heat and lava upflow (Fig. 10.13).

To sum up. Most of the surface of Venus is flat and relatively young, modified by volcanism during the past few hundred million years. Global plate tectonics has *not* taken place, but some local activity has occurred.

Evolution of Venus

Comparative planetology includes study of the evolution of planets as well as their current physical properties. The internal evolution of terrestrial planets primarily hangs on size and mass. The larger, more massive planets undergo more activity for a longer time. We expect the earth and Venus to be very similar, yet in most vital respects, they are not. How did the differences develop,

(a)

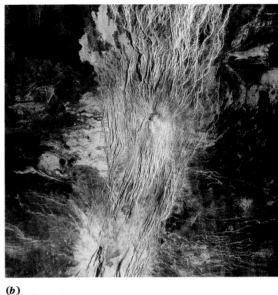

(b)

(c)

FIGURE 10.12 Magellan images of volcanism and tectonic movements. (*a*) Running across the central part of this image is part of a lava channel some 1200 km long and 20 km wide. Very fluid lavas carved this channel in the Lada Terra region at 51° S latitude. The image covers about 150 km across. (*b*) Devana chasm at the equator is a rift zone with multiple faulting and deep fractures. It is part of a wider zone of surface deformation. The lava flows in the center are about 70 km by 280 km and flow down one flank of the rift valley. (*c*) Maat Mons, an 8-km-high volcano, with foreground lava flows. The vertical relief is greatly exaggerated. (All courtesy of JPL and NASA.)

given that the two planets have about the same size, density, mass, interior composition, and structure? Probably from its proximity to the sun and consequent lack of liquid water on the surface. On earth, the oceans and carbonate rock from the remains of ocean life contain a large amount of the carbon dioxide. On Venus, the carbon dioxide cannot be trapped in this way. It stays in the atmosphere, keeping up the severe greenhouse effect.

Although Venus has a high degree of overall volcanism and local outbreaks, it does *not* show the global tectonic pattern as on the earth. The crust of Venus has experienced some localized crustal

FIGURE 10.13 Global view of Venus, center on longitude 270°, from a mosaic of Magellan images. Note the extensive lava flows. (Courtesy of JPL and NASA.)

plate movement. Venus has a solid, mostly unbroken crust without the planetwide structure of plates found on the earth. That's a real puzzle.

We can speculate about the history of Venus using the earth as a guide. We infer that with the other terrestrial planets, Venus formed by accretion about 4.6 billion years ago (Section 8.5). Venus' interior differentiated as a result of internal radioactive and impact heating. During the first 500 million years, a crust formed and solidified. About 3 to 4 billion years ago, large masses bombarded the surface (as they did the earth) and fractured the crust. Volcanoes erupted. A heavy bombardment by smaller bodies from space cratered the surface; this activity ended about 3 billion years ago. That ancient surface is gone.

Since the initial bombardment, huge volcanoes vented through cracks in the surface; their cones form the shield volcanoes visible today (Fig. 10.14). Lava flows have crossed hundreds of kilometers of terrain. Volcanism has reworked the surface almost completely in the last 400 million years.

Venus may well resemble the earth at an early age, from 4.5 to 2.5 billion years ago. The Venus of today appears like a view of the earth's distant past. Our current notion is that Venus seems to have evolved in a sequence similar to that of the earth, but more slowly and for unknown reasons.

■ Why is it surprising that unlike our earth, Venus shows no sign of global tectonic activity?

10.5 Mars: General Characteristics

Mars once appeared to be the most likely home of extraterrestrial life in the solar system, as portrayed in the fantasies of science fiction writers. But the Viking landers have dashed this optimistic attitude: Mars now is barren of life as we know it (more in Chapter 22). Viking and other spacecraft have described a new Mars (Fig. 10.15): a planet with plentiful ancient craters, giant canyons, and huge volcanoes.

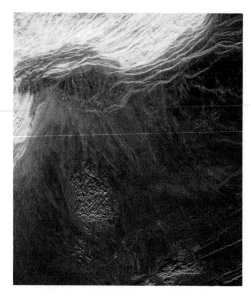

FIGURE 10.14 Magellan image of the western region of the volcano Maxwell Montes and the Lakshmi plateau. The view is about 300 km across. (Courtesy of JPL and NASA.)

Orbit and Day

Mars orbits the sun at an average distance of 1.52 AU. But Mars' distance from the sun varies considerably (by about 9 percent) because its orbit is fairly eccentric. So the distance between Mars and the earth at opposition, when both planets lie on the same side of the sun, ranges widely from less than 56 million to more than 101 million kilometers.

Surface markings visible through telescopes make the Martian rotation rate easy to measure. In 1659 the Dutch physicist and astronomer Christian Huygens (1629–1695) observed the rotation rate to be close to 24 hours. Modern measurements place the value at 24 hours, 37 minutes; a day on Mars lasts only a bit longer than on the earth. The rotation axis of Mars is tilted at almost the same angle as the earth's, with the result that the seasonal variations are similar (but Mars is much colder than the earth overall).

Size, Mass, and Density

Mars has a diameter of some 6800 km, only 53 percent the earth's size. So Mars is about half the earth's size. Mars has moons, so we can use Newton's form of Kepler's third law (Focus 4.3) to find its mass from their orbits: it is only 6.4×10^{23} kg, only 11 percent of the earth's mass.

Mars' bulk density is 3900 kg/m^3, only a bit higher than the moon's (3300

FIGURE 10.15 Wide-angle view of the southern hemisphere of Mars. Note the many shallow craters, especially at bottom. (Courtesy of NASA.)

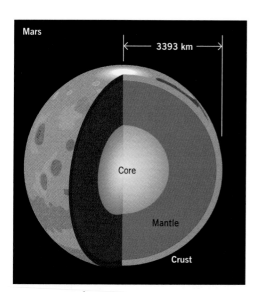

FIGURE 10.16 One model of the Martian interior; note the small size of the metallic core (iron and iron sulfide) relative to the rocky (olivine) mantle. (Based on a model by J. S. Lewis.)

kg/m³) and much less than the earth's (5500 kg/m³). This comparatively low density implies that Mars' interior (Fig. 10.16) differs from the earth's. Its core is smaller and probably consists of a mixture of iron and iron sulfide, having a lower density than the materials in the earth's core. The Martian mantle probably has the same density as the earth's; it could be made of olivine (an iron–magnesium silicate), iron oxide, and some water.

These physical properties show that Mars is an intermediate case between the larger (earth and Venus) and smaller (moon and Mercury) terrestrial planets. So we expect Mars to provide key clues about a middle stage of evolution, and it does!

Magnetic Field

Mars has an extremely weak planetwide magnetic field, only 0.002 times the strength of the earth's. A value that small presents a puzzle if the dynamo model correctly describes planetary magnetic fields. Mars rotates as fast as the earth. Though the Martian core is smaller, it should contain an ample amount of metals. We have no direct proof that the core is liquid, but the evidence for past volcanic activity implies a hot mantle, hence probably, a hot, liquid core. So Mars

should have a moderately strong magnetic field, but it does not.

Perhaps, as for Venus, we are viewing Mars in the middle of a magnetic field reversal. But it is unlikely that we would catch both planets in the process of changing polarity. Again, the dynamo model is called into question by the weak field of another earthlike planet.

Atmosphere and Surface Temperature

Mars has a thin atmosphere. The Viking landers found average surface pressures of roughly 0.005 times the earth's surface pressure. (You'd have to travel 40 km up in the earth's atmosphere to experience pressure that low.) This thin atmosphere consists of 95 percent carbon dioxide, 0.1 to 0.4 percent molecular oxygen, 2 to 3 percent molecular nitrogen, and about 1 to 2 percent argon—very similar in relative composition to the atmosphere of Venus (but nowhere near as dense).

The Viking orbiters measured the water vapor in the atmosphere and found the greatest amounts in the high northern latitudes in the summer. If all the water vapor in the Martian atmosphere were condensed as liquid, it would form an "ocean" only 3μm deep! On earth, the atmospheric water vapor is typically several centimeters of precipitable water, and, of course, the oceans are several kilometers thick. Mars is a very dry planet now, compared with the earth.

It cannot rain on Mars today because of the low surface pressure. Only in the deepest canyons, where the atmospheric pressure is higher, could water be a liquid on the surface. However, water ice is common on Mars, either on the surface or in the clouds (Fig. 10.17). Some evidence suggests the presence of water in a permanent frost layer beneath the surface in most regions.

Although the atmosphere contains mostly carbon dioxide, its low density limits its greenhouse effect. At the Martian equator, when Mars is closest to the sun, the difference between noon and midnight amounts to almost 100 K. The summer tropical high of 310 K (37 °C) is exceptional. For a period of two Martian

Mars

At a Glance

Equatorial radius	3397 km
Mass	6.42×10^{23} kg
Bulk density	3940 kg/m³
Escape speed	5.0 km/s
Atmosphere	Carbon dioxide, nitrogen

FIGURE 10.17 Canyons on Mars filled with water ice in the morning in the region of Noctic Labyrinthus: Viking 1 image, October 1976. (Courtesy of NASA.)

months, the surface temperature remains below the freezing point of water both day and night. The layer of water ice that coats the rocks and soil in the winter is extremely thin, less than a millimeter.

■ In what way does Mars call into question the dynamo model for planetary magnetic fields?

10.6 Remote Sensing of the Martian Surface

Because an astronomer must peer through two atmospheres, the view of Mars is usually blurry. A small telescope shows the main surface features: the white polar caps, light reddish-orange regions, and dark areas. The HST has provided a clearer view (Fig. 10.18). Spacecraft visits, though, gave us a new vision of Mars.

The Sands of Mars

The large surface features on Mars are dark, apparently greenish-gray areas, in contrast to the overall reddish-orange. Some early observers thought that these features were vegetation. The greenness, however, turns out to be an illusion caused by the contrast of the light and dark areas. The dark regions are not really green: they are red, just not as red as the lighter regions.

The light reddish regions make up almost 70 percent of the Martian surface to give Mars its striking ruddy appearance. In 1934 the American astronomer Rupert Wildt (1905–1976) suggested that these areas contain ferric oxide— rusted iron! Iron oxides come in many forms on the earth; all are characteristically brown, yellow, and orange. The surface of Mars contains substantial rusted iron combined with water. Perhaps as much as 1 percent of the surface is water bound up with minerals—a rusty sand (Fig. 10.19).

Global Dust Storms

The reddish sand—finer than that on the earth's beaches—is blown up by fierce winds, greater than 100 km/h, that create planetwide dust storms. These occur most violently when Mars is closest to

the sun. Then the dust clouds (Fig. 10.20), whipped up to heights of 50 km, shroud the entire planet in a yellowish haze for a month or two. It takes many months for the fine dust to completely settle back to the surface. These global storms sandblast the surface and mix it up so much that the surface composition over the planet is basically the same.

This wind-driven dust caused most of the changes in Martian surface features seen in the past. For example, the windstorms blow dust into dunes or deposit it in streaks around mountains and craters.

Canals and the Polar Caps

In 1877 Schiaparelli recorded Martian surface features in great detail. He

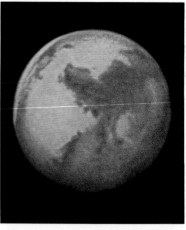

FIGURE 10.18 Hubble Space Telescope (HST) color image of Mars taken when Mars was 85 million kilometers from the earth on December 13, 1990. The resolution is 50 km. (Courtesy of NASA/JPL; image by P. James, University of Toledo.)

FIGURE 10.19 Candor chasm, Mars. This image has been processed from Viking orbiter photos and color enhanced to bring out the ground features. The colors are *not* the true colors. Candor chasm is one part of the great Valles Marineris. The surface has been shaped by wind erosion and collapse of the sides of the valley walls. (Courtesy of USGS; image by A. S. McEwen, Arizona State University and the USGS, from NASA photographs.)

charted a number of dark, almost straight features, which he called *canali,* Italian for "channels." The word was translated into English as *canals,* however, which carries the misleading implication that they were artificial structures.

The so-called canals ignited the curiosity of the American astronomer Percival Lowell (1855–1916). To pursue his interest in Mars, Lowell (Fig. 10.21*a*) in 1894 founded an observatory near Flagstaff, Arizona, to take advantage of the excellent observing conditions there. Shortly afterward, he published Martian maps showing a mosaic of more than 500 canals (Fig. 10.21*b*). In a series of popular books, Lowell argued that the canals were artificial waterways, con-

FIGURE 10.20 The beginning stages of a dust storm in the southern hemisphere of Mars in February 1977. The dust clouds are about 20 km high. (Courtesy of USGS; image by Jody Swann and A. S. McEwen from NASA photographs.)

(a)

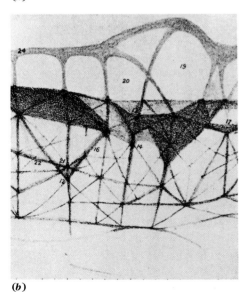

(b)

FIGURE 10.21 Lowell and his canals on Mars. (*a*) Percival Lowell: ". . . the solidarity of the Martian land system points to an efficient government. . . ." (Courtesy of Yerkes Observatory.) (*b*) Surface features on Mars, including some "canals," drawn by Percival Lowell in 1896–1897. (Courtesy of Lowell Observatory.)

structed by Martians to carry water from the **polar caps** to irrigate arid regions for farming. Lowell believed that the polar caps were water ice and that the dark regions were areas of vegetation that displayed seasonal growth, prompted by water from the polar caps.

We know now that the polar caps *do* consist mostly of water ice, especially the residual cap left in the summer, which ranges in thickness from year to year from a meter to a kilometer (Fig. 10.22). The outer reaches of the caps, prominent in winter, consist of carbon dioxide ice, which condenses at a lower temperature than water ice. (At Martian surface pressures, water ice condenses at 190 K, carbon dioxide ice at 150 K.)

Water *does* exist on Mars—but it does not flow freely on the surface because both the temperature and the pressure are too low. (If all the water in the polar caps could cover the surface as a liquid, it would form a layer only about 10 m deep.) As the carbon dioxide cap vaporizes in each hemisphere, the released gas flows toward the opposite hemisphere, transporting the pole cap material throughout the atmosphere.

Lowell was wrong about the vegetation and the canals. Windblown dust deposits may have created temporary features that were seen as the largest and fuzziest of the canals. The smallest ones were likely optical illusions enhanced by wishful thinking. A comparison of Lowell's canal maps with orbiter photos in-

FIGURE 10.22 The south polar ice cap of Mars in the summer of 1977. The residual cap visible here is about 400 km across and contains mostly dry ice. Water ice may exist in the underlying layered terrain. (Courtesy of USGS and NASA; image by T. Becker.)

dicates that only one real feature (part of Valles Marineris) corresponds to any of the so-called canals.

■ **What are the reasons for the absence from Mars today of rivers, lakes, and ponds of water?**

10.7 The Martian Surface Close Up

Martians never had a chance to invade the earth—they don't exist. But we've invaded Mars—not to conquer, but to learn. We have found out that the red planet was once an active world geologically but now is a calm, cold desert.

General Surface Features

Mariner 9 showed spectacular features of a geologically active planet. The two Martian hemispheres have different topological characteristics: the southern hemisphere is relatively flat, older, and heavily cratered; the northern hemisphere is younger, with extensive lava flows, collapsed depressions, and huge volcanoes. Near the equator, separating the two hemispheres, lies a huge canyon, called *Valles Marineris* (Fig. 10.23). This chasm is 5000 km long (about the length of the United States) and some 500 km wide in places. It is likely a rift valley—the only one on Mars.

The Viking landers touched down on Mars in 1976. Seen close up, the Martian surface is bleak and dry (Fig. 10.24a). Large rock boulders are strewn about, amid gravel, sand, and silt. The boulders are basaltic. Some contain small holes (Fig. 10.24b) from which gas has escaped; the holes make the rock look spongy. On earth, such basalts originate in frothy, gas-filled lava; the Martian rocks probably had a similar origin.

Both landers uncovered indirect evidence for once-flowing Martian surface water. The region around Viking 1 is a floodplain, where water sorted the smaller rocks into gravel, sand, and silt. The ground there also resembles the hardened soil of the earth's deserts. Such soil forms when underground water percolates upward and evaporates at the surface. Upon evaporating, the water leaves behind minerals that harden the soil.

Arroyos and Outflow Channels

The Mariner 9 mission discovered, and the Viking orbiters confirmed, a number of sinuous *outflow channels* that appear to have been cut in the surface by running water (Fig. 10.25a). The largest ones have lengths up to 1500 km and widths as great as 100 km. (These channels are *not* the canals seen by Lowell and others; they are too small to be visible from the earth.) These channels resemble the arroyos commonly found in the Southwest of the United States. An **arroyo** is a channel in which water flows only occasionally (Fig. 10.25b). They provide direct evidence of water erosion.

What makes us think that the Martian channels were actually cut by flowing water? The evidence is pretty strong by analogy to the land forms cut by water on earth: the flow direction is downhill; the flow patterns meander;

FIGURE 10.23 Mosaic of Viking images showing Valles Marineris (across the middle and center) and the Tharsis ridge, with its huge shield volcanoes (at the left). (Courtesy of USGS; image by A. S. McEwen, Arizona State University and USGS from NASA photographs.)

(a)

(b)

FIGURE 10.24 Viking views of the Martian surface. (*a*) Viking 1 panorama in 1977; the view is to the north of the lander. Note the thick reddish dust covering the surface. (*b*) Viking 2 lander view of the surface of Mars. Many of the rocks have small holes, which indicates a volcanic origin. The rocks are 10 to 25 cm in size. The sandy soil, blown by the wind, piles up between the rocks. (Courtesy of NASA.)

tributary structures appear; and sandbars are cut by smaller flow channels, as is commonly found in arroyos on the earth. Most such channels originate within the cratered highlands, just north of Valles Marineris. Some sections of the ancient terrain in the southern hemisphere show networks of small valleys, which resemble terrestrial runoff channels.

The formation of the Martian arroyos required extensive running water for at least a short period of time. Since Mars does not have liquid surface water now, different conditions occurred in the past—for example, a warmer climate and a denser atmosphere (Section 22.7). Brief water and mud flows during this period could have cut out the arroyos and valleys (Fig. 10.26) we see today.

Volcanoes

By far the most awesome Martian features are the shield volcanoes clustered on and near the Tharsis ridge—clear evidence of Martian volcanism. The largest is *Olympus Mons,* some 600 km across at its base (Fig. 10.27). The cone reaches 25 km above the surrounding plain, and its base would span the islands of Hawaii, which are made of several volcanoes. If put down in California, Olympus Mons would cover the area from San Francisco to Los Angeles (Fig. 10.28). It soars above sea level more than 2.5 times the height of Mt. Everest! Indirect evidence from the nearby lava flows implies that the volcanoes are about a billion years old.

The huge mass of Olympus Mons requires that the Martian crust beneath it

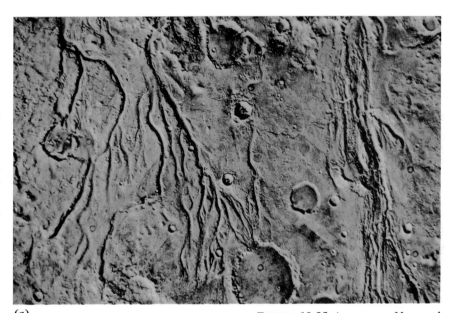

(a)

(b)

FIGURE 10.25 Arroyos on Mars and the earth. (*a*) Most Martian outflow channels originate from regions of collapsed and disrupted terrain that result from the catastrophic release of groundwater. Here the terrain has a slope that drops about 3 km over the span of the area shown, some 1300 km by 900 km. Note how some arroyos cut through craters, an indication that they formed after the craters. (Courtesy of NASA.) (*b*) Arroyo on the earth. Note the wavy patterns in the sand created by sudden and brief flows of water during summer thunderstorms in New Mexico. (Photo by M. Zeilik.)

be thicker than the crust beneath smaller such volcanoes on the earth. Geologists estimate the thickness of the Martian crust to be some 120 km, about twice that of the earth's. This thick crust may well explain why Mars never developed a global plate tectonic system.

FIGURE 10.26 Evidence for subsurface water on Mars. Underground water flows sapped the terrain here to create a landscape of valleys in the eastern region of Mangala Valles. (Courtesy of NASA.)

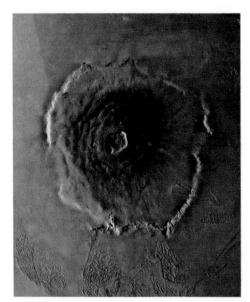

FIGURE 10.27 Olympus Mons. This mosaic made from Viking photographs shows the volcano rising above the surrounding terrain; the base spans some 600 km, the ridge around the base is 6 km high; and the summit lies 27 km above the plain. The central depression at the summit is about 3 km deep and 25 km across. Old lava flows drape over the sides of the volcano. (Courtesy of A. Allison and A. S. McEwen, USGS.)

Olympus Mons crowns a string of volcanoes situated on the Tharsis ridge. Occasionally, thin ice clouds decorate the tops of the volcanoes there. These clouds might result from erratic spurts of outgassing. On earth, volcanic activity spews forth gases (including water vapor) from the earth's mantle. Such outgassing by the giant Martian volcanoes in the past may have added significantly to the Martian atmosphere.

The Tharsis ridge labels Mars' northern hemisphere, which differs so dramatically from the southern one. The ridge rises about 10 km above the average surface height for the planet and contains numerous volcanic structures. Over time, impact craters have been wiped out by volcanic resurfacing, and very few are visible now. In contrast, the southern hemisphere is basically a desert, pockmarked by old, eroded craters. The geologic inference from this difference is that about 3 billion years ago in the northern hemisphere, a huge mass of lava oozed out from under the surface, creating the volcanic plains and the volcanoes over a long period of time. This initial flow destroyed the older deserts and craters in this region. Other flows probably followed.

Cratered Southern Hemisphere

The southern hemisphere of Mars has a cratered terrain that resembles the ancient highlands of the moon or the plains of Mercury. The landscape contains impact craters that range in size from huge, lava-filled basins down to some only a few meters across (Fig. 10.29). The Martian craters come in the same varieties as lunar and Mercurian ones, some with central peaks that mark their impact origin. In general, the Martian craters are shallower than the ones on the moon and Mercury. Many are filled with wind-blown dust. Wind scours the craters and piles the dust in dunes within the craters' bowls. So we know that wind erosion has worked for a long time on Mars.

The Martian craters do *not* usually have rims of ejected material common to craters on the moon and Mercury. Instead, many Martian craters have bulges that protrude from their rims. We think that the original impact melted frozen ice

FIGURE 10.28 The size of Olympus Mons compared with the states of California and Nevada.

in the Martian soil. This melted water quickly turned the ground into mud, which flowed in bulges away from the crater.

The Evolution of Mars

Putting together the mass of information available today to infer the Martian past is a tough task. Again, we rely on the earth for comparison, along with the moon. Mars has ancient impact craters, somewhat eroded, in the southern hemisphere. The northern shows evidence of extensive volcanism. Between them lies the giant rift valley of Valles Marineris—the only suggestion of local tectonics. Mars lacks mountain ranges, and so any evidence for plate tectonics. Because the Martian surface-to-volume ratio lies between those of the earth and the moon, we expect that the planet lost internal energy at an intermediate rate.

A Martian planetary history may have followed a sequence like the early part of the earth's. First, after the formation of Mars by accretion, impact craters covered the surface. Shortly afterward, the planet differentiated (as did the earth; Section 8.5) to form a crust, a mantle, and a core. Regions of thicker crust rose to higher elevations. In the second phase, thin regions of the crust fractured, and the Tharsis ridge uplifted, cracking the surface around it.

During this time, a primitive atmosphere, denser and warmer than at present, held large amounts of water vapor from the volcanic outgassing. Rainfall may have eroded the surface in furrows and then percolated to a depth of a few kilometers. Decreasing temperatures formed ice at shallow depths. When heated (perhaps by volcanic activity), this ice melted, leading to the formation of collapse and flow features. Planetwide water erosion carved the surface.

In the next phase, extensive volcanic activity occurred, especially in the northern hemisphere. The Tharsis region continued to uplift, generating more faults. Valles Marineris formed at this time from expansion of the crust. Finally, volcanism—most of it concentrated on the Tharsis ridge—broke the surface and spewed out great flows of lava.

FIGURE 10.29 Argyre Planitia in the Martian southern hemisphere. Argyre is a large impact basin (left of center). Many relatively uneroded craters are on the surface. (Courtesy of NASA.)

Since that last time of great eruptions, wind erosion has mainly sculpted the Martian surface. A few small impact craters probably have formed from time to time. Strange, small craft from the earth have touched down on the surface, finding a windy, dusty, and rocky desert.

As noted earlier, we have scant indication for plate tectonics on Mars. Whereas on earth volcanoes tend to form in chains, in which one crustal plate encounters another, on Mars volcanoes are pretty much clustered in one highland region. On earth, plates also show up as continental masses with basins in between; Mars does not have such crustal configurations. The implication is that because of its small mass and size, Mars cooled quickly and developed a thick, inactive crust and mantle. So it never experienced the large convective motions in the mantle that drive the plate tectonics on the earth. Because Mars cooled off faster, as a planet, it did not evolve as much as ours.

In summary. The evolution of terrestrial planets involves changes to their interiors, surfaces, and atmospheres. These changes are driven mostly by a planet's internal heat: the more massive the planet, the more internal heat it generates (from radioactive decay); the longer it retains this heat, and the greater its evolution.

Surfaces are modified by several processes: impacts of interplanetary debris, outflow of internal heat, volcanism,

Comparative Evolution

At a Glance

Moon and Mercury	Ended 2.5 to 3.2 billion years ago
Mars	Most ended 2 billion years ago; few regions 600 million
Venus	Surface about 400 million years old; continuing volcanism?
Earth	90% surface less than 600 million years old; active volcanism and plate tectonics

TABLE 10.1 MAIN STAGES IN THE EVOLUTION OF TERRESTRIAL PLANETS	
STAGE	PROCESSES
I	Formation by accretion, heating of crust and interior, crust formation
II	Crust solidification, intense impacts, cratering of surface
III	Basin formation and flooding, lowlands formation
IV	Low-intensity impacts, atmosphere formation by outgassing
V	Volcanoes, crustal movement, continent formation

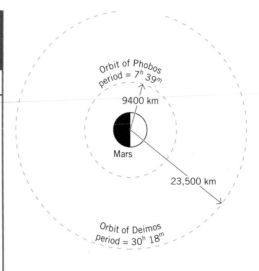

FIGURE 10.30 Orbits of Phobos and Deimos around Mars. Note that Phobos revolves around Mars faster than the parent planet rotates.

erosion by wind and water, and crustal movements (if the mantle is hot). The atmospheres change from interaction with sunlight, escape into space, degassing (from a hot mantle), and life (if it exists).

Earthlike worlds have five general stages of evolution (Table 10.1). Mars has evolved the least, to stage IV. Venus has changed somewhat more, into the beginning of stage V. Only the earth has made it through stage V and continues to evolve.

In September 1992, NASA launched the Mars Observer spacecraft on a return trip to the red planet. The spacecraft should arrive in August 1993. Once there, the Mars Observer will have the capability to see details as small as 1.5 meters from its orbit! However, such resolution will be reached for only small surface regions, about 1.5 km across. This effort will boost our information of the surface and refine our models for the evolution of Mars.

■ Since Mars shows evidence of once having had abundant water, what do you suppose happened to it all?

10.8 The Moons of Mars

Two moons encircle the planet Mars; appropriately, they are named Phobos and Deimos ("Fear" and "Panic"), after the companions of the god Mars. Asaph Hall (1829–1907) at the U.S. Naval Observatory discovered the two moons in 1877. *Deimos* and *Phobos* both lie close to Mars and orbit the planet rapidly (Fig. 10.30). Deimos, the outer moon, circles Mars in 30.3 hours; Phobos, the inner moon, takes a mere 7.3 hours. In fact, Phobos is one of only two moons (the other is Jupiter's innermost satellite) that orbit the parent planet faster than that body spins. So while Deimos rises in the east and sets in the west, as our moon does, Phobos rises in the west and sets in the east! Like the earth's moon, the Martian moons keep the same face to the planet.

Spacecraft observations indicate that Deimos and Phobos have the same general shape: an ellipsoid with three axes. Phobos, the larger, has axes about 27, 21, and 19 km long; Deimos's axes are only 15, 12, and 11 km. Photographs also show that Phobos (Fig. 10.31) and Deimos (Fig. 10.32) have cratered surfaces. The sizes and numbers of these

(a)

(b)

FIGURE 10.31 Phobos close up. (*a*) Phobos photographed from 612 km by the Viking 1 Orbiter. This side of Phobos is the one that faces Mars. The largest crater is Stickney, 10 km in size. (*b*) Phobos viewed from a distance of 120 km. The smallest craters are about 10 meters in diameter. Note the many grooves; their origin has not been explained. (Both courtesy of NASA.)

craters suggest that the surfaces of these satellites are at least 2 billion years old.

The satellite surfaces are very dark, ranking among the blackest objects in the solar system. They reflect only 2 percent of the light that strikes them—much less than our moon, which reflects about 7 percent. Deimos and Phobos are un-evolved, primitive bodies made early in the solar system's history. Their surfaces are dark and cratered. Any internal heat was lost abruptly. You will see (Chapter 12) that such bodies are common and resemble a certain type of asteroid. Also, you will find that coal-black material is widespread in the solar system beyond Mars.

■ Do the surfaces of Deimos and Phobos resemble that of the lunar highlands?

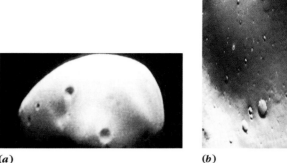
(a)

(b)

FIGURE 10.32 Deimos. (*a*) Overall view. The largest crater is 1.3 km in size. The illuminated part of the moon is about 12 km by 8 km. (*b*) Close-up view of a region 1.2 km by 1.5 km bearing features as small as 3 meters. The boulders are about the size of a house. (Both courtesy of NASA.)

Key Concepts

1. Venus and Mars, the terrestrial planets most like the earth (Table 10.2), have bulk density close to that of earth, and so have somewhat similar interior structures (Venus more so than Mars).

2. The atmospheres of Venus and Mars have essentially the same composition: almost all carbon dioxide. Both atmospheres are also very dry. But the Martian atmosphere is very thin compared to that of Venus. Even so, wind erosion takes place on Mars.

3. The average surface temperature of Venus is very high (740 K) compared to the earth (290 K); in contrast, Mars is quite cold (225 K). Venus is so hot because of an extreme greenhouse effect from its thick carbon dioxide atmosphere. The Martian atmosphere is so thin that it provides little greenhouse effect.

4. The surface of Venus has mountains, volcanoes, upland plateaus, rolling plains, large valleys, and impact craters. Most of the surface is very flat and lacks large impact basins. Almost the entire surface has been remolded by volcanism, so it is relatively young (though older than the earth's youngest surface regions).

5. Venus, because it has almost the same mass as the earth, went through a similar evolutionary sequence; Mars, with less mass, did not evolve

TABLE 10.2 RELATIVE COMPARISON OF THE TERRESTRIAL PLANETS				
PLANET	DIAMETER (earth = 1)	MASS (earth = 1)	DENSITY (water = 1)	SURFACE PRESSURE (earth = 1)
Mercury	0.38	0.055	5.4	$\sim 10^{-15}$
Venus	0.95	0.82	5.2	90
Earth	1.0	1.0	5.5	1.0
Mars	0.53	0.11	3.9	0.01
Moon	0.27	0.012	3.3	$\sim 10^{-15}$

nearly as much. Neither planet developed global tectonics, though both show evidence of volcanism, Venus more than Mars.

6. Mars is a desert world, with much less water than the earth; most of the water is probably ice in the cores of the polar caps and under the surface. Venus is a hot place, and also very dry compared to the earth.

7. The surface of Mars has giant volcanoes (now extinct), arroyos cut by water (in the past), a huge rift valley system, and impact craters. The northern hemisphere is volcanic and younger than the southern.

8. The arroyos and eroded craters indicate that in the past Mars had an atmosphere extensive enough to permit liquid water to exist on the surface (now

both the surface pressure and the temperature are too low).

9. Terrestrial planets go through similar evolutionary sequences, with mass mostly determining how far a planet will evolve (distance from the sun plays a secondary role): the less mass, the faster a planet loses internal heat and its atmosphere, and the less it evolves. Volcanism and tectonics play the major role in the evolution of the surface. Wind and water erosion are secondary processes.

10. The Doppler shift occurs for light and radio waves when a receiver and a source are moving relative to each other. For planets, the Doppler shift can be used to determine rotation rates, especially for planets whose surfaces we cannot see visually.

Key Terms

arroyos
blueshift
Doppler shift
outgassing
polar caps
radial velocity
redshift
retrograde rotation
shield volcano

Study Exercises

1. How is the Martian surface similar to that of Venus? (Objective 10-2)

2. In what one respect is Venus most like the earth? Most different from the earth? (Objectives 10-1, 10-2, and 10-3)

3. How is Mars most like the earth? Most different from the earth? (Objectives 10-1, 10-2, and 10-3)

4. Suppose you are kidnapped by an evil alien creature who threatens to drop you on Venus or Mars with a limited amount of supplies. Which planet would you prefer, and what are the reasons for your choice? (Objective 10-7)

5. Under what conditions could the Martian arroyos have formed? (Objectives 10-2, 10-5, and 10-9)

6. Using your knowledge of terrestrial volcanoes, outline how the Martian volcanoes may have affected the evolution of the Martian atmosphere. (Objectives 10-8 and 10-9)

7. In what major respects are the atmospheres of Mars and Venus similar? Different? (Objectives 10-2, 10-3, and 10-11)

8. Compare the contents of the Martian polar caps in summer and in winter. (Objective 10-2)

9. What one surface feature can be used most effectively to compare the relative ages of the earth, Mars, and Venus? (Objectives 10-5 and 10-10)

10. Would you expect very many or very strong quakes on Mars now? Why or why not? (Objectives 10-2 and 10-4)

11. Of the terrestrial planets, which one would show the greatest Doppler shift from a reflected radar signal? (Objective 10-12)

12. If you bounced a radar signal off a planet, from which region would you expect to find *no* Doppler shift (a Doppler shift of zero)? (Objective 10-12)

13. What is the most significant difference between the global magnetic fields of Venus and Mars compared to the earth? How does this difference create a problem for the dynamo model for planetary magnetic fields? (Objective 10-12)

14. Do the surfaces of Deimos and Phobos resemble that of our moon in any ways? (Objective 10-13)

Problems & Activities

1. Suppose a radar signal at a frequency of 10 GHz is bounced off Mercury. Upon its reception at the earth, what would you expect to be the amount of its Doppler shift from Mercury's rotation?

2. Compare the escape speed from Mars to that of the earth. Of Venus.

3. Use the orbital information about Deimos or Phobos and Newton's form of Kepler's third law to calculate the mass of Mars.

4. Use the properties of ellipses and the orbital characteristics of Mars and the earth to show that the closest the two planets can come at opposition is 51 million km.

5. What is the angular diameter of Mars when viewed at closest opposition, as given in Problem 4?

6. Compare the surface gravity of Venus to that of the earth. Of Mars.

7. In the 1993 opposition, Mars reached a maximum angular diameter of 15 arcsec. What was its size-to-distance ratio? How far was it from earth at this opposition? How does this value compare to the closest opposition possible (Problem 4)?

See for Yourself!

You will need a good small telescope (at least a 15-cm objective) to observe even the major surface features of Mars. The best time to make such observations is at opposition, when the earth and Mars are closest and so has its largest angular size. These opportunities take place while Mars is undergoing its retrograde motion, which occurs in intervals of 780 days. At the January 7, 1993, opposition, Mars had an angular diameter of 15 arcsec. The next opposition of Mars is in February 1995 (see Fig. 1.8). Then Mars will come within 0.68 AU of the earth, not as close as the 1993 distance of 0.63 AU.

The key trick for observing Mars is to do so on a night of good seeing, when star images are very steady and have an angular size of about 1 arcsec. You then need to choose an eyepiece with the highest magnification possible for the conditions and your telescope's aperture: about 250 to 300 power. Don't expect surface features to jump out at you! After all, you are observing through two atmospheres. Look patiently for moments of good seeing, when the image will become unusually steady. You should have no trouble seeing the polar cap of the winter hemisphere. Other markings will take more luck. ◾

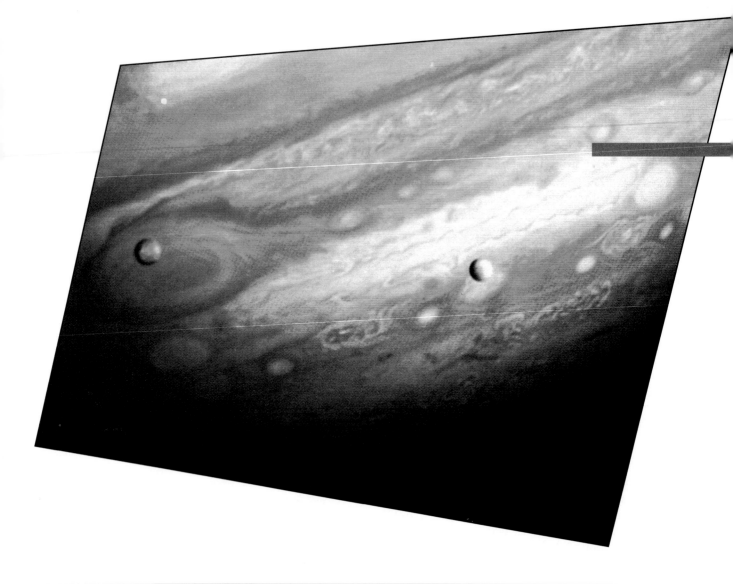

After studying this chapter, you should be able to:

11-1 Compare and contrast the Jovian planets as a group to the terrestrial planets, emphasizing the greatest differences.

11-2 Contrast the Jovian planets to one another in terms of their relative sizes, relative masses, bulk densities, atmospheric compositions, internal structures, and unique features.

11-3 Compare and contrast the interior and composition of Jupiter to those of the earth.

11-4 Compare the rings of Saturn with those of Uranus, Neptune, and Jupiter in terms of size, shape, and possible composition.

11-5 Present the unique characteristics of Pluto that make it neither a Jovian nor a terrestrial planet.

11-6 Describe the general properties of the Pluto–Charon double-planet system.

11-7 Compare and contrast the general characteristics, surface features, and evolution of the Galilean satellites of Jupiter: Io, Europa, Ganymede, and Callisto.

11-8 Compare and contrast the Galilean satellites to the earth's moon and to Pluto and Charon, especially in terms of evolutionary processes.

11-9 Compare the magnetic fields and magnetospheres of Jupiter, Uranus, Neptune, and Saturn to those of the earth, and apply a dynamo model to them.

11-10 Contrast Saturn's largest moon, Titan, with the other moons of Saturn, the Galilean moons of Jupiter, the largest moons of Uranus, and the earth's moon.

11-11 Argue that the Jovian planets have evolved since their formation by heat flow from their interiors, but, in contrast with the terrestrial planets, they leave behind no record of their changes.

Central Question

How do the Jovian planets differ from the terrestrial ones, and what do these differences imply about different evolutions?

The Jovian Planets: Primitive Worlds

Mephistopheles: Nor are the Names of Saturn or Jupiter feign'd, but are erring stars.

CHRISTOPHER MARLOWE, *The Tragical History of Doctor Faustus*

Jupiter and Saturn are impressive, awesome planets—two giant worlds languidly circling the sun. Banded Jupiter drags along its coterie of satellites. Saturn is set in its rings like a prize gem of space. Uranus and Neptune seem like twins at a distance but differ greatly when viewed close up. You can barely grasp the weird environments of these giant worlds: planets many times the diameter of the earth, mostly gases and liquids, and without breathable atmospheres.

The realm of the Jovian planets—Jupiter, Saturn, Uranus, and Neptune—marks a region completely different from that of the terrestrial planets. The difference includes not only the physical properties of the planets but also their many moons and graceful rings. The satellites here are worlds in their own right, some as large as the smaller terrestrial planets. These moons have surface environments that differ remarkably from those in the inner solar system, so their evolutionary histories were different—a point that illuminates another facet of the early solar system. And the double-planet system of Pluto and Charon may have more in common with these moons than with the Jovian planets.

This chapter investigates the features of the Jovian planets that set them apart from the terrestrial ones. The major differences between the Jovian and terrestrial planets furnish additional clues about the evolution of the planets and the formation of the solar system. The key point is this: the Jovian planets are primitive worlds, looking very much the same today as when they were formed.

11.1 Jupiter: Lord of the Heavens

The **Jovian planets** are those whose physical properties resemble those of their prototype—Jupiter (Fig. 11.1). Jupiter ranks as the largest and most massive body in the solar system (after the sun). Jupiter's total mass is about 2.5 times that of all other planets put together (and more than 300 times the earth's). Eleven earths placed edge to edge would stretch across Jupiter's visible disk,

FIGURE 11.1 Giant planets. The Jovian planets, shown to scale relative to the earth. Note the banded atmospheres of Jupiter (far left) and Saturn. Uranus, to the right of Saturn, and Neptune both have methane gas in the atmosphere that absorbs red light and so gives an overall bluish color. The rings of Jupiter, Uranus, and Neptune are not visible. (Courtesy of NASA/JPL.)

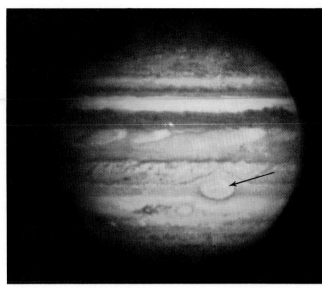

FIGURE 11.2 Jupiter and its Great Red Spot photographed by the Hubble Space Telescope in May 1991. The colors here are true colors. Note the banded structure of the clouds in the upper atmosphere caused by strong jet streams. The clouds contain small crystals of frozen ammonia and traces of colorful chemical compounds. The Great Red Spot (arrow) is a giant storm; below and to its left is a white oval. (Courtesy of NASA/ESA.)

and more than a thousand earths would be needed to fill its volume—truly a giant planet!

Physical Characteristics

Just as Jupiter's mass is the largest among the planets, so is its diameter—more than 140,000 km. Yet for all this size, Jupiter's material is much less concentrated than the earth's, for Jupiter's density is only some 1300 kg/m³—not much more than water!

All the Jovian planets have low densities compared with the terrestrial planets—one key difference between the two classes. These low densities imply that the Jovian planets are made of quite different stuff. The terrestrial planets are basically rocks and metals, made of elements such as iron, aluminum, oxygen, and silicon. Jupiter, in contrast, is made mostly of hydrogen and helium in a liquid and gaseous state.

This hydrogen and helium came together when Jupiter formed, and the giant planet has lost little, if any, since then. Why not? Jupiter's huge mass means that its escape speed (Section 4.5) is large, about 60 km/s. Remote from the sun, Jupiter's upper atmosphere is cold, only about 130 K, and hydrogen mole-

cules there move at about 1 km/s. So even hydrogen molecules do not have escape speed. If the hydrogen cannot escape from the upper atmosphere, more massive atoms and molecules also cannot get away. Jupiter has retained its atmosphere for eons and will hold it for eons to come. What you see now is basically the atmosphere and mass with which Jupiter was born. (The earth and other terrestrial planets lost atmospheric materials early in their histories.)

Atmospheric Features and Composition

The visible disk of Jupiter (Fig. 11.2) is *not* the planet's surface but its upper atmosphere, which shows alternating strips of light and dark regions that run parallel to the equator. The light regions, called **zones,** have lower temperatures than the dark regions, called **belts.** The lower temperatures imply that the zones are higher than the belts because, in general, the temperature in a planet's atmosphere decreases with altitude. These differences in temperature suggest that the zones flag the tops of rising regions of high pressure, and the belts mark the descending areas of low pressure (Fig. 11.3). This convective atmospheric flow

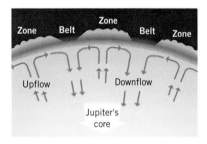

FIGURE 11.3 The circulation in Jupiter's upper atmosphere. Rising air creates high-pressure regions (called *zones*) and the downflow makes low-pressure areas (called *belts*). The zones generally appear lighter in color than the belts because they are higher up in the atmosphere. The convection is driven by heat flowing out of the planet's interior.

transports energy out to space from the planet's interior and so implies that its interior is hot.

Markings in the clouds allow measurement of Jupiter's rate of rotation. (It can also be measured from the Doppler shift of the light emitted by the approaching and receding edges of Jupiter.) The rate varies with latitude: Jupiter spins in 9 hours and 50 minutes at its equator and in 9 hours 55 minutes at its poles. Such a variation of rotation rate is called **differential rotation;** it indicates that the body is fluid. (A solid body like the earth rotates so that each point in its surface has the same rotational period.) This rapid rotation and Jupiter's large radius produce an equatorial speed in excess of 43,000 km/h and makes the planet fairly oblate.

Such an enormous rotation speed drives the circulation in Jupiter's atmosphere. It causes the permanent high-pressure zones and low-pressure belts to stretch out completely around the planet. Jet streams zip along at the boundaries between the belts and zones, creating atmospheric disturbances. Typical wind speeds are 100 m/s—about three times faster than the earth's jet streams. The Voyager missions zoomed in on these complex streams and swirls of Jupiter's upper cloud layer, showing

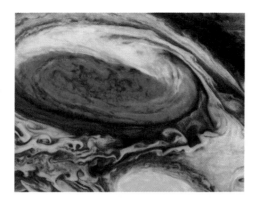

FIGURE 11.5 Close-up of the Red Spot and the turbulent region near it. The smallest details visible are about 60 km across; the image has been computer-processed to enhance these features. (Courtesy of NASA.)

the turbulent atmospheric flow in detail (Fig. 11.4).

The most permanent and famous atmospheric disturbance is the *Great Red Spot* (Fig. 11.5), first observed by the Englishman Robert Hooke in 1630. The Red Spot changes in size; it is some 14,000 km wide and up to 40,000 km long—it could easily swallow the earth! The Red Spot is a few degrees cooler than the surrounding zone and pokes about a few kilometers above it, so it is a rising region of high pressure. The Red Spot rotates counterclockwise, just as expected from a high-pressure zone in Jupiter's southern hemisphere. It turns once in 7 days, a huge vortex pushed by the surrounding atmospheric flow. In turn, the Red Spot deflects nearby clouds and forces them around it like leaves in a whirlpool. Behind the Red Spot runs a region of turbulent flow (Fig. 11.5) from the atmosphere flowing past it. The Red Spot stands out as a long-lived (hundreds of years) atmospheric eddy.

Why is the Red Spot red? (Many other spots on Jupiter are white.) It is likely that colored compounds are made by chemical reactions, driven by the outflowing heat. In general, the colors seen in Jupiter's upper atmosphere (blue, red, and yellow) probably result from various chemical compounds formed there.

Jupiter's upper atmosphere contains by mass about 82 percent hydrogen, 18 percent helium, and traces of all other elements—essentially the same composition as the sun. Most of this material is in the form of molecules. Examples are: methane, ammonia, molecular hydrogen, and water. The visible clouds at the tops of the zones are most likely ammonia ice crystals. Below them may float liquid ammonia and water ice

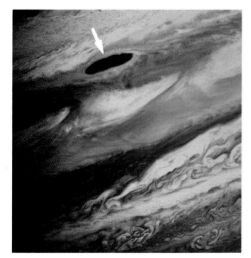

FIGURE 11.4 Close-up of a brown oval (arrow) in Jupiter's atmosphere photographed by Voyager 1. Such features last one or two years; this one has a length about equal to the earth's diameter. They appear to be clear regions that allow us a view down to lower levels. Winds in the belt above the oval flow at about 100 m/s. (Courtesy of NASA.)

Jupiter	
At a Glance	
Equatorial radius	71,492 km
Mass	18.99×10^{27} kg
Bulk density	1330 kg/m^3
Escape speed	59.6 km/s
Atmosphere	Hydrogen, helium

clouds. The entire atmosphere may be 1000 km thick. In fact, no distinct boundary lies between atmosphere and interior. The atmosphere gets denser and hotter farther in, gradually merging into the liquid interior.

A Model of the Interior

We can infer Jupiter's internal structure from physical models that include two key pieces of information: (1) Jupiter's low density and its atmospheric composition imply a solar mix of material throughout, and (2) Jupiter radiates into space more energy than it receives from the sun (about twice as much), so it must be *hot* inside. The internal heat is probably left over from Jupiter's formation. (Large, massive planets lose heat very slowly.)

Models of Jupiter, assuming a solar composition and internal heat, come up with a differentiated interior (Fig. 11.6). The atmosphere covers the planet like a thin skin and consists mostly of molecular hydrogen. As one dives into the planet, the density, temperature, and pressure increase, so the hydrogen exists in a *liquid* state. At a pressure of about 2 million atmospheres, the hydrogen is squeezed so tightly that the molecules are separated into protons and electrons that move around freely and can conduct electricity. This state, called **metallic hydrogen,** has recently been observed in a laboratory on the earth. It continues to within about 14,000 km of the planet's center. Here, perhaps, if Jupiter does have a solar composition, lies a core of heavy elements (perhaps mostly rocky materials).

Most of Jupiter is hydrogen, and most of that hydrogen is liquid—quite a contrast to the earth's interior (Section 8.2) and that of the other terrestrial planets. The core temperature may be about ten times hotter than the earth's core. The flow of the heat outward from the core, along with the rapid rotation, drives the circulation to produce the beautiful banded atmosphere.

Magnetic Field

Jupiter has a magnetic field some ten times as strong as the earth's. At the

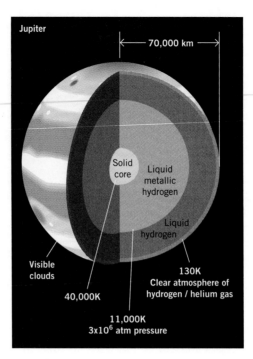

FIGURE 11.6 Model for Jupiter's interior structure. Below the atmosphere is a thick layer of liquid molecular hydrogen; below that, a region of metallic hydrogen because of the high pressure (above 3×10^6 atm). The core is a dense material, perhaps silicates, that may be molten because it is quite hot, perhaps 40,000 K. The core may have a mass about ten to twenty times that of the earth. (Adapted from a NASA diagram.)

cloud tops, the field strength is about 4×10^{-4} T. Recall that the earth's magnetic field at the surface is about 0.5×10^{-5} T. (The tesla, T, is the SI unit of magnetic field strength: see Appendix A.) The magnetic field axis is tilted about 10° with respect to the rotation axis. The magnetic field rotates with Jupiter. Its period, 9 hours 55 minutes 30 seconds, is taken as the rotation of Jupiter as a body. This technique is used to infer the actual rotation periods of the Jovian planets.

Jupiter's strong magnetic field produces a magnetosphere much larger than that of the earth, though the formative processes are basically the same. As the solar wind plows into this intense field, it creates an enormous shock wave, called a *bow shock,* that enshrouds the planet (Fig. 11.7). The magnetic field traps plasma particles from the sun in belts, similar to the Van Allen belts around the earth, close to Jupiter. The sun side field acts as a buffer to deflect the solar wind around Jupiter. On the night side, a magnetic tail stretches out to a length of a few astronomical units and may reach as far as Saturn! Electrons moving close to the speed of light in the inner regions of the magnetosphere generate radio emission that has been mapped from earth (Fig. 11.8).

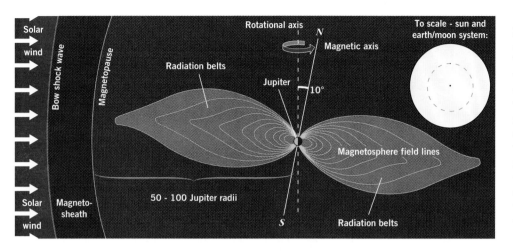

FIGURE 11.7 Model for Jupiter's magnetosphere based on spacecraft data. The size and shape vary according to the strength of the solar wind. The solar wind forms a bow shock wave where it interacts with the magnetic field; behind it lies the magnetosheath, a region of relative calm. Note that the tilt of the magnetic axis to the rotation axis (10°) is about the same as that for the earth. The size of the sun, and within it, the earth and moon system, are shown on the same scale at the upper right. (Adapted from a NASA diagram.)

How does Jupiter generate such an intense magnetic field? Recall that a dynamo model (Section 8.3) of the earth's magnetic field pictures currents in the liquid metal core generating the magnetic field like an electromagnet. For a large part of Jupiter's interior, liquid metallic hydrogen can conduct electric current. So the conditions for a planetary dynamo to operate are available: a fluid able to conduct electricity, filled with convective currents driven by heat and rapid rotation. A dynamo in the metallic hydrogen zone could produce Jupiter's magnetic field—the model seems to work well here.

Auroras

Voyager and *HST* pictures of Jupiter's night side showed polar **auroras** (Fig. 11.9), which were expected because of Jupiter's large magnetosphere. We presume that these auroras have the same causes on Jupiter as on earth: excitation of the upper atmosphere by energetic charged particles pouring in at the north and south magnetic poles. These particles flow from the sun and are trapped by Jupiter's magnetic field in its magnetosphere. Magnetic reconnection events (Section 8.3) then drive the plasma into Jupiter's atmosphere.

■ What does the differential rotation indicate about Jupiter's physical state?

11.2 The Many Moons and Rings of Jupiter

Jupiter possesses an entourage of at least sixteen moons (Appendix B, Table B.7).

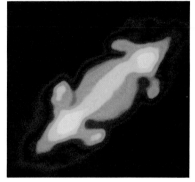

FIGURE 11.8 Jupiter's inner magnetosphere. These observations with the VLA at a wavelength of 21 cm show the radio emission from high-speed electrons spiraling along the magnetic field lines of the inner part of the Jovian magnetosphere. The structure is similar to the Van Allen radiation belts of the earth's magnetosphere. The gas in the atmosphere of the planet also produces radio emission, which shows up here. (VLA observations by I. de Pater; courtesy of NRAO/AUI.)

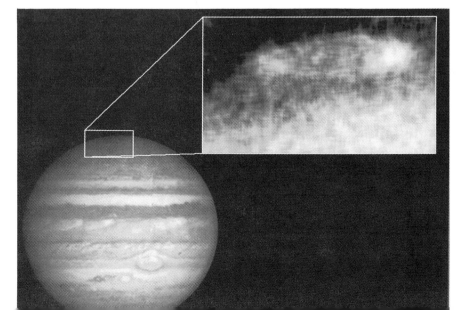

FIGURE 11.9 An aurora over Jupiter's north polar region in February 1992. At left is an image of Jupiter's full disk taken in visible light. The white box encloses the region shown at the upper right. This image was taken in the ultraviolet and shows the oval of the aurora (compare to the earth's: Figs. 8.11 and 8.12). (Courtesy of John Caldwell, Institute for Space and Terrestrial Science and York University; NASA/ESA.)

The brightest and largest, whose orbits lie within 3° of Jupiter's equatorial plane, were observed and reported by Galileo. These huge **Galilean moons** orbit within 2 million km of Jupiter; from closest outward, they are *Io, Europa, Ganymede,* and *Callisto.* Like our moon, each is locked in synchronous rotation and so keeps one face toward Jupiter. These moons are among the largest in the solar system, along with Saturn's Titan and Neptune's Triton (Fig. 11.10).

Each Galilean moon is a world of its own; together, they exhibit dramatic differences in surface features and internal structures. Key differences are hinted at by the bulk densities of the moons: Io, 3500 kg/m^3; Europa, 3000 kg/m^3; Ganymede, 1900 kg/m^3; and Callisto, 1800 kg/m^3. This list is in order of increasing distance from Jupiter, and you can see the pattern: the density of the moons decreases with increasing distance from Jupiter. Such density differences show that the compositions of Io and Europa resemble that of our moon—mostly rock, with perhaps a little icy material. In contrast, Ganymede and Callisto contain substantial amounts of water ice or other low-density icy materials, and much less rock.

Io

Io has a thin atmosphere, composed mainly of sulfur dioxide. (Only two other satellites, Saturn's Titan and Neptune's Triton, are known to have an atmosphere.) At the surface, the atmospheric pressure is about 10^{-10} atm. Io's atmosphere has a peculiar property: it gives off a bright, continuous glow of emission from sodium atoms. This sodium glow surrounds Io like a yellow halo out to a distance of about 30,000 km.

What produces Io's sodium cloud? Volcanic eruptions, at least in part. Io has at least eleven active volcanoes; erupting volcanoes and fuming lava lakes cover its surface (Fig. 11.11*a*). This activity implies that the interior is hot; sulfur and sulfur compounds melt and are vented in volcanic outflows. The colors of the surface indicate that the outflows contain liquid sulfur.

Io's volcanoes eject plumes of gas (containing sulfur dioxide) and dust up

FIGURE 11.10 Large moons in the solar system. Using the earth's moon (center) for scale, this image shows the Galilean moons of Jupiter (Io, Europa, Ganymede, and Callisto); Titan, moon of Saturn; and Triton, moon of Neptune. (Artwork prepared by Stephen P. Meszaros for NASA.)

to heights of 250 km at velocities of up to 1000 m/s. (In contrast, the earth's large volcanoes spit out material at about 50 m/s.) On a nearly airless body like Io, the volcanic gas and dust crest, like a fountain plume, in several minutes and then spread and fall in a dome shape (Fig. 11.11*b*). Io's escape velocity (2.5 km/s) is greater than the speed with which the volcanic dust and gases erupt; so there is little direct loss of the material to space.

Io's volcanoes have a different shape from those commonly found on the earth, Venus, and Mars. Few appear as cones or shields; rather, they resemble collapsed volcanic craters. Lava simply pours out of a crater vent and spreads outward for hundreds of kilometers (Fig. 11.12). So dark-colored lava lakes, including one about the size of the island of Hawaii, surround many of Io's volcanoes. The temperature of these lava lakes is about 400 K. (The melting point of sulfur is 385 K.)

Why is Io's interior so hot? The gravitational effects of the other Galilean moons force Io into a slight perturbation of its circular orbit. The tidal forces of Jupiter (which are large because Io is close) cause Io to flex. These tidal stresses continually act on Io. Its interior heats up from the recurring push and pull of the tidal forces, much like continued squeezing of a rubber ball will heat it. This heating drives the volcanism. The surface of Io must be very young, since volcanic activity continually alters it. No impact craters appear on Io; volcanic flows have covered them up. Io's surface seems to be the youngest in the solar system, probably less than 1 million years old.

Europa

Europa's surface shows bright areas of water ice among darker, orange-brown areas. It is crisscrossed by stripes and bands that may be filled fractures in the moon's icy crust (Fig. 11.13). The dark markings crisscross its face, making it look like a cracked eggshell (Fig. 11.14). Some of these shallow cracks extend for thousands of kilometers, split to widths up to 200 km, but they reach depths of only 100 meters or so. Europa's surface is really incredibly smooth. In relation to its size, its dark markings are no deeper than the thickness of ink drawn on a Ping-Pong ball. This moon has the smoothest surface we've seen so far in the solar system.

Europa's surface is almost devoid of impact craters. So Europa's surface cannot be a primitive one; it has been resurfaced. The crust may have been warm and soft enough some time after formation to wipe out early impact craters. Europa's cracked surface indicates that its solid, icy crust is thin and its interior was once hot. As Europa cooled, its crust turned to smooth, glassy ice that later cracked. Beneath the ice veneer may now lie a layer of liquid water.

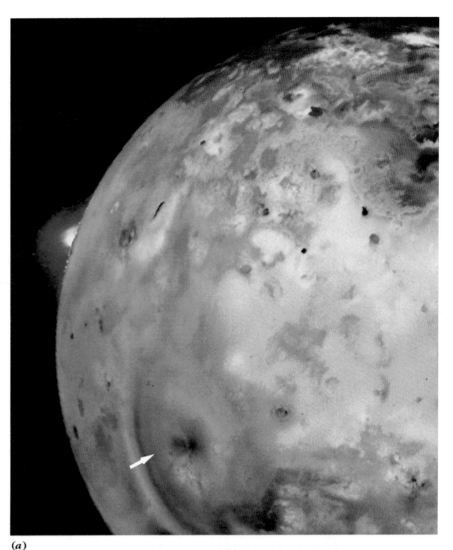

(a)

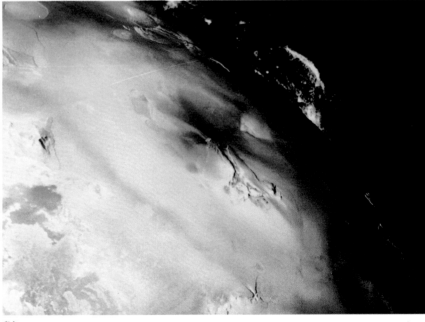

(b)

FIGURE 11.11 Io, an active moon. (*a*) Pele and Loki volcanic plumes on Io. These two active volcanoes are visible in this specially processed image. The plume from Loki appears at the upper left, against the background of space. Pele (arrow) is below it and to the right. Note that its plume spreads out like a dark, circular umbrella. (Image created by A. S. McEwen, Arizona State University and the U.S. Geological Survey; courtesy of JPL and NASA.) (*b*) Another view of an outburst from Pele. The plume here covered an area equal to that of Alaska and rose more than 300 km above the surface. Pele is at the center of the photo, appearing as a complex of hills with a central valley. (Image created by A. S. McEwen, Arizona State University and the U.S. Geological Survey; courtesy of NASA/JPL.)

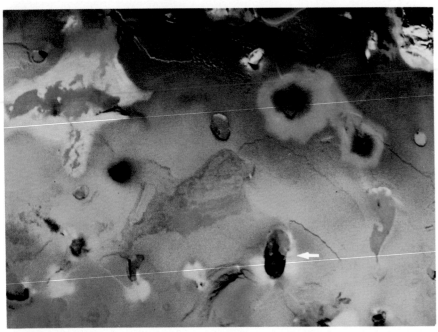

FIGURE 11.12 Close-up view of Io's surface in its northern hemisphere. Dark spots (such as the one with the arrow) with the irregular radiating patterns are volcanic vents with lava flows. (Image created by A. S. McEwen, Arizona State University and the U.S. Geological Survey; courtesy of NASA/JPL.)

FIGURE 11.13 Voyager view of Europa with a resolution of 10 km. Note the dark cracks along the surface and the relatively few craters. This false-color image was designed to bring out variations in the surface features; the colors are much different from the true ones. (Courtesy of NASA.)

Ganymede

Largest moon of Jupiter, Ganymede has two basic types of terrain (Fig. 11.15): *cratered* and *grooved*. Craters up to 150 km in size densely mark the surface of the cratered terrain. Their abundance indicates that the cratered terrain is old, some 4 billion years. Compared with those on the moon and Mercury, the craters are shallow for their diameter—an indication of an icy rather than rocky surface. Very bright rays extend from many of the craters on Ganymede (Fig. 11.15*a*). These features attest impacts on an icy surface at the time of formation.

The grooved terrain separates the cratered terrain into polygonal segments (Fig. 11.15*b*). It consists of a mosaic of light ridges and darker grooves where the ground has slid, sheared, and torn apart. Long cracks, where the ground has moved sideways for hundreds of kilometers, also cover the surface.

No large mountainous regions or large basins relieve the terrain on Ganymede; in fact, no relief is greater than about 1 km. This suggests that the crust of Ganymede is somewhat plastic, probably because of the large amount of water ice (perhaps 90 percent) in the

crust. Ganymede's bulk density is only about 1900 kg/m^3, so it contains about half water and half rock overall. Occasional stresses on the water–rock crust have created the fracture patterns. Some ridges and grooves overlie others, an indication that there may have been many episodes of crustal deformation in the

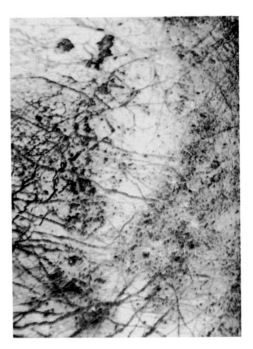

FIGURE 11.14 Close-up view of Europa's surface covering an area about 600 km by 800 km along the day-night line, which highlights the surface features. The bright ridges are about 5 to 10 km wide and 100 km long. The dark bands are 20 to 40 km wide and thousands of kilometers long. (Courtesy of NASA.)

past. Basically, Ganymede has been geologically inactive for 4 billion years.

Callisto

Farthest out of the Galilean moons, Callisto most resembles a terrestrial planet (Fig. 11.16). It has a surface riddled with craters of a wide range of sizes. Some have bright ice rays; others are filled with ice. Callisto's craters are shallow—several hundred meters deep or less. Why? Because the surface is a mixture of ice and rock; the surface slowly flows, flattening out the land.

Callisto has a huge, multiringed basin called *Valhalla* (Fig. 11.16). Its central floor is 600 km in diameter; twenty to thirty mountainous rings that have diameters of up to 3000 km surround it like a bull's-eye. The rings look like a series of frozen waves. They were formed in a stupendous collision that melted subsurface ice, causing the water to spread in waves that quickly froze in the 95 K surface temperature. The ripples are preserved as rings. The central floor has fewer craters than the rest of the terrain, an indication that the impact occurred after much of the initial cratering had been accomplished.

Asteroidal Moons

Jupiter's other moons are asteroidlike bodies (Section 12.1), and we expect that they are indeed captured asteroids. (This is likely, for, after all, Jupiter lies just outside the asteroid belt.) There are two groups of four, one group at a distance of about 12 million km that orbit counterclockwise, and another at about 23 million km that orbit clockwise.

We have observed one moon, Amalthea, which may resemble these outermost moons. Only 181,000 km out from Jupiter, it whizzes around once every 12 hours. It is elongated, 270 km by 155 km (Fig. 11.17); the surface is dark red. This moon's irregular shape, small size, and carbon-rich, cratered surface show its asteroidlike character.

The Rings of Jupiter

Jupiter has millions of moons—tiny ones that make up its **ring system,** discovered by Voyager 1. The rings are so

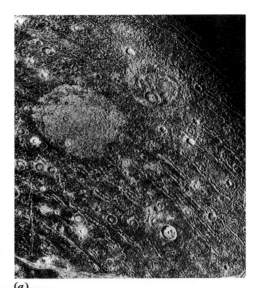
(a)

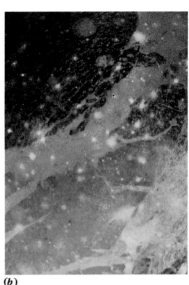
(b)

FIGURE 11.15 Ganymede. (*a*) The older surface of Ganymede. The smallest features are about 3 km across. The impact craters that have bright rays are younger than those that do not. The curved troughs and ridges mark part of a huge impact basin. (*b*) Grooved and ridged terrain, which make up the newer parts of Ganymede's surface. The resolution is 6 km. (Both courtesy of NASA.)

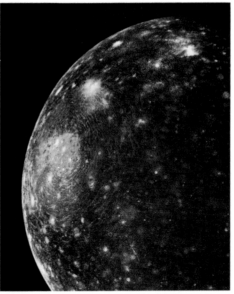

FIGURE 11.16 The giant ringed basin on Callisto (upper right). The bright spot at the basin's center is about 600 km across; the outer rings, 2600 km. Note the lack of ridges or mountains surrounding this impact basin, which is called Valhalla. Also note the many impact craters with bright rays. (Courtesy of NASA.)

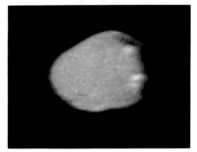

FIGURE 11.17 Amalthea, Jupiter's innermost moon. The indentations at top and bottom are craters. Note the overall reddish color. The long axis is about 270 km. (Courtesy of NASA.)

thin (less than 30 km thick) that they are essentially transparent. They are most visible when viewed edge-on; then the particles scatter light well. To possess this property, the particles must be small (about 10 μm). We do not yet know what they are made of.

Dramatic pictures of the back-lit rings (Fig. 11.18) show that the rings have a definite structure. The outer, brightest part is 800 km wide and lies about 128,500 km from Jupiter's center. Within it is a broader ring some 6000 km wide. And within that ring lies a faint sheet of material that extends from 119,000 km out from Jupiter's center down to the cloud tops. A faint, outer ring surrounds the whole system and reaches out over 200,000 km, almost to the orbit of Amalthea.

■ Of the four Galilean moons, Io is unique. Why?

11.3 Saturn: Jewel of the Solar System

Saturn bears a marked affinity to Jupiter, but its fantastic ring system outranks in splendor that of the larger planet (Fig. 11.19). Saturn is only slightly smaller, but it has only about a third the mass of Jupiter. Why such a difference? Because Saturn has the lowest density of any of the planets—about 700 kg/m^3, which is much less than that of water. Saturn could float! It must have an interior rich in light elements and lacking in rocky materials.

Atmosphere and Interior

The atmospheric structure of Saturn resembles that of Jupiter: belts running parallel to the equator, driven by its rapid rotation. Saturn's rotational period is 10 hours, 14 minutes at the equator and varies with latitude; it, too, shows *differential rotation*. Disturbances in the belts rarely occur compared with the frequency of such events on Jupiter. Mighty storms do occur at intervals of about 30 years, roughly at midsummer in the northern hemisphere (Fig. 11.20). A warming of the atmosphere may trigger these storms, which last for several weeks before the fierce atmospheric winds blow them apart.

The atmosphere of Saturn has roughly the same composition as that of Jupiter: mostly hydrogen and helium. Spacecraft observations indicated about half the percentage of helium found in

FIGURE 11.18 Jupiter's ring system. The sun, which is to the rear of this image, is backlighting the particles of the rings. Note the bright, sharp edge and diffuse inner region. The rings are very thin and narrow. (Courtesy of NASA.)

FIGURE 11.19 Saturn viewed by the *HST* in August 1990, when the planet was about 1.4×10^9 km from the earth. The resolution is about 700 km. Note that the bands in the atmosphere are much less prominent than those of Jupiter. The overall ring system structure is visible. The outer ring is the A ring; within it is a small gap called the Encke division. The wider Cassini division separates the A ring from the B ring. Inside the B ring lies the faint C ring. (Courtesy of NASA.)

Jupiter's upper atmosphere. Methane, water vapor, and ammonia make up a minority of the gases.

Saturn's clouds appear far less colorful than those of Jupiter—mostly a faint yellow and orange. Because of the low temperatures on Saturn compared to

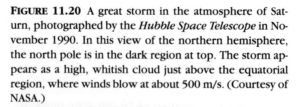

FIGURE 11.20 A great storm in the atmosphere of Saturn, photographed by the *Hubble Space Telescope* in November 1990. In this view of the northern hemisphere, the north pole is in the dark region at top. The storm appears as a high, whitish cloud just above the equatorial region, where winds blow at about 500 m/s. (Courtesy of NASA.)

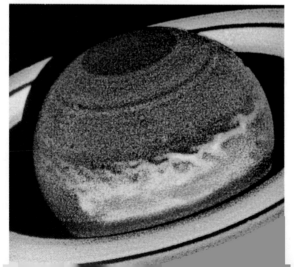

Jupiter, the clouds lie lower in the atmosphere, and a high-altitude haze subdues our view. However, the Voyager photos showed much of the same complexity of cloud patterns seen on Jupiter (Fig. 11.21), with wind speeds much higher, up to 500 m/s near the equator. Weather patterns can change weekly, while large storm systems persist for longer times—on the order of a few years.

Saturn's interior (Fig. 11.22) probably reflects Jupiter's composition—roughly the same as that of the sun. Saturn may have a small, rocky core some 20,000 km in diameter and a mass of about 20 earth masses (or it may have no such core at all!). It probably has a large zone of liquid hydrogen and a smaller one of metallic liquid hydrogen. So, like Jupiter, much of Saturn's interior is in a liquid state.

Similarities to Jupiter

Saturn resembles Jupiter in two other important respects. First, infrared observations show that Saturn emits more energy, as infrared radiation, than it receives from the sun—about twice as much. As with Jupiter, this excess heat may be left over from the planet's formation.

Second, radio and spacecraft observations of Saturn show that it, too, has a strong magnetic field and so a large magnetosphere. The magnetic axis aligns within one degree of Saturn's rotation axis. The magnetic field is probably produced by a dynamo effect in the liquid metallic hydrogen zone of Saturn, in the same way it is presumably produced in Jupiter. The magnetic field creates Van Allen-like belts around Saturn, which trap charged particles from the sun.

■ **What may comprise the core of Saturn?**

■■■

11.4 The Moons and Rings of Saturn

Saturn's clique of moons totals at least eighteen (Appendix B, Table B.8). With two exceptions (Phoebe and Iapetus), all the moons stick close to Saturn's equatorial plane. Masses for some of the

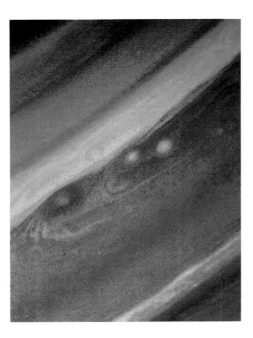

FIGURE 11.21 Voyager 2 photo of the northern hemisphere of Saturn, showing jet streams and turbulent wind flows; winds can reach speeds of 100 m/s. (Courtesy of NASA.)

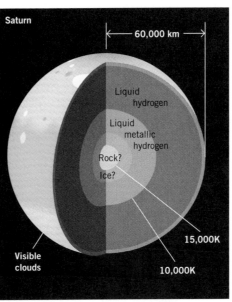

FIGURE 11.22 Model for the interior structure of Saturn. Note that the metallic hydrogen zone is smaller compared to that of Jupiter. The inner, rocky core may have a mass of about twenty times that of the earth.

smaller moons were determined from their gravitational attraction on spacecraft. The densities range from 1200 kg/m³ for Tethys and a few others to 1900 for Titan, similar to the densities of the outer, low-density Galilean moons of Jupiter.

The moons of Saturn fall into three groups: big Titan by itself; the six large icy moons (Mimas, Enceladus, Tethys, Dione, Rhea, and Iapetus, in order outward from Saturn); and the small moons (Phoebe, Hyperion, and the rest). Overall, their densities are less than 2000 kg/m³, which implies that the moons are mostly ice (60 to 70 percent) with some rock (30 to 40 percent). In contrast to the Galilean moons, there is no trend of densities with distance from Saturn. Like Jupiter's moons, all except one (Phoebe) keep the same face toward Saturn.

Most of the moons are cratered. Some cratered terrain has been modified on the larger moons, which implies internal heating to melt parts of the icy surfaces. In contrast, the smaller moons, which are also densely cratered, show no changes—they still have their original surfaces.

With these basics in mind, let's look at selected moons in some detail. This investigation will reveal a key aspect of the outer solar system: many bodies contain ices (frozen volatiles of various kinds) and a dark, carbon-rich rocky material. These light and dark substances dominate the cold regions of the solar system. They are very likely primitive materials from the era of planetary formation.

Titan

Titan, Saturn's largest moon, has a mass of 1.35×10^{23} kg and a diameter of 5120 km. Its density is about 1900 kg/m³, which implies a 50-50 composition of ice and rock.

Titan was the first moon found to have an actual atmosphere. Voyager 1 showed that it consists mostly of molecular nitrogen (some 80 percent), with about 1 percent methane and perhaps a trace of argon. Several hydrocarbons other than methane have also been detected, including ethane, acetylene, and ethylene. The surface temperature is 94 K; the atmosphere's surface pressure, about 1.5 atm. That's an incredibly thick atmosphere for its size!

Voyager flyby photos (Fig. 11.23) showed a layer of orange smog as well as a blue color along Titan's edge. This hue indicates that the atmosphere varies in composition. The surface was completely obscured from Voyager's view. But the pressure and temperature data, along with the spectroscopic detections of nitrogen and hydrocarbons, have led to models of a surface covered with a frigid ocean of ethane, methane, and nitrogen up to a depth of 1 km, beneath which may reside a layer of acetylene.

Other Moons

Saturn's four largest moons, after Titan, are Iapetus, Rhea, Dione, and Tethys, with diameters ranging from some 1100 km to 1500 km (Mimas and Enceladus are much smaller than these four.) Their surfaces are heavily cratered (Fig. 11.24). Except for Iapetus, they have bright, icy surfaces.

Iapetus is a strange moon (Fig. 11.25*a*) with the most extremes of surface cover. The hemisphere leading in its orbit is only one-sixth as bright as that following (like the difference between a blackboard and a field of snow). The leading surface is covered with dark debris picked up during the journey around the planet; the source of this may be dust knocked off of Phoebe by meteorite impacts. Here we see the stark contrast between bright ices (albedo about 5 percent) and sooty stuff (albedo ranging from 30 to 60 percent) on the same body.

Only Enceladus does not have a surface thick with craters (Fig. 11.25*b*). It also shows some linear grooves and fractured regions, a sure sign of recent modification of the surface. A hot interior can melt the icy surface or cause volcanism of water and ice. The interior may be heated by tidal stresses from other nearby moons (as Io is heated).

The rest of the moons are all small, heavily cratered bodies, 300 km or less in diameter. The largest is Hyperion, which has an odd shape (Fig. 11.26), like a thick hamburger. The other moons are also

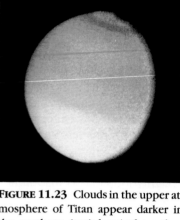

FIGURE 11.23 Clouds in the upper atmosphere of Titan appear darker in the northern (top) hemisphere than in the southern; this difference shows that clouds in the two hemispheres differ in thickness. A well-defined band marks the equator. (Courtesy of NASA.)

Titan

At a Glance

Equatorial radius	2560 km
Mass	1.35×10^{23} kg
Bulk density	1880 kg/m³
Escape speed	1.5 km/s
Atmosphere	Nitrogen, methane

FIGURE 11.24 The surface of Dione. Note the many impact craters. The sinuous valleys are old surface faults in the icy crust eroded by impacts. (Courtesy of NASA.)

cratered but much smaller. We presume that all these bodies are basically ice, as are the larger moons (except for Titan).

The Ring System

In 1659 Huygens observed that Saturn "is surrounded by a thin, flat ring" that does not touch the body of the planet. Further observations by Giovanni Cassini uncovered a gap in Huygens' single ring; this gap is known as *Cassini's division*. The rings lie tipped about 26° to the orbital plane. Because of their tilt, they change their appearance as viewed from the earth during the course of Saturn's revolution about the sun.

The near disappearance of the edge-on rings indicates that they are very thin, no more than few kilometers thick. At their edges, the rings are only a few hundred *meters* in thickness. Although thin, the rings are wide; the three main rings visible from the earth (called A, B, and C) reach from 71,000 to 140,000 km from Saturn's center (Fig. 11.27).

The Voyager photographs revealed spectacular detail in the ring system. Although the A ring is relatively smooth, the B and C rings break up into numerous small **ringlets** (Fig. 11.28), like grooves on a phonograph record. Many hundreds, perhaps a thousand, light and dark ringlets surround the planet, with widths as small as 2 km (and possibly smaller), the best resolution of Voyager. Even the Cassini division, apparently empty as seen from earth, was found to contain at least twenty ringlets.

Pioneer 11 discovered a new ring out beyond those heretofore known: the F ring lies 3500 km outside the edge of the rings visible from the earth; it appears to be some 300 km wide and a mere 3 km thick. Voyager 1 photos resolved this ring into a complex system of knots and a braided structure of at least three strands. Nine months later, Voyager 2 photos showed that the braiding had disappeared! This change indicated an unstable disturbance in the rings. Another such very narrow ring (the G ring) is 10,000 km farther out.

Two other extremely faint rings are known. The E ring extends beyond the F ring to at least 6.5 Saturn radii (400,000 km). A ring inside the C ring, called the D ring, extends at least halfway to the surface of Saturn.

The rotation rate of the rings varies in a regular way. The speeds range from 16 km/s at the outer boundary of the A ring to 20 km/s at the inner boundary of the B ring. These speeds, measured by the Doppler shift (Section 10.2), agree with those expected from Kepler's third law for individual masses placed at the ring distances from Saturn. This agreement indicates that separate particles comprise the rings and orbit Saturn just as the planets orbit the sun. The same is true for all the rings of the Jovian planets. (If the rings were solid like a record, the rotational speed of the outer edge would be higher than that of the inner edge.)

Infrared observations show that Saturn's rings are made of particles of water ice or rocky particles coated with water ice. The ice does not evaporate because the surface temperature of the particles is only about 70 K. Observations of radio signals from Voyager reflected by the rings indicate that the particles are about one meter in diameter but range in size from centimeters to tens of meters— from ice lumps the size of golf balls to boulders as big as a house.

The gravitational effects of the satellite just outside the A ring (Atlas) and the two that straddle the F ring (Prometheus and Pandora) play important roles in the dynamics of the rings. The F-ring

(a)

(b)

FIGURE 11.25 Medium-sized moons of Saturn. (*a*) Iapetus, showing details as small as 20 km. Note the impact craters. The dark region at bottom covers the icy crust of the hemisphere that faces in the direction in which Iapetus orbits. (*b*) Enceladus. The grooves and linear features (tens of kilometers long) on the surface imply that the crust was deformed by internal heat. The largest craters visible are about 35 km in diameter. Resolution is about 2 km. (Both courtesy of NASA.)

FIGURE 11.26 Three views of Hyperion, showing its unusual shape as it rotates. Note the impact craters and variation of light and dark regions on the surface. (Courtesy of NASA.)

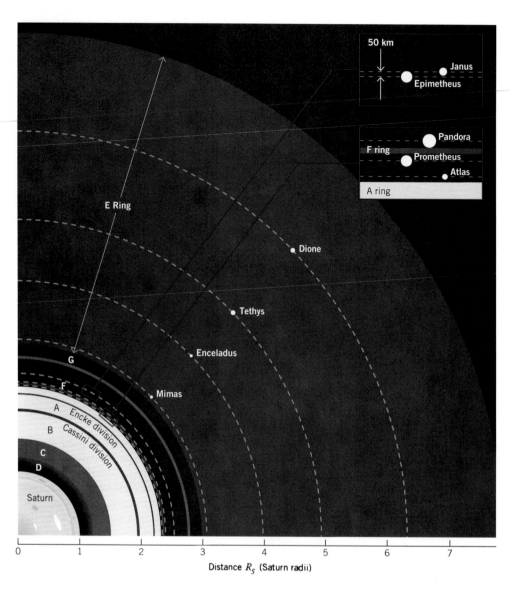

FIGURE 11.27 Schematic diagram of Saturn's ring system and the orbits of some of its moons as seen from above Saturn's north pole. The co-orbital satellites, Janus and Epimetheus, essentially share the same orbit. Prometheus and Pandor act as shepherd satellites for the F ring. Distances are given in units of the planet's radius (R_S). (Adapted from a NASA diagram.)

moons, in particular, are called the **shepherd satellites** because they keep the ring particles in a narrow range of orbits. The inner moon accelerates the inner ring particles as it passes them (as expected from Kepler's third law, it has a shorter orbital period). These particles spiral outward to larger orbits just as the tidal force of the earth on the moon forces the moon into a larger orbit. In a physically similar way, the more slowly moving outer moon decelerates outer-ring particles as they pass by, with the result that they spiral inward. The balance of these interactions constrains the particles' motions and preserves the narrowness of the F ring. Likewise, the A-ring shepherd causes the sharp outer edge of the A ring. Because tidal forces tend to spread rings out, shepherd satellites work to preserve the sharp edges. The

FIGURE 11.28 Voyager 2 false-color image of Saturn's C and B rings, taken from a distance of 2.7×10^6 km. The image processing emphasizes the difference in color between the B ring (yellow) and the C ring (blue). These colors are not the actual colors of the particles in the rings. Rather, the differences indicate that the surface compositions of the particles in these two rings must be different. Note the many ringlets in both rings. (Courtesy of NASA.)

kind of herding of ring particles seems common for the rings of the Jovian planets.

Although deceptively solid in appearance and covering a large area of space, the rings have a total mass estimated to be just 10^{16} kg, only about 10^{-6} the mass of our moon, and a mere 10^{-10} the mass of Saturn. The rings as we see them also have short lives—perhaps only 100 million years. Somehow their material is continually replenished.

■ In addition to its enormous size as a moon, in what way is Titan unique in the solar system?

11.5 Uranus: The First New World

On March 13, 1781, the then unknown amateur astronomer William Herschel perceived a star "visibly larger than the rest" in the constellation of Gemini and "suspected it to be a comet." Observations later in March and in April proved that the object's orbit was not like that of a comet. Herschel concluded that he had found a new planet—the seventh in the solar system and the first to be discovered with a telescope. He named it Uranus after the mythological father of Saturn.

Atmospheric and Physical Features

At an average distance of 19.2 AU, it takes Uranus 84 terrestrial years to journey around the sun. Far from the sun, the upper atmosphere must be very cold—58 K. Like Jupiter and Saturn, the atmosphere contains mostly molecular hydrogen and helium. Uranus has a distinctive bluish-green color, which comes from sunlight that penetrates deep into the planet's atmosphere; some red light is absorbed in the atmosphere, and much of the blue and green is reflected back into space. This color is expected from an atmosphere that contains methane gas.

The Voyager 2 flyby in January 1986 provided the first detailed views of Uranus (Fig. 11.29). They showed ammonia clouds lying below a thick layer of haze. The clouds have a delicately banded structure. Winds blow the clouds

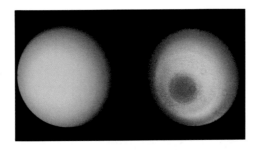

in the same direction as the planet rotates. Occasional plumes of clouds appeared in the upper atmosphere, perhaps produced by violent upflows. Voyager 2 found that the rotation at the cloud tops varied from 17 hours at the equator to 15 hours near the poles. Uranus' rotation axis lies almost in the plane of its orbit! Uranus spins on its side. Journeying around the sun in this lopsided manner (Fig. 11.30), Uranus exposes each pole to sunlight for 42 years at a time; night at the opposite pole lasts equally long.

The low bulk density of Uranus, 1300 kg/m³, implies that it contains mostly lightweight elements (Fig. 11.31). Uranus is thought to consist of roughly 15 percent hydrogen and helium, 60 percent icy materials (water, methane, and ammonia) and 25 percent earthy materials (silicates and iron). Its internal structure differs greatly from that of Jupiter or Saturn. It may have a very small icy-rock core, encased in an icy or watery mantle (perhaps with some rocky material).

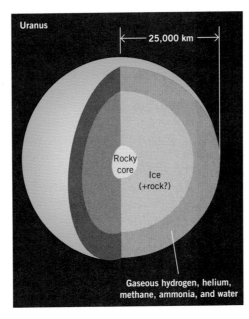

FIGURE 11.29 Uranus from Voyager 2, taken at a distance of 9×10^6 km in January, 1986. The image at left is processed to show the planet as it would appear to the eye; the bluish-green color results from the absorption of red light by methane in the atmosphere. The right-hand image uses false colors to bring out details in the structure of the upper atmosphere. Note the dark hood over the south pole with a series of concentric rings. (Courtesy of NASA/JPL.)

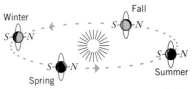

FIGURE 11.30 The orientation of the spin axis of Uranus as the planet orbits the sun, and the seasons in the southern hemisphere. Note how the axis of rotation lies in the plane of the orbit.

Uranus

At a Glance

Equatorial radius	25,559 km
Mass	8.66×10^{25} kg
Bulk density	1270 kg/m³
Escape speed	21.3 km/s
Atmosphere	Hydrogen, helium, methane

FIGURE 11.31 Model for the interior of Uranus, based on Voyager data. The interior consists of three main regions: a very small rocky core (density about 8000 kg/m³); an icy (or watery) region with perhaps some rocky material (density from 5000 to 1000 kg/m³); and a gas layer, composed mostly of hydrogen and helium, with enhanced concentrations of methane, ammonia, and water (density about 300 kg/m³). The rocky core, if it exists, may have a mass about equal to that of the earth.

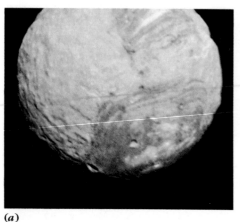

(a)   *(b)*

FIGURE 11.32 Miranda imaged by Voyager 2. (*a*) Overall view of the surface, showing ridges, valleys, and scarps. Resolution is about 3 km. The largest impact craters visible are about 30 km in diameter. (*b*) Close-up of the surface, showing the heavily modified grooved and mottled terrain. Note the craters in both regions. (Both courtesy of NASA.)

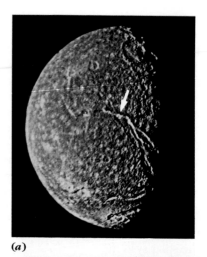

(a)

(b)

Moons and Rings

Five moons were known for Uranus before the Voyager mission: Miranda, Ariel, Umbriel, Titania, and Oberon. They all move in the planet's equatorial plane and revolve in the direction of the planet's rotation. Because the moons lie in the same plane as Uranus' equator, their orbits as seen from the earth are alternately edge-on, then fully open, every 21 years; in 1966 they appeared edge-on, but in 1987 they appeared face-on.

Miranda is the smallest moon and the closest to Uranus. The others range in diameter from some 1100 km to 1600 km. Their surfaces appear to be made of a dirty ice, very much like that of Saturn's Hyperion. The bulk densities range from about 1400 to 1700 kg/m³, which implies these are bodies made of rock and ice.

Voyager 2 revealed the surfaces of these moons. Miranda has the most complex surface, with many different types of terrain jumbled together, including grooved regions, faults, and 5-km-high cliffs—certainly the strangest surface seen to date in the solar system (Fig. 11.32). How could such a small moon be so deformed? At some time, it must have had a hot interior, but the source of that heating has not yet been figured out. Tidal stress is a possible culprit.

Oberon has a dense cover of impact craters (an ancient surface) and at least one mountain (possibly volcanic) about 5 km high. Scarps and faults also spit the surface. Titania also has a surface

plastered with many impact craters, strewn with valleys 100 km wide and hundreds of kilometers long (Fig. 11.33*a*). One valley cuts across the entire surface! Also visible are multiringed impact basins. Ariel has impact craters, large fractures, and valleys (Fig. 11.33*b*). Some regions appear to be resurfaced by ice floes. Finally, Umbriel has the least dramatic surface, with overlapping impact craters. The heavy cratering is more evidence of an era of torrential impacts early in the solar system's history.

Voyager also discovered a slew of new moons—ten that have been named after other characters of Shakespeare. The largest (called Puck) is only 170 km in diameter and orbits inside the already-known moons. Six moons (Appendix B, Table B.9) orbit between Uranus and Puck. These inner moons have diameters between 40 and 80 km. The other moons are near the edge of the ring system and have dark surfaces with low albedos. These serve as shepherd satellites to keep the rings stable (Fig. 11.34).

Uranus' ring system was discovered accidentally in 1977 when astronomers watched Uranus pass in front of a faint star and were surprised to see the star momentarily dimmed a few times before Uranus covered it. Rings blocking out the starlight had caused the dimmings. Observations of Uranus passing in front of another star in 1978 showed nine rings; Voyager added two more, for a total of at least eleven. The smallest

FIGURE 11.33 Titania and Ariel. (*a*) Titania shows many impact craters, which indicates that the surface is old and unevolved. Note the large fault valleys (arrow); they are about 75 km wide and 1500 km long. At the upper right is a basinlike structure. Note the overall gray color, typical of the moons of Uranus. (*b*) Ariel also has valleys (arrow), probably caused by faulting of the surface. Craters are newer and light in color. (Both courtesy of NASA.)

rings have widths of only a few kilometers; the largest is about 100 km wide.

Why didn't astronomers detect the rings of Uranus earlier? (After all, we've known of Saturn's rings for about 300 years.) For one reason, the rings of Uranus are not very wide. More important, the material that makes up Uranus' rings is extremely black—the albedo is 5 percent. In contrast, because they are covered with water ice, the particles in the rings of Saturn have an albedo of more than 80 percent. So the particles in the rings of Uranus probably are bare of ice and likely are made of dark carbon materials. Most of the particles in the rings are a few meters across, but many smaller particles are distributed throughout the Uranian ring system.

Magnetic Field

The Voyager 2 mission showed that the magnetic field is tilted 59° with respect to the rotational axis (Fig. 11.35), with the north magnetic pole closest to the south geographic one. The magnetic field turns once in about 17 hours, 20 minutes, and that has been taken as the planet's rotation period. (Why the magnetic field has such a large tilt is a puzzle. Moreover, the field does not center on the core of Uranus!)

The magnetic field of Uranus is strong enough to develop a stable magnetosphere. Uranus' magnetosphere contains more particles with higher kinetic energies than Saturn's. You would expect frequent auroras in this case, and, indeed, they were observed on the planet's night side by Voyager.

■ If you were on Uranus, how long would the summer and winter last?

11.6 Neptune: Guardian of the Deep

Frankly, we didn't know very much about the cold world of Neptune until the Voyager 2 flyby in 1989 (Fig. 11.36). The eighth planet from the sun (usually—it is right now ninth, as Pluto has moved closer), Neptune was discovered in 1846 because of its gravitational effects on the orbit of Uranus (Section 4.5).

Calculations of Neptune's orbit show that it should have been very close to Jupiter in the sky in January 1613. Galileo's journals have entries showing that he observed an object in the vicinity of Jupiter near Neptune's predicted position on December 27, 1612, and again on January 28, 1613, when he detected a small motion of Neptune with respect to a nearby star. Inexplicably, Galileo never followed up on this discovery and so failed to recognize the object as a new planet.

Physical Properties

Far from the sun, Neptune revolves once in 165 years. Like Uranus, Neptune shows off with a light bluish color (from methane in the atmosphere). The main constituents are molecular hydrogen and helium; methane makes up a minor amount. The upper atmosphere displays distinct cloud bands.

Infrared observations show that

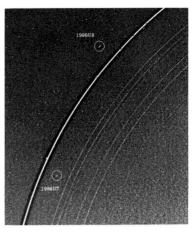

FIGURE 11.34 Rings and small moons of Uranus. The two moons (initially called 1986U7 and 1986U8; now named Cordelia and Ophelia) discovered by Voyager orbit inside and outside the epsilon ring (outlined in black). They act as shepherd satellites to keep the ring narrow and stable. (Courtesy of NASA.)

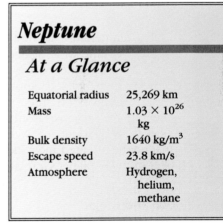

Neptune

At a Glance

Equatorial radius	25,269 km
Mass	1.03×10^{26} kg
Bulk density	1640 kg/m^3
Escape speed	23.8 km/s
Atmosphere	Hydrogen, helium, methane

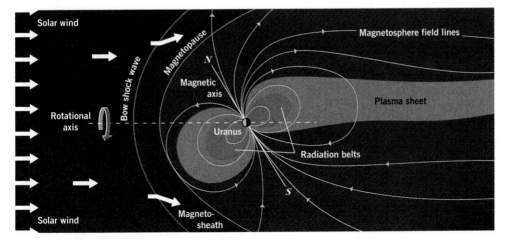

FIGURE 11.35 The magnetosphere of Uranus. Note the large tilt with respect to the rotational axis. Plasma in the solar wind is picked up by the magnetosphere to form a long plasma tail downstream and also to form into Van Allen-type belts of trapped, charged particles. Note the formation of the bow shock where the solar wind encounters the magnetic field; behind the bow shock lies the magnetopause.

FIGURE 11.36 Neptune imaged by Voyager 2. The planet's bluish color is caused by the presence of methane in its atmosphere. (Courtesy of NASA/JPL.)

FIGURE 11.37 Neptune's moon Proteus, the second largest (after Triton). Note the rough, dark surface. (Courtesy of NASA/JPL.)

Neptune's temperature is about 60 K, whereas if Neptune were heated by the sun alone, we would expect a value of 44 K. So Neptune, unlike Uranus, has internal heat. It gives off three times as much energy as it receives from the sun; the heat flow is about 0.3 W/m². The thermal energy emitted most likely was left over from the planet's formation.

Moons and Rings

The only moons of Neptune known before Voyager were Triton and Nereid. The orbit of the latter is extremely eccentric, and its distance from Neptune ranges from about 1.4 million to 9.6 million km. The larger satellite, Triton, revolves with a period of 5.9 days in a retrograde (east-to-west) orbit that is inclined 23° to the plane of Neptune's equator. Its period of rotation is the same; so it is another moon in synchronous rotation. Triton has a diameter of about 2700 km, based on Voyager observations. Nereid has a size of only some 340 km. Triton is one of the few moons with an atmosphere—a very thin one that contains mostly nitrogen with traces of methane.

Voyager captured images of six small moons. They range in diameter from 50 km to 400 km and exhibit the usual rugged surfaces of Jovian moons (Fig. 11.37). Their albedos are all low, about 5 percent or so. Two of them lie close to the outer and middle rings, and so may well act as shepherding moons. Except for the smallest and innermost moon (called Naiad), all lie within the plane of Neptune's equator.

FIGURE 11.38 The rings of Neptune; note how fuzzy the inner one appears. The outermost ring shows the denser segments along it. The black band down the center blocks out the light from the disk of Neptune. (Courtesy of NASA/JPL.)

Voyager showed clearly for the first time Neptune's somewhat mysterious ring system, which contains four individual rings (Fig. 11.38). The two brightest, outer rings have radii of 53,000 and 62,000 km with a tenuous ring spread between them. Closer in lies a wide ring some 2000 km across. The outer ring is clearly clumpy, with three brighter segments ordered along a fainter but complete ring (like sausages on a string). This structure explains the earlier ground-based observations that were interpreted as a series of arcs around the planet. Three rings have been named Galle, Leverrier, and Adams (Fig. 11.39). We now know that Saturn's ring system is the only exception among the Jovian planets in terms of its broad extent (Fig. 11.39).

One other point to note about the ring systems of the Jovian planets. All of them lie within the Roche limit for the parent planets. So it is unlikely that any of the rings' pieces will accrete to make a larger mass. We then have a few scenarios for the origin of the rings. One, they could be material left over from the planet's formation. Since they lie within the Roche limit, they would not accrete to make satellites. Two, they could result from a body or bodies that passed with the Roche limit and were tidally ripped into pieces. Of these two notions, I feel that the first is the most workable.

Atmospheric Features

In August 1989, Voyager 2 skirted a mere 5000 km above Neptune's clouds and took a close view of conspicuous markings in the upper atmosphere of Neptune. The most striking became known as the Great Dark Spot (Fig. 11.40), which is a storm some 30,000 km across, rotating counterclockwise in a few days. A region of high pressure, the Dark Spot lacks the typical atmospheric methane; here, we are looking deep into Neptune's atmosphere.

Bright, cirruslike clouds accompany the Dark Spot (which is actually bluish) and also appear in some other latitude bands. Most of these clouds change size or shape from one rotation to the next. Believed to be condensed methane, the clouds lie about 50 km above the general cloud layer, which

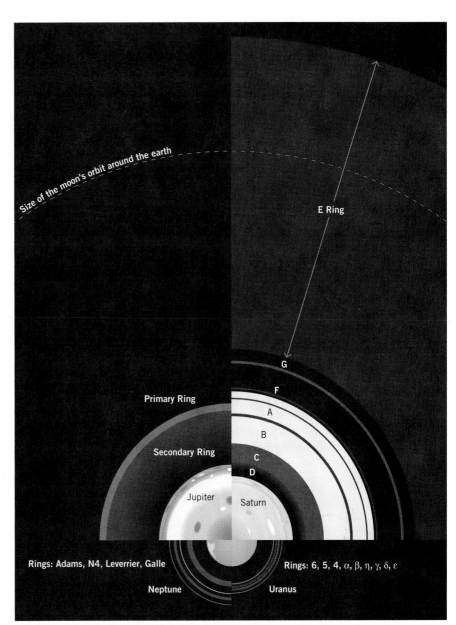

FIGURE 11.39 Comparison of the ring systems of the Jovian planets, drawn to relative scale. Most of Neptune's rings now have been named: N1 is *Adams*, N2 is *Leverrier*, and N3 is *Galle*. (Adapted from a NASA diagram.)

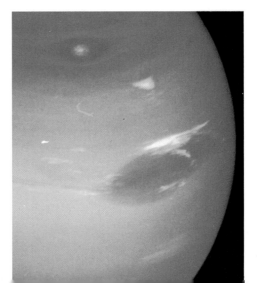

FIGURE 11.40 The Great Dark Spot, accompanied by white, high-altitude clouds. These clouds change their appearances in times as short as 2 hours. Note the lack of a general banded structure in the atmosphere. Relative to the planet's core, the Dark Spot moves westward. (Courtesy of NASA/ JPL.)

consists of hydrogen sulfide. Compared to that on bland Uranus, the atmospheric activity on Neptune came as a surprise. It is likely driven by the outflow of Neptune's internal heat. A few other dark and bright spots are present, but the complex swirls and banded structure seen on both Jupiter and Saturn seem to be absent on Neptune.

Magnetic Field

Voyager picked up radio signals from Neptune's magnetosphere. Tracking these signals gave an accurate measure of the rotation period: 16 hours, 3 minutes. A surprise was the discovery that the magnetic axis is tilted about 47° from Neptune's axis of rotation, almost as much as the tilt of Uranus' magnetic axis. The reason for these large tilts is not yet known. The magnetic field strength is about one-fifth that of the earth's. The dipole is strangely offset about four-tenths of a Neptune radius toward the south pole from the planet's center (Fig. 11.41). The magnetosphere has a very low density of trapped charged particles.

Triton

Triton (Fig. 11.42) displays a fascinating pink and blue face. The cratering here is not too heavy, which means that the sur-

face must be relatively young and recently modified—subject to meltings and refreezings. The overall temperature is low, a mere 37 K. On parts of the surface lie frozen ice lakes, shaped like lunar maria. Some are stepped, which suggests a series of meltings and freezings. But in general the surface relief is quite low—less that 200 meters. Triton's atmosphere is about 800 km thick; it contains mostly nitrogen with a trace of methane.

Near Triton's south pole (which is now in a summer season), the surface ice, consisting of methane and nitrogen, appears to have evaporated in spots. In other regions, small flows have filled valleys and fissures—slow-moving glaciers of methane and nitrogen. In other sections, the icy surface appears to have melted and collapsed. Dark, elongated streaks tens of kilometers across seem to be the trails of ice volcanoes. Just 30 meters below the surface, the pressure is high enough to liquefy nitrogen ice. When the surface cracks, the liquid can burst out and turn to gas, which shoots up several kilometers above the surface. The vapor condenses and falls back down, creating a thin, dark layer on the surface. One such geyser was seen actively erupting during the Voyager encounter.

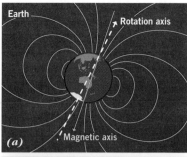

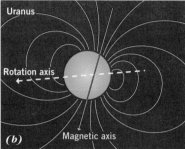

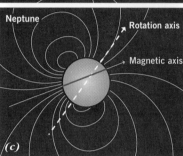

FIGURE 11.41 Comparison of the magnetic fields of the earth, Uranus, and Neptune. The fields are modeled as if a bar magnet were set in the planet's interior, with the angle and position inferred from the magnetosphere. Note how the fields of Uranus and Neptune are offset from the centers of the planets and grossly misaligned with the rotation axes.

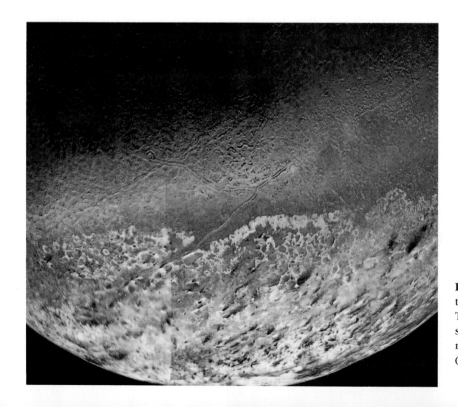

FIGURE 11.42 High-resolution mosaic of the portion of Triton's hemisphere that faces Neptune. The southern pole cap (at bottom, center) is slightly pink. From the ragged edge of the pole cap northward (up), the surface is redder and darker. (Courtesy NASA/JPL.)

Triton's bulk density turns out to be 2100 kg/m³ (about the same as Pluto: see Section 11.7). That implies a rocky core surrounded by a mantle of methane and water ice. The icy surface, which is primarily molecular nitrogen (N_2) ice, results in an average albedo of roughly 70 percent. Neptune's orbit results in a seasonal cycle of 165 years; summer in the southern hemisphere has already lasted some 30 years. Still, so far from the sun, the surface remains so cold that regions of nitrogen ice are still visible.

■ Despite its vast distance from the sun, Neptune has an atmosphere that is surprisingly active. How is this possible?

11.7 Pluto and Charon: Guardians of the Dark

Early in the twentieth century, Percival Lowell became fascinated with the problem of a planet beyond Neptune (called *Planet X*) and initiated a search program at Lowell Observatory. When Lowell died, in 1916, the search for Planet X was terminated pending the completion of a new telescope, which occurred in 1929.

Clyde W. Tombaugh (Fig. 11.43) worked at the new search, which started on April 1, 1929. Bearing in mind the way Neptune was discovered (Section 4.5), most astronomers had assumed that Planet X would be similar to Neptune because of irregularities in Neptune's orbit. So they had searched for a visible disk of a new planet. Instead of concentrating on a disk, however, Tombaugh looked for Planet X's motion relative to the background stars.

The photographic search was tedious, but on February 18, 1930, Tombaugh noted two images on different photographs, in the area near a star in Gemini, that had shifted slightly (Fig. 11.44). The shift was such that the object had to be a body orbiting the sun. The discovery was announced within a month, on March 13 — Lowell's birthdate. The name Pluto was officially accepted by Lowell Observatory.

Orbital and Physical Properties

Pluto's average distance from the sun is 39.44 AU. Since it has a highly eccentric orbit, it ranges from 29.7 to 49.3 AU from the sun, so it is never closer to the earth than 28.7 AU at closest approach (opposition). On January 21, 1979, Pluto edged closer to the sun than Neptune. It will orbit closer to the sun than Neptune until March 1999. However, because of the high inclination of its orbit, Pluto is actually well above Neptune's orbital plane. Besides, Neptune is now about 60° around its orbit from Pluto, so there is no danger of a collision!

Methane ice coats Pluto's surface, which means that the surface is bitter cold, no more than 60 K even during the daytime. Recent observations show nitrogen and carbon monoxide ices as well as methane. The overall albedo is about 50 percent. Observations of Pluto's brightness have uncovered a cycle (because of patches of the ice) about every 6.4 days. This variation is generally accepted as Pluto's rotation period.

On September 5, 1989, Pluto reached perihelion in its orbit. A little more than a year earlier, on June 9, 1988, Pluto had passed in front of a star as observed in Australasia. Coordinated observations from this region confirmed that Pluto has an atmosphere, which stretches over 600 km from the planet's surface. This atmosphere probably con-

FIGURE 11.43 Clyde W. Tombaugh, who discovered Pluto. (Photo by M. Zeilik.)

FIGURE 11.44 Discovery photos of Pluto (arrow). Note the change in the planet's position relative to the stars between these two photos. (Courtesy of Lowell Observatory.)

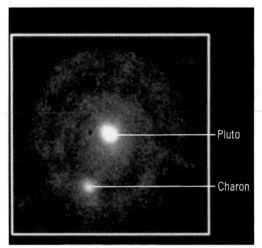

FIGURE 11.45 Pluto and Charon. At the time of this *Hubble Space Telescope* observation done with the Faint Object Camera (FOC), Charon (lower left) was about 0.9 arcsec from Pluto. (Courtesy of NASA/ESA.)

sists of nitrogen, carbon monoxide, and methane gas (with a surface pressure of a mere 10^{-8} atm or so) that has been released from the ice on the surface as the planet is heated by its closest approach to the sun in 248 years.

Charon: Pluto's Companion Planet

In June 1978, James Christy of the U.S. Naval Observatory in Flagstaff, Arizona, noticed what appeared to be a bump on Pluto's image in a photo. Checking older photos, Christy found seven showing the same bump, always oriented approximately north–south. He proposed that the bump was the faint image of Pluto's moon partially merged with the image of the planet. Christy named this moon *Charon*, after the mythological boatman who ferried souls across the river Styx to Hades, where Pluto sat in judgment. The Hubble Space Telescope took an image that clearly showed Pluto and Charon (Fig. 11.45) but could not reveal any details on their surfaces.

A few years after this discovery, Pluto and Charon eclipsed each other as seen from earth. (Such alignments occur only twice each 248 years.) These eclipses (Fig. 11.46) indicate that Pluto has a diameter of some 2300 km; Charon's diameter is roughly 1200 km (Table 11.1 and Fig. 11.47). The observations of

Charon show a revolution period of 6.4 days (the same as Pluto's rotation period, so it is in synchronous rotation) at a distance of 19,100 km from Pluto. That the revolution period of Charon is the same as Pluto's rotation indicates that the two bodies are tidally locked. We can use Charon to find Pluto's mass by Kepler's third law. The result: Pluto has a mass about 0.002 that of the earth or one-fifth that of our moon. Charon's mass is about one-fifth that of Pluto. This mass ratio of 5 to 1 for Pluto and Charon means that the center of mass of their binary system lies *outside* the body of Pluto. So they really do make up a double-planet system! (In contrast, the center of mass of the earth–moon system lies within the earth.)

Together, Pluto and Charon have an average density of some 2100 kg/m^3—much higher than earlier estimates. Pluto may contain as much as 75 percent rocky material and may have a density of 2000 kg/m^3. Contrast the Jovian moons, which have a larger percentage of water ice. Charon's density might fall as low as 1200 kg/m^3 (or even less); recent infrared observations revealed water frost on its surface. So Charon, like moons of Saturn, may well consist mostly of icy materials. The eclipses indicate that Charon has a dark gray surface in contrast to Pluto's reddish one. In addition, the equatorial region of Pluto is darker than its polar caps.

The idea that Pluto is a small, rocky-ice planet fits with a speculation that Pluto is actually an escaped moon of Neptune. Why? The orbit of Pluto is so eccentric that it sometimes comes within Neptune's orbit, which means that at

TABLE 11.1 THE PLUTO–CHARON SYSTEM

PROPERTY	PLUTO	CHARON
Mass (earth = 1)	0.0022	0.0004
Diameter	2240 km	1120 km
Density	$\approx 2000\ kg/m^3$	$\approx 1200\ kg/m^3$?
Rotation period	6.39 days	6.39 days

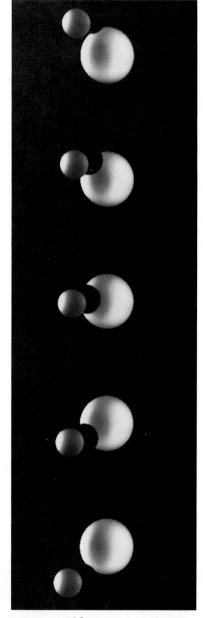

FIGURE 11.46 Computer simulation of the eclipses of Pluto and Charon. The sequence at the left, showing Charon passing in front of Pluto as seen from the earth, recurs 3 days later in the form of Charon slipping behind Pluto. (Courtesy of K. Horne, M. Bule, and D. Tholen.)

times Pluto is actually closer to the sun than Neptune. Recall that Neptune's Triton is a large satellite with an unusual retrograde orbit. If Pluto originally had been a moon of Neptune along with Triton, a close encounter between the two might have caused Triton to reverse its orbital motion, throwing Pluto free of Neptune, as well. Most astronomers consider this model to be an intriguing speculation; one support for it is that Triton and Pluto are about the same in density and diameter.

In another scenario, a collision with a large body stripped Pluto of some of its icy materials and changed the rock-to-ice ratio from that of a lower-density Jovian moon. (The analogy here is to a giant impact stripping the rocky mantle of Mercury, leaving behind a denser object.)

The search at Lowell Observatory for other planets beyond Neptune ended in 1945 with no further positive results. Tombaugh's search, which continued for 13 years, would have uncovered a planet like Neptune as far as 100 AU from the sun. Later searches were also negative.

■ In what sense does Pluto *not* resemble the Jovian planets?

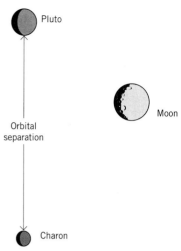

FIGURE 11.47 The Pluto – Charon system drawn to scale, with the earth's moon shown on the same scale. (Adapted from a NASA diagram.)

Key Concepts

1. As a group, the Jovian planets (Table 11.2) differ most from the terrestrial ones on the basis of their lower densities, greater diameters, and greater masses. Also, in chemical composition they are more like the sun than the earth. Jupiter and Saturn are mostly fluids in their differentiated interiors.

2. The Jovian planets are large masses that consist mainly of hydrogen and helium, with common molecules of methane, hydrogen, ammonia, and water. Their visible disks show the tops of their atmospheres; they have no solid surfaces. The large escape speeds and low temperatures of the Jovian planets imply that their atmospheres have changed little since their time of formation.

3. Jupiter, Saturn, and Neptune emit into space more heat than the energy they receive from incoming sunlight, so their interiors must be hotter than would be the case if the sun were the sole source of their heating. Their internal heat was probably generated during their formation. The outflow of this heat drives the evolution of these planets. However, because they do not have solid surfaces, evidence of an evolutionary history has not been preserved.

4. Jupiter, Saturn, Uranus, and Neptune have strong planetary magnetic fields, which implies that they have liquid, conducting interiors—but their compositions rule out a metal core like the earth's. For Jupiter and Saturn, dynamos likely arise from a zone of liquid metallic hydrogen, with organized flows set up by these planets' rapid rotations. For Uranus and Neptune, the source of the circulating electrical currents is closer to the cloudtops.

5. Jupiter's largest moons decrease in density outward from the planet, an indication that at some time in the past Jupiter was hot enough to affect them. These Galilean moons also have undergone different amounts of crustal evolution, as indicated by the extent of their unmodified impact cratering—Io the most and Callisto the least. Io, the most volcanically active body in the solar system, has a hot interior. Its surface is very new, modified by volcanic eruptions.

6. The rings of Jupiter are thin and consist of small particles; those of Saturn are wide, consist of larger, icy particles, and contain many small ringlets; those of Uranus and Neptune are thin and narrow and consist of very dark, small particles. The particles in these rings orbit the planet following Kepler's laws. Small moons in orbits near the rings may help to keep the rings stable.

7. Except for Titan, Saturn's largest moons are basically ice. They have evolved somewhat since the time of their formation, as indicated by their valleys and

TABLE 11.2 A COMPARISON OF THE JOVIAN PLANETS			
	DIAMETER (earth = 1)	MASS (earth = 1)	BULK DENSITY (water = 1)
Jupiter	11.0	318	1.3
Saturn	9.5	95	0.7
Uranus	4.1	15	1.2
Neptune	3.9	17	1.7
Pluto	0.17	0.002	2.1

modified craters. The smallest moons, also of dirty ice, are heavily cratered and unevolved. Titan is the only moon in the solar system with a thick atmosphere.

8. The atmospheres of Uranus and Neptune are essentially the same (hydrogen and helium); but their interior structures are different, as implied by the difference in their densities. Neptune has more violent weather, driven by its heat outflow. Uranus has no or little outflow of heat.

9. Except for Miranda and Triton, the moons of Uranus and Neptune are dark, icy bodies with little evidence of crustal evolution. Their surfaces are heavily cratered. Miranda has a weirdly modified surface, and Triton shows signs of volcanism.

10. Pluto is a small icy world, with a methane-coated surface and a very thin atmosphere. Its moon, Charon, is about half Pluto's size and one-fifth its mass. Pluto and Charon orbit a common center of mass as a double-planet system. Akin to the nuclei of comets, in their physical characteristics, Pluto and Charon are nevertheless much larger in size.

11. The solid materials in the outer solar system are largely ices and carbon-rich rocks. For the most part, the solid bodies were modified 4 billion years ago by an intense bombardment of solid debris. Minor heating and surface alterations have taken place in a few instances. Volcanism appears on Io and Triton. None of these solid bodies shows evidence of plate tectonics.

Key Terms

auroras	Galilean moons	ring systems	zones
belts	Jovian planets	ringlets	
differential rotation	metallic hydrogen	shepherd satellites	

Study Exercises

1. In what significant respect is Jupiter most different from the other Jovian planets? (Objective 11-1)

2. Suppose you flew very close by Jupiter. What outstanding features would you see in the atmosphere? (Objectives 11-2, 11-4, and 11-7)

3. How do we know the bulk density of Pluto? (Objectives 11-2 and 11-5)

4. How do we know that the rings of Saturn are thin? (Objective 11-2)

5. What fact makes it relatively easy to find the masses of the Jovian planets? (Objective 11-2)

6. In one word, state the greatest difference between the Jovian and terrestrial planets. (Objective 11-1)

7. In two sentences, compare the rings of Saturn to those of Jupiter and to those of Uranus. (Objective 11-4)

8. How is Jupiter's magnetic field similar to the earth's? How is it different? Answer the same questions for Saturn. (Objective 11-9)

9. What features does Pluto have in common with the Galilean moons? (Objective 11-8)

10. In one short sentence, describe the interior compositions of the moons of the Jovian planets. (Objectives 11-7, 11-8, and 11-10)

11. What, in general, does a heavily cratered surface indicate about a body's evolution and age? (Objective 11-11)

12. In what way is the magnetic field of Uranus different from that of Jupiter or Saturn? (Objective 11-9)

13. In what way is the magnetic field of Neptune very different from that of Jupiter or Saturn? (Objective 11-9)

14. What is the main difference in the composition of the interior of the earth compared to the interior of Jupiter? What is the major difference in the physical state of the materials? (Objective 11-3)

15. In what sense can Pluto and Charon be considered a double-planet system? (Objective 11-6)

Problems & Activities

1. Although Jupiter does not have a solid surface, you can calculate its "surface" gravity at the cloud tops. Compare it to the surface gravity of the earth. *Hint:* For any spherical mass, its surface gravity, *g,* is g = GM/R^2 where *R* is the radius.

2. Apply Newton's form of Kepler's third law to Pluto and Charon and find the sum of the masses of these two bodies. What additional information do you need to find the individual masses?

3. Use the orbital properties of any of the Galilean moons to calculate the mass of Jupiter, using Newton's form of Kepler's third law.

4. At closest approach to the earth, Jupiter has an angular diameter of about 47 arcsec. What is its distance? Pluto has an angular diameter of a mere 3 arcsec at closest approach to the earth. What is its distance then? In both cases, what is the alignment of these planets with the earth?

5. Compare the escape speeds of Jupiter and Pluto to that of the earth.

6. What is the total heat output, in watts, from Neptune? (*Hint:* What is the total surface area of Neptune?)

See for Yourself!

You'll find that Saturn and its rings are the smash hit of telescopic views in the solar system! Any small telescope with a magnifying power of 30 or greater will reveal the main body of the rings. At 100 to 200 power with a 6- to 8-inch telescope, you can easily see the A, B, and C rings from the outside inward. The dark Cassini's division appears as a gap between the A and B rings; actually, it too contains many small particles. The best time to observe Saturn is during oppositions, which occur at intervals of 378 days. At favorable oppositions, Saturn's disk can have an angular diameter as large as 20 arcsec; the rings, about twice that size. Among the moons, Titan is easy to spot, since it is bright and a fair distance from the rings.

It's fairly hard to see any features of Saturn's upper atmosphere with a small telescope. Patience and good seeing help! The great storm of 1990, which appears to be one in a regular series, was readily visible in telescopes of 4-inch apertures or larger. The next such storm will occur in 2019. ■

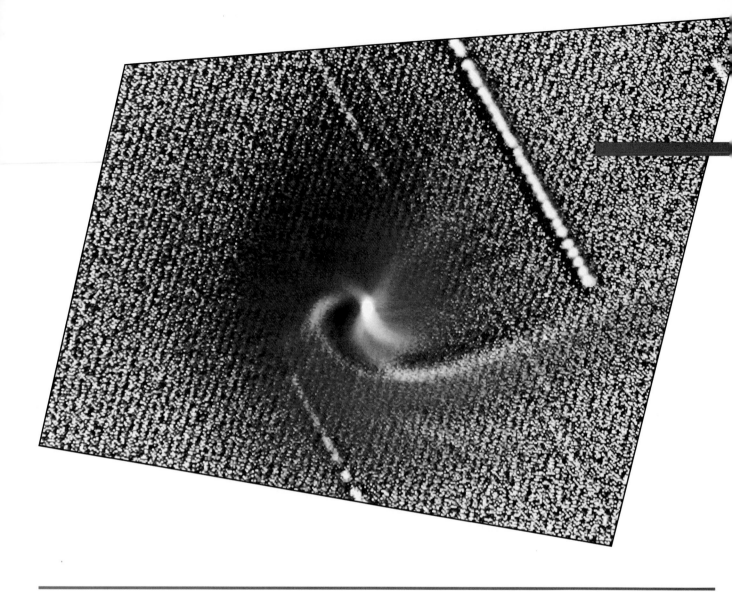

LEARNING OBJECTIVES

After studying this chapter, you should be able to:

12-1 Identify at least two dynamic and two chemical properties of the solar system that any model of origin must explain.

12-2 Describe and compare the general physical properties of comets, asteroids, meteoroids, and meteorites, and state what the radioactive dating of meteorites implies for the dating of the formation of the solar system.

12-3 Specify what clues asteroids, comets, and meteorites provide about the formation of the solar system, with special emphasis on the composition of each.

12-4 Describe the concepts of angular momentum and the conservation of angular momentum; explain what is meant by the *angular momentum problem* for nebular models.

12-5 Describe one possible way out of the angular momentum problem in modern nebular models.

12-6 Describe briefly the chemical condensation sequence, using one Jovian and one terrestrial planet to illustrate its use.

12-7 Describe the role of accretion in the formation and initial heating of the planets and the role of impacts in the subsequent intense bombardment.

12-8 Compare and contrast a model for the formation of Jupiter and Saturn to that of the terrestrial planets.

12-9 Describe the general process of gravitational contraction of a spinning cloud of gas and dust, applying the concept of the conservation of energy and angular momentum.

12-10 Sketch a modern scenario for the formation of the solar system, and evaluate how well it explains the known chemical and dynamic properties listed for Objective 12-1.

Central Question

What physical processes resulted in the formation of the planets from a cloud of gas and dust?

The Origin and Evolution of the Solar System

But indeed the whole story of Comets and Planets and the Production of the World, is founded upon such poor and trifling Grounds, that I have often wondered how ingenious Men could spend all pains in making such fancies hang together.

CHRISTIAN HUYGENS

How did the solar system originate? Because the puzzle of the origin of the solar system remains unresolved, it still arouses astronomers' curiosity. In fact, it has challenged astronomers for centuries! Many models have been proposed. None has been completely successful. One scientific justification of the space program rested on finding information to support or refute theoretical ideas of the solar system's formation. And data from spacecraft have provided some key clues.

To unravel the puzzle of the solar system's genesis requires more than astronomy. It demands the interplay of astronomy with physics, chemistry, and geology. The question refuses to be answered simply. This chapter investigates in some detail the current approach, in which the formation of the solar system is seen as a natural result of the formation of the sun from an interstellar cloud of gas and dust.

The general outlines of this process reasonably explain the basic features of the solar system. Many details remain to be resolved; no one model yet fits them all together. We have painted the broad strokes of the big picture, and it has a profound implication: that other solar systems should be fairly common.

12.1 Debris Between the Planets: Asteroids

The space between the planets is *not* empty. Comets, meteoroids, and asteroids are found there. These bodies, along with gas and dust, make up the interplanetary debris. Scooped up, the total mass would amount to less than the mass of the earth. Yet, as archaeologists have discovered, a trash heap holds valuable clues to a city's history, even though it contains much less mass than the city. Likewise, the solar system's debris is a fruitful hunting ground for clues to its past.

This section deals with debris in the form of asteroids; the next two cover comets and meteoroids. You will see again the trend that cropped up with the satellites of the Jovian planets. The makeup consists of light-colored ices and dark-colored silicates.

Asteroids: Minor Planets

An **asteroid** is an irregular, rocky hunk, small both in size and in mass compared to a planet. Ceres, the largest known asteroid, has a diameter of only about 1000 km and a mass of perhaps 10^{21} kg. (That's about one-third the size of the moon and $\frac{1}{100}$ its mass.) Its density is about 3000 kg/m^3. All the asteroids added up may amount to a mass of a few percent that of the moon.

Most asteroids orbit the sun in a belt between Mars and Jupiter, at an average distance of 2.8 AU (Fig. 12.1); these bodies make up the **asteroid belt.** Their orbits are fairly eccentric, typically 0.1 to 0.3. A few stray asteroids, in highly eccentric orbits, veer widely from this region. Icarus, for example, skirts closer to the sun at perihelion than Mercury does. On June 15, 1968, Icarus careened past the earth with a closest approach of 6.4 million kilometers, just twenty times farther away than the moon. Some 5500 asteroids have been discovered so far, and many await future sightings.

Most asteroids fluctuate in brightness from their rotation—proof that they have irregular surfaces or shapes (or both). The Galileo spacecraft, on its way to Jupiter, captured an image of the asteroid Gaspra (Fig. 12.2). It has a lumpy shape, some 20 km by 12 km by 11 km, and a cratered surface. The smallest craters are 300 meters wide. Gaspra rotates once every 7 hours. Large chunks of the asteroid have been struck off in past collisions—probably typical events in the life of an asteroid. Deimos and Phobos, the moons of Mars (Section 10.8), resemble Gaspra—strengthening the idea that they are captured asteroids. We expect other asteroids to look like these, showing rough, crater-pitted surfaces.

A controversy has developed about whether binary asteroids might exist, loosely held together by gravity against tidal forces of the sun and planets. In ac-

cord with Kepler's laws, each asteroid in a pair would revolve around a common center of mass. Confirmatory observations of such binary systems have been scant until now.

In August 1989, by chance, a newly discovered asteroid flew close to the earth (within 6×10^6 km) and through the field of view of the Arecibo radio telescope. High-resolution observations (300 m) clearly show a double-lobed shape that might be either a single object like a dumbbell or two asteroids, each about one kilometer in diameter, orbiting each other in about 4 hours. More recent radar observations of an asteroid named Toutatis clearly showed two big chunks of rock, held together (temporarily) by gravity (Fig. 12.3). Toutatis is

Asteroids At a Glance	
Orbits	Most in asteroid belt
Sizes	Small, less than 1000 km diameter
Shapes	Irregular
Surfaces	Cratered and cracked
Composition	Rocks and metals

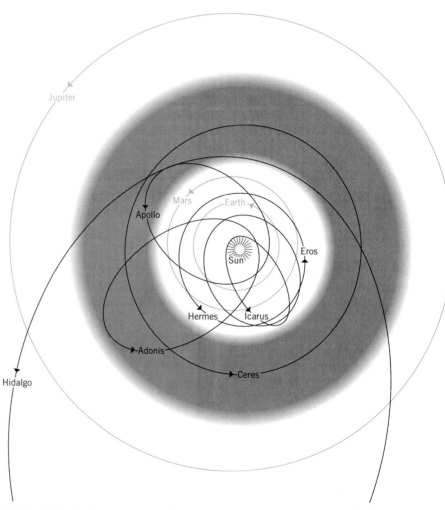

FIGURE 12.1 Orbits of asteroids. Most orbits are located in the zone called the asteroid belt between Mars and Jupiter. Some are found outside the belt, such as Icarus, which comes very close to the sun (and the earth). Note how eccentric the asteroidal orbits are compared to those of the planets; a fair number, such as Icarus and the Apollo asteroids, actually cross the orbit of the earth.

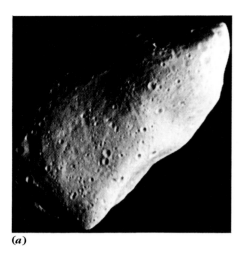

(a)   (b)

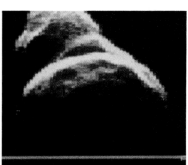

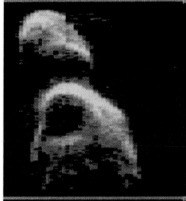

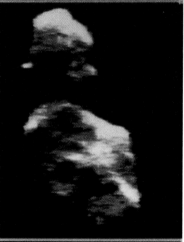

FIGURE 12.2 Asteroids. (*a*) The asteroid Gaspra, imaged from a distance of 5,300 km by the Galileo spacecraft. The illuminated part of the asteroid is about 19 km by 12 km by 11 km. The smallest craters visible here are about 300 meters across. More than 600 craters are visible. The north pole is at the upper left. Gaspra rotates once in about 7 hours. (*b*) A montage of Gaspra (top) and the moons of Mars, Deimos (lower left) and Phobos (lower right) shown to the same scale. Note all have irregular shapes but different surfaces. Phobos and Deimos may well be captured asteroids. (Both courtesy of NASA.)

one of the largest of a group of asteroids that approaches close to the earth. In December 1992, it passed a mere 3.5 million kilometers from us!

These observations support the notion of compound or binary asteroids systems. Frequent collisions and fragmentation may create such systems, which have short lifetimes (maybe 10^5 years).

Albedos and Composition

On the basis of their albedos, asteroids fall mainly into two compositional classes. Some are relatively bright (albedos of about 15 percent) and others are much darker (albedos of 2 to 5 percent), an indication that they contain a substantial percentage of carbon compounds. Those in the lighter class are dubbed **S-type** (stony) asteroids; the darker ones have been christened **C-type** (carbon). (Ceres is a C-type asteroid.) The S-type, in addition to having higher albedos, show spectral features indicative of silicate materials. A third class, called **M-type** (metallic), has attributes that suggest the presence of metallic substances. They have albedos of about 10 percent. Only 5 percent of the total number of asteroids belong to this class.

Based on albedos, compositions in the asteroid belt vary with distance from

the sun. Near the orbit of Mars, almost all asteroids have S-type characteristics. Farther out, we find fewer high-albedo asteroids and more that are dark. At the outer edge of the belt, 3 AU from the sun, some 80 percent of the asteroids are C types. In fact, the C type may well be the most abundant kind of asteroid overall.

■ **What facet of our knowledge about asteroids has strengthened opinions about the origins of Phobos and Deimos?**

12.2 Comets: Snowballs in Space

You probably associate comets with long, graceful tails (Fig. 12.4). In fact, not all comets exhibit tails, even at perihelion; and when far from the sun, comets do not have visible tails. When first sighted telescopically, a comet typically appears as a small, hazy dot. This bright head of the comet is called the **coma;** the tail develops from the coma. Sometimes the coma contains a small, starlike point called the **nucleus.** We think that cometary nuclei are smaller than 10 km across; none was ever viewed as more than a point of light until the spacecraft missions to Halley's Comet (discussed

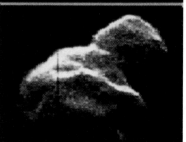

FIGURE 12.3 The asteroid Toutatis. In these radar images, you can see two separate pieces revolving around each other. (Courtesy of NASA/JPL.)

FIGURE 12.4 Comet West in March 1976, before perihelion passage. Note the separation of the tail into a gas (plasma) part and a dust one, which spreads in a wide fan. The gas tail appears bluish from its emission; the dust tail reflects sunlight and so appears yellowish. The plasma tail is blown back by the solar wind; the dust particles fall behind in their own orbits around the sun. (Courtesy of Dennis Di Cicco, *Sky and Telescope* magazine.)

later). Those flybys confirmed a size of roughly 10 km.

During a comet's rush toward perihelion, it grows brighter and sprouts a tail as its material is heated by the sun and vaporizes. A comet's tail may stretch for millions of kilometers, and the gaseous part of it points away from the sun because it is blown by the solar wind. Note that the ion tail points *ahead* of the comet in the outbound half of each orbit.

Composition

Comets can have two types of tail: gas and dust (Fig. 12.4). The physical differences between the two show up in their spectra. The spectrum of the gas or **ion tail** has emission lines, indicating that it is ionized. So the ion tail is a plasma. The **dust tail** shows a spectrum of sunlight, reflected from dust expelled out of the coma. The pressure from sunlight de-

taches dust from the coma to form a tail. (Recall that light can be thought of as particles—photons. Photons reflected off a dust particle apply a pressure to the surface and push a particle if it has a low enough mass.)

The ion tail's spectrum shows gases such as carbon monoxide, carbon dioxide, and molecular nitrogen. These tails consist of ions carrying magnetic fields (Section 8.3) at high speeds through interplanetary space. The magnetic fields interact with the plasma of the gas tail and impart to a comet its distinctive cast.

At great distances from the sun, the coma also shows a reflected solar spectrum. At about 1 AU from the sun, the coma exhibits emission of molecules. As the comet speeds nearer to the sun, emission lines of silicon, calcium, sodium, potassium, and nickel appear. The coma may reach a diameter of a million kilometers. Ground-based infrared observations confirm that comets contain considerable amounts of silicate dust.

For all their stunning length against the sky, comets have very small masses. Halley's Comet, one of the largest, has an estimated mass of only about 10^{14} kg (roughly 10^{-8} that of our moon) and releases about 10^{11} kg during each perihelion passage, only about 0.001 of its total mass. With so little mass, a comet achieves its spectacular display only by spreading itself very thin.

The mass expelled from a comet, mostly in the form of gas, flies off into space. The nucleus supplies this mate-

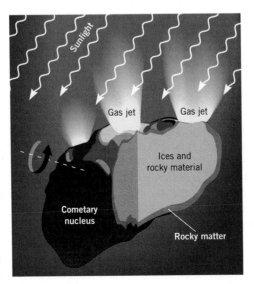

FIGURE 12.5 The dirty snowball cometary model (simplified!). The nucleus is a pudding of ices and rocky material encrusted by a thin, rocky shell. Sunlight heats the nucleus and vaporizes the ices. Gas streams out to make the coma and tail. The nucleus has an irregular shape and rotates about its center of mass.

8 MAR 76   12 MAR 76   14 MAR 76   18 MAR 76   24 MAR 76

rial, but what is the nature of the nucleus? In 1950 American astronomer Fred L. Whipple developed the **dirty snowball comet model** (Fig. 12.5). It pictures cometary nuclei as compact, solid bodies made of frozen ices of water, ammonia, and methane, embedded within rocky material. When a comet nears the sun, the icy material vaporizes, forming the coma—a cloud of mostly gas and some dust. Continued vaporizing enlarges the coma and creates the tail. As the ice evaporates, a thin coating of rocky material remains to form a solid, but fragile, crust on the nucleus. Jets burst through this crust as ices vaporize and expand. From the missions to Halley's Comet, we now know that this idea is basically correct. Most comets are snow (water ice) and dust.

As a comet rounds the sun, the nucleus can break up. Comet Ikeya–Seki, the great sun-grazing comet of 1965, passed within 470,000 km of the sun and survived. But some comets passing close to the sun are not so lucky. Comet West (1976) split into at least four pieces after its perihelion passage (Fig. 12.6). Such a breakup leaves debris with a cometary composition. A few comets have actually collided with the sun (Fig. 12.7), resulting in their destruction.

Orbits

Comets that make regular returns to the sun have elliptical orbits and are called **periodic comets.** (Halley's Comet is the most famous periodic comet.) The periodic comets fall into two groups: long period and short period, with the dividing line at orbital periods of about 200 years. The *short-period comets* tend to orbit in the same direction as the planets, to stay in the plane of the solar system, and to have smaller orbital eccentricities.

In contrast, *long-period comets* swing around the sun in highly eccentric orbits, cutting into the plane of the solar system from all angles.

A periodic comet eventually (in a million years or less) suffers one of three fates: it dissipates or breaks up; it collides with a planet; or it is ejected from its elliptical orbit and out of the solar system by a close encounter with a planet. So the periodic comets we see, which are eliminated rapidly compared to the age of the solar system, must be very recent additions to the solar system. Hence, some reservoir continues to supply them, as explained below in connection with the *Oort Cloud.*

Not all comets have elliptical orbits. Some pass the sun once and never return—unless the planets perturb their open orbits into elliptical ones. These comets are seen on their first trip to the inner solar system, and then never again. They are one-shot celestial visitors.

The Comet Cloud

Many comets have been observed in the past, but where are they all now? What is the origin of these flimsy bodies? Their orbits suggest that all comets are attached gravitationally to the sun, except for those thrown into escape paths by encounters with the planets. From the observed orbits of long-period comets, we find that the average semimajor axis is about 50,000 AU; the corresponding orbital period is about 10 million years. The eccentricity of these orbits is close to 1, and the aphelion distances are about 100,000 AU. These comets, in accord with Kepler's second law, travel very slowly, only a few thousand kilometers per day, at aphelion. They spend most of their time coasting through the depths of space, 100,000 AU from the sun.

FIGURE 12.6 The breakup of Comet West in March 1976. After its perihelion passage, Comet West's nucleus split into at least four pieces, as shown in this sequence over 16 days. (Courtesy of New Mexico State University Observatory, photos by C. Knuckles and A. Murrell.)

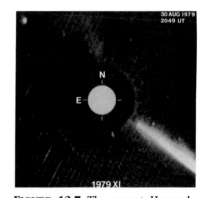

FIGURE 12.7 The comet Howard–Koomen–Michels approaches the sun in August 1979. Its perihelion distance was less than one solar radius. Shortly after this satellite photo was taken, the comet impacted dense regions of the sun's atmosphere and completely vaporized. (Courtesy of U.S. Naval Research Laboratory.)

If you could step back, far from the sun, and take a wide-angle photograph at any given time, you would find a large number of comets hovering in a cloud far beyond Pluto. This cometary cloud, proposed by the Dutch astronomer Jan H. Oort (1900–1992) in 1950, and sometimes known as the **Oort Cloud,** makes up the solar system's cometary reservoir.

According to Oort's model, most comets never come near the sun and we never see them directly. But occasionally, the gravitational action of passing stars or giant interstellar clouds of gas and dust (Section 15.1) pushes a comet into an orbit that moves it closer to the sun. Comet Kohoutek, seen in 1973, was a first-timer in the neighborhood of the sun. In passing by Jupiter or Saturn, an infalling comet may be captured into a periodic comet. All comets are eventually lost because of vaporization of the nucleus. The comet supply must be replenished. To ensure sufficient input to make up the losses, Oort's picture requires at least 10^{11}—perhaps as many as 10^{12}—comets to be clustered in the cloud.

Where did the Oort Cloud come from? The comets in it formed along with the rest of the solar system. It's unlikely that they formed at the present distance of the Oort Cloud, because the density of material there at the time of the solar system's formation would have been very low. Some astronomers suggest a possible origin in the asteroid belt between Mars and Jupiter; others think they formed just beyond the orbit of Pluto, about 100 AU from the sun. Both these suggestions run into the same difficulty: How did the comets get from their place of origin to their present location in the cloud? According to one idea, the gravitational effects of Jupiter kick comets out of an asteroid belt into long-period orbits. Another is that a compact cloud of comets lies just beyond the rim of the orbit of Pluto. This inner cloud would survive encounters that prune the Oort Cloud; cometary nuclei stored here might drift into the outer cloud.

Interactions with the Solar Wind

In 1951 the German astronomer Ludwig Biermann (1907–1986) suggested that the solar wind interacts with the coma to generate a comet's ion tail. This idea has been confirmed by spacecraft observations that the magnetic field carried by the wind can indeed drag ions from the coma. (Recall from Section 8.3 that a plasma such as the solar wind can carry along a magnetic field.)

Comets are now viewed as solar wind telltales, with magnetic fields playing a major role in their structure. The two plasmas, that of the solar wind and that of the tail, interact by magnetic fields to create a comet's shape. Without such magnetic fields and plasma interactions, we wouldn't have comets!

Halley's Comet

When Edward the Confessor died in January 1066, he left no direct heir to the English throne, and the nobles chose Harold as their king. In the same year, a bright comet traveled across the sky—the commonly accepted sign of a ruler's death and misfortunes to follow. Meanwhile, across the Channel, William of Normandy cleverly interpreted the comet as a sign portending his victory (Fig. 12.8). With this psychological edge for his men, he sailed to the British Isles and conquered the dispirited Saxon armies near Hastings. By the end of the year, William the Conqueror was crowned king. The comet of 1066 turned out to be one that returned on a regular basis.

But cometary orbits remained unknown until Edmond Halley (1656–1742), the Astronomer Royal, calculated the orbits of the comets of 1531, 1607, and 1682 by a method devised by his friend Isaac Newton. Halley found, to his amazement, that the orbits were almost identical. Noting that the comets appeared at intervals of approximately 75 or 76 years, Halley concluded that these several comets were in fact one and predicted that it would return around 1758. He was right. The comet that is named in his honor was sighted on Christmas night in 1758, after Halley had died. **Halley's Comet** was the first comet to be recognized as a permanent member of the solar system, with an elliptical, periodic orbit (Fig. 12.9). The spectacular comet of 1066 was none other than Halley's Comet.

Comets

At a Glance

Orbits	Elliptical for periodic ones
Size of nucleus	Very small, less than tens of kilometers
Shapes	Irregular
Surfaces	Craters and vents; small hills
Composition	Rocky with carbon, embedded ices (water)

FIGURE 12.8 Halley's Comet in A.D. 1066, from a section of the Bayeux tapestry. (Courtesy of the Picture Collection, The Branch Libraries/New York Public Library.)

Halley's Comet is the granddaddy of all known comets. Its passages have been recorded at least twenty-nine times, as far back as 239 B.C. Halley's Comet moves in an extremely elongated ellipse. Having passed aphelion (farthest distance from the sun) beyond Neptune's orbit in 1948, it returned to the earth's neighborhood in 1985–1986, when it reached perihelion on February 9, 1986, at a distance of 0.59 AU. After rounding the sun, the comet passed by closest to the earth (0.42 AU) on April 11, 1986.

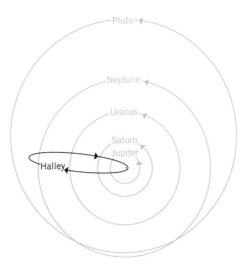

FIGURE 12.9 The orbit of Halley's Comet relative to those of the outer planets. Note how eccentric the orbit is, even compared to that of Pluto.

The 1986 Encounter with Halley's Comet

Unfortunately, as you may recall, this return of Halley's Comet presented the worst viewing in the past 2000 years! Even from the Southern Hemisphere, the sight wasn't very spectacular—except on the night of April 24, 1986, when a total eclipse of the moon took place. I viewed the comet from a mile-high mountain in New Zealand. With the moon's light dimmed in the eclipse, the southern Milky Way stood out as a broad flow of stars. Between the Milky Way and the moon, Halley's Comet hung fastened to the sky, its tail spread in a ghostly fan.

Though spacecraft data have taken the spotlight, ground-based work built up the foundation to understand Halley's and other comets. Sequences of photos enabled astronomers to study the plasma flowing from the sun and its interaction with the gases from the comet. A comet orbiting close to the sun plows through the solar wind, and its tail acts like a windsock in space—indicating the direction (and speed) of the solar wind. Occasionally, the plasma tail pinches off and is swept away by the solar wind. As the old tail leaves, a new plasma tail grows in its place—a snakelike shedding of gas (Fig. 12.10). Magnetic reconnection works here as it does in the earth's magnetosphere (Section 8.3).

The most spectacular results of the spacecraft missions came from the European Space Agency's Giotto, which swooped within 600 km of the core of Halley's Comet on March 14, 1986. Well before that closest approach, Giotto sampled the ionized zone. Most of the ions are related to water; in fact, most of the

(a)

(b)

FIGURE 12.10 Magnetic reconnection event in the tail of Halley's Comet photographed (*a*) on March 9, 1986, and (*b*) on March 10, 1986. During this time, a part of the gaseous coma is blown down the tail when the comet crosses a change in the magnetic polarity in the solar wind. (Photos by E. Moore; courtesy of Joint Observatory for Cometary Research, New Mexico Institute of Mining and Technology/NASA.)

comet's gas is water. What was not expected was the presence of fair amounts of sulfur and carbon compounds—the simplest chemicals of life as we know it (Chapter 22).

Giotto's last image, from a distance of about 1700 km, revealed a dusty realm (Fig. 12.11). The nucleus itself was peanut-shaped, about 8 km wide and 16 km long. And its surface was dark—black as velvet—with an albedo of only a few percent. The surface also appeared rough, with at least one hill a few hundred meters high and about a kilometer wide at its base. Craterlike structures about a kilometer in size were noted in some regions (Fig. 12.12). These structures showed the nucleus rotating once every two days.

Most dramatic were the dust jets. The sunlit side of the comet ejects heated materials—ice that vaporizes to gas, and the dust with it. These gases and particles blow off in jets only a few kilometers wide—at least nine were visible during the Giotto close encounter. The surface sources of the jets were less than a kilometer in diameter, and the outspray contained about 80 percent water vapor and 20 percent dust. And that dust was almost all carbon (like the lead in a pencil), with a small amount of sandlike material mixed in. No one had predicted before the mission that the dust would contain so much carbon—and it is the black carbon that makes the surface of the nucleus so dark, like that of a C-type asteroid.

Although most of the expelled gas is water vapor, about 4 percent is carbon dioxide, a small percentage carbon monoxide, and a very small amount a chain of carbon atoms—a simple organic molecule.

As expected from the dirty snowball model and confirmed by observations, Halley's Comet sheds a layer a few meters deep from its nucleus on each passage. So the comet is not forever, but it will survive for a good number of future orbits.

■ Why do you think Halley's Comet is so famous?

12.3 Meteors and Meteorites

A **meteor** is the flash of light from the vaporization of a solid particle in the earth's atmosphere (Fig. 12.13). As it plunges through the air, the particle is burnt up by air friction, leaving behind it a bright trail, a glowing column of light. Before the particle enters the atmosphere, it is called a **meteoroid:** a solid object traveling through interplanetary space—destined eventually to strike another object. Of course, other objects (comets, asteroids, and planets) also travel through the interplanetary void; a meteoroid differs from those chiefly in its small size, no more than a few meters in diameter, usually much less.

If a meteoroid survives its plunge through the atmosphere and strikes the earth's surface, the body is then called a

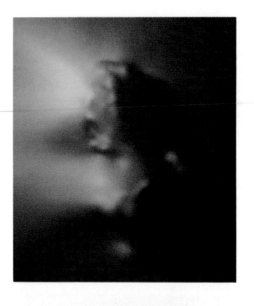

FIGURE 12.11 Composite, detailed image of the nucleus from the Halley Multicolour Camera exposures on March 13, 1986. (Courtesy of H. U. Keller and copyright 1986, Max-Planck Institut für Aeronomie, Lindau/Harz, Germany.)

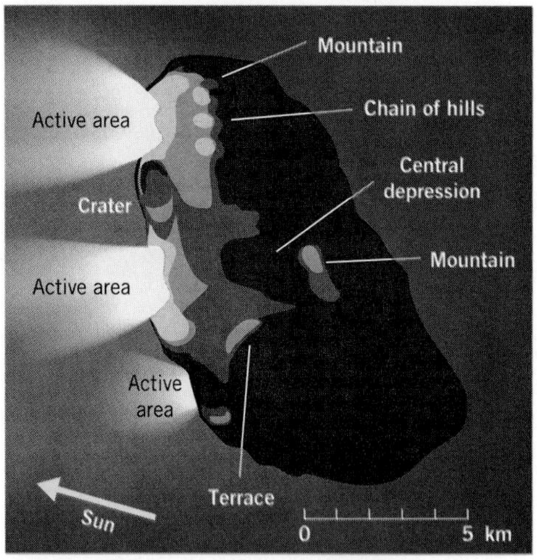

FIGURE 12.12 Schematic drawing of the major surface features of the nucleus.

FIGURE 12.13 A bright fireball of the Perseid meteor shower in 1988. (Courtesy of Steve Traudt, Synergistic Visions.)

meteorite. Most meteoroids are fragile, delicate particles (Fig. 12.14) that crumble quickly in their contact with the air. A meteoroid is typically a flimsy dust speck whose demise is its only remarkable aspect. Few hit the earth; most fall into the sun.

The dirty snowball comet model explains the intimate connection between comets and meteors. During a comet's passage by the sun, solar heating causes a continual loss of icy material from the cometary nucleus. So the dust and solid particles interspersed in the ices flake off and scatter in an untidy array around the comet. This solid debris has a low density. The older the comet and the greater its number of passages by the sun, the greater the loss of icy material and release of meteoroid material. About 99 percent of meteors are of cometary origin; few of these leave meteoritic debris. (Those that do probably are associated with asteroids.)

The cometary debris tends to follow the orbit of the original comet; at certain times of the year, the earth crosses the path of the debris and a *meteor shower* takes place. During a shower, many meteors are visible over a limited period of time. The most reliable of these is the Perseid shower, visible in August. The name of a shower identifies the constellation from which the meteors appear to radiate. The best time to observe most showers is in the early morning, although not all the showers peak then.

Comet Swift–Tuttle, which causes the Perseid shower, had not been observed since the nineteenth century. It was rediscovered by accident in 1992 (Fig. 12.15). American astronomer Brian Marsden recalculated its orbit, which has a period of about 130 years, and at first concluded that the comet had one chance in 10,000 of bashing the earth in the year 2126! Later calculations, based on new observations and a better fix on the orbit, caused Marsden to cancel the alert—for the next thousand years.

Types of Meteorites

In terms of physical and chemical composition, meteorites fall into three broad classifications: irons, stones, and stony-irons. The **irons,** which are generally about 90 percent iron and 9 percent

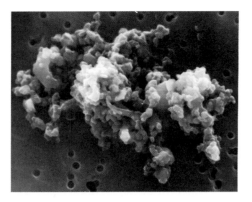

FIGURE 12.14 A particle of interplanetary dust about 10 μm in size. Note its open, fragile structure. This dust is probably material from a comet. (Courtesy of D. Brownlee, University of Washington.)

FIGURE 12.15 Comet Swift–Tuttle, which was rediscovered in 1992. Its debris produces the Perseid meteor shower. (Courtesy of the U.S. Naval Observatory.)

nickel, with a trace of other elements, are the most commonly found. They are easy to identify because of their high density and melted appearance. The **stones** are composed of low-density silicate materials similar to the earth's crustal rocks. When examined under a microscope, many stones are seen to contain silicate spheres, called **chondrules,** embedded in a smooth matrix. These stones are known as **chondrites** (Fig. 12.16*a*). Finally, **stony-iron** meteorites represent a crossbreed between the irons and the stones and commonly exhibit small stone pieces set in iron.

One of the most curious kinds of chondrite is the **carbonaceous chondrite** (Fig. 12.16*b*). The chondrules in such meteorites are embedded in material that contains a large fraction of carbon compared with other stony chondrites—typically about 2 percent carbon compared to the total mass. Their carbon content gives these meteorites a dark appearance.

Carbonaceous chondrites also contain significant fractions of water (about 10 percent) and volatile materials. In addition, if some gas were extracted from the sun and cooled, the condensed elements would be chemically very similar to carbonaceous chondrites. This similarity suggests that carbonaceous chondrites formed out of the same primordial material as the sun and have suffered no major heating or changes since that time. The pristine nature of chondrites makes them a major focus in attempts to study the origin of the solar system.

Origin of Meteorites

Most meteorites are too dense to have derived from comet-related meteoroids; that's why they survived the plunge to the ground. They resemble the inferred physical characteristics of asteroids. Orbits of meteorites that have been determined prove to be like those of asteroids rather than of comets. Collisions between asteroids fragment them, and these small pieces provide a source of meteorites.

What about the origin of iron meteorites? An important clue appears when a polished surface of an iron meteorite is etched with acid. Large crystalline patterns, called **Widmanstätten figures,** become visible (Fig. 12.17) upon such treatment. Terrestrial iron does *not* show these patterns when etched.

A nickel–iron mixture, when cooled slowly under low pressures from a melting temperature of about 1600 K, forms large crystals. The key point is that the cooling must be very gradual (about 1 K every million years). But metals conduct heat well, and in the cold of space, a molten mass of nickel and iron would cool rapidly without forming large crystals. So how could nickel–iron meteorites grow Widmanstätten figures? They need protection from the cold. It's likely that nickel–iron meteorite material solidified inside small bodies, termed **parent bodies.** To allow a cooling of only 1 K every million years, these objects must have been at least 100 km in diameter.

Most parent bodies were only a few hundred kilometers across. Once formed, they could be heated by the radioactive decay of short-lived isotopes, which would cause them to differentiate—the densest material sinking to the center, and the least dense frothing to the surface. Such an object would end up with a core of metals and a cover of rocky material, which would cool to form a crust. Insulated by the crust, the molten metals would be able to cool slowly and form Widmanstätten figures. If, later, the parent bodies were to collide and fragment, pieces from the outer crust would make stony meteorites, pieces from farther down would become stony-iron meteorites, and the core would produce the iron meteorites. (Based on the same reasoning, we expect the earth's core to be nickel–iron.)

Are any parent bodies around now? Yes—as asteroids. Recall the three main types of asteroid: the dark C-type, containing much carbon; the lighter S-type, composed of silicate materials; and the intermediate M-type, with metallic characteristics. These types probably are related to the carbonaceous chondrites, the stony meteorites, and the irons, respectively.

The view of meteorites outlined here implies that their parent bodies were among the first solid objects to form in the origin of the solar system. So the ages of meteorites should provide a direct indication of the age of the solar system. Meteorites can be dated by using radioactive decay techniques (Focus 8.1). Such methods give ages very close to 4.6 billion years. Meteorites provide

(a)

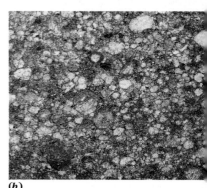

(b)

FIGURE 12.16 Chondritic meteorites. (*a*) Carbonaceous chondrite. Note the dark overall appearance resulting from the high percentage of carbon. (Courtesy of O. Richard Norton/Science Graphics.) (*b*) Detailed view of an ordinary chondrite from New South Wales, Australia. This close-up shows some chondrules a few millimeters in diameter, embedded in a finer matrix. (Courtesy of O. Richard Norton/Science Graphics.)

FIGURE 12.17 Widmanstätten figures (crystal patterns) in a nickel–iron meteorite from Toluca, Mexico. (Courtesy of O. Richard Norton/Science Graphics.)

Meteor Showers

At a Glance

Name	Date of Maximum Visibility	Hourly Rate of Meteors	Associated Comet
Perseid	Aug 12	40	Swift–Tuttle
Aquarid	May 4	5	Halley
Orionids	Oct 22	13	Halley
Taurid	Nov 1	5	Encke
Leonid	Nov 17	6	Tempel 1
Geminid	Dec 14	55	Ikeya

TABLE 12.1 INTERPLANETARY DEBRIS

MATERIAL	LOCATION	COMPOSITION	ORIGIN
Asteroids	Generally found between Mars and Jupiter in the ecliptic plane.	Rocky material (silicates), iron, nickel	Uncollected debris from formation of the solar system
Comets	Short period: elliptical orbits in ecliptic plane with aphelia near Jupiter and Saturn. Long period: orbits not confined to the ecliptic plane.	Ices of water, methane, ammonia, carbon dioxide, and rocky material (silicates)	Perhaps uncoalesced building blocks of Jovian planets and fragments from collisions between them
Meteors and meteoroids	Swarms or streams in elliptical orbits generally in ecliptic plane.	Flaky silicate materials, possibly some ices	Remains of exhausted comets (if dirty snowball model is correct)
Meteorites	On earth's surface.	Silicates, iron, nickel	Parent meteor bodies, asteroids
Interplanetary gas and dust	Throughout the solar system, mostly in ecliptic plane.	Gas: mostly hydrogen and helium Dust: silicates, graphite	Solar wind, fragmentation of comets, asteroids, and meteoroids

astronomers with a direct, reliable estimate of *when* the solar system formed. The real issue is *how*.

Table 12.1 summarizes our understanding of interstellar debris. Now that we have investigated the remains of the solar system's formation, let's turn to the puzzle of how it came about.

■ **The most dramatic meteor showers are the Perseids of August. Which comet is responsible for creating these displays?**

12.4 Pieces and Puzzles of the Solar System

The dynamic and chemical properties of the solar system impose crucial limitations on any model of its formation. These features serve as broad templates for shaping more specific questions.

Chemical

Chemically, the solar system falls into three broad categories of material: gaseous, rocky, and icy (Table 12.2). Each group is distinguished by its melting point. The gaseous and icy materials together, sometimes called **volatiles,** are generally gaseous under the conditions expected during the solar system's formation. The bodies of the solar system are composed of various combinations of the three groups. Though one class of materials may dominate, any solar system body contains some of each group.

For example, the sun contains mainly gaseous materials and also icy/rocky materials, but as gases (because the sun is hotter than 2000 K), not solids. The terrestrial planets and the asteroids are mainly rocky/metallic; Jupiter and Saturn mostly gaseous; Uranus, Neptune, Pluto and Charon, and comets mostly icy.

Dynamic

Dynamically, the solar system displays a regular structure in terms of its motions. Viewed from above the sun's north pole, the solar system shows the following regularities:

1. The planets revolve *counterclockwise* around the sun; the sun rotates in the same direction.
2. The major planets, except Mercury and Pluto, have orbital planes that are only slightly inclined with the plane of the ecliptic (the plane of the

Meteorites
At a Glance

Sizes	Centimeters to meters
Shapes	Irregular
Composition	Rocks (some with high carbon content) and metals
Sources	Comets and asteroids

TABLE 12.2 GENERAL CLASSES OF SOLAR SYSTEM MATERIALS

CLASS	EXAMPLES OF MATERIALS	SOLIDIFICATION TEMPERATURE (K)
Rocky	Iron, iron sulfide, silicates	500–2000
Icy	Water, methane, ammonia	100–200
Gaseous	Hydrogen, helium, neon, argon	0–50

earth's orbit); that is, the orbits are *coplanar*.

3. Except for Mercury and Pluto, the planets move in orbits that are very nearly *circular*.

4. Except for Venus, Uranus, and Pluto, the planets rotate *counterclockwise*, in the same direction as their orbital motion.

5. The planets' orbital distances from the sun follow a *regular spacing;* roughly, each planet lies twice as far out as the preceding one.

6. Most satellites revolve in the *same* direction as their parent planets rotate and lie *close* to their planets' equatorial planes.

7. Some satellites' orbital distances follow a *regular spacing rule*.

8. The planets together contain *more* angular momentum than the sun (99.5 percent versus 0.5 percent).

9. Long-period comets have orbits that come in from *all directions and angles,* in contrast to the *coplanar* orbits (in the same plane) of the planets, satellites, asteroids, and short-period comets.

10. All the Jovian planets are known to have *rings*.

Models of Origin

A successful model must explain as many of the foregoing dynamic and chemical properties as possible. A good model must account for the greatest number of the listed characteristics, and explain them in some internally consistent, simple fashion (Section 2.1). That is what is meant by the requirement that a model be "aesthetically pleasing."

The sun contains most of the mass of the solar system (99.9 percent versus 0.1 percent). The rest lies close to the plane of the solar system. In terms of the layout of mass, the solar system is quite thin. If it were the size of an average pancake, the solar system would be only a centimeter thick, with the orbits of the planets contained inside it.

Finally, a successful model must deal not only with the dynamic and chemical regularities of the system but also with the interplanetary debris: comets, asteroids, and meteoroids. Contem-

porary models consider these bodies to be important relics of the solar system's early history.

Most models today are variations of **nebular models,** in which the sun condenses from an interstellar cloud of gas and dust that also forms a disk, a **solar nebula,** out of which the planets condense. The nebular models view the solar system as a natural outcome of the sun's formation and, perhaps, of any star's formation. If nebular models are correct, planetary systems are very common in our Galaxy and in other galaxies.

■ **The sun represents nearly all the mass of the solar system. Does it also contain the lion's share of angular momentum?**

12.5 Basics of Nebular Models

The essential feature of nebular models is that the sun and then the planets form from a cloud of interstellar material. The sun's formation takes place in the center of a flattened cloud. The planets grow from the disk of the cloud. That's the basic picture and the nub of the problem, as developed in this section.

We know that the solar system is now basically flat, with the sun in the center. Examining the rings of the Jovian planets, we see that they are flat and consist of small particles orbiting their parent planet. You can imagine that if somehow the small particles could be stuck together, Jupiter would end up with another moon. So the problem has two parts: how to make a flat solar system, and how to get the planets to grow out of the cloud.

Angular Momentum

To tackle the problem of making a flat solar system, we need to consider a basic physical idea: the *conservation of angular momentum* (Focus 9.1). The basic point is this: once a body has started spinning, it will keep on spinning as long as no outside influence affects it. The amount of angular momentum depends on how much mass the body has and how much it is spread out. If, by itself, the body changes size—for instance, if it contracts gravitationally—it will naturally spin faster to keep its angular momentum the same. (A mass contracts gravitationally when it pulls itself together by the gravitational forces between the particles that make it up.)

A familiar example of the conservation of angular momentum is the spin-up of an ice skater. The skater goes into the spin with arms outstretched. As the skater pulls his or her arms in, the rate of spin increases. Why? Because to conserve angular momentum, the skater must rotate faster *around* the body's spin axis as mass is brought closer to it.

Consider now a large spherical cloud of gas and dust particles, slowly spinning. Imagine that it pulls itself inward by its gravity (Fig. 12.18). What happens? It will spin faster and collapse down along the spin axis to make a flat disk with a fat center. That's just what we need to form the solar system. As a natural result, we get the planets' orbits aligned in a thin disk, and the sun rotates in the same direction as the revolution of the planets.

With this neat solution comes a serious problem: the present distribution of angular momentum. Although the sun holds 99 percent of the system's mass, it contains less than 1 percent of the angular momentum. The outer planets have the most, 99 percent of the total. If all the planets with their present angular momenta were dumped into the sun, it would spin once every few hours rather than once a month. In the cloud's collapse as suggested in this model, the sun forms from the central part. So it should be spinning very rapidly. Actually, the sun spins 400 times more slowly than this rate. The angular momentum is there, but not in the right place! To adopt a nebular model, we must account for the present distribution of angular momentum, a problem still being worked out.

One solution involves the interaction of magnetic fields and charged particles to rearrange the distribution of angular momentum. Basically, the spin of the central part of the nebula must be decreased and transferred to the outer regions. Charged particles and magnetic

fields interact so that the particles spiral along the magnetic lines of force. Such interactions provide a way to transfer spin from the young sun to the outer parts of the nebula.

Here's an analogy to the process. Imagine standing in a swimming pool, up to your neck in water, with your arms extended. Spin around as fast as you can. Your arms will drive the water around and force it to swirl. In response, you'll feel a drag of the water on your arms. If you didn't keep yourself spinning, you'd rapidly slow down as angular momentum was transferred from you to the water.

As the sun forms, it heats the interior regions of the nebula. Here the gas is ionized, and magnetic field lines trap these particles. As the sun rotates, it carries its magnetic field lines with it; these drag along the charged particles, which in turn unite with and drag along the rest of the gas and dust. So the magnetic field spins around the material in the nebula near the sun. At the same time, the mass of the nebula resists the rotation. This drag on the magnetic field lines stretches them into a spiral shape (Fig. 12.19). The magnetic field links the material in the nebula to the sun's rotation. So the nebular material gains rotation (and angular momentum) and in the process causes a drag on the sun's rotation, which slows it down.

Whatever process transferred the angular momentum, the transfer must have taken place *before* large solid ob-

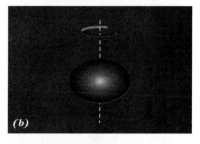
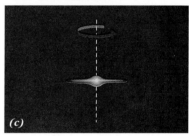

(*a*)

(*b*)

(*c*)

FIGURE 12.18 Gravitational contraction of a spinning cloud. (*a*) The process begins with the cloud slowly spinning. As it contracts (*b*), material falls in along the spin axis to form a disk with a central bulge (*c*). The cloud ends up spinning faster than at the start of the contraction because of the isolated cloud's conservation of angular momentum.

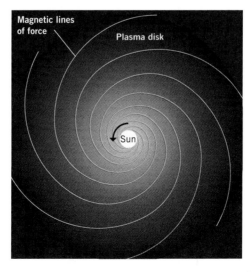

Magnetic lines of force

Plasma disk

Sun

FIGURE 12.19 Top view of a possible primeval solar magnetic field. One end of the field lines rotates with the sun; the other end is dragged by the ionized material (plasma) in the disk. Hence the spin of the central regions is transferred to the outer parts.

jects formed in the nebula. The transfer mechanism just described works effectively *only* on plasmas.

Heating of the Nebula

Finally, one other fundamental process occurs: that of *gravitational contraction*. Whenever a mass pulls itself together by its own gravity, it gets hotter, as a result of the transformation of energy from one kind (gravitational potential energy) into other kinds (heat and light). The basics are just like those for accretion, but a larger number of smaller particles is involved.

Here's how. Consider a ball held above the earth's surface (Fig. 12.20*a*). Its speed is zero and it has no kinetic energy. But it does have potential energy, as you can tell when you drop it (Fig. 12.20*b*). The ball falls, and as it does, its speed increases—it accelerates. So its kinetic energy increases the more it falls.

Instead of a ball, consider a cloud of gas containing a large number of particles (Fig. 12.21). Think of each particle as a small ball. Imagine that the cloud contracts gravitationally, from the combined attraction of all the particles. As it contracts, the particles gain speed. Though at the start the velocities are directed inward, collisions will soon distribute them in random directions, with only a slow net motion inward. (The collisions slow down the gravitational contraction.) So the average kinetic energy of the particles increases. Temperature measures the average kinetic energy of the particles involved, so the temperature of the gas increases.

Meanwhile, the density also increases. So the particles collide more often from the combined effects of the increases in velocity and density. The collisions excite some atoms; these emit photons. So the net result of the gravitational contraction is that some of the initial gravitational energy is converted to heat (a higher temperature means greater average speeds of the particles of the gas) and some ends up as photons. In fact, half goes into raising the temperature and half into light. The key point is this: *as the cloud contracts, it gets hotter.*

With angular momentum and gravitational contraction in mind, let's turn to

FIGURE 12.20 Conversion of gravitational energy into kinetic energy by the fall of a mass. (*a*) The mass has zero speed when it is first dropped. (*b*) As it falls, its speed and so its kinetic energy increase.

the next basic problem: how to form the planets.

■ In a nebular model, does the magnetic transfer mechanism of angular momenta work better on plasmas or solid bodies?

12.6 The Formation of the Planets

A successful nebular model must account in some detail for four important stages in the solar system's evolution: the formation of the nebula out of which the planets and sun originate, the formation of the original planetary bodies, the subsequent evolution of the planets, and the dissipation of leftover gas and dust. Modern nebular models (there are more than one!) give tentative explanations for these stages, but many details are lacking. No one model today is entirely satisfactory.

Making Planets

How can planets grow? There are three main methods: gravitational contraction,

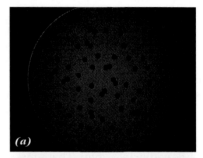
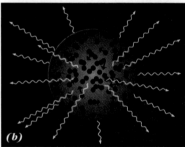

FIGURE 12.21 Heat and light produced by the gravitational contraction of a cloud of material. (*a*) Consider a spherical cloud of gas particles, starting out with zero speed. (*b*) As the cloud contracts gravitationally, the particles gain speed and so kinetic energy (heat); they also collide, excite each other, and so radiate light.

Nebular Models

At a Glance

Origin	Interstellar cloud of gas and dust
Processes	Gravitational contraction, conservation of angular momentum
Sun	Forms at center of nebula, reflects original composition
Planets	Form in disk of nebula, reflect temperature locally
Formation processes	Condensation, accretion

accretion, and condensation. **Gravitational contraction** works if regions in the nebula have enough mass to be able to contract by their own gravity to form a planet. **Accretion** occurs when small particles collide and stick together to form larger masses that eventually grow into planets. (An example: as snowflakes fall through the air, they may collide and stick to form clusters of snowflakes.) **Condensation** involves the growth of small particles by the sticking together of atoms and molecules. (An example: water molecules combine in clouds to form raindrops.)

The formation of the planets was a multistep process; first small bodies formed, followed by larger ones, which eventually evolved into the planets. Accretion was probably the main process in building most of the planets (and their satellites). The first fairly large bodies, from a few kilometers to a few hundred kilometers in size, are called **planetesimals.** Then, the planetesimals collide and accrete to make planet-sized masses (Fig. 12.22), called **protoplanets.** These objects evolved over billions of years into the planets of today.

How to get from small dust grains to large protoplanets? Grains collide and accrete to form larger, pebble-sized objects, which quickly fall into the plane of the nebula. The pebbles then accumulate into planetesimals by gravitational attraction. Whatever materials happen to be available at a certain distance from the center of the nebula make up a planetesimal. So the planetesimals reflect the local compositions of material.

Once the first planetesimals have formed, they may gather into larger bodies, perhaps almost as large as the moon. Somehow these objects end up in a few protoplanets. Here gravity helps. By the time the planetesimals have gathered (in a few tens of thousands of years) into several somewhat larger bodies (500 km in size), they will have enough mass to help pull in other smaller masses from a distance. So a growing planet will sweep clear a zone of the nebula to feed its mass. For the terrestrial planets to grow to their present sizes, calculations indicate a time of roughly 100 million years for the protoplanets to form by the accumulation of smaller masses. This clearing

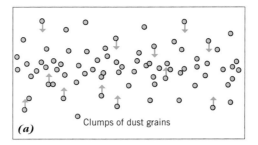

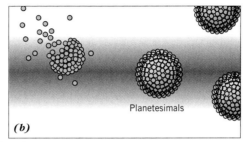

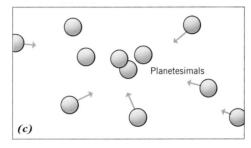

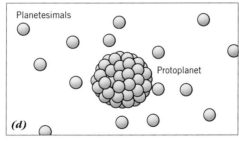

FIGURE 12.22 Possible scenario for the formation of the planets. (*a*) Dust grains collide and stick to form small objects that fall into a thin disk. (*b*) Gravity pulls these together to make asteroidal-sized bodies, the planetesimals. (*c*) The planetesimals collect in clusters to form the cores of the protoplanets (*d*), which evolve into the planets.

out of a region around each protoplanet has led to the spacings of the planets that we see now.

Chemistry and Origin

How did the protoplanets acquire differences in chemical composition? The basic concept is that of a **condensation sequence.** The nebula's center must have been hot, a few thousand kelvins. Here solid grains, even iron compounds and silicates, could not condense. Elsewhere, the identity of the materials that would condense as new grains depended on the temperature (Fig. 12.23). Just below 2000 K, grains made of terrestrial materials would condense (Table

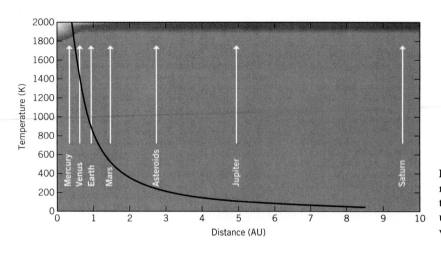

FIGURE 12.23 How the temperature in the solar nebula might have decreased with increasing distance from the sun: a cross section of the solar nebula and a possible distribution of temperatures within it at the beginning of planetary formation.

12.2); below 273 K, grains could form of both terrestrial and icy materials.

The exact sequence in which compounds could condense from a heated nebula is crucial to this idea. At different temperatures, the gases available and the solids present react chemically to produce a variety of compounds. The key result of the condensation sequence: the densities and compositions of the planets can be well explained with the condensation sequence *if* the temperature in the nebula drops *rapidly* from the center outward. Under this assumption, then, it is feasible to suggest that at different distances from the sun, different temperatures allowed different chemical compounds to condense and form grains that eventually made up the protoplanets. If a material could not condense because the temperature was too high, it would not end up in the protoplanet.

Unlike the Jovian planets, which are much like the original nebula in composition, the terrestrial planets contain little in the way of icy and gaseous materials. At the earth's distance, the temperature was about 600 K. Here, silicates of both iron and magnesium could condense, along with iron oxide and iron sulfide. So the model predicts a planet with a smaller nickel–iron core than Mercury's, and a larger mantle, rich in silicates and oxides of iron and magnesium. The earth has such an interior (Fig. 8.2). What about Mars? Here the temperature was lower still. The resulting core would consist mostly of iron sulfide, and the mantle would be olivine, rich in iron oxide and water. We don't know the interior of Mars very well, but the predic-

tion from the condensation sequence is compatible with our general model about the Martian interior.

Caution. It is how *low* the temperature falls that determines the chemical reactions to produce condensates. In general, the condensation sequence requires that a certain ***minimum*** temperature be reached to account for the known chemical composition of the planets. Roughly, these temperatures are 1400 K for Mercury, 900 K for Venus, 600 K for the earth, 400 K for Mars, and 200 K for Jupiter.

When the central regions of the nebula finally form the sun, solar energy evaporates any icy material in the nebula and around the inner protoplanets or planetesimals. Intense solar radiation pressure and a strong solar wind (a solar gale!) push leftover gases out of the solar system.

Leftover planetesimals bombard the planetary surfaces, leaving remnant craters on the terrestrial planets. This bombardment heats the surfaces of the planets. Radioactive decay heats the interiors and melts them. Dense elements, such as iron and nickel, sink to form a core. Less dense materials, such as silicates, float to form surface froth, which cools to become the crust. So the planets become differentiated—the first step in their evolution.

Not all planetesimals end up in larger bodies. Some rocky/metallic ones remain; they become asteroids. The icy ones hang around as the nuclei of comets. Most are tossed into the outer solar system by the gravitational influences of large masses, such as Jupiter. The Oort Cloud now contains these primordial icy bodies.

The condensation sequence implies that myriad small bodies rich in volatiles and carbon formed in the cold reaches beyond Mars—the moons of the Jovian planets and even Pluto and Charon are samples. The icy-carbon objects were the original supplies of the solar system. Those that ran into each other accreted or fragmented. Those that plunged toward the sun heated and in part vaporized. Yet, they have evolved less than the planets. The survivors today make up the interplanetary debris.

Jupiter and Saturn: A Different Story

Jupiter and Saturn may not have formed in accordance with the planetesimal accretion model just described. Rather, they may have condensed gravitationally from single large blobs of material in the nebula. The internal heat they gain comes from the conversion of gravitational potential energy into heat during gravitational contraction.

Theoretical models of the evolution of Jupiter start with a mass sixteen times Jupiter's present size and a surface temperature of about 1000 K, as well as a total energy output per second of almost 0.01 the sun's. The early high power of Jupiter may explain, using condensation sequence again, why the Galilean satellites decrease in density going outward from Jupiter (Section 11.2). If Jupiter were hot at the time of the satellites' formation, the inner ones would not have accreted as much icy materials as the outer ones. So the Galilean moons may mimic the condensation and accretion of the terrestrial planets. Saturn may have progressed through a similar sequence.

In fact, the evolution of Jupiter and Saturn may follow that of the solar system generally. Their rings may be unaccreted material.

Evaluation of the Nebular Model

By and large, the nebular model makes a firm basis for exploring the origin of the solar system. Here's how it deals with the ten points listed in Section 12.4.

1. *The planets' revolution and sun's rotation:* well explained from the original rotation of the nebula before collapse.

2. *Coplanar orbits:* well explained by the conservation of angular momentum applied to the nebula's rotation during the collapse.

3. *Circular orbits:* explained fairly well by the interactions of planetesimals.

4. *Planets' rotation:* expected from the spin of the nebula; exceptions may be the result of violent collisions.

5. *Planets' spacings:* fairly well explained by the sweeping out of zones of planetesimals by the gravitational attraction between the protoplanets.

6. *Satellite systems:* nebular processes on a smaller scale; giant impacts may cause exceptions.

7. *Satellite spacings:* sweeping up of material during early accretion of protosatellites.

8. *Angular momentum distribution:* involves magnetic fields and plasmas early on, to shift distribution from center to disk of nebula.

9. *Comets:* explained fairly well as icy planetesimals thrown out of orbits near Jupiter, Saturn, Uranus, and Neptune.

10. *Planetary ring systems:* explained vaguely as unaccumulated debris, perhaps some as the result of giant impacts.

■ How does the formation of protoplanets from planetesimals help in elucidating the process of the nebular model?

Key Concepts

1. Any model for the origin of the solar system must account for its general chemical and dynamic properties in a unified way.

2. The key dynamic properties include the following: the system is flat; most of the mass is in the center (the sun); most of the angular momentum is in the planets; the planets' orbital directions are all the same and also most of their rotational directions; and the planets are regularly spaced.

3. The key chemical properties include the general division of the terrestrial and Jovian planets, the differences among planets in each group, and the compositions of asteroids, meteoroids, and comets.

4. Interplanetary debris provides important clues about the origin of the solar system, especially from the physical properties of comets (dirty snowballs), asteroids (rocks and metals), and meteorites (rocks and metals).

5. The crystalline patterns found in iron meteorites indicate that their material cooled slowly from a molten state, which implies that they formed in larger parent bodies.

6. A spinning cloud of interstellar material will flatten out (as a result of the conservation of angular momentum) and also heat up as it contracts gravitationally.

7. The condensation sequence works if the temperature in the solar nebula decreases rapidly from center to edge. Which materials vaporize and condense depends on how high and then how low the temperature gets at a certain distance from the sun.

8. Magnetic fields can transfer the spin of the sun to the rest of the solar nebula (before the planets form) to account for the distribution of angular momentum now.

9. The general process of planet formation may have been condensation of grains, followed by accretion of grains into planetesimals, and clumping of planetesimals into protoplanets, culminating in the evolution of the protoplanets into the bodies we see today.

10. Jupiter and Saturn may have been formed by the gravitational contraction of large blobs of material rather than by planetesimal accretion.

11. Comets and asteroids are very likely leftover planetesimals formed in different regions of the solar system.

12. Comets interact with the solar wind to generate long, magnetized plasma tails; their icy nuclei are irregular in shape and very dusty.

Key Terms

accretion

asteroid

asteroid belt

carbonaceous chondrites

chondrites

chondrules

coma

condensation

condensation sequence

dirty snowball comet model

dust tail

gravitational contraction

ion tail

irons

meteorites

meteoroids

meteors

nebular models

nucleus (of a comet)

Oort Cloud

parent bodies

periodic comets

planetesimals

protoplanets

S-, C-, and M-type asteroids

solar nebula

stones

stony-irons

volatiles

Widmänstatten figures

Study Exercises

1. Suppose you hopped in your spaceship on Saturday night and flew to an asteroid. What would you see? (Objective 12-2)

2. Then you speed off to a comet. What would you see? How would the comet's appearance differ from the asteroid you just visited? (Objective 12-2)

3. After many perihelion passages of the sun, what happens to a comet in the dirty snowball comet model? (Objective 12-2)

4. How can you tell an iron meteorite from a piece of terrestrial iron and nickel? (Objective 12-2)

5. What is the major weakness of the nebular model with respect to the dynamic properties of the solar system? (Objectives 12-4 and 12-5)

6. What one planet fits least well with the general chemical and dynamic properties of the solar system? (Objectives 12-1 and 12-10)

7. Use the chemical condensation sequence to explain the general chemical differences between the earth and Jupiter. (Objective 12-6)

8. How is it that the solar system is flat rather than spherical? (Objectives 12-4 and 12-10)

9. In what ways did the Giotto mission confirm the dirty snowball model of cometary nuclei? (Objectives 12-2)

10. Suppose you found a dense, irregular fragment on the ground. How could you tell that it was or was not a meteorite? (Objective 12-2)

11. What is the possible source of material for the intense bombardment of planetary surfaces some 4 billion years ago? (Objective 12-7)

12. How did the protoplanets acquire their original internal thermal energy? (Objective 12-7)

13. In what key way may the formation of Jupiter have differed from that of the earth? (Objective 12-8)

14. By what process does a contracting cloud of gas and dust heat up? (Objective 12-9)

15. If an initially spherical, spinning cloud of gas and dust contracts gravitationally, into what shape will it finally end up? (Objective 12-9)

Problems & Activities

1. Suppose that the asteroid Ceres is made entirely of rocky material and has a spherical shape. What would its mass be? Its escape speed?

2. Suppose that the nucleus of a comet were composed entirely of icy materials and had a radius of 10 km. What would its mass be? Its escape speed? Compare these values to those for Ceres.

3. Imagine a double-asteroid system with an orbital period of 4 hours and a separation of 100 meters. What would be the sum of the mass of the two asteroids?

4. If icy planetesimals were perturbed by Jupiter out to the Oort Cloud (with a typical aphelion of 100,000 AU), what would be their orbital periods? Assume that they now orbit the sun.

5. The asteroid Toutatis was observed at a distance of 2.5 million km from the earth. What was the resolving power needed to show that it was two separate objects?

6. When Ceres and the earth are closest together (Ceres in opposition to the earth), what is Ceres' angular diameter?

See for Yourself!

Meteor showers can be surprising celestial fireworks! All it takes to observe them well is a dark site, a comfortable (reclining!) chair, and a lot of patience. And dress properly for the time of year and location. It also helps if you have a small group of observers; then each one can concentrate on a certain region of the sky. People can call out sightings and one person can record them by the hour.

Showers are named after the constellation out of which they appear to radiate. So the Perseids appear to come from the direction of Perseus, the Orionids from Orion, and so on. The general rule is to watch for a specific shower *after* midnight from a location that is as far from any light pollution as possible. This rule varies a bit depending on the shower: for the Perseids and Orionids, 4:00 A.M. is optimal; for the Geminids, 2:00 A.M. is better. Start the watch about midnight and work until about 6:00 A.M., as if you were on shift.

Keep in mind the meteors orbit the sun in a *stream* of debris from a comet. So major annual showers actually begin a few days before and continue a few days after their date of maximum. For the Perseids, August 9 to 14 is the range; for the Orionids, October 20 to 25; and for the Geminids, December 12 to 15. If you become an avid fan of meteors, you will probably cover spans of a few days such as these. ■

PART TWO *Epilogue*

This part has shown you the new richness in our knowledge of the solar system. But don't let this wealth of information deflect you from the main point: What do the present properties of the solar system reveal about its origin and evolution?

By now you realize that the models we have developed to answer this question strike a very tentative chord. But we do have a pretty good idea of the general themes. The formation of the solar system naturally accompanied the birth of the sun from a cloud of interstellar materials. The closeness in age of the sun, moon, earth, and meteorites implies that the process of formation lasted a short time—at most a few hundreds of millions of years—compared to 4.6 billion years, which is the age of these bodies. The fast formation contrasts with the slow evolution that follows.

To glean the conditions in the solar system's early years, we must look at the fossil objects in it: the interplanetary debris of comets, asteroids, and meteoroids. These objects have evolved little (compared to the planets) and so give us the most direct clues to the past.

That past involves a nebular model for the solar system's formation. The start of the process requires the gravitational contraction of a huge interstellar cloud. What prompts such a contraction? We don't really know yet. But we do have evidence for the existence of enormous clouds between the stars. We can also see very young stars that have been born from such clouds. And that starbirth may well involve the formation of planets. So the fact that stars are common implies that planets are common, too. Stars and planets are connected in starbirth; that is one of the key themes of this book.

Since their birth, the planets have evolved, some more than others. The Jovian ones have changed little since their formation; they have essentially retained their youthful appearances. In contrast, the terrestrial planets have reshaped their interiors and surfaces and transformed their atmospheres. And the earth, a restless planet, has changed the most; its structure now differs greatly from its protoplanetary state.

I can't help wondering how many other planets circle other stars in the sky, especially when I observe on a winter night. How many monsters such as Jupiter? How many jeweled Saturns? How many barren spheres like Mercury? And how many like the earth, where perhaps my counterpart also dreams of inhabited worlds? ■

The Unifying View

ASTRONOMICAL CONCEPTS

Building planets, using the earth as a model:
  Need mass and size to find bulk density, which offers clues to chemical composition (metals, rocks, and ices).
  A planet's mass can be found from a natural or artificial satellite and Kepler's third law.

A planet's size can be found from its angular diameter and distance.

Evolution of planets is driven by heat flow from hot interiors, which can appear as volcanism and plate tectonics on the surface.
  If planets are fluid or partially fluid early in their lives, gravity causes their interiors to differentiate by density.

Energy may be left over from the planet's original accretion from debris and/or from radioactive decay.

Building a solar system, using the local one as a model:
  The solar system contains four general forms of material: the sun, the terrestrial planets, the Jovian planets, and debris.

INFORMATION CONCEPTS We perceive the world through our senses, sometimes aided by instruments:
Flyby missions to the planets and their moons, to Halley's Comet, and to an asteroid, as well as the exploration of the moon's surface, have provided new information and insights about the origin of the solar system.

Model building is the essential process by which science discovers knowledge and reveals information:
Current models have been tested fairly well for the interiors of the earth and moon, but not for the other planets.
Current nebular models for the formation of the solar system explain fairly well the main features but tend to be vague about details.

ENERGY CONCEPTS The total energy of a system is conserved; it has a constant value.
A cloud of gas and dust contracting by its own energy has a constant total energy.

Transformations occur among the various forms of energy, and energy can be transferred from one location to another.
As a cloud of particles contracts, gravitational potential energy is converted into kinetic, thermal, and radiative energy.

Protoplanets are heated by accretion.
The heat flowing from the hot interior of a planet eventually is emitted into space as infrared radiation.

MOTION AND FORCE CONCEPTS Newton's laws of motion, plus his law of gravitation, explain and predict the motion of the masses in the solar system in a unified way.
Gravity naturally brings together a great enough mass of material.

The total angular momentum of an isolated system is conserved.
A spinning cloud of particles naturally contracts into a disk, with an increase in the rate at which it spins.

QUANTUM CONCEPTS Electromagnetic energy has both wave and quantum properties.
The Doppler shift results from a change in the observed wavelengths for light emitted by a source moving away from or toward an observer.

Improve Your Night Vision

AUTUMN

Fred Schaaf

The stars of the traditional autumn constellations are dimmer than those of any other season. Some of the fairly bright autumn constellations, mostly in the north or high in the south—Pegasus, Andromeda, Perseus, Cassiopeia, and the dimmer Cepheus and Cetus—are related by being named after characters in one of the greatest of all Greek myths.

But there is more of interest than just starlore among the autumn constellations. For instance, we can consider the brightest examples of the three most important types of variable star.

Algol is the brightest of the *eclipsing binaries*. Instructions for how to observe it and record its dramatic brightness changes appear in "See for Yourself" at the end of Chapter 14.

A star that requires more patience than Algol—but also can give more reward—is Mira, the brightest of the *long-period variables*. The name Mira itself means "wonder," and it is deserved. Although at some times Mira is faint in binoculars (much too faint for the naked eye), at other times physical changes in the star itself cause it to become a quite conspicuous naked-eye object. This star,

whose Greek letter designation is Omicron Ceti, has an average period between one maximum and the next of about 330 days. But neither the period nor the amplitude (brightness range) of Mira is ever exactly the same! So an amateur astronomer who studies this star can provide original information on it every year. In some years, the excitement is particularly great when Mira has an unusually bright maximum (once, back in 1779, it almost rivaled the star Aldebaran!).

To estimate the brightness of a star like Mira, you need detailed star charts giving the magnitudes of "comparison stars" (nearby stars of unvarying brightness, which can be used as comparisons as Mira brightens or fades). If you are interested in obtaining such a chart, or possibly following the changes of many variable stars, you should write to the American Association of Variable Star Observers (AAVSO), at 25 Birch Street, Cambridge, MA 02138.

What is the third important type of variable star? The *cepheids*. You can find the prototypical example of the class—Delta Cephei—high in the north or northwest on autumn evenings. It is precisely predictable and does not vary greatly in brightness, but you will want to observe it

because of the importance to astronomy of the cepheid class.

Perhaps the single most interesting celestial object of autumn is the farthest thing most people will ever see with the naked eye. Almost right overhead at midnorthern latitudes in the middle of November evenings is M31, the Great Galaxy in Andromeda. To the naked eye in the country or with binocular-aided vision in the suburbs, M31 appears as a greatly elongated smudge of light—but this is actually a spiral galaxy roughly twice the size of our own giant Milky Way. The light you see reaching you from M31 tonight left there well over 2 million years ago.

Autumn is the season that features four of the year's eight or nine most interesting "meteor showers." *Meteor* is the name given to the streak of light caused by a piece of space debris burning up from the friction of its immense speed as it enters our upper atmosphere. *Meteor showers* are increased numbers of these meteors, all appearing to diverge from a certain point in the heavens—the *radiant*—as a result of perspective (just as parallel railroad rails appear to diverge from a spot on the horizon). The poet Milton wrote "Swift as a shooting star in autumn thwarts the night."

The autumn showers include the Orionids (about October 20–25), the Taurids (often best in first half of November), the Leonids (November 16–17—with the next of the rare Leonid "meteor storms" of thousands of meteor per hour visible for some observers in 1998 and 1999); and the Geminids (December 12–15—a rival of the Perseids for sheer numbers of meteors and variety of colors, trails, and behavior). You can narrate into a tape recorder special features of different meteors and the times at which the meteors appeared. Or you can keep simple counts in your mind, making sure to maintain a separate tally for yourself (don't combine different observers' sightings) and a sepa-

rate tally of the meteors from the shower and those that come from elsewhere in the sky. (More advice and information on observing meteor showers can be found in "See for Yourself" at the end of Chapter 12.)

Planetary nebulas—clouds of gas thrown off by certain dying stars—are properly visible only in telescopes, preferably far from city lights. Many of these nebulas look something like the blue-green planets Uranus and Neptune, hence the name. Several of the most fascinating for your small telescope are the Ring Nebula in Lyra and the Dumbbell Nebula in Vulpecula (both still visible in the west among the departing summer constellations).

In the 1990s Saturn itself is found among the autumn constellations. Usually a small telescope shows its glorious rings easily—in 1995 and 1996, however, they will be elusive but captivating as they go through their rare "edgewise" presentations to the earth and sun. Will you see Saturn well in your small telescope? (You can identify it as a bright golden point of untwinkling light with the naked eye.) Although cold fronts bring nights of exceptional clarity—excellent atmospheric "transparency"—to much of the United States and Canada in September and October, the turbulence associated with these weather systems may give poor "seeing" (refer to Summer essay at the end of Part One).

The least subtle of autumn's special sky phenomena may be the huge rising moon at autumn dusks. This moon illusion is still not fully understood and apparently results from a number of factors. Does such a moon still look unusually big if you observe it through a tube or cupped hands that separate it from surrounding landscape in your field of view? Judge how many times bigger than usual the moon looks for a number of nights around harvest moon (the full moon nearest to autumn equinox) and hunter's moon (the next full moon).

PART THREE

THE UNIVERSE OF STARS

T HE UNIVERSE EVOLVES — THE whole universe and all it contains. That's what I mean by *cosmic evolution*. The universe has a history, and its parts do, too. Unraveling all these stories brings us insight into the grand scheme of things and our place within it: the cosmic connections that link us all.

Stars form the nexus of cosmic evolution. They make up the crucial links in the chain of cosmic connections. With their births, planets form. With their light, life survives. With their deaths, new elements are forged. Without the lives and deaths of stars, the flow of cosmic evolution would cease.

All this from what appear to be simply points of light in the sky — except for one star, the sun. The realization that stars are actually other suns marked a crucial transformation in astronomers' picture of the cosmos. Because we orbit a mere 8.3 light minutes from it, the sun is exposed to our close and careful view — a vision impossible for us to achieve with the stars. That view permits us to make the solar–stellar connection, using the sun as our basic model for stars.

Chapter 13 presents the sun in all its glory. Then Chapter 14 focuses on the stars as we understand them as suns. With the basic physical properties revealed, Chapter 15 turns to the birth of stars from the matter between stars. Once born, stars evolve through a life cycle, described in Chapter 16. We can basically understand this cycle, even though we have not seen any one star proceed from birth through evolution. Finally, stars die, often in violent ways, and leave behind bizarre remains to mark their deaths. Chapter 17 tells the stories of their ends — and how these may connect to new beginnings. ∎

PART THREE OBJECTIVE: To be able to specify how stars of different types evolve, and how their evolution involves and changes the interstellar medium.

13-1 Outline a contemporary method to find the precise distance to the sun.

13-2 Describe current methods for finding out the sun's mass, size, density, luminosity, and surface temperature.

13-3 Define a *blackbody,* describe the characteristics of blackbody radiation, and apply these to the sun or to describe astronomical situations.

13-4 Describe the physical meaning of the term *opacity* and tell how this property affects the flow of radiation through the sun.

13-5 Describe the appearance of the sun's spectrum and outline the atomic processes that produce this spectrum.

13-6 List the sun's two most abundant elements and describe how these and others have been found.

13-7 Briefly, in a sentence or two, explain the source of the sun's energy.

13-8 State the specific thermonuclear reactions that are supposed to produce the sun's energy and describe the conditions needed for them to take place.

13-9 Discuss the results and possible consequences of the solar neutrino experiments.

13-10 Trace the flow of energy from the sun's core to the earth and describe how the features of the quiet sun (photo-

sphere, chromosphere, corona, and solar wind) result from this energy flow.

13-11 Select one feature of the active sun, describe its observed characteristics, and explain the main features in terms of energy flow and magnetic fields.

13-12 Describe the model for an ideal gas and apply this model to the sun, to other stars, and to other appropriate astronomical situations.

Central Question
How does the sun produce its life-giving energy, and how do its energy production and flow affect its physical characteristics?

13

Our Sun: Local Star

Now this day,
My sun father,
Now that you have come out standing to your sacred place,
That from which we draw the water of life . . .

ZUNI PRAYER AT SUNRISE, TRANSLATED BY RUTH BUNZEL

What is the sun? Basically, a hot, huge globe of gas with fiery fusion reactions in its core. There, deep in the heart of sun, the energy bound up in matter is unleashed. Slowly, over about two hundred thousand years, that energy (as light) flows to the sun's surface. Free of the sun's material, it flies out to space. And 8.3 minutes later, a very small part of that light strikes the earth, and miracles occur.

The sun serves as our local link to the stars. Our sun has the same basic structure as the other stars in the sky, which we can see directly only as pinpoints of light. How to find out the physical characteristics of these distant lights? By using our sun as a guide. The sun serves astronomers as the local laboratory for testing ideas about stars in general; our understanding of stars hinges on what we understand about our sun. That analogy marks the essence of the solar–stellar connection.

13.1 A Solar Physical Checkup

To infer the physical properties of nearby stars, we learn about the nature of the sun (Fig. 13.1). The crucial properties are mass, energy output, radius, surface temperature, and chemical composition. To tackle some of these, we need an essential fact: the distance of the earth from the sun. (You will find throughout this book that a critical question—perhaps *the* critical question in all astronomy—is how to find the distances to celestial objects.) An analysis of the sun's light provides other facts without requiring knowledge of the distance. Let's see what we know more or less directly about the sun.

How Far?

The earth's orbit about the sun is elliptical (Section 3.5 and Focus 3.2). The semimajor axis of its orbit, which is the average earth–sun distance, has a length called an *astronomical unit (AU)*. So the earth is 1 AU from the sun. How large is the AU in basic physical units, such as kilometers?

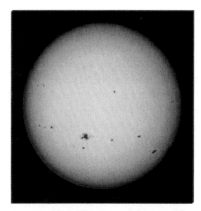

FIGURE 13.1 The face of the sun. The sun's photosphere in visible light. Note that the photosphere has a granular look and fades out at the edge. The dark regions are sunspots. (Courtesy of the Observatories of the Carnegie Institution of Washington.)

275

Recall that simple angular measurements establish the scale of the solar system in terms of the AU (Section 3.2) and that Kepler's laws (Section 3.5) completely describe the orbital motions of the planets, with the AU as the basic unit. So at any time you can draw up a scale map of the solar system with the planets' positions all neatly laid out in AU.

To work out the distance scale in kilometers requires only one piece, already known in AU, be measured in kilometers. An analogy: suppose you're given a map of your region of the country with no distance scale in kilometers but with all locations laid out correctly relative to one another. Hop in your car and drive between two points on the map while keeping close track of the distance in kilometers between these two points, which are separated on the map by so many centimeters. You've found the distance scale for the map.

To do the same for the solar system, we use the known speed of light. Radar signals, which travel at light speed, are bounced off Venus, and the time between transmission and reception is accurately measured and converted into kilometers. (This is analogous to driving the car between two locations on a map.) Kepler's laws give the earth–Venus distance in AU at the time of the radar measurement. (This is analogous to having a distance scale on a map in centimeters.) The radar measurement is usually made when the earth and Venus are close together on the same side of the sun. The distance at that time is some fraction of an AU (about 0.3 AU). So we know a fraction of an AU in kilometers, hence one AU in kilometers: 1.496×10^8 km. [Remember it by rounding off to 150 million (150×10^6) km.]

Why bother bouncing radar off a planet? Why not bounce it off the sun and measure the AU directly? Mainly because the sun, which is a hot gas and thus lacks a solid surface, does not reflect radar signals very well. Also, the sun generates its own radio noise, which makes it hard to detect a returning signal.

How Big?

Viewed from the earth, the average angular diameter of the sun is 32 arcmin.

The Sun

At a Glance

Distance from the earth	1 AU = 1.496×10^8 km
Radius	6.966×10^8 m
Mass	1.991×10^{30} kg
Luminosity	3.86×10^{26} W
Flux at earth	1370 W/m^2
Surface temperature (photosphere)	5780 K
Average density	1410 kg/m^3
Age	4.5 to 5 billion years
Composition (surface) by mass	Hydrogen, 74%; helium, 25%; other elements, 1%

(That's about 0.5°, or half the tip of your finger at arm's length.) Because we know the AU in kilometers, we get the sun's actual size from its angular size (which provides the size-to-distance ratio). For 0.5°, that's about 1/110. So the sun's diameter is roughly $1/110 \times (150 \times 10^6) \approx 1.4$ million kilometers, or 109 times the earth's diameter (Fig. 13.2). Imagine the earth the size of a dime. The sun would be about 2 meters in size and 200 meters from the coin-sized earth.

How Massive?

To find the sun's mass, you need the earth–sun distance, the earth's average orbital velocity, Newton's second law of motion, and Newton's law of gravitation

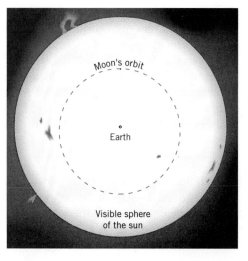

FIGURE 13.2 The relative sizes of the sun (its photosphere) and the moon's orbit drawn to scale.

(Section 4.4). Or you can use Newton's form of Kepler's third law (Focus 4.3), knowing the earth's distance and orbital period. The sun's mass comes out to be about 2×10^{30} kg, or more than 300,000 times the earth's mass.

How Dense?

The sun is a big, massive object. Yet its density (the mass divided by the volume) is low: 1400 kg/m³, only 40 percent denser than water! This low average density, along with its hotness, implies that the sun is a gas. In fact, you will see that the sun is a gas throughout, even in its dense core.

Caution. Don't confuse *density* with *mass* or *size*! For example, the sun is much more massive than the earth, and larger, yet it has a much lower average density. Why? The sun is a gas, while the earth is solid rock and metals.

■ Why can't radar signals be bounced off the sun to measure the AU distance in kilometers?

13.2 Ordinary Gases

Because stars are gaseous, to know what happens inside the sun (and other stars), you must have some idea of how gases behave. Let me describe a simple model for ordinary gases that you can apply in most situations. (You will come across extraordinary gases in Chapter 17.)

Temperature

A gas consists of particles—atoms, molecules, or ions, or perhaps all three. Simplify the situation in a real gas by imagining all the particles to be the same, and like hard spheres. Picture these spheres trapped in a small box. Once set into motion, the spheres keep moving, bounding off the walls of the box and colliding with one another (Fig. 13.3).

These collisions push any one sphere at times faster or slower, but over time the sphere has a definite average speed. Also, there will be collisions over time, with the result that all the spheres in the box will have the same average speed. We use this average speed of par-

ticles in a gas as a measure of average kinetic energy of its particles—a measure of its *temperature*. If all the particles were motionless, a gas's temperature would be zero—absolute zero on the kelvin temperature scale (Appendix A).

As already mentioned, this book uses the *kelvin,* or absolute, temperature scale. One kelvin degree is the same amount of a temperature difference as one degree in the centigrade or Celsius scale. The zero points, however, are different. The kelvin scale uses absolute zero; the Celsius scale uses the freezing point of water. On the kelvin temperature scale, the freezing point of water is 273.15 K. At room temperature, about 300 K, the average speed in a gas of hydrogen particles is about 3 km/s, or about 11,000 km/h! Higher temperatures mean higher average speeds of the particles.

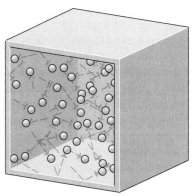

FIGURE 13.3 Hard-sphere model of a gas in a box. The temperature of the gas is a measure of the average kinetic energy of the particles. The internal pressure is created by many collisions among the particles.

Pressure

Consider putting a partition in the gas container. The spheres ram and bounce off both sides of this partition; each collision exerts a small force on it. The combined force of all collisions is the *pressure* of the gas. This book uses the pressure of the atmosphere at sea level as the basic unit of pressure; it's called *one atmosphere* (atm). The SI unit for pressure is the *pascal,* abbreviated *Pa* (see Appendix A). One atmosphere amounts to about 100 kPa. For comparison, the pressure of the tires on my truck is about 200 kPa. The compression pressure in the gasoline engine is roughly 1000 kPa, about 10 atm.

Imagine that you increase the temperature of the gas. The average speed of the particles increases, so they collide with one another and strike the partition more frequently and with greater force. The pressure increases. How, in relation to the increase in temperature? For ordinary gases, the increase is a direct one. So if you double the temperature (in kelvins), the pressure doubles.

What happens to the pressure if you increase the number of particles in the box? (You've done this if you've pumped up a bicycle tube.) Adding more particles to the same space increases the density of particles—on the average, each cubic meter contains more particles. For a gas, the number of particles in a unit volume is called the *number density*.

Suppose you increase the number density four times without changing the temperature. Each cubic meter now contains four times as many particles. So four times as many collisions occur, on the average, against the partition, and each collision has the same average force as before. The pressure increases; it is four times greater. Increasing the number density directly increases the pressure of an ordinary gas.

This hard-sphere model shows that gas pressure depends directly on *both* the number density and the temperature. With these basics about simple gases in mind, let's turn to that huge hot ball of gas that is the sun and infer some of its important properties.

■ **If the temperature of an ordinary gas increases, what happens to its pressure?**

13.3 The Sun's Continuous Spectrum

Most of what we know about the sun and other stars comes by way of light. Decoding the message of sunlight and starlight takes up much of the time, energy, and ingenuity of astronomers. Let's look at how the message is read.

Luminosity

The sun's **luminosity** (or *power*) is its total output in radiative energy each sec-

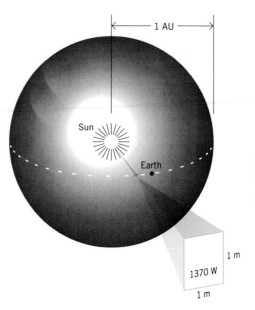

FIGURE 13.4 Measuring the sun's luminosity from the earth from the sun's flux at the earth and the earth–sun distance. An imaginary sphere with a radius of 1 AU would capture all the sun's radiative energy output. At the earth's distance, then, one square meter captures only a small part of this total energy output, or about 1370 watts.

ond. By the time the light reaches the earth, it has spread out over a large region of space—in fact, over a sphere whose radius is the earth–sun distance. So you cannot *directly* measure this radiative energy, the sun's luminosity, all at once.

How to find the sun's luminosity? To avoid having the earth's atmosphere absorb some of the sunlight, place a special detector in a satellite orbiting the earth. Point the detector directly at the sun. Measure the radiant energy absorbed each second by the detector: it amounts to 1370 watts for each square meter of the detector's area (Fig. 13.4). So on a surface 1 AU in radius (an imaginary sphere surrounding the sun), every square meter catches this amount of power, which could run an electric hair dryer if converted to electricity.

Enrichment Focus 13.1

THE SUN'S LUMINOSITY

The sun's *luminosity,* or *power,* is its total output of radiative energy each second. How can we determine it? By measuring the solar *flux* at the earth. The flux, f, equals about 1370 W/m². This amount of energy crosses *each* square meter on an imaginary sphere of radius d surrounding the sun, where d is the earth–sun distance in meters.

We must know this distance to calculate the luminosity! Since the area of this sphere is $4\pi d^2$, the luminosity L of the sun is given by the relation

$$L = 4\pi d^2 f$$

Putting in the numbers, we write

$$L = 4\pi (1.50 \times 10^{11} \text{ m})^2$$
$$(1.370 \times 10^3 \text{ W/m}^2)$$
$$= 3.9 \times 10^{26} \text{ W}$$

The earth intercepts only about 4.5×10^{-10} of this total, but that's still a large amount, almost 2×10^{17} W!

Totaled up over the entire surface, the energy per second amounts to roughly 3.8×10^{26} W. (See Focus 13.1.) That's enough power to light 3.8×10^{24} 100-watt light bulbs all at once! The earth intercepts only a small part of this power, only about 10^{-10} of the total, but that still amounts to a lot. In one year, the entire earth's surface catches about 10^{18} kilowatt-hours (kWh).

Astronomers use a special name for the amount of energy passing though one given unit of area (such as a square meter) each second: **flux.** The sun's flux at the earth, 1370 W/m^2, serves as the standard of comparison for the flux from other celestial objects.

Surface Temperature and Blackbody Radiators

The sun's color (yellowish white) provides a clue to its **surface temperature.** We can assign a temperature to the sun's surface by examining its continuous spectrum, ignoring the lines for the moment. (Recall, Section 5.1, that the sun has an *absorption-line spectrum.*)

To see how, consider heating a piece of metal in a very hot flame. At first the metal gives off infrared radiation before it emits any visible light. Eventually, like the coils of an electric stove at its highest setting, it emits a dull red light. Then, as it gets hotter, it glows more brightly reddish, then orange, yellowish, yellowish-green, white, and, finally, bluish white. The overall color change in the visible region of the spectrum corresponds to the temperature of the metal; a metal's continuous spectrum changes with temperature in a special way.

How are color and continuous spectra related? Let's look at the metal's emission in some detail (Fig. 13.5). Measure the brightness at a range of wavelengths (from ultraviolet to infrared) for the metal at different temperatures. Then plot the metal's spectrum of emitted flux versus wavelength. Note three features in these spectra: the emission *peaks* at some wavelength, the peak shifts to *shorter* wavelengths as the metal gets hotter, and the metal emits *more* intensely at *all* wavelengths at *higher* temperatures.

Your eye, because it responds only

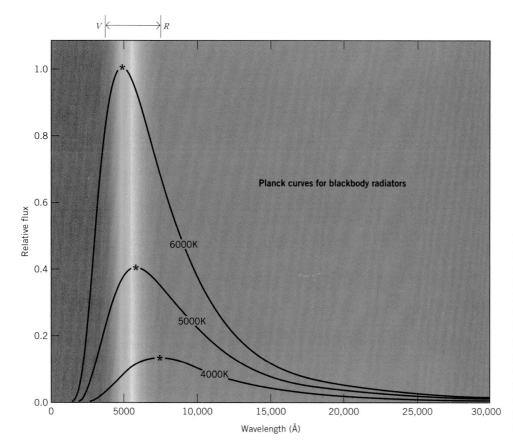

FIGURE 13.5 The spectrum of a hot metal (or opaque solid, liquid, or gas) as it varies with temperature (3000 to 5000 K). The visible region (from violet, V, to red, R) is indicated; the long-wavelength end of the curve goes out into the infrared. The asterisks mark the peak of the emission for each temperature; note that peaks occur at shorter wavelengths for higher temperatures. The shapes of these curves, called *Planck curves,* are typical of the continuous emission from blackbody radiators.

to the visible region of the spectrum (violet to red), sees only a small portion of the complete spectrum. (For example, your eye doesn't sense any infrared. When an electric stove is on low, you don't see the coils glow, yet your hand can detect the infrared emitted by them.) For the temperatures considered here, however, your eye does reasonably discern the shift in the balance of colors at the peak emission.

An object whose spectrum has this characteristic shape (Fig. 13.5) and the variation with temperature just described is called a **blackbody radiator,** or simply a **blackbody** (Focus 13.2). Its spectrum is sometimes called a **Planck curve,** in honor of Max Planck, a pioneer in quantum physics. The spectrum of a blackbody depends *only* on its temperature and not on other properties (such as composition).

Perfect blackbodies do not actually exist, but the sun and similar stars emit radiation enough like an ideal blackbody to make the model worth using. (This is especially true for stars with surface temperatures about the same as the sun's.) The wavelength at which a star's contin-

uous spectrum peaks is related directly to its temperature: the *higher* the temperature, the *shorter* the peak wavelength.

A common question about blackbodies: Why are they called *black* when they give off *light*? Blackbodies are so-named for their *light-absorbing* abilities. They absorb light at any wavelength completely and reflect none. When a blackbody absorbs radiative energy, it heats up and emits at all wavelengths, even though the peak of emission may not be visible to our eyes. Physically, a good absorber, when heated, is a good radiator. And to conserve energy, it must emit as much energy per second as is put in. When that balance is achieved, the blackbody will have a certain temperature. Note that any opaque and hot solid, liquid, or gas will produce a blackbody spectrum, as you would expect from Kirchhoff's rules (Section 5.2). The key word here is *opaque*.

Let's examine the sun's continuous spectrum (Fig. 13.6) as measured at the earth's surface. Note that when the spectrum is measured at sea level, the earth's atmosphere absorbs parts of it—especially the ultraviolet (absorbed by ozone)

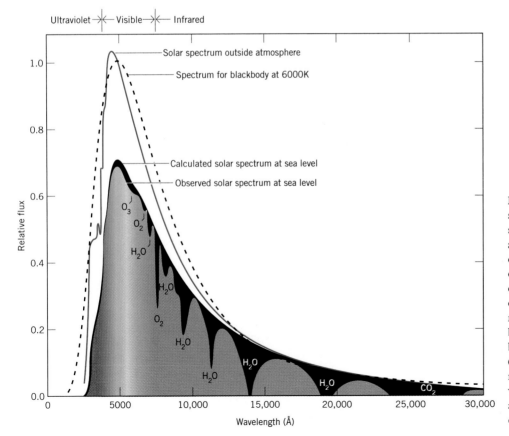

FIGURE 13.6 The absorption of the sun's radiation by the earth's atmosphere. The sun's radiation measured above the earth's atmosphere is indicated by the solid line. (Note how close this curve comes to a blackbody curve at a temperature of 6000 K, indicated by the dashed line.) The actual measured spectrum at sea level is labeled according to whether the carbon dioxide (CO_2), water vapor (H_2O), oxygen (O_2), or ozone (O_3) is responsible for the absorption. Water vapor is responsible for most of the absorption in the infrared—the major cause of the greenhouse effect.

Enrichment Focus 13.2

RADIATION FROM BLACKBODIES

A blackbody is an object that absorbs completely any radiation that falls on it. The radiation from a blackbody has a characteristic shape (spectrum) that depends only on the temperature of the body and not on any other property, such as composition.

A blackbody has a number of special characteristics. First, a blackbody with a temperature greater than absolute zero emits *some* light at *all* wavelengths. Second, a hotter blackbody emits *more* light at *all* wavelengths than does a cooler one. Third, a hotter blackbody emits a *greater proportion* of its radiation at *shorter* wavelengths than a cooler one. Finally, the amount of radiation emitted by the surface of a blackbody depends on the *fourth power* of its temperature (in kelvins). So if one blackbody is twice as hot as another of the same size, it emits $2^4 = 2 \times 2 \times 2 \times 2$, or 16 times as much power in total. The hotter of them also emits 16 times as much power as the cooler one from each square meter of surface.

Two of the foregoing properties of blackbodies are described by simple equations. First, the amount of power emitted for every square meter of a blackbody at temperature T kelvins is

$$F = 5.67 \times 10^{-8} \, T^4 \, \text{W/m}^2$$

This relation is called the **Stefan–Boltzmann law.** For example, every square meter of the sun's photosphere (5800 K) emits about 6.3×10^7 W. Note that F is the flux emitted at the *surface* of the blackbody. Explicitly,

$$F = 5.67 \times 10^{-8}(5780)^4$$
$$= 5.67 \times 10^{-8}(1.12 \times 10^{15})$$
$$= 6.33 \times 10^7 \, \text{W/m}^2$$

Second, the wavelength at which the energy output from a blackbody peaks is

$$\lambda_{\text{max}} = \frac{2.90 \times 10^{-3}}{T}$$

where T is the temperature in kelvins and λ_{max} is the wavelength, in meters, at which the peak output occurs. This relation is called **Wien's law.** For the sun, the peak for a temperature of 5800 K is about 4.9×10^{-7} m (0.49 μm, or 4900 Å).

Note that both the Stefan–Boltzmann and the Wien equations can be used to infer the sun's surface temperature. To use the first, you need to find out how much power each square meter of the surface puts out. For the second, you have to determine the wavelength at which the sun's spectrum peaks. These methods of determining temperature can be applied to other stars, if they also radiate roughly like blackbodies.

Note that if a star is spherical and assumed to radiate like a blackbody, we can find its luminosity from the radius and the surface temperature. The luminosity is the total surface area ($4\pi R^2$ for a sphere) times the flux, or

$$L = 4\pi R^2 F$$
$$= 4\pi R^2 (5.67 \times 10^{-8} T^4)$$

---

and the infrared (absorbed by water vapor and carbon dioxide). This absorption in the infrared causes the greenhouse effect (Section 8.4) and also forces astronomers to place infrared telescopes above the earth's atmosphere (Section 6.3). But very little of the visible light is absorbed. You see that the spectrum peaks at about 5000 Å. Without the atmospheric absorption (Fig. 13.6), the sun's spectrum follows a more or less continuous curve with this one peak in the yellow-green part of the spectrum. To produce this peak requires a surface temperature of about 5800 K.

Opacity

If the sun's continuous spectrum has a blackbody shape, the emitting region must also be a good absorber of light.

Recall (Section 5.4) that atoms can absorb light by electron transitions to higher energy levels. What specific transitions make the sun a good absorber? To see this point, you first need to understand the concept of **opacity.** On a clear day, you can see through the air to far distances (many kilometers). The air is *transparent* to light; its opacity is low. On a very foggy day, you can't see far at all (only a few meters in San Francisco!). The air is *opaque* to light.

Interactions of light with atoms and electrons make a gas opaque. When a photon is absorbed, it no longer exists and so can't carry energy any farther. The photon's energy is *not* destroyed; it is transferred to an electron. When the electron loses the energy, it emits a photon. But—and this is the key point—that photon can be emitted in *any* direction, including back in the direction from which it originated. It heads off and moves only a short distance before it is reabsorbed. When another photon is reemitted, it probably zips off in a different direction. So photons in an opaque gas travel very short distances before they are absorbed. In a gas of lower opacity, they travel greater distances. When a gas is opaque, photons bounce around, hence travel slowly through it; when it is transparent, the photons fly straight through it.

If we define the visible surface of the sun as the layer at which the gases become visibly transparent, that layer is only a few hundred kilometers thick and has an average temperature of 5800 K. This region essentially defines the sun's **photosphere**—the place at which photons can escape from rather than bounce through the sun's atmosphere. Photons bouncing out from the center reach this region and can then fly straight into space.

■ How can we assign a temperature to sun's surface?

13.4 The Solar Absorption-Line Spectrum

Recall (Section 5.2) that a spectroscope shows the solar spectrum to consist of a continuous spectrum crossed with dark lines. How is this absorption-line spectrum produced in the sun's atmosphere?

The Formation of Absorption Lines

Use the concept of opacity again—for an absorption line. Consider the transition of an electron that produces the 4383-Å line of iron (Fig. 13.7). When we measure the absorption or emission of this line, we find that it is *not* perfectly sharp; that is, it is not all centered in an infinitely narrow band. It has a finite width in terms of wavelength, centered on 4383 Å, where the iron atom absorbs very well. At somewhat shorter or longer wavelengths, say 4379 Å or 4387 Å, the iron atom can still absorb light, but not as well. In other words, a gas containing iron has a much higher opacity for 4383-

Å photons (at the line's center) than for those with wavelengths shorter or longer than the central wavelength.

Caution. Absorption lines are *not* perfectly black and they do contain *some* energy. They are dark only relative to continuous emission at neighboring wavelengths.

In the photosphere, the temperature rises quickly as you go down into it: a few thousand kelvins in a few hundred kilometers. This sharp temperature change results in the sun's dark-line spectrum. In effect, you can consider the atmosphere as overlying a hot surface, just the situation needed, according to Kirchhoff's third rule (Section 5.2), to produce an absorption-line spectrum (Fig. 13.8).

The Chemical Composition of the Photosphere

Astronomers have analyzed tens of thousands of absorption lines in the solar spectrum to find the chemical composition of the atmosphere. The intensity of the dark lines from a particular chemical element is related to the amount of that element found in the atmosphere. The line's intensity also depends on the temperature and density in the photosphere. Iron produces most of the absorption lines; other strong lines come from hydrogen, calcium, and sodium (Section 5.4).

Identifying particular elements requires matching their "fingerprint" patterns. But it's a harder task to find out an element's actual abundance. To determine the abundance, you need to know

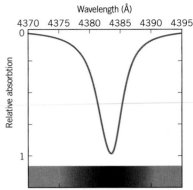

FIGURE 13.7 The absorption profile of the 4383-Å line of iron in the photosphere of the sun. This profile results from the ability of the iron atom to absorb photons at wavelengths near 4383 Å: that is, the relative opacity. The opacity is the greatest at the center of the line. Note how rapidly it decreases away from the center of the line.

FIGURE 13.8 The sun's dark-line (absorption) spectrum. Note the many absorption lines with different intensities (darknesses), each appearing at a specific color. This spectrum covers the complete range from the red to the violet. Note that it is folded around. Start at the upper right (which looks black) and move across the page horizontally to the left. Then drop down to the next band and move horizontally to the right. Continue this pattern to follow the wavelengths from longest to shortest. (Courtesy of NSO, Sacramento Peak.)

how atoms of a particular element absorb and emit light. Then you must make a model of the sun's photosphere. You can construct this model from theoretical calculations and basic physical concepts; it consists of a list of temperatures, densities, and pressures at different depths in the photosphere. Then you add some amount of the element, and you can calculate how intense its absorption lines must be. You try to match the calculated intensity to the observed intensity by playing with different values of the abundance. A match between observed and calculated intensities gives the correct abundance.

Most of the mass is hydrogen (74 percent) and helium (25 percent). All other elements (loosely called *metals* or *heavy elements* by astronomers) make up a mere 1 percent. These abundances are relative to the total *mass* in the photosphere.

Caution. This procedure gives only the abundance in the *photosphere.* It does not tell you the abundance in the *interior.*

If an element's absorption lines don't show up in the visible spectrum, does it mean that this element does not exist in the sun? Not necessarily. Perhaps so little of the element exists that it does not produce detectable absorption lines. Or the strongest absorption lines may be in a region of the sun's spectrum unobservable through the earth's atmosphere. Or the temperature, pressure, and density of the sun's atmosphere may inhibit the formation of the element's spectral lines. If the conditions are too hot or too cool, the lines do not form.

Consider helium, for example. If you examine carefully the sun's visible spectrum, you won't find any absorption lines of helium. Why not? In the photosphere, the temperature is too low to excite helium to levels that can absorb photons of visible wavelength. Above the photosphere, the temperature rises to 40,000 K, which is high enough to excite helium atoms, and during an eclipse you can find bright lines from helium. In contrast to the continuous spectrum, however, too few excited helium atoms exist in this hot region to produce emission lines strong enough to be seen directly.

■ What happens to the temperature of the photosphere as you move down through its layers?

13.5 Energy Flow in the Sun

The thermal energy flow, from the hotter core to the cooler surface and beyond the atmosphere, controls the internal structure and environment of the sun. The sun that we see directly consists of the sun's outer layers, together known as the atmosphere. Because the sun's atmosphere is a gas, it does not have sharp boundaries. But we have been able to discover that the atmosphere has three different zones: the *photosphere,* the *chromosphere,* and the *corona.* These regions are the visible expression of the sun's energy outflow. Before I deal with each one, let's look at how energy can be transported from place to place.

Conduction, Convection, Radiation

In general, energy is carried by **conduction, convection,** or **radiation.** If you've ever relaxed in front of a roaring fire, you've experienced all these processes. You are directly warmed by the fire's heat, infrared radiation that travels directly from the fire to you to be absorbed by your skin. That's energy transported by *radiation.*

Much of the energy from the fire, however, is wasted up the chimney. The

The Sun's Structure

At a Glance

Core	Site of fusion reactions; temperature 8×10^6 to 16×10^6 K; energy transport by radiation
Convection zone	Energy transport by convection; temperature below 500,000 K; outer 0.3 of the sun's radius, below photosphere
Photosphere	Origin of continuous and absorption-line spectra; temperature from 6400 to 4200 K; energy transport by radiation
Chromosphere	Energy transport by radiation and magnetic fields; temperature from 4200 to 10^6 K
Corona	Energy transport by radiation and magnetic fields; temperature from 1 to 2×10^6 K; source of the solar wind

fire heats the air just around it. Air is gas, which expands and becomes less dense when heated. Cooler, denser air flows in and pushes the hotter air up and out of the house, where it cools off. This transport of energy by mass motions of a gas (or liquid) is *convection.*

Finally, you may have left the poker in the fire. If you grab the handle without thinking, you will find it very hot. This is a result of *conduction:* the poker's electrons that were actually in the fire were heated and so moved around at high speeds. They banged into their neighbors, agitating them. These particles collided with their neighbors, and so on throughout the poker, from one end to the other. The kinetic energy was transferred by direct collisions along the poker.

In ordinary stars like the sun, radiation or convection can transport energy. (Conduction does not play an important role because the sun's material is a gas and not very dense. However, conduction does become important in extremely dense stars, where matter is in a state different from the ordinary: Chapter 17.)

Which process is at work? That depends on the local conditions in the gas. The general rule is this: the transport process that works most efficiently is the one that operates. This efficiency depends on how well radiation transports energy through a material. And that rests on *opacity.* Basically, opacity measures how efficiently some material absorbs photons.

In the sun's interior, photons travel about a millimeter between absorptions. The gas here is fairly opaque, but radiation is still more efficient than convection for transporting the energy. As photons journey toward the surface, they run into a region having a temperature low enough to make the gas extremely opaque. The photons are bouncing around so much that they make little progress outward, and so do not transport energy very well. Convection by circulation of the gas can transport energy more efficiently, and a convection zone forms as a series of convective cells. The cells make up the final 0.3 of the sun's radius.

Photosphere

The sun's photosphere has a bubbly look, like the surface of a pot of boiling oatmeal (Fig. 13.9). Each bubble has an irregular shape about 2000 km across (half the size of the United States!) and lasts for about 10 minutes. This phenomenon is called *photospheric granulation.* You can visualize the photospheric granulation as the top layer of a seething zone where hot blobs of gas spurt to the surface, radiate energy, cool, and flow downward.

Just below the boiling photosphere lies a region of convection. Here the outward flow of energy heats the gas and makes it rise. The situation resembles that of the formation of cumulus clouds in the earth's atmosphere. Here on earth hot gases rise, carrying moisture. As they rise, they cool, and eventually the water condenses to form clouds. The photosphere marks the top of the sun's convective zone. The base of the convection zone is at about 0.7 solar radius.

Chromosphere

Just before and after totality in a solar eclipse, a bright, pink flash appears above the edge of the photosphere. This is the **chromosphere,** the solar atmosphere just above the photosphere. The pink color comes from the emission of Hα, the first line of the Balmer series of hydrogen (Section 5.3), in the red region of the spectrum. A spectroscope can be tuned to the Hα line. An image taken of the sun through such a device shows the emission from the chromosphere at any time, not just during eclipses (Fig. 13.10). The temperature, density, and pressure in the chromosphere determine the intensities of its emission lines. These line intensities provide clues to the physical conditions there.

The chromosphere is an active place. It is constantly pierced by jets of gas, called *spicules,* that spurt up to heights of 10,000 km and die out in a few minutes (Fig. 13.11). Because of the fountains of spicules, the chromosphere is not a uniform layer, but a jagged one.

The chromosphere begins a few hundred kilometers above the top of the

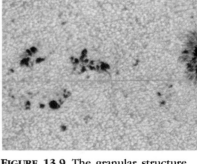

FIGURE 13.9 The granular structure of the photosphere. Individual granules, each about 1500 km across, mark the tops of hot, rising bubbles of gas. The darker regions between the granules are the places of downflow of the cooler, denser gas. The darkest objects in this image are sunspots, visible in June 1989. Note how they appear as breaks in the granular pattern. (Courtesy of NOAO/Kitt Peak.)

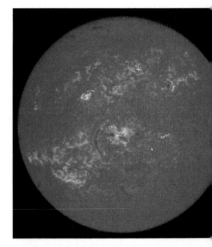

FIGURE 13.10 The chromosphere, visible in Hα light. The bright regions mark areas of magnetic activity and are associated with the sunspots visible in the photosphere. The dark filaments are clouds of hydrogen gas supported by magnetic fields in the chromosphere. (Courtesy of NOAO/Kitt Peak.)

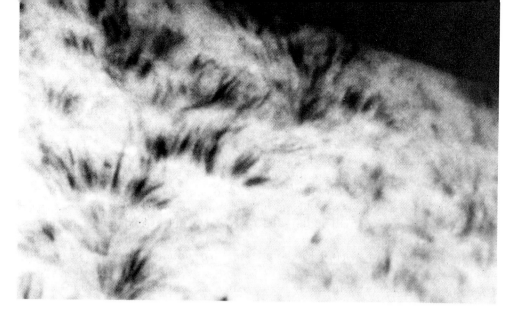

photosphere and extends only about 2000 km higher, where it merges into the corona. The chromosphere has a density about a thousand times less than the photosphere but, surprisingly, it gets much hotter. The cleanest way to distinguish these different regions of the atmosphere is by their temperatures (Fig. 13.12). The temperature rises from 4300 K to above 400,000 K in the first 2500 km of the chromosphere above the photosphere. This rise to high temperatures produces the emission lines from this region.

Why don't we see emission lines from the chromosphere when we look through it down to the photosphere? The answer is that the chromosphere has such a low density that it is *transparent* to the light passing through it. The photospheric spectrum (continuous and absorption line) makes it through the chromosphere, which adds only a little emission.

Why is the chromosphere generally hotter than the photosphere? If photons heated it, you'd expect that it would be cooler than the photosphere—certainly not hotter. Some other energy source must do the heating. The secret probably lies with energy bound up in the magnetic fields that project up from the photosphere.

You know that magnetic fields store energy if you've ever pulled two magnets apart: you have to put out an effort—work—to separate them. The trick, then, is to learn how the magnetic

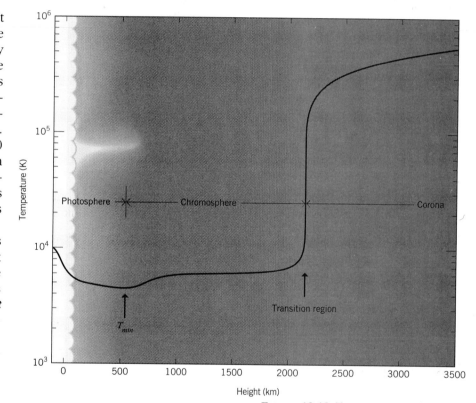

FIGURE 13.12 Temperature profile of the sun's atmosphere. The height is taken as the distance from the base of the photosphere. Note how the temperature *decreases* in the photosphere and drops to a minimum value. It then rises fairly gradually in the chromosphere. A sharp rise marks the transition between the chromosphere and the corona. (Adapted from a diagram by Eugene H. Avrett.)

energy moves upward from the photosphere. Magnetic fields can transport energy by moving material or by generating electric currents that move through conducting materials. Currents flowing upward from the photosphere to the corona can leave some energy in the chromosphere (more in Section 13.7).

Corona

You can see the sun's splendid **corona** directly during a total solar eclipse (Fig.

13.13). Although as bright as the full moon, the corona is normally obscured by the photospheric light scattered in the earth's atmosphere. During a total eclipse, when the photosphere is blocked out by the moon, the sky becomes dark enough for the corona to be visible.

Spectroscopes reveal that the corona emits bright lines, whioch were a mystery for many years because their patterns did not match any known elements. We now know that highly ionized atoms of iron, nickel, neon, and calcium, rather than a strange element, produce the emission lines. (Recall that when an atom loses an electron, its energy levels change.) Because it takes large amounts of energy to rip many electrons from an atom, the corona must be very hot. For example, to strip iron of 16 of its normal 26 electrons requires a temperature greater than 2 million kelvins, and temperatures this high occur in parts of the corona.

What makes the corona so hot? Space telescopes have shown that most of the corona's emission lines, from highly ionized elements, show up in the ultraviolet. These lines indicate that at the base of the corona, the temperature rises rapidly, roughly 500,000 K in a few hundred kilometers, in a thin transition region between the chromosphere and corona (Fig. 13.12). This sharp rise may result from the same physical process as in the chromosphere: transport of energy by magnetic fields (Section 13.7).

Solar Wind

The solar wind permeates the solar system. Its source is the sun's corona, which extends into space to distances greater than those indicated by its visual appearance. The corona moves; it flows out from the sun into space to create the **solar wind.** The outflow of material occurs because the high temperature of the coronal gas causes it to exert a large outward pressure—a greater force than the inward pull of gravity. As a result, the gas from the corona streams away from the sun with greater than escape speed. The gas stream becomes the solar wind, so named because of its high speed. The wind is a plasma flow, carrying trapped magnetic fields along with it.

At the earth's orbit, the solar wind typically whips by at 500 km/s. The speed varies as the solar wind blows in gusts. The particles in the solar wind (mostly protons and electrons) travel the distance from the sun to the earth in about 5 days. (Remember light takes 8.3 minutes.) The earth swims through the solar spray and catches some of the solar wind particles in its magnetic field to create its magnetosphere (Section 8.3).

The plasma of the solar wind carries along a remnant of the sun's magnetic field. This transported field makes up the magnetic field between the planets. As the sun rotates, it twists the interplanetary magnetic field into a spiral (Fig. 13.14). This field interacts with

(a)

(b)

FIGURE 13.13 The sun's corona. (*a*) Wide-angle view during the March 1988 solar eclipse. A special camera was used to bring out the details in the corona both close to and far from the sun (out to several hundred thousand kilometers). Note the streaming structures (arrow) and asymmetry in the outer regions. (Courtesy of NCAR.) (*b*) Close-up view of the base of the corona during the March 1991 eclipse. (Courtesy of NSO, Sacramento Peak.)

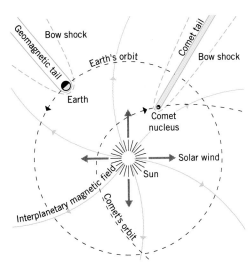

FIGURE 13.14 The spiral shape of the sun's magnetic field as it is carried by the solar wind into interplanetary space. The outflow of the solar wind carries a magnetic field from the sun. As the sun rotates, the magnetic field twists into a spiral that makes up the interplanetary magnetic field. This field and the plasma of the solar wind interact with planetary magnetic fields (such as the earth's) and those of comets.

planets (to make their magnetospheres) and with comets (to make their tails). The wind travels as far as 100 AU, well beyond the orbit of Pluto.

■ **Why doesn't conduction play a major role in the energy transport within the sun?**

13.6 The Solar Interior

The interior of the sun contains the bulk of the sun's mass and the furnace that generates its energy. All the features of the sun, from photosphere to solar wind, result from the energy flow out from the interior. How does the sun generate its energy?

Energy Sources

During the nineteenth century, when the earth was found to be very old, attempts to explain the source of the sun's power became embarrassingly difficult. The crux of the problem is not the rate of energy production but its longevity. The sun has been at roughly the same luminosity for at least 3.5 billion years, according to present geologic evidence. Ordinary chemical reactions, such as burning, could not provide the amount of energy necessary for so long a history. If the sun were composed entirely of oxygen and coal, it would have burned to a dark cinder in a few thousand years. The sun could not be a coal-burning furnace; it wouldn't last long enough.

In the middle of the nineteenth century, both Hermann von Helmholtz (1821–1894) and William Thomson, Lord Kelvin (1824–1907), proposed that the sun shone because it was releasing gravitational energy by shrinking. That is, gravitational contraction converted gravitational potential energy to radiative energy (Section 12.5); half goes into heat energy and the other half ends up as light.

Because of the sun's substantial mass, a contraction rate of only 40 meters a year would liberate the required energy. Gravitational energy stored in the sun would last for about 50 million years, far longer than the earth's age as set in the nineteenth century. But when later geologic investigations expanded the earth's age to billions of years, the sun's resources of gravitational energy no longer sufficed to account for the sun's shining. What does?

At the beginning of this century, Albert Einstein provided the key idea explaining the sun's energy. In the special theory of relativity, mass and energy are related by the equation

$$E = mc^2$$

where E is the energy (in joules) released in the conversion of mass m (in kilograms), and c is the speed of light (in m/s). Because c^2 is a large number (about 9×10^{16}), a minute mass stores enormous energy. For example, the conversion of 1000 kg of matter into energy unleashes about 10^{13} kWh (9×10^{19} J), roughly equal to the total energy consumption of the United States in one year!

Nuclear Transformations

How is matter changed into energy? Two operations in nature unleash the energy frozen in the nucleus of an atom: *nuclear fission* and *nuclear fusion*. In the process of **nuclear fission,** the nucleus of an atom of a heavy element (such as uranium or plutonium) splits into two lighter nuclei. The mass of the remnants adds up to less than that of the original nucleus. The deficit in mass is released as energy. In **nuclear fusion,** the nuclei of atoms of lighter elements are fused together to create a heavier nucleus. However, the product has less mass than the original particles. The missing mass has been converted to energy.

The sun produces energy by fusion. Hydrogen is the most abundant element and also has the smallest nuclear charge (one proton). The fusion of hydrogen nuclei results in the production of helium. To make a helium nucleus from hydrogen nuclei releases 4.2×10^{-12} J of energy. That's a minuscule amount—lighting a match produces 10^3 J! Many hydrogen nuclei must react each second to supply the sun's power.

Fusion Reactions

Two sets of fusion reactions are possible to transform hydrogen to helium: the **proton–proton chain** (*PP chain* for short) and the **carbon–nitrogen–oxygen cycle** (*CNO cycle*). The CNO cycle contributes a minor amount to the energy of the sun, but acts as a key source in more massive stars (Chapter 16).

Let's look at the steps in the PP chain. A collision between two protons starts it off (Fig. 13.15); if these nuclei collide with enough energy (a temperature of at least 8×10^6 K), the protons stick together. A heavy hydrogen nucleus (^{2}H) forms, consisting of a proton and a neutron, n. The other positive charge breaks away as a positron, e^+. (A **positron** is the antiparticle of the electron and carries a positive instead of a negative charge.) Almost lost in the shuffle is a neutral, supposedly massless particle called a **neutrino,** ν, which zips away at light speed. The dense solar interior offers no barrier to the neutrino's travel. In about 2 seconds, the neutrino escapes into space, carrying away some energy.

In the solar interior, the positron quickly collides with an electron, e^-, and the two antiparticles are annihilated to form two gamma rays, γ. (When matter and antimatter collide, they are trans-

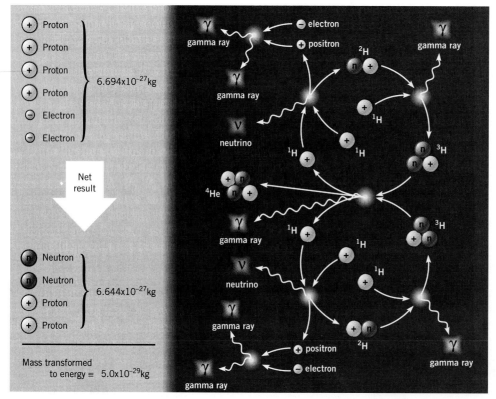

FIGURE 13.15 The primary proton–proton reaction that occurs in the core of the sun. Net result is the formation of one helium nucleus that has a mass of about 5.0×10^{-29} kg less than the unbound particles that go into the reaction. The missing mass has been released as energy in these fusion reactions.

formed from matter to light.) Meanwhile, the heavy hydrogen crashes into another proton and forms light helium (^{3}He) and a gamma ray. This sequence is repeated. Another light helium nucleus is created; it collides with the one just formed. In this final reaction of the PP chain, the usual result is ordinary helium (^{4}He) plus two protons and another gamma ray.

Keep in mind that very high temperatures (at least 8×10^6 K) and high densities are needed for protons to collide not only with enough energy to fuse but frequently enough to generate the sun's energy. This requirement restricts the energy production to the sun's **core,** about the inner 25 percent of its radius (Fig. 13.16). Only here does the temperature hit at least 8×10^6 K; it reaches almost 16×10^6 K at the very center of the core.

Each completed PP chain unlocks about 4.2×10^{-12} J. To account for the sun's luminosity, 1.4×10^{17} kg of matter must be converted to energy each year. That amounts to 4.5×10^6 metric tons of matter transformed each second, or about the mass of a million automobiles!

To sum up. The input for the PP chain is six protons and two electrons, and the usual output is one helium nucleus, two protons, two neutrinos, and miscellaneous gamma rays. Two protons come out, so only four are involved in the net result. Note that the net result of this nuclear cooking is the creation of helium and energy.

The neutrinos fly right out into space with about 2 percent of the energy, but the gamma rays, which carry off the bulk of the energy, find the sun's interior opaque. Photons bounce along random paths as they are absorbed and reemitted. Slowly, the photons creep out toward the sun's surface, from regions of higher to lower temperature (Fig. 13.17). The energy is transported by radiation through a large fraction of the interior, which is called the *radiative zone.* As a consequence of the temperature decline, an average photon's energy declines. The original gamma rays are degraded, by interaction with the sun's material, into many lower-energy photons. In two hundred thousand years, the photons

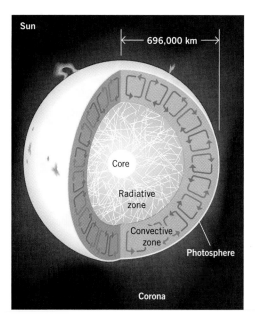

FIGURE 13.16 Model of the general interior structure of the sun. The fusion reactions take place in the core; the energy released here flows as radiation through the radiative zone. When the opacity becomes high enough, the energy transport occurs by convection. Above the convective zone lies the photosphere. The corona marks the outer limit of the sun's atmosphere.

reach the convective zone and then finally break out of the photosphere in the form of less energetic, visible radiation.

The sun is a fusion furnace, forging helium from hydrogen in its hot core. Slowly, the core's helium abundance increases, and its hydrogen abundance decreases. The energy you see now was produced in the core an average of some 200,000 years ago.

How long can the sun survive at this rate? The sun has enormous hydrogen supplies. However, only the hydrogen in and close to the core can burn. This amounts to about 10 percent of the total. Also, the PP chain transforms only 0.7 percent of the mass into radiant energy. Even with these restrictions, the sun's fusion energy can last about 10 billion years; it is now about 5 billion years old. Fusion solves the energy problem for the sun.

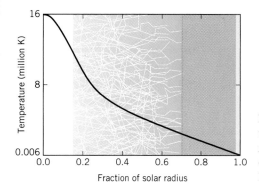

FIGURE 13.17 Temperature profile of the sun's interior. Note the sharp drop from the core to the photosphere. (Based on a theoretical model by John Bahcall and R. Ulrich.)

Solar Neutrino Problem

How can we visualize the sun's interior when we can't see below its surface? We have some idea of the interior conditions because astrophysics can make theoretical models of the sun. We do this by applying simple physical laws to a ball of hot gas, mostly hydrogen and helium, at whose center proton–proton reactions take place. Electronic computers calculate the models, which match the observed characteristics of the sun. These models show, for example, how the sun's temperature increases from edge to center.

These models of the sun's interior have been constructed as carefully as possible. However, an ongoing experiment casts some doubt on them. Remember the neutrinos produced in the PP chain? These particles fly directly out of the sun's core. So if we could detect these neutrinos, we could see into the sun's interior. (During the time you've read this sentence, *trillions* of neutrinos have passed through your body!)

Raymond Davis and his colleagues developed one strange telescope to catch the solar neutrinos. This instrument consists of about 378,000 liters of a chlorine compound placed in a huge tank located about 1.6 km below the earth's surface in an old gold mine in Lead, South Dakota. How does this "telescope" work? A chlorine atom can absorb a neutrino and convert it to argon. By a very delicate procedure, the argon gas can be flushed out of the tank and the amount measured. Thus the number of neutrinos captured during a given time span is gauged.

Nuclear theory (based on other experiments) and models of the solar interior predict how many argon atoms should be produced from the incorporation of neutrinos each day. The prediction is about one atom per day. The experimental results? Since 1970 the detected solar neutrinos amount, on average, to only *one-third* the number predicted from standard models of the sun's interior and our current knowledge of the nuclear reactions involved! To put the result another way, the experimental result implies that PP reactions take place now in the core at one-third the

---

Proton–Proton Fusion

At a Glance

Minimum temperature needed	8×10^6 K
Input	Four protons and two electrons
Output	One helium nucleus, two neutrinos, gamma rays
Energy produced	4.2×10^{-12} J
Site	Sun's core

---

rate needed to account for the sun's luminosity.

What's wrong? First, the experimental equipment may harbor undetected problems (though this is unlikely because the experiment ran for many years). The chlorine capture experiment is the only one to have run for a long time, however, and its results need confirmation from other, independent experiments. About ten are now in preparation or progress.

One such investigation, the Kamiokande II experiment (see Section 17.4), has reported a neutrino flux half that predicted by standard models of the interior. Another, called the Gallex experiment, has a detector of 30 tons of gallium located under a 2.9-km mountain in the Appennines north of Rome. Its initial result was about 60 percent of the predicted rate. So the deficit of neutrinos seems to be real, though the exact amount is still uncertain.

Second, if the results are sound, they are telling us that the standard solar model is incorrect or that something yet unknown happens to neutrinos. So the experiments may provide new information on the properties of the sun's interior or on the properties of neutrinos. Perhaps solar models made with much lower abundances of heavy elements produce far fewer neutrinos—but they contradict our general picture of the manufacture of elements in stars. Solar models can be rigged to agree with the experiment, but then they contradict other astrophysical data or the earth's climatic history. (The neutrinos observed from Supernova 1987A, Section 17.4, confirmed that some of the expected reactions *do* produce neutrinos.)

Among all the proposals (literally, hundreds!), no really satisfactory explanation has cropped up. We are still stumped. And we don't even know whether the problem lies with the solar models (and so the "physicists are right") or with the properties of neutrinos (and so the "astronomers are right")! That's the essence of the solar neutrino problem.

Solar Vibrations

A way to solve the puzzle of the nature of the sun's interior comes from a surprising source: sound waves. Wave motions (somewhat like the seismic waves in the earth; Section 8.2) resonate in the sun's interior and appear on the surface. The sun rings like a bell at different frequencies, just as the earth's interior rings from earthquakes. In analogy to the seismology of the earth, the study of such phenomena in the sun is called **helioseismology**—the analysis of solar vibrations to infer the sun's interior properties. The vibrations cause the gases of the photosphere to move up and down (Fig. 13.18)—motions visible as Doppler shifts of the photospheric gas.

The vibration first observed by solar astronomers involves 5-minute oscillations of the photosphere, which make the gas rise and fall at a speed of about 0.4 km/s. They grow and decay over about half an hour. Such vibrations are excited by the turbulence in the convective zone. They involve millions of tones turned on by the noise of convection, and they allow us to probe the physical properties of the convection zone. To date, astronomers have measured thousands of frequencies to accuracies close to 0.01 percent. The results indicate that the convective zone goes somewhat deeper than we had believed—the outer 30 percent of the radius rather than 20 percent. Meanwhile, long-period vibrations penetrate deep into the sun before they return to the surface; they can tell us about the core. Results to date indicate a somewhat smaller core.

With the appropriate computer models of the vibrations, we can infer

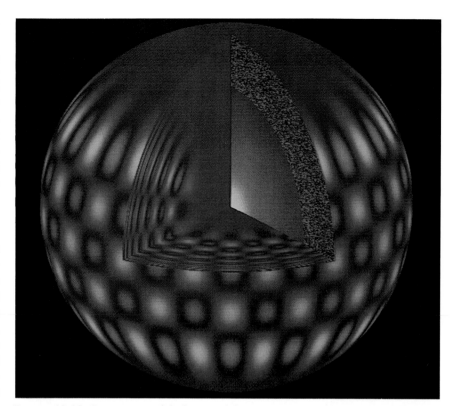

FIGURE 13.18 Computer model of some of the resonant tones of the solar interior. Blue represents expanding regions; red, contracting ones. The pattern on the photosphere results from the resonances of waves in the sun's interior. The patterned region just below the surface represents the sun's convection zone. (Courtesy of NOAO.)

some of the sun's interior properties. For example, the helium abundance of the interior affects the tuning of the sun's vibrations. The best match to the observations requires about 29 percent helium—higher than that inferred from the atmosphere. But, this high-helium model requires an even higher production rate of solar neutrinos, aggravating that problem. The results pretty much rule out a rapidly rotating solar core or extremely strong internal magnetic fields—two options that could decrease the neutrino flux.

■ What type of nuclear reaction occurs in the sun's core?

13.7 The Active Sun

Events of the active sun are localized, short-lived phenomena on or near the solar surface. The most important of these—sunspots and flares—are all associated with the locally intense solar magnetic field. In general, these areas of

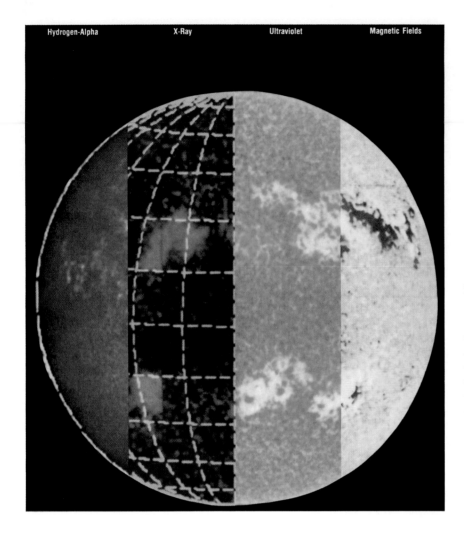

Hydrogen-Alpha   X-Ray   Ultraviolet   Magnetic Fields

FIGURE 13.19 Active regions on the sun. The sun viewed by four different techniques at the same time to show different regions affected by magnetic fields. From the left to right, as it appears in the Hα line, x-rays, ultraviolet, and regions of strong magnetic fields. Each image relates to different levels in the sun: strong magnetic fields in the photosphere (white and black are opposite magnetic polarities); Hα marks the upper photosphere and lower chromosphere; ultraviolet, the upper chromosphere; x-ray, the corona. The effects of magnetic fields are visible at each of these levels usually by the brighter areas of emission in the levels above the photosphere. (Courtesy of NASA/GSFC; image by J. Gurman.)

solar activity are called **active regions** (Fig. 13.19).

One other solar property contributes to active-sun phenomena: the sun's nonuniform rotation. Because the sun is a gas, it spins faster at the equator (one rotation every 25 days) than at the poles (one rotation every 31 days) and therefore shows **differential rotation.** So as you travel from a pole toward the equator on the sun, the photosphere's rotation rate increases. Differential rotation can distort solar magnetic field lines and drive the development of active regions.

Sunspots

Sunspots appear as dark blotches on the solar disk (Fig. 13.20). With a temperature of about 4200 K, a sunspot is relatively cooler than the photosphere and so appears dark in contrast. In fact, a sunspot is almost four times fainter than

the photosphere (it emits one-fourth the flux).

Sunspots have a strong tendency to form in groups. They are born in an active region where photospheric granules separate, and a tiny spot appears between them as a dark pore. Such pores have magnetic field strengths averaging 0.25 T—thousands of times stronger than the sun's overall magnetic field. Usually more pores soon become visible and coalesce over a period of several hours to form a sunspot. A large, single sunspot group may contain a hundred individual spots, persisting for two or more solar rotations.

Sunspot Cycle

In 1843 Heinrich Schwabe (1789–1875), a German amateur astronomer, noted that the sunspot numbers vary periodically. On the average about 11 years pass

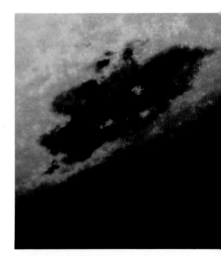

FIGURE 13.20 Sunspot group near the edge of the sun in March 1989. Many flares were associated with it. (Courtesy of NOAO.)

between sunspot *maxima* (times of greatest numbers of sunspots) and sunspot *minima* (times of least numbers of sunspots). Later the British astronomer Walter Maunder (1851–1928) discovered not only that sunspot numbers vary during a cycle but also that the sunspots' positions change. Sunspots usually appear only in the zone between the solar equator and 35° north or south latitude. At the start of a sunspot cycle, a few spots emerge at high latitudes. As the cycle progresses, the sunspot latitude zone migrates toward the equator; new spots appear in this band. As the survivors of one cycle expire near the equator (about 5° north or south latitude), new spots from the next cycle form at the higher latitudes.

In 1908 George Ellery Hale (1868–1938) detected intense magnetic fields associated with sunspots. The strongest magnetic sunspots have field strengths exceeding 0.4 T. (That's about 8000 times stronger than the earth's field at its surface.) Hale also noticed that sunspot groups contain spots of opposite magnetic polarity. For example, the east and west spots of a twin group might have north and south polarity. If this arrangement holds at a given time in the northern hemisphere of the sun, the situation at the same time in the southern hemisphere is reversed (Fig. 13.21). Later Hale discovered that the magnetic polarities for the west and east spots have a 22-year average cycle, just double the 11-year average cycle of sunspot numbers. That is, if a spot group has the west spot with a south polarity, in the next sunspot cycle the west spot will have a north polarity (Fig. 13.21), and in the following cycle, south polarity again.

Current investigations indicate that there is little historical evidence of an 11-year cycle in sunspot activity before 1700. Sunspots were discovered by Galileo with his new telescope in about

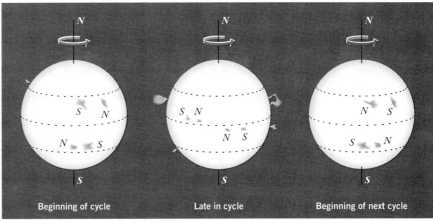

FIGURE 13.21 The magnetic cycle for sunspots. At the beginning of a sunspot cycle, if spots with north polarity are leading groups in the sun's northern hemisphere, then south polarity leads groups in the southern hemisphere. At the beginning of the next cycle, the polarity of the leading spots is reversed for each hemisphere.

1613, but, as first noted by Maunder, hardly any sunspots were seen in the 70-year period from 1645 to 1715 (Fig. 13.22). This period corresponded to an unusual cold spell sometimes called the Little Ice Age (the average temperature of the earth dipped about 0.5 K) that extended from the sixteenth to the eighteenth century.

The relative consistency of the cycle in modern times may be a brief phase that recurs over longer times. Evidence in the layering of Australian rocks some 700 million years old hints at an activity cycle very much like that recorded since A.D. 1700. Overall, solar activity behavior may be much more complex than we have inferred from the limited time spans investigated so far. And it may well affect the earth's climate, albeit in complex ways.

Physical Nature of Sunspots

Sunspot magnetic fields are generated by enormous electric currents, much as the field of an electromagnet is produced. Exactly how this formation happens is unclear. One possibility is that the sunspots' magnetic fields suppress the hot gas rising from the convective zone. The hot gas runs into a magnetic thicket and has trouble breaking through the surface

FIGURE 13.22 Sunspot cycles from 1600 to the 1980s, plotted in terms of the number of observed sunspots monthly during a year. The peaks (sunspot maxima) come roughly 11 years apart. Note how few sunspots were seen prior to 1700, compared to modern times. (Adapted from a diagram by J. Eddy.)

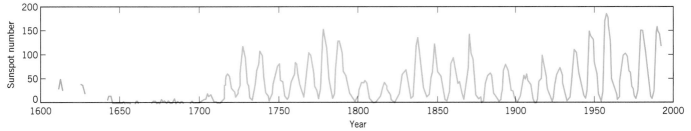

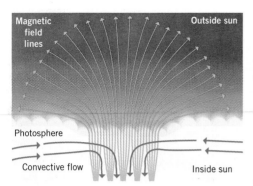

FIGURE 13.23 One model of the distribution of magnetic field lines in a sunspot driven by the convective upflow and downflow below the photosphere. The convective upflow of the sun's plasma pushes the magnetic field lines together. (Adapted from a figure by E. N. Parker.)

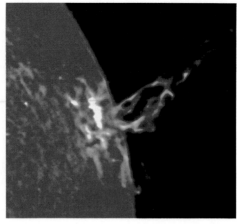

FIGURE 13.24 Eruptive prominence, June 20, 1989. The hydrogen gas in the upper parts may escape into space. That in the lower region streams back down to the sun. (Courtesy of NSO, Sacramento Peak.)

in the sunspot's center (Fig. 13.23). A convective downflow may draw away some of the hot gas, removing heat from the region. As a result, the sunspot is cooler and darker than the surrounding photosphere.

Prominences

Magnetic fields also play a major role in the production of **prominences,** huge clouds of hydrogen gas above the photosphere. When viewed along the sun's edge, prominences often loop and surge up into the corona (Fig. 13.24). When seen against the photosphere, a prominence resembles a dark snake winding across the solar disk. Because they are cool clouds of coronal gas, prominences absorb some photospheric light and appear dark in contrast.

Prominences are almost always found near a sunspot group in an active region. Frequently, a young prominence may disappear, only to reappear a few days later, relatively unchanged. This observation suggests that the basic underlying structure of the prominence is the magnetic field, made visible when gas is at the right temperature and density to radiate visible light.

Flares

Sunspots are floating islands of electromagnetic storms. Associated with them in active regions are short-lived, violent discharges of energy called **solar flares.** These energetic bursts appear with sunspots and sometimes bridge the gap between two close spots. Near large sunspots, about a hundred small flares occur each day. The elapsed time between the birth of a flare and its rise to peak intensity is only a few minutes. The decay time is about an hour. Emitting myriad forms of energy—x-rays, ultraviolet and visible radiation, high-speed protons, and electrons—a large flare blows off about 10^{25} J (Fig. 13.25), the equivalent of the energy released by a bomb of 2 billion megatons! (A **megaton,** in turn, is the energy equivalent to that of an explosion of a million tons of TNT.)

On March 6, 1989, the strongest solar flare in 20 years blasted detectors on satellites. At its height, the temperature in the flare's plasma reached 10 million kelvins (Fig. 13.26). A week later, storms in the earth's magnetosphere ignited auroras, disrupted radio communications, and resulted in power surges that blacked out lights for 6 million people in Quebec, Canada. The total energy expelled by the flare was estimated at 10^{30} J. This violence marked the prelude to the activity cycle that peaked in July 1989. We are now in the downside of the cycle.

Because large active regions are the most frequent locations for large solar flares, astronomers think that the concentration of energy arises from local twists and kinks in the magnetic field loops (Fig. 13.27). The energy accumulates until its release is triggered. The flare starts in the corona and strikes down into the chromosphere in the form of high-speed electrons.

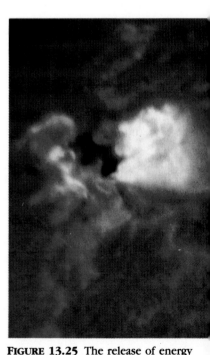

FIGURE 13.25 The release of energy in an energetic solar flare, viewed in $H\alpha$ light, March 9, 1989. The onset of the flare takes place in the magnetic loops high in the corona. The result shows up as a blast of $H\alpha$ emission from the bases of the magnetic loops down in the chromosphere. (Courtesy of NOAO.)

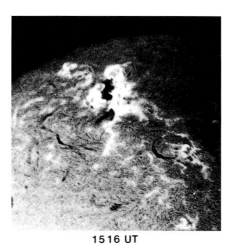

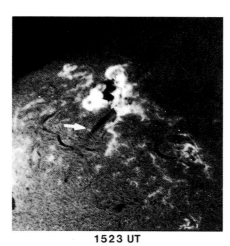

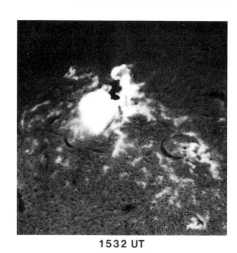

| 1516 UT | 1523 UT | 1532 UT |

FIGURE 13.26 The enormous solar flare of March 1989. This sequence of photos taken in Hα light shows the development of the flare. The first shows an early stage, the second the ejection of charged particles, visible as the dark, spraylike structure at the lower left (arrow), and the third the flare at maximum, when it grew to the strongest optical flare ever recorded. (Courtesy of NSO, Sacramento Peak.)

FIGURE 13.27 Magnetic loops in the sun's lower corona, as viewed in ultraviolet light. The plasma follows along the magnetic field lines, which reach out of the sun at one polarity and return with the opposite polarity. (Courtesy of NASA/JPL.)

Flares blast energetic particles, mostly protons and electrons, into space. A majority of the flare's energy, however, escapes as x-ray and ultraviolet light. Arriving at the earth about 8.3 minutes later, the flare's radiation rips through the upper atmosphere and tears electrons from neutral atoms. The increase in ionization disrupts long-distance radio communication. A few days later, the slower protons and electrons approach the earth, but they are usually trapped by the earth's magnetic field. These particles are called **solar cosmic rays**. They can injure unshielded humans in space (or on the moon). Luckily, the earth's atmosphere and magnetosphere guard us from the hazards of such outbursts.

Occasionally the earth's magnetic reservoirs overflow with charged particles, particularly at times of sunspot

FIGURE 13.28 Model for the physical process that creates a solar flare. (*a*) Magnetic loops extend into the corona from active regions of north and south polarity. A *neutral line* between the north and south regions indicates where the field has zero strength. (*b*) Magnetic reconnection occurs above the photosphere, joining field lines of opposite polarity. (*c*) The point of reconnection moves up the field lines into the corona, driving plasma outward and downward as the field lines, like stretched rubber bands, release energy. (Adapted from a figure by R. Noyes.)

One model for how a flare works involves *magnetic reconnection* (Section 8.3) of magnetic loops in the corona (Fig. 13.28). Such loops connect the north and south polarities of an active region; at the top of the loops, field lines extend away from the sun. Stresses on the magnetic field, perhaps from turbulence in the convective zone, cause reconnection at the top of the loop. This change forces some charged particles down into the chromosphere and some out into space—an action that we see as a flare.

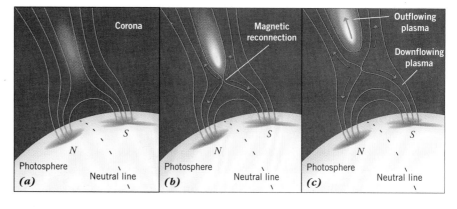

maxima when flare activity is at its peak. As the particles spill into the earth's upper atmosphere, the swift electrons bump into atmospheric atoms and excite them. When the atoms de-excite, they emit visible radiation, causing in the sky the glow called the *aurora*.

Coronal Loops and Holes

Because the coronal gas is so hot, it emits low-energy x-rays and has an irregular distribution above and around the sun (Fig. 13.29). The large loop structures indicate where the ionized gas flows along magnetic fields that arch high above the sun's surface and return to it. The hot gas is trapped in these magnetic loops. Solar physicists now view the corona as consisting primarily of such loops.

Some regions of the corona appear dark, especially at the poles and down the middle part of the sun. Here the coronal gas is much less dense and less hot than usual—these regions are called **coronal holes.** The coronal holes at the poles do not appear to change very much, but those above other regions are related to solar activity.

What makes a coronal hole? Solar astronomers believe that coronal holes mark areas in which fields from the sun continue outward into space rather than flowing back to the sun in loops. So the coronal gas, not tied down in these regions, can flow away from the sun out of the coronal holes; this flow makes the solar wind (Fig. 13.30).

The coronal gas does *not* follow the differential rotation of the photosphere. Rather, it rotates at the same angular speed at all latitudes (as the earth does). This fact implies that the bottoms of the magnetic loops are anchored deep below the photosphere, perhaps at the very bottom of the convective zone, where such fields might be generated by a solar dynamo, where organized flows of electric currents generate magnetic fields.

Now back to why the corona (like the chromosphere) is so hot. The magnetic loops, rising from the photosphere up to 400,000 km into the corona, play the key role. The magnetic field in a loop is twisted by photospheric motions at its base. Slow twisting generates electric fields that then heat the coronal gas. This

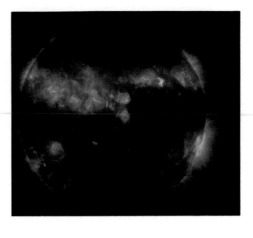

FIGURE 13.29 X-ray image of the sun (October 1989) showing fine structure in the coronal loops, where the plasma temperature is one million kelvins. The active regions are strung out parallel to the equator. (Courtesy of Leon Golub and IBM Research/SAO.)

doesn't take much energy, because the coronal gas is so thin that it holds little heat.

Model of the Solar Activity Cycle

Magnetic fields drive the sun's activity over a 22-year average cycle. We believe, although all the details are not worked out, that the cycle occurs from a coupling of the sun's overall magnetic field, which is driven by convection, with its differential rotation. As the name implies, in the **dynamo model,** the sun has an electromagnetic dynamo in its convection zone (Fig. 13.31). Interior convection and the differential rotation drive this dynamo to generate magnetic fields.

Consider an activity cycle starting with the sun's magnetic field aligned mostly north–south. Differential rotation stretches these field lines east–west along the equator. This activity produces tubes of magnetic fields parallel to the equator. Active regions develop where

FIGURE 13.30 Streamers in the sun's corona during a total eclipse in March 1988. This is a composite of a white-light and x-ray image (look back at Fig. 13.13 (*a*). The coronal photograph in white light shows the streamers where the sun's magnetic field extends into space and is the source of the solar wind. The x-rays (red) at 170 Å show the corona a few hours before the eclipse (Courtesy of NCAR/University Corp. for Atmospheric Research/NSF.)

the tubes burst out of the surface, starting at midlatitudes. The differential rotation mixes up the polarities so that the field reverses after 11 years and the cycle repeats after 22 years. Cyclonic motions in the convective zone help in the regeneration of the north–south field at the start of each cycle.

The main point here is that convection and differential rotation drive the solar dynamo and produce the magnetic activity cycle. We believe that the same processes create magnetic fields and activity cycles in other stars like the sun (Section 14.7).

■ Why do sunspots appear darker on the photosphere?

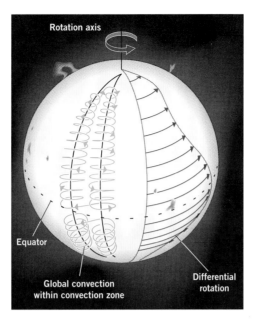

FIGURE 13.31 Convection and the solar magnetic field; one model for the solar dynamo. The global convection (in the convection zone) and the differential rotation (because the sun is a gas) drive the solar dynamo that generates the sun's global magnetic field. (Adapted from a diagram by P. Gilman.)

Key Concepts

1. The astronomical unit (AU), the average earth–sun distance, is about 150×10^6 km. It's found with precision by bouncing radar signals off Venus and measuring their round-trip travel time when Venus and the earth are separated a known distance in AU.

2. With the distance known, the sun's size is found from its angular size, its mass from the earth's orbital period and Newton's law of gravitation and second law, and its low bulk density from its volume and mass. The sun is an ordinary gas throughout that obeys simple laws relating pressure, temperature, and density.

3. To find the sun's luminosity (total radiative energy output per second) requires a knowledge of the earth–sun distance in kilometers and a measurement of the sun's flux at the earth. We measure the flux above the atmosphere with satellite detectors.

4. The sun's surface temperature can be found from its color or the wavelength at which its continuous spectrum hits a peak intensity, if we assume that the sun radiates like a blackbody. (It does, more or less.) Blackbody radiators are hot, opaque bodies whose continuous emission depends only on their temperatures.

5. The solar gases become transparent to light at the sun's surface—its photosphere. The continuous spectrum forms in this region; the absorption lines form in higher, cooler layers. Where the lines form depends on the opacity of the material at different levels.

6. The sun's absorption lines indicate what elements are in the photosphere. The photosphere contains by mass about 74 percent hydrogen, 25 percent helium, and 1 percent all other elements (generally called by astronomers *metals*).

7. Energy flows from the sun's hotter core to its cooler photosphere. For most of the interior, radiation transports the energy. The photosphere has a bubbly structure, indicating that a zone of convection exists beneath it. This convection carries energy to the surface to be radiated into space. The chromosphere above the photosphere is hotter and visible by its emission lines during an eclipse. The corona is even hotter than the chromosphere, as indicated by emission lines of highly ionized atoms.

8. The sun's energy comes from fusion reactions in which four protons are fused into a helium nucleus with the release of energy. Only in the sun's core is the temperature high enough for these fusion reactions to occur. The sun has enough hydrogen to fuel these reactions for billions of years. Gradually, as its hydrogen decreases, the sun's core increases in its percentage of helium.

9. Solar neutrino telescopes have found far fewer neutrinos than predicted by models of the sun and nuclear reactions. This result, if it stands, implies either that our models of the interior are wrong or that we have something new to learn about neutrinos.

10. The sun undergoes cycles of activity in which active regions—where sunspots, flares, and prominences tend to be found—become more common and then less so; this cycle lasts an average of 22 years, as indicated by the polarity of sunspots. The magnetic

field and activity cycle are driven by the sun's convection and differential rotation.

11. Active regions result from the development of strong (thousands of times higher than the average) local magnetic fields, which produce sunspots (dense concentrations of magnetic field lines) and solar flares, which emit electromagnetic radiation (especially x-ray, ultraviolet, and radio) and high-speed particles that cause auroras when they reach the earth. The solar wind leaves the sun through coronal holes, cool regions in the corona from which the sun's magnetic field streams into space.

12. The sun serves as the local model on which we base our understanding of other stars. In particular, we expect them to be massive, gaseous bodies (made mostly of hydrogen and helium) whose radiance comes from fusion reactions in their dense, hot cores.

Key Terms

active region
blackbody (radiator)
carbon-nitrogen-oxygen (CNO) cycle
chromosphere
conduction
convection
core (of the sun)
corona

coronal holes
differential rotation
dynamo model (solar)
flux
helioseismology
luminosity
megaton
neutrino
nuclear fission

nuclear fusion
opacity
photosphere
Planck curve
positron
prominences
proton-proton (PP) chain
radiation

solar cosmic rays
solar flares
solar wind
Stefan-Boltzmann law
sunspots
surface temperature
Wien's law

Study Exercises

1. Why do astronomers bounce radar signals off Venus to find the distance to the sun? (Objective 13-1)

2. What *measurements* must be made to find out the sun's luminosity? Which of these is *least* accurate? (Objective 13-2)

3. Suppose you examined sunlight with a spectroscope. Describe, in general, what you would see. (Objective 13-5)

4. Explain the appearance of the sun's spectrum by simple atomic processes. (Objective 13-5)

5. Why do you *not* see helium absorption lines in the sun's visible spectrum, even though helium is the sun's second most abundant element? (Objective 13-5)

6. Should the chemical composition of the sun's core differ from that of its photosphere? Why or why not? (Objectives 13-6, 13-8, and 13-10)

7. Describe how to estimate the sun's surface temperature from its continuous spectrum. (Objective 13-2)

8. Use the concept of opacity to explain why it takes photons hundreds of thousands of years to walk out of the sun's core to the surface. (Objective 13-4)

9. In one sentence, describe the source of the sun's energy. (Objectives 13-7 and 13-8)

10. Describe a method by which you could infer the temperature of a sunspot from its continuous spectrum. (Objectives 13-3 and 13-11)

11. We can see spectral lines of helium from the sun's chromosphere but not the photosphere. Why? (Objectives 13-5 and 13-6)

12. How do we know that the photosphere marks the top of the sun's convective zone? (Objectives 13-4 and 13-10)

13. In what sense do neutrinos allow us to "see" the sun's core directly? What have been the results of the solar neutrino experiments to date? (Objective 13-9)

14. The deeper you go into the sun, the higher the density and temperature. What do you expect, since the sun is a gas, to happen to the pressure? (Objective 13-12)

15. Consider an ideal gas. Suppose you double its density and triple its temperature. By how much will its pressure change? (Objective 13-12)

Problems & Activities

1. The sun's luminosity comes from fusion reactions in the core that convert matter to energy. Calculate the number of kilograms converted each second.

2. The sun has been using fusion reactions for close to 5 billion years. If its luminosity has been constant over this time, calculate the amount of matter converted to date.

3. Solar energy heats the earth, and the earth's surface radiates this energy back to space. The earth acts like a blackbody with a temperature close to 300 K. Calculate the peak wavelength of emission and the flux emitted from the earth.

4. Your body also emits like a blackbody. Estimate its temperature and calculate the peak of emission and the energy emitted per square meter. Compare that to the energy per square meter emitted by the sun.

5. What is the wavelength at which the continuous emission from a sunspot has a peak in the flux? *Hint:* Assume that a sunspot radiates like a blackbody!

6. Calculate the fraction of the sun's luminosity that is intercepted by the earth. *Hint:* Consider the earth a disk, whose area is πR^2, where R is the earth's radius.

See for Yourself!

You can actually measure the size of the sun by hand! Get a piece of cardboard about 8 inches by 10 inches, white paper, a ruler, a pencil, a pen, and a sunny day. Take a ballpoint pen (or the pencil) and make a small hole in the center of the cardboard. Line up the card and the paper *parallel to each other* and project a round image of the sun through the pinhole onto the paper. Have a fellow student mark with the pencil around the edge of the sun's image. Also have your companion measure the distance from the pinhole to the image in *centimeters*. Then measure the diameter of the sun's image, also in *centimeters*.

From the position of the pinhole, the angular size of the sun's image on the paper and the sun in the sky are the same. So the ratio of the sun's actual diameter to its distance from the earth is the same as the ratio of the sun's image to the distance between the paper and pinhole (Fig. A.3):

$$\frac{\text{diameter of sun}}{\text{distance earth-sun}} = \frac{\text{diameter of image}}{\text{distance hole-paper}}$$

Let S be the sun's diameter, D the earth-sun distance, x the image diameter, and d the distance from the pinhole to the paper; then

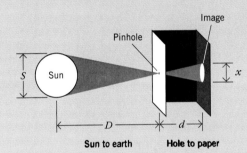

FIGURE A.3 Geometry for finding the sun's diameter from a measurement of its image projected by a pinhole.

$$\frac{S}{D} = \frac{x}{d}$$

and solving for the sun's diameter, we have

$$S = \frac{D \times x}{d}$$

You have measured x and d; now you just need D. Use 150×10^6 km; the result will then come out in kilometers for the sun's diameter. ■

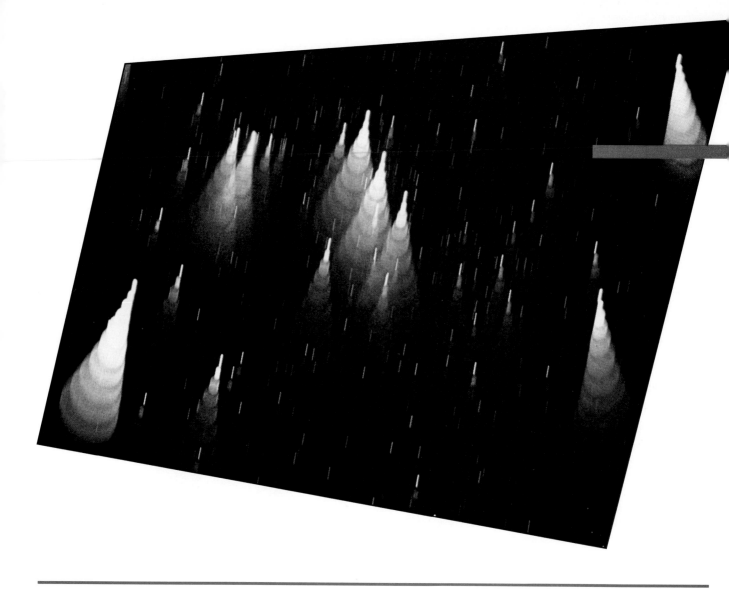

After studying this chapter, you should be able to:

14-1 Outline the methods astronomers use to find the following physical properties of stars: surface temperature, chemical composition, size (radius or diameter), mass, luminosity, and density.

14-2 Describe the relationship between a star's color and its surface temperature.

14-3 Describe the relationship of a star's luminosity, surface (effective) temperature, and size, assuming that it radiates like a blackbody.

14-4 Explain the difference between flux and luminosity and relate flux, luminosity, and distance.

14-5 Describe what is meant by the *inverse-square law for light* and apply this relation to astronomical situations.

14-6 Show by a simple diagram the relationship between a star's distance and its parallax, noting the limitations imposed by the size of the earth–sun distance.

14-7 Sketch a Hertzsprung–Russell diagram for stars, indicating the positions of sun, the main sequence, giants, supergiants, and white dwarfs.

14-8 Use the Hertzsprung–Russell diagram to infer the relative luminosities, surface temperatures, and sizes of stars represented on it.

14-9 Outline the steps by which astron-omers determine the masses of binary stars and apply the Doppler shift to binary systems.

14-10 Sketch and make use of the mass–luminosity relation for main-sequence stars; from it, argue that stars have finite lifetimes.

14-11 Give two observational indications that stars have magnetic activity like the sun's.

14-12 Use simple algebraic forms of the inverse-square law for light, stellar parallax, and the mass–luminosity relation.

Central Question
How do astronomers determine the physical properties of stars?

The Stars as Suns

I have made the stars!
I have made the stars!
Above the earth I threw them.
All things above I made
And placed them to illumine.

A PIMA INDIAN CREATION SONG BY
EARTH DOCTOR

Astronomers know what the stars are: enormous fusion reactors in space, like our sun. That view is recent, developed in full in this century. Until the end of nineteenth century, astronomers really didn't pay much attention to the inner workings of stars. And the vast distances of stars from the sun kept them aloof from human understanding.

With the development of spectroscopes and atomic theory (Chapter 5), we finally acquired in the 1930s and 1940s the crucial tools to analyze starlight and compare the stars to the sun. By finding distances to nearby stars, we were able to infer basic physical properties, which we then applied to more distant stars. These revelations reinforced the connection between the sun and the stars, *the* essential link to the rest of the cosmos. The stars are suns; the sun is a star. These aspects mark both sides of the solar–stellar connection.

14.1 Some Messages of Starlight

Go out on a clear January night to view the constellations. Face south. Orion (refer back to Fig. 1.1) immediately catches your eye. Two stars in Orion shine the brightest: Betelgeuse, which looks reddish, and Rigel, blazing bluish white. To the south and east of Orion lies Canis Major, the Great Dog, Orion's hunting companion. You can easily notice Sirius, the jewel of Canis Major, as the brightest star in the sky. (See the winter star map in Appendix G for these constellations.)

The differences in these stars typify the astronomers' dilemma: What can we know about these distant stars? How do these stars compare with our sun? Are they larger? More luminous? Hotter? How to find out these physical properties? In what ways do these stars resemble the sun? How do they differ?

Starlight carries information about the physical properties of stars (Fig. 14.1). By deciphering starlight and comparing it to sunlight, we can infer the nature of stars. Their physical properties include chemical composition, surface temperature, radius, luminosity, and mass. For some of these properties, we do not need to know the distance to the stars; for others, we do. In all cases, we observe, measure, and analyze the light from the stars as was done for the sun.

Brightness and Flux

If you have even casually looked at the night sky, you know that stars do not appear to have the same brightness. You make this judgment with your eyes, which work essentially like a refracting telescope (Section 6.2): the lens of each eye focuses light onto its retina, which detects the light by transforming the radiative energy into electrochemical impulses that travel to your brain. How bright a star appears to you depends on how much energy, each second, strikes your retina. The more energy received, the greater brightness you perceive.

Similarly, the objective of a telescope (Fig. 14.2) gathers light to a focus, and a detector senses the radiative energy. How bright a star appears in a telescope depends on how much energy, each second, arrives at the earth from the

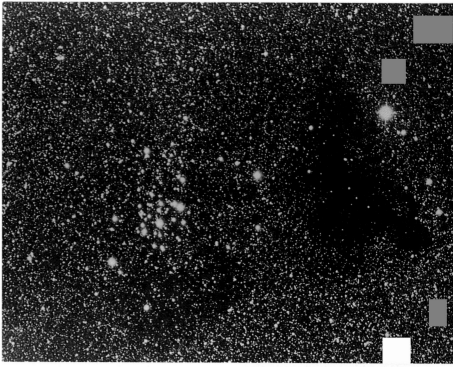

FIGURE 14.1 New General Catalog 6520 (NGC 6520), a star cluster (to left of center) in the southern Milky Way. Note that the stars have various colors: white, reddish, yellowish, and bluish. To the right of middle is a cloud of dust blocking out light from stars behind it. (Courtesy of Anglo-Australian Observatory; photo by D. Malin.)

star and how large the objective of the telescope is. Just to determine the brightness of the star, then, we need to get rid of the dependence on aperture size. We do this trick by considering only the flux crossing the telescope's objective.

Recall that *flux* is the name given to the amount of energy passing through each unit of area in a second. Since watts (Appendix A) are a measure of energy per second, flux can have units such as watts per square meter. For example, the flux at the earth from the sun is 1370 watts per square meter. Compared to the sun, the brightest star, Sirius, sends a tiny flux to the earth—a mere ten-millionth of a watt per square meter!

Caution. Brightness actually means flux. Brightness is a more common word but lacks the specific physical meaning of flux. (Astronomers have a special way to measure flux at the earth; see Focus 14.1.) Also, do *not* confuse the flux emitted at the *surface* of a star with that measured at the earth.

Flux and Luminosity

Remember how to work out the sun's luminosity (Section 13.3)? Measure the sun's flux, find the earth–sun distance in

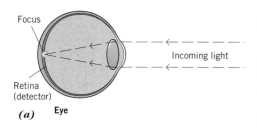

(a) **Eye**

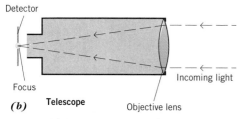

(b) **Telescope**

FIGURE 14.2 Comparison of the optics of the human eye and a telescope. Both focus an image on a detector. In the case of the eye, the retina serves as the detector surface.

kilometers, construct an imaginary sphere with a 1-AU radius, and total up all the energy hitting that sphere. We can find a star's luminosity by the same procedure (Fig. 14.3). Measure the star's flux at the earth. Then, knowing its distance, find the area of a sphere surrounding the star, and add up the flux over the total area. Simple enough—but we need to know the distance to the star, and that may not be easy to come by (Section 14.2).

Why worry about a star's luminosity? Because it tells how much radiative energy the star emits each second, and that's fundamental to understanding stars as suns.

Caution. To find a star's total luminosity requires the measurement of its flux over the *complete* range of the electromagnetic spectrum. That might not be possible. First, our detector may work over only a limited range. Second, the earth's atmosphere absorbs some of the energy. Third, matter in interstellar space may also do some absorbing. If the flux is measured over a limited range of the spectrum, say the visual, astronomers report this fact by using the term *visual flux.* Then the luminosity calculated from it is only the *visual luminosity* of the star. Even sunlike stars emit only 40 to 50 percent of their light in the visual part of the spectrum, so their visual luminosity is only about half their total luminosity. The luminosities for stars given in this book are almost all visual luminosities to have a consistent basis of comparison. But beware that these visual luminosities will be *less* than the total luminosity over all wavelengths.

The Inverse-Square Law for Light
You may have noted that the *distance* plays a key role in relating flux to luminosity, because the brightness of a light source relates in a very specific way to your distance from it.

Consider the following experiment. Put a bare light bulb in a socket and turn it on. Take a light meter—a device used to measure the intensity of light, commonly used with automatic exposure cameras—and place it 1 meter from the bulb (Fig. 14.4). Note the reading on the

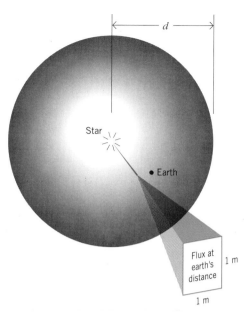

FIGURE 14.3 Determining the luminosity of a star from its distance and the flux at the earth. When it reaches the earth, a star's light has spread out over a sphere having a radius that is the distance between the earth and the star; its area is $4\pi d^2$. The flux received at the earth is a small fraction of the star's luminosity because the light has spread over a huge area in space.

light meter, and call that one unit. Now move the light meter 2 meters from the bulb; its reading will be one-fourth that at the 1-meter position. Move the light meter to a distance of 3 meters; the reading will now be one-ninth that at 1 meter. Note that the light intensity decreases as the *inverse square* of the distance. So that the flux is *inversely proportional to the distance squared.* (What would the reading be at 4 m? Right—1/16 that at 1 m.)

How does this inverse-square relation happen? Imagine a bulb placed in the center of two transparent spherical

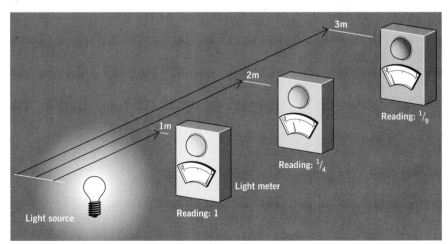

FIGURE 14.4 Flux and distance. As a light meter is moved away from a light bulb, the measured flux decreases. The greater the distance, the lower the reading, so that at 2 meters, the flux is only 1/4 that at 1 meter; at 3 meters, 1/9 as much.

FLUX AND MAGNITUDE

For reasons of history and convention, astronomers generally use a quirky way to talk about stellar brightness. It's called _apparent magnitude_. Flux and apparent magnitude are two different methods used to describe the same property. I do _not_ use magnitudes in this book because I think they are confusing to the novice and unnecessary for a basic understanding of the material. However, you may come across the concept elsewhere, so I'll discuss it briefly here.

The magnitude scale on which stars are rated evolved from a convention established by Hipparchus (160–127 B.C.) and has now become traditional. In his catalog of stars, Hipparchus classified apparent magnitudes by rating the brightest star he could see as magnitude 1 and the faintest as magnitude 6. As this system was used, some stars were found to be brighter than magnitude 1; for example, Vega is magnitude 0, and Sirius is magnitude −1.4. The first peculiarity to note about the magnitude scale is that the _larger_ the magnitude of a star (the more positive), the _fainter_ it is; but the smaller the magnitude (the more negative), the brighter it is. That is, a star of magnitude 6.5 is fainter than one of magnitude 4.2, and one of magnitude −1.3 is fainter than one of magnitude −2.1. But a star of magnitude 12.5 is brighter than one of magnitude 17.9, and one of magnitude −1.8 is brighter than one of magnitude −0.9. Confusing? Perhaps it helps to think of magnitude as measuring the amount of _faintness;_ larger numbers mean fainter stars (Fig. F.10).

Most star catalogs or finding charts give the apparent magnitudes of the objects. If you look at Appendices D and E, you will find the apparent _visual_ magnitude listed. Here "visual" refers historically to the magnitude as sensed by the eye. Astronomers today use a special filter in the middle of the visual spectrum.

On Hipparchus' scale, stars of first magnitude were about a hundred times brighter than stars of sixth magnitude. The modern system therefore _defines_ a difference of 5 magnitudes as corresponding to a brightness ratio of 100. A difference of one magnitude then amounts to a brightness ratio of 2.512. This strange number pops up because $2.512 \times 2.512 \times 2.512 \times 2.512 \times 2.512 = 2.512^5 = 100$, another way of stating that a difference of 5 magnitudes equals a ratio of 100 in brightness. Table F.2 will help you keep straight the magnitude differences and brightness ratios. Note that

TABLE F.2 MAGNITUDE DIFFERENCES AND FLUX RATIOS

A MAGNITUDE DIFFERENCE OF:	EQUALS A FLUX RATIO OF:
0.0	1.0
0.2	1.2
1.0	2.5
1.5	4.0
2.0	6.3
2.5	10.0
4.0	40.0
5.0	100.0
7.5	1,000.0
10.0	10,000.0

differences in apparent magnitude provide a way of comparing fluxes if you convert to brightness ratios.

Astronomers also use a system called **absolute magnitude** to talk about luminosities in the form of magnitudes. Astronomers do compare the luminosities of stars, in an imaginary way, by setting them at a standard distance of 10 parsecs (pc), which is 32.6 light years.

Imagine that you could transport the stars in the sky, including the sun, to 10 pc from the earth. Stars that are in reality closer than this distance would appear fainter; those that are farther would get brighter, as expected from the inverse-square law. The term _absolute magnitude_ refers to how bright a star would appear at 10 pc from us.

Compare the absolute magnitudes of these stars: the sun, 4.8; Sirius, 1.4; Polaris, −4.6. Note that Polaris is actually the most luminous, followed by Sirius, with the sun last.

When we know a star's absolute magnitude, we can compute its luminosity by comparing its absolute magnitude to that of the sun. The trick here is to convert magnitude differences into brightness ratios (Table F.2). Let's compare the sun with Sirius. In the visual range, the sun's absolute magnitude is 4.83, that of Sirius 1.41. The difference is roughly 3.4 magnitudes, so the brightness ratio is about $(2.512)^{3.4} \approx 22$.

If you know a star's distance and apparent magnitude, you can work out its absolute magnitude. In fact, if you know any two of the three properties, you can find the third. Algebraically, the relationship is

$$m - M = 5 \log d - 5$$

where m is the apparent magnitude, M the absolute magnitude, and d the distance in parsecs.

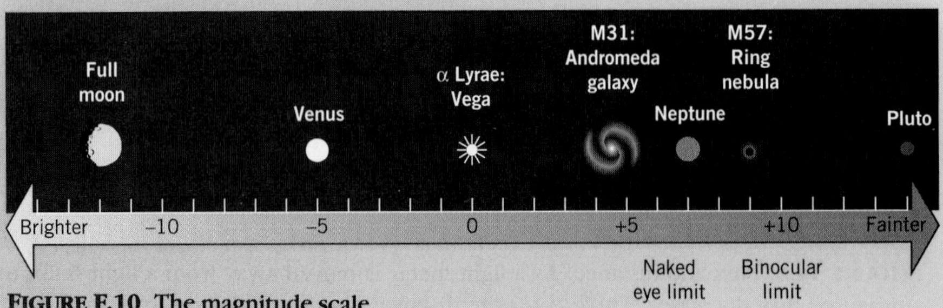

FIGURE F.10 The magnitude scale.

shells, one having twice the radius of the other (Fig. 14.5). Light the bulb briefly so that it emits a pulse of light. As the light moves away from the bulb, it expands in a sphere in all directions. But the total amount of energy in the pulse remains the same, no matter how large the sphere. As the light passes through the first sphere, it covers a certain area. As it goes through the second, it covers a larger area. How much larger? The area of a sphere is directly related to the radius squared; so the larger sphere has four times the area of the smaller one, and the light spreads out four times as much.

Now pick any square meter of surface for both spheres. Because of the radiation's dilution over a larger area, the small patch you select on the larger sphere has only one-fourth as much light striking it as the patch on the smaller sphere. If the ratio of radii were increased to 3, the decrease in brightness would be by 9; if increased to 4, the decrease would be by 16. This is the **inverse-square law for light** (or any electromagnetic radiation)—a crucial concept for astronomy.

Suppose you observe that the flux of one star is a hundred times that of another. If you assume that both stars have the same luminosity, how do their distances compare? According to the inverse-square law, the brighter one must be ten times as close as the fainter one.

■ **What does luminosity tell us about a star?**

14.2 Stellar Distances: Parallaxes

The key to finding a star's luminosity from its flux is its distance. For nearby stars, we have a direct method to find distances: triangulation, similar to the procedure used by surveyors on the earth. Stellar triangulation is called **heliocentric** or **trigonometric parallax.**

You actually observe parallaxes all the time, although you are usually unaware of it. Put your hand out at arm's length with a pencil held upright (Fig. 14.6). Now alternately open and close each eye. The pencil will appear to jump back and forth relative to the distant background. This angular shift in the pencil's position is the parallax of the pencil. Now if you measure the amount (in angular measure), by which the pencil appears to have shifted and also the distance between your eyes, you can calculate how far the pencil is from your head.

Imagine that your pencil is a nearby star, the background more distant stars, and your eyes the sighting positions of the earth in orbit around the sun separated by a time of 6 months. From these two positions in the earth's orbit (separated by 2 AU), the nearby star appears to shift in position (its parallax) compared with the more distant stars (Fig. 14.7). Measure the angular size of that shift. Half of this angular shift is defined as the *astronomical parallax* of the star. Because you know the diameter of the earth's orbit, you can calculate the distance to the star. Note that the *farther away a star is, the smaller its parallax will be.* (Focus 14.2 shows this concept in more detail.)

You perceive parallax, then, when you view a relatively close object from each of two ends of a baseline. As the earth moves from one place in its orbit to another, a nearby star's position seems to change relative to the more distant stars. The maximum shift occurs when your viewings of the star are 6 months apart—that is, when you are sighting from opposite sides of the earth's orbit. Half the total angular shift is the star's *parallax.* Measure the amount of that shift. Then, because you know the diameter of the earth's orbit (2 AU), you can calculate the distance to the star. In essence, the parallax angle gives the size-to-distance ratio, with "size" in this case the size of the baseline and "distance" the distance to the star in units of the baseline.

The parallax of Sirius is 0.38 arcsec. The distance to Sirius is about 546,000 times the distance to the sun (546,000 AU), or a distance of 8.6 ly. From this information, we can calculate the luminosity of Sirius compared to the sun's luminosity from Sirius' flux. We measure the flux from Sirius at the earth. With the

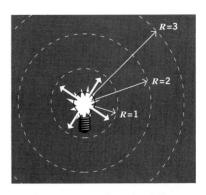

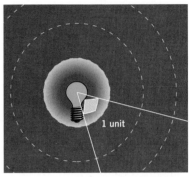

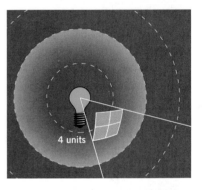

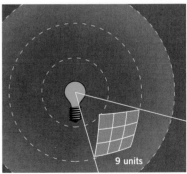

FIGURE 14.5 Geometry for the inverse-square law for light. Place a light bulb in a series of concentric, transparent shells. Since each larger shell has a greater area than its predecessor, light that covers one unit of area in the first shell covers 4 units in the second and 9 in the third. So the flux at the second shell is 1/4 the first; at the third, 1/9. The spreading out of the light dilutes the flux.

star's distance known, the inverse-square law gives us the rate at which we would receive energy from the star, *if* it were 1 AU from us. We compare that figure to the amount we do receive from the sun. From the comparison, we know how luminous the star is compared with the sun. For example, if we could move Sirius up to the sun, it would appear 23 times brighter than the sun. So Sirius is 23 times more luminous than the sun.

Heliocentric parallax works accurately only for close stars. Note that the parallax of Sirius, a very close star, is less than 1 arcsec, about 0.0005 the angular diameter of the moon! That's the size of a United States quarter at a distance of a little more than 5 km. Because parallaxes are so small, it should not surprise you that it was not until 1838 that the German astronomer F. W. Bessel (1784–1846) succeeded in measuring the first stellar parallax. People who measure parallaxes with ground-based telescopes have refined their techniques to an accuracy of about 0.001 arcsec, so they can get accurate distances out to about 300 ly.

Earth-based observers have a natural limitation in finding parallaxes: the size of the AU. A larger AU would make it possible, given the same techniques, to measure more distant stars reliably. An astronomer on Mars, for instance, has a 50 percent larger baseline to use and could go 50 percent farther out in distance. (Of course, the measurements would take longer because Mars has a larger orbital period than the earth.)

Many stars lie more distant than 300 ly. How are their distances measured? By ingenious, indirect methods. A spectroscopic procedure comes later in this chapter (Section 14.5). A special space telescope, called *Hipparchos,* is now measuring parallaxes of many stars with an accuracy about ten times better than can be achieved from the ground today. Astronomers expect that when the satellite's mission is done, we will know the parallaxes of some 22,000 stars with an error of about 0.0003 arcsec and of 120,000 or so with an error around 0.002 arcsec.

■ As the distance of a star increases, what happens to the size of its parallax angle?

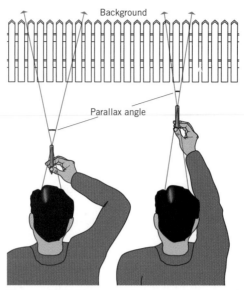

FIGURE 14.6 Observing the parallax of a pencil held at arm's length. As you alternately open and close each eye, the pencil appears to jump in position relative to the background markers. The angle of the apparent movement is the *parallax angle.* The farther away you hold the pencil, the smaller the parallax angle. The baseline for the shift is the distance between your eyes.

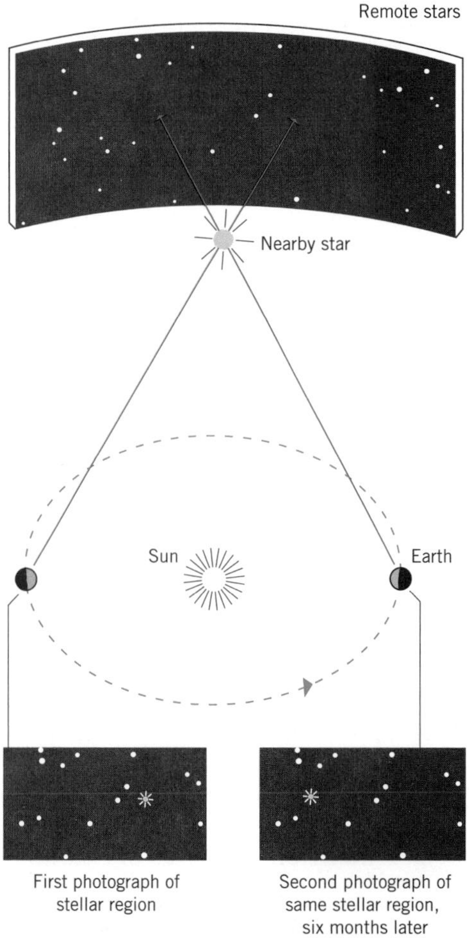

FIGURE 14.7 Heliocentric stellar parallax. As the earth goes around the sun, the positions of nearby stars shift relative to the positions of more distant stars in 6 months. The nearby star returns to its original relative position in a year. The closer the star, the greater the observable shift. The shift has a one-year cycle, the earth's period of revolution. The baseline for the shift is the diameter of the earth's orbit. Note that a larger baseline results in a larger shift for a star at the same distance.

Remote stars

Nearby star

Sun   Earth

First photograph of stellar region

Second photograph of same stellar region, six months later

Stars	
At a Glance	
Composition	Hydrogen, helium
Energy sources	Fusion reactions, gravitational contraction
Core	Site of fusion reactions, temperatures 10^7 K and higher
Luminosity	Total amount of energy produced each second
Photosphere	Source of stellar spectra

Enrichment Focus 14.2

HELIOCENTRIC (TRIGONOMETRIC) STELLAR PARALLAX

Suppose you travel out into space and look back at the earth–sun separation (1 AU). You keep going until the angular size of the AU is 1 arcsec. Now you station a star at this point and hurry back to the earth. You observe this star for one year and measure its shift. Half of the total shift for one year would equal 1 arcsec. We say the star is 1 **parsec** from the sun—that is, the distance at which the *par*allax is one *sec*ond of arc—a distance roughly equal to 3.26 light years. Suppose the star were twice as far away; it would have half the parallax, 0.5 arcsec. At half the distance, its parallax would be double, that is, 2 arcsec. Note this simple inverse relationship of parallax and distance. Algebraically,

$$d(\mathrm{pc}) = \frac{1}{p\,(\mathrm{arcsec})}$$

where d is the distance and p is the parallax.

An example: the star 40 Eridani has a parallax of 0.2 arcsec. How far is the star from the sun? Since the star's parallax is one-fifth that for a star 1 parsec away, 40 Eridani must be 5 pc distant. Explicitly,

$$d = \frac{1}{0.2\ \mathrm{arcsec}}$$

$$= 5\ \mathrm{pc}$$

Here's a simple geometric explanation for the parallax formula. Travel out in space to some star. Then draw an imaginary circle (Fig. F.11) centered on the star and through the sun *(S)* so that the circle's radius is d. Note that the parallax angle, p (in degrees), is some fraction of the total circle (360°). Also, R, the earth–sun distance (1 AU), corresponds closely to an arc on the

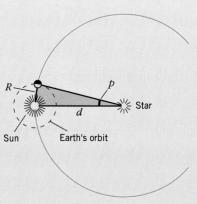

FIGURE F.11 Geometry for heliocentric parallax.

circumference of the circle; its fraction of the circumference is the same as the fraction p is of 360°. So

$$\frac{R}{2\pi d} = \frac{p\,(°)}{360°}$$

or

$$d = \frac{360°}{2\pi}\frac{4}{p\,(°)}$$

Convert 360° to seconds of arc:

$$d = \frac{360 \times 60 \times 60}{2\pi}\frac{R}{p\,(\mathrm{arcsec})}$$

$$= 206{,}265\ \frac{R}{p\,(\mathrm{arcsec})}$$

Now R is 1 AU; define 1 pc as 206,265 AU. Then

$$d(\mathrm{pc}) = 206{,}265\ \frac{(1/206{,}265\ \mathrm{pc})}{p\,(\mathrm{arcsec})}$$

or

$$d(\mathrm{pc}) = \frac{1}{p\,(\mathrm{arcsec})}$$

Using light years as the unit of distance, we have

$$d(\mathrm{ly}) = \frac{3.26}{p\,(\mathrm{arcsec})}$$

14.3 Stellar Colors, Temperatures, and Sizes

Let's return to the stars in the winter sky. I've dealt with one observable property: their fluxes. Another property you can observe is color. Betelgeuse looks flaming reddish, Rigel and Sirius appear bluish white, and Capella shines yellowish white. What can we learn from these differences?

Color and Temperature

The different colors in Betelgeuse, Rigel, Sirius, and Capella suggest different **stellar surface temperatures.** (Recall Section 13.3 about the sun's surface temperature.) Rigel is the hottest of the four; it is

so hot that it emits more blue light than any of the other visible colors. In contrast, Betelgeuse is the coolest, for it radiates mostly red light. The colors of stars are a direct function of their surface temperatures.

Caution. When we look at starlight with our eyes, we see only the visible part of its entire spectrum. Stars, in fact, radiate very much like blackbodies (Focus 13.2), so their continuous spectra, from shortest to longest wavelengths, have the characteristic blackbody shape, called a *Planck curve.* The visible range covers but a small portion of a blackbody's spectrum.

One way to measure a star's temperature is to find the peak of its Planck

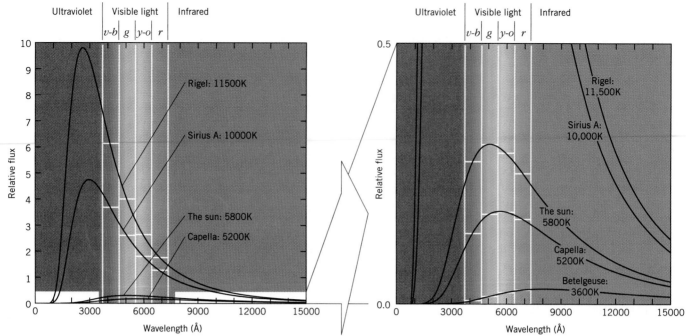

FIGURE 14.8 Continuous blackbody spectra. The fluxes of blackbodies with temperatures equivalent to those of Rigel, Sirius, the sun, Capella, and Betelgeuse compared in the color bands violet-blue (v-b), green (g), yellow-orange (y-o), and red (r). These are compared to continuous spectra of different temperatures (that match closely that of each star). The blackbody spectra have the distinctive shape known as a Planck curve. Note that the visible range is only a small part of the total spectrum; the color of a star depends on the eye's perception of a star's continuous spectrum in the visible range. The relative flux scale is much larger for the part of the figure on the left.

curve (Fig. 14.8). The hotter the star, the shorter the wavelength of the peak (Focus 13.1). Although simple, this technique has two drawbacks. First, we need to measure a wide range of the spectrum to find the peak. Second, at the ground, the earth's atmosphere may absorb the radiation at the wavelength of the peak. For example, the hottest stars have spectra that peak in the ultraviolet, which doesn't make it through the atmosphere to a ground-based telescope.

But the temperature of a blackbody can also be worked out by measuring the *relative* brightness at any *two* wavelengths—that's the meaning of *color* (Fig. 14.9). A reddish star does not emit red light only, but the relative amount of the longer (red) wavelengths is greater than that at the shorter (blue) wavelengths. All stars emit light over the entire visible range, but each shows relatively different amounts of red and blue light.

Temperature and Radius

One reason I've focused on color is that the stellar temperatures obtained from the colors are related to the sizes of stars, if the stars radiate at least somewhat like blackbodies.

How can this estimate be done? In the constellation Scorpius, you can find the bright reddish star Antares (see the summer star chart in Appendix G). If you

view Antares through a telescope, you'll find it has a faint bluish-white companion. Antares is a binary star (more on this in Section 14.6); it and the bluish-white companion revolve around each other, bound by gravity. The red star is called Antares A, the bluish-white companion Antares B. Their surface temperatures turn out to be roughly 3000 and 15,000 K, respectively.

We receive about forty times more flux from Antares A than Antares B. Now both stars lie at the same distance from the earth, so the difference in flux cannot arise from different distances; it must come from differences in the radiative output at the stars' surfaces. So we conclude that Antares A is forty times more luminous than Antares B. How can a

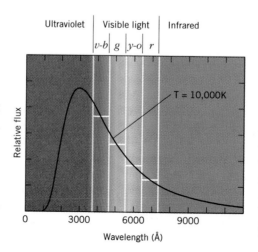

FIGURE 14.9 For a 10,000 K blackbody curve, note that the flux in the blue range is greater than that in the red. The ratio of these fluxes, measured accurately, defines the *color* of the star. This measurement can be applied to other stars of different temperatures.

cool star be so much more luminous than a hotter one (especially in light of the fact that hotter stars emit more power per surface areas than cooler ones)?

Recall (Focus 13.2) that the amount of power emitted by *each unit of area* of a blackbody's surface depends only on its temperature—in fact, on the fourth power of the temperature. For example, Antares B is about five times hotter than Antares A, so each square meter of Antares B emits 5^4, or 625 times the power of Antares A. If Antares A had 625 times the surface area of Antares B, both stars would have the same luminosity. Because Antares A has forty times the luminosity of Antares B, it must have 40×625, or 25,000 times the surface area.

To sum up. A star's luminosity is related to its surface temperature (which determines how much power each square meter emits) *and* its surface area (which determines the total number of square meters doing the emitting).

Now you have a way to infer the radius of a star from its luminosity and surface temperature. As an example, take the star Capella. Its surface temperature is 5200 K, almost the same as the sun's. But its luminosity is 130 times that of the sun. If Capella radiates like a blackbody, its radius is about fourteen times larger than that of our sun.

Direct Measurement of Diameters

If we could measure a star's angular diameter and its distance, we could calculate its actual diameter directly. Because stars are so far away, however, their disks have very small angular diameters—on the order of a few milliarcseconds (one *milliarcsecond* is 10^{-3} arcsec), far too small to resolve with earth-based telescopes. (Stellar images rarely have angular diameters less than about one arcsecond, even during excellent seeing conditions. Lunar-based optical telescopes will be able to achieve such resolutions.)

One method of obtaining the high angular resolution required involves *lunar occultations*. The moon *occults* a star when it passes in front of the star as viewed from the earth. Occultation observers make very rapid measurements of a star's light as the moon occults it. Because we know well the moon's angular speed in the sky, we can find out a star's angular diameter from accurate timing of the occultation. From the angular diameter, we can then find the actual diameter if we know the distance to the star.

One star measured this way is Aldebaran, the eye of Taurus the Bull. The results: Aldebaran's radius is 3.3×10^7 km, or about fifty times larger than the sun's radius. That's about 20 percent smaller than the size inferred from temperature and luminosity (and the assumption that the star radiates like a blackbody). The angular measurement is likely the better value because it is a direct measurement,

■ **If two stars have the same radius, and one has a higher surface temperature, which one is more luminous?**

14.4 Spectral Classification of Stars

The absorption (or dark-line) spectra of many stars resemble the spectrum of the sun. From the sun's spectrum, astronomers infer its photospheric composition. The same procedure can be applied to stars. But there's one trick: a star's spectrum is affected by its *surface temperature* as well as by its composition.

Temperature and the Balmer Lines

We now know that most stars have pretty much the same composition, but their spectra are *not* all alike. Consider, for example, the Balmer lines of hydrogen (Fig. 14.10). Stars cooler than the sun have spectra with weaker (less intense) Balmer lines. Stars somewhat hotter than the sun show stronger (darker) Balmer lines. But stars *much* hotter than the sun again have spectra with weak Balmer lines.

How to understand this? Recall (Section 5.3) that the Balmer series arises from transitions between the second energy level of hydrogen and any level

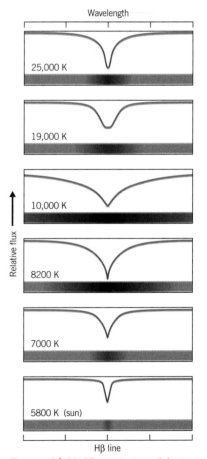

FIGURE 14.10 The variation of the intensities (darkness) of one Balmer line of hydrogen (the second line in the series, the Hβ line) in stars with different temperatures. The profile of the line in the sun is at the bottom. Note how the intensity (total area) of the line increases from higher temperatures until it reaches a maximum at 10,000 K stars; it then decreases.

above it; these transitions run upward for absorption and downward for emission. To absorb a Balmer line, a hydrogen atom must be excited to level 2. In the sun, only one of every 10^8 atoms is excited; not enough energetic collisions occur to kick up many electrons from the ground level. So the Balmer lines in the sun's spectrum are not very strong. In a cooler star, fewer energetic collisions occur, and fewer hydrogen atoms are excited. The Balmer lines are even much weaker. In a hotter star with more energetic collisions, more atoms are excited, about one out of every million. So the Balmer lines are more intense (darker) in such a star. If the star is sufficiently hot, collisions are so violent that many electrons are knocked out of the atom entirely, leaving hydrogen *ions* (protons) behind. So with fewer excited atoms, the Balmer lines are weaker (less dark).

The key point. Balmer lines of hydrogen are produced by atoms with an electron in the second energy level; that is, the atom must be *already* excited. If for some reason a star doesn't have very many hydrogen atoms excited to the second level, the Balmer lines in that star will be weak. This can happen if the star has a very high temperature, so that virtually all its hydrogen is ionized, or if the star has a relatively low temperature, so that even though there is much neutral hydrogen, there are very few excited atoms in the second level.

Spectral Classification

At the turn of this century, workers at Harvard College Observatory classified stellar spectra by using absorption lines, especially hydrogen Balmer lines. The results were codified in the *Henry Draper Catalog*. This assignment first fell to Williamnia Fleming (1857–1910), who selected 10,351 stars for the classification scheme, and supervised a staff of women to bring the catalog into final form.

Much of the final task of developing an expanded catalog in 1924 was done by Annie Jump Cannon (1863–1941; Fig. 14.11), who single-handedly classified the spectra of more than

250,000 stars! The original classification scheme was set up strictly on the basis of the strength of various lines (Fig. 14.12), well before there was any understanding of the effects produced by different temperatures. The Balmer lines played a vital role in this scheme: stars with the strongest Balmer lines were called class A, those with slightly weaker lines class B, and so on. Some classes were later dropped because they contained too few stars or only very peculiar ones, and the order was rearranged to one of decreasing temperature, once the explanation of the line intensity in terms of temperature was understood.

The **stellar spectra sequence** from hotter to cooler in the Harvard classification now runs O-B-A-F-G-K-M. The standard mnemonic for the sequence is: *Oh, Be A Fine Girl* (or *Guy*), *Kiss Me!* Almost all stellar spectra fit into this sequence. The O stars have spectra with weak Balmer lines of hydrogen and lines of ionized helium. The A stars have the strongest Balmer lines. In F stars the Balmer lines fade and many other lines appear, mostly of metals. The sequence from type O to M, looking at the continuous spectra from the stars, is also a color sequence. O stars appear bluish white, G stars yellowish, and M stars reddish. Astronomers further divide each class into subclasses from hotter to cooler, labeled from 0 to 9 (for example: G0, G1, G2, G3, G4, G5, G6, G7, G8, G9). Each subclass is distinguished by slightly different intensities of specific absorption lines.

The strengths of the Balmer lines suggest that the differences in stellar spectra reflect primarily differences in *temperatures,* not in *abundances.* These temperature differences result in different degrees of ionization and excitation of the atoms in the star. The strength of the atom's spectral lines is determined by the number of atoms that are excited and the number that are ionized. This revelation in twentieth-century astrophysics occurred because of the work of an Indian physicist, Meghnad N. Saha (1893–1956), who proposed in 1920 that differences in temperature dramatically controlled the ionization conditions for elements in stellar atmospheres. In 1925

FIGURE 14.11 Annie Jump Cannon (left) and Henrietta Swan Leavitt (right) did fundamental work in spectroscopic astrophysics at Harvard College Observatory. (Harvard College Observatory photograph.)

Temperature
(Type)

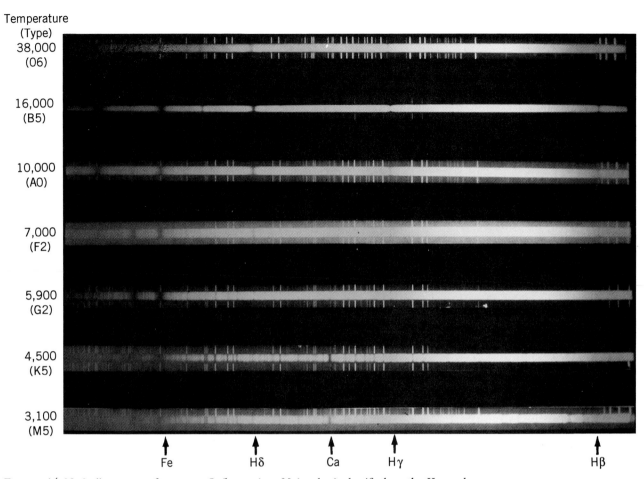

FIGURE 14.12 Stellar spectra from type O (hottest) to M (coolest) classified on the Harvard system. Note these differences among the spectra: (1) the intensities of the hydrogen Balmer lines, and (2) the intensities of elements such as calcium and iron, which appear strong in some spectra and do not appear at all in others. (Courtesy of Palomar Observatory.)

in her Ph.D. thesis, Cecilia Payne-Gaposhkin (1900–1979) applied Saha's ideas to the spectra in the Harvard collection. She came up with the basic temperature classification sequence (described above) from O to M stars, thereby confirming that temperature plays the key role in the spectra. Payne-Gaposhkin determined that stars were made mostly of hydrogen and helium—contrary to the notions of the time that stars were made of stuff like the earth.

Consider first O stars, which have the hottest surface temperatures, 30,000 K and higher (Fig. 14.12). At these high temperatures, atoms collide violently. The energies in such collisions can rip the electrons from hydrogen atoms so that most of the hydrogen is ionized. Very few neutral atoms remain to absorb

at wavelengths corresponding to the Balmer series. Because most of the hydrogen is ionized in O stars, the Balmer lines are weak. This is also true for B stars, whose temperatures range from 11,000 to 30,000 K (Fig. 14.13).

Now turn to A stars (Figs. 14.12 and 14.13), whose surface temperatures range from 8000 to 11,000 K. Atoms and ions collide less violently, and most of the hydrogen is neutral. To absorb Balmer lines, the hydrogen atom must have its electron in the *second* energy level. Although the collisions are not strong enough to ionize the hydrogen, they do possess the energy to *excite* the electrons out of the lowest energy level. In A stars, many hydrogen atoms are excited by collisions with other atoms, so their electrons are in the second level. These excited atoms readily absorb light at the Balmer wavelengths, and the lines appear strongest.

In F and G stars (Figs. 14.12 and 14.13), surface temperatures are still lower, typically 6000 K. Very few hydrogen atoms are ionized. In addition, the impacts between atoms do not have enough energy to excite very many hydrogen atoms; so most have their electrons in the lowest energy state. Such atoms cannot absorb at Balmer wavelengths. As a result, the Balmer lines disappear almost completely. However, the lines from singly ionized calcium become stronger.

The variation in Balmer line intensities arises from collisions that excite and ionize atoms. How much the collisions ionize or excite depends on temperature. So each spectral type corresponds to a restricted range of surface temperatures, which are listed in Table 14.1.

Other lines from other elements can be analyzed in a fashion similar to that for the Balmer lines (Table 14.1). You can see that the observation of stellar spectra coupled with an understanding of the atom gives astronomers information about the physical conditions in the atmosphere of a star. An analysis of spectral lines based on atomic theory

also provides information about the abundance of elements in stars in the same way as for the sun. Astronomers have found that, just like the sun, stars consist mostly of hydrogen and helium. (In Chapter 16, you'll see that stars do differ somewhat in composition relative to their abundance of heavy elements,

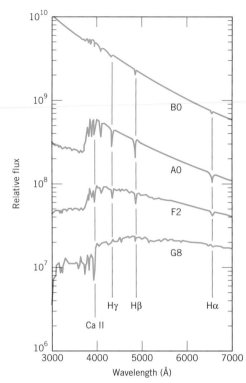

FIGURE 14.13 Comparison of the spectra for stars of spectral types B0 (top; surface temperatures = 30,000 K), A0 (surface temperature = 10,000 K), F2 (surface temperature = 7000 K), and G8 (surface temperature = 5600 K). The hydrogen Balmer series is shown, as well as the lines from singly ionized calcium. (Adapted from a diagram by A. Hernden and R. Kaitchuck.)

TABLE 14.1 FEATURES OF THE STELLAR SPECTRAL CLASSES

SPECTRAL CLASS	COLOR	SURFACE TEMPERATURE (K)	PRINCIPAL FEATURES	EXAMPLES
O	Bluish-white	30,000	Relatively few absorption lines. Lines of ionized helium and other lines of highly ionized atoms. Hydrogen lines appear only weakly.	Naos
B	Bluish-white	11,000–30,000	Lines of neutral helium. Hydrogen lines more pronounced than in O-type stars.	Rigel, Spica
A	Bluish-white	7,500–11,000	Strong lines of hydrogen. Also lines of singly ionized magnesium, silicon, iron, titanium, calcium, and others. Lines of some neutral metals show weakly.	Sirius, Vega
F	Bluish-white to white	6,000–7,500	Hydrogen lines are weaker than in A-type star but still conspicuous. Lines of singly ionized metals are present, as are lines of other neutral metals.	Canopus, Procyon
G	White to yellowish-white	5,000–6,000	Lines of ionized calcium are the most conspicuous spectral features. Many lines of ionized and neutral metals are present. Hydrogen lines are weaker even than in F-type stars.	Sun, Capella
K	Yellowish-orange	3,500–5,000	Lines of neutral metals predominate.	Arcturus, Aldebaran
M	Reddish	3,500	Strong lines of neutral metals and molecules.	Betelgeuse, Antares

TABLE 14.2 PROPERTIES OF KEY STARS IN THE WINTER SKY

STAR	SPECTRAL CLASS	SURFACE TEM-PERATURE (K)	RADIUS (sun = 1.0)	DISTANCE (light years)	LUMINOSITY (sun = 1.0)[b]
Epsilon Orionis	B0	24,800	37	1600?	470,000?
Rigel	B8	11,550	74	880?	90,000?
Regulus	B7	12,210	3.6	69	270
Sirius A	A1	9,970	1.7	8.7	23
Procyon A	F5	6,510	2.1	11	7
Sun	G2	5,780	1.00	1.6×10^{-5}	1.00
Capella[a]	G5	5,200	14	41	130
Epsilon Eridani	K2	5,000	0.7	11	0.28
Aldebaran	K5	3,780	61	60	700
Betelgeuse	M2	3,600	1200	1400?	21,000?
Sirius B	White dwarf	30,000	0.0073	8.7	0.003

[a]Capella is a double star. The temperature, radius, and luminosity are those of the brighter and cooler component.

[b]The luminosities and radii of Epsilon Orionis, Rigel, and Betelgeuse are only approximate because their distances are estimated.

but hydrogen and helium still make up the bulk of their mass.)

■ **What produces the Balmer lines of hydrogen?**

14.5 The Hertzsprung–Russell Diagram

Table 14.2 summarizes the astronomical and physical properties of key stars in the winter sky. Examine this table for a minute. Can you find any pattern in it? Note the wide range of luminosities and sizes and the smaller range of temperatures. How are these properties related to the internal anatomy of the stars? Again we face the astronomer's dilemma: how to find out vital information from points of light in the sky. The solution lies in the spectroscopy of these stars.

Temperature Versus Luminosity

Let's turn the table into a picture: a temperature–luminosity diagram (Fig. 14.14) based on the spectral types of the stars. Such diagrams were independently set up early this century by Ejnar Hertzsprung (1873–1967) and Henry N. Russell (1877–1957). In their honor, such a plot is called a **Hertzsprung–Russell diagram** (commonly abbreviated *H–R diagram*). Note that astronomers follow

the strange convention of plotting temperature so that it increases to the *left* on the horizontal (temperature) axis of the H–R diagram.

Examine Figure 14.14. Notice Sirius B in a corner by itself and Betelgeuse alone in the upper right-hand region. Regulus, Sirius A, Procyon, the sun, and Epsilon Eridani fall close together along a sloping line. No obvious patterns, though, appear in this plot.

Now inspect a different H–R diagram, one for the nearest stars, all those within 20 ly of the sun (Fig. 14.15). Notice that most of the stars are less luminous and cooler than the sun. (In fact, we can see them only because they are so close to us.) The star *Alpha Centauri A* has almost the same luminosity and

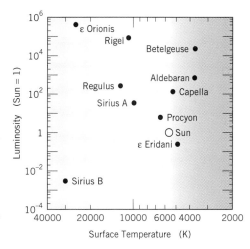

FIGURE 14.14 Temperature–luminosity diagram for important stars in the winter sky (see Table 14.2). The sun is included for comparison. The vertical axis is luminosity (increasing upward) in solar units; the horizontal axis is surface temperature, which increases to the left rather than to the right as you might expect. This is a historic convention that astronomers have not yet changed.

temperature as the sun. This star is the sun's twin and is also the star nearest to us. (Alpha Centauri A is one member of a triple-star system with Alpha Centauri B and Proxima Centauri, and in this sense is unlike the sun.)

Finally, a pattern! The stars' properties clearly do *not* fall in a random scatter in Figure 14.15. Rather, there is a trend; if you draw a line through the points from luminous, hot Sirius A to the coolest, faintest star in the lower right-hand corner, you have identified the **main sequence.** Most nearby stars fall on the narrow strip of the main sequence in the H–R diagram. The few stars in the lower left-hand corner, Sirius B included, have very high surface temperatures but low luminosities, so they must be very small. Such a peculiar star is called a **white dwarf** star.

Consider another H–R diagram (Fig. 14.16), one for the brightest stars you can see in the sky. Compare it with the preceding plot. What a difference! Almost all these stars have a much higher luminosity than the sun. And many of them are also much hotter stars (of spectral classes O and B). The main sequence no longer appears so obvious. Only a handful of stars show up in both diagrams.

What are the physical differences among these stars? Take Betelgeuse, whose properties put it in the upper right-hand corner of the H–R diagram. Here is a star whose surface is much cooler than the sun's. So if Betelgeuse were the same size as the sun, it would be much less luminous. But Betelgeuse has a luminosity 21,000 times that of the sun. To be so much cooler and more luminous than the sun, Betelgeuse must be very much larger (Section 14.3), about 1200 times the size of the sun—a star so big that it could swallow up Mars if it were placed in the center of our solar system! Astronomers call Betelgeuse a *supergiant* star.

Here is the reason that the H–R diagram for the nearest stars (Fig. 14.15) differs from that for the brightest stars (Fig. 14.16): the first diagram contains ordinary stars with sizes like that of the sun; no giants are among the nearest stars, for they are very rare. The sun is a main-sequence star. Most of the stars in

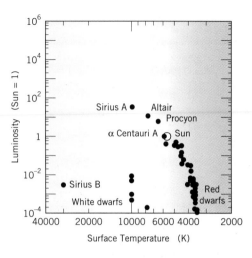

FIGURE 14.15 A Hertzsprung–Russell (luminosity–temperature) diagram for the stars nearest to the sun. Note that the vertical axis is luminosity in solar units; the sun's luminosity is 1.0. The horizontal axis is surface temperature in kelvins. The sun and Alpha Centauri A have essentially the same values of the luminosity and temperature, so their points overlap. Except for the white dwarfs, the stars' points fall along a narrow band called the *main sequence.* See Appendix D for information on these stars.

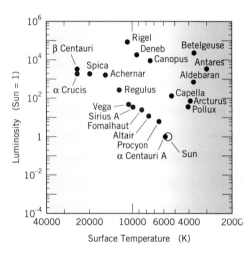

FIGURE 14.16 A Hertzsprung–Russell (luminosity–temperature) diagram for the brightest stars in the sky. Axes are the same as for Figure 14.14. Again, the sun and Alpha Centauri A have almost the same luminosity and surface temperature, and so they overlap on this diagram. The stars split into two groups: one rises to the left and the other to the right. Rigel and Betelgeuse, both in Orion (Fig. 1.1) fall at the tops of these ranges. Appendix E contains information on these stars.

the sun's vicinity are also main-sequence stars, of spectral class M. So they are cool, not very luminous, and of lower mass. The second diagram contains many **giant** and **supergiant** stars, still visible among the closer stars because of their high luminosity, even though they are scattered widely through space.

Now piece these two diagrams together. Most stars fall on the gentle curve of the main sequence (as in Fig. 14.17). A scattering of stars cuts across the tip of the diagram; these are the very luminous supergiants. A group of luminous stars extends off the main sequence, these are the giants. Finally, note the white dwarf stars at the bottom of the diagram.

You might have heard that the sun is a typical or average star. In some sense this is true, for the sun is neither the largest nor the smallest kind of star. However, it is *not* true that most stars in the Galaxy are like the sun. In the solar

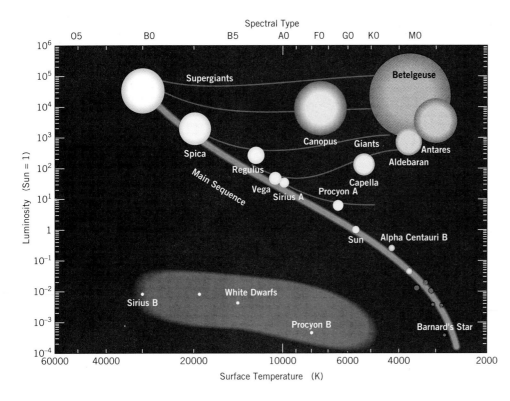

FIGURE 14.17 True-color representation of the Hertzsprung–Russell diagram. The horizontal axis is temperature; the vertical one, luminosity, in units of the sun's luminosity. The solid white lines show where stars of different luminosity classes fall on the diagram: Ia supergiants at the very top, Ib supergiants below, III giants just below them, and finally V main-sequence stars. The relative sizes of the stars are shown correctly within each luminosity class, but not between them. The colors are those as perceived by the eye looking at these stars through a telescope.

neighborhood, G stars like the sun are relatively rare. The most common kind of star in our immediate vicinity, and in fact in the Galaxy in general, is a main-sequence M star, a cool, reddish star of very low luminosity, on the lower end of the main sequence (Fig. 14.18).

Spectroscopic Distances

Once we have an H–R diagram for many stars, we can use it to infer approximate distances to stars. How? First, determine the spectral type of a star. Suppose it's an M star. Then look at the H–R diagram to find the luminosity. Measure the flux, and use the inverse-square law for light (Section 14.1) to find the distance from the luminosity and the flux.

But a problem arises: you can see that M stars have a wide *range* of luminosities, from 10^{-5} solar luminosity for main-sequence M stars to about 10^5 solar luminosities for supergiant ones. How do we decide what luminosity an M star has over a great range of 10^{10}?

Fortunately, we can tell from a star's spectrum. Recall that the strengths of dark lines relate to a star's temperature. The energy of the collisions depends on the temperature in a star's atmosphere. But for gases at the same

temperature, the *rate* of collisions depends on the density of the gases. So in a denser gas, collisions are more frequent than in a less dense gas at the same temperature.

Giant stars, because they are so huge, typically have atmospheres of lower density than main-sequence stars. The more frequent collisions in main-sequence stars make certain absorption lines in their spectra appear to be broader than the same lines in spectra of giant or supergiant stars (Fig. 14.19). So a star's size, hence its luminosity, is given indirectly by the *widths* of certain absorption lines when comparing the spectra of stars of the same spectral type. (Different lines are used for different spectral types.)

Such an analysis reveals that stars fall into **luminosity classes.** The recognized luminosity classes (Fig. 14.20) are Ia, most luminous supergiants; Ib, less luminous supergiants; II, luminous giants; III, normal giants; IV, subgiants, and V, main-sequence stars. The sun falls into luminosity class V. (White dwarfs do not have luminosity classes because they are all essentially the same size.)

So a star's spectrum allows it to be classified by spectral type *and* luminosity class. For a given spectral type, you

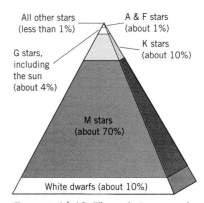

FIGURE 14.18 The relative populations of stars in the Milky Way Galaxy. This pyramid, which shows the relative numbers of common stars, illustrates that the sun is not an "average" or "typical" star.

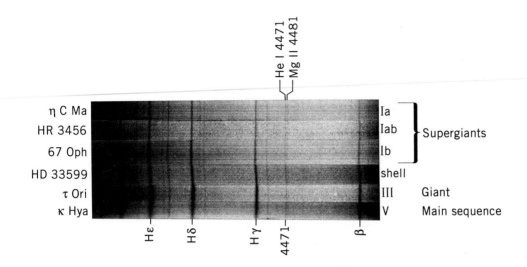

FIGURE 14.19 Luminosity class and spectra of stars all having the same surface temperature (spectral class B5). The star names are listed to the left of each spectrum. The top three spectra are those of supergiant stars. (Type Iab lies between types Ia and Ib.) The next is a peculiar type of star that has a shell of material around it. Below this one is a giant, and on the bottom a main-sequence star. Note the different intensities of the dark lines, especially those of the hydrogen Balmer series. These lines become broader and strong from top to bottom, and so allow astronomers to infer the luminosity class of the stars. (Courtesy of Nancy Houk; from *An Atlas of Objective-Prism Spectra* (1984), by N. Houk, N. J. Irvine, and D. Rosenbush.)

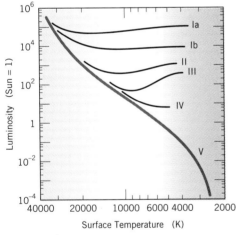

FIGURE 14.20 Luminosity classes of stars on the H–R diagram. The classes run from I, the largest supergiants, to V, stars like the sun. Classes Ia and Ib are supergiants; class II, luminous giants; class III, normal giants; class IV, subgiants; and class V, main-sequence stars. The lines in this diagram represent the center of the range for each class.

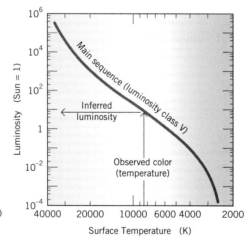

FIGURE 14.21 Using spectral class and luminosity class (in this case V, main sequence) to infer the luminosity and so the distance to a star, if its flux is known. From a star's color (or spectrum), estimate its temperature on the horizontal axis. Move straight up until you hit the main sequence, and then move horizontally to the left to find the star's luminosity. Comparing the luminosity to the measured flux allows an inference of the distance, because the flux varies as the inverse-square law for light.

can estimate the luminosity (Fig. 14.21) within a range of probable error (the width of a luminosity class on the H–R diagram). If you know a star's luminosity and its flux, you can calculate its distance from the inverse-square law for light.

The procedure for working out distances from spectra is called **spectroscopic distances.**

In summary. To make an H–R diagram, first find stars close enough to the sun to measure their distances reliably by parallax. Then calculate their luminosities from their distances and fluxes. Next, take spectra of the stars to find out their spectral class. From the spectra, you determine how hot the stars are. Then plot the luminosity and spectral type.

The result is a calibrated H–R diagram—calibrated in the sense that you have the *luminosities* of the stars plotted against their *surface temperatures*. The H–R diagram graphically summarizes some of the important physical properties of stars. It also serves as a visual sorting aid to bring to light different classes of stars. And it can be used as a tool for obtaining the distances of other stars.

■ What type of star has a high temperature but a low luminosity?

14.6 Weighing and Sizing Stars: Binary Systems

You have seen how stars differ in properties such as luminosity and size. What about mass? Some stars are larger than the sun, some smaller. But sizes do not tell us directly whether a star is more or less massive than the sun. How to find a star's mass?

Binary Stars

We have no direct way of knowing the mass of an isolated star. To find masses, we examine the gravitational effects of one object on another. Recall how we find the sun's mass (Section 13.1): we look at the acceleration of the earth as it orbits the sun. Similarly, we use the accelerations of two stars orbiting one another to find their masses. Two stars, bound by their mutual gravity and revolving around a common center of mass, are called **binary stars.**

If both stars are visible, we can trace out their orbital motion by observing them over a long time, which gives us the angular size of the orbit and the orbital period. But that's not enough to get the masses! In addition, we need to find the distance to the binary system so that we can convert their angular separation into a physical one. Next, it is likely that the plane of a star's orbit is tilted from a direct face-on view; this orbital tilt needs to be accounted for. Then we have enough information to find the *sum* of the masses from Newton's revised form of Kepler's third law (Focus 4.3). To find the *individual* masses, we need one more piece of information: we must know how far each star is from the

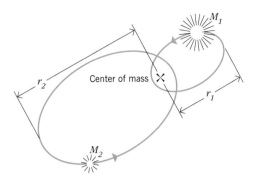

center of mass of the system. (The *center of mass* is the balancing point between the stars, as if they were on two ends of a seesaw. Refer back to Section 4.5.)

In a binary system each star orbits the center of mass at a distance inversely proportional to its mass. The more massive star lies closer to the center of mass (Fig. 14.22). For instance, in Figure 14.22, M_2 lies 4.5 times farther from the center of mass than does M_1. That means that M_1 has 4.5 times the mass of M_2. As the system travels through space, and so across our line of sight, the center of mass traces a straight line (Fig. 14.23), while the two stars spiral around it. This corkscrew motion identifies the stars as a binary system and locates the center of mass.

Remarkably enough, *most* stars are in binary or multiple star systems. For nearby sunlike stars, about 45 percent are probably single, while some 55 percent are known to be double, triple, and even quadruple! For example, Alpha Centauri is a triple system. Two of the stars, Alpha Centauri A and B orbit each other with a separation of about 20 AU. The third star, called Alpha Centauri C (or Proxima Centauri), orbits a few thousand AU from them.

FIGURE 14.22 Center of mass in a binary star system. Both stars (M_1 and M_2) move in elliptical orbits (r_1 and r_2) around the center of mass. The more massive star is the one closer to the center of mass: in this case, star M_1.

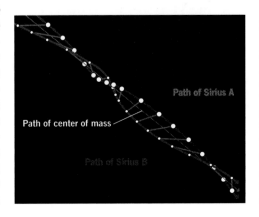

FIGURE 14.23 Motions of the binary star system consisting of Sirius A, and its companion, Sirius B, showing the motions of Sirius A and its companion about their center of mass, and that of the center of mass of the two stars relative to background stars, as the center of mass moves through space. The two stars trace a corkscrew path against the sky. Since, however, Sirius A is much brighter than Sirius B, its motion around of mass is easier to observe.

Types of Binary System

All binary systems are physically the same, but we observe them in different ways. This fact of astronomical life has prompted astronomers to divide binaries into three general classes: *visual, spectroscopic,* and *eclipsing.*

In a **visual binary,** a telescope clearly shows both stars. With enough observations, we can trace the orbital path of the fainter star (called the *secondary*) around the brighter one (the *primary*). Because a supposed pair may be only an accidental line-of-sight juxtaposition of two physically separate stars, we must sometimes wait many years to confirm that the two stars really are bound by gravity. As you'd expect from Kepler's third law, binary stars with large orbital separations have long orbital periods.

Suppose two stars are so close together that we cannot resolve them with a telescope, and their orbital periods are so short that the stars move quickly in their orbits. We can identify this binary by looking for two sets of lines in the spectrum (one from each star) and measuring the Doppler shifts (Section 10.2 and Focus 10.1) produced by the orbital motion. This pair of stars is a **spectroscopic binary.** Note we can observe the Doppler shifts as long as the orbit is *not* face-on as we view it.

The Doppler Shift for Binary Stars

Let's apply the Doppler shift to binary stars. Imagine the more massive star to be stationary, with the secondary revolving about it. As the secondary recedes from the earth, you see its spectral lines redshifted compared with those of the primary; as the secondary approaches, you see its lines blueshifted (Fig. 14.24). At the intermediate points, when the secondary travels across the line of sight, you see no shift.

If the two stars do not differ greatly in luminosity, both spectra can be observed, especially the cycle of shifts of the secondary with respect to the primary. (The smaller-mass star will have the higher velocity, because it is farther from the center of mass.) Sometimes the spectrum of the secondary is too faint to be seen. We then use the Doppler shift in the primary's spectrum alone (but we get less information).

These wavelength shifts can be turned directly into radial velocity shifts by using the Doppler effect (Focus 10.1). We use the radial velocities and the period to get the circumference of the orbit. Then we work out the radius of the orbit and so the separation of the two stars. So a spectroscopic binary gives us direct information on the system's orbit.

Note that we can measure the *actual* velocity of the stars in kilometers

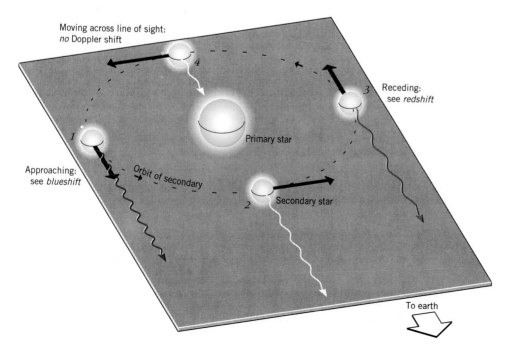

FIGURE 14.24 The cycle of Doppler shifts from a spectroscopic binary system. Consider the primary star to be fixed (not revolving around the center of mass). When the secondary is at position 1, it is moving toward the earth (negative radial velocity); so its spectral lines appear blueshifted. At 2, the secondary moves across our line of sight, so we see no shift. The same is true at 4. Finally, at position 3, the secondary moves away from us (positive radial velocity), so we see a redshift. So from 1 to 2, we see a decreasing blueshift, from 2 to 3 an increasing redshift, from 3 to 4 a decreasing redshift, and so on.

per second; we get the *actual* radius of the orbit, in kilometers, for each star, relative to the center of mass. We can use this information, along with Kepler's third law, to determine the stars' *individual* masses.

Eclipsing Binary Stars

The basic difficulty is that we cannot get the stars' individual masses unless we know the *tilt* of the orbit with respect to our line of sight. Why? A system with a tilted orbit has its stars moving toward us and across the sky at the same time, and so the Doppler shift doesn't give us the total velocity, just a part of it. Generally, we don't know the system's tilt with respect to our line of sight. But in a few cases, the orbits are tilted just so that one star passes in front of the other, producing an eclipse. That's an **eclipsing binary system.**

Algol (Arabic for "demon star"), in the constellation Perseus, is the prototype of eclipsing binaries. It has a period of 2.87 days and, in mideclipse, plummets sharply in brightness (Fig. 14.25*a*). Algol A, the brighter star (260 solar luminosities) is a B8 V (surface temperature 12,500 K); its companion, Algol B, at 5 solar luminosities, is a fainter K2 IV star (surface temperature, 4500 K; see Fig. 14.25*b*). From the measured radial velocities and the orbital period, Algol A has a mass 4.6 times that of Algol B. From Kepler's third law, the total (combined) mass of the system is 4.5 solar masses. So Algol A is 3.7 solar masses and Algol B 0.8 solar mass.

Eclipsing binaries give us another property of the stars directly: their diameters. When Algol B passes in front of Algol A (primary eclipse, during which the less luminous star passes in front of the more luminous one), the duration of the eclipse depends on the diameter of B relative to A and the relative orbital speeds. When Algol A swings in front of Algol B (secondary eclipse, during which the more luminous star crosses in front of the less luminous one), the duration depends on the diameter of A relative to B and the relative orbital speeds. So we can find the radius of each star: 2.9 solar radii for Algol A and 3.5 solar radii for Algol B.

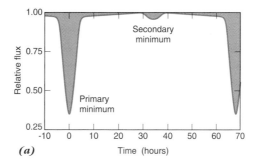

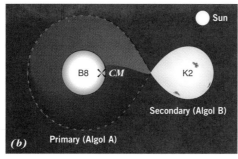

(a)

(b) Primary (Algol A) — Secondary (Algol B) — B8 — CM — K2 — Sun

Sirius: A Binary System

Finding stellar masses from binary stars is so important I will do another system in detail. Sirius is a binary star; the main star, called Sirius A, is the one you see in the sky. Its companion, Sirius B, is much fainter, and you need a moderately large telescope to see it. The orbital motion of Sirius B is in Figure 14.26 (plotted with Sirius A fixed), which shows that the orbital period is close to 50 years. The distance to the stars is 8.64 ly. Then the separation (semimajor axis) is about 20 AU. Use Kepler's third law with this information to find that Sirius A has about twice the sun's mass, while Sirius B has about the same mass as the sun.

A careful analysis shows that Sirius A has a mass of 2.1 solar masses, a radius of 1.7 solar radii, and an effective temperature of 9970 K. In contrast, Sirius B has a mass of 1.05 solar masses, a radius of 0.0073 solar radius (*much* smaller than the sun), and an effective temperature of 30,000 K. Sirius B turns out to be a white dwarf (Section 17.1).

The Mass–Luminosity Relation for Main-Sequence Stars

Binary systems act like scales that allow us to weigh the two stars. In most cases, the luminosities of the two stars can also be determined. When the luminosities of

FIGURE 14.25 Algol is an eclipsing binary system: because the orbits of the two main stars lie almost edge-on as seen from the earth, eclipses can occur. Algol A, the primary, is the brighter star. Its dimmer companion is called Algol B. (*a*) The eclipse of Algol A by its cool companion produces the largest dip (*primary minimum*) in the light curve, a plot of how the observed flux changes with time. When the companion circles around Algol A, the loss of light is less, so a smaller dip (*secondary minimum*) occurs in the light curve. (*b*) Top view of the Algol system. Algol A, the primary, is a B-type star (temperature of 12,500 K) with a diameter about three times that of the sun. Its dimmer companion, Algol B, is cooler; it is a K-type star with a surface temperature about 4500 K and a radius about 3.5 times the sun's. The center of mass (*CM*) is located just inside the primary star. The shape of the secondary is distorted from a sphere by the gravitational force of the primary. The dashed line around the primary indicates its zone of strong gravitational influence. (Adapted from a diagram by Mercedes Richards.)

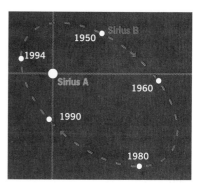

FIGURE 14.26 Orbit of Sirius B relative to Sirius A, plotted with Sirius A as fixed (not moving around the center of mass). Note that Sirius B was closest to Sirius A in 1944 and again in 1994. The orbital period is about 50 years.

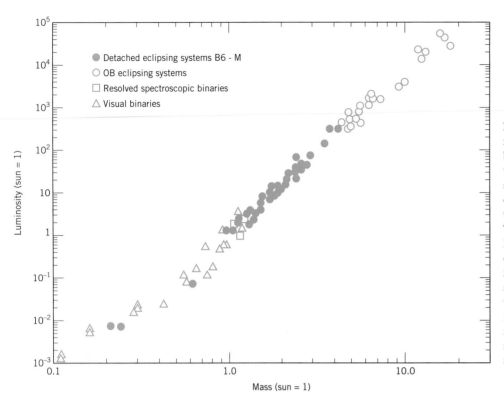

FIGURE 14.27 The mass–luminosity relation for stars, as determined from binary systems, in which the individual masses can be determined. These stars are also those whose distances can be measured, which means that their luminosities can be calculated. The points marked with solid circles have the smallest errors; these are eclipsing binaries with spectral types from middle B to M. The open circles are eclipsing binaries with spectral types O and B; the observational errors here are larger. The squares represent spectroscopic binaries; the triangles, visual binaries. (Adapted from a diagram by R. C. Smith.)

main-sequence stars are plotted against the stars' masses, the points fall into a definite pattern (Fig. 14.27). For main-sequence stars, the mass determines the luminosity, and the resulting correlation is called the **mass–luminosity relation** (sometimes abbreviated *M–L relation*).

Basically, the mass–luminosity relation shows that a star's luminosity is *roughly* proportional to the *fourth power* of its mass for stars with a mass *greater than* 0.4 solar mass. For example, a star with a mass 10^1 times that of the sun has about 10^4 times the sun's luminosity. Main-sequence stars follow the mass–luminosity relation fairly well; hence, the upward swing in luminosity of the main

sequence from M to O stars reflects an *increase* in the stars' masses. So main-sequence O stars are more massive than main-sequence M stars Table 14.3). Astrophysicists had predicted the mass–luminosity relation theoretically, and its confirmation came from investigation of binary stars. (In Chapter 16 you will see the importance of the mass–luminosity relation for stellar evolution theory.) Giant and supergiant stars also follow mass–luminosity laws, but they differ from that for main-sequence stars.

The mass–luminosity diagram tells us that the masses of other stars do not differ widely from the sun's mass. According to theoretical considerations,

TABLE 14.3 TYPICAL PROPERTIES OF MAIN-SEQUENCE STARS

SPECTRAL CLASS	SURFACE TEMPERATURE (K)	MASS (sun = 1)	LUMINOSITY (sun = 1)	RADIUS (sun = 1)	APPROXIMATE LIFETIME (years)
O5	45,000	60.0	800,000	12	8×10^5
B5	15,400	6.0	830	4.0	7×10^7
A5	8,100	2.0	40	1.7	5×10^8
F5	6,500	1.3	17	1.3	8×10^8
G5	5,800	0.92	0.79	0.92	12×10^9
K5	4,600	0.67	0.15	0.72	45×10^9
M5	3,200	0.21	0.011	0.27	20×10^{11}

stars with masses greater than about 100 solar masses are unstable, and bodies with masses less than roughly 0.1 solar mass cannot become hot enough to start nuclear reactions and become stars. The predicted narrow range of stellar masses is borne out by the H–R diagram and the mass–luminosity relation.

Stellar Densities

The mass–luminosity relation says that the masses of stars do not vary over a severely wide range. Yet their sizes do, as indicated by the existence of supergiant and giant stars. So stellar densities vary widely. Remember that density of an object is its mass divided by its volume.

The sun's average density is 1400 kg/m^3. Let's compare the sun to Sirius B, a white dwarf. Sirius B is about *3 million* times denser than water. This fact tells you that Sirius B—and other white dwarfs—cannot be ordinary stars and cannot have an internal structure like that of the sun.

Stellar Lifetimes

The mass–luminosity law immediately provides a way of comparing **stellar lifetimes.** The argument goes like this. The total amount of energy available to a star from the conversion of hydrogen to helium is directly proportional to its mass (its fuel supply). So a star with more mass can produce energy for a longer time than one with less mass, *if* both stars give off energy at the *same* rate. But the rate at which a star loses energy is given by its luminosity. A greater luminosity means that a star is producing energy faster, using up its mass more quickly. Because luminosity increases as the fourth power of the mass, more massive stars use up their mass faster and have shorter lifetimes. For example, a star of 10 solar masses has a lifetime of about 20 million years, in contrast to 10 billion years for the sun (the representative G star in Table 14.3).

Warning. Do not confuse the *age* of a star with its *lifetime!* The lifetime is the total span of active life from fusion reactions. The age is the amount of time that has elapsed since fusion reactions

began. The sun's expected lifetime is about 10 billion years (true for all solar-mass stars), while its estimated age is 5 billion years. Other solar-mass stars can have different ages. When we say a star is "young" or "old," we are comparing it to stars with similar *lifetimes*.

■ **What do we need to know to determine the individual masses of binary stars?**

14.7 Stellar Magnetic Activity

If stars are suns, do they also show the aspects of the active sun (Section 13.7)? Do they have coronas, starspots, and flares? Remember that these phenomena occur in *active regions,* which result from strong magnetic fields.

Because stars are far away, we cannot directly see their coronas (if they have them). How to observe them? The hot plasma in the sun's corona shows up in x-ray photos as bright regions; in contrast, the photosphere and chromosphere appear dark. So x-ray observations of stars provide an indirect view of their coronas—if *stellar coronas* resemble the sun's in general ways.

The *Einstein X-Ray Observatory* examined a number of nearby stars for x-ray emission. From these observations, we find that stars of *all* spectral classes have x-ray emission and, by deduction, coronas. The quiet sun seems to be one of the weaker x-ray emitters, below 10^{20} W. In contrast, the active sun puts out some 10^{22} W. Other G stars emit upward of 10^{24} W—about 100 times more than the active sun. The implication is that such stars have more extensive and hotter coronas (temperatures up to 10 million kelvins), connected with extensive active regions (Fig. 14.28). Coronas appear common in other stars.

The sun's corona is the source of the solar wind. With coronas, many stars have outflows as stellar winds. Most O and B stars have hot, rapid winds that blow off 10^{-6} solar mass per year at speeds of thousands of kilometers per second. Cool giant and supergiant stars also have strong winds, but they are

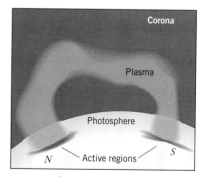

FIGURE 14.28 Schematic view of coronal loops above active regions in a star like the sun. The magnetic field lines extending from the north and south poles from the photosphere in an active region contain the hot plasma in a loop up in the corona.

cooler and less speedy than those from hot stars.

How could we observe **starspots?** During times of sunspot maxima, sunspots cover only about 0.1 percent of the sun's total surface. If you observed the sun as a star from light years away, you'd need extremely sensitive equipment to note any variation in the sun's light as it rotated and sunspots passed into and out of view. So you might expect that it would be useless to try to search for starspots by the same kind of observations of changes in light output as stars rotate.

Luckily, this solar analogy proves right in concept but wrong in scale. Some stars, even G stars, turn out to be hyperactive, with enormous concentrations of starspots compared to the sun— starspots covering such a large fraction (up to 50 percent) of the surface that we can actually see them indirectly! As the star rotates, we alternately view the spotted and clean sides. When few spots face us, we see the most light from the star. When the spotted hemisphere turns our way, there is a decrease in the amount of light we see.

Clearly, many stars must be observed over long times (decades) to confirm activity cycles that parallel the sun's. Such a long series of observations of sunlike stars was begun in 1976 at Mt. Wilson Observatory and continues today. The project so far shows that some sunlike stars exhibit activity cycles. A few last longer than decades; others on the order of decades; and some show no sign of activity. Basically, it appears that all stars of spectral type F or cooler undergo some form of magnetic activity at some times in their lives.

On the sun, we know that flares occur in active regions. The same appears to be the case for some other stars, but the flares burst forth with much more energy than in the case of the sun. *Stellar flares* can be observed at many different wavelengths, including those in the radio (with a radio telescope). Some cool stars explode with enormous radio flares—millions of times stronger than the most energetic solar radio flare ever recorded. Such flares last for about a day, and in that time, the energy emitted in just the radio part of the spectrum amounts to 10 times as much energy as the sun puts out at all wavelengths in one second! Others flare optically and give off a total of some 10^{27} W, or a thousand times more than a similar solar flare.

To sum up. Other stars have characteristics of the active sun, so strong magnetic fields, probably twisted and tangled, lie at the roots of this activity.

■ **What observational evidence do we have of magnetic activity on the other stars?**

Key Concepts

1. Stars are fusion reactors, more or less like our sun (Table 14.3), with similar physical properties, which we can infer by analyzing their light and measuring (or guessing) their distances. The most important physical properties are mass, luminosity, size, surface temperature, and chemical composition. The latter two can be found from spectroscopic analysis, which does not require that we know the distances.

2. Astronomers measure the brightness of stars as seen in the sky by their flux, or the amount of energy from a star that reaches the earth each second over a given area (such as a square meter). A telescope is used to measure the fluxes of stars, which are much smaller than the flux from the sun.

3. Light intensity changes with distance; the flux goes as the inverse-square of the distance. This law applies to electromagnetic waves of all kinds. Hence, distance plays a key role in the fluxes we receive at the earth from stars.

4. Only the distances of the closest stars can be measured directly by a triangulation technique called heliocentric parallax; the closer the star, the larger its parallax angle. We use the earth's orbit as the baseline for this measurement.

5. If a star's distance is known, we can find its luminosity from its flux by applying the inverse-square law for light. A star's luminosity tells us how much energy it emits per second.

6. Assuming that stars radiate like blackbodies, their colors indicate their surface temperatures. Hotter stars are bluer, and cooler stars are redder. We can also infer their sizes from their surface temperatures and luminosities, which allow an estimate of surface areas.

7. The angular diameters of stars are generally too small to measure from the earth with current instruments. In a few cases, specialized equipment or techniques allow us to measure stellar angular diameters. Then if we know the distances, we can calculate the actual diameters.

8. The Hertzsprung–Russell diagram is a graph of the surface temperatures and luminosities of stars. On it, stars fall into distinct groups: main sequence, giants, supergiants, and white dwarfs. Sizes of stars are indicated on the H-R diagram by their luminosity classes. Once an H–R diagram has been made for a large number of stars, we can use it to estimate distances by spectroscopic distances using spectral types and luminosity classes.

9. Binary stars are two stars locked by gravity in orbits around a common center of mass. We can apply Kepler's third law to find the masses of the stars if we know their distance, their orbital period, and the location of each star relative to the center of mass.

10. Binary stars provide the only means of finding directly the masses of stars. For main-sequence stars, we find that the masses and luminosities are related (the more massive stars are more luminous): the luminosities are directly proportional to the fourth power of the mass. We infer that more massive stars have shorter lives than less massive ones.

11. We are beginning to recognize that stars have magnetic activity, both erratic and cyclic, similar to that of the active sun. These include phenomena starspots and flares, driven by magnetic fields.

Key Terms

absolute magnitude

binary star

eclipsing binary system

giant

heliocentric parallax

Hertzsprung–Russell diagram

inverse-square law for light

luminosity class

main sequence

mass–luminosity relation

parsec

spectroscopic binary

spectroscopic distances

starspots

stellar lifetimes

stellar spectral sequence

stellar surface temperatures

supergiant

heliocentric (trigonometric) parallax

visual binary

white dwarf

Study Exercises

1. In the winter sky, you see the following stars: Capella (yellowish), Betelgeuse (reddish), and Sirius (bluish). List these stars in order of increasing surface temperature. Estimate the surface temperature of Betelgeuse and of Sirius. Do they differ much? (Objectives 14-1 and 14-2)

2. Consider stars with the following spectral types: M I, G III, and A V. Which star is the largest? Which the most luminous? (Objectives 14-3 and 14-10)

3. What limits the accuracy of heliocentric parallax measurements? (Objective 14-6)

4. Refer to Figure 14.14 to answer the following questions. (a) Capella and the sun have roughly the same surface temperature. Which star is larger? (b) Regulus and Capella have about the same luminosity. Which star is larger? (c) Vega and Sirius have about the same surface temperature. Which star is more luminous? (d) Which star would appear redder, Vega or Pollux? (Objectives 14-2 and 14-8)

5. Since we can't see the disks of other stars directly, how do we know that some of them have coronas like the sun? (Objective 14-11)

6. What procedure does an astronomer follow to find out a star's density? *Hint:* Divide mass by volume to get density. (Objectives 14-1, 14-8, and 14-9)

7. Consider a binary star system that does not eclipse and in which one star is much brighter than the other. Then the absorption lines from the fainter star do not appear in the spectrum, but those of the brighter one do. Describe how the Doppler shift would appear from the orbital motion of the stars. (Objective 14-9)

8. The star Regulus has a mass about five times that of the sun. Use Figure 14.27 to estimate the luminosity of Regulus. (Objective 14-10)

9. Jupiter is about 5 times as far from the sun as the earth is ($\approx$ 5 AUs compared to 1 AU). By how much less is the sun's flux at Jupiter compared to that of the earth? (Objective 14-5)

10. Imagine that you observe two stars with the same color. What can you infer about their surface temperatures? (Objective 14-2)

11. In what region of the H–R diagram do you find supergiant stars? (Objective 14-7)

Problems & Activities

1. Imagine that the sun were moved to one parsec from the earth. What would its flux be then? *Hint:* How many AUs make up a parsec (Focus 14.2)?

2. Consider a star whose flux at the earth is one unit of stellar flux. Imagine that the star is moved to five times its distance. How would the flux change?

3. Imagine measuring the parallax of Sirius from Mars. What size would the parallax angle be?

4. Suppose the sun were placed at a distance of 10 ly from the earth. What would its angular diameter then be?

5. Compare the power per second per square meter of photosphere emitted by the sun to that of Sirius. To Betelgeuse. *Hint:* Assume that these stars radiate like blackbodies.

6. For the sun, Sirius, and Betelgeuse, compare the wavelengths at which each body's continuum emission peaks. *Hint:* Same in Problem 5!

7. Sirius has a mass 2.1 times that of the sun. According to the mass–luminosity relation, how luminous should Sirius be compared to the sun? How does this estimate compare to its measured value (Table 14.2)?

8. Recall (Focus 13.2) that the *surface flux* of a blackbody (power emitted per unit area) depends only on the temperature. Then the total power or luminosity of a star is simply its *surface flux times its surface area*. If Sirius has a surface temperature of 10,000 K, what is its surface flux? If its radius is 1.7 solar radii, what is its luminosity? *Hint:* The area of a sphere is $4\pi R^2$, where R is the radius.

See for Yourself!

You can, with a little practice, observe Algol's light variation from most locations in the Northern Hemisphere. The monthly magazine *Sky and Telescope* provides the dates and times of primary eclipses. Look them up in the "Celestial Calendar" section. You will find the dates and times are given in Universal Time rather than local time. Instructions are given on how to convert from Universal Time to the time at your viewing location. In North America, you subtract 5, 6, 7, and 8 hours to go from Universal Time to Eastern Standard, Central Standard, Mountain Standard, and Pacific Standard, respectively.

Algol remains at its minimum for about 2 hours. The drop into the minimum and rise to maximum is fairly sharp, taking about 4 hours. So you can observe the entire eclipse on a good night when the center of the eclipse occurs near midnight. Start your observations about 5 or 6 hours *before* the predicted time of mideclipse.

The winter star chart in Appendix G shows the location of Algol in Perseus. Algol is the second bright star in Perseus, so it is fairly easy to spot. Figure A.4 gives a closer view of this region of the sky. Note that the brightest star in Perseus, called Mirphak, is just a little brighter than Algol (outside of its eclipses). You can use Mirphak as a comparative reference, since its luminosity is constant. It is located about 9° (one fist) away from Algol.

At mideclipse, Algol emits only 30 percent of the flux that it gives off outside of the eclipses (when you see the light from both stars combined). On the magnitude scale (Focus 14.1), the decrease is 1.3 magnitudes, which should be very apparent during the span of your observations. ■

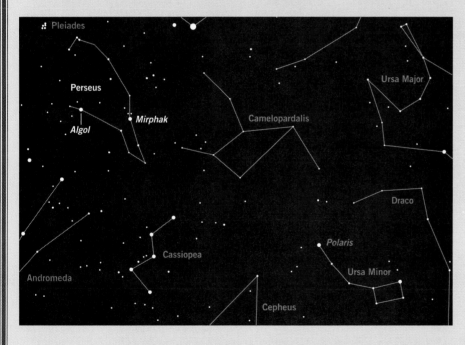

FIGURE A.4 Sky chart for locating the positions of Algol and Mirphak in the constellation Perseus.

Starbirth and Interstellar Matter

It's lovely to live on a raft. We had the sky, up there, all speckled with stars, and we used to lay on our backs and look up at them, and discuss about whether they was made, or only just happened . . .

MARK TWAIN, *Huckleberry Finn*

Stars have finite lives—long by human standards, but limited nevertheless. How long a star lives depends on its mass. A star like the sun will survive for some 10 billion years. More massive stars live scant millions of years. The fact that we observe stars more massive than the sun means not only that starbirth occurred in the past but also that it is going on now. Otherwise we would see fewer stars, especially the massive ones, which die off relatively quickly.

How are stars born? In clouds of gas and dust between the stars. Contrary to first impressions, interstellar space is *not* empty. It contains gas and dust, both thinly spread out and in clumps. Hydrogen makes up most of the gas, which outnumbers the dust particles in a typical volume enormously. The interstellar gas contains more than neutral atoms: some of it is ionized and some is bound up in molecules. A large fraction of the interstellar gas is locked up in short-lived clouds. From these interstellar clouds, stars are born.

The gas and dust between the stars, called the interstellar medium, marks the locus of star-forming action in the Galaxy. Here, stars form in vast stellar nurseries. Some will become massive stars, some sunlike; there will be many binaries, and some will perhaps develop with planetary systems.

15.1 The Interstellar Medium: Gas

The **interstellar medium** is the gas and dust between the stars. What do we know, in general, about the **interstellar gas?** First, it is made mostly of hydrogen. Second, it tends to clump in clouds. Third, a hot, dilute gas exists between the clouds. Fourth, the gas in different locations contains neutral atoms, ionized atoms, free electrons, and molecules. Fifth, on the average the interstellar gas is very tenuous. The distance between interstellar atoms is roughly 100 million

The Interstellar Medium

At a Glance

Contents	Gas and dust
Dust	Small grains of silicates and ices
Gas	Atomic, molecular, and ionized
Structure	Clumpy, with a thin gas between clouds
Source of	Stars and planets

times larger than the size of the atoms themselves. If two people were separated by a proportional distance relative to their size, they would be about 100 million meters apart, about the distance between the earth and moon! Sparse as it is, the interstellar gas occasionally clumps and forms stars.

Bright Nebulas

On a winter night, you can easily spot the constellation Orion (look back at Fig. 1.1). Dangling from Orion's belt is a short sword; if you look closely, the middle star appears fuzzy. A small telescope pointed at this fuzzy patch shows you a diffuse, convoluted cloud surrounding a small cluster of stars. This bright cloud is called the *Orion Nebula*. (*Nebula* is the Latin word for "cloud," and the use of **nebula** is an astronomical holdover from the nineteenth century.) Another name for the Orion Nebula is *Messier 42* (or M42), the forty-second object in the catalog compiled by the French astronomer Charles Messier in the eighteenth century.

Only 1500 ly from the earth, the Orion Nebula, roughly 20 ly in diameter, is typical of **emission** or **bright nebulas** (Fig. 15.1), which emit light. At the end of the nineteenth century, spectroscopic analysis demonstrated that these bright nebulas show spectra of emission lines, hence consist of hot gas. Recall from Kirchhoff's rules for spectra (Section 5.2) that an emission-line spectrum indicates a hot, diffuse gas. The hot Orion Nebula has lines of hydrogen, helium, and oxygen predominating in its bright-line spectrum.

Nebulas such as Orion do *not* shine by their own light. Instead, they borrow the energy from hot O or B stars located in or near them. The essential physical process is this: the gas absorbs high-energy ultraviolet photons given off by the central star (or stars) and gives off photons in emission lines at lower energies. For a hydrogen gas, the visible light comes out mainly in the red region of the spectrum (Fig. 15.2). (Focus 15.1 has more on this emission process.)

The star (or stars) in a bright nebula is quite hot, about 30,000 K, and emits

FIGURE 15.1 Close-up of the Orion Nebula (Messier 42), a typical emission nebula of ionized gas; a smaller nebula appears above it. The colors of the emission depend on the elements that have been excited by collisions. The dark regions within the nebula are those bearing concentrations of dust. This image has been processed to bring out the fine detail in the structure of the nebula. The filaments and sheets of gas hint that the nebula is a dynamic place. (Courtesy of David Malin, AATB.)

FIGURE 15.2 The Lagoon Nebula, Messier 8, in Sagittarius. The gas shows a complex structure. Note the cluster of stars embedded in the gas. (Courtesy of NOAO/Kitt Peak.)

many photons with enough energy to ionize hydrogen. Most photons are absorbed by the gas surrounding the star so that, out to a considerable distance (a few tens of light years), the gas is almost totally ionized. This zone of ionized hydrogen is called an **H II region** (Fig. 15.3). (H I stands for neutral hydrogen and H II for ionized hydrogen. In general, a neutral atom is labeled I, an atom with one electron removed is labeled II, with two electrons removed, III, and so on.)

Interstellar Atoms

Bright nebulas are composed almost entirely of hydrogen. So the neutral gas ionized to form them must also have contained mostly hydrogen, as a neutral atom (designated **H I region**). Astronomers surmised that hydrogen atoms populated interstellar space, but they did not observe the H I gas until 1951.

21-cm Emission from Hydrogen

Recall (Section 5.1) that the hydrogen atom has one proton in the nucleus and one electron in orbit around it. Both the proton and the electron have angular momentum. You can imagine them spinning like miniature tops. According to the rules of quantum physics, the electron and the proton can be oriented in the atom so that the two spins either align or oppose each other. If the spins oppose, the total energy of the atom is just a bit less than if the spins align. As usual, the atom prefers to be in the lower energy state. Suppose the spins are aligned. Eventually the electron flips over and emits a low-energy photon—energy that corresponds to a wavelength of 21.11 cm (Fig. 15.4).

How are the protons and electrons in hydrogen atoms aligned in the first place? By collisions with electrons and other atoms. The gas in interstellar space is very sparse, and collisions between two atoms occur only once every few million years. On the other hand, once the spins in a hydrogen atom have become aligned, about 10 million years on the average passes before the proton flips and the atom drops to its lowest energy state. It emits a 21-cm photon. This

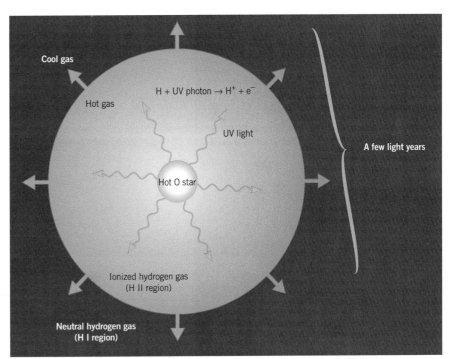

FIGURE 15.3 Schematic diagram of an idealized H II region. A hot O star (or stars) emits ultraviolet light that can ionize hydrogen for a few parsecs around. The absorption of the ultraviolet heats the ionized gas to a temperature of about 10,000 K. The hot, ionized gas expands into the cooler, neutral gas surrounding it. Real H II regions are not perfectly spherical, as drawn here.

is a rare event for any one atom. But because so many hydrogen atoms exist in interstellar space, enough are emitting 21-cm radiation at any given time that the interstellar gas radiates strongly at this wavelength and can be detected with radio telescopes (Fig. 15.5).

Surveys at 21 cm find that most neutral hydrogen is concentrated in the plane of the Milky Way (look back at Figure 6.23b). On the average, the hydrogen atoms have a temperature of 70 K and a density of 3×10^5 in a cubic meter. So in a volume of space equivalent to the volume of your body, you'd find only 3×10^4 hydrogen atoms, whereas in fact your body contains some 10^{27} atoms.

Other gases also exist in interstellar space. Even before atomic hydrogen was observed with radio telescopes, optical observations had revealed the presence of atoms of several other kinds (and the first molecules). Superimposed on the spectra of some stars, astronomers found sharp, dark lines of elements such as so-

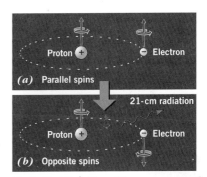

FIGURE 15.4 The 21-cm emission from hydrogen atoms. (*a*) Collisions can line up the spins of the proton and electron so that they are parallel. (*b*) After some time, if the atom does not hit another, the electron flips so that the spins are opposed, and the atom emits a 21-cm photon.

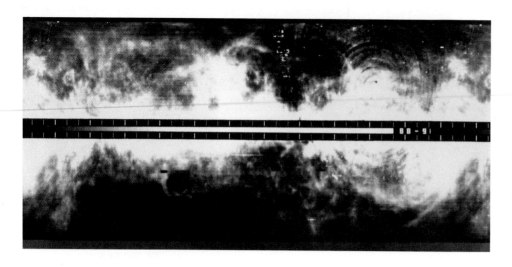

FIGURE 15.5 A 21-cm map of one section along the plane of the Milky Way. The bar through the center marks and plane and is an artifact of the image processing. The view extends about 60° above and below the plane. The 21-cm emission, displayed here in white, shows large filaments and shells. (Courtesy of Carl Heiles.)

dium (Fig. 15.6). These narrow absorption lines are produced when starlight passes through cool regions of the interstellar gas, as expected from Kirchhoff's rules (Section 5.2). Such optical observations uncovered atoms of sodium, potassium, ionized calcium, and iron in the interstellar medium.

By far the most abundant atom in the interstellar gas is hydrogen. Because space is so empty and the temperature of the gas is so low, most hydrogen atoms remain in their lowest energy state, from which they can absorb only ultraviolet

Enrichment Focus 15.1

EMISSION NEBULAS: FORBIDDEN LINES

Let me describe in more detail the process by which a bright nebula emits light. Virtually all the atoms in an H II region are ionized. Most of these atoms are hydrogen. But other elements, such as oxygen and nitrogen, are also in the gas and are ionized. In the low-density conditions of H II regions, some ions emit unusual lines called *forbidden lines.* Let's see how.

The ordinary electron transitions described so far (such as the transition from level 3 to level 2 in hydrogen that produces the Hα line) occur very quickly, in about 10^{-8} s. Some excited states, though, have much longer durations—as long as a few minutes. Now, under most conditions, such as in the sun's photosphere, an ion in a iong-lived excited state is a sitting duck for a collision with another particle in the gas. This collision will knock the electron out of the excited

state before it can drop down to a lower one. No photon will be emitted.

But in a bright **emission nebula,** gas densities are quite low— roughly 10^9 particles in a cubic meter. Collisions are relatively rare, so an electron in a long-lived excited state can fall to lower energy levels before a collision. Lines from such states are called *forbidden lines* because the electrons stay in the excited states for a very long time compared to the usual 10^{-8} s; some last as long as minutes!

Singly ionized oxygen (O II) has two possible forbidden transitions. This ion has two such sublevels above the ground level (Fig. F.12). A collision can bump an electron into either level. If it goes up to the higher level and then drops back down to the ground state, it emits a photon at 3726 Å; from the lower level, de-excitation results in a 3729-Å photon. Such lines are in the

violet region of the visible spectrum. Other lines in other regions of the spectrum are emitted by other ions. For example, doubly ionized oxygen (O III) can emit two lines at 4959 and 5007 Å, in the green region. Emission lines from H II regions are a mixture of forbidden lines and ordinary emission lines, such as the Hα line.

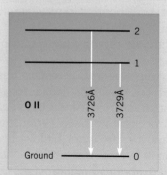

FIGURE F.12 Energy-level diagram for the forbidden-line emission from singly ionized oxygen (O II).

light. Since ultraviolet cannot penetrate the earth's atmosphere, these absorption lines cannot be observed with ground-based telescopes. With the advent of earth-orbiting ultraviolet telescopes, we have finally observed cold hydrogen.

Clouds and Intercloud Gas

The interstellar absorption lines were the first direct indication of the existence of a pervasive interstellar medium. They led to the idea that cool regions in the medium exist in the form of small clouds. Ultraviolet and radio observations show that the neutral hydrogen gas has a very patchy distribution in clouds with diameters from tenths to tens of light years. The average density of neutral hydrogen is somewhat less than 1 million atoms per cubic meter. However, the observations also show that in certain directions in space, the neutral hydrogen density is ten times less. (For comparison, the density of air at the earth's surface is about 10^{25} particles per cubic meter.)

The space between interstellar clouds also contains gas. And, as you'd expect, it consists mostly of hydrogen. Is it ionized or neutral? Well, it seems to be both—a thin, neutral gas and an even thinner, hotter (about 10,000 K) ionized gas between the clouds.

Ultraviolet observations provide direct evidence for a *very* hot gas of oxygen, stripped of five electrons, permeating the intercloud regions. (The symbol for five-times-ionized oxygen is O VI.) To rip so many electrons from an oxygen atom requires a very high temperature—about 1 million kelvins. Because this hot gas has about the same temperature as the sun's corona, it is called the **coronal interstellar gas.** Evidence for it comes also from the x-rays it emits (look back at Fig. 6.23*d*).

Interstellar Molecules

So much for hot stuff. What about cold material? You might expect it to be in the form of simple molecules (such as water) because atoms tend to combine into molecules at cold temperatures. Optical astronomers in the 1930s made the first discoveries of some molecules (such as CH and CN).

But the optical part of the spectrum

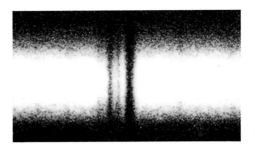

is not the most fruitful region to search. A molecule consists of atoms linked together in particular arrangements by electron bonds. It can have different energy states according to how the atoms vibrate or the molecule spins (Section 5.3). As with changes in electronic states in an atom, when a molecule changes its vibrational or rotational state, it can emit or absorb a photon. For changes in vibrational states, the photons are infrared ones; for rotational states, radio ones. In the cold regions where molecules can exist in interstellar space, occasional collisions between molecules (or perhaps with atoms) kick the molecules and get them spinning. These excited molecules emit radio photons that can be observed as an emission *line,* generally at millimeter wavelengths.

The radio search for polyatomic molecules began in earnest in the 1960s. More than 80 molecules have been found so far. Table 15.1 lists some key ones. Carbon monoxide (CO) is one of the most common molecules in space (Fig. 15.7). It is easily excited by collisions in places where the densities are a mere 10^8 per cubic meter. Also common are ethyl alcohol, which gives the kick to beer and wine, and water. Note that many of the molecules are *organic*—that is, compounds in which carbon plays a central role. The most abundant atoms in these molecules (carbon, hydrogen, nitrogen, and oxygen) are also the most abundant in living creatures on the earth. In addition, astronomers have observed sulfur and phosphorus, which also play biogenic roles (more in Chapter 22).

By far the most abundant molecule is molecular hydrogen. But molecular hydrogen does not emit or absorb at radio wavelengths. Instead, the hydrogen molecule absorbs and emits ultraviolet and infrared wavelengths. Infrared

FIGURE 15.6 Interstellar absorption lines of the D lines of sodium. In cool gas clouds this element absorbs some of the starlight that passes through it. The view is through the interstellar medium in the direction of the star Epsilon Orionis in the constellation Orion. Each of the narrow D lines is split up into five components, each one corresponding to a cloud with a different radial velocity and so a different Doppler shift for the line. (Courtesy of Palomar Observatory, California Institute of Technology.)

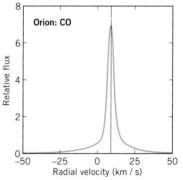

FIGURE 15.7 Microwave emission line from carbon monoxide (CO) from the region of the Orion Nebula. The molecular emission comes from a cloud behind the nebula as we see it in the sky. Note that the emission line has a width. This widening occurs because different regions of the gas are moving with different radial velocities (horizontal axis). In fact, the central peak lies at a velocity of about 10 km/s. The vertical axis gives the relative flux, as measured by the radio telescope. (Adapted from observations by John Bally and Charles J. Lada.)

emission lines of molecular hydrogen have been observed from heated interstellar clouds (at a temperature of about 2000 K, the hydrogen molecules are excited and emit infrared lines), and ultraviolet absorption lines from cool clouds in front of hot stars have been detected in the spectra of these stars. Overall, about half the hydrogen in the Milky Way appears to be in the form of molecules.

Molecular Clouds

Most interstellar molecules are localized in dark, dense, cold conglomerates called *molecular clouds.* These clouds often lie near H II regions; one of the closest molecular clouds sits behind the Orion Nebula, as we view it from earth (Fig. 15.8). The molecular cloud here consists of two parts: a large, low-density cloud surrounding a dense, small core. The low-density cloud has an enormous extent: it is at least 30 ly across, and contains at least 10^4 solar masses of material. The core is only 0.5 ly in size and has a mass of only 5 solar masses.

The Orion region presents an excellent example of a **giant molecular cloud.** Observations so far indicate that the bulk of the material of the interstellar medium is bound up in complexes of giant molecular clouds (Fig. 15.9). These immense globs of molecules, held together by gravity, have the following typical properties.

1. They consist mostly of molecular hydrogen; many other molecules are present, but these make up only a small fraction of the mass.
2. The cloud complexes have average densities of a few hundred million molecules per cubic meter; the individual clouds are slightly denser, with a few billion molecules every cubic meter.
3. They have sizes of a few tens of light years.
4. The total masses of the complexes range from 10^4 to 10^7 solar masses; 10^5 solar masses is typical. Masses of individual clouds are about 10^3 solar masses.

TABLE 15.1 SOME IMPORTANT INTERSTELLAR MOLECULES

COMPLEXITY	MOLECULE SYMBOL	MOLECULE NAME
Diatomic	H_2	Hydrogen
	OH	Hydroxyl radical
	CN	Cyanogen radical
	CO	Carbon monoxide
Triatomic	H_2O	Water
	HCN	Hydrogen cyanide
	H_2S	Hydrogen sulfide
	SO_2	Sulfur dioxide
Tetratomic	NH_3	Ammonia
	H_2CO	Formaldehyde
	HNCO	Hydrocyanic acid
	HC_2H	Acetylene
Pentatomic	CH_4	Methane
	HCOOH	Formic acid
	HC_3N	Cyanoacetylene
Hexatomic	CH_3OH	Methyl alcohol
	CH_3CN	Methyl cyanide
	$HCONH_2$	Formamide
Heptatomic	CH_3NH_2	Methylamine
	CH_3C_2H	Methylacetylene
	$HCOCH_3$	Acetaldehyde
Octatomic	$HCOOCH_3$	Methyl formate
	CH_3C_3N	Methyl cyanoacetylene
Nonatomic	CH_3CH_2OH	Ethyl alcohol
	CH_3CH_2CN	Ethyl cyanide

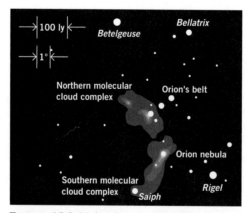

FIGURE 15.8 Molecular clouds and regions of starbirth in Orion. A carbon monoxide flux map of the emission from giant molecular cloud associated with the Orion Nebula is superimposed on the stars. This map covers about 15 degrees on the sky. The scale for one degree and 100 ly on this map is shown in the upper left. The hottest, densest part of the cloud lies near the Orion Nebula; it is called OMC1 for Orion Molecular Cloud 1. It is part of the Southern Molecular Cloud Complex. Another complex lies just to the north and east of the last star in Orion's Belt. This map was made by a millimeter-wave antenna at Columbia University.

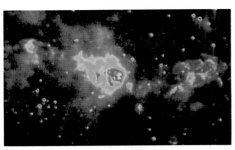

FIGURE 15.9 Infrared map of emission from hydrogen molecules in the constellation of Cepheus. In this composite of many individual images taken at a wavelength of 2.12 μm, the false colors are dark blue for the weakest emission and red for the strongest. The emission comes from hydrogen molecules at a temperature of 2000 K; the molecules have been heated by shock waves generated by newborn stars. These giant molecular clouds are about 2400 ly away; the area covered is 3.3 ly by 1.7 ly. (Courtesy of Adair P. Lane and NOAO/Kitt Peak.)

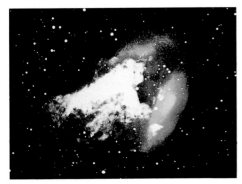

FIGURE 15.10 Messier 17 and its adjacent molecular cloud. This image combines an optical photo of the H II region (black and white; the emission shows as white) with infrared observations (false colors). The color was constructed to mimic emission at a wavelength of 40 μm. This infrared emission comes from the dust in the molecular cloud to the right of M17; the color changes from violet to red to illustrate the energy flow here. (Courtesy of A. Myers, Ames Research Center, NASA.)

The cores of these clouds are unusual places compared with the average interstellar medium. Here the temperatures are a frigid 10 K, and the densities get as high as 10^{12} molecules per cubic meter. That's an immense concentration by interstellar standards, yet it is only 10^{-13} times the density of molecules in the air at the earth's surface!

Giant H II regions, which surround young, massive stars, are always found near molecular cloud complexes (Fig. 15.10). This proximity suggests that giant molecular clouds play a key role in the process of star formation. Chemically, clouds with active star formation appear to be very different from those in which none is now taking place.

Table 15.2 summarizes the main components of the interstellar gas.

■ **What is the most abundant element comprising the interstellar gas?**

15.2 The Interstellar Medium: Dust

Dust also occupies interstellar space. There's not much out there; on the average, there's one dust particle in every million cubic meters—that's roughly a cube with sides one football field in size! The dust amounts to about one percent of the total mass of interstellar matter, but it can cut out light from distant objects or from those shrouded in dense clouds. Piercing the veil of **interstellar dust** has been an important goal of radio and infrared astronomers. That breakthrough has been critical in revealing the process of starbirth.

Interstellar Gas

At a Glance

Found in	Clouds, space between clouds
Detected by	Optical, radio, infrared, x-rays
Made of	Atoms, ions, molecules
Composition	Mostly hydrogen, some helium

TABLE 15.2 MAJOR PARTS OF THE INTERSTELLAR GAS

COMPONENT	INDICATOR	TEMPERATURE (K)	DENSITY (number/m³)	FRACTION BY MASS (percent)[a]
Molecular clouds	CO	10–50	10^8 to 10^{15}	40
H I regions	21-cm radiation	50–100	10^6 to 5×10^7	40
Intercloud gas	21-cm radiation	7,000–10,000	2×10^5 to 3×10^5	20
Intercloud coronal gas	O VI	1,000,000	10^2 to 10^3	0.1
H II regions	Hα, continuous radio	10,000	10^7 to 10^{10}	Little

[a]The fractions by mass are very approximate and should be taken only as a guide.

Cosmic Dust

Direct observations hint at dust between the stars. Dark nebulas, such as the famed Horsehead Nebula in Orion (Fig. 15.11), display dramatic cutoffs of light due to dust. Astronomers give the name **dark cloud** to an interstellar cloud that contains so much dust that it blots out the light of stars within it and behind it. The dark rifts and lanes in the Milky Way, once attributed to the lack of stars, are actually regions heavily obscured by dust. In some regions of the sky, dusty, swirling clouds blot out the light of the stars and H II regions behind them (Fig. 15.12).

Some bright nebulas are *not* emission regions but clouds of dust reflecting light from nearby stars. An example is the nebula that surrounds the Pleiades (Fig. 15.13). The spectrum of this nebula does not exhibit the bright emission lines. It shows simply the absorption-line spectrum of the Pleiades stars—light reflected by dust. A bright nebula that arises from the reflection of starlight by dust is called a **reflection nebula,** and it has a bluish color. Let's see why.

Generally, interstellar dust makes itself known in two ways: by **extinction,** the dimming of starlight, and by **reddening,** the scattering of the blue wavelengths more than the longer wavelengths. Let's look at each of these processes by recalling what happens to sunlight when the sun is near the horizon (Section 8.4). One, the sun is dimmer. Two, the sun appears redder when near the horizon because of the preferential scattering of blue light through much more air than when the sun is overhead.

Imagine starlight traveling through a dust cloud. The particles can absorb some of this light as it comes through. The dust particles can also scatter the starlight, so it goes off in a different direction from the original one. In either case, less light exits the dust cloud than enters it. Astronomers call this dimming of starlight *extinction.*

When starlight is extinguished, blue light is more strongly scattered than red. So red light penetrates the dust cloud more readily than blue. When you observe a star through the dust cloud,

FIGURE 15.11 The Horsehead Nebula in Orion in visible light. The red emission nebula contrasts with the famous dark, dusty cloud called the Horsehead from its silhouette. To the upper right is a bluish reflection nebula. (Courtesy of David Malin, AATB.)

more of its red light reaches your eye than does blue. The star appears redder than it actually is: astronomers call this process *reddening*. What about the blue light? The part that is scattered bounces around the dust cloud until it finally exits. So the cloud, a reflection nebula, appears bluish.

Reddening, since it is a color effect, is much easier to measure than extinction. We can estimate the quantity of the dust from the amount of reddening. From the H–R diagram we know that a certain spectral classification of a star corresponds to a certain color. A star's spectral class can be determined (from the strength of absorption lines), even if its light is reddened. We measure the star's color compared with that expected for its spectral class. The difference in color, the reddening, tells how much dust lies along the line of sight to the star.

Dust and Infrared Observations

Interstellar dust blocks out visible light. This fact makes optical astronomers unhappy, for it limits their view of distant

FIGURE 15.12 The Cone Nebula in the constellation Monoceros (east of Orion). This close-up view shows many dark, dusty lanes in the H II region. A young star, visible in the infrared, lies at the tip of the cone. (Courtesy of David Malin, AATB.)

FIGURE 15.13 The Pleiades star cluster (Messier 45) immersed in a bright reflection nebula. Dust surrounding the stars reflects their light preferentially in the blue. (Courtesy of David Malin, AATB.)

FIGURE 15.14 The core of the Orion Nebula. The small group of stars in the center is the Trapezium cluster (arrow). Note the three stars in a line at the left. (Courtesy of Jim Riffle, copyright 1990, Astro Works Corporation.)

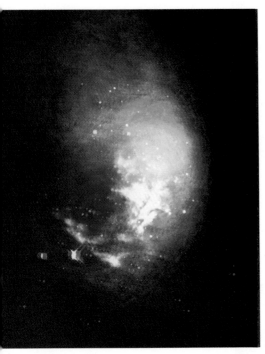

FIGURE 15.15 An infrared false-color map of the central part of the Orion Nebula. The colors represent infrared emission from 30 to 100 μm and are superimposed on a black and white photo that shows the visible stars. The strongest part of the infrared emission comes from the core of the molecular cloud that lies behind the visible H II region. Here, newborn stars have formed out of the molecular cloud. The young Trapezium cluster, the center of the H II region, lies near the center (see Figure 15.14). (Courtesy of NASA.)

stars and galaxies. Infrared astronomers fare better, for infrared radiation penetrates dust. In fact, some infrared radiation from space comes from the dust itself. So infrared astronomers can both *see* dust and *see through* dust!

How does dust emit infrared? The dust is made of small, solid particles called *grains*. Basically, dust grains act roughly like (very small) blackbody radiators. If the grain has a temperature of around 100 K, its emission will peak in the infrared.

The Orion Nebula marks a region studded with strong infrared sources. Optically, the core of the nebula (Fig. 15.14) is the densest part: hot gas (mostly hydrogen) is ionized and excited by the O stars there. The brightest stars form a trapezoid figure (easily seen in a small telescope) called the *Trapezium cluster.*

Now let's look at an infrared map (Fig. 15.15) of the core region. Quite a difference! What we are seeing is the infrared emission from cold dust (about 70 K) located somewhere at or near the center of the molecular cloud—dust heated by something capable of putting out 70,000 solar luminosities. The visible Orion Nebula, illuminated and sustained by the Trapezium stars, lies in front of the molecular cloud like a hot bubble (Fig. 15.16). This model provides some key pieces to the process of starbirth.

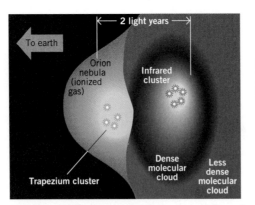

FIGURE 15.16 One model of the association of the Orion Nebula, its molecular cloud, and infrared sources. The Orion Nebula is a hot gas bubble, expanding outward, on the front of the molecular cloud. The infrared sources are embedded in the molecular cloud. (Based on a model by B. Zuckerman.)

Interstellar Dust

At a Glance

Found in	Interstellar gas clouds
Detected by	Extinction, reddening
Structure	Small core with larger mantle
Radius	0.1 to 1.0 μm
Composition	Silicate core with icy mantle; tarry crust
Formation site	Stellar winds of cool stars

The Nature of Interstellar Dust

What is the interstellar dust made of? One indirect clue comes from the cosmic abundances of candidate elements for dust grains: only elements that make up an appreciable fraction of the interstellar material can contribute in a large part to the dust grains. These include (in order) hydrogen, oxygen, carbon, nitrogen, and silicon.

The most abundant elements make up rather common substances: hydrogen and oxygen for water, carbon and hydrogen for methane, carbon and oxygen for carbon dioxide, nitrogen and hydrogen for ammonia, silicon and oxygen plus metals for silicates (compounds of Si and O commonly found in earth rocks). Compounds like water, methane, and carbon dioxide are loosely called *icy materials* because they are solids at temperatures below about 100 K. (Recall that they make up the bulk of the nucleus of a comet: Section 12.2.)

Now the dust grains themselves. To account for the properties of interstellar extinction, astronomers have developed models featuring **core–mantle grains** (Fig. 15.17). The small core, about 0.05 μm in radius, can consist of silicates, iron, or graphite; silicates are most likely. The mantles, about 0.5 μm in radius, are made of icy materials, likely some mixture of them all. When grains drift into hot regions, such as an H II region, their mantles evaporate, leaving behind a bare core. Infrared observations bolster the idea that silicates and ices (at least water ice) form parts of interstellar grains. Their spectra show absorption bands like those of silicates and water ice.

The icy materials in the mantles may be processed into organic compounds. Laboratory simulations of the conditions in interstellar space show that dirty ice mantles can absorb ultraviolet light (which permeates the dust-free regions of interstellar space). Ultraviolet photons have enough energy to break up chemical compounds and promote the creation of new ones, such as formaldehyde (H_2CO). That should sound familiar, for it is one of the molecules found in the interstellar gas. The experiments have also created residues that can survive much higher temperatures than ordinary ices—up to 500 K. The

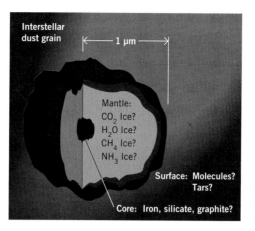

FIGURE 15.17 Simplified model of an interstellar dust grain; the composition is not really known. The dust may consist of any of the materials listed or some combination of them. A small core covered with a larger mantle is the main feature. The surface may be coated with a tarry substance.

compositions of these organic materials are unclear, but one of them, a yellow tar, probably makes up a large part of processed grain mantles.

To sum up. Interstellar dust contains elements common to the interstellar gas in the general forms of ices, silicates, graphite, and metals (such as iron). Some of the mantles of grains contain organic compounds.

Dust and the Formation of Molecules

Dust and molecules are intimately allied in space: wherever you find a molecular cloud, you usually find lots of dust. The reverse is also true. If you see a dark nebula, you have also likely found a molecular cloud. This association implies that grains play a role in the formation of molecules.

Consider the problem of forming an interstellar molecule. You have to get widely separated atoms together and chemically bound—no easy task in the dilute gas of interstellar clouds. Solution? Use cold dust grains as the sticking and forming surfaces. If a hydrogen atom hits a cold grain, it will stick. Add another hydrogen atom to the same grain, and a hydrogen molecule forms, which does not stick as well as atomic hydrogen. It eventually pops off into space. Grain surface formation seems to work for hydrogen. For other small molecules, chemical reactions in the gas alone seem to explain the formation of molecules up to those containing four atoms.

What about more complex molecules? They may form in the icy mantles of grains, where ultraviolet light provides the energy for chemical processing of dirty ice to more complex molecules. Some of these compounds are unstable and can explode, hurling the molecules into the interstellar gas.

To sum up. Hydrogen molecules form on interstellar grains, and molecules of up to four atoms form in the interstellar gas. The formation of more complex molecules may be the product of chemical reactions, driven by ultraviolet light, in grain mantles.

Formation of Cosmic Dust

Where does the dust come from? Interstellar grains have cores of dense solids (such as silicates and graphite), which solidify at a few thousand kelvins (as in the solar nebula).

How and where are these grains made? The denser grains are probably made in the atmospheres of *cool* supergiant stars. Such stars blow mass into space at rates of about 10^{-5} solar mass per year. As gaseous material streams outward, its temperature drops, and solids can condense out of the vapor. In fact, spectra of some supergiant stars show silicates, indicating that such dust exists around them. In a rarer class of stars, in which carbon is somewhat more abundant than oxygen, graphite particles and particles made of silicon carbide form in the outflowing material.

Cool giant stars, which are more common than supergiants, also lose mass—at about 10^{-6} solar mass per year. These giants add dust to the interstellar medium. For both giants and supergiants, some stars of spectral class M are enveloped in so-called *circumstellar material,* which contains both gas and dust. These stars have the highest mass loss rates of all, as great as 10^{-4} solar mass a year. They are large contributors to interstellar dust.

The ices that make up grain mantles likely condense on cores in the deep interiors of dense molecular clouds where they are protected from ultraviolet radiation. Here the temperatures are low and the gas densities high, so bare grains can grow crusts of ices. A core grows a new mantle once every 10^8 years or so,

since grains will lose their icy mantles in an environment where temperatures range above a few hundred kelvins.

■ **What is the major benefit of using infrared astronomy to view interstellar dust?**

15.3 Starbirth: Theoretical Ideas

Let's deal now with how stars form. Here's the big picture: stars are born out of interstellar molecular clouds by gravitational collapse. Because these clouds contain many times the mass of a single star, they fragment into much smaller pieces during the process of star formation. To try to cast a clear light on our present understanding of starbirth, I have divided the topic into two broad parts: theoretical ideas (this section) and observational clues (Section 15.4).

Collapse Models

Newton first recognized the basic process of star formation: *gravitational collapse.* A cloud with enough mass and a low temperature will naturally contract from its own gravity. As gravitational potential energy becomes kinetic energy, the material in the cloud heats up (Section 12.5). Eventually the temperature (and density) builds up to a point that the outward pressure brings down the rapid collapse to a slower contraction. Greater pressure halts the collapse. And the temperature in the center finally reaches the kindling temperature of fusion reactions. At that moment, a star is born. Prior to the ignition of fusion reactions, the cloud is called a **protostar.** As a protostar evolves, its luminosity, size, and surface temperature change.

Protostars of different masses evolve in physically different ways. Despite differences because of mass, theoretical models display some common stages.

1. The collapse occurs fast because it starts out in free-fall, that is, controlled only by gravity.

2. The central regions collapse more rapidly than the outer parts, and a

small condensation forms at the center; this core will become a star.

3. Once the core has formed, it accretes material from the infalling envelope of material surrounding the core.

4. The star becomes visible either by accreting all the surrounding material onto itself or by somehow dissipating the shroud of dust.

With these general features in mind, let's look in a little detail at the formation of protostars.

Protostar Formation

Imagine a nonrotating interstellar cloud of gas, mostly molecular hydrogen, and dust. Assume that this cloud has sufficient density to contract gravitationally. Its initial diameter is a few light years. As the collapse proceeds, material at the cloud's center increases in density faster than at the edge. Because of the increase in density, the collapse at the center speeds up (as expected from Newton's law of gravitation). It collapses faster, grows denser, and so collapses still faster. The rest of the cloud's mass is left behind in a more slowly contracting envelope.

With the rapid fall of material in the core, at some point the density of the core reaches a critical value at which the cloud becomes opaque. The core heats up to a few hundred kelvins, its pressure increases, and the core's collapse slows to a contraction and then eventually stops. A protostar forms (Fig. 15.18).

Meanwhile, the envelope continues merrily falling inward, showering mass on the core. The protostar slowly contracts; its size is twice that of the sun, its luminosity, a few times the sun's. The total time from the start of collapse to this stage is about one million years. The infant star hides in its womb, with dust cutting out the protostar's light. However, the light absorbed by the dust heats it so that it gives off infrared radiation. An observational sign of protostars should be small, intense sources of infrared radiation in or near known dense clouds of gas and dust.

Eventually, the star rids itself of its cloaking cloud. It may fall entirely onto

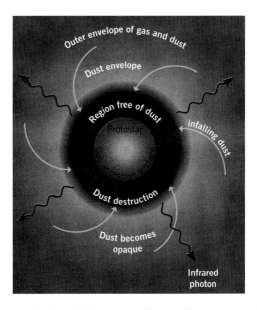

FIGURE 15.18 Schematic drawing of a protostar and its surrounding environment. Once the central mass has stabilized, material from the envelope continues to fall inward to accrete on the star's surface. Just around the star is a zone free of dust, which has been vaporized by the high temperatures there. Infrared photons are emitted by the dusty shell where the dust is thick enough to be opaque to photons emitted by the protostar.

the young star, or it may be blown away by a strong stellar wind. As the cloud dissipates, we see a **pre-main-sequence star,** one that is larger and cooler than it will be in its final state on the main sequence. The total time elapsed from the onset of collapse to reaching the main sequence is about 50 million years for a solar-mass star.

Collapse with Rotation

The models just described lack at least one fact of astronomical life: rotation. It is likely that interstellar clouds rotate at least a bit. Any rotating mass has *angular momentum* (Focus 9.1). An isolated, rotating mass must conserve angular momentum. So as a spinning, spherical mass collapses gravitationally, it must spin faster. It will eventually collapse *along* its rotation axis into a disk, as described in Section 12.6 for the formation of the planets.

The addition of spin to theoretical models of protostar collapse makes the calculations much tougher and the results less conclusive. To give you the flavor of results: in some cases, a ring or bar of material results. These rings and bars turn out to be unstable in some instances; they break up into two or three blobs, which sometimes coalesce. The smallest mass of a fragment is about 0.01 solar mass; even these small fragments end up rotating rapidly, although the cloud originally may have had very slow rotation.

(a)

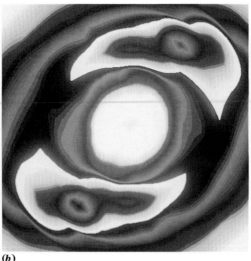

(b)

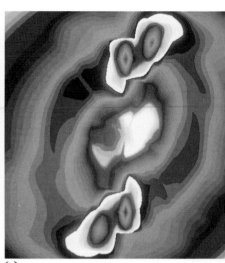

(c)

FIGURE 15.19 Computer model of the birth of protostars from a spinning cloud of gas. The calculation is designed to conserve angular momentum. These false-color images show the density of the gas, with blue and black the densest regions. The sequence in time from *(a)* to *(c)* spans about a thousand years. The cloud fragments into blobs whose masses are appropriate for protostar cores. (Courtesy of A. P. Boss, Carnegie Institution of Washington.)

Recent three-dimensional computer models (Fig. 15.19) have probed this process more extensively. The cloud begins with a simple pancake structure (Fig. 15.19*a*), but its rapid rotation throws off two spiral arms (Fig. 15.19*b*). The central region does not split cleanly into two pieces; instead, the fragments crash together to form a bar. Most of the material in the spiral arms wraps up into a ring (Fig. 15.19*c*).

The important point to remember is this: with spin added to the clouds, the calculations develop rings or bars that fragment into a few blobs. If each blob eventually becomes a star, we then have a natural explanation for the finding that most stars in the Milky Way are in binary or multiple systems. Or, if the masses are much less and end up in rings, they can result in planetary systems.

■ How does a massive protostar create the dusty cloak that conceals it?

15.4 Starbirth: Observational Clues

Let's look at the real world to see how models relate to observations. We have uncovered more information about massive starbirth than about the birth of solar-mass stars. There's a good reason: massive protostars have greater luminosities than solar-mass ones, and, when they reach "stardom," massive stars ionize the gas around them. Radio telescopes can then detect the ionized gas.

Since all this action is cloaked by dust, infrared and radio observations permit us to inspect stellar wombs.

Signposts for the Birth of Massive Stars

Let me outline what models predict should be the hallmarks, in the radio and infrared, of the birth of a massive star. With this outline as a guide, we'll then look at the observations.

First, because stars condense from molecular clouds, we need to find a molecular cloud; molecules emit at millimeter wavelengths. Second, the initial free-fall collapse warms the dust to roughly 30 K. This dust emits infrared radiation that peaks at roughly 100 μm. Third, as the protostar forms, the interior dust reaches about 300 K and so emits with a peak at 10 μm. Fourth, as the protostar reaches the main sequence, it ionizes the hydrogen gas, and a compact H II region develops, observable at centimeter wavelengths. Fifth, as the hot, ionized gas expands, the radio and infrared intensity decreases. Finally, the H II region expands enough to blow off its dusty cloak, and the star appears to optical view.

With this scenario in mind, let's return to our old friend the Orion Nebula (Fig. 15.1). The H II region around the Trapezium marks the oldest (most evolved in an evolutionary sense) part of the region. The Trapezium cluster consists of a few hundred stars. The O and B stars here, which are no more than a mil-

Starbirth	
At a Glance	
Where	Molecular clouds, large and small
When	Now, about ten stars per year in Milky Way
How	Gravitational contraction, fragmentation
What	Protostellar sources, visible in the infrared

lion years old, ionize the gas. The distance between the Trapezium stars and the front edge of the molecular cloud is roughly one light year. In an evolutionary sense, the molecular cloud core that lies behind the Orion Nebula is the youngest (least evolved) part of the region. Where is starbirth happening?

Infrared images reveal that the core of the Orion Nebula contains a dense cluster of young stars (Fig. 15.20). More than 500 stars appear, most visible only in the infrared; these make up the densest young cluster known. Radio observations show that the gas here hits a density of 10^{14} molecules per cubic meter. Infrared observations of hot hydrogen molecules in this region show that the gas moves along at high speeds—some 100 to 150 km/s—probably powered by a wind from one or more protostars. So the star formation here now occurs in the densest regions, stirred by the swift outflows from the newborn stars. This action results in an intricate structure of the Orion molecular cloud: dense filaments, hot sheets and streams of gas, and cool clumps interspersed with bubbles and cavities (Fig. 15.21).

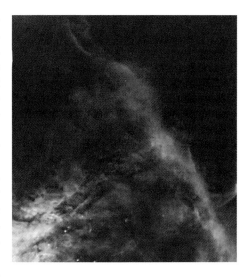

FIGURE 15.21 Gas plume (at lower right) in the Orion Nebula, as imaged by *HST*. The true-color representation uses emission detected from hydrogen (red), oxygen (blue), and nitrogen (green). Note the complexity of the structures in the gas, driven by new stars. The area shown is a little more than a light year across. (Courtesy of C. R. O'Dell, Rice University, and NASA.)

The Birth of Massive Stars

The foregoing observations add up to a picture of a sequence of massive-star formation from giant molecular clouds. The hot stars have heated the gas around them, ionizing it and destroying the molecules there. The hot gas of the H II region slowly expands and runs into the cold, dense molecular cloud. Here a shock wave forms. The shock wave, moving at about 10 km/s, prompts gravitational collapse and star formation out of the molecular cloud.

Moving more into the fragmented molecular cloud, the shock wave triggers more star formation, and each group—of OB stars—will develop an H II region and another shock wave (Fig. 15.22). So the molecular cloud will finally self-destruct in an orgy of star formation.

Once the star formation has started at an end of a molecular cloud, it propagates through it in a chain reaction. But what triggers the initial burst of star formation? The answer to that question is not yet clear.

FIGURE 15.20 Young stars in the Orion region. This infrared image shows at least 500 stars, most of which are invisible at optical wavelengths. In the false-color coding, the hottest sources are blue and the cooler ones red. In the reddish region to the upper right lie two bright, yellowish sources, which are probably active regions of star formation. The angular area covered is about 5 arcmin on a side, centered on the Trapezium. (Courtesy of Goddard Space Flight Center, NASA.)

The Birth of Solar-Mass Stars

The current observational evidence for the formation of stars like the sun is skimpy and doesn't hold together in a inviting way. One point is clear: like massive stars, solar-mass stars are born from molecular clouds. The questions are, In which clouds, and how?

Clusters of stars are thought to be young if they are close to gas and dust in the form of dark clouds. Dark clouds are one type of small molecular cloud. They typically have temperatures of 10 K, densities of 10^8 atoms in a cubic meter, and masses up to a few hundred solar masses.

A prime example of a dark cloud with active star formation is *Rho Ophiuchi* (in the constellation of Ophiuchus; see the star map for summer in Appendix G). The central region of this cloud emits far infrared. Near-infrared observations (Fig. 15.23) show many possible **young stellar objects** *(YSOs),* the generic name for all the stellar objects in early stages of formation. The basic observational evidence for starbirth is infrared pointlike (starlike) sources in dark clouds.

So according to one view, solar-mass stars are born in fairly massive dark clouds or perhaps in giant molecular clouds along with massive stars. They form from fragments throughout the cloud, rather than at the edges as massive stars do. The birth of the massive stars sweeps away the gas and dust to reveal the stars. In this picture, most starbirth takes place in dark, massive clouds, out of which OB groups form. So the sun may have been born in an OB group, such as the one we described in Orion.

From observations so far, we suspect that in general, stars are born from molecular clouds, massive stars from massive clouds, and less massive stars (the majority of those in the Galaxy) from less massive molecular clouds. Observations by *IRAS* (Section 6.3) appear to support this general notion (Fig. 15.24). That survey found many warm (temperatures from 70 to 200 K) point sources of infrared emission in the cores of dark, molecular clouds. These may be protostars, heating up the encircling dust. Whatever the details, we estimate that about ten new stars are born each year in the Milky Way.

Among YSOs are one type called **T-Tauri stars** (named after their prototype, *T Tauri,* in the constellation of Taurus). Most T-Tauri stars are low-mass (0.2 to 2.0 solar masses), pre-main-sequence stars. So they are young, aged from 10^5 to 10^8 years; they are surrounded by hot, dense envelopes; and they are losing mass by swift stellar winds. T-Tauri stars

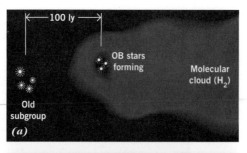

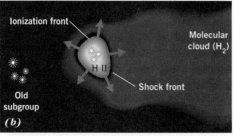

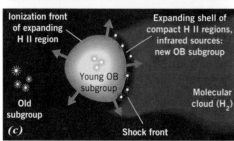

FIGURE 15.22 Schematic drawing for a sequential model of massive-star formation from a giant molecular cloud. (*a*) The formation of a small cluster of hot stars in a molecular cloud (*b*) creates an expanding H II region and shock wave that (*c*) drives the collapse of more of the molecular cloud to make another small cluster of stars. Each episode of star formation leaves behind an OB subgroup as the formation process moves through the elongated shape of a giant molecular cloud.

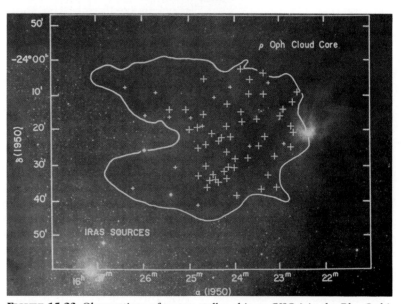

FIGURE 15.23 Observations of young stellar objects (YSOs) in the Rho Ophiuchi dark cloud. Infrared observations at 2 μm show YSOs (marked by crosses) in the darkest, densest parts of the clouds. About 20 new stars have been born here in the past million years. This region is a smaller part of the area shown in Figure 15.24. (Courtesy of C. J. Lada and L. Wilking.)

FIGURE 15.24 Protostar formation in the Rho Ophiuchi dark cloud. The largest image shows an optical view of the region near Rho Ophiuchi, just north of Antares. The dark cloud lies in the center; this is the region of the infrared sources shown in Figure 15.23. The small insert is a false-color map from *IRAS*, in which 12 μm emission is blue, 60 μm green, and 100 μm red. It reveals a group of young stars called the *IR Cluster*. The object called IRAS 1629A, about 500 ly away, appears to be a collapsing cloud forming a protostar. (Courtesy of the University of Arizona; photograph by the Anglo-Australian Observatory.)

almost always appear within dark clouds, where star formation is most active. Some T-Tauri stars have clear evidence of thin disks of circumstellar material with diameters of a few hundred AU. These might well mark sites of planetary formation.

Molecular Outflows and Starbirths

Observations of molecular emission around protostars have discovered high-speed (up to 100 km/s; typically 50 km/s) flows of gas. Doppler shift measurements show that these flows tend to be *bipolar*: two streams moving in op-posite directions (one has blueshifts, the other redshifts). The flows carry considerable mass (many times that of the sun) and can span a few light years; so enormous amounts of energy push them along (some 10^{40} J). Such **bipolar outflows** do appear to be associated with the birth of a star. They probably last only a short time—no more than 10,000 years.

A simple model for the outflows envisions a young massive star putting out a strong stellar wind. Surrounding the star is a dense disk of gas and dust. This disk would naturally channel the flow of the stellar wind, causing it to stream out along the thin axis of the disk, making two streams. When these two streams push enough material outward, two opposing lobes of gas should form. The early stages of this process are buried in gas and dust, but later stages are visible to optical and radio telescopes.

Models for these bipolar flows basically involve a disk of material around the protostar, threaded with a magnetic field (Fig. 15.25). The disk forms from gravitational contraction with angular momentum. The magnetic field resides in the surrounding molecular cloud. Ionized gas that falls into the disk drags the magnetic field with it. The hot disk generates both a molecular flow (from the outer regions) and an ionized flow from its inner parts. The ionized outflow is channeled by the magnetic field. Essentially, the rotating disk acts like a flywheel that stores energy that is released in the outflow.

The bipolar flow stage marks a *very* brief act in the process of starbirth and the evolution of YSOs. The sequence goes like this: a star forms in the gravitational collapse of part of a rotating molecular cloud (Fig. 15.26*a*); the central part gathers into the protostar, which appears as an infrared source surrounded by a dense disk (Fig. 15.26*b*). Next, the star develops a powerful wind that breaks through the disk in opposite directions (Fig. 15.26*c*). The wind carries along clumps of molecular material and strikes the cloud, producing shock waves. Cavities carved by the outward flow enlarge and push away more of the dark cloud (Fig. 15.26*d*), eventually revealing the star.

Planetary Systems?

The discovery of the bipolar flows strongly hints that disks of material typically form around massive stars during their formation. From such disks planetary systems might form. So we have a clue that the nebular model (Chapter 12) for planetary formation might actually operate elsewhere in the Galaxy.

In fact, we are just finding observational clues that planetary formation may be happening now. An image of a star called Beta Pictoris (Fig. 15.27) shows a possible planetary disk. At a distance of

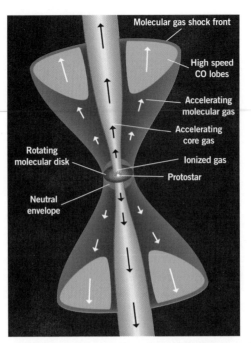

FIGURE 15.25 Schematic model for the source of the bipolar outflows from a young stellar object. The key feature here is the presence around the star of a disk of material formed by accretion. It is hot enough to produce ultraviolet photons to ionize some of the gas nearby. If the star has formed with a magnetic field, the ionized outflow follows along the magnetic field lines. The outflow creates a shock front that backs up the material in two lobes. (Adapted from a diagram by R. E. Pudritz and C. A. Norman.)

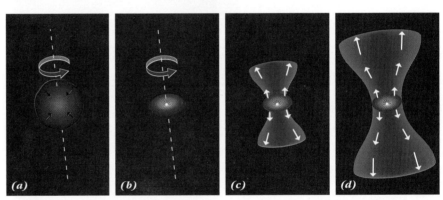

FIGURE 15.26 Evolutionary sequence in the formation of a bipolar outflow. (*a*) Gravitational collapse of a rotating interstellar cloud. (*b*) Formation of a disk with a protostar at the center. (*c*) Start of the outflow along the axis of the disk. (*d*) Development of two large lobes from the outflow.

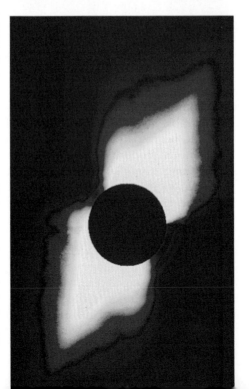

FIGURE 15.27 Circumstellar disk around the star Beta Pictoris. This CCD image shows a disk of material on both sides of the star, whose image is artificially blocked. The circle of the blocking material has a diameter of 80 AU at the star's distance. The disk around the star appears nearly edge-on and extends out some 400 AU. (Courtesy of D. A. Golimowski, S. T. Durrance, and M. Clampin.)

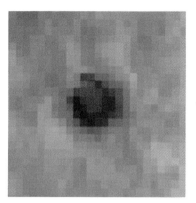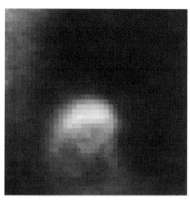

FIGURE 15.28 Protoplanetary disks around young stellar objects. These *HST* images show newly formed stars, less than about a million years old. The original spin of the star-forming cloud has spread the material out into a broad disk. Each disk has a hole in the center where the star is located. Material boils off the disk's surface and is blown by stellar winds. These images are composites of ones taken in hydrogen and oxygen. Each image is only 12 light *days* across! (Courtesy of C. R. O'Dell, Rice University, and NASA.)

50 ly from the sun, Beta Pictoris is a fairly nearby star. The disk of dusty material extends 60×10^9 km from the star. It shows up edge-on, and calculations indicate that planets may already have formed here. Spectroscopic observations show Doppler shifts that hint at a rapidly rotating, clumpy cloud of gas making up the inner regions of the disk.

The *IRAS* telescope has also gathered some strong, indirect evidence for planetary systems. It found a cool (some 90 K) cloud of solid particles around the nearby star Vega. These dust grains are about a thousand times larger than those found in the interstellar medium, and they total about 0.01 earth mass. The cloud has a diameter of roughly 170 AU, about twice the diameter of our solar system. Vega is a young star (about one-fifth the sun's age), so the observations suggest the possibility of a site of planetary formation.

Observations of bipolar outflows from T-Tauri stars, which have about the same mass as the sun, imply that many of these bodies have disks of materials with sizes on the order of 100 AU—just what you would expect for nebulas out of which planetary systems can form.

Finally, the *Hubble Space Telescope* has taken images in the Orion Nebula region that show disks around some YSOs (Fig. 15.28). These disks appear as thick disks with a hole in the middle where the star is located. The radiation from the hot star boils off material from the disk's surface. The material is then blown into a cometlike tail by the wind from the YSO. So the formation of such disks may commonly accompany the birth of stars.

■ Approximately how many new stars are born in the Milky Way each year?

Key Concepts

1. The interstellar medium contains both gas and dust. The gas, mostly hydrogen, comes in a variety of forms: molecules, atoms, and ions. The dust is in the form of small grains. Interstellar dust is far less abundant than the gas but does make up about 1 percent of the total mass of the interstellar medium.

2. The gas clumps in clouds of various sizes, ranging from small clouds of atoms to the giant molecular clouds; between the clouds is a hotter intercloud gas. A wide variety of molecules has been found in molecular clouds; some of these are organic compounds. But the most common element is hydrogen, as atoms, ions, and molecules. Much of the interstellar dust is in molecular clouds.

3. Dust makes itself known by the reddening and extinction of light and also by infrared emission when

it heats up. The dust is made of grains about a micrometer in size; the core of the grains contains silicates, graphite, or iron. The mantles are made of icy materials and organic compounds. The grains are formed in the outflow of material from cool, supergiant stars. Dust grains are associated with molecular clouds and aid the formation of simple molecules.

4. Stars are born out of molecular clouds by the process of gravitational collapse and contraction. A protostar forming in a cloud gets its energy from the conversion of gravitational potential energy to kinetic energy. The process of starbirth is hidden from our direct view, but we can infer its execution by infrared and radio observations. Somehow the surrounding material is dissipated to reveal the star.

5. Interstellar clouds likely have initial spin and so angular momentum. That angular momentum must be conserved. As they collapse, they naturally form a disk with a few large blobs of material. The gravitational collapse of rotating clouds naturally results in the formation of planetary or multiple star systems.

6. Infrared and radio observations imply that massive stars are formed in small groups (about ten) out of giant molecular clouds in a chain-reaction sequence in which one groups triggers the birth of the next. The process may be started by a supernova remnant hitting the giant molecular cloud. Starbirth for massive stars occurs quickly.

7. For solar-mass stars, starbirth happens more slowly than for massive stars. Solar-mass stars may be formed out of small molecular clouds or as a spin-off of the birth of massive stars from giant molecular clouds.

8. The formation of stars involves a bipolar outflow of gas (both molecular and ionized), collimated in part by a magnetized disk or ring around the young star. Many young stellar objects display such outflows, which blow away the surrounding cloud. This outflow marks a very brief phase in the making of a star.

9. Starbirth is taking place now in the Galaxy, as gravity shapes interstellar material into stars. The rate is estimated at about ten stars per year, mostly less massive ones. The observational evidence is the detection of pointlike sources of infrared emission in molecular clouds.

Key Terms

bipolar outflows	emission nebula	interstellar dust	protostar
bright nebula	extinction	interstellar gas	reddening
core–mantle grains	giant molecular clouds	interstellar medium	reflection nebula
coronal interstellar gas	H I region	nebula	T-Tauri stars
dark clouds	H II region	pre-main-sequence star	young stellar objects

Study Exercises

1. Describe one way in which astronomers observe each of the following: (a) interstellar H I, (b) interstellar H II, (c) the coronal interstellar gas, and (d) interstellar molecules. (Objectives 15-1 and 15-2)

2. Outline *two* ways in which astronomers "see" interstellar dust. (Objectives 15-1 and 15-3)

3. What if any evidence do we have that dense materials make up part of the interstellar dust? And that icy materials do? (Objective 15-4)

4. List the observational evidence that leads to the guilt-by-association argument for the formation of massive stars in small groups from giant molecular clouds. (Objective 15-7)

5. What powers a protostar? (Objectives 15-6 and 15-9)

6. In a theoretical picture for the formation of a massive star, how is it that we never see a massive star directly (optically) until it reaches the main sequence? (Objective 15-9)

7. Once it has reached the main sequence, how does a massive star influence its parent cloud? (Objective 15-11)

8. What *observational* evidence do we have that solar-mass stars form from dark clouds? (Objective 15-8)

9. What kind of observation tells us that the outflows

from star-forming regions take place in two opposite directions? (Objective 15-11)

10. In one sentence, argue that starbirth must be occurring now. (Objective 15-10)

11. In the sequential model for massive-star formation from giant molecular clouds, what mechanism promotes the formation of a new group of stars, once the process has begun? (Objective 15-9)

12. What is one possible source of interstellar dust grains? (Objective 15-5)

Problems & Activities

1. A protostar can have a temperature of 600 K and a luminosity 1000 times that of the sun. At what wavelength is its peak emission? What is its radius? *Hint:* Assume that the protostar radiates like a blackbody.

2. Compare the pressure in the core of a molecular cloud to that in an H II region. Which is greater? *Note:* The pressure is directly proportional to the number density and the temperature.

3. A fast bipolar outflow moves at about 100 km/s. How long would it take such a stream of gas to cover a distance of 1 AU?

4. If the dust particles around Vega act like blackbody radiators, at what wavelength does their emission peak?

5. Assume that your body is made of water molecules. Estimate the number of water molecules in the mass of your body. If these were spread out in a density equivalent to that at the core of a molecular cloud, how much volume would they fill? Assume that the cloud is spherical!

See for Yourself!

Without radio or infrared eyes, you cannot directly see regions of star formation. But with just a good pair of binoculars you can view the hot gas associated with the formation of massive stars. And you also can spot star clusters that once formed from gas and dust clouds. Keep in mind that the field of view of 7 × 50 binoculars is 7°.

In the winter sky, the easiest H II region to pick out is the Orion Nebula (Fig. 1.1) just below the belt stars. You should be able to discern some structure, but nowhere near the detail visible in this book's photos. Then swing west and north to Taurus (Appendix G) past its bright star Aldebaran until you come to the Pleiades, a young star cluster some 100 million years old. Count how many stars you can spy with just binoculars. Now move down back to Orion and over to Sirius, the brightest star in the sky. Right below it by about 4° is the star cluster called Messier 41 (M41). This cluster is estimated to have an age of about 200 million years. It is about 20 ly across and 2200 ly from the earth.

The summer sky is a glory for nebulas and star clusters. You can find Scorpius with bright Antares. Sagittarius is just to the east and a bit north. Find the star at the top of the spout of the "teapot" of Sagittarius. About 6° north of it is the Lagoon Nebula, Messier 8. You should see the young star cluster within it. Just north of the Lagoon is the Triffid Nebula, Messier 20. You probably won't be able to see more than a faint patch of light. Slowly scan the Milky Way from Sagittarius north to Altair. This starwalk should amaze you! ■

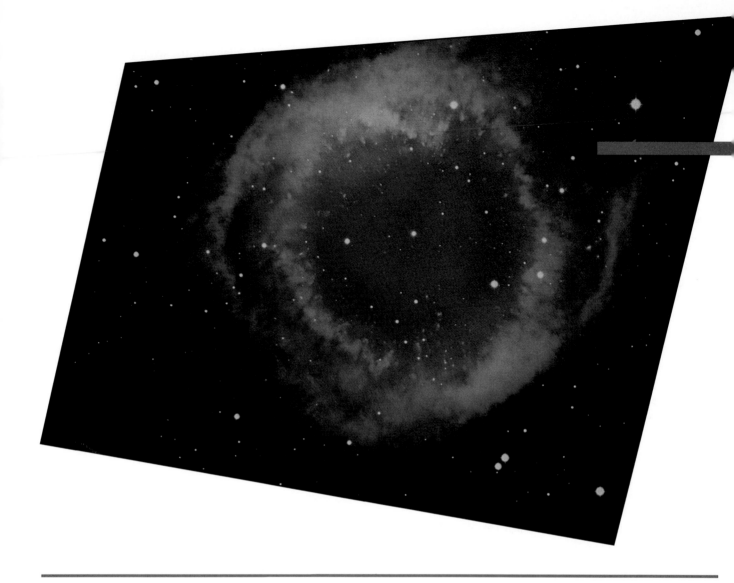

Star Lives

Not from the stars do I my judgment pluck,
And yet methinks I have astronomy....

WILLIAM SHAKESPEARE, SONNET 14

Gravity controls the history of newborn stars. A star survives as long as it can counteract the relentless gravitational crunch. The story of this battle against gravity runs like the aging of a person from birth to death, but it takes millions to billions of years. As long as a star shines, it lives. When its radiance fades, it dies.

To put this span into a human perspective, imagine time speeded up so that one year passes in one-fifth of a second. Then the sun would live only 65 years or so. The sun's birth would be quick; only 4 months would pass from the start of the collapse of the sun's embryonic cloud to its establishment as an immature but full-fledged star. For about the next 60 years the sun would shine calmly as it passed through middle age. Old age would gradually fossilize the energy production of the sun. In about 5 years the elderly sun would slowly expand to almost a hundred times its present size; it would become a bloated red giant. Then a sudden burst would blow off the sun's atmosphere, leaving behind a hot core that would cool quickly.

How a star lives depends mostly on how much mass it has at birth. The important stages that mark its life and the duration of each stage are related directly to a star's mass. Stars that are much more massive than the sun have short, frenetic lives.

This chapter presents, first, some theories of stellar evolution, contrasting the lives of stars like the sun to those of more massive stars. For most stars, the themes are the same; rapid birth, long middle age fusing hydrogen to helium in the core, and a slow aging to a red giant, violence, and death. Second, it offers some observational evidence in support of these theoretical concoctions. You will see how stars burn, falter, and flame out. Our sun, too, will eventually die— and the earth with it.

16.1 Stellar Evolution and the Hertzsprung–Russell Diagram

Stars evolve because they shine. As a star loses energy to space, it must change. While it lives, a star does not cool off. Most objects cool off (their temperature goes down) when they lose energy to their environment. It's natural to think that a star does, too, but it doesn't. Nuclear reactions generally replenish the energy a star radiates into space. When one fuel in the core (say hydrogen) runs out, another (say helium) can ignite, at a temperature made higher by the compression of matter by gravity. Only when a star cannot fuse matter any more will it cool off. Then the star has died (Chapter 17).

I'll approach stellar evolution from both theoretical and observational points of view. This chapter presents mostly *theoretical* ideas, because observations of a star's evolution do not come easily. Why? The sun's anticipated lifetime is more than 100 million human lifetimes. So there is no way you could watch a single star, like the sun, evolve. But you can see many different stars at one time. You can organize these stars on the Hertzsprung–Russell diagram if you know their luminosities and spectral types. Then you can use the H–R diagram of many stars to guess at the evolution of one star. Let's see how.

Classifying Objects

Suppose you asked 18- and 19-year-old adults in the United States for their weight and height and plotted the data (Fig. 16.1). Note the trend: the points tend to fall along a line that shows that weight generally increases with height. That result shouldn't surprise you.

Suppose you recorded the height and weight of every person you encountered randomly. Plot the data again as weight versus height (Fig. 16.2). You still have a general trend, rather than a random scatter of points. But you also have other groups that don't follow the first set. Why the difference? The first graph includes people of the *same* age; the second, people at *different* ages.

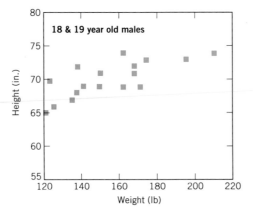

FIGURE 16.1 Height-weight diagram for a sample of 18- and 19-year-old adults from an introductory astronomy class at the University of California, Berkeley. Units are inches and pounds.

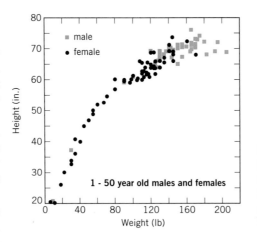

FIGURE 16.2 Height-weight diagram for a sample of U.S. people, ranging in age from 1 to 50 years. Units are inches and pounds. (Data collected by an introductory astronomy class at the University of New Mexico.)

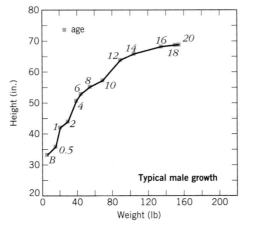

FIGURE 16.3 Height-weight-time diagram for a typical U.S. male from birth (*B*) to 20 years of age. Units are inches and pounds. (Data from the National Center for Health Statistics Growth Chart.)

Now here's a third graph (Fig. 16.3). It's a plot of weight versus height for an average U.S. male at different times in his life, as indicated by points representing the person's age from birth to 20 years. The line connecting the points shows how a *single* person's height and weight change. You can correctly interpret the graph for many people in evolutionary terms *if,* and *only if,* you know how one person evolves. Time was implicit in the second graph

(Fig. 16.2). In the same way, time and age implicitly play a role in an H–R diagram, in which are plotted two essential properties of stars: surface temperature and luminosity. These external changes displayed by stars with aging are analogous to height and weight in human beings.

Time and the H–R Diagram

How does the H–R diagram tell you about an evolutionary sequence of a star? Imagine that you have a large family. They have gathered together, and you take a snapshot of the group, from the newest-born to the great-grandparents. Most of the people in the picture are in their middle age (20 to 60 years old); you have a few infants, some children and teenagers, and a few old people. You have so many middle-aged people because most of your life you will be middle-aged; you spend relatively less time as an infant, teenager, or old person. So the relative numbers of people in these stages of their lives reflects the relative duration of each stage. Each person, of course, has his or her unique life. This snapshot just happens to have caught all the family members together at one time.

Now suppose you have a collection of any objects that evolve in a specific evolutionary sequence. Each follows its own special history but passes through similar stages. You can estimate the relative time spent in any evolutionary stage by the relative numbers you find at that stage compared with others. (This argument holds true only if birth and death go on at a constant rate; if no more people are born, eventually you will see only old people—and then none.)

Recall the H–R diagram for many stars, which shows the luminosity and surface temperature of the stars. Most of the stars fall on the main sequence, so stars found here are going through the longest, most stable stage in their evolution. The main sequence, from O to M stars, represents a sequence from higher to lower masses. Here's the evolutionary meaning of the main sequence: *it marks stars at the stage of converting hydrogen to helium in their cores; stars remain at this stage for the greatest part of their lives*. Any star fusing hydrogen to helium in its core is a main-sequence star; it falls on the main sequence of the H–R diagram. That's why we see so many main-sequence stars now. Other stages, such as becoming a red giant, must be shorter. How do we know? Because we see far fewer red giants now than main-sequence stars.

A star's mass (hence the pressure and temperature in its core) determines how the star will evolve. So you must examine the H–R diagram to find how stars of different masses evolve. But what is the correct interpretation of the H–R diagram for the evolution of stars? You need some hints from the physical nature of stars and theoretical calculations to make up the star models.

■ **What is the sun's expected lifetime in human terms?**

Stellar Lives

At a Glance

Mass	Mainly determines the lifetime of a star
Chemical composition	Secondary effect on star lives
Energy sources	Gravitational contraction and fusion
Lifetimes	From tens of millions to tens of billions of years
Evolution	Changes in radius, luminosity, surface temperature, and chemical composition

16.2 Stellar Anatomy

What is a star? A huge, hot ball of gas, mostly hydrogen, heated by thermonuclear reactions in its core. You can imagine a star as a naturally controlled hydrogen bomb. Why doesn't it blow up? Because gravity persistently pulls it together. All its life, a star must withstand the inward squeeze of gravity. How does it do it? By producing an outward pressure, usually from fusion reactions.

Pressure and Energy Balance

For most of its life, the outward pressure exerted by a star results because the star is hot. A star consists of gas; a hot gas has a high pressure, and the outward force balances the inward gravitational force. This balance must hold true at every level throughout the star (Fig. 16.4); otherwise it would be unstable. So the first physical requirement for a model of a stable star is that it be in balance, neither expanding nor contracting.

Second, the star must generate energy internally. For most of a star's life, thermonuclear fusion reactions operate as the internal furnace (details to come in Section 16.4). In general, the rate of energy produced inside the star equals the rate at which the energy radiates away at its surface. This balance holds not only overall but also at each layer within the star. Otherwise the star will be unstable and will expand or contract.

As an analogy, consider an assembly line that carries cars as they pass through different assembly stages. At each stage, the rate at which the cars come in must be equal to the rate at which they go out. Otherwise the cars will pile up at one location. If the flow of heat through a star were uneven, the temperature of various layers would change. These temperature changes would result in pressure changes capable of causing the star to expand or contract. So, generally, a star meets the condition of energy loss equaling energy production. And its energy flows from hotter to cooler regions.

Energy Transport

Third, we need to look at how a star transports energy from its core to the surface (Section 13.5). Basically, three methods are available: *conduction, convection,* and *radiation.* Which of these occurs depends on the star's opacity (Section 13.3). The *opacity* of a star's material directly affects its radiative energy transport. The more opaque a star's material, the slower the flow of radiation through it.

You can think of opacity as acting like insulation in a house in winter. The furnace is generating heat. The greater the house's insulation, the slower the flow of the heat to the cold outside. A slow flow keeps the exterior of the house cold and the inside hot. In a poorly insulated house, the outside is warmer and the inside colder than in a well-insulated house.

Likewise, if the opacity of a star

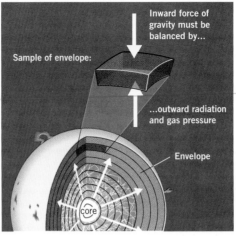

FIGURE 16.4 The balance of gas pressure and gravity in a star. For the star to be stable, the outward pressure from the internal heat (generated by the fusion reactions in the core) must balance the inward pull of gravity. This balance must hold at every level in the star; otherwise, the star will either expand (pressure is higher) or contract (gravity is greater).

were suddenly lowered, radiation would escape more easily and the star would grow more luminous, but its inside would become cooler. So the internal pressure would drop, and the star would contract until balance was regained. The opposite occurs if the opacity increases: the radiation is dammed up, the star becomes hotter, and it expands.

Whether convection occurs at any region in a star depends on the opacity of the material there. If the material is transparent, energy will flow more easily by radiation. If it is so opaque that the radiative energy flow gets bottled up, convective transport will take command and operate instead. Generally, a star's opacity depends on its chemical composition, density, and temperature. A greater density or abundance of metals increases the opacity; a greater temperature decreases it.

Gravity keeps a star from expanding and just balances the outward pressure of the hot gases. At the same time, the outflow of energy is just balanced by its production. This balance cannot continue forever. A star loses energy to space—it shines! Thermonuclear fusion reactions supply the energy to keep it hot. Eventually the star runs out of fuel, and the pressure can no longer balance its gravity. Then it contracts. A star can survive as long as it can find a means to produce energy, to stay hot, hence to withstand gravity. When it fails to do this, the star dies. How stars survive marks the central theme of this chapter.

■ **What are two main physical requirements for a model of a star?**

16.3 Star Models

A theoretical model of a star incorporates all the physical conditions just described. A star must produce energy; gravity and pressure must be in balance; and energy must be transported evenly from core to surface. In addition, we must know how the matter in a star behaves. Luckily, for most stars for most of their lives, that behavior is the same as for an ideal gas (Section 13.2). In some massive stars, internal pressure from radiation adds to the internal pressure from the gas. (Remember that light has particle properties, and so, in a sense behaves like the physical particles in a gas. That is, it exerts *radiation pressure.*)

We need to know exactly what thermonuclear processes produce energy, as well as the mass and chemical composition of the model star. Then a few complicated equations formulate the problem, and these are solved in a consistent way to find the physical conditions in the star (temperature, pressure, and density, for instance) from the center to the surface. This catalog of values for important physical properties for a specified mass and composition is called a **star model.** In general, we expect density, temperature, and pressure to decrease as we go from the center of the star to its surface (Fig. 16.5). One pair of numbers of this catalog, the luminosity and temperature at the surface, specifies a point on an H−R diagram, the *theoretical* point for a model star of given mass and chemical composition. This point on the H−R depends on the specifics of the stellar model.

Goals of Models

In practice, the construction of a star model requires very tedious calculation. High-speed, modern computers can, however, complete the calculations for one instant in a star model's life in a few minutes. The evolutionary sequence of one star, using different models for different stages in a star's life, may take hours of computer time. And they depend on the specific models used.

Why all this fuss? Because the study of stellar evolution rests on the construction of physically reasonable models of stellar interiors. We rely on the models because we cannot directly observe the interiors of stars and we cannot wait millions to billions of years to watch individual stars evolve. *Models reveal how time enters into the H−R diagram for real stars.*

Almost everything said in this chapter rides on the validity of stellar models. And one fly swims in the ointment: solar neutrino experiments, which cast some doubt on our models for the sun (Section 13.6). Most of us ignore that problem, hoping that the missing solar neutrinos can be explained without requiring the conclusion that present stellar models are wrong.

General Results

What do star models tell us? They show that as a star evolves, its radius, temperature, and luminosity change in complicated ways. Such changes result in the change in a star's position on the H−R diagram. A series of these points at various places on an H−R diagram makes up a star's **evolutionary track.**

For a simple example, if a star's surface temperature increases (but not its

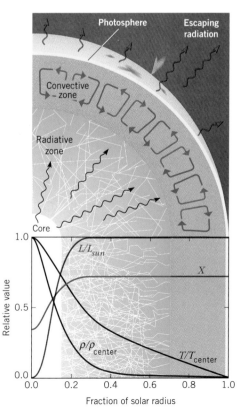

FIGURE 16.5 Model for the sun's interior at its current age (4.6 billion years). Shown are the temperature, T, density, ρ, accumulated luminosity, L, and hydrogen abundance (by fraction of total mass) X, as these change with radius. Note that the luminosity is divided by the sun's luminosity; the temperature and density by the central temperature and density. (Based on calculations by J. Bahcall and R. Ulrich.)

luminosity), the star's position on the H–R diagram moves horizontally from right to left, with no vertical change (line A in Fig. 16.6). If the luminosity increases (but not the surface temperature), the star's point on the H–R diagram moves vertically from bottom to top, with no horizontal change (line B in Fig. 16.6). Both motions *on the H–R diagram* represent changes in the physical properties of the star. The path is the evolutionary track. Each star model of a different pair of physical characteristics (such as mass and chemical composition) traces out a different evolutionary track.

Note in the first case that to keep the luminosity constant at a higher temperature requires that the star's surface area, and so its radius, decrease. In the second case, with no change in temperature, the star's luminosity can go up only if its surface area increases, and so its radius. You see that a star's surface temperature, luminosity, and radius are all related and change with time.

Caution. What moves around on the H–R diagram is *not* the star in space but a point *representing* the changes in luminosity and temperature of the star.

The goal of studying stellar evolution is to understand the *physical* causes behind a star's evolutionary track on the H–R diagram.

■ Why do astronomers devise star models?

16.4 Energy Generation and the Chemical Compositions of Stars

What makes a star evolve? The answer lies in the heart of the star. Here thermonuclear reactions cook lightweight elements into more complex ones by fusion. This continual change in chemical composition and its effects on a star's structure mark a secondary theme of stellar evolution.

Hydrogen Burning
The sun generates energy by the proton–proton (PP) reaction (Section 13.6). Four hydrogen nuclei combine to form one helium nucleus and release a certain amount of energy. A minor part of the sun's energy comes from another reaction sequence, called the carbon–nitrogen–oxygen cycle (CNO cycle). In more massive stars, this cycle becomes more important than the PP cycle. If a star's central temperature is greater than about 20×10^6 K, the CNO cycle produces more energy than the PP reaction. (Both can go on at the same time.) The net result of the CNO cycle is the same as that of the PP reaction: four hydrogen nuclei are converted to one helium nucleus, with the release of energy. The CNO cycle predominates in stars with masses greater than about 1.5 solar masses.

Let's look at the CNO cycle in a little detail (Fig. 16.7). The complete cycle has six reaction steps. First, a proton (^{1}H) collides with a carbon nucleus (^{12}C) to convert it to a radioactive nitrogen isotope (^{13}N). The nitrogen nucleus emits a positron and a neutrino to become a carbon isotope (^{13}C). A proton blasts into this particle and, with the emission of a gamma ray, turns it into stable nitrogen (^{14}N). Then a proton bangs into the nitrogen nucleus to form the radioactive oxygen isotope (^{15}O) and a gamma ray. The oxygen isotope decays into the nitrogen isotope (^{15}N), a positron, and a neutrino. Finally, the nitrogen isotope gets knocked by a proton and splits into ordinary carbon (^{12}C) and helium (^{4}He). The carbon comes back unscathed; it acts only as a catalyst, helping to glue four protons together to make helium. Total energy released to the star is 4.0×10^{-12} J, some as neutrinos.

These reactions take place only in the star's core, where temperatures are high enough to keep them going. Be-

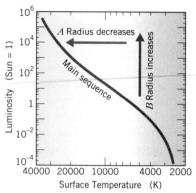

FIGURE 16.6 Motion of a star's point on an H–R diagram as its size changes. A star that stays at a constant luminosity but increases in temperature must be decreasing in size (path A). If a star stays at a constant surface temperature but increases in luminosity, its size must increase (path B). Generally, a star's changes in luminosity, temperature, and radius are more complex than shown here for these two special cases.

Stellar Models

At a Glance

Based on	Initial mass, chemical composition
Balance	Gravity and internal pressure
Energy flow	Convection and radiation
Results	Evolutionary track on H–R diagram

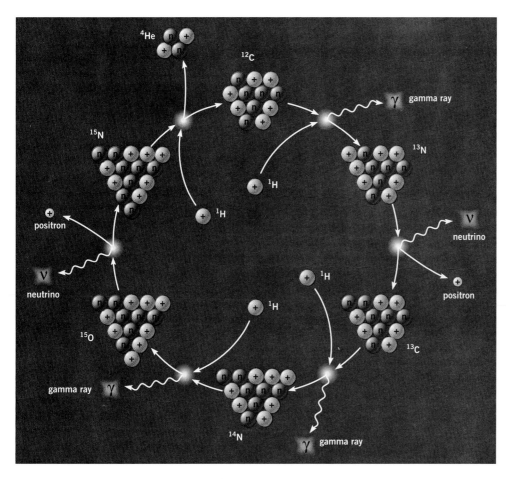

FIGURE 16.7 The carbon–nitrogen–oxygen (CNO) cycle, a hydrogen-burning fusion process. Note that the net result is the conversion of four hydrogen nuclei to one helium nucleus; the carbon entering the first step returns in the last.

cause both involve the fusion of hydrogen as a fuel, they are given the generic name of *hydrogen burning.* Because a star has only a limited amount of hydrogen to burn, the core is eventually converted almost entirely to helium, and the CNO and PP reactions cease. Fusion reactions no longer supply energy to keep the temperature and pressure high.

Helium and Carbon Burning

What next? The core contracts and heats up (remember, the self-contraction converts gravitational potential energy to kinetic energy). When the core's temperature gets up to roughly 100×10^6 K, another reaction can take place: the **triple-alpha reaction,** so named because three helium nuclei (also known as alpha particles) fuse to form one carbon nucleus, with the release of energy (Fig. 16.8). This process is generally called *helium burning.*

What happens when the helium runs out? The core contracts and heats up again. If the temperature increases

enough, carbon can be fused into heavier elements in the processes of *carbon burning.* Such processes require extreme temperatures, at least 600×10^6 K. Iron, the most stable of all nuclei, ends the sequence of nuclear fusion. To form elements heavier than iron by fusion reactions, energy must be added and absorbed. The steady climb from hydrogen to iron in fusion reactions in stellar cores is one type of **nucleosynthesis,** the nuclear cementing process that makes heavy elements in a universe that otherwise would consist only of hydrogen

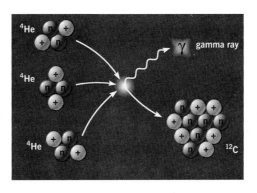

FIGURE 16.8 The triple-alpha process. This helium-burning fusion reaction converts three helium nuclei into one carbon nucleus and produces energy.

and helium made in its Big Bang origin (Chapter 21).

You may be wondering why a star's temperature goes *up* when fusion reactions *stop*. The cause is gravitational contraction. A star is always losing energy to space by radiation. If that energy is not replaced by nuclear reactions, it must come from somewhere else—namely, from gravitational contraction. As the star contracts, some of the gravitational potential energy of its particles transforms into kinetic energy and some is radiated away (Section 12.5). So the temperature goes up until the ignition temperature of the next set of fusion reactions is reached. Fusion turns on again, the pressure increases, and the core contraction stops.

During a star's life, short periods of gravitational contraction alternate with long spells of fusion burning. (Note that just the *core* needs to contract, not the star as a whole.)

■ **Which nuclear reaction is the predominant source of energy in our sun during its main-sequence phase?**

16.5 Theoretical Evolution of a 1-Solar-Mass Star

Gravity instigates the birth of a star. The shape of the subsequent evolution are controlled by the star's mass. Let's look at the evolution of a star like the sun, a 1-solar-mass star, having the sun's chemical composition.

Evolution to the Main Sequence

Let me briefly review a 1-solar-mass star's pre-main-sequence evolution (Section 15.3). The key point to recall is that a protostar gets its energy from gravitational energy, *not* fusion reactions.

The protostar forms by the gravitational contraction of an interstellar cloud. Once the dense core of the cloud has formed (point 1 of the evolutionary track in Fig. 16.9), the rest of the cloud accretes upon it (points 1 through 4 in Fig. 16.9). For a while, a protostar has a larger radius than it will have as a main-sequence star, and the surface temperature is lower (point 2 in Fig. 16.9). However, the protostar has a higher luminosity than it will have when it reaches the main sequence, about thirty times larger after 100,000 years has elapsed. How can this be, if it is cooler? Because it is larger and so has more surface area to radiate energy.

At this stage, the star's temperature is so low that its opacity is relatively high (even though its density is low). Convection transports the energy outward. So a protostar is completely convective—a huge, bubbling ball of gas. The efficient transport of energy makes the star very luminous (points 2 to 3 in Fig. 16.9).

As the protostar shrinks, its luminosity decreases (points 3 to 4 in Fig. 16.9). Meanwhile, the core continues to heat up. Eventually the core hits eight million kelvins, high enough to start thermonuclear reactions. When the protostar gets most of its energy from thermonuclear reactions (PP reactions in the case of the sun) rather than gravitational contraction, it achieves full-fledged stardom (point 5 in Fig. 16.9). The star is now called a **zero-age main-sequence (ZAMS)** star. The total time elapsed from

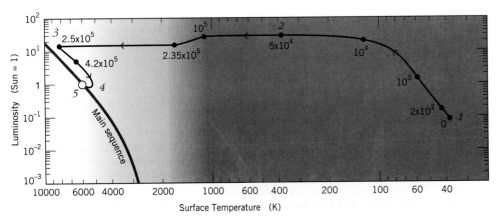

FIGURE 16.9 The evolutionary track on an H–R diagram of a pre-main-sequence, solar-mass star. The labels indicate the approximate time in years since the formation of a core of an interstellar cloud at point 1. The zero-age main sequence (ZAMS) is indicated at the left. (Adapted from a diagram by K.-H. Winkler and M. J. Norman.)

initial collapse to arrival as a star on the main sequence is only 50 million years (from point 1 to point 5 in Fig. 16.9).

Evolution on the Main Sequence

Where the star ends up on the main sequence depends mainly on its mass. The more massive the star, the hotter and more luminous it is; the less massive the star, the cooler and less luminous it is. The main sequence, on the H–R diagram, is a series of stars of decreasing mass (but similar chemical composition), from the upper left-hand corner (O stars with high mass) to the lower right-hand corner (M stars with low mass). A sunlike star spends about 80 percent of its total lifetime on the main sequence, a period called its **main-sequence lifetime.** For this star, energy is transported through its interior mostly by radiation. Convection works in a zone just below the surface.

How the star evolves further also depends on its mass. Because massive stars have higher luminosities, the hydrogen in them burns faster than in low-mass stars. For example, main-sequence lifetime for a 15-solar-mass star lasts only 5 *million* years. Why? Because of its greater mass, the 15-solar-mass star has a greater luminosity, about 30,000 times greater than the sun. (Recall the mass–luminosity law, Section 14.6.) So its fusion reactions must go on at a much faster rate than in the sun—also 30,000 times faster. So even though the 15-solar-mass star has more hydrogen to fuse, it does so at incredibly fast rates.

The main-sequence phase ends when almost all the hydrogen in the core has been converted to helium. During this time the temperature in the core increases gradually. This results in a greater flow of energy to the surface, and the star's luminosity increases (point 1 to point 2 on the evolutionary track in Fig. 16.10; see also Table 16.1). The star is now poised to become a *red giant.*

Evolution off the Main Sequence

When the hydrogen in the core is used up, the thermonuclear reactions cease

there. However, they keep going in a shell around the core, where fresh hydrogen still exists. At the end of fusion reactions in the core, gravity takes over and the core contracts. This heats up the shell of burning hydrogen, so the reactions go faster and produce more energy. The luminosity increases. The radius of the star increases, and its surface temperature decreases. The temperature drop increases the opacity, and convection carries the energy outward in the star's envelope. The star now becomes a red giant as it expands in size.

Its position on the H–R diagram moves to the region of lower surface temperatures and higher luminosities (point 3 to point 4 in Fig. 16.10). The sun as a red giant would engulf Mercury!

Gravity has compressed the red giant core to such a high density that it no longer behaves like an ordinary gas. The core becomes a **degenerate electron gas** (Focus 16.1). In this state, the electrons produce a **degenerate gas pressure,** which depends only on den-

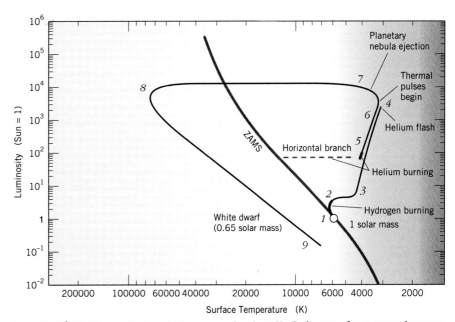

FIGURE 16.10 Theoretical evolutionary track on an H–R diagram for a one-solar-mass star. The evolution begins here at the ZAMS (point 1) and ends with the formation of a white dwarf from the core of a red giant star. As hydrogen is depleted in the core, the star's luminosity increases (1 to 2). When the core runs out of hydrogen, the star burns hydrogen in a shell around the core (2 to 4); it becomes a red giant. Gravitational contraction heats the core until the temperature becomes high enough to start the ignition of helium in the core (4); for a short time, helium fusion occurs in the core (5). When the star's core is depleted of helium, the star burns both hydrogen and helium in a shell and becomes a red giant again (5 to 7). Finally, the star throws off its out layers (7) to expose its core (8) and becomes a white dwarf (9).

TABLE 16.1 EVOLUTIONARY PHASES OF A SOLAR-MASS STAR

H−R POSITION (Fig. 16.10)	PHYSICAL PROCESSES
1	Hydrogen core burning begins (main sequence)
2	Hydrogen core burning ceases; hydrogen shell burning begins
3	Hydrogen shell burning continues; convection dominates energy transport
4	Helium flash; helium core burning starts (red giant)
5	Hydrogen shell burning accompanies helium core burning
6	Fusion reactions stop; hydrogen and helium shell burning continues; thermal pulses
7	Expulsion of outer layers
8	Planetary nebula with central star
9	All thermonuclear reactions cease; white dwarf; slow cooling

sity, not temperature, and this enables the core to attain and preserve a balance even though no fusion reactions are going on. Degenerate gases are also good conductors of heat, so they have the same temperature throughout. (You will encounter degenerate gases again in Chapter 17.)

As the bloated star attains its red giant status (point 4 in Fig. 16.10), the core temperature—which has been steadily increasing as the core contracts—hits the minimum to start helium burning by the triple-alpha process. When part of the degenerate core ignites, the heat generated by the fusion spreads rapidly throughout the core. The rest of the core quickly ignites. The increased temperature runs up the rate of the triple-alpha process, generating more energy, further increasing the temperature, and so on. This out-of-control process in the core is called the **helium flash.** The whole process of helium core ignition in the helium flash takes place in a very short time—perhaps only a few minutes! We will never see this helium flash in a star, for the action takes place deep in the core.

The star adjusts to this event by decreasing its radius and luminosity a bit. The star quietly burns helium in the core and hydrogen in a layer around the core (point 5 in Fig. 16.10). This phase is the helium-core-burning analogue to the star's main-sequence phase (in which the hydrogen core burns).

Evolution to the End

Eventually the triple-alpha process converts the core to carbon. The reaction stops in the core but continues in a shell around it. This situation—core shut down but the thermonuclear reactions going on in a shell—resembles that of the star when it first evolves off the main sequence. The burning shell makes the star expand. It again becomes a red giant (point 6 in Fig. 16.10).

Because the rate of the triple-alpha reaction changes greatly in response to small changes in temperature, the helium-burning shell causes the star to become unstable. Bursts of triple-alpha energy production cause thermonuclear explosions in the shell. They have the prosaic name of **thermal pulses.** The explosions occur about every few thousand years and cause the luminosity of the star to rise and dip rapidly by up to 50 percent. "Rapidly" here means in a few years or tens of years! Each blast generates an assault of energy. To move it out efficiently, the region becomes convective. The bubbling gases carry outward heavy elements fused in each explosion.

Meanwhile, the star has developed a very strong outflow of mass from its surface, sometimes called a *superwind* to distinguish it from the normal stellar wind of a red giant. The superwind blows in gusts that quickly (in about 1000 years) rip off the envelope of the

Enrichment Focus 16.1

DEGENERATE GASES

When matter is packed to very high densities (greater than 10^8 kg/m^3), it no longer behaves in ordinary ways. In a normal gas, the particles are widely separated and rush helter-skelter into one another and rebound away (Section 13.2). In highly compressed material, little space exists between particles. The matter is so jammed together that the electrons on the outside of the atoms are, in a sense, touching one another. The nuclei can no longer hold electrons in their usual energy levels, and the electrons move among the nuclei. But there's not much space for moving about.

Electrons abide by a quantum property called the *Pauli exclusion principle*. It states that no two electrons can be together in exactly the same energy state. Picture a small box containing one electron in some energy state. Imagine adding another electron to our box. The electron already in it probably occupies the lowest energy state possible. So the next electron must occupy the second level available, and so on. In contrast to a low-density state, in which energy levels are available for occupation, a very dense gas has far fewer levels, which get filled up quickly. (This is a simplified explanation because electrons have spin, so two electrons, one with its spin axis up and the other with its spin axis down, can occupy the same energy state.)

What happens if we try to cool this dense gas by letting electrons give up kinetic energy? They can't lose very much kinetic energy, for only high energy levels are open. So no heat can be extracted from these electrons (in a sense, their temperature is zero). Yet they exert a great pressure because they move with high speeds.

A gas in this state is called a *degenerate electron gas,* and the pressure from the uncertainty principle in action is called the *degenerate gas pressure.* Unlike an ordinary gas, for which the pressure is directly proportional to temperature (Section 13.2), degenerate gas pressure is nearly independent of the temperature.

Electrons become degenerate at densities of about 10^8 kg/m^3. Such densities occur in stellar cores after main-sequence hydrogen burning and also in white dwarfs. When electrons are degenerate, they conduct heat very efficiently, and temperature variations are quickly smoothed out. (In a sense, a degenerate gas acts like a metal; it can conduct current and heat well.) So degenerate cores have the same temperature throughout. A high enough temperature can relieve the electrons of their degenerate condition: "high enough" is at least a few hundred million kelvins for electrons.

star (point 7 in Fig. 16.10). A hot core is left behind. The expelled material forms an expanding shell of gas heated by the hot core. Astronomers call this a **planetary nebula** for historic reasons. (It looks like a planetary disk when viewed with a small telescope. This description was first given by William Herschel, who discovered Uranus.) The hot core appears as the central star of a planetary nebula (Fig. 16.11). A star may shed its envelope in a series of bursts, each releasing a shell of material. This gas expands at a typical speed of 20 km/s.

The nebula keeps expanding until it dissipates in the interstellar medium in only a few thousand to a few hundreds of thousands of years. What happens to the core? For a star of roughly 1 solar mass or less, the core never reaches the ignition temperature of carbon burning. Why not? Because the core has become degenerate and cannot contract and heat up to ignite carbon burning. In about 75,000 years it forms a white dwarf star,

composed mostly of carbon (point 8 to point 9 in Fig. 16.10). Without energy sources, the white dwarf cools to a **black dwarf,** the dark culmination of a 10-billion-year biography. The cooling down occurs over billions of years while the degenerate gas pressure continues to support the star.

The Fate of the Earth

What will happen to the earth when the sun dies? The sun will expand to a size of about 1.1 AU, so that the earth will orbit in its outer atmosphere. The earth's orbital velocity of 30 km/s corresponds to a supersonic speed within a red giant envelope. The earth's air will be ripped off and the mantle vaporized. The drag will cause the earth's orbital radius to decrease, and in less than 200 years, the earth will fall into the sun's core to vaporize completely. Puff—our planet will be gone!

Lower-Mass Stars

The evolution of stars of lower mass than the sun resembles that for the sun, with one exception. Few stars less massive than the sun have had time to evolve off the main sequence. The universe simply isn't old enough. A 0.74-solar-mass star, for example, has a main-sequence lifetime of 20 billion years, longer than most estimates for the age of the universe (Focus 7.1). Below 0.08 solar mass, the object never becomes hot enough to ignite fusion reactions.

Chemical Composition and Evolution

A star's mass most strongly directs its evolution. Its chemical composition plays a secondary role. If a star has a much smaller percentage of heavy elements than the sun, 0.01 percent instead of 1 percent, would its evolutionary track make *much* different gyrations?

The answer is *no*. The general trends stay the same: the star becomes a red giant, undergoes a helium flash, sits for a while during helium core burning, returns to red giant status, undergoes a series of thermal pulses, and finally expels its outer layers to make a planetary nebula leaving a white dwarf.

However, low-metal stars do differ significantly with respect to where they bide their time on the H–R diagram during their stage of burning helium in the core. Basically, stars with smaller percentages of heavy elements and less mass than the sun make a horizontal line on the H–R diagram, called the **horizontal branch** (Fig. 16.10), during the phase of burning helium in their cores. These stars have a lower mass than when they were main-sequence stars. They have lost some mass as a result of stellar winds during their first red giant phase (before the helium flash).

To sum up. The evolution of low-mass stars like the sun proceeds through the stages of protostar, pre-main-sequence star, main sequence, red giant, planetary nebula, white dwarf, black dwarf.

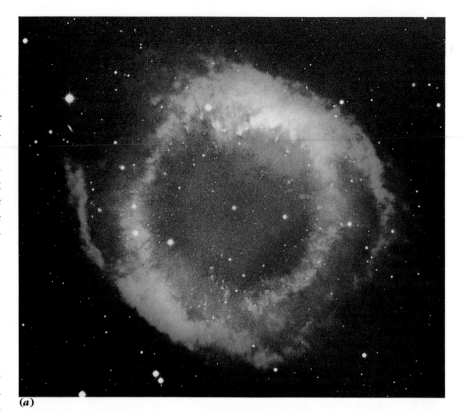
(a)

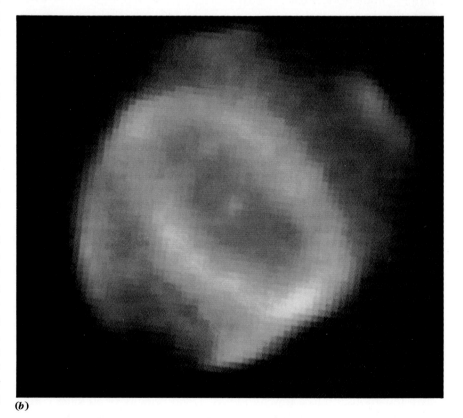
(b)

FIGURE 16.11 Planetary nebulas. (*a*) The Helix planetary nebula (New General Catalog 7293). The central star was once the core of a red giant. The nebula, which looks like a ring, actually forms a series of shells; it was the envelope of the red giant. (Courtesy of David Malin, AATB.) (*b*) Young planetary nebula, imaged by the *HST*. The expanding cloud of gas, mostly in a ring, is tilted at about 35°. Smaller clumps and wisps show the turbulence in the cloud. Observations over the past few decades have shown that this star has changed from an ordinary star to a young planetary nebula. (Courtesy of Matt Bobrowsky, CTA Incorporated, and NASA.)

■ When the sun runs out of hydrogen in its core, what type of star will it become?

16.6 Theoretical Evolution of Massive Stars

Now to examine the history of stars much more massive than the sun but with the same chemical composition. Regulus, the bright star that marks the heart of Leo the Lion, is a 5-solar-mass star (see the spring star chart in Appendix G for the location of Regulus in the sky). The evolution of stars such as Regulus shows the trends for *middle-mass stars*—those that range from about 5 to 10 solar masses. Such stars differ in their evolution from less massive stars because they can reach higher temperatures in their cores and throughout their interiors. Radiation pressure plays a larger role in providing the internal pressure supporting the stars. (In fact, a massive star cannot exceed about 100 solar masses because radiation pressure blows it apart.) The greater temperatures have other notable outcomes for the general evolutionary trends:

1. While on the main sequence, the star burns hydrogen by the CNO cycle.
2. The star's *main-sequence lifetime* is shorter.
3. The higher temperatures kindle fusion of carbon and heavier elements in the core.
4. The helium-rich core does not become degenerate.
5. Energy is transported through the interior mostly by convection rather than radiation.

All this evolution happens about a hundred times faster than for a solar-mass star.

Evolution of a 5-Solar-Mass Star

The PP reaction first ignites in the core as the star reaches the ZAMS. When the core's temperature rises to about 20 million K, the CNO cycle produces more en-

ergy each second than the PP reaction. The CNO cycle uses up the core's hydrogen quickly. Note that a 5-solar-mass star on the main sequence is more luminous and hotter than a 1-solar-mass star (Fig. 16.12).

When the central hydrogen fusion fires are exhausted, the star contracts. New hydrogen falls to the inner regions and ignites in a shell around the burnt-out core. The luminosity increases, but the radius expands, so the surface temperature drops. The star becomes a red giant (Fig. 16.12).

Meanwhile the core contracts until it gets hot enough to ignite the triple-alpha process. In contrast to a 1-solar-mass star, the core has *not* become degenerate. So no helium flash occurs in the core, just a relatively gentle triple-alpha ignition. (Stars with masses greater than 2.25 solar masses do not develop degenerate helium cores because helium ignition takes place before the stars can acquire sufficient density.)

The star burns helium in its core for about 10 million years (Fig. 16.12). When the helium runs out, the star again contracts. Now fresh helium falls into the core to make a thin, helium-burning shell around the core. The star becomes a red giant again. (Note that whenever fusion reactions take place in a shell rather than in the core, the star expands.) During the second time as a red giant, thermal pulses ignite in a shell around

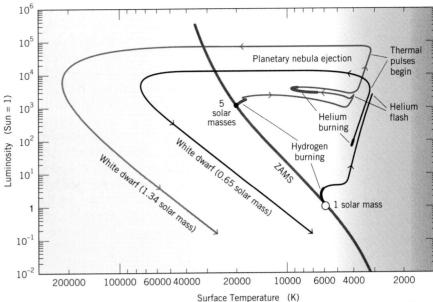

FIGURE 16.12 Theoretical evolutionary tracks off the main sequence (ZAMS) for stars of 1 and 5 solar masses. The thicker parts of the evolutionary tracks indicate phases of long-term fusion reactions in the cores. This diagram shows both stars going through a planetary nebula phase before they become white dwarfs; note the large percentage of mass lost by each. Note the overall similarity of the sequence of evolutionary stages. The thermal pulses occur in a helium-burning shell after helium burning has ceased in the core. (Based on a diagram by I. Iben, Jr.)

Red Giants

At a Glance

Mass	Less than 1 up to 8 solar masses
Size	Tens of solar radii
Energy transport	Convection
Phase	Hydrogen shell burning
Signatures	Thermal pulses, stellar winds

the core (Fig. 16.12), driven by the triple-alpha process. The star flashes and pulsates.

What next? We're really not sure. One possibility is that superwind develops and blows off the outer layers to leave behind a hot core surrounded by a planetary nebula (Fig. 16.11). The core becomes a white dwarf.

More Massive Stars

A somewhat more massive star can develop a degenerate carbon core of about 1.4 solar masses. When the core gets hot enough to ignite carbon burning, it should do so in a flash fire. This reaction may generate so much energy that the star blows apart. Such a cataclysmic explosion may result in a *supernova*. The outer layers of the star blast into space. The core is crushed to immense densities. (More in Section 17.4.)

We are not certain what main-sequence mass marks the transition to stars that certainly will explode as supernovas. Such a demise seems sure for *high-mass stars*—those with masses of 20 solar masses and greater. These stars are O stars during their main-sequence phase. Their evolutionary tracks off the main sequence are very complex, in part because these stars shed mass with strong stellar winds. No details here! (See Chapter 17.) These stars turn into supergiants after they leave the main sequence, and soon afterward they self-destruct in a fierce supernova.

■ Does a 5-solar-mass star become a red giant?

16.7 Observational Evidence for Stellar Evolution

The description of stellar evolution given in the preceding sections is based on theoretical calculations. How well do these ideas connect with the real astronomical world? To find out, we look for actual stars at various stages in their life cycles, and with different masses and compositions, that are like the stars predicted by our models of stars. The way we do so is to examine stars in clusters.

Stars in Groups

Most of the stars in the Milky Way are in multiple systems, which are isolated from other stars by great distances. But we do find that some stars come in groups held together by their own gravity, at least for some time. These groups are called *star clusters*. The Milky Way contains star systems of two contrasting types: **open clusters** (sometimes called **galactic clusters**) and **globular clusters.** How do they differ?

The Pleiades in the constellation Taurus (see Appendix G for winter) is a good example of an open cluster, with a loose array of stars. The Pleiades lie about 400 ly from the sun. The cluster contains about 100 stars within a diameter of 10 ly for an average density of about 0.1 star in a cubic light year. These statistics are pretty typical of open clusters: they contain from fewer than a hundred up to a thousand stars in a space a few tens of light years in size. So their star densities are not more than a few stars per cubic light year. Astronomers have catalogued some thousand open clusters to date, and perhaps 20,000 inhabit the Milky Way (Fig. 16.13).

One key characteristic of open clusters is their H–R diagrams. The one for the Pleiades is pretty typical (Fig. 16.14). Note that the stars below a surface temperature of 10,000 K (spectral type A0) fall squarely on the main se-

FIGURE 16.13 An open cluster in the Milky Way (New General Catalog 188), which lies about 5000 ly away. Note how spread out the stars appear; they span about 20 ly. The stars in this rather old cluster (some 10 billion years) have had some time to move apart. (Courtesy of L. C. Coombs.)

quence. Above 10,000 K, the stars lie above and to the right of the main sequence. Most open clusters show the same properties: the lower-mass stars fall on the main sequence, but at some point the higher-mass stars turn off it (Fig. 16.15).

Globular Clusters

Globular clusters contrast dramatically with open clusters (Table 16.2). As the name implies, a globular cluster has a distinct spherical shape (Fig. 16.16). Binoculars show it as a miniature, fuzzy sphere, like a cotton ball. A small telescope reveals some of the brightest stars.

The stars at the center of a globular cluster are crammed as high as a hundred per cubic light year (Fig. 16.17)! If the earth orbited a star in the core of a globular cluster, its nearest neighbors would be a few light *months* away. You would see thousands of bright stars scattered over the sky. In all, a globular cluster contains up to a million stars of roughly 1 solar mass or less jammed into a space only 100 ly or so in diameter.

An H–R diagram of a globular cluster (Fig. 16.18) clearly shows its major difference from an open cluster in terms of its stellar type. The main sequence turns off to the red giant branch, and the upper end of the main sequence has disappeared. The more massive stars have evolved off the main sequence. A horizontal branch of stars returns from the giant region to the region of the absent upper main sequence. This slash across the H–R diagram, called the *horizontal branch,* is the special signature of a globular cluster. These features indicate that globular clusters are very old, with an age of 15 billion years!

Recall that a horizontal branch contains low-mass stars, burning helium in their cores and having chemical compositions that differ significantly from that of the sun. The observation that globular clusters have a horizontal branch strongly implies that the stars there are less massive and have fewer heavy elements than the sun. They are poor in metals because they are so old, and the gas and dust from which they were born contained only a small percentage of heavy elements. However, that means

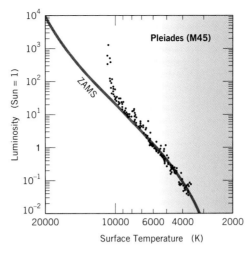

FIGURE 16.14 H–R diagram for the Pleiades. The line indicates the zero-age main sequence; the dots are the observed points for the stars. The luminosities are given for the visual range of the spectrum, with the sun's visual luminosity equal to 1.0. (From *An Atlas of Open Cluster Colour-Magnitude Diagrams,* by G. L. Hagen, David Dunlop Observatory.)

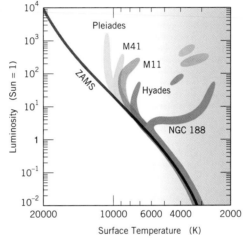

FIGURE 16.15 Schematic H–R diagram for selected open clusters whose distances are known. Each cluster turns off the main sequence at a different point. The stars above the turnoff point have evolved off the main sequence to become red giants. All clusters have stars that occupy the lower part of the main sequence below the turnoff points. Note that NGC 188 is relatively very old, and the Pleiades are relatively very young.

FIGURE 16.16 A globular cluster (Messier 13 in Hercules), which lies about 22,000 ly away. Compare the distribution of stars to the open cluster NGC 188 in Figure 16.13; note the great concentration in the center of M13. (Courtesy of NASA.)

that at least some heavy elements were in the cloud from which the cluster formed, so it seems that every globular cluster survived at least one supernova explosion very early in its life.

Stellar Populations

The striking difference between the H–R diagrams of open and globular clusters implies that the stellar types in the two kinds of cluster are quite different. Astronomers call stars of the kinds found in open clusters **Population I stars** and those in globulars **Population II stars.** The brightest Population I stars are bluish-white; the brightest Population II stars, reddish. Also the brightest Population I stars (O stars) have about one hundred times the luminosity of the brightest Population II stars (red giants). The luminous Population I stars must be relatively young, since they are O and B supergiants.

A really crucial distinction *cannot* be seen directly in the H–R diagram: chemical composition. Spectroscopic observations show that Population I stars have essentially the same chemical composition as the sun—1 to 2 percent, by mass, of heavy elements (which are all the elements more massive than hydrogen and helium). Population II stars contain about 0.01 this amount, only 0.01 to 0.02 percent of the mass.

From the description of stellar evolution so far, you should see that Population I stars must be *younger* than Population II stars, in the sense that they were born *later.* According to the Big Bang model for the origin of the cosmos (Chapter 21), the original stuff of the cosmos was mostly hydrogen and helium—essentially no heavy elements. The heavy elements must be made in stars. When a massive star dies, it spews a lot of material back into the interstellar medium. This blown-off material has been enriched in heavy elements; from it, new stars will be born. So we know that Population II stars are *older* (formed *earlier*) than Population I stars because they have fewer heavy elements. Population I stars have been formed out of enriched, recycled material.

TABLE 16.2 COMPARISON OF STAR GROUPS

CHARACTERISTIC	OPEN CLUSTERS	GLOBULAR CLUSTERS
Mass (solar masses)	10^2-10^3	10^4-10^5
Diameter (ly)	6–50	60–300
Color of brightest stars	Reddish to bluish-white	Reddish
Density of stars (solar masses/ly^3)	0.1–10	1.0–100
Examples	Pleiades (Fig. 15.13)	Hercules (M13) (Fig. 16.16)

Caution. Not all Population I stars are luminous bluish-white stars. In fact, many Population I stars are stars like the sun, and the vast majority are faint red dwarf stars (the lower right-hand end of the main sequence). Bluish-white Population I stars are the most luminous and so the easiest to spot. Also, not all red giants are Population II stars; but in a group of Population II stars, the most luminous ones will be red giants. And, Population I stars have a range of ages,

FIGURE 16.17 The dense, core region of globular cluster M13. This false-color CCD image, which has the brightest colors at the most intense emission, covers an angular extent about 5 arcmin square. The core itself is only some 10 ly in diameter! (Courtesy of I. R. King, © 1983 California Institute of Technology.)

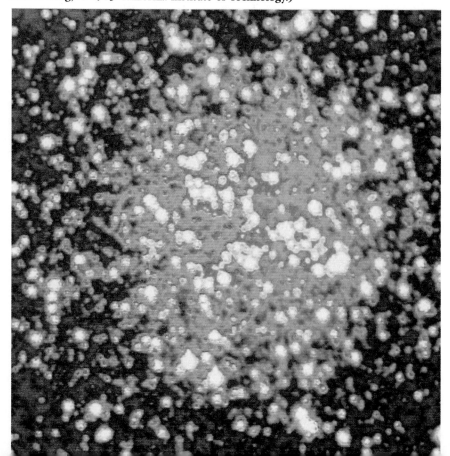

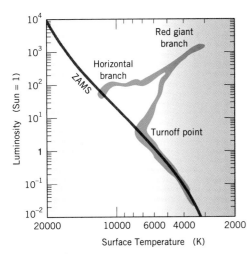

FIGURE 16.18 Schematic H–R diagram for a globular cluster. The stars along the lower part of the main sequence have masses less than 1 solar mass. The horizontal branch indicates the points for stars that have metal abundances less than that of the sun; on it to the left, the masses are smaller. Note the well-defined red giant branch.

from a few tens of millions of years old to perhaps 10 billion years old.

Comparison with the H–R Diagram of Clusters

When a star cluster forms (galactic or globular), its stars are born at essentially the same time with the same chemical composition, but the masses vary. The more massive stars evolve more rapidly than the less massive stars. So the more luminous stars evolve more quickly to becoming red giants. Each star of different mass follows a different evolutionary track. As a cluster ages, stars of lower mass will evolve off the main sequence.

At the beginning of its life, a galactic cluster's H–R diagram resembles that of a young open cluster (Fig. 16.19*a*); a little later, that of a middle-aged open cluster, such as the Pleiades (Fig. 16.19*b*). Much later, the H–R diagram is similar to that of an old open cluster (Fig. 16.19*c*). Note that the **turnoff point** away from the main sequence moves down to lower-mass stars as the cluster ages; *a cluster's turnoff point indicates its age.*

An example: the old galactic cluster Messier 67. Figure 16.20 shows an H–R diagram for the stars in M67. Note the

theoretical line that corresponds to stars with masses ranging from 0.7 to 3 solar masses after an evolution of 5×10^9 years. The evolutionary line fits the turnoff point, so we infer that the cluster has this age.

Let's line up the H–R diagrams for a variety of galactic and globular clusters (Fig. 16.21). The stars in the galactic clusters do not peel off the main sequence at the same point. In some clusters (such as the Pleiades), only a few stars at the upper end of the main sequence have evolved away from it. In others, the turnoff point lies farther down the main sequence, but none is below a luminosity of about that of the sun. In contrast to galactic clusters, all globular clusters (such as Messier 3 in Fig. 16.21) have remarkably similar H–R diagrams, with roughly the same turnoff points.

What are the implications of this comparison for stellar evolution? First, it says that globular clusters are older than open clusters. (M3 *is* actually older than M67; it appears higher up on the H–R diagram because of its Population II composition.) Second, it implies that the stars in globular clusters are roughly the same age. Calibrated with theoretical models, the turnoff points for globular clusters indicate that they are 15 billion years old, give or take about 2 billion years. M3, for instance, has an estimated age of about 13 billion years. The stars in the globular clusters are the oldest stars known. Third, open clusters have a large range of ages. The ages of open clusters can also be estimated from their turnoff points by comparing them to those found from theoretical calculations. For example, M67 has an approximate age of 5 billion years (Fig. 16.22), and the Pleiades, 100 million years. The typical age of a galactic cluster is some 350 *million* years, with a large spread.

To sum up. As a cluster ages, the main sequence will gradually shorten as the stars peel off, in order of mass, and evolve over into the red giant region. This turnoff point gives the age of a cluster, when compared to the evolutionary tracks of theoretical models. The comparison also confirms the general validity of the star models.

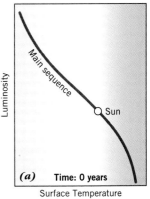

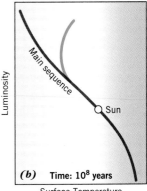

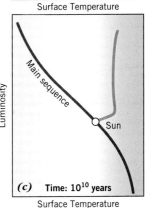

FIGURE 16.19 Theoretical H–R diagrams of a cluster of stars with the same chemical composition, born at the same time (*a*), but having different ages. "Sun" indicates the position of a solar-mass star. Note how the turnoff point moves down the main sequence as the cluster ages (*b* and *c*).

Variable Stars

A star's luminosity varies little and slowly during its main-sequence sojourn. Later in life, a star's luminosity varies dramatically, over short periods of a few hours to a few years. Some of these **variable stars** (so called because their luminosities change rapidly with time) lie above the main sequence on the H–R diagram (Fig. 16.22). What is their evolutionary status? We know from model calculations that these variables are post-main-sequence stars and that those that vary regularly are in the helium-core-burning stage.

There's a staggering array of variable stars; I'll mention just three: *RR Lyrae variables, cepheid variables,* and *red variables* (Fig. 16.22). The RR Lyrae stars and Population I and II cepheids are **periodic** or **regular variables;** their change in luminosity with time follows a regular cycle over a certain period.

RR Lyrae stars (named after their prototype, RR Lyrae, whose period is 13.6 hours) vary in luminosity with periods of typically 12 hours. They are Population II stars and have about one hundred times the sun's luminosity. About 5000 RR Lyrae stars are known. **Cepheid variables** (named after their prototype Delta Cephei) fall into two groups: Population I, called *classical cepheids,* and Population II. Classical cepheids have periods of light variation typically 5 to 10 days in duration. Population II cepheids vary in periods from typically 12 to 20 days.

In contrast, the **red variables** have irregular cycles of light variation that range from 100 to 2000 days. They are red giant and supergiant stars, both Population I and II, with luminosities roughly one hundred times that of the sun.

These variables have a key physical characteristic in common: they *pulsate.* Doppler shift observations of their spectra show that as they vary, these stars expand and contract. (When the star is expanding, its surface moves toward us, producing a blueshift in the lines of the spectrum; when it contracts, the surface moves away, producing a redshift in the lines of its spectrum.) Population I cepheids, for instance, expand and contract at speeds of about 30 km/s.

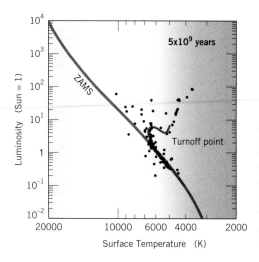

FIGURE 16.20 Comparison of theoretical models (solid line) and observations of the stars in the cluster Messier 67 (solid dots) to find the turnoff point from the ZAMS and so the age of the cluster (roughly 5 billion years, which means that it is a very old open cluster). (Based on calculations by D. A. VandenBerg.)

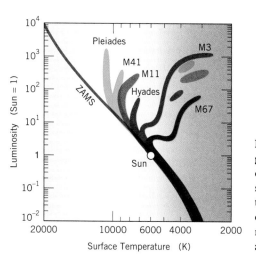

FIGURE 16.21 Composite H–R diagram for some galactic and globular clusters. The further down the main sequence the turnoff point occurs, the older the cluster. The parts of the clusters found to the right are those made of red giant stars. (Adapted from a diagram by A. Sandage.)

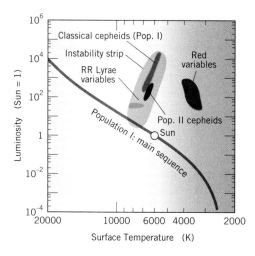

FIGURE 16.22 The positions of selected variable stars on the H–R diagram. The Population I and II cepheids and the RR Lyrae stars occupy a region known as the instability strip; here stars are burning helium in their cores and are giants. The red variables, which are undergoing helium shell burning, are red giant and supergiant stars. (Adapted from a diagram by J. P. Cox.)

Cepheids and RR Lyrae stars lie in a region of the H–R diagram called the **instability strip.** Evolutionary tracks of low-mass (0.5 to 0.7 solar mass) Population II stars and high mass (3 to 20 solar masses) Polulation I stars transverse this strip during their *helium-core-burning* phase. During this phase, a star's outer layers become unstable and pulsate with a regular period.

The red variables, on the other hand, fall in a region of the H–R diagram characterized by *shell burning* rather than core burning. Helium shell burning is unstable and results in thermal pulses, which create convective zones in a region of the star that is rich in helium. Such zones may drive the star to pulsate.

So theoretical calculations indicate where on the H–R diagram we expect instability and variations. Observations validate the theoretical work and confirm basics of the models.

Central Stars of Planetary Nebulas

In the scenario for the evolution of a 1-solar-mass star, when a red giant blows off its outer layers, it leaves behind a hot, dense core. This cinder cools to form a white dwarf. We expect the same to be true of low-mass and some middle-mass stars.

If this picture is correct, you'd expect the central stars of planetary nebulas to fall along the evolutionary track (points 8 to 9 in Fig. 16.10) after the red giant stage. Well, they do (Fig. 16.23)! Some central stars are extremely hot and luminous. One recently measured by the *HST* suggests a surface temperature of 200,000 K—the hottest star known to date! Others are not so hot. Their positions on the H–R diagram fall neatly above and to the left of those for white dwarfs. So the stars of planetary nebulas mark a transition between the core of a red giant and a white dwarf. This observation nicely confirms that stars evolve from red giants to white dwarfs.

■ **What is the main difference between Population I stars and Population II stars?**

16.8 The Synthesis of Elements in Stars

Gravitational potential energy furnishes the initial heat to get fusion reactions going. To survive, a star must fuse lighter elements into heavier ones to generate energy. The more mass a star has, the greater the central temperature produced by gravitational contraction and the heavier the elements it can eventually fuse. From the ignition temperatures needed for fusion reactions, we can set limits on the heaviest elements that a star of a certain mass can fuse (Table 16.3). For example, our sun can burn helium to carbon in its core but will never get hot enough to fuse carbon.

Table 16.3 summarizes the principal stages of nuclear energy generation and nucleosynthesis in stars. Note that the products (or ashes) of one set of reactions usually become the fuel for the next set of reactions. What a beautiful scheme for energy production in the universe!

Also note that only very massive stars can produce elements heavier than oxygen, neon, sodium, and magnesium in their cores. Few stars have this much mass, so most stars come to the end of their nuclear evolution without having manufactured some important elements. This fact emphasizes the importance of

Star Clusters

At a Glance

Main types	Open, globular
Star types	Population I (open), Population II (globular)
General ages	Globulars older, open younger
Ages found by	Turnoff point of main sequence

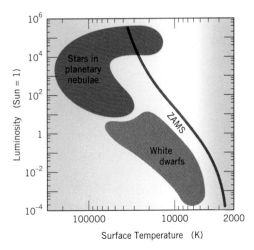

FIGURE 16.23 Schematic diagram for the location of central stars of planetary nebulas on the H–R diagram. These stars are the hottest known; note the very high surface temperatures. The close fit of the region of planetary nebula central stars to the region for white dwarfs supports the idea that the two are linked in evolutionary sequence. By the time the star has cooled to the white dwarf region, its planetary nebula has expanded and faded away.

TABLE 16.3 STAGES OF THERMONUCLEAR ENERGY GENERATION IN STARS

PROCESS	FUEL	MAJOR PRODUCTS	APPROXIMATE TEMPERATURE (K)	APPROXIMATE MINIMUM MASS (solar masses)
Hydrogen burning	Hydrogen	Helium	2×10^7	0.1
Helium burning	Helium	Carbon, oxygen	2×10^8	1
Carbon burning	Carbon	Oxygen, neon, sodium, magnesium	8×10^8	1.4
Neon burning	Neon	Oxygen, magnesium	1.5×10^9	5
Oxygen burning	Oxygen	Magnesium to sulfur	2×10^9	10
Silicon burning	Magnesium to sulfur	Elements near iron	3×10^9	20

Source: Adapted from a table by A. G. W. Cameron.

massive stars in the scheme of cosmic evolution—they create heavy elements *and* throw some back into the interstellar medium.

Nucleosynthesis in Red Giant Stars

Red giant stars play a vital role in the scheme of cosmic nucleosynthesis. That view, based on theoretical models, sees the thermal pulses in a helium-burning shell as the site for producing certain isotopes, especially those that are rich in neutrons. (A neutron-rich isotope is one that has more neutrons than protons in the nucleus.) This process can occur in two stages for stars of low and middle mass.

One stage takes place when a star first becomes a red giant. Thermal pulses, sparked by the triple-alpha process, burn helium to carbon, which is then transformed to oxygen. The convective zone that develops as a result of the pulses reaches down to the star's core and pulls up elements that have been made with hydrogen burning. This whole process, which occurs in *every star* as it becomes a red giant for the first time, is called the *first* **dredge-up.**

For medium-mass stars, such as one of 5 solar masses, a second phase of nucleosynthesis takes place after a star has burned the helium in its core (Fig. 16.12). Thermal pulses can then convert helium to carbon, carbon to oxygen, nitrogen to magnesium, and iron to certain neutron-rich isotopes of heavier elements. The convective zone brings these

to the surface, a process called the *second* **dredge-up.**

All these processes would have no effect on the rest of the cosmos except for one crucial fact: red giant stars have strong stellar winds (Fig. 16.24). These blow off material from the surfaces of the stars, with the result that the processed material from the first and second dredge-ups is sent out into the interstellar medium, from which future stars will form.

You now see where part of the periodic table of the elements (Appendix F) comes from. What about the rest? You'll find out in Chapter 17.

■ How do elements synthesized in red giants make it out to the interstellar medium?

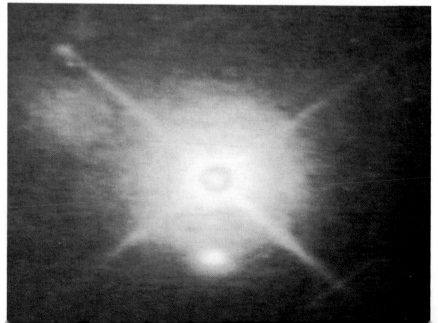

FIGURE 16.24 Mass loss from the red supergiant star Betelgeuse. This image shows a gaseous halo of potassium around the star; the potassium serves to trace out the extent of the gas cloud. The material blown off from Betelgeuse also contains dust grains of graphite. (Courtesy of NASA.)

Key Concepts

1. A Hertzsprung–Russell diagram provides a picture of many different stars at different stages of their lives. Time is implicit in an H–R diagram; neither axis is time. We make time explicit by constructing theoretical models of stars at different stages of their lives. A series of stages for any one star traces out its evolutionary track on an H–R diagram.

2. Stars maintain a balance between gravity and internal pressure generated by the heat from fusion reactions. To strike this balance, a star must produce energy inside. The loss of this energy to space requires that the star evolve. If stars did not shine, they would not evolve.

3. Fusion reactions in cores normally generate a star's energy. When fusion reactions stop, gravitational contraction of the core can produce heat and light and raise the temperature high enough to ignite the next stage of fusion reaction. As fusion reactions use up fuel, the ashes produced can become the next fuel, if the temperature gets high enough. So fusion products are the sources of new fuel in the sequence of stellar nucleosynthesis.

4. Star models are based on the physical properties of stars. These models assume that a star's internal pressure balances its gravity; that an energy source (usually fusion) is in the core; that energy flows out from hotter to cooler regions by radiation or convection; that a star is a gas throughout; and that equal amounts of energy are radiated from the interior and produced within it. Specific models depend on the initial mass and chemical composition.

5. A newly born star shines from energy produced by gravitational contraction. When fusion fires ignite in the core, the star finally achieves the main-sequence stage of its life. The star first appears on the zero-age main sequence. Stars on the main sequence in the H–R diagram are fusing hydrogen to helium in their cores.

6. More massive stars (upper part of the main sequence) use up their fuel very fast, hence evolve faster than less massive stars (lower part of the main sequence). The more massive stars have higher fuel requirements because they use the CNO cycle and because they have higher core temperatures than stars of lower mass.

7. When fusion reactions are occurring in a star's shell (or shells) around the core and not in the core, the star expands in size. This shell burning usually means that the star becomes a red giant. Our sun (like other low-mass stars) will evolve to a red giant and will blow off its outer layers to make a planetary nebula. The former red giant core will become a white dwarf, then cool to a black dwarf. Some middle-mass stars follow the same sequence.

9. Other medium-mass stars become red giants a number of times, then die in a supernova explosion. High-mass stars (greater than 10 to 20 solar masses) become supergiants after their main-sequence phase. They also die in supernovas.

10. A comparison of the H–R diagrams for clusters of stars confirms our basic ideas about stellar evolution because these stars were formed about the same time, with the same initial chemical compositions. Only their masses differ. The more massive stars evolve first off the main sequence, so the main-sequence turnoff point indicates the age of a star cluster. It is calibrated by theoretical models of stars.

11. Massive stars can fuse elements up to iron in their cores in their normal lives. Red giants can fuse some heavy elements in convective helium-burning shells. These are brought to the star's surface by convection and blown off by stellar winds into the interstellar medium. So the percentage of heavy elements in the interstellar medium increases as time passes.

12. A star's mass determines how it develops and how fast it lives (Table 16.4); more massive stars have shorter lives than less massive ones. Initial chemical composition plays a secondary role in determining the history of a star.

TABLE 16.4 COMPARISON OF THE GENERAL EVOLUTION OF STARS OF DIFFERENT MASSES

MASS	EVOLUTIONARY SEQUENCE
Low (< 1 solar mass)	Protostar → pre-main sequence → main sequence → red giant → planetary nebula → white dwarf → black dwarf
Middle (≈ 5 to 10 solar masses)	Protostar → main sequence → red giant → planetary nebula or supernova
High (> 20 solar masses)	Protostar → main sequence → supergiant → supernova

Key Terms

black dwarf

cepheid variables

degenerate electron gas

degenerate gas pressure

dredge-up (first and
    second)

evolutionary track

globular clusters

helium flash

horizontal branch

instability strip

main-sequence lifetime

nucleosynthesis

open clusters

periodic (regular)
    variables

planetary nebula

Population I stars

Population II stars

red variables

RR Lyrae stars

star model

thermal pulses

triple-alpha reaction

turnoff point

variable stars

zero-age main sequence (ZAMS)

Study Exercises

1. A star like the sun consists completely of an ordinary gas. Why doesn't it suddenly collapse gravitationally? (Objective 16-2)

2. Present calculations indicate that a solar-mass protostar is much more luminous than the sun. Yet it's much cooler at the surface. How can the protosun be much cooler and yet more luminous than the present sun? (Objective 16-4)

3. How can you tell from an H–R diagram that the stars in the Pleiades cluster are younger than those in the Hyades? (Objective 16-6)

4. Why are massive stars able to fuse heavier elements than less massive stars? (Objective 16-11)

5. What evidence do we have that red giant stars become white dwarfs? (Objective 16-7)

6. What is a main difference between the evolution of a 1-solar-mass star and a 5-solar-mass star? (Objectives 16-5 and 16-10)

7. Compare and contrast the stellar *types* in open and globular clusters. (Objectives 16-7 and 16-9)

8. What kind of star is our sun likely to become, and what kind of corpse will it eventually leave? (Objective 16-3)

9. Outline the evolution of a cluster of stars containing half 5-solar-mass stars and half 1-solar-mass stars. (Objective 16-6)

10. What kind of fusion reaction can produce carbon in a star? What minimum mass is needed for this reaction to occur? (Objectives 16-8 and 16-11)

11. In what sense is time implicit in an H–R diagram? How do astronomers make it explicit? (Objectives 16-1 and 16-9)

12. Why do stars evolve? (Objective 16-4)

13. How does the method of energy transport differ in high-mass stars compared to solar-mass ones? (Objective 16-12)

Problems & Activities

1. How many kilograms of matter are converted to energy each second in a 5-solar-mass, main-sequence star?

2. When the sun was a protostar, it had a maximum luminosity about 1000 times its present luminosity and a surface temperature of roughly 1000 K. What was its radius? *Hint:* Assume that it radiates like a blackbody.

3. Let's draw a few points of a small range of the sun's evolutionary track on an H–R diagram. Use a piece of graph paper with the horizontal axis for temperature (increasing to the left) and the vertical axis for luminosity. For a luminosity equal to 1.0 and a surface temperature of roughly 6000 K, place a point for the sun now. The radius equals 1.0. Label this point with an age of 4.6 billion years. Then add points from the table below. Note you will need to calculate the surface temperatures at different times, assuming a blackbody radiator.

TIME (billions of years)	LUMINOSITY (sun [now] = 1.0)	RADIUS (sun [now] = 1.0)	SURFACE TEMPERATURE (kelvins)
3.0	0.95	0.95	
6.6	1.2	1.08	
7.7	1.3	1.14	
9.8	1.8	1.36	

These calculations pretty much give the sun's evolution from the pre-main-sequence phase to the end of its main-sequence phase. (Table from calculations by S. Turck-Chièze, S. Cahen, M. Cassé, and C. Doom.)

4. A typical planetary nebula will have an angular diameter of 20 arcsec. What is its size-to-distance ratio? A typical distance to a planetary nebula is 500 ly. What is the nebula's physical diameter?

5. The smallest angular diameter of a galactic cluster is 0.5 arcmin. The largest angular diameter is 50 arcmin. What is the size-to-distance ratio of each? Assume that all galactic clusters have the same physical diameter. Then what is the range of distance for such clusters?

See for Yourself!

Globular clusters and open clusters are fun objects to view. The larger and brighter of the open clusters are easy to pick out with binoculars. Almost all globular clusters, though, need the light-gathering power of a reasonable telescope, with an objective of some 20 cm or so.

In the winter sky, the Pleiades (Messier 45) and the Hyades are pretty sights. Between Gemini and Leo lies the faint constellation Cancer. Right in its center is Messier 44, the Praesepe ("Beehive") cluster. I usually picture it as a set of triangular arrangements of stars. In the north, the clusters h and Chi Persei lie close together just away from the brighter stars in Perseus. In the summer sky, scanning with binoculars up the Milky Way will reveal many open clusters in the region.

Globular clusters take more effort—and a telescope. Messier 13 in Hercules is your best first shot in the summer sky. Another good globular is Messier 22 in Sagittarius, just north and west of the "teapot." In the fall, Messier 15 in Pegasus is bright, with a very compact core. You can use the sky charts in *Astronomy* and *Sky and Telescope* magazines to find the exact positions of these clusters. Remember that when you look at them, you are seeing the oldest stars in the Milky Way Galaxy. ■

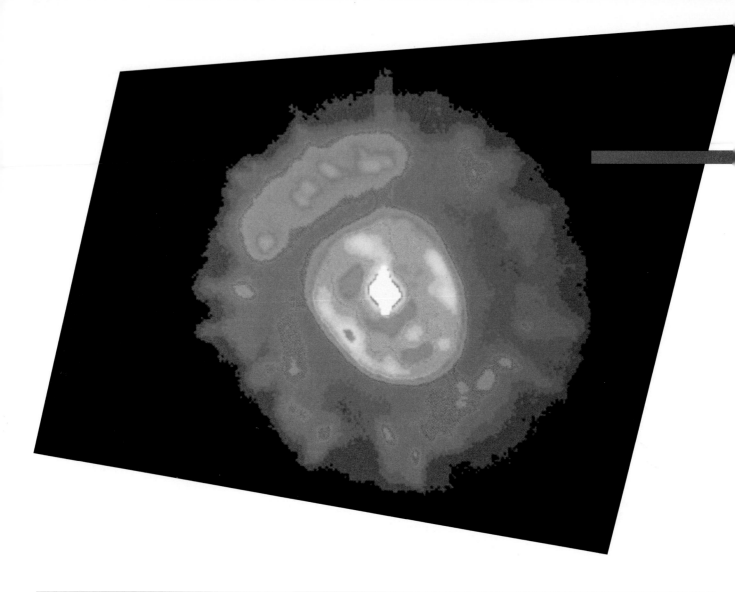

LEARNING OBJECTIVES

After studying this chapter, you should be able to:

17-1 Compare the physical natures of white dwarfs, brown dwarfs, and neutron stars and describe the place of each in stellar evolution.

17-2 Describe the basic physical properties of a degenerate star in contrast to an ordinary star.

17-3 Argue, with observational support, that pulsars are rapidly rotating, highly magnetic, neutron stars.

17-4 Compare and contrast the observed features of a nova and a supernova.

17-5 Outline a possible model for a nova explosion that involves a binary star system, and connect the model to observations.

17-6 Outline possible models for supernova explosions (using supernova 1987A as an example of one type) and describe the effects of the aftermath of this explosion on the interstellar medium.

17-7 Cite observational evidence that the Crab Nebula is a supernova remnant and describe the effect of the pulsar on the nebula now.

17-8 Describe how synchrotron radiation is emitted, identify its observed properties, and apply this concept to appropriate astrophysical situations.

17-9 Describe a black hole in terms of escape speed and the speed of light, and tell what happens to an observer falling into a black hole from the viewpoints of the infalling observer and an outside observer far from the black hole.

17-10 Describe how nucleosynthesis can occur in a supernova and identify possible products of such nuclear reactions.

17-11 Describe the properties of cosmic rays, and discuss their possible relationship to supernovas.

17-12 State and evaluate the observational evidence for the existence of black holes.

Central Question
How do stars die, and what corpses do they leave behind?

Stardeath

I believe a leaf of grass is no less than the journey-work of the stars.

WALT WHITMAN, *Song of Myself*

Many stars die violently! Imagine that one year equals one-fifth of a second, so the sun would live 65 years. In this speeded-up time, about twenty-five stars in the Milky Way wink out every second. These deaths are signaled by blasts of the superwinds that eject a star's outer layers. The discarded shells replenish the interstellar medium, which has been depleted by the formation of stars and planets. And they add heavy elements to it.

The dead star's remnant core cools. Locked tight by gravity, it forms a cinder in space. In some instances, the core develops into a neutron star, a smooth, spinning sphere of nuclear matter. In others, the core disappears through a warp in spacetime as a black hole. In still others, the burned-out core becomes a white dwarf star, a solid carbon crystal—the eventual fate of our sun.

Observations imply that almost all stars throw off mass before they meet their ends. A supernova is the most destructive example of mass loss. But a supernova is constructive too—in its immense explosion, many heavy elements of the universe are made and thrown to the currents of space. Supernovas spice the interstellar medium and provide the impulse to the birth of new stars.

17.1 White and Brown Dwarf Stars: Common Corpses

The evolution of a 1-solar-mass star (Section 16.5) illustrates the constant battle between pressure and gravitational forces. Because unlike thermonuclear reactions, gravity never lapses, the final state of any star depends on the physical properties of matter at high densities and the total mass of the star. When all the thermonuclear reactions cease, what pressure can support the star? What will the star become at the time of its death?

Physics of Dense Gases

The conventional model of the atom surrounds the nucleus with a cloud of electrons. The distance from the nucleus to the first electron shell is about a thousand times the diameter of the nucleus, so an atom is mostly empty space. At the high temperatures in stars, atoms are ionized, and the electrons run around free of the nuclei. As a star is crushed to higher densities in its evolution, the electrons form a *degenerate electron gas* (Focus 16.1).

In an ordinary gas, the pressure depends directly on the temperature. But in a degenerate gas the pressure is *not* related to the temperature; it depends only on density. The density at which a gas switches to a degenerate state is about 10^8 kg/m^3. A degenerate star of 1 solar mass has an average density of about 10^9 kg/m^3. If the sun were compressed to this density, it would be roughly 7000 km in radius, about the size of the earth (Fig. 17.1).

White Dwarfs in Theory

In 1935 the Indian astronomer Subrahmanyan Chandrasekhar applied the physics of a degenerate electron gas to the model of a star. He found that the pressure exerted by the electrons could resist the force of gravity only for stars of less than 1.4 solar masses and that such stars would have a density of about 10^9 kg/m^3. Such a star, at the end point of its thermonuclear history, is a **white dwarf.** No heavier elements are fused, and no new energy produced. The stored thermal energy flows to the surface by conduction because of the great density.

How does a white dwarf fend off gravity? By the outward pressure from the degenerate electron gas, called the *electron degeneracy pressure.* What about the nuclei? They form a crystal structure embedded in the degenerate gas. If the expulsion of the outer layers of the star left a carbon core, the white dwarf, once it had cooled enough, would be a solid carbon crystal. Many white dwarfs probably contain a mixture of largely carbon and oxygen.

Chandrasekhar also discovered that the more massive the white dwarf, the *smaller* its radius. (In contrast, the more mass a main-sequence star has, the larger it is.) How does this feature come about? More mass means more gravity. To balance gravity requires internal pressure. In a white dwarf, the pressure

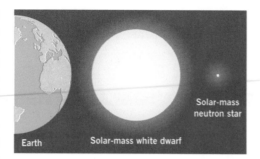

arises from the nature of a degenerate gas, where greater pressures are a response to greater density. So the *more massive a white dwarf, the smaller its size.*

A crucial point arrives when the mass of the white dwarf is about 1.4 solar masses. Such a star has the highest density and smallest radius possible. Add a bit more mass and the gravitational forces overwhelm the degenerate electron gas pressure. The star collapses; it cannot be stable. This amount of mass, 1.4 solar masses, is called the **Chandrasekhar limit** and signals the point at which degenerate electron matter is crushed by gravity.

Observations of White Dwarfs

In 1862 the American optician Alvan Clark (1804–1887) observed **Sirius B** (Fig. 17.2), the faint companion to Sirius (Section 14.6). Later this star was found to be a white dwarf with an average density of about *3 billion* kilograms per cubic meter! So, on the earth, a spoonful of white dwarf stuff would weigh about as much as a car.

Some white dwarfs seen so far are actually white, but a few are yellowish, and some are reddish with surface temperatures 4000 K and lower. A few have very high surface temperatures—up to 70,000 K! These hotter ones (temperatures from 25,000 K and up) emit x-rays. Their luminosities simply result from the outflow of internal thermal energy left over from nuclear reactions in the past. Very slowly (over billions of years), their stored internal energy radiates into space. Eventually, a white dwarf becomes a *black dwarf* (not to be confused with a black hole!).

FIGURE 17.1 Comparison of the relative sizes of the earth, a solar-mass white dwarf, and a solar-mass neutron star drawn to the same relative size scale.

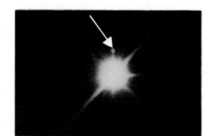

FIGURE 17.2 Sirius B (arrow), a white dwarf star, companion to Sirius A (brighter image to right). The sizes of these images relate only to the brightnesses of the stars, not to their physical diameters. (Courtesy of Lick Observatory.)

So far, about 1000 stars have been identified as white dwarfs. Because of their low luminosities, white dwarfs are hard to see. We can find only those within about 1000 ly. This makes it hard to estimate the total number of white dwarfs in the Galaxy, but they may make up 10 percent of all stars.

To sum up. A white dwarf is a stellar corpse with roughly the mass of the sun and the size of the earth. A degenerate electron gas supports it against the crush of gravity. Our sun will become a white dwarf, then a black dwarf—a cold corpse in space.

Brown Dwarfs

White dwarfs are the corpses of stars that started their main-sequence lives with masses ranging from 0.08 to (perhaps) 5 solar masses. What becomes of objects born with less mass?

Because they do not reach high enough temperatures for hydrogen fusion, such masses develop into huge, Jovian-type bodies. Gravitational energy released by contraction provides their power; they have the promise of protostars, but never reach stardom. Astronomers call such masses **brown dwarfs**— a misnomer as bad as "white dwarf" because they are not really brown (and have never really been stars), and some are not all that small (larger than Jupiter)! Perhaps they are best called stars below the *minimum hydrogen-burning limit,* but I find that phrase awkward.

Such masses have very cool (and decreasing) surface temperatures—less than 3000 K—so that most of their emission lies in the red and infrared part of the spectrum. These cool temperatures and expected low luminosities (roughly a hundred-thousandths the sun's luminosity) make brown dwarfs hard to detect. Infrared observations could permit detection, and searches are being conducted in regions of dark clouds with young stellar objects. But to date we haven't yet found a really convincing candidate.

■ **What kind of gas comprises a white dwarf?**

17.2 Neutron Stars: Compact Corpses

What happens to stars that have more than 1.4 solar masses of core material at the end of their evolution? Strange things! Degenerate electron gas pressure cannot support them, and self gravity crushes them to higher and higher densities.

Degenerate Neutron Gases

At about 10^{13} kg/m^3, the inward pressure forces **inverse beta decay** to occur (Fig. 17.3). In this process, an electron (sometimes called a *beta particle*) and a proton join together to form a neutron and a neutrino. This happens both to free protons and to protons in the nucleus of a heavy element. At around 10^{15} kg/m^3, the neutrons are no longer bound to the nuclei and begin to form a separate gas. At 10^{17} kg/m^3, the nuclei suddenly fall apart into a gas with 80 percent neutrons. A spoonful of this material on the earth would weigh about a *billion* tons.

At this density, the neutrons become degenerate in the same manner that electrons do at white dwarf densities. The neutrons provide a degenerate gas pressure, called the *neutron degeneracy pressure,* and so balance the inward pull of gravity. This pressure allows the formation of a stable **neutron star,** a star composed mainly of neutrons. Its diameter will be about 10 to 20 km, depending on its mass. (As with white dwarfs, the *greater* the mass of a neutron star, the *smaller* its radius, because it is made of a degenerate gas.)

Physical Properties

A neutron star is a weird beast compared with an ordinary star. The exact structure of a model depends on assumptions made about the behavior of matter at very high densities. In a one model for a 1.4-solar-mass neutron star, the radius is about 16 km (Fig. 17.4). The inner 11 km is a fluid core, consisting mostly of neutrons. The next 4 km out makes an inner crust, a neutron-rich fluid or perhaps a solid lattice. The outer crust, about 1 km thick, is a crystalline solid similar to the

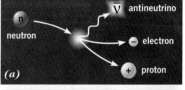
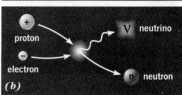

FIGURE 17.3 Beta decay and its reverse process, inverse beta decay. (An electron is sometimes called a *beta particle* for historic reasons.) (*a*) In beta decay, a neutron decomposes into a proton, an electron, and an antineutrino (an antimatter neutrino), $\bar{v}$. (*b*) In inverse beta decay, a proton plus an electron transforms into a neutron and a neutrino.

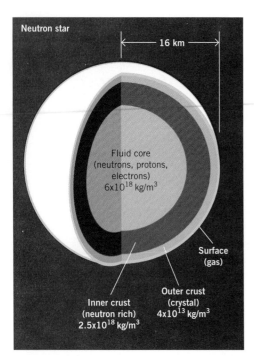

FIGURE 17.4 Theoretical model for the cross section of a 1.4-solar-mass neutron star. Note that the model shows three distinct interior layers. The exact details of any model depend on the known and assumed properties of the behavior of matter at high densities. (Adapted from a diagram by F. K. Lamb.)

interior structure of a white dwarf. In the outer few meters, where the density falls quickly, the neutron star has an atmosphere of atoms, electrons, and protons. The atoms are mostly iron.

Because a neutron star is so dense, it has an enormous surface gravity. For example, a 1-solar-mass neutron star with a radius of 12 km has a surface gravity 10^{11} times greater than that at the earth's surface! This enormous pull means that there are no mountains on a neutron star—elevations reach a few centimeters at most. This intense gravitational field also results in a huge escape speed, as much as about 80 percent the speed of light. Objects falling onto a neutron star from a great distance have at least the escape speed when they hit. That means that even a small mass carries a fantastic amount of kinetic energy. For example, a marshmallow dropped onto a neutron star from a few AU out will slam into the surface with megatons of kinetic energy!

An ordinary star, having at the *end* of its evolution a mass equal to or greater than 1.4 solar masses, probably ends up as a neutron star. Theoretically, a stable neutron star with a mass of less than 1.4 solar masses can also form. These low-mass neutron stars could be made in the pile-driver compression of a supernova explosion in which the interior of the star is crushed to the very high densities of neutron stars. The lowest mass possible is about 0.1 solar mass.

In an analogy to the Chandrasekhar limit, a mass limit for neutron stars is reached when the gravitational forces overwhelm the degenerate neutron gas pressure. This limit—not known exactly, but estimated at about 2 solar masses (it might be as high as 3)—signals the next crushing point of matter by gravity.

To sum up. A star with a mass between 1.4 and (roughly) 2 to 3 solar masses at the time of its death naturally forms into a degenerate neutron gas; hence the name "neutron star." Neutron stars lack any fusion reactions.

■ As the mass of a neutron star increases, what happens to its radius?

Neutron Stars	
At a Glance	
Mass	Up to 3? solar masses
Size	About 10 km in radius
Density	Averages 10^{17} kg/m³
Internal pressure	Degenerate neutron gas
Energy source	Stored internal heat; rapid rotation

17.3 Novas: Mild Stellar Explosions

Aristotle asserted that the heavens were unchanging. But in 1572 Tycho Brahe observed a new star, or *stella nova* (a Latin phrase usually contracted to simply *nova*). Kepler kept a close watch on another nova that burst into view in the constellation Ophiuchus in 1604.

With the advent of photography and large telescopes in the nineteenth century, astronomers discovered many novas scattered throughout the sky. By the beginning of this century, a nova was no longer considered to be an actual new star but rather the sudden eruption of light from an existing star (Fig. 17.5). Also in this century, astronomers recognized that some of these outbursts took place with extraordinary violence. These special flare-ups are now called *supernovas* to distinguish them from ordinary novas. The novas of 1572 and 1604 are now known to have been supernovas.

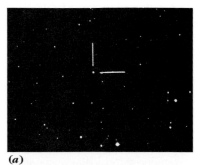

(a)

(b)

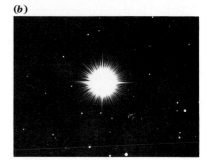

FIGURE 17.5 Nova Herculis 1934. These photos show the star (*a*) before (marked by the white lines) and (*b*) during its outburst in 1934. (Courtesy of Lick Observatory.)

Both novas and supernovas represent explosions of stars. For ordinary novas, only the outer layers of the star participate in the explosion. For supernovas, the interior regions are also involved.

Ordinary Novas

In a typical nova outburst, a star in just a few days increases in brightness up to 50,000 times (sometimes up to a million). It stays at peak brilliance for several hours. Then slowly, in a few hundred days, the nova's light declines to an inconspicuous level—usually brighter, however, than the star's prenova level. A plot of a nova's rise and fall in brightness (or luminosity) is called its *light curve* (Fig. 17.6). Most novas have the same general shape for their light curves: a sharp rise and a gradual decline.

A typical nova hits a peak at greater than 100,000 solar luminosities. All told, a nova emits during its flare-up and demise some 10^{38} J; thus in a few hundred days, a nova emits as much energy as the sun generates in about 100,000 years! We estimate that about 100 novas take place each year in the Milky Way.

A nova's spectrum undergoes pronounced and complicated changes during its outburst. Generally, the prenova star's spectrum has broad, dark absorption lines with weak or no bright emission lines. At maximum, the nova has absorption lines like a supergiant star. Some time after maximum, the nova's spectrum develops emission lines similar to those from H II regions (Section 15.1), such as the Orion Nebula. The Doppler shifts of these lines range from a few hundred to a few thousand kilometers per second.

What does the evolution of a nova's spectrum say about the outburst? First, the star's photosphere expands dramatically in size, to 100 to 300 solar radii. Second, the photosphere then collapses back onto the star. Third, a shell of material is blown off the star and rapidly expands away from it (Fig. 17.7). Overall, a nova spurts off about 10^{26} kg of material, about 10^{-4} solar mass. If the prenova star has about 1 solar mass, the ejected material makes up only a small fraction of the total.

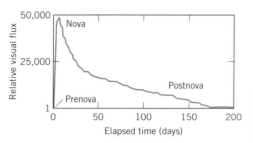

FIGURE 17.6 Light curve for a typical nova. Note the sharp rise from the prenova luminosity and the more gradual decline over a span of a few hundred days. (Adapted from a diagram by C. Payne-Gaposhkin.)

A Nova Model

What prompts a nova to explode? One major clue: observational studies indicate that almost all novas occur in close binary systems, that is, binary stars with short orbital periods—typically about 4 hours. In such systems the two stars are so close together (about 1 solar diameter) that matter may flow between them. For instance, Nova Herculis 1934 is an eclipsing binary system, now called DQ Herculis. From its eclipses, we know that its orbital period is 4.6 hours and that the primary star is about 0.6 solar mass, and the secondary 0.4 solar mass. A nova occurs on the white dwarf companion of about a solar mass in such binary systems.

Here's one scenario. Consider a binary star that originally consists of a main-sequence star and a less massive companion. At the end of its life, the main-sequence star has become a white dwarf (Fig. 17.8). In a close binary star system, hydrogen can come from the companion star to the white dwarf. How?

Around each star lies a region of space in which its gravitational force dominates. This region is called the **Roche lobe** (Fig. 17.8*a*). Any matter within the Roche lobe is gravitationally bound to that star and cannot escape to the other star. But where in a binary system the gravitational pull from one star just cancels that from the other, the two Roche lobes touch and join. This means that a gravitational highway exists for the flow of matter between the stars. For the mass exchange to happen, material from the companion star has to get out to the Roche lobe. How? When it becomes a red giant (Fig. 17.8*b*), its atmosphere swells up and reaches its Roche lobe boundary. Material then flows from the red giant to the white dwarf.

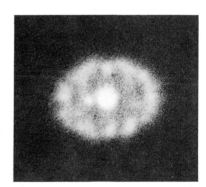

FIGURE 17.7 Nova Herculis 1934, photographed 40 years after the outburst. The shell of blown-off material, which is expanding into the interstellar medium, has a radius of about 0.05 ly. Figure 17.5 shows the actual explosion; many years had to elapse before the shell had expanded to a large enough size to be visible. The star that supplied the fresh fuel for the outburst is at the center; the white dwarf in this binary system is too faint to show up here. (Courtesy of Stewart Observatory, University of Arizona.)

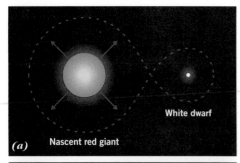

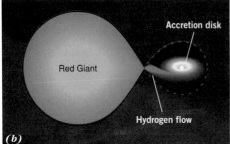

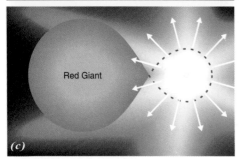

FIGURE 17.8 One model for the evolution of a binary system to produce a nova outburst. (*a*) One star has evolved to a white dwarf; its companion expands as it evolves off the main sequence to become a red giant. (*b*) The companion is now a red giant and matter flows through the Roche lobe into an accretion disk around the white dwarf. (*c*) The fresh hydrogen falls onto the white dwarf and fuels an explosion within the surface layers of the white dwarf.

The donor star does not have to be a red giant. In very close binary systems, the Roche lobe around a low-mass, main-sequence companion star can be about the size of the star. The white dwarf tidally distorts its companion, and mass transfer takes place.

As the matter falls toward the white dwarf, the gas forms into an **accretion disk** (Fig. 17.9). The material gathers in a disk from the conservation of angular momentum (Focus 9.1). The matter spirals into the white dwarf's surface from the accretion disk. Fresh hydrogen gradually accretes on the white dwarf's surface, forming a new envelope of fuel.

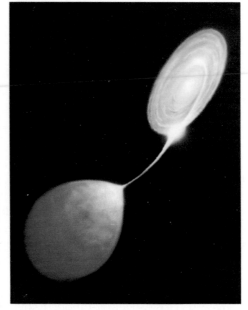

FIGURE 17.9 Artist's conception of a red giant–white dwarf binary system with matter flowing from the red giant into an accretion disk around the white dwarf.

Additional material piles on, compressing and heating the disk.

When the temperature at the bottom of the accreted layer has reached a few million kelvins, hydrogen fusion reactions ignite. Because the gas here is degenerate, the ignition is explosive (Fig. 17.8*c*). The runaway fusion reactions heat the entire layer to some hundred million kelvins. Then the layer expands and rapidly results in an envelope about the size of a red giant! This material then blows into space at speeds of hundreds of kilometers per second.

In any case, the ejected material will have more heavy elements from the thermonuclear processing. Observations of the expanding nebular material around novas do show enhanced (compared to the sun) abundances of elements such as carbon, nitrogen, oxygen, and neon.

In summary. Novas occur when material accretes onto a white dwarf. This infall heats the material, igniting the runaway thermonuclear reactions that blow off the outer layers. Most novas are members of short-period binary systems, which contain a white dwarf companion.

■ **In what type of stellar system do most novas occur?**

FIGURE 17.10 Supernova explosion in the elliptical galaxy Centarus A: (*a*) The galaxy before the explosion and (*b*) the galaxy with the supernova at its peak (arrowhead). (Courtesy of NOAO/CTIO.)

TABLE 17.1 SUPERNOVAS OBSERVED IN THE GALAXY BY THE NAKED EYE

DATE (A.D.)	CONSTELLATION	BRIGHTNESS	DISTANCE (kly)	OBSERVERS
185	Centaurus	Brighter than Venus	8.1	Chinese
369	Cassiopeia	Brighter than Mars or Jupiter	33	Chinese
1006	Lupus	Brighter than Venus	11	Chinese, Japanese, Korean, European, and Arabian
1054	Taurus (Crab Nebula)	Brighter than Venus	6.5	Chinese, Southwestern Native American, and Arabian
1572	Cassiopeia	Nearly as bright as Venus	16	Tycho and many others
1604	Ophiuchus	Between Sirius and Jupiter in brightness	20	Kepler, Galileo, and many others

Source: Adapted from a table compiled by W. C. Straka.

17.4 Supernovas: Cataclysmic Explosions

As violent as novas may appear, they cannot match the fierce destruction of a star in a **supernova.** These cataclysmic explosions spew out energy in extraordinary amounts, about 10 billion times the sun's luminosity at their peak (more than 100,000 times that of a nova). A supernova usually signals the death of a massive star.

The name *supernova* was coined by Fritz Zwicky (1898–1974) and Walter Baade (1893–1960) for the extraordinary novas discovered in our own and other galaxies. More than 300 supernovas have been found in other galaxies (Fig. 17.10). Both the new stars observed by Tycho and Kepler were supernovas. Since the supernova of 1604, no such grand explosion has been seen in the Milky Way. (However, one exploded in the Large Magellanic Cloud in February 1987; see below.) Only six have been confirmed in the Milky Way during recorded history (Table 17.1). (Over the age of the Galaxy, however, hundreds of millions of supernova explosions have occurred.)

Chinese astronomers used the term *guest stars* for temporary celestial objects, such as novas or comets. One guest star flared in the sky on July 4, 1054. Close study of Chinese and Japanese accounts of this visitor confirms that the star remained visible to the unaided eye

FIGURE 17.11 The Crab Nebula, a supernova remnant. (See the winter star chart in Appendix G for the location of the Crab Nebula in Taurus.) This 1973 photograph shows the true colors and emphasizes the filamentary structure in the expanding material. (Courtesy of NOAO/Kitt Peak; image by W. Schoening and N. Sharp.)

TABLE 17.2 GENERAL PROPERTIES OF SUPERNOVAS

PROPERTY	TYPE I	TYPE II
Origin	Binary system with one star a white dwarf	Old massive star
Ejected mass (solar masses)	≈ 1	≈ 5
Velocity of ejected mass (km/s)	10,000	5000
Total kinetic energy (J)	$\approx 10^{44}$	$\approx 10^{44}$
Visual radiated energy (J)	$\approx$ few $\times 10^{42}$	$\approx 10^{42}$
Spectra	No hydrogen lines	Hydrogen lines
Corpse	None?	Neutron star
Frequency	1 in 60 years	1 in 40 years

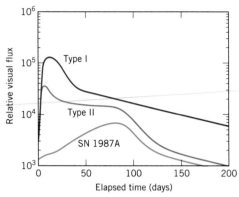

FIGURE 17.12 Typical light curves for Type I and Type II supernovas. Both have a sharp rise to maximum, but their declines in brightness differ in character: Type I supernovas generally have higher peaks and are brighter overall during the declining stage. The light curve for supernova 1987A (SN 1987A) is shown for comparison.

for more than 650 days in the night sky and was visible in daylight for 23 days! The position noted by the Oriental astronomers placed the event in the constellation Taurus. In the twentieth century, the Crab Nebula became the first firmly identified supernova remnant in our Galaxy (Fig. 17.11).

Classifying Supernovas

Astronomers classify supernovas into two general categories (Table 17.2) by their spectra and the shape of their light curves (Fig. 17.12). **Type I** light curves exhibit a sharp maximum (reaching about 10 billion solar luminosities) and die off gradually. **Type II** have a broader peak at maximum (emitting about 1 billion solar luminosities) and die away more rapidly. Studies of other galaxies have revealed that Type II supernovas occur in association with Population I stars. In confusing contrast, Type I supernovas occur in association with *both* Population I and II stars (Section 16.7). Recall that Population I stars are younger and rich in metals; Population II are older and metal-poor.

The spectra of Type I and II supernovas can contain both emission and absorption lines, which change as the properties of the ejected material evolves. The basic difference is that Type II spectra show *strong hydrogen lines*

and those of Type I do *not*. This indicates that Type I explosions involve stars that lack hydrogen—bodies that are highly evolved. The features of the spectra are the best way to differentiate the type of supernova explosion.

The total energy output (excluding neutrinos) of a supernova is stupendous: 10^{44} J, or approximately the entire energy output of the sun during its projected lifetime of 10 billion years. At its brightest, a supernova shines with a light of *10 billion* suns—about that of a typical galaxy! But most of the energy output is in the form of neutrinos (10^{46} J!) rather than electromagnetic radiation.

How often this kind of cosmic violence takes place is still debated. The rate of occurrence in any one galaxy is low in human terms, but the vast number of visible galaxies ensures that a few supernovas will be observed every year. From the rate in other galaxies, we estimate that a Type I supernova bursts forth in a galaxy, on the average, once in 60 years. Type II supernovas may be more frequent: one explosion in a galaxy roughly every 40 years. Some astronomers argue that the true frequency is greater because we do not observe all supernova events.

One wonderful supernova went off in spring 1993 in a nearby galaxy, located between the Big Dipper and Polaris. (The galaxy is called Messier 81 and lies about 12 million ly distant.) Called supernova 1993J (SN1993J, the tenth one to be observed in 1993), it was the brightest supernova visible from the northern hemisphere in decades. Because of its position near the celestial north pole, some telescopes (such as radio telescopes) could observe it continuously. After its discovery at the end of March by Francisco García, an amateur astronomer in Spain, word went out electronically to astronomers around the world. Even space telescopes, such as *ROSAT,* took aim at this special opportunity—and x-rays were detected, the first time ever right after a supernova. Radio telescopes, such as the VLA, picked up the early radio emission. By measuring an exact position, astronomers determined that a red supergiant star—as expected from theoretical models of Type II supernovas—had blown up.

The Origin of Supernovas

What kinds of star become supernovas? Because Type I supernovas often occur in Population II stars, these stars may have a mass about that of the sun. In contrast, Type II supernovas are thought to be stars much more massive than the sun—stars that live their normal lives as O and B stars. These stars probably explode once they have evolved off the main sequence to become red supergiants and have formed an iron core (the next section gives details). They blow off a few solar masses of material and leave a remnant mass behind.

Type I supernovas are more puzzling. How can a 1-solar-mass star detonate as violently as a supernova? A widely held idea resembles that for binary novas (Fig. 17.8). Imagine a binary system containing a white dwarf and a red giant star. Assume that the white dwarf has a mass close to the Chandrasekhar limit (1.4 solar masses). As material flows onto the white dwarf, it builds up enough to push the star's mass over the Chandrasekar limit. The star collapses because it cannot support the increased gravitational forces. The collapse heats the material and ignites the carbon of the white dwarf. The carbon burns swiftly into nickel, cobalt, and iron—elements that are seen in the spectra of Type I supernovas. These stars probably blow themselves to bits and leave *no* remnant mass.

Supernova 1987A

It finally happened—the astronomer's dream. On the night of February 24, 1987, Ian Shelton of the University of Toronto was photographing the Large Magellanic Cloud (LMC; see Section 19.6) from Las Campānas Observatory in Chile. On the photo was a new, bright star in the LMC. By luck, Shelton had photographed the same region 25 hours earlier. A comparison of the photos dramatically showed the star—the brightest supernova since Kepler's in A.D. 1604. The supernova (Fig. 17.13) was given the name SN 1987A ("A" for the first supernova discovered in 1987.)

Probably the first person to see the supernova directly and recognize its significance was Albert Jones of New Zealand. Jones is one of the most active visual observers of variable stars in the world. One evening, finishing up his observing as clouds were rolling in, Jones spotted a "bright blue object" in the LMC. Later that night the skies cleared, and Jones could confirm the sighting, which was announced worldwide.

The supernova in the LMC exploded some 170,000 years ago, at the time when glaciers enveloped the earth. When its flash finally reached us, current astronomical instruments permitted the *first* opportunity to study a supernova in detail. These observations have confirmed our basic model of a Type II supernova.

The foremost observation was the discovery of neutrinos from the explosion—a prediction first made for supernovas in 1966 by Stirling Colgate and collaborators. In Kamioka, Japan, a joint project of the United States and Japan, called the Kamiokande II, detected a

FIGURE 17.13 Supernova 1987A (arrow) in the Large Magellanic Cloud (LMC). To the upper left of the supernova is the Tarantula Nebula, a large, complex H II region. (Courtesy of ESO.)

burst of neutrinos about one day before the supernova burst into visibility. Neutrino events were also observed from detectors in Ohio by the Irvine–Michigan–Brookhaven (IMB) experiment (Fig. 17.14). The detection of neutrinos strongly implies a Type II supernova—the death of a massive star. For the first time, observations caught the actual formation of a neutron star!

In the models for Type II supernovas, a star with a mass of 10 or more solar masses develops an iron core. When the core has grown to a mass greater than 1.4 solar masses, it suddenly collapses. In less than one second, the electrons in the core are forced to merge with protons to form neutrons. The infalling material bounces off this neutron core, sending out a shock wave that creates the optical supernova explosion, which takes a day or so to become visible. In contrast, the neutrinos that are produced (one for each combination of an electron and proton) zip right out of the core. These neutrinos, in fact, carry away most of the energy, some 10^{46} J. But only a few (about 20!) were detected of the estimated 10^{15} per square meter in the burst that arrived at the earth from SN 1987A.

Most observations backed up the idea that this was a Type II explosion, but with a twist. The supernova did not get as bright as expected. It was about a hundred times fainter than a Type II should be at maximum. (Fig. 17.15). That is *not* typical behavior for a Type II (see the light curve in Fig. 17.12).

Initial spectra showed strong, broad hydrogen lines, as expected from the hydrogen in the envelope around the exploding core. Doppler shift measurements indicated an expansion rate of about 17,000 km/s, as expected for a Type II event. The evolution of the spectrum was typical. It first showed very strong lines of hydrogen alone. Later, strong emission lines of heavier elements (some formed in the explosion) dominated the spectrum (Fig. 17.16).

The star that blew up was a blue supergiant (called Sanduleak −69° 202), which was located in a region of active star formation in the LMC (Fig. 17.17). The star was estimated to have had a mass of 20 solar masses, a luminosity of

10^5 times the sun's, a surface temperature of 16,000 K, and a radius about 50 times that of the sun. It's still a puzzle that this star was a blue rather than a red supergiant before it blew up.

By mid-July, the light echoes of the supernova appeared as two concentric rings (Fig. 17.18). The echoes arise from thin sheets of material 470 and 1300 ly in front of the supernova; these reflected the light and delayed the signal's arrival to the earth. These reflections took a longer time to reach the earth because the light traveled a longer path. In 1990, an *HST* image of the region revealed some of the material expelled in the explosion and some discharged before the explosion (Fig. 17.19).

Supernova 1987A marked the supernova of a lifetime—the first time we have seen the core collapse of a massive

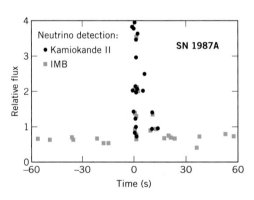

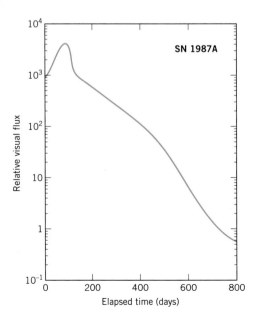

FIGURE 17.14 Detection of neutrinos from SN 1987A. Time zero on the horizontal axis indicates the moment of core collapse. Two sets of data are plotted here: one from the Kamiokande II experiment and the other from the Irvine–Michigan–Brookhaven (IMB) detector. The vertical axis gives the energy of the neutrinos in units of millions of electron volts (MeV), where $1 \text{ eV} = 1.6 \times 10^{-19}$ J. The points near and below 8 MeV are background noise. The neutrinos from the supernova all arrived within about 12 seconds (peak).

FIGURE 17.15 Light curve of SN 1987A from the first day of the visible outburst until May 16, 1989, some 800 days later. These observations were made at visual wavelengths, which only carried a small fraction of the energy of the blast. (Courtesy of NOAO/ CTIO.)

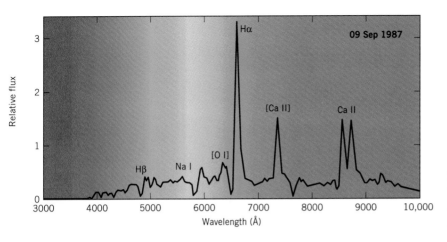
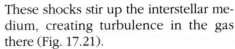

FIGURE 17.16 Spectrum of SN 1987A in September, 1987 which showed strong emission lines of oxygen, calcium, and sodium along with hydrogen. (Courtesy of NOAO/CTIO.)

These shocks stir up the interstellar medium, creating turbulence in the gas there (Fig. 17.21).

Radio astronomers have one up on optical astronomers in the hunt for galactic supernova remnants: they can observe low-density excited gas that has no detectable optical emission. For example, the radio astronomers were the first to detect the supernova remnant Cassiopeia A (Fig. 17.22), which took place about A.D. 1680. Later work with optical

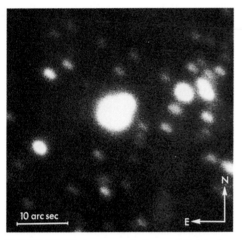

FIGURE 17.17 A pre-explosion red light photograph of Sanduleak −69° 202, the progenitor of SN 1987A, taken in 1979. Note the bulge to the upper right, showing that the star is double. (Courtesy of ESO, © 1987.)

star. It provided a number of crucial tests for supernova models—tests that were passed with flying colors.

Supernova Remnants

A supernova bangs out a blast wave into the interstellar medium. Traveling at supersonic velocities, the shell of material creates a *shock wave* that plows through the interstellar gas and dust. The shock wave's collisions with the cool clouds of the interstellar medium can excite the interstellar material, making it glow and produce emission lines. This luminous material marks a **supernova remnant.** The Loop Nebula in Cygnus (Fig. 17.20) is such a remnant. Note that it looks spherical—a shell produced by the interaction between the interstellar medium and a supernova shock wave.

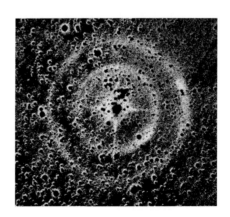

FIGURE 17.18 Light echoes in July 1988 from the SN 1987A. The image has been processed to bring out the rings, so the star images come out black. The rings arise from dusty sheets of gas between us and the supernova. The dust scatters the light to make the rings. The cross is an artifact introduced by the supports for the secondary mirror of the telescope. (Courtesy of the Anglo-Australian Observatory, photo by D. Malin.)

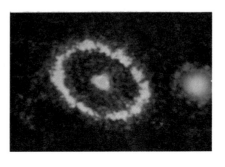

FIGURE 17.19 False-color *HST* image of SN 1987A. The material thrown off by the blast is the central, reddish region. It is about 0.1 ly across, or 100 times the diameter of the solar system. The outer, yellowish ring is 1.3 ly across. It is made of material discharged by the star prior to the explosion by a strong stellar wind. (Courtesy of NASA/ESA.)

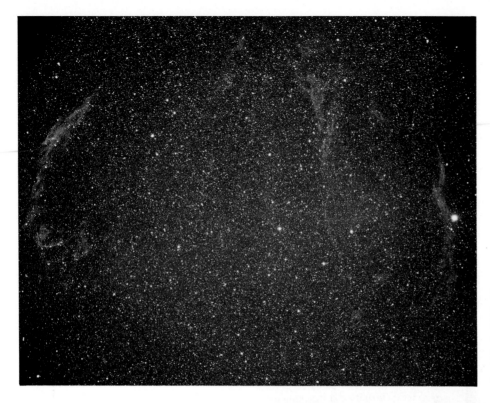

telescopes revealed faint patches of nebulosity at the same location—debris from the supernova. The explosion that formed this remnant involved a 15-solar-mass star. The material now spans almost 20 ly and moves outward at some 8000 km/s.

Radio astronomers recognize supernova remnants by a special property of their radio emission. A plot of intensity versus frequency displays a *nonthermal spectrum* (Focus 17.1). If a radio source is observed at a variety of frequencies (Fig. 17.23), the shape of its spectrum distinguishes between a possible supernova remnant (nonthermal) and an ordinary H II region (thermal; a hot gas). The

Supernova 1987A

At a Glance

Type	Variation of Type II
Location	The Large Magellanic Cloud
Progenitor star	B3 supergiant
Core collapse duration	10 milliseconds
Expected corpse	1.4-solar-mass neutron star
Total energy released	10^{46} J (mostly neutrinos)

FIGURE 17.21 An *HST* image of a small portion of the Cygnus loop. It is a combination of three colors: oxygen (blue), hydrogen (green), and sulfur (red). As the shock waves form a diagonal across the image from upper left to lower right, their structure is clear. The blue ribbon stretching across the center may be a knot of gas ejected at high speeds by the supernova. (Courtesy of J. J. Hester, Arizona State University/NASA.)

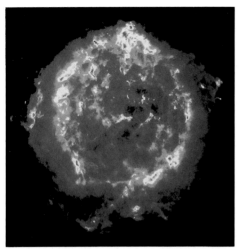

FIGURE 17.22 False-color radio image of the supernova remnant Cassiopeia A. This VLA map, made at a wavelength of 6 cm, was processed on a Cray supercomputer; red and yellow show the strongest emission and blue the weakest. The supernova that created this remnant exploded in A.D. 1680. The outer layers of the star were ejected at a high velocity into the interstellar medium. The expanding shell has decelerated enough by now to permit the more slowly expanding material from deeper within the star to break through the shell from the inside. The passage of the ejecta through the shell creates the conical extensions that leave craterlike structures in their wake along the rim. (Courtesy of NRAO/AUI; VLA observations by P. E. Angerhofer, R. Braun, S. F. Gull, R. A. Perley, and R. J. Tuffs.)

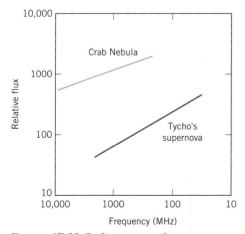

FIGURE 17.23 Radio spectra of two supernova remnants. Plotted here are the overall spectra at frequencies from 10 to 10,000 MHz of the Crab Nebula and Tycho's supernova. The flux decreases at higher frequencies (shorter wavelengths)—a typical trend for the nonthermal spectra of supernova remnants.

Planck curve of a blackbody radiator is the prototype of thermal emission.

A nonthermal spectrum indicates that the radio emission comes from very-high-speed electrons accelerating in magnetic fields, producing **synchrotron radiation.** The intensity and wavelength range of this radiation depend on the intensity of the magnetic field and the kinetic energy of the electrons. Because supernova remnants display nonthermal spectra, we know that the synchrotron process generates their radio emission, which means that the remnants contain magnetic fields and some high-energy particles.

X-ray astronomers can observe young supernova remnants directly; more than thirty have been observed so far. The huge shock waves generated by the blast plow through the interstellar medium at speeds of hundreds of kilometers per second. They compress and heat the interstellar gas to temperatures of a few million kelvins in the zone just behind the blast wave. This hot gas shows up in x-ray photos (Figs. 17.24 and 17.25).

The Crab Nebula: A Supernova Remnant

Eleventh-century Chinese and Japanese astronomers carefully watched the supernova that produced the **Crab Nebula.** This contrasts sharply with the lack of comment by European astronomers at the time. Although ignored in Europe, the supernova may have been observed and recorded in the southwestern part of North America. One good example is a rock painting in Chaco Canyon, New Mexico, which may represent the predawn conjunction of the waning crescent moon and the supernova (Fig. 17.26). The Anasazi (Section 1.6), who lived in Chaco then, may have made this painting to commemorate the event.

The star that exploded may have had a mass of 9 solar masses. The material blown off in the explosion is still expanding. Indeed, Doppler shift measurements of the optically visible filaments in the Crab Nebula demonstrate that the gas *is* expanding. The Doppler shift of the expanding filaments implies—if the expansion rate has been constant at 1450

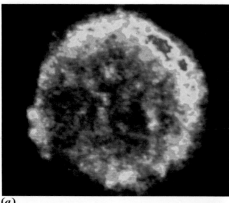

(a)

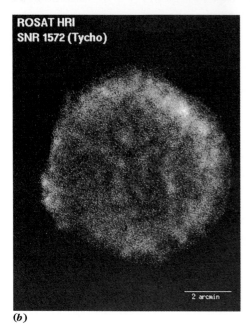

ROSAT HRI
SNR 1572 (Tycho)

2 arcmin

(b)

FIGURE 17.24 Tycho's supernova in x-rays. (*a*) X-ray, false-color image of Tycho's supernova remnant, taken by the *Einstein X-Ray Observatory*. Note the overall spherical shape. Red and yellow represent the strongest emission; blue the weakest. (Courtesy of Einstein Data Bank, Harvard–Smithsonian Center for Astrophysics.) (*b*) *ROSAT* x-ray image of the Tycho remnant. This image has a higher angular resolution and measures a region of the x-ray spectrum different from that of the *Einstein* image. (Courtesy of *ROSAT*; image by H. Fink, Max-Planck Institute for Extraterrestrial Physics.)

km/s—that they began expanding at around A.D. 1132. That's pretty close to the actual date of the explosion observed from earth. (The supernova's distance is 6500 ly, so the explosion actually preceded human observation by thousands of years.) What has powered the nebula since the explosion expelled the outer layers of the star?

In 1953 the Ukrainian astronomer Josef Shklovsky (1916–1985) solved the source of the Crab Nebula's radio and

optical emission when he suggested that the synchrotron process produced it (Focus 17.1). Shklovsky's argument was clinched when Soviet astronomers found that the optical emission was strongly polarized (Fig. 17.27). Later observations showed that the X-ray emission was also polarized.

What does it mean for electromagnetic radiation to be *polarized?* You already know if you wear polarizing sunglasses (with Polaroid lenses). Look at sunlight reflected off a road or a car through such sunglasses in their normal orientation. Then twist them 90°. You'll see that the brightness of the reflected sunlight changes. That's because reflected light is polarized, and the polarizing lenses when correctly oriented, cut down the glare from the reflected light. (Figure 17.27 shows the orientation of the polarizing material.)

Light has wave properties and vibrates up and down. Waves are said to

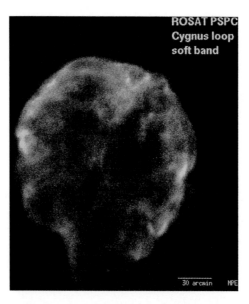

ROSAT PSPC
Cygnus loop
soft band

30 arcmin   MPE

FIGURE 17.25 *ROSAT* x-ray image of the Cygnus Loop supernova remnant. Note the filamentary structure. (Courtesy of *ROSAT*, image by B. Aschenbach, Max Planck Institute for Extraterrestrial Physics.)

FIGURE 17.26 Pre-Columbian painting in Chaco Canyon, New Mexico, that may represent the A.D. 1054 supernova. The view is looking up at the painting, which is on the underside of a rock ledge. A starlike symbol lies next to a crescent symbol. The hand indicates that something extraordinary is shown. On the morning of July 5, 1054, the supernova, brighter than Venus, rose with the waning crescent moon. (Photo by M. Zeilik.)

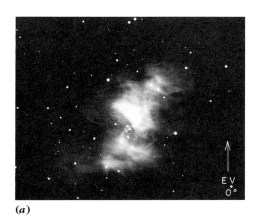

(a)

(b)

FIGURE 17.27 The Crab Nebula viewed through polarizing filters. The arrows indicate the orientation of the polarization of the filters. The differences in images (*a*) and (*b*) show that the light is polarized because different amounts are passed through the filters in their two orientations. (Courtesy of Palomar Observatory, California Institute of Technology.)

undergo **polarization** if the planes of their vibrating motion tend to be oriented in some direction—for instance, the plane of this page of paper. Light that is unpolarized has no preferred orientation: the planes of wave vibration occur in all directions in equal amounts. Synchrotron emission is usually strongly polarized.

The solution to one puzzle posed another question even more vexing: What is the source of the energetic electrons? As these electrons spiral through the magnetic field emitting synchrotron radiation, they lose energy rapidly. For the electrons producing the optical emission, half their energy would be drained off in only 70 years. So the supply of electrons must be continuously replenished. The problem of the electrons' source became even more acute when x-ray emission was discovered in 1963. The electrons that produce synchrotron x-ray emission have higher energies than those that produce optical emission. They also deplete their energy faster, losing half in only a few years. The Crab Nebula emits about 100 times more energy in the form of x-rays (Fig. 17.28) than as radio or optical emission, so a large amount of energy must be added to the nebula over a time of only a few years.

The energy problem disappeared in 1968 with the discovery of a pulsar in the Crab Nebula (Section 17.6).

■ Why can it be argued that the frequency of supernova occurrences in our Galaxy may actually be higher than has been thought?

17.5 The Manufacture of Heavy Elements

A massive star dies in a supernova. The blast can leave behind a neutron star or a black hole. So a massive star digs its own grave. But it also blasts newly synthesized elements into interstellar space.

Nucleosynthesis in Stars

Massive stars (those of 8 solar masses or greater) burn heavier elements in their sequence of thermonuclear energy generation (look back at Table 16.4). Each step up in the fusion chain requires higher temperatures to overcome the greater repulsive electrical force of nuclei with more protons. These fusion reactions can fire only in the cores of massive stars. Compared to the main-sequence hydrogen burning, they have very short durations—a thousand years or less.

The end to the fusion chain comes with iron, the most tightly bound of the normal nuclei. To split iron into lighter elements takes an input of energy. To fuse it to heavier ones also requires additional energy. So nuclear reactions nat-

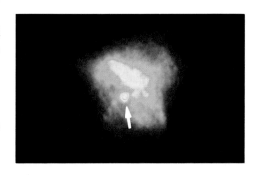

FIGURE 17.28 X-ray image of the Crab Nebula, taken by the *Einstein X-Ray Observatory*. The bright patch at the center is the pulsar (arrow). The diffuse glow around it is synchrotron emission from the expanding material. (Courtesy of Einstein Data Bank and Harvard – Smithsonian Center for Astrophysics.)

Enrichment Focus 17.1

THERMAL AND NONTHERMAL (SYNCHROTRON) EMISSION

The spectrum of blackbody radiation has a characteristic shape. The energy output increases from longer to shorter wavelengths, peaks at a wavelength that depends on the blackbody's temperature, and then decreases at shorter wavelengths until it hits zero. The spectrum of blackbody emission is the archetype of thermal emission, which arises basically from the motions of the particles involved. The greater the motions, the higher the output of radiation (and the hotter the source).

Nonthermal emission does not follow the characteristic signature of blackbody radiation. In general, the nonthermal spectrum increases in intensity at longer wavelengths (Fig. F.13). Synchrotron emission is a frequently found example of nonthermal emission; it arises from the acceleration of charged particles (usually electrons) in a magnetic field. Moving charged particles interact with magnetic fields so that they spiral around the magnetic field lines rather than traveling across them (Fig. F.14). The spiral paths are curved, so the particle is continually accelerated and thus emits electromagnetic radiation. The frequency of emission is directly related to how fast the particle spirals; the faster the spiral, the higher the frequency. Increasing the magnetic field strength tightens the spiral and so increases the frequency.

The velocity of the charged particle also affects the frequency directly, so more energetic particles can produce higher frequency emission, but they also require strong fields to keep

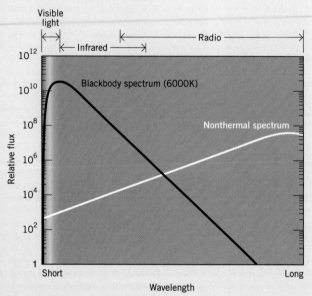

FIGURE F.13 Comparison of the spectra of a thermal (blackbody) source at 6000 K and a nonthermal (synchrotron) source. Note that the wavelengths run from short to long on the horizontal axis.

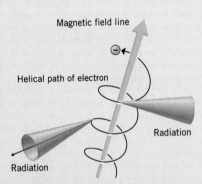

FIGURE F.14 Schematic diagram of synchrotron emission from an electron spiraling along a magnetic field line. The emission occurs in a narrow beam in the plane of the electron's spiral path.

them in a tight spiral. As the particle radiates, it loses energy and generates lower-energy (longer-wavelength) radiation. So a synchrotron source needs a continually replenished supply of

electrons to keep emitting at relatively short wavelengths (high-energy photons).

One important point about synchrotron emission: it is *polarized*. Thermal emission, such as from the sun, is not polarized. Synchrotron-emitting electrons, when viewed side-on in their spiral motion, appear to be moving back and forth along almost straight lines. Their synchrotron emission has its waves more or less aligned in the same plane. So synchrotron radiation is polarized and, at visible wavelengths, can be observed as such with Polaroid filters.

We use the term *synchrotron emission* because such radition was first observed from the General Electric synchrotron, in which magnetic fields contained electrons that were accelerated to high energies.

urally stop at iron; no other rearrangement of nuclei can generate any more energy.

But the elements that we know do not end at iron (Appendix F); many are heavier. Where and when are they manufactured in the course of cosmic evolu-

tion? A supernova explosion acts as nature's special workshop for forging some elements heavier than iron. In the most important process, high-energy neutrons bombard various nuclei. Because they have no charge, the neutrons have an easy time penetrating a nucleus. This nu-

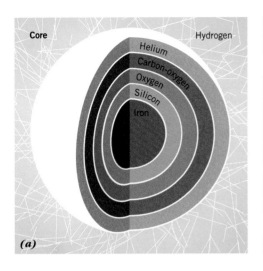

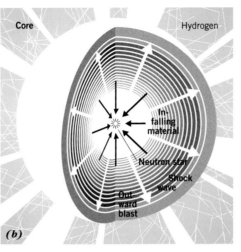

FIGURE 17.29 One model of the interior of a massive star undergoing a supernova explosion. Here we are looking at the core of a Population I star at the end of its life. Fusion reactions in shells give layers of different elements (*a*). At high enough temperatures, the iron core disintegrates into helium. This process soaks up heat from the core, and it collapses rapidly (*b*), sparking explosive ignition of fusion reactions. This explosion produces a burst of neutrinos. (Adapted from a diagram by J. C. Wheeler.)

clear capture of neutrons leads to a buildup of heavier nuclei.

Two processes take part in this buildup. A neutron under normal conditions of low density will disintegrate (in about 1000 s) into a proton, an electron, and an antineutrino. This is called **beta decay** (Fig. 17.3*a*). The rate of beta decay naturally divides neutron nucleosynthesis into two processes. In one process, neutrons are captured at a rate *faster* than the beta-decay rate, so neutron-rich nuclei are formed. This process is called the **rapid process** (*r-process* for short). In the other, nuclei capture neutrons *slower* than the beta-decay rate, so proton-rich nuclei are made. This process is called the **slow process** (or *s-process*). A combination of both these processes leads to the manufacture of most of the elements and isotopes heavier than iron. In supernovas, the time scales are short, and it is mainly the r-process that is effective.

Nucleosynthesis in a Supernova

Here's one model of how a Type II supernova happens (Fig. 17.29*a*). Imagine a very massive star with a core of iron. Its interior temperature decreases from the core outward, leading you to expect the star's interior to be layered like an onion. Around the iron core is a silicon layer; here temperatures do not get high enough to fuse silicon to iron. Around that layer is one of oxygen; here temperatures are too low to fuse oxygen to silicon. It has taken the star all its life to get to this stage, but at an accelerating pace.

For a 25-solar-mass star, for instance, the hydrogen-burning phase in the core lasts 7 million years; helium, 500,000 years; carbon, 600 years; neon, one year; oxygen, 6 months, and silicon, *one day*—a rapid dive to disaster!

When the core ends up as iron, its fusion reactions stop. Relentlessly, gravity squeezes the core to higher temperatures and densities. When the core's temperature gets to about 5×10^9 K, the photons there have so much energy that they penetrate the iron nuclei and break them down into helium. As the iron disintegrates into helium, large amounts of heat are used up. Photons that normally would provide the radiation pressure to support the star are splitting iron nuclei instead. The pressure in the core no longer supports the star, and it collapses suddenly. The gravitational collapse rapidly pumps heat into the material. This collapse is *fast*—speeds in the outer part of the core hit 70,000 km/s. In one second, a core the size of the earth drops to a radius of only 50 km and its central density rises to several times that of an atom's nucleus!

Three important events occur in this collapse. First, protons and neutrons released by the disintegration of nuclei in the core pelt and penetrate remaining nuclei. If these particles capture neutrons, they can be transformed to heavier elements. Second, the layers above the core plummet inward toward the core and heat up. Suddenly, ignition temperatures of many fusion reactions are reached. They turn on explosively; neutrinos are produced.

Third, the inner core's collapse creates a neutron star; its degenerate pressure halts the collapse, and the material rebounds outward. As the infalling matter from the outer core crashes in the rebounding inner core, a shock wave forms; it bullies its way outward from the core in just tens of milliseconds (Fig. 17.29*b*). It carries material with it into the interstellar medium—material enriched with heavy elements. About 99 percent of the energy released in this explosion comes out in the form of neutrinos.

Other Sites of Nucleosynthesis

Because I've emphasized nucleosynthesis in supernovas, you might be thinking that *only* supernovas make elements heavier than iron. That's not the case. Red giants also manufacture some heavy elements by the s-process. While a red giant undergoes helium shell burning and thermal pulses, neutrons are produced and added slowly to iron (and other elements) to fuse heavier elements in isotopes that are rich in neutrons (Section 16.8). Although this process is indeed slow, the red giant stage with thermal pulses lasts long enough to synthesize an appreciable amount of heavy elements. These are then dredged up to the surface by convection. A red giant can blow off s-process materials by its stellar wind.

In general, elements made in red giants complement those made in supernovas to fill up the periodic table. Nucleosynthesis in red giants makes many of the elements lighter than lead but heavier than iron. Supernovas synthesize elements heavier than lead, such as uranium and thorium. (When you use electricity generated by a nuclear power plant, you're turning on your light by the fossil remains of a massive star!) Old stars and dying stars spice up the composition of the universe. From this enriched material, new stars and planetary systems form. And so we live on a planet rich in metals from stars.

■ When nuclear reactions in a massive star stop, what is the last element synthesized before a supernova?

17.6 Pulsars: Neutron Stars in Rotation

Models of supernovas suggest that a neutron star may remain as the corpse of the exploded star. A neutron star found in a supernova remnant would clinch this argument. But how would a neutron star be visible? In a way not anticipated by astronomers: as a **pulsar,** a member of a category of objects accidentally discovered in the summer of 1967 by an English radio astronomy group headed by Anthony Hewish.

The Hewish group was mapping the sky at radio wavelengths to observe quasars (Section 20.5). Jocelyn Bell Burnell, then a graduate student in charge of the data analysis, noticed a strange signal that suddenly disappeared, only to reappear 3 months later. The Hewish group focused on this strange signal and found radio pulses occurring at a very regular rate, once every 1.33730113 seconds. Flushed with excitement, they searched the sky for any similar signals and discovered three more objects emitting radio bursts at different rates. The Hewish group concluded that the objects must be natural phenomena and named them *pulsars.*

Observed Characteristics

To date, more than 350 pulsars have been studied in detail. For a given pulsar, the period between pulses repeats with very high accuracy, better than 1 part in 100 million. The amount of energy in a pulse, however, varies considerably; sometimes complete pulses are missing from the sequence. Although the flux and shape vary from pulse to pulse, the average of many pulses from the same pulsar defines a unique shape (Fig. 17.30). The average pulse typically lasts for a few tens of milliseconds. A *millisecond* is 10^{-3} second.

For the well-studied pulsars, the periods (time interval between pulses) range from 0.0016 to 4.0 seconds, with an average value of 0.65 second. (Only a few pulsars are known so far to have millisecond pulses; see the end of this section.) When accurate radio observations have been made, periods have been seen to increase. The rates of slowing down have typical values of about 10^{-8} s/y. Such small increases can be measured only with atomic clocks, whose stability is better than 10^{-10} s/y. (Your digital wristwatch is off by about 100 seconds in a year.)

These observed general properties of pulsars provide clues to the possible physical properties of these precise cosmic clocks. The pulse duration indicates the largest size of the bodies emitting the radiation. How? Suppose that you could switch off the sun. Because of the finite speed of light, it would take slightly more than 2 seconds for the entire solar disk to appear dark, and then another 2 seconds to regain its original brightness if quickly turned on again. The part of the sun that we see when we look at the edge of the visible disk is at a greater distance (halfway around the sun) than the part at the center of the disk. So if the sun is turned off, we won't know that the edge is dark until about 2 seconds after we see the center become dark.

Clearly an object the size of the sun, or any ordinary star, is too large to be a pulsar, which has pulse durations amounting to a few hundred milliseconds or less. The typical pulse durations of a few tens of milliseconds imply sizes of roughly 3000 km or less. What stellar objects exist at this size or smaller? And, what objects can contain the large quantities of energy that pulsars emit? A dense object in some way moving rapidly acts as a storehouse of kinetic energy. This clue points to degenerate stars as the candidates for pulsars.

Clock Mechanism

Here's the basic issue: pulsar models must account for the precise clock mechanism of pulsars, that is, the extremely regular repetition of pulses. Basically, one clock mechanism works the best: *rotation.*

Consider one rotation period equal to the period between pulses. Then what kinds of objects are rotating? Here's the basic idea: if a spherical mass rotates too rapidly, its gravity will not be able to hold it together. Mass will fly off tangent to its equator. Neutron stars have average densities of about 10^{17} kg/m^3. So

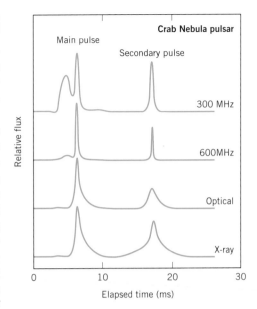

FIGURE 17.30 Average pulse profiles from the Crab Nebula pulsar observed at a variety of frequencies, from radio (300 and 600 MHz) to optical and x-rays. This pulsar, called PSR 0531 + 21, emits a strong main pulse and a weaker secondary one. Each pulse from a pulsar has a special shape at a given frequency of observation. Averaged over many pulses at that frequency, every pulsar has a characteristic shape to its pulses. Note how these differ at different frequencies. (Adapted from a diagram by F. G. Smith.)

they can rotate once every millisecond without losing mass—fast enough for even the fastest pulsar. (White dwarfs cannot do so.) Of course, they can rotate more slowly, too.

To conclude. A rotating neutron star can provide the clock mechanism for pulsars.

Pulsars and Supernovas

Are supernovas or their remnants connected to pulsars? Yes! We have two good examples so far: a pulsar in the Crab Nebula and another in the Gum Nebula, both supernova remnants.

The Crab Nebula pulsar (Fig. 17.31) is called PSR 0531 + l21. (PSR stands for

FIGURE 17.31 The Crab pulsar in blue light. The white rectangle shows the location of the pulsar and nearby stars. This area is displayed in the sequence of 33 photos at the right, which run from top to bottom and from left to right. Here is a complete sequence in the pulse cycle shown with very high time resolution—1 ms. Thus the series of photographs shows the light variation over one complete cycle of 33 ms. Note the two pulses. (Courtesy of NOAO/Kitt Peak, image by Nigel Sharp.)

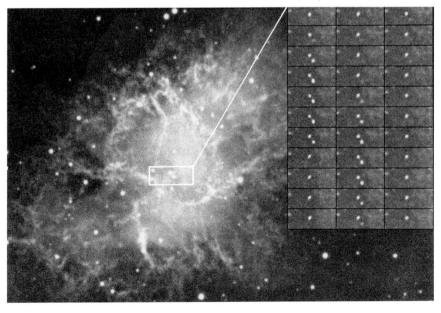

pulsar, and the numbers give the pulsar's position in the sky.) PSR 0531 + 121 has a fast period: 0.033 second, or 30 pulses per second! Two key features of the Crab pulsar. First, it is the only pulsar discovered so far to pulse not only at radio and optical wavelengths, but also in the infrared, x-ray, and gamma-ray regions of the spectrum. The total energy emitted in the pulses is about 10^{28} W, which is about 100 times more than the total luminosity of the sun. Second, the Crab pulsar was one of the first to exhibit a definite slowdown in pulse period, at a rate of about 10^{-5} s/y.

The discovery of the Crab pulsar solves the energy problem of the Crab Nebula. At all wavelengths, the Crab Nebula emits about 10^{31} W, or about 10^{5} times the sun's luminosity. What is the source of this energy? If the pulsar is a rotating neutron star, its slowdown in period gives a change in rotational energy of about 5×10^{31} W.

That's enough to power the nebula—if the rotational kinetic energy of the neutron star can somehow be converted to the kinetic and radiative energy of the nebula. In other words, the light we see now from the nebula ultimately derives from the pulsar.

If the Crab pulsar were the only one associated with a known supernova remnant, it might be written off as a chance coincidence. But astronomers know of another one: the pulsar called PSR 0833−45, in the constellation Vela in the Gum Nebula (Fig. 17.32). In 1976 astronomers finally observed optical pulses from the Vela pulsar. The pulses come every 80 ms and have two peaks, separated by about 22 ms. Also, gamma-ray telescopes have detected pulses from Vela. So this pulsar resembles the Crab pulsar in key ways: both are rapid, both emit pulses over a wide range of the electromagnetic spectrum, and their gamma-ray pulse profiles are very similar.

Of course, astronomers are desperately seeking the pulsar from SN 1987A. No luck yet!

A Lighthouse Model for Pulsars

Now to tie these observations together in the accepted basic model for pulsars—a rotating, magnetic neutron star—otherwise known as the **lighthouse model.** The model has two key components: the neutron star, whose great density and fast rotation ensure a large amount of rotational energy, and a dipole magnetic field that transforms the rotational energy to electromagnetic energy. Needed is a very strong magnetic field, some 10^{8} T.

The region close to the neutron star, where the magnetic field directly and strongly affects the motions of charged particles, is called the pulsar's *magnetosphere* (in analogy to planetary magnetospheres). Here all the energy conversion action takes place. One model pictures the magnetic axis as tilted

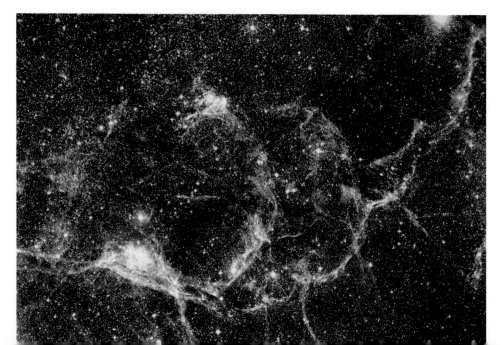

FIGURE 17.32 The Vela supernova remnant, called the Gum Nebula. The Vela supernova explosion blasted material outward at thousands of kilometers per second. These shock waves radiate; this photograph shows green light given off by oxygen atoms in the gas. Note how the filaments form circular arcs. The entire nebula is more than 2000 ly in diameter. (Courtesy of the SPL/Photo Researchers.)

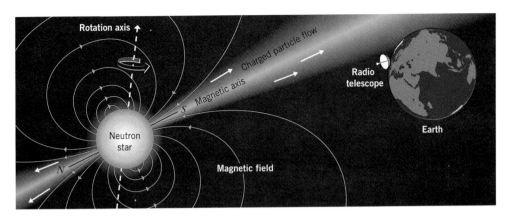

FIGURE 17.33 The lighthouse model of a pulsar. The magnetic axis is tilted with respect to the spin axis. Electrons from the neutron star's surface flow out along the magnetic field lines and so escape mostly at the north and south magnetic poles. These emit synchrotron radiation, which we see as pulses when a pole spins across our line of sight. The rotation of the neutron star provides the clock mechanism for the pulsar.

with respect to the rotational axis (Fig. 17.33). As the pulsar spins, its strong magnetic field spins with it, and this spin induces an enormous electric field at its surface like an electric dynamo. This electric field pulls charged particles— mostly electrons—off the solid crust of iron nuclei and electrons. The electrons flow into the magnetosphere, where they are accelerated by the rotating magnetic field lines. The accelerated electrons emit synchrotron radiation (Focus 17.1) in a tight beam more or less along the field lines.

You can now see how a pulsar pulses without actually pulsating. (Recall that the cepheid variables, in contrast, vary by actual expansion and contraction.) If the magnetic axis falls within our line of sight, each time a pole swings around to our view (like the spinning light of a lighthouse), we see a burst of synchrotron emission. The time between pulses is the rotation period. The duration of the pulses depends on the size of the radiating region. As the pulsar generates electromagnetic radiation, the torque (Focus 9.1) from accelerating particles in its magnetic field slows its rotation. This slowdown is observed.

For us to see a pulsar in the lighthouse model, the neutron star's magnetic axis must be oriented just right for the pulses to beam at the earth and give a blinking effect. The Milky Way contains many pulsars that we cannot see because of unfavorable orientations.

Binary Radio Pulsars

Most stars in the Milky Way are members of binary or multiple star systems. Even after one member of a binary becomes a supernova, the system usually remains intact. Radio pulsars have also been observed in binary systems.

The first one found, called PSR 1913 + 16 (in the constellation of Aquila; see the seasonal star chart for summer: Appendix G), has a pulse period of only 0.059 second. Surprisingly, its period goes through a large cyclic change in only 7.75 hours. Such regular changes would naturally come about in a binary system of the pulsar and a companion with an orbital period of 7.75 hours. What is seen is a Doppler shift in the signal produced by the orbital motion of the system. When the pulsar is moving away from us, its pulses are spread out and come at longer intervals. When it is moving toward us, the pulses are pushed together and come at shorter intervals.

PSR 1913 + 16 lies about 15,000 ly away. Visual and x-ray observations have so far failed to detect the pulsar. From radio observations alone, we know that the pulsar and its companion have an orbital semimajor axis of only a few times the sun's radius! Their combined masses are 2.8 solar masses of about equal mass, so perhaps both are neutron stars. There are other pulsars in binaries, seen with x-ray telescopes (Section 17.8).

PSR 1913 + 16 confirms a prediction that Einstein made in 1915 from his general theory: an accelerating mass should radiate energy in the form of gravitational waves. (In analogy, accelerating charged particles emit electromagnetic waves.) Such gravitational waves interact feebly with matter, so it is hard to detect them directly. PSR 1913 + 16 serves as a laboratory for an indirect test: the pair moves at such high speeds (up to 400 km/s) in their orbits, and are so close together, that their strong accelerations should produce gravitational radiation. If so, the orbit of the pulsar should decrease in size, and indeed, such a decrease has been accurately measured using the pulses as the ticks of a local clock. The orbit decreases about 3.5 meters per year, decreasing the orbital period, at a rate just about that predicted by general relativity.

Very Fast Pulsars!

Because of observational limitations, astronomers prior to 1982 had no luck in finding pulsars with periods shorter than that of the Crab pulsar. Then, while investigating a peculiar radio source in the constellation of Vulpecula, radio astronomers homed in an extremely fast pulsar—a period of 1.558 *milliseconds* (Fig. 17.34). If the lighthouse model is applicable to this pulsar, called PSR 1937 + 214, it must spin 642 times per second (twenty times faster than the Crab pulsar) so that its surface rotates at roughly one-tenth the speed of light, very close to its breakup speed. Such pulsars are called **millisecond pulsars** by astronomers. Some 20 are known to date. Many have been found in globular clusters, which implies that they are very old.

One of the curious features of millisecond pulsars is the great stability of their rotation rates. For instance, PSR 1937 + 214 loses only 3.2×10^{-12} ms in a year, compared with an ideal clock. So this pulsar provides the best time standard available today, even beating out atomic clocks (which are accurate to a few microseconds in a year; a *microsecond* is 10^{-6} s). That contrasts to the very fast spindown rates of ordinary pulsars. One explanation is that the millisecond pulsars have very weak magnetic fields, perhaps a thousand times weaker than typical. Then how does such a pulsar become observable?

The scenario proposed is that of a pulsar resurrected in a binary system. Millisecond pulsars may have aged gracefully after formation in a supernova. Then, billions of years later, their low-mass companions evolved finally to red giants and matter flowed from them into a disk around the dead pulsar. This material makes a rapidly spinning accre-

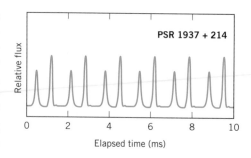

FIGURE 17.34 Radio pulses from a millisecond pulsar (called PSR 1937 + 214) observed at a frequency of 1412 MHz. Like the Crab pulsar, this one shows a large main pulse and a smaller secondary pulse. Note the time scale at the bottom spans a mere 10 ms. (Based on observations by D. C. Backer, S. R. Kulkarni, C. Heiles, M. M. Davis, and W. M. Goss.)

tion disk around the neutron star. The magnetic field of the pulsar is entwined with the disk; this linkage spins up the pulsar so that it lives again.

One millisecond pulsar has gained notoriety as the "black widow" pulsar. Its official astronomical name is PSR 1957 + 20; its special signature is a glowing, outward-streaming nebula. This pulsar generates a hot, high-speed wind, with particles traveling close to the speed of light. This wind does damage not only to the surrounding medium but also to the pulsar's companion star. This star, when it went through its red giant phase, resurrected the old pulsar by mass transfer and spinup. The pulsar's wind now rams into the companion star, creating a shock wave, heating one side of the star to 5000 K, and stripping off the star's surface material. So in a symbolic sense, the pulsar is gobbling up the star that gave it birth—and all millisecond pulsars may do the same.

Pulsars with Planets?

Millisecond pulsars are thought to be old (many billion years), companions to stars in some kind of binary systems. Radio astronomers can time their pulses with great precision and detect even small changes in them. One millisecond pulsar, called PSR 1257 + 12, showed two periods in the small variations of its timing compiled over a time span of more than a year: one around 67 days and the other around 98 days. (These changes were so small that they amounted to a 0.7 *meter* per second wobble in the pulsar.) What could cause them? Small masses orbiting the neutron star!

Alexander Wolszczan and Dale A. Frail have analyzed the results as the effect of at least two (and possibly three) planet-sized masses orbiting a 1.4-solar-mass neutron star. One, which may have a mass of about 3.9 earth masses, is thought to swing around the neutron star in 67 days at a distance of 0.36 AU. The other may have a mass of 3.2 earth masses and orbit in 98 days at a distance of 0.47 AU. The third and much less massive (and less likely!) body would orbit at 1 AU with a period of about one year.

How might such objects form? Millisecond pulsars are apt to have an accretion disk around them made of matter from their stellar companion. Somehow planet-sized masses form in this disk, maybe in a manner similar to the formation of the protoplanets in the solar nebula (Section 12.6). For now, this is a speculative idea. But, if planets do form under conditions that involve a mass as different from the sun as a neutron star, it may well be plausible that planetary systems are much more common than now believed.

■ **What is the rotation rate of the Crab pulsar?**

17.7 Black Holes: The Ultimate Corpses

Many stars in the Milky Way will die with masses much greater than the neutron star limit of about 3 solar masses. Will there be any barrier to the collapse of this material? No. The crush of gravity overwhelms all outward forces, including the repulsive forces between particles with the same charge and degenerate pressure. No material can withstand this final crushing point of matter. The collapse cannot be halted; the volume of the star will continue to decrease until it reaches zero. The density of the star will increase until it becomes infinite.

Before a mass reaches total collapse, bizarre events occur near it. As its density increases, the paths of light rays emitted from the star are bent more and more away from straight lines going away from the star's surface. Eventually the density reaches such a high value that the light rays are wrapped around the star and do not leave. The photons are trapped by the intense gravitational field in an orbit around the star. The es-

cape velocity from the star is then greater than the speed of light. Any additional photons emitted after the star attains this critical density can never reach an outside observer. The star becomes a **black hole** (Fig. 17.35).

The Schwarzschild Radius
Let's consider the meaning of *black* in *black hole* in terms of escape speed (Section 4.5 and Focus 4.4). The escape speed from the earth is about 11 km/s. Imagine squeezing the earth so that it became smaller and denser. The escape speed from its surface would increase. Imagine the earth compressed until its escape speed equaled the speed of light. Then nothing emitted at its surface, *not even light,* could escape into space. Nothing could get away, so to an outside observer, the earth would appear to be black.

How small must an object become to be dense enough to trap light? Einstein's general theory of relativity provides an answer. Just after the publication of the general theory, the German astrophysicist Karl Schwarzschild (1873–1916) calculated this critical size, now called the **Schwarzschild radius.** For the sun, the Schwarzschild radius is about 3 km—smaller than the typical sunspot! The mass of any object gives its Schwarzschild radius directly. For 1 solar mass, it's 3 km; for 2 solar masses, 6 km; for 10 solar masses, 30 km; and so on. (The earth's Schwarzschild radius is a mere centimeter!)

How to get a star as small as its Schwarzschild radius? Two ways are possible. First, runaway gravitational collapse. If you put together more mass than about 3 solar masses, it must naturally squeeze itself into a black hole. Nothing we know about—not even the hardness of matter itself—can stop this final crushing. Second, a supernova. A star's self-destruction can slam matter into a size smaller than its Schwarzschild radius (and make black holes with masses less than 3 solar masses).

Note. An object of *any* mass can be made into a black hole if a force compresses it enough. But a stellar corpse with a mass greater than the neutron star

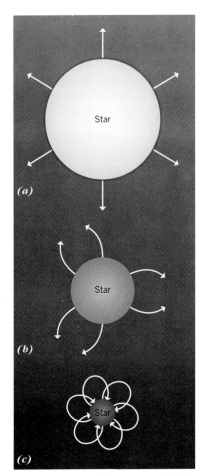

FIGURE 17.35 The trapping of light by the collapse of a mass into a black hole. Imagine a person standing on the star's surface with a flashlight. The star's mass, when much larger than its black hole size, has a gravity low enough to permit light to leave on straight-line paths (*a*). As the star contracts, the gravity at the surface increases, and light paths are bent. When the star is a little larger than its black hole size (*b*), all light paths, except for the one straight up, fall back to the surface. When the star is smaller than its black hole radius (*c*), even the straight-up photon returns. (Adapted from a diagram by W. Kaufmann.)

limit of 3 solar masses *must* become a black hole after thermonuclear reactions have ceased, for there is no known source of pressure that can support it. This limit applies only to the fate of stars.

The Singularity

We've seen how black holes may form. Once a black hole has formed, what happens to the matter that makes it? Einstein's general theory predicts that the matter will keep collapsing gravitationally until it has *no volume*. But it still has mass, so its density is infinite. This theoretical end to runaway gravitational collapse is called a **singularity.** Matter has literally squeezed itself so small that it occupies no space. Yet it's still there. What a paradox! How can matter *not* take up space? The general theory of relativity points to the formation of a singularity, cloaked in the center of a black hole, as the natural end of gravitational collapse.

Journey into a Black Hole

Let me describe the theoretical properties of a black hole, both inside and outside. A person falling into a black hole meets a fate an outside observer cannot even find out about—unless the outsider drops in too. Let's take an imaginary journey into a black hole to follow the adventures of a crazy astronaut who takes the plunge. We'll compare this trip with what an outside observer would see of it. You and a friend start out in a spaceship orbiting a few AU from a 10-solar-mass black hole. Nothing peculiar here. The ship orbits the black hole in accordance with Kepler's laws, as it would any ordinary mass. In fact, Kepler's third law and the spaceship's orbit permit you to measure the hole's mass.

Your friend volunteers to hop in. She takes with her a laser light and an electronic watch. You and she synchronize watches. Once a second, according to her watch, she will send a laser flash back to you.

Down she goes! For a long time as she falls toward the black hole, nothing strange happens. But as she gets closer, stronger and stronger tidal gravitational forces (Section 9.1) stretch her out (if she falls feet first) from head to toes. Also,

another tidal force squeezes her together, mostly at the shoulders. (You feel such tidal forces on the earth, but the forces are so weak that they don't bother you.) Near a black hole, tidal forces grow enormously. An ordinary human being would be ripped apart about 3000 km from a 10-solar-mass black hole. Let's suppose your friend is indestructible, so she can continue her trip. As she drops, the tidal forces get stronger fast. But nothing else seems strange to her. Every second on the dot she sends out a blast of laser light. Peering down, she can just make out a small black region in the sky. (Recall that a 10-solar-mass black hole has a radius of only 30 km.)

Then she crosses the Schwarzschild radius! But nothing new happens to her. No solid substance, no signs mark the edge of the black hole. However, no amount of energy can push her out of the black hole; she has crossed a one-way gate in spacetime. The trip now swiftly ends for your unfortunate friend. Quickly—in about 10^{-5} second (by her watch) after crossing the Schwarzschild radius—she crashes into a singularity (if it exists!). Crushed to zero volume, she vanishes. Even if a singularity does not lie in the black hole's center, the mass that made the black hole probably does, and your friend would smash into it.

But what of your view, back in the spaceship, of your friend's adventure? You would *never* see the end of her journey; in fact, you'd not even see her fall into the black hole. As she dropped closer to the black hole, you'd notice that the light from her laser was redshifted, with the shift increasing as she fell closer to the black hole. (The light had to work against gravity to get to you, so it lost energy and increased in wavelength.) Also, you'd find that the time between laser flashes increased. What had happened? Compared with your watch, your friend's watch appeared to slow down as she traveled into regions of stronger gravity. So your watch and hers disagreed about how long it took her to reach to the black hole.

As she came closer to the Schwarzschild radius, the watches got more and more out of sync. The times between your reception of her flashes stretched out. In fact, a laser burst sent out just as she crossed the Schwarzschild radius would take an *infinite* time to reach you. It also would suffer an *infinite* redshift. To you, her fall would seem to grow slower and slower as she got closer to the black hole, but she would never appear to fall into it. Your measurement would show time slowing down so much near a black hole that it seemed to be frozen. In addition, the light would get more and more redshifted until you no longer could detect it.

A black hole practices cosmic censorship. It prevents you from seeing your friend even fall into it, and you cannot know what happens to your friend, once inside.

Caution. Black holes are *not* cosmic vacuum cleaners. Many people think that they have infinitely powerful gravitational fields that suck up everything that gets near them, scouring out the universe. Not so! Suppose you'd never heard anything about the strange properties of black holes. Just thinking about Newtonian gravitation, what do you think would happen to the orbit of the earth around the sun if the sun suddenly shrank down to a ball 6 km across? That's right, nothing! The masses of the sun and earth haven't changed, and neither has the distance from the earth to the center of the sun. So the force of gravity on the earth hasn't changed, and neither has its orbit.

You can think of black holes as ghosts. The Schwarzschild radius is *not* a physical object. It's a region of spacetime so severely curved that weird things happen near it. Likewise, the singularity predicted to be in the center of a black hole cannot be a real object. It's a region where space and time end in this universe. What a strange corpse from the death of a massive star!

■ What are the two ways in which a star can reach its Schwarzschild radius?

17.8 Observing Black Holes

How to actually observe a black hole? With difficulty! Light emitted inside cannot get out. Light sent out close by is strongly redshifted, so it's hard to detect. In addition, a black hole is small, only a

few kilometers in size. You'll have a hard time seeing an isolated black hole.

But a black hole near any mass might be observable. Matter falling toward a black hole gains energy and heats up. (It's also squeezed by tidal forces.) Heated enough, the atoms are ionized. Gravity accelerates the ionized gas, and it emits electromagnetic radiation. If heated to a few million kelvins or so, the material gives off x-rays. Before it is trapped in the gravitational gulf, infalling material can send x-rays into space. So a black hole passing through an interstellar cloud or close to a star can sweep up material, which can radiate before it crosses the Schwarzschild radius. Hence, x-ray sources are good candidates for black holes.

The *Uhuru* satellite, launched in 1970 and designed to observe x-ray sources, detected about 160 strong x-ray objects. Some of these x-ray sources are prime candidates for black holes because they are binary—the x-ray source and a normal star (a potential source of infalling material) orbit a common center of mass.

Binary X-Ray Sources

Why are *binary* x-ray sources most suspect? Imagine a black hole orbiting a supergiant star. If they are very close together, their orbital period is a few days or so. The star has a huge, distended atmosphere—and material from this atmosphere can fall to the black hole from a Roche lobe (Fig. 17.36).

Falling toward the black hole, the material gains kinetic energy, heats up to a million kelvins or so, and emits x-rays. Only a small region around the black hole gives off x-rays, and since the material may fall in sporadically, you might expect the intensity of the x-rays to vary quickly. Also, imagine the black hole and star with their orbital plane in our line of sight. When the black hole goes behind the star, its x-rays will be cut off. In this case we would see an eclipsing x-ray binary system.

So a sign of a possible black hole is a rapidly varying x-ray source, which may be eclipsed at regular intervals in a binary system. Have we seen such variable x-ray sources? Yes! Each of these ob-

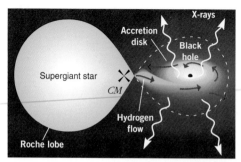

FIGURE 17.36 One model for a black hole as an x-ray source. In a binary system, the black hole is coupled with a supergiant star (which fills its Roche lobe). Material from the supergiant star flows to the black hole, where it first falls into an accretion disk. As it falls, the gas heats up to about a million kelvins and emits x-rays from the accretion disk, before it ultimately enters the black hole. The arrows indicate the flow of material from the supergiant star. "CM" marks the center of mass of the system. (Adapted from a diagram by H. Gurskey.)

jects (Table 17.3) is believed to be a main-sequence or post-main-sequence star swinging around an x-ray source. The objects have x-ray luminosities in the range of 10^{29} to 10^{31} W. (That's 200 to 20,000 times the luminosity of the sun, and all in x-rays.) Three of the sources (Hercules X-1, Centaurus X-3, and Small Magellanic Cloud X-1) have short-period x-ray pulses; they are x-ray pulsars.

Five of the systems exhibit x-ray eclipses; the x-ray source passes behind the normal star as we view the system. Using spectroscopic analysis of the light from the visible star, we can observe the changes in Doppler shift, which permits us to find out the orbital periods—they are typically a few days. These short periods indicate that the orbits are only a few times larger than the primary stars. If we can determine the separation of the two objects, we can ascertain—from Kepler's third law—the sum of the masses (normal star plus x-ray source). If we can get an idea of the mass of the normal star from its luminosity, then we can also determine the mass of the x-ray source. And if that mass turns out to be large enough (greater than 2 to 3 solar masses, the upper limit for a neutron star), the x-ray source must be a black hole.

Is Cygnus X-1 a Black Hole?

To prove the reality of black holes we need to observe one. We have only a few good candidates. So far, the most likely is Cygnus X-1, a strong x-ray source in the constellation Cygnus (see the fall star chart in Appendix G). Cygnus X-1 emits about 4×10^{30} W in x-rays. Astronomers have identified Cygnus X-1 (Fig. 17.37) with an O supergiant star called HDE 226868.

Optical observations show that the dark lines in the spectrum of the blue supergiant go through periodic Doppler shifts in 5.6 days (Fig. 17.38). So the supergiant orbits with the x-ray source about a common center of mass every 5.6 days. The supergiant has a massive

FIGURE 17.37 The blue supergiant star HDE 226868 (arrow) about which Cygnus X-1 orbits. See the summer constellation chart in Appendix G for the location of Cygnus X-1 in Cygnus. (Courtesy of J. Kristian, the Carnegie Observatories.)

TABLE 17.3 SOME BINARY X-RAY SOURCES

NAME	DISTANCE (kly)	BINARY PERIOD (days)	CHARACTERISTICS OF X-RAYS	CHARACTERISTICS OF VISIBLE STAR/ COMPANION
Cygnus X-1	8.3	5.6	Vary in duration from 0.001 to 1 second	Blue supergiant of about 16 solar masses; 9 (?)-solar-mass black hole
Centaurus X-3	20	2.087	Eclipses with 0.488-day duration; pulses every 0.72 second	Blue giant of about 19 solar masses; 1.1-solar-mass neutron star
Small Magellanic Cloud X-1	215	3.89	Eclipses with 0.6-day duration; pulses every 0.72 second	Blue supergiant of about 17 solar masses; 1.1-solar-mass neutron star
Vela X-1	4.6	8.96	Eclipses with 1.7-day duration; flares lasting a few hours	Blue supergiant of about 25 solar masses?
Hercules X-1	16	1.70	Eclipses with 0.24-day duration; pulses every 1.24 seconds	Companion HZ Her, about 2.4 solar masses; 1.5-solar-mass neutron star
Scorpio X-1	2.2	0.787	Vary in duration down to 1 second; no eclipses	Less than 2 solar masses; orbital period 0.78 days

but optically invisible companion: Cygnus X-1.

Recall that only for binaries can we find the masses of stars directly. But we need to know the separation of the stars, their distances from the center of the mass, and the orbital tilt. In two regards of these, the mass of Cygnus X-1 is hard to find out. We can observe the Doppler shift in the spectrum of the visible companion, but we cannot obtain the velocity of the x-ray source. And because Cygnus X-1 does *not* eclipse, we can't pin down its orbital inclination. So we don't have enough information to determine well both *individual* masses.

But we can make some reasonable estimates. The mass of a blue supergiant star is typically 15 to 40 solar masses. The orbital period and velocity of the supergiant give us a relation between the masses of the supergiant and the x-ray source; this relation, however, is uncertain by the amount of orbital tilt. X-ray observations imply a tilt between 36° and 67°. These values suggest that Cygnus X-1 has a mass of about 9 solar masses; it is likely above 6. If so, and if the limit for a neutron star is 3 solar masses, Cygnus X-1 is a black hole (Fig.17.39*a*).

Another strong candidate has the prosaic name of A0620-00. It is located in Monoceros, near Orion. This binary system contains an ordinary K star orbiting the center of mass every 7.75 hours at an orbital speed of 430 km/s (about 14 times the orbital speed of the earth around the sun). The distance of the K star from the center of mass has been measured: it is 0.014 AU, and we can infer its mass as 0.7 solar mass. (The system is about 3200 ly from earth.)

What is the companion? It is not seen, but emission lines from the hot gas in an accretion disk surrounding it have been recorded. The Doppler shift in these lines indicates that the companion has at least 3.2 solar masses—just enough to be a black hole. Again, we do not know the orbital tilt of the system; the 3.2-solar-mass result is a *minimum* mass that comes from assuming that we observe the system edge-on. A plausible estimate for the companion is 8 solar masses (Fig. 17.39*b*). So here, too, we have a pretty firm case for a black hole.

Not all binary x-ray sources contain

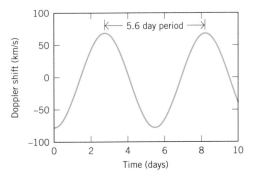

FIGURE 17.38 The orbital period of Cygnus X-1 inferred from the Doppler shift in the visual spectral lines of the blue supergiant star about which Cygnus X-1 orbits. These lines show a cycle of red- and blueshifts that indicate a period of 5.6 days. (Based on observations by C. Bolton.)

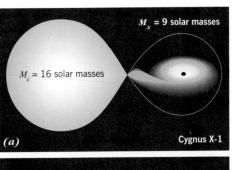

FIGURE 17.39 Possible black hole binary systems shown schematically. The Roche lobes are drawn, with the companion star (M_C) assumed to fill its Roche lobe. (*a*) Cygnus X-1, based on a 16-solar-mass companion and a reasonable orbital inclination. (*b*) A0620-00, again based on a reasonable value for the orbital inclination. (Adapted from a diagram by J. E. McClintock.)

black holes. The x-ray emission could arise from accretion disks around neutron stars. An example is Centaurus X-3 (abbreviated Cen X-3). Its x-ray emission pulses every 4.8 seconds, and x-ray eclipses take place every 2.1 days. Since Cen X-3 eclipses, we can pin down its mass to about 1.1 solar masses—so it is not a black hole, but a low-mass neutron star formed in a supernova. It now orbits a star of some 20 solar masses.

The fact that Cen X-3 is an x-ray pulsar also supports a neutron star model, in analogy with the model of radio pulsars as magnetic neutron stars. A flow of material from a companion (from a Roche lobe) forms an accretion disk. The x-ray pulses might arise from accreting matter channeled into the magnetic polar regions by the intense magnetic fields.

■ Why do we look to strong binary x-ray sources as potential sites for a black hole?

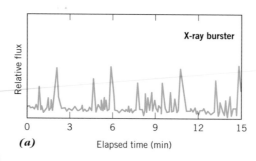

(a)

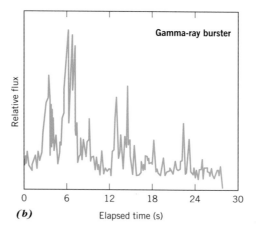

(b)

FIGURE 17.40 High-energy celestial bursters. (*a*) X-ray bursts from the rapid x-ray burster called MXB 1730-335. These observations cover a time span of about 15 minutes; the interval between bursts can be as short as 10 seconds. (Adapted from a diagram by W. Lewin and colleagues.) (*b*) A gamma-ray burster was observed in 1978 by the Pioneer Venus Orbiter. Note the irregularity in the spacing and flux of the outbursts. (Based on observations by R. W. Klebesdal, W. D. Evans, E. E. Fenimore, J. G. Laros, and J. Terrell.)

17.9 High-Energy Astrophysics

The bizarre objects presented so far have one aspect in common: all emit x-rays vigorously at some stage in their lives. They make up part of the field of *high-energy astrophysics*. Other players include gamma-ray emitters and also energetic particles known as cosmic rays. This section deals with two aspects of a high-energy face of the universe.

X-Ray and Gamma-Ray Bursters

The x-ray variability I've discussed so far has been rather gentle. But some x-ray sources display powerful bursts, increasing one hundred times in luminosity (Fig. 17.40*a*). These are called **x-ray bursters.** Some occur in regular intervals of a few hours or days. Others fire off in rapid sequence—thousands of bursts in a day in a staccato pattern. Some bursters appear to reside in globular clusters, which implies that they involve solar mass or smaller stars. Most bursters lie in the plane of the Milky Way, which implies that they are objects within the Galaxy.

One common type of burst seems to be well explained by a thermonuclear flash on the surface of a neutron star. In these bursts, the x-rays get a hundred times stronger. The bursts rise to a peak in about 1 second and last for about 10 seconds. The width of the radiating area is about 10 km, just that of the entire surface of a neutron star. The model for these bursts pictures hydrogen and helium accreting onto the surface of the star; this gas ignites explosively under conditions of high pressure and density. The source of the accreted material is probably a companion in a binary system.

More spectacular and bewildering than the x-ray bursters are the **gamma-ray bursters** (Fig. 17.40*b*). When quiet, they are so faint that they are difficult to observe in any part of the spectrum. Yet an active gamma-ray burst, which lasts for a few seconds, will outshine all other gamma-ray sources in the sky—including the sun!

What causes these bursts is really unclear, though they are likely associated with neutron stars. Ideas have included synchrotron emission from near-light-speed electrons in very strong magnetic fields (some 10^8 T), which might be found near the surfaces of magnetized neutron stars.

Cosmic Rays

Back to the interstellar medium. Through it zip **cosmic rays:** particles that travel

Death of Stars

At a Glance

MAIN-SEQUENCE MASS (SOLAR MASSES)	NORMAL LIFE	DEATH AND FINAL CORPSE
0.1–0.5	M stars	White dwarf
0.5–8	K–B0 stars	Planetary nebula, white dwarf
8–20	B0–B5 stars	Supernova, neutron star, or black hole
20–60	O stars	Supernova, black hole

close to the speed of light. When these particles crash into the earth, they make our only direct connection to matter outside the solar system. While traveling through space, they carry a large fraction of the total energy in the interstellar medium. And their origin probably is related to stardeaths.

For historic reasons, cosmic rays are misnamed. They are *not* electromagnetic radiation. They are charged particles—usually protons. In order of relative abundance (after hydrogen), cosmic rays are made of helium nuclei, electrons, light nuclei (nuclear charges from 3 to 5), medium-mass nuclei (charges from 6 to 9), and heavy nuclei (charges greater than 10).

Observations indicate that cosmic rays, except for those expelled from solar flares on the sun, stream to the earth *uniformly* from all directions in space. This fact implies that the high-energy particles from sources within the Milky Way get mixed up before they arrive at the earth. How? Recall that magnetic fields force the motion of charged particles into spirals. The radius of this corkscrew motion depends on the particle's speed, its charge, and the strength of the magnetic field: the stronger the field: the tighter the spiral.

The Milky Way's interstellar magnetic field averages some 10^{-10} T. A proton in such a field arcs on a curve with a radius of about one light year. Whatever the original direction of the proton, by the time it has traversed a light year, its direction has changed. So all the particles within the Milky Way have their directions homogenized.

Cosmic rays carry energies far beyond that possible in the most powerful particle accelerators on the earth. How do these particles get accelerated to such high speeds? A number of ways are possible. One, supernova explosions can accelerate particles to near light speeds. Two, supernova remnants and neutron stars (as pulsars) can generate cosmic rays. Three, some cosmic rays may come from sources outside of the Milky Way. Four, the intergalactic magnetic fields and supernova shock waves can reaccelerate particles from lower to high energies.

Anywhere that the magnetic field lines converge can be a place of particle reacceleration to bring charged particles back up to speed. Shock waves mark regions in which magnetic fields are compressed, say in supernova remnants. So the deaths of stars probably are the sources of most of the cosmic rays in the Milky Way.

■ **Are cosmic rays energy or matter?**

Key Concepts

1. The mass of a star at the time of its *death* determines the corpse it leaves behind: white dwarf, up to 1.4 solar masses; neutron star, 1.4 to perhaps 3 solar masses; black hole, greater than 3 solar masses. (The actual neutron star limit may be as low as a little more than 2 solar masses.)

2. White dwarfs and neutron stars are supported against gravity by the degenerate gas pressure from electrons in a white dwarf and from neutrons in a neutron star. Gravity will never crush them any smaller; they will exist forever.

3. Novas occur in close binary systems when hydrogen from the companion falls onto the surface of a white dwarf from an accretion disk. This fresh fuel ignites explosively on the surface to produce a nova outburst.

4. Supernovas (Type II) are the explosions of massive stars (greater than 5 to 10 solar masses) that have evolved iron cores. Type I supernovas are the explosions of solar-mass stars in a binary system, perhaps involving white dwarfs close to the Chandrasekhar limit.

5. Supernova explosions make many of the elements heavier than iron and blast them into the interstellar medium (along with elements made in their normal lives). These newly minted nuclei enrich the matter between the stars.

6. The Crab Nebula is a supernova remnant, emitting light now by the synchrotron process. Its current emission is powered by a pulsar in the center; that pulsar was formed in the supernova explosion of A.D. 1054, which involved a star of 10 solar masses.

7. Pulsars are rapidly rotating neutron stars. This idea is inferred most strongly from the regular timing of the fastest pulsars; only neutron stars are small enough and dense enough to rotate so rapidly. Pulsar emission comes from high-speed electrons in a neutron star's intense magnetic field—one example of synchrotron emission.

8. A black hole forms when a mass becomes so compacted that its escape velocity is greater than the velocity of light. Black holes are small: a 1-solar-mass black hole has a radius of 3 km; 2 solar masses, 6 km; 10 solar masses, 30 km; and so on.

9. Einstein's theory of general relativity predicts that time appears frozen near a black hole and that a singularity resides in its center. Black holes can be seen only by their interaction with visible matter; an especially good circumstance would be a black hole–ordinary star binary system.

10. Cygnus X-1, in a binary system, is the good candidate for a black hole. It emits x-rays from a hot accretion disk around the suspected black hole, which has a mass of at least 6 solar masses. Most other binary x-ray sources contain neutron stars. Very few others are candidates for black holes.

11. Supernova explosions, remnants, and pulsars may be the sources of cosmic rays. X-ray and gamma-ray bursters probably involve binary systems with at least one neutron star.

Key Terms

accretion disk	gamma-ray burster	rapid process	synchrotron radiation
beta decay	inverse beta decay	Roche lobe	Type I supernova
black hole	lighthouse model	Schwarzschild radius	Type II supernova
brown dwarf	millisecond pulsar	singularity	white dwarf
Chandrasekhar limit	neutron star	Sirius B	x-ray bursters
cosmic rays	polarization	slow process	
Crab Nebula	pulsar	supernova remnant	

Study Exercises

1. In a short paragraph, describe the primary characteristics of a white dwarf. (Objective 17-1)

2. In a short paragraph, describe to a friend who has not studied astronomy the chief features of a neutron star. (Objective 17-1)

3. What observational evidence do we have for the actual existence of neutron stars and white dwarfs? (Objectives 17-1, 17-2, and 17-3)

4. Look around you. Of the items you see, what would not be there if supernovas didn't occur? (Objectives 17-6 and 17-10)

5. Assuming no loss of mass, what will be the final form of (a) a 0.5-solar-mass star and (b) a 2-solar-mass star? (Objective 17-1)

6. Make a list of the observational evidence that supports the idea of the Crab Nebula as a supernova remnant. (Objectives 17-6 and 17-7)

7. In what way does a black hole practice censorship? (Objectives 17-9 and 17-12)

8. If a black hole is really black, how can it be an x-ray source? (Objective 17-9)

9. Why can't you find out what happens inside a black hole? (Objective 17-9)

10. What one feature of pulsars links them most strongly to rapidly rotating neutron stars? (Objective 17-3)

11. What is the source of the electromagnetic radiation in the pulses of pulsars? (Objective 17-3)

12. How can you tell whether cosmic rays originate outside the solar system? (Objective 17-11)

Problems & Activities

1. Calculate the escape speed from the surface of a solar-mass neutron star with a radius of 10 km.
2. Imagine a brown dwarf with a temperature of 1500 K and a radius a tenth of the sun's radius. If it radiates like a blackbody, what would its luminosity be?
3. The Cassiopeia A supernova remnant has an angular diameter of 6.5 arcmin. What is its size-to-distance ratio? Its distance is some 9000 ly. What is its actual diameter?
4. Tycho's supernova expands at a radial speed of about 2000 km/s. This expansion is visible as a motion of the material of 0.2 arcsec/y. What is the distance to the remnant?
5. If our sun were reduced to the size of a neutron star, what would its average density be?
6. If *your* mass were reduced to its Schwarzschild radius, how large would it be?

See for Yourself!

You will need a medium-sized telescope (20 cm or so objective) and a dark sky to see the only visible supernova remnant in the northern sky: the Crab Nebula (Messier 1) in Taurus. It's actually fairly easy to locate, just above the star at the end of the lower horn of the Bull (Fig. A.5). By eye, you can easily see Aldebaran in Taurus, which is seen within the Hyades, although it is not a member of this open cluster. (In fact, it lies about halfway between us and the cluster.) Moving east and north of Aldebaran about 20°, you will find the end star of the horn. It is called Zeta Tauri. The Crab Nebula is located about one degree above it. Its overall angular extent is 4.5 arcmin by 7.0 arcmin.

The gas of the nebula is currently expanding at about 1500 km/s. Its radius across its widest dimension is about 14 ly. If the velocity of expansion has been constant, how far in the past (in years) did the expansion begin? What date does that give you for the supernova explosion? How does your estimate compare to the actual date? If they are not the same within about a hundred years, how might you explain the difference? ■

FIGURE A.5 The location of the Crab Nebula in Taurus. (Star chart made by Voyager software for the Macintosh™.)

PART THREE *Epilogue*

This part has described contemporary concepts in our understanding of the births, lives, and deaths of the stars. Here briefly are the major themes.

A star's evolution depends mainly on its mass; more massive stars have higher core temperatures and, so higher luminosities. They evolve faster and live shorter lives. Their higher core temperatures allow them to fuse heavier elements. They are likely to die in supernova explosions. Stars like the sun will evolve to become red giants, white dwarfs, and then black dwarfs. Most stars will go through this evolution because most stars in the Milky Way contain 1 solar mass or less of material.

During its life, a star struggles constantly against the relentless force of gravity. The star resists gravitational collapse by pressure from heat in its interior. Fusion reactions in the core provide this heat; they are first ignited by the heat from gravitational collapse. To withstand gravity, a star must fuse heavier and heavier elements. Eventually it runs out of fuel, and gravity wins.

Stars recycle some material back into the birthplace of stars, the interstellar medium, but the medium's composition has been changed by the addition of heavier elements. Some of the material stays locked in a star's corpse, never to participate again in cosmic evolution.

How a star dies depends on its mass at its time of death. If its mass is less than 1.4 solar masses, its corpse takes the form of a white dwarf. If it is greater than 1.4 but less than roughly 3 solar masses, gravity naturally crushes it to a neutron star. In either case, gravity does not defeat the star completely. But if it has greater than about 3 solar masses, gravity wins out absolutely and forms a black hole. But keep in mind that *any* mass can become a black hole.

The stars form the crucial evolutionary links in the chain of cosmic evolution. They produce light and warmth vital to life on any planet around them. They create the elements out of which planets are made. How? By fusion reactions. What ignites these reactions? Gravitational contraction. Gravity, the driving force of the astronomical universe, squeezes matter into heavier elements in the hearts of stars.

Violence marks the death of stars. For stars of about 1 solar mass, the death rattle involves only a small fraction of the star's mass. For massive stars, almost the entire star participates in the cataclysm of a supernova. ■

The Unifying View

ASTRONOMICAL CONCEPTS

Measuring distances is crucial to all aspects of astronomy:
Heliocentric parallax is the only direct way to find the distances to stars.

Spectroscopy provides information about the physical natures of stars independent of knowing distances:
The Hertzsprung–Russell diagram is a key tool for sorting stars by type, and spectroscopy plays the main role in the sorting process.

The sun and other stars shine from fusion reactions or the conversion of gravitational potential energy:
The most important property of star is its mass, which determines how it is born and how it will live and die.
A star's internal physical conditions change as it evolves; these changes can be traced out on a Hertzsprung–Russell diagram.

Stars (and planetary systems) are formed by the gravitational contraction of the matter between the stars:
Interstellar matter is made of dust grains and gases (in various forms); most of this matter is contained in small and large clouds.

Many stars are in binary systems:
We can measure the masses of stars directly only in binary systems.
Common novas occur in binary systems.
One type of supernova takes place in binary systems.
Many x-ray-emitting stars are in binary systems.
We can search for black holes in binary systems.

INFORMATION CONCEPTS

We perceived the world through our senses, sometimes aided by instruments:
Spectroscopy is a key method in the investigation of stars.

Model building is the essential process by which science discovers knowledge and reveals information:

Models of stars are based on the sun and physical processes of hot gases.

Stellar models must match as closely as possible the observed properties of stars, such as mass, luminosity, radius, and chemical composition.

ENERGY CONCEPTS

Energy is the ability to do work and transform matter:

Stars shine by transforming matter to energy.

Energy comes in many forms such as potential, kinetic, thermal, radiative:

Gravitational potential energy ignites stars.

Stars emit radiative energy carried by photons.

Mass is energy ($E = mc^2$); thermonuclear reactions in the cores of stars release energy from matter.

Transformations occur among the various forms of energy, and energy can be transferred from one location to another:

Atoms and molecules absorb and emit light, usually at unique energies to produce stellar spectra.

Thermal energy flows from hotter to cooler regions; in most stars, this flow takes place by radiation or convection; in degenerate stars, by conduction.

Mass and energy warp spacetime; a black hole is a severely curved region of spacetime.

MOTION AND FORCE CONCEPTS

Newton's laws of motion plus his law of gravitation explain and predict the motion of the masses in the stellar systems in a unified way:

Kepler's laws are an alternative expression of forces and gravitation; they properly describe the motions of binary stars.

Einstein unified space and time in the cosmos:

The general theory of relativity predicts the existence of black holes with singularities inside.

To an outside observer, time slows down near a black hole.

QUANTUM CONCEPTS

On a very small scale, matter is made of discrete particles:

Matter at very high densities (degenerate matter) behaves in strange ways that apply to white dwarfs and neutron stars.

Thermonuclear reactions transform nuclei and release neutrinos.

On a very small scale, energy comes in discrete units:

Photons are released by thermonuclear reactions in stars.

Physical phenomena are governed by a few basic interactions of matter and energy, often at the quantum level:

Stellar spectra usually contain absorption lines.

Continuous stellar spectra resemble the spectra of blackbody radiators.

Synchrotron emission occurs when charged particles travel at high speed through magnetic fields.

Interstellar gas can emit by visible light (lines), continuous radio, millimeter and centimeter lines, infrared and ultraviolet lines, and by synchrotron radiation.

Improve Your Night Vision

WINTER

Fred Schaaf

Many people think that skies are clearer in winter because the stars look so bright. Actually, the winter constellations just happen to have the stars of greatest apparent brightness—five of the seven brightest visible from 40° N latitude on earth appear in one relatively compact group of these constellations.

The center of the winter host of bright star patterns is Orion, with his belt of three almost equally bright, equally spaced stars in a row. If we start with the easternmost of the three belt stars and work our way out on a spiral that includes most of winter's brightest stars, we can in appropriate order visit examples of each spectral type.

"Spectral type" designates the categorization of stars according to the appearance of the chemical spectrum of their light. The progression of types is usually given from hottest to coolest stars, and the types in this order are represented by the following sequence of letters: O, B, A, F, G, K, M.

The easternmost star in Orion's belt is Alnitak (Zeta Orionis), a star of spectral type O. The color is white or bluish-white, the surface temperature as hot as that of almost any star. To the southwest of Alnitak is much brighter Rigel, the most famous of *blue supergiants,* but slightly less hot and classified as a type B star. Orion is so bright because most of its naked-eye stars are members of an *OB association,* a vast collection of very young, very hot, very bright O- and B-type stars.

From Rigel we curve out to Sirius, by far the brightest star in all of the heavens. White, with a less noticeable tinge of blue (some say none), Sirius is a type A star, much cooler than Rigel. Follow the spiral to Procyon (a white or slightly yellow-white type F), then to Capella (yellow or more accurately yellow-white type G, like the sun), then to Aldebaran (orange or light orange type K), and finally back to Orion and Betelgeuse (red or deep orange type M). Betelgeuse is the most famous, the biggest, and the brightest of the *red supergiants,* with a surface temperature of only about 3600 K, little more than half that of the sun and less than a tenth that of type O stars). Betelgeuse is known to vary in brightness, not very regularly, over periods of

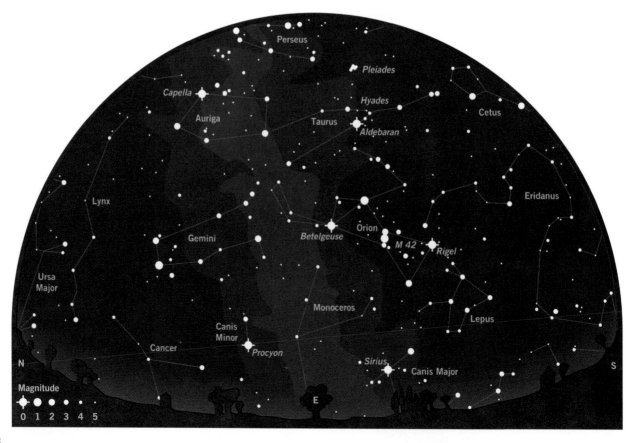

months and years—how bright does it look compared to Aldebaran, Procyon, or Rigel tonight?

We still can't tear our gaze away from the center of the winter host, the glorious Orion. Around one of the stars in the roughly north–south line of them just below the belt—a line marking the sword of Orion—a careful naked-eye look reveals a hazy glow. Binoculars begin to show a luminous tuft. Medium-sized amateur telescopes reveal a big fan of radiance gleaming mostly green, structured with wisps and filaments of light, sprinkled with numerous stars, extended out from around a tight knot of four easily visible bright stars. The entire structure of glorious gas is called the Great Nebula in Orion.

A *nebula* is a cloud of gas and dust in interstellar space, but there are various kinds. This one happens to be our most splendid example of a diffuse nebula, a birthing place of stars. The O and B stars of Orion were born very recently, as astronomical reckoning goes, out of nebulosity like that of the Great Nebula. We now have strong evidence that starbirth is going on inside the Great Nebula even as we watch.

Before leaving Orion with its mighty young blue supergiants and supreme but rapidly aging red supergiant, we should note that both Sirius and Procyon have companion stars that are among the smallest stellar corpses— *white dwarfs*. Red giants eventually lose their outer layers (sometimes explosively, as a *supernova*) and eventually become either *black holes, neutron stars,* or *white dwarfs*. The companion of Sirius, only about the width of earth (it's roughly the mass of the sun!), is fairly bright but is overwhelmed by the nearby glow of Sirius itself, so few amateurs have gotten to see it in their telescopes.

About the only way to escape from enthrallment by Orion and Sirius is to cast your gaze over at the two great *open star clusters* of Taurus, the Hyades and the Pleiades. The open clusters (also called galactic clusters) are far less tightly condensed than the older globular clusters and contain far fewer stars—usually only hundreds or dozens, not hundreds of thousands or millions. But there is no globular cluster within 5000 ly of earth, whereas innumerable open clusters are closer (partly because they are so much more common in the Galaxy), and the Hyades and Pleiades are about 130 and 410 ly away. No wonder then that we can see the brightest of their individual stars easily with the naked eye, and the clusters themselves sparkle like gorgeous mini-constellations or asterisms of their own.

The Hyades form a V shape or arrowhead of stars with the much brighter star Aldebaran. Aldebaran, which lies about twice as close to us, is not a member of the cluster. Scan through the Hyades with a pair of good binoculars and note what colors you see. Do you notice many blue stars? Do you think that the Hyades is a very young cluster?

Trying the same binocular scan with the Pleiades will show you almost no stars that are not bluish-white or white. In fact, the Pleiades are young enough ("young" might be 100 million years old or somewhat less) for careful telescopic observers to detect lingering traces of the nebula that produced them. Of course, the most famous observational test with the Pleiades is counting how many of them you can see with the naked eye on various nights. As we will find in our essay for spring, there is now a special and very practical new reason for performing this test.

PART FOUR

GALAXIES AND COSMIC EVOLUTION

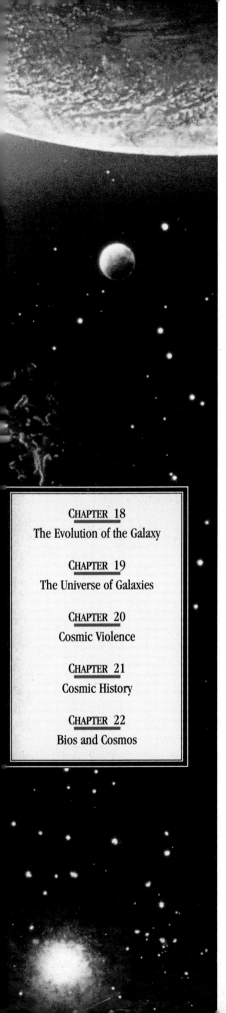

How are the stars arranged in space? Astronomers have found in this century that the sun resides in a vast pinwheel of stars—more than 100 billion of them. Exceeding 120,000 ly in diameter, this vast system is the Milky Way Galaxy. Chapter 18 deals with our Galaxy, based on our understanding of the physical nature and evolution of stars.

Beyond the Galaxy, billions of other galaxies inhabit the remote reaches of space—the universe is truly a universe of galaxies, as described in Chapter 19. New telescopes have revealed that many galaxies, especially in their cores, generate energy violently. Some appear to be blasting matter outward in narrow jets that contain particles traveling close to the speed of light. This violent face of the universe (Chapter 20) is an important subtheme of cosmic evolution.

Chapter 21 investigates the physical evolution of the universe since the Big Bang. As the cosmos expanded and cooled, various kinds of matter froze out in forms that we can see today—or may not have seen yet! Here we find the closest connection between the smallest pieces of the cosmos—elementary particles—and the universe as a whole. That connection is made at the Big Bang. Then another look at the future of the universe, and the evidence so far that indicates that it will expand, more and more slowly, forever. The cosmos started with a bang but may end with a whimper. However, the cosmos may be filled with a large fraction of dark matter that would force the cosmos to collapse in a Big Crunch!

The evolution of the universe, stars, and galaxies sets the stage for the chemical and biological evolution (Chapter 22) that resulted in ourselves and other living creatures on the earth. Natural processes bind simple chemicals, abundant in the cosmos, into more complex substances that serve as the foundation of life. From those complex molecules, simple life formed early on in the history of the earth. If that process crowns the course of cosmic evolution, then other planets in our Galaxy should have life on them, too. ■

PART OBJECTIVE: To be able to tie together the major themes of cosmic evolution in discussing the origin and history of the cosmos and of life on the earth.

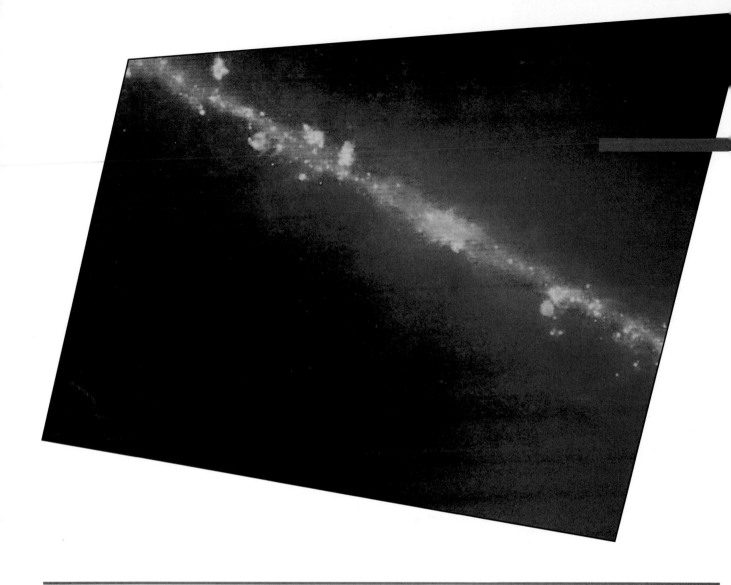

The Evolution of the Galaxy

A broad and ample road, whose dust is gold,
And pavement stars, as stars to thee appear
Seen in the galaxy, that milky way....

JOHN MILTON, *Paradise Lost*

Like a majestic cosmic pinwheel, our Milky Way Galaxy spins slowly in space. Our sun orbits the center of it along with more than *100 billion* other stars. Because the sun is about 30,000 ly from the nucleus, it completes one revolution roughly every 250 million years.

What is the three-dimensional structure of this enormous system of stars? Optical astronomers probe the structure near the sun, and radio astronomers study regions farther away. The details are not yet in, but astronomers have been able to establish the broad outlines of the Galaxy's structure—a remarkable achievement, considering that we are buried within the Galaxy. These investigations show that the Galaxy does have a spiral pattern.

This overall spiral pattern of the Galaxy evolves. Theoretical work sees the grand design as driven by waves; it links many of the galactic entities in the disk by the process of galactic evolution. The Galaxy's structure evolves—but at a rate so slow that we won't see any changes in our lifetimes.

The Galaxy looked very different in the past, especially just after its formation. The evolution of stars, interstellar matter, and the nucleus has driven significant changes. This chapter takes a brief excursion into galactic history to see how our spiral Galaxy evolved.

18.1 The Galaxy's Overall Structure

Astronomers study the structure of the Galaxy to infer its past and predict its future. Our current model is pretty good. It pictures the Galaxy in three main parts: a central region called the nuclear bulge, a disk, and a halo (Fig. 18.1). The **disk,** the main body of the Galaxy, has a diameter of some 120,000 ly and a thickness of about 1000 ly. Population I stars and interstellar clouds of gas and dust inhabit the disk, the gas extending out farther than the stars. The sun resides

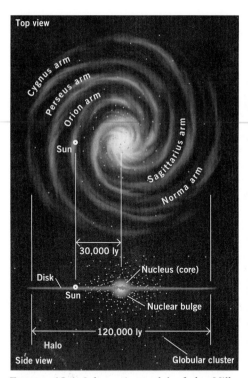

FIGURE 18.1 Schematic model of the Milky Way Galaxy, showing its main features: nucleus, halo, and disk. The sun lies in the disk, about 30,000 ly from the center, on the inner edge of a spiral arm. The realm of the globular clusters defines the halo, shown here only in part. The nuclear bulge in the center surrounds the core.

FIGURE 18.2 Wide-angle view of the central bulge and the disk of the Galaxy. This infrared image was taken by the *Cosmic Background Explorer* (*COBE*). The false colors display the emission at 1.2 μm (blue), 2.2 μm (green), and 3.4 μm (red). Stars in our Galaxy dominate the emission at these wavelengths. The region shown spans 168° across, or almost half the sky. (Courtesy of NASA/Goddard Space Flight Center.)

a little above the disk at a distance of approximately 30,000 ly from the Galaxy's center.

The **nuclear bulge** encases the central regions of the Galaxy, including the mysterious **nucleus,** the very heart of the Galaxy. The bulge is about 12,000 ly in diameter and 10,000 ly thick; it contains old Population I stars. Infrared observations from space clearly disclose the nuclear bulge (Fig. 18.2). The spherical **halo** encircles the nuclear bulge and the disk. Globular clusters, containing Population II stars, make up the most obvious material in the halo, which has a diameter of at least 120,000 ly.

Stars of different types (Section 16.7) inhabit each part of the Galaxy: young, metal-rich Population I stars in the disk; old, metal-rich Population I stars in the bulge, and very old, metal-poor Population II stars in the halo. You will find this distribution a key clue about the Galaxy's history.

We are in a bad position to observe the Galaxy's structure, for we reside in the Galaxy's dusty disk. Imagine, for example, that you are watching the half-time show at a football game. The band has set up an elaborate formation, and from the stands, you can easily tell what the formation represents. But suppose you were down on the field instead, at the edge of the formation. It would at first appear to be a jumble! You could eventually figure out the shape if you could find the distances to all the band players. You could make a map of their positions by plotting the distances and directions of each person. But suppose you had to do this mapping in a fog so dense that only the closest people were visible. Then you would need some other method of estimating distances—perhaps by the intensity of the sounds of the instruments that came through the fog.

We believe that the disk of the Galaxy contains a few spiral arms. A **spiral arm** extends for thousands of light years and contains many O and B stars, gas, and dust with a density higher than in the region between arms. The spiral arms contain most of the gas, dust, and young stars in the Galaxy. Near the sun, for example, about half the matter is in gas and dust, the rest in stars. In contrast, for the

Galaxy as a whole, only a few percent of all the material is in the form of gas and dust.

Optical astronomers, who try to find the distances to spiral arms, find their view blocked by interstellar dust. Radio waves get through. So radio astronomers can pick up the radio emission from clouds of gas that probably mark spiral arms. But they have more difficulty in determining distances than the optical astronomers do. Although the results from the two techniques do not agree in all details, the following structure has been uncovered. We infer that our Galaxy also has two major arms (interior to the sun's orbit) and perhaps four arms (exterior to it) wound around the nucleus. The coherent, spiral pattern is clearest in these outer parts. Other galaxies exhibit a spiral design (Fig. 18.3).

■ Which general types of star are located in the Milky Way's halo?

18.2 Galactic Rotation: Matter in Motion

The sun and its nearest neighboring stars move in a variety of directions at a variety of speeds. In relation to *nearby* stars, the sun travels at a speed of about 20 km/s. These stars and the sun have roughly similar orbits. But how to find out what motion the sun and nearby stars share in relation to the center of the Galaxy? And what about the orbital motions for matter in other parts of the Galaxy?

The Sun's Speed Around the Galaxy

We use a variety of indirect approaches to find out how fast the sun moves around the Galaxy. One method utilizes the motions of globular clusters. The globulars seem to orbit the Galaxy in random orbits, with a roughly spherical distribution around the nucleus (Fig. 18.4). With respect to the nucleus, the average motion of all globular clusters is roughly zero. In other words, the system of globulars has no overall rotation about the galactic center (although individual clusters move rapidly, some in one direction, some in another).

FIGURE 18.3 A spiral galaxy with well-defined spiral structure (NGC 2997). Note the differences between the colors in the nucleus in those in the spiral arms. The bluish color of the arms comes from concentrations of young, hot stars. (Courtesy of David Malin; AATB.)

A study of the radial velocities of the globular clusters by Doppler shifts enables us to find the sun's motion with respect to the system of globulars. Now, the system of globulars has *no* rotational motion with respect to the nucleus. So the sun's motion relative to the system of globulars is its rotational motion with respect to the Galaxy's center. Such an analysis shows that the sun revolves at about 250 km/s.

The exact value of the sun's speed around the Galaxy is very hard to determine and is subject to large errors. The value usually agreed upon is 250 km/s. But recent observations indicate that it is probably smaller, perhaps 220 km/s; this book uses 220 km/s as the sun's orbital speed.

The Galaxy

At a Glance

Diameter of disk	120,000 ly
Diameter of halo	300,000 ly
Sun's distance from center	30,000 ly
Total mass	About 10^{12} solar masses
Age	About 15×10^9 years

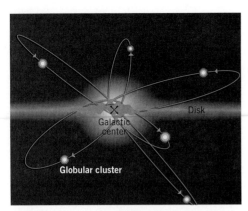

FIGURE 18.4 Schematic picture of the orbits of globular clusters around the galactic center. Each globular moves on a highly eccentric orbit, so that most of its time it lies far from the nucleus, as expected from Kepler's second law. The far ends of these orbits define the spherical halo of the Galaxy.

The Sun's Distance from the Center

How far is the sun from the Galaxy's center? That's also a tough question to tackle because we cannot see the center optically. However, we can look above and below the galactic plane, where the obscuration is less, to observe objects thought to be symmetrical about the galactic center.

For example, we can use the *RR Lyrae variables* found in globular clusters. These are Population II stars; they have periods of light variation that typically are about a half-day in duration (Section 16.7). No matter what their period, RR Lyrae stars have essentially the

same mean luminosity: 50 solar luminosities. So once an RR Lyrae star has been identified by the shape of its light curve, we can easily calculate its distance from a measurement of its flux and the inverse-square law for light. By this technique, recent work has found the sun's distance from the galactic center to be approximately 28,000 ly, within a range of 24,000 to 33,000 ly. I'll use 30,000 ly; but you must realize that the measured distance to the center is somewhat uncertain. Other observations indicate a distance perhaps as small as 23,000 ly.

This technique of using RR Lyrae variable stars in globular clusters to find the distance of the sun from the Galaxy's center was developed by Harlow Shapley (1885–1972) around 1915. It marked a crucial step in our understanding of the Galaxy and the sun's location within it. Shapley's argument rests on the observation of the distribution of globular clusters, which concentrate in the southern sky (Figs. 18.5 and 18.6). From what vantage point do we, whirling around the sun, view these groups of stars? Shapley knew that our Galaxy has the shape of a flattened disk. He assumed that the globular clusters had a uniform distribution around the Galaxy's nucleus in a huge sphere that outlines the halo.

Now to infer the size of the Galaxy and the sun's location. Suppose the sun were located in the center of the Galaxy (Fig. 18.7*a*). Trace lines of sight in a number of directions. Because we have assumed a central vantage point in a uniform distribution of objects, every line of

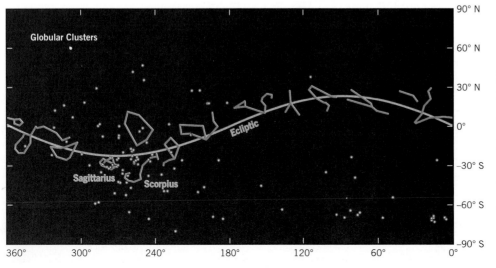

FIGURE 18.5 The concentration of globular clusters in the region of the sky toward the center of the Milky Way. This full-sky view shows the locations of the brightest globular clusters from north to south around the entire sky. Each dot marks one globular cluster; also indicated are the zodiacal constellations and ecliptic. Note the concentration near Scorpius and Sagittarius. (Diagram generated by *Voyager* software for the Macintosh, Carina Software.)

FIGURE 18.6 The concentration of globular clusters in the region of the sky toward the constellation Sagittarius. Each white circle marks a globular cluster. About one-third of all known globulars are in this region, which is only 2 percent of the entire sky. In other directions, the number of globulars is much less for same-sized areas of the sky. (Courtesy of Harvard College Observatory.)

sight you chose should intercept the same number of globular clusters. So, the expected distribution should be uniform over the sky. But this uniformity does not in fact exist.

Shapley therefore chose to give up the central location of the sun in the Galaxy (Fig. 18.7b). Now some lines of sight cut longer distances than others through the globular clusters, so more clusters are seen in these directions than along other lines of sight. Thus instead of being symmetrical, the expected distribution is most heavily concentrated *in the direction of the galactic center.* So with the sun away from the center, the predicted result matched the observational one.

The distances to the globular clusters can be worked out from RR Lyrae stars. Then the size of the sphere of globulars marks the extent of the halo (and so the size of the Galaxy), and the distance of the sun from the center of this sphere indicates the distance of the sun from the Galaxy's center.

Rotation Curve and the Galaxy's Mass

Knowing the sun's velocity and distance, we apply Kepler's third law to deduce

the mass of the Galaxy (Focus 18.1). The result, about 10^{11} solar masses, refers *only to the mass interior to the sun's orbit.* What about the mass outside it?

With the sun's orbital distance and velocity known, we find the **galactic rotation curve**—of how fast an object some distance from the galactic center revolves around it. The rotation curve tells us the overall distribution of matter in the Galaxy, because gravity controls those orbital motions.

Imagine that the Galaxy's mass is mostly concentrated in the nucleus—a setup that resembles the solar system. Although the planets exert mutual attraction, each orbits about the sun following Kepler's laws (Fig. 18.8). The stars in the galactic disk should revolve about the nucleus of the Galaxy much as the planets revolve around the sun. The stellar motions should follow Kepler's laws (Section 3.5), and the orbital speeds of the stars should decrease with increasing distance from the Galaxy's center.

In fact, the orbital speeds don't follow Kepler's laws well (Fig. 18.9), and that tells us an important fact about the Galaxy: the major part of the mass is *not* concentrated at the center! From close to the center out to 1000 ly, the curve rises steeply, then drops, bottoming out at about 10,000 ly. It then rises slowly out to the position of the sun, where it seems to drop again. In the outer parts of the Galaxy, we know from carbon monoxide observations of molecular cloud complexes that the curve rises beyond the sun's orbit. It then appears to flatten out at a distance of 50,000 ly from the galactic center. If the motions followed Kepler's laws, we would expect the

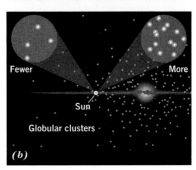

FIGURE 18.7 The position of the sun in the Galaxy inferred from the observed distribution of globular clusters in the sky. Assume that the globulars have a uniform distribution around the center of the Galaxy. (*a*) The situation if the sun were in the center of the Galaxy: you would see roughly the same number of globulars in every direction in the sky. (*b*) The actual situation, with the sun away from the center: more globulars are visible in the direction of the center than in other directions.

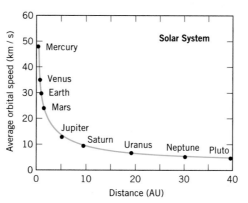

FIGURE 18.8 Solar system rotation curve for the planets. This is Keplerian rotation, inasmuch as almost all the mass of the solar system is concentrated in its center—the sun. Note how rapidly the orbital speeds decrease with distance.

Enrichment Focus 18.1

THE MASS OF THE GALAXY

The sun swings around the Galaxy at 220 km/s at a distance of 30,000 ly from the center. Let's use this information to estimate the mass of the Galaxy. Assume that the sun moves in a circular orbit. Apply Kepler's third law in the form used for binary star systems (where $k = 1$) to the Galaxy:

$$M_1 + M_2 = \frac{R^3}{P^2}$$

where R must be in AU and P in years. The mass then comes out in solar masses. At 220 km/s it takes the sun about 2.8×10^8 years to complete a circuit of the Galaxy. Now R is roughly 30,000 ly and 1 ly = 6.32×10^4 AU; so 30,000 ly = 1.9×10^9 AU. Then

$$M_1 + M_2 = \frac{(1.9 \times 10^9)^3}{(2.8 \times 10^8)^2}$$

$$= \frac{6.9 \times 10^{27}}{7.8 \times 10^{16}}$$

$$= 0.9 \times 10^{11} \text{ solar masses}$$

What is $M_1 + M_2$? It is the mass of the sun plus the mass of the Galaxy. The mass of the sun is so small compared with that of the Galaxy (just look at the result!) that we ignore it. So the Galaxy's mass *interior* to the sun's orbit is about 10^{11} solar masses. The flat rotation curve implies that much more lies beyond the sun's orbit. We need to observe farther out until we see a decline in the rotation curve. Then we can accurately calculate the total mass of the Galaxy.

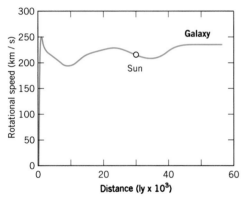

FIGURE 18.9 Galactic rotation curve, based on a combination of carbon monoxide and hydrogen observations, assuming that the sun's speed is 220 km/s and its distance from the galactic center is 30,000 ly. Note that the rotational speeds do not follow a downward trend outside the sun's orbit. (Adapted from a diagram by D. P. Clemens.)

rotation curve to decline beyond the sun's orbit, and it does *not*.

What does this rotation curve say? *Much of the Galaxy's material must lie out beyond the sun's orbit.* From the rotation curve out to 60,000 ly, the Galaxy's mass is 3.4×10^{11} solar masses. Other estimates give about twice this value, close to 10^{12} solar masses. So at least as much mass lies exterior to the sun as interior to it.

■ **How are RR Lyrae stars used to find the sun's distance from the center of the Galaxy?**

18.3 Galactic Structure from Optical Observations

Photos in blue light taken of nearby spiral galaxies, such as Messier 31, the Andromeda Galaxy, show the bluish spiral arms distinctly. This appearance results from bluish O and B supergiants, H II regions, and Population I cepheids that cluster in the spiral arms (Fig. 18.10). These objects are called **spiral tracers.**

Spiral Tracers and Spiral Structure

How to apply these spiral tracers to our Galaxy to determine the layout of its spiral arms near the sun? It's not a simple operation! First, it requires an accurate and reliable technique for measuring the distance to each of the tracers. Second, optical observations are restricted by the blotting out of starlight by dust: Since most of the interstellar dust lies concentrated in the galactic plane, the sun sits in the thick of the interstellar smog. Third, the sun's location in the plane gives us a poor vantage point for seeing the Galaxy's spiral structure because we are forced to observe it edge-on rather than face-on.

Let's see, for example, how cepheid variables have been used to delineate spiral features. The cepheids have the advantage that their distances are easy to determine by the **period–lumi-**

FIGURE 18.10 Optical spiral arm tracers. Spiral arms of Messier 31, the Andromeda Galaxy. Bright O and B stars define the spiral arms; dark, dusty regions line the inner edges of the arms. These serve as spiral-arm tracers. (Courtesy of Jim Riffle, Astro Works Corporation.)

nosity relationship. This crucial distance-measuring technique is so important that you should know how it works. (It will appear in later chapters.)

A variable star is one whose luminosity (and so flux) changes with time. A *light curve* is a graph of the change in a star's flux with time. A star whose light varies in a regular fashion is known as a *periodic variable;* Delta Cephei sets the standard for one such class of variables—called *cepheid variables* or *cepheids* (Section 16.7)—by the special shape of their light curves (Fig. 18.11). From Doppler shift observations, we know that cepheids actually expand and contract as they vary in luminosity.

For the cepheid variables, we find that a special relationship, the *period–luminosity relationship,* connects the luminosity of a cepheid variable to its period (Fig. 18.12). Basically, it states that the *longer the period of light variation of a cepheid, the more luminous it is.*

Here's how we can use the relationship to find distances to cepheids.

1. Find a cepheid (identifying it by its light curve).

2. Measure its period of light variation, from peak to peak.

3. Find the star's average luminosity from the period–luminosity relationship.

4. Measure the star's flux by telescopic observations.

5. Calculate its distance from the inverse-square law for light.

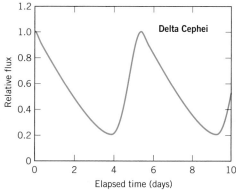

FIGURE 18.11 A light curve for a typical cepheid variable star—Delta Cephei, the prototype of the class—shows how the star's brightness varies with time. Note that the rise in brightness is steep, but the decline less so. The period of this wavelike brightness variation is about 5.4 days.

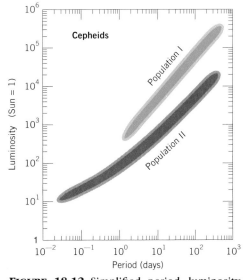

FIGURE 18.12 Simplified period-luminosity relation for cepheids, which fall into two groups—Types I and II, based on stellar population. In both cases, the general trend is the same: the longer the period, the more luminous the star. By comparing the observed flux to the luminosity estimated from the period, you can use the inverse-square law for light to find the distance to the cepheid.

Today we know that the stars that used to be lumped together as cepheid variables are actually *three* types: Type I (classical) cepheids, Type II cepheids, and RR Lyrae stars. Type I cepheids are Population I stars, Type II belong to Population II. RR Lyrae stars, also Population II, are commonly found in globular clusters. (Shapley used these and the technique described here to get the distances to globulars: Section 18.2.)

Optical Maps of Spiral Structure

The optical maps of spiral structure must be viewed with caution because interstellar dust restricts where and how far we can look from the sun and how well we can estimate distances to optical tracers. If dust is in the way, the measured flux is decreased, and we would estimate a *larger* distance than the actual distance. Although the outline of spiral structure is likely correct, observations of other galaxies show that irregularities commonly occur. It's futile to draw a master diagram from optical data alone, which works out to distances of only some 15,000 ly.

Despite disagreements about the details, most optical astronomers concur that their investigations have disclosed at least three major arm segments spaced about 7000 ly apart. The Galaxy appears to have a spiral structure with much irregularity in the general pattern, which may consist of two or four spiral arms—but we cannot tell for certain yet which is correct.

■ What is the physical activity of cepheids that results in their variability?

18.4 Exploring Galactic Structure by Radio

The prime drawback to optical mapping of the Galaxy is obscuration from interstellar dust. Radio observations do not have this handicap because dust does not easily stop radio waves. Radio astronomers can reach far beyond the restricted range of the optical astronomers. The 21-cm line from hydrogen (Section 15.1), which comes from the concentrations of neutral hydrogen clouds in the spiral arms, is useful to radio astronomers as a spiral tracer; the best radio tracer seems to be the carbon monoxide millimeter-line emission from giant molecular clouds.

Radial Velocities and Rotation

To distinguish among spiral arms, we look in different directions and at different *velocities* in the same direction. The 21-cm radiation from H I clouds arrives at the earth Doppler-shifted to different wavelengths because of the different velocities of the hydrogen gas clouds. These differences in velocities come mostly from the rotation of the Galaxy. So *if* we know how the Galaxy rotates (and that's the tricky part), we can translate 21-cm observations into a map of spiral structure.

What technique do radio astronomers use? They look for 21-cm radiation from specific H I clouds. Because of galactic rotation, the clouds along the line of sight have different radial velocities, hence different Doppler shifts, so their signals are received at slightly different wavelengths.

Here's a case looking outward from the sun (Fig. 18.13). Our line of sight intercepts two nearby clouds at succes-

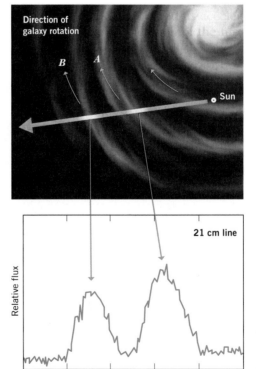

FIGURE 18.13 Using 21-cm emission to trace out spiral structure. Consider looking out of the Galaxy through two spiral arms (*A*, and *B*). The sun is overtaking A and B. So the radial velocity difference between us and *A* is smaller than that between us and *B*, and the blueshift from *A* is less than that from *B*. The emission from both is blueshifted to wavelengths shorter than 21.11cm.

sively greater distances. The inner cloud travels faster than the outer one. In its revolution around the galaxy, the sun is approaching both clouds. So the difference between the sun's velocity and the innermost cloud's velocity is the least, and its Doppler shift is the least. In contrast, the difference between the sun's velocity and velocity of the outer cloud is the most, so it has the greatest Doppler shift. If each of these clouds corresponds to a piece of a spiral arm, how to arrange them at different distances from the sun?

Assume nearly circular orbits. We know the sun's speed around the Galaxy. When we observe 21-cm emission, we know the direction in which we are looking and can measure a radial velocity. From that radial velocity, we infer the rotational speed of the cloud. We then look up this speed on the galactic rotation curve to find the distance to which it corresponds.

In essence, the galactic rotation curve tells us that at a given position in the sky, a certain radial velocity corresponds to a specific distance from the sun. Scanning around the plane of the Galaxy, we make a series of 21-cm observations. These can be connected to trace a spiral arm and so outline the Galaxy's structure. The same technique applies to the millimeter emission from molecular clouds.

Radio Maps of Spiral Structure

Conflicting radio maps have been drawn by different investigators. The heart of the problem is that the neutral gas clouds don't follow the simple scheme of circular rotation; in addition to their circular motion, they have their own random motions. Unfortunately, such noncircular motions lead to incorrect distances and so disrupt the unity of the spiral-arm map.

Despite such problems, the radio maps do hint at a large-scale spiral structure. At distances greater than 15,000 ly, radio astronomers probe regions inaccessible to optical astronomers. The two techniques complement each other. And the structure of the Galaxy derived from them shows through in its broad outline, although it is messy in the region interior to the sun's orbit.

Beyond the sun, the spiral-arm pattern appears more clearly, thanks to observations of giant molecular clouds at millimeter wavelengths. And the emission lines from molecules make Doppler shift observations possible. Using the same technique as for the 21-cm neutral hydrogen line, and complementing those observations, recent work shows four arms as overall structure (Fig. 18.14).

This technique has clarified the spiral-arm picture using giant molecular clouds with masses greater than 10^5 solar masses (Fig. 18.15). The Carina arm stands out most clearly. It stretches for more than 80,000 ly, with a giant molec-

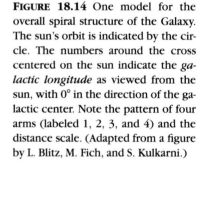

Spiral Tracers

At a Glance

Optical	OB stars, H II regions, cepheid variables
Radio	21-cm emissions from H I regions, millimeter from CO
Results	Four main spiral arms
Caution	Limited areas mapped so far

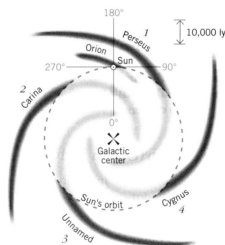

FIGURE 18.14 One model for the overall spiral structure of the Galaxy. The sun's orbit is indicated by the circle. The numbers around the cross centered on the sun indicate the *galactic longitude* as viewed from the sun, with 0° in the direction of the galactic center. Note the pattern of four arms (labeled 1, 2, 3, and 4) and the distance scale. (Adapted from a figure by L. Blitz, M. Fich, and S. Kulkarni.)

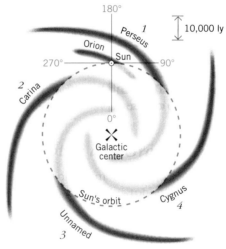

FIGURE 18.15 Giant molecular clouds used to outline spiral structure. Radio observations at 2.6 mm were used to infer the positions of giant molecular clouds from their Doppler shifts. The sizes of the circles indicate the masses of the cloud complex (see legend at bottom). Note how well these clouds delineate the Carina, Sagittarius, and Perseus arms. The empty areas have basically not yet been surveyed. (Observations by R. S. Cohen and colleagues.)

ular cloud roughly every 2000 ly along the segment. The Perseus and Sagittarius arms are also delineated but overall contain far less mass in molecular clouds. Overall, the maps from the giant molecular clouds probably provide the best portrayal of the Galaxy's spiral pattern.

■ At what distance in the Milky Way does our visual observing ability cease?

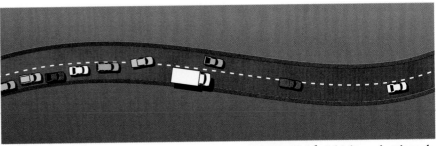

FIGURE 18.16 A highway bottleneck as an analogy to a density wave. The slow-moving truck has cars jammed up behind it waiting to pass. The blockage consists of different cars at different times, but it is always behind the truck. Viewed from above, the jammed-up area moves more slowly than the average speed of the cars; it is a density wave.

18.5 The Evolution of Spiral Structure

For many years astronomers supposed that the spiral arms in our Galaxy and others were *material* arms, a coherent bunch of objects—stars, nebulas, gas, dust—somehow physically held together. Such a point of view faced two questions: What holds the material in an arm together? How does an arm persist for a long time?

The Windup Problem

The persistence of an arm is hard to explain because it should wind up; some parts of the arm rotate more slowly than others. So after a few rotations of the Galaxy, the arms should have disappeared. The Galaxy has turned about twenty times since the origin of the solar system, yet the arms are still there!

Here's an analogy to this windup problem. Imagine you and two friends are going to run around a track. You station yourself in the middle, one friend a few meters in from you, closer to the track's center, and the other a few meters out from you. You start running, lined up. Now insist that the friend inside run around faster than you and the one outside slower. You can guess what will happen after one or two laps: the lineup will be disrupted. If spiral arms were material arms, the same thing would happen to them after a few rotations.

Astronomers have been struck by the persistence of spiral arms. We can't tell this from our Galaxy alone, for it could be that we are observing at a very special time, soon after the formation of the arms. But that is not likely to be true

for all galaxies. Of the brightest galaxies in the sky, more than 60 percent have coherent spiral arms (Fig. 18.3).

How does this tell us that spiral arms are persistent phenomena? Recall that as we look out in space, we look back in time. Suppose that a galaxy's spiral structure did in fact disappear by winding up after a few rotations. Assume, too, that all galaxies formed at the same time. Then we should *not* see spiral structure in nearby galaxies, but we should see it in more distant galaxies. That's not the case! Both nearby and distant galaxies exhibit spiral structure. Seeing so many galaxies as spiral in shape implies that the structure lasts for at least some few billions of years or that it is regularly renewed.

The Density-Wave Model

How to explain this persistence? The view today pictures spiral arms not as material arms at all but rather as the result of spiral waves of higher density moving through the Galaxy's disk. These waves produce all the signposts of a spiral arm—young stars, H II regions, lanes of dust. None of these objects lasts very long. As they die and the density wave moves on, new spiral-arm tracers are born from the interstellar medium at the new location of the wave. So a spiral arm always contains objects of the *same kinds,* but not the *same objects.* Any particular arm is a transient phenomenon. Individual objects revolve at the speed determined by gravitational forces appropriate for their distance from the center, but the *wave* pattern rotates with a constant angular speed and does not wind up. This approach is called the **density-wave model** of spiral structure. Of the models proposed to date to explain spiral structure, it best describes the overall scheme.

What's a density wave? A sound wave is a density wave. Clap your hands. You push against air molecules to compress them together. This first group of molecules bangs into adjacent ones in the direction of their motion, which transfers the compression to the next bunch of molecules. As this compression (density wave) travels forward, it leaves behind a trough of lower density. Two important points here: a sound wave requires a source to start it, and the high-density part of the wave persists even though the specific particles that make it up change at different points in the medium.

Here's an analogy: suppose you are driving on a heavily traveled mountain road (Fig. 18.16). Everyone moves along happily at the speed limit. Ahead, an overloaded truck can go about half the maximum speed. Cars jam up just behind the slow-moving truck as the drivers wait for a clear road ahead in order to pass. When they do pass, they resume moving along at the speed limit, leaving the poor trucker behind. Imagine that you watched this situation from the air and concentrated on the motion of the cars. You'd see a denser region of traffic just behind the truck, where the cars pile up for a short time; the truck moves down the road at half the average speed of the cars. You would also note that the jam-up persists, even though it does not contain the same cars. New cars get caught up in it as other cars move out.

You can think of the cars as the stars, the interstellar medium in the galactic plane, and the jam-up as the visible effect of a moving density wave (the truck). The jam-up creates a region in the disk of increased density of stars and gas—that's a spiral arm.

This density-wave idea *assumes* that a two- or four-armed spiral density wave sweeps through the galactic plane. Although the origin of this wave is not explained, once formed, it persists for a billion years or so. The gas in the disk piles up at the back of the wave. The compression squeezes small molecular clouds together to form giant molecular cloud complexes, which in turn form young stars and H II regions. The compression of the interstellar medium by the density wave forms the features associated with a spiral arm.

Caution. Just because I've been emphasizing luminous O and B stars (because they are good tracers of spiral arms), don't be misled into thinking that a density wave initiates the formation of *only* O and B stars. It probably prompts the formation of stars with all possible masses. The massive ones die quickly compared with the less massive ones, which are far more numerous. The less massive stars remain between the spiral arms after the density wave has moved on, but they are less conspicuous because of their low luminosities.

During the short lifetimes of the newly formed O and B stars, a density wave moves only a short distance. So these stars, while they last, mark the spiral arm clearly. As a density wave moves on, it provokes the formation of more stars. These take the place of the ones that had rapidly faded out. So the spiral arms are maintained by density waves. They generate a spiral wave pattern by making regions of higher gravitational forces and so higher densities of matter in the Galaxy's disk.

Status of the Density-Wave Model

How well does the density-wave model describe the observed spiral structure? First, it outlines the grand scheme of an overall spiral pattern that we observe in other galaxies and in our own. (About half of all spirals show this grand design.) Second, it explains the persistence of the spiral arms (over a duration of about a billion years) in spite of galactic rotation. Third, it predicts the general features of a spiral arm. So the density-wave model succeeds fairly well in explaining the prominent features of spiral structure.

However, the model so far falls flat on a few points. It does not explain the origin of the density waves. Nor does it clearly work out what sustains them. As the density waves ripple through the interstellar medium, they lose energy and should dissipate in about a billion years. But as evidenced by the abundance of spiral galaxies, the density waves—if they are the correct explanation—last longer than a billion years or are renewed several times as the old waves fade out. Some mechanism keeps supplying energy to maintain them or to trigger a new series of density waves.

To sum up. Observations and computer calculations leave little doubt that spiral density waves exist and indeed are fairly common, but we don't yet know exactly how they come into being. This is a puzzle to be solved in the study of our Galaxy and others.

■ Why do we need a density-wave model to explain spiral structure?

18.6 The Center of the Galaxy

Although dust largely obscures the center of the Galaxy, a few regions of low absorption open up optical glimpses of the nuclear region (Fig. 18.17). In addition, we can surmise the nature of the stars in the nucleus from observations of the nuclei of other galaxies. Observations of these two kinds imply that the nucleus contains mostly old Population I stars densely packed together. So jammed are these stars that if you lived in the nucleus, the nighttime sky would be as bright as twilight on the earth. In fact, the total power output of the nucleus is about one *billion* solar luminosities!

Radio, infrared, and x-ray observations can all probe the nucleus. They show that the heart of the Galaxy is a bizarre and active place. Motions of gas here suggest a high concentration of mass at the very center—perhaps a black hole. In the very center of the Galaxy lies a radio source less than 13 AU (about 2 light *hours*) in size.

Radio Observations

Let's look first at the continuous emission of the galactic center (Fig. 18.18). An intense radio source lies smack in the direction of the center. It is called *Sagittar-*

FIGURE 18.17 The galactic center region. The area spanned covers about 40°. Note the dark lanes of dust cutting through the central region of the Milky Way's plane. (Courtesy of NOAO/KPNO.)

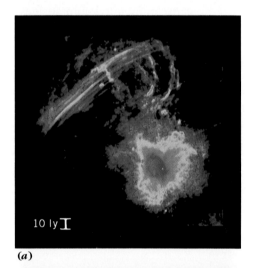

(a)

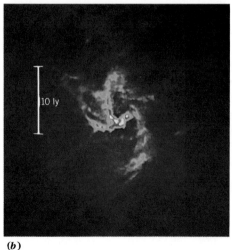

(b)

FIGURE 18.18 Radio views of the galactic center. Each of these false-color radio maps of the radio emission shows a smaller region. Blue represents the regions of lowest radio emission. (*a*) A 20-cm radio map of Sgr A (lower right), showing the strongest emission (red) from the center. Note the arc of filaments extending from the upper right to the left; each filament has a length of a few light years. The galactic plane crosses the image from upper left to the lower right. The filaments are perpendicular to the plane and parallel to each other. The halo-like emission (blue) around Sgr A is about 75 ly in diameter. The area is about 250 ly by 250 ly. (Observations with the VLA by F. Yusef-Zadeh, M. Morris, and D. Chance; courtesy of NRAO/AUI.) (*b*) High-resolution flux map of the Sgr A source made from VLA observations at a wavelength of 6 cm. White represents the strongest emission. Note the spiral-like structure of Sgr A. This map, which shows the thermal emission from this region, reveals details not visible in the Sgr A emission in (*a*). The area here covers about 30 ly by 30 ly. (Observations by K. Y. Lo and M. J. Claussen; courtesy of NRAO/AUI.)

ius A (*Sgr A* for short). Clustered around Sgr A—and all lying more or less along the galactic plane—are radio sources in a string. When investigated at different radio wavelengths, these sources appear to have characteristics of H II regions: hot ionized gas around young OB stars. The total extent of this region is about 300 ly by 850 ly, and the ultraviolet energy output from the OB stars needed to keep the region ionized is at least 5 million solar luminosities.

The Sgr A complex combines different radio structures. Some radio emission here is from ionized gas. But some also comes from high-energy electrons traveling through a magnetic field—synchrotron emission (Focus 17.1). High-resolution radio maps show that Sgr A actually consists of two separate radio sources: one, called *Sgr A East,* emits by the synchrotron process; the other, *Sgr A West,* seems to be more like a giant H II region. Sgr A West is associated with an agglomeration of infrared sources (to be explained shortly). Within Sgr A West lies a compact radio source, smaller than 13 AU, which appears to mark the actual core of the Galaxy; it is called *Sgr A*.* It shows synchrotron emission with a variable power of some 10^{28} W. Overall, the ionized gas here, which amounts to a few million solar masses of material, rotates at about a few hundred kilometers per second.

Observations of a larger region here show that the ionized gas loops in a filamentary bent arc (Fig. 18.18*a*). The arc may consist of material extending from Sgr A and guided in loops by local magnetic fields. A high-resolution map of the galactic center region at 6 cm (Fig. 18.18*b*) shows the thermal emission from a hot, ionized gas from the inner 10 ly of the Galaxy. A curious aspect of the emission is that it has a spiral shape (not to be confused with spiral arms in the disk!).

Sgr A also gives off radio-line emission from molecules such as CO. The observations indicate that this molecular cloud, which may contain as much as a million solar masses of material, is in front of Sgr A East, so it is not right at the center of the Galaxy. Also, the gas and dust appear to form a ring about 15 ly in diameter.

FIGURE 18.19 Infrared (2.2 μm), wide-angle image of the central 150 ly of the galactic center. The bright, white region near the center of the image is the nucleus. (Observations by R. Joyce and R. Probst at Kitt Peak National Observatory; courtesy of NOAO.)

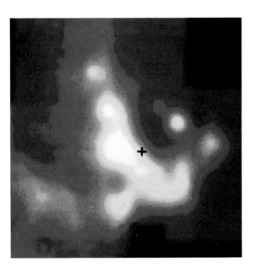

FIGURE 18.20 Close-up image of the galactic center at the infrared wavelength of 8.3 μm. The angular size of the area covered is 30 arcsec by 30 arcsec or about 5 ly by 5 ly. Note individual sources and the extended emission. This false-color map has white as the strongest emission and red the weakest. The cross marks the position of Sgr A*; note the lack of any strong emission there. (Courtesy of NASA/Goddard Space Flight Center.)

Infrared Observations

Early infrared observations showed that the galactic center region emits strongly at 2.2 μm (Fig. 18.19). The most intense part of this emission coincides with Sgr A. What is the source of this radiation? Simply the combined 2.2-μm emission from all the old Population I stars that inhabit the galactic nucleus. High-resolution observations have revealed many sources packed together around the nucleus (Fig. 18.20).

Observations of the same region near 8 μm look quite different; they show the infrared emission from dust that is heated by the radiation from stars (Fig. 18.20). Some of the heating radiation comes from the old Population I stars. But some also derives from high-luminosity O stars; the condensations in these maps are probably the locations of newly formed O stars. The combined lu-

The Galactic Center

At a Glance

Size	Inner 1000 ly
Luminosity	10^9 solar luminosities
Mass in gas	7×10^7 solar masses
Mass in stars	7×10^9 solar masses

minosity from them, in the range from 2 to 20 μm, is roughly a million times that of the sun. Over the entire infrared range, the galactic center emits about 100 million times the sun's luminosity!

X-Rays

The galactic center emits x-rays in the form of an extended x-ray source about 2° in size and pointlike, short-lived sources very close to the galactic center. Some of these sources give off bursts of x-rays, with the amount of energy in each burst comparable to that from the short-lived sources. To date it is not clear how or indeed whether these sources relate to each other.

Images from the *Einstein X-Ray Observatory* show the galactic center in its full, high-energy glory (Fig. 18.21). Within 300 ly of the center, the x-ray emission is modest in strength. It consists of a complex of weak sources (covering some 200 ly by 150 ly) embedded in a halo of weaker, diffuse emission. About twelve sources are here. One coincides with Sgr A West; it has an x-ray luminosity of 10^{28} W. The other sources lie along the ridge in the same location as the cluster of infrared sources.

The Inner 30 Light Years

We now have a pretty good overall idea of the distribution of gas and dust within 30 ly of the Galaxy's center (Fig. 18.22). The bulk of the continuous radio emission comes from Sgr A West in a filamentary structure that shows a western arc and a northern arm centered on the radio point source Sgr A*. This source does not quite coincide with an infrared source; it lies very close to or exactly at the galactic center. The far-infrared observations point to a disk of neutral gas and dust some 30 ly across. The inner few light years of this disk have a lower density than its average. The dusty disk lies pretty much in the plane of the Galaxy. The two lobes that mark the inner edge of this disk (Fig. 18.22) appear on either side of the ionized region. The disk, which contains hydrogen molecules at some 2000 K heated by shocks, rotates with its axis lined up to that of the general galactic rotation.

Does a Black Hole Lurk in the Core?

A puzzle generated by the radio and infrared line observations arises from the rapid rotational motions near the Galaxy's core—the rotational velocities *increase* closer to the core as shown by their Doppler shifts.

Why is that a problem? Well, the rotational velocities are so high that a huge concentration of mass is needed to hold the speedy gas together. For example, if the Galaxy's core simply contained a cluster of stars, you would expect the rotational velocities to *decrease* toward the core, because as you get closer in, you have less and less mass to bind the moving materials gravitationally. To account for the rapid rotation requires a mass in the core of 4 million solar masses—all lumped together in a region only 0.1 ly in diameter!

What form might this mass have? One possibility is that it is locked up in a supermassive black hole. If it were in the form of, say, a cluster of solar-mass stars, these stars would be located, on the average, only 1 to 2 AU from each other. Some million solar masses of material

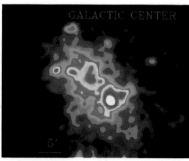

FIGURE 18.21 Computer-generated, false-color image of the x-ray emission from the galactic center, observed by the *Einstein X-Ray Observatory*. Note the angular scale for 5 arcmin at bottom. White represents the strongest emission, dark blue the weakest. (Courtesy of Einstein Data Bank, Harvard–Smithsonian Center for Astrophysics.)

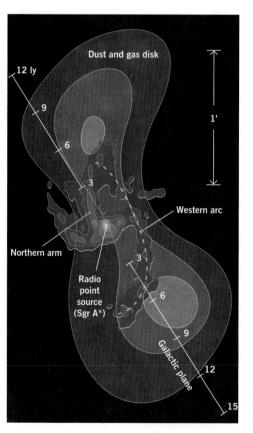

FIGURE 18.22 Schematic diagram of the inner 3 ly of the Galaxy. The galactic plane is marked by the diagonal line; the scale for 1 arcmin lies vertically at the right. The radio contours show the point source at the nucleus (Sgr A*), the northern arm, and the western arc of Sgr A. The dashed lines outline the disk of gas and dust, which is tilted to our line of sight. (Adapted from a diagram by M. K. Crawford and colleagues.)

could be jammed in such a star cluster in the nucleus—and the more in the form of stars, the less mass needed in a black hole. The dynamics of the nucleus do not *require* a supermassive black hole; one of moderate mass, say a thousand solar masses, could also explain the motions there.

The idea that a supermassive black hole lurks in the heart of the Galaxy has yet to be confirmed. Indirect support for the idea comes from observations that a few other galaxies may have a similar mass concentration in their nuclear region. (More on this in Chapter 20.)

To recap. Infrared, radio, and x-ray telescopes can probe the galactic center directly. They have shown that although the nucleus is very small, it emits enormous amounts of energy, about 10 percent of the total from the Galaxy. Within it, material orbits the center at a rapid rate. At the very center lies a massive object, perhaps a black hole. If it does exist, such a supermassive black hole can explain the observed radio and infrared emission and the total luminosity of the galactic center region.

■ **What is the source of the intense radio emission at the galactic center?**

18.7 The Halo of the Galaxy

The globular clusters outline the halo around the Galaxy (Fig. 18.23). Little else is visible in the halo. Stray stars are sometimes seen here, as well as some gas that is hot and ionized. The halo may also contain as-yet undetected objects, such as very faint low-mass stars, and it extends far beyond the edge of the disk.

Globular Clusters

You encountered the physical characteristics of globular clusters in Section 16.7. To review briefly: a globular cluster has a spherical shape (some tens to hundreds of light years in diameter) and contains up to a million Population II stars, each with a little less than 1 solar mass. The globular clusters form a spherical distribution around the Galaxy's center.

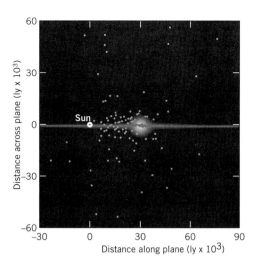

FIGURE 18.23 The distribution of globular clusters about the plane of the Galaxy. The view is edge-on. The sun's position is indicated. (Adapted from a diagram by W. E. Harris.)

Some elliptical orbits bring them out to extreme distances of 300,000 ly from the Galaxy's nucleus. The clusters orbit at speeds about 100 km/s, diving into and shooting out of the disk.

The outer halo of the Galaxy has no exact boundary. Observations of the placement of globular clusters indicate that the halo extends out to at least 300,000 ly, far beyond the limits of the Galaxy's disk, to the Magellanic Clouds, two companion galaxies that are gravitationally bound to the Galaxy (Section 19.6).

Other Material in the Halo

The rotation curve (Fig. 18.9) discloses that the halo actually contains considerable material, perhaps much more mass than is in the rest of the Galaxy. This conclusion rests on the fact that the rotation curve flattens out at distances far from the center of the Galaxy. So more mass is out there, but its form is *not* obvious. Using other spiral galaxies as a guide, the mass in the halo may be four or five times greater than the total mass known so far. What might it be?

The halo may contain a large number of low-mass, faint red stars that are difficult to observe directly. Observations of a few other nearby galaxies like the Milky Way imply that they may have extensive, massive halos of faint red stars.

The halo also contains gas, but much less than the disk of the Galaxy. Observations at 21 cm show hydrogen clouds traveling with high speeds above and below the galactic plane. Most of the halo's gas, however, is probably ionized hydrogen. This gas could come from the disk, blown out by supernova explosions, expanding H II regions, and stellar winds. So the halo may be fairly hot and expanding into intergalactic space.

But the halo may also contain other objects, currently unobservable. Astronomers give the generic name **dark matter** to this invisible material. Low-mass stars would be very hard to detect. Smaller objects, similar to planets, may also exist. Recall that hydrogen masses with less than 0.08 solar mass never get hot enough to become stars. Some astronomers have suggested numerous, low-mass black holes. But these would be impossible to confirm observationally if not in binary systems.

Finally, a massive halo can help out the density-wave model for the spiral structure in the disk. The mass in the halo could stabilize the arms gravitationally and ensure that the density waves propagate for long periods of time (a billion years). A massive, invisible halo may well be the necessary ingredient to lock in a well-defined spiral structure in a galaxy.

■ **What makes up the Galaxy's halo?**

18.8 A History of Our Galaxy

We have a pretty fair idea of the architecture of the Milky Way Galaxy. What clues does this information provide about the birth and evolution of the Galaxy? The crucial clues come from two sources: the chemical composition of galactic material and its dynamics.

The process of galactic evolution links the chemistry with the dynamics. This linkage marks an important theme of cosmic evolution, because the evolution of the Galaxy results from the evolution of all the stuff that makes it up.

Populations and Positions

Population I and Population II stars differ considerably in their abundances of heavy elements. In general, Population II stars contain about 1 percent of the metal abundance of Population I stars.

In fact, however, we do not find a stark and simple division of metal abundances into just two groups. Rather, we find a continuous range of abundances, from about 3 percent to less than 0.1 percent for the ratio of metals to hydrogen. So, though the division into two populations is a useful tool, a continuous range of populations actually exists. And when these populations are catalogued by metal abundance, we find a striking correlation with average distance from the Galaxy's disk. *The lower the metal abundance, the farther the objects are found from the plane of the disk.*

To interpret this key observation, we rely on basic concepts of starbirth, stardeath, and the recycling of the interstellar medium (Chapters 15 and 17). First, stars are born from clouds in the interstellar medium. Their atmospheric elemental abundance reflects that of the gas from which they formed. Second, stars inherit the orbital motions about the Galaxy of their parent gas and dust clouds. Third, massive stars evolve quickly and spew back into the interstellar medium material that is enriched by heavy elements. As long as new stars—especially massive ones—are born, the abundance of heavy elements in the interstellar medium gradually increases as the Galaxy ages.

With these basics in mind, let's try to make sense of the observations, which show that the youngest objects (highest heavy-element abundances) hug close to the disk, whereas the oldest objects (lowest heavy-element abundances) range far from the disk. Other objects fall in between these extremes. So the halo of the Galaxy is its oldest part, and the spiral arms its youngest.

We can estimate the Galaxy's age by finding the oldest stars in the halo. A comparison of theoretical models for globular cluster stars with H−R diagrams for them (Section 16.7) indicates an age of 15 billion years (with an error range of about 2 billion years). That's *when* the Galaxy formed. Let's see *how* it formed.

The Birth of the Galaxy

Because globular clusters contain the oldest stars associated with the Galaxy, the halo marks the fossil remains of the Galaxy's birth. Within it, globulars orbit the Galaxy on extremely elongated elliptical paths. Most of the time, the globulars move slowly through the halo at the outer extremes of their orbits; only briefly do they whip in and around the nucleus. These stars exhibit the motions of the cloud from which they were formed. So the Galaxy must have been born from a gas cloud that was initially huge—at least 300,000 ly in radius.

Here's one model of the Galaxy's birth. Imagine a tremendous, ragged cloud of gas roughly twice as big as the Galaxy's halo today (Fig. 18.24). Its density is low. This proto-Galaxy cloud probably is turbulent, swirling around with random churning currents. Slowly at first, the cloud's self-gravity pulls it together, with its central regions getting denser faster than its outer parts. Throughout the cloud, turbulent eddies of different sizes form, break up, and die away. Eventually, the eddies become dense enough to contain sufficient mass to hold themselves together. These might be hundreds of light years in size—incipient globular clusters. Each blob then splits up to form individual stars—all born at about the same time (15 billion years ago).

Meanwhile, the gas contracted more and fell slowly into a disk. (Sound familiar? Look back at Section 12.5 on the formation of the solar nebula.) Why a disk? Because the original cloud had a little spin, and the conservation of angular momentum (Focus 9.1) requires that it spin faster around its rotational axis as it contracts (Fig. 18.24). The kinetic energy of the cloud slowly decreases, as gas clouds collide and heat is radiated away. The disk rapidly flattens.

As the disk forms, its density increases, and more stars form. Each burst of starbirth leaves behind representative stars at different distances from the present disk. Finally, the remaining gas and dust settle into the narrow layer we see today. Somehow density waves appear and drive the formation of spiral arms.

During this time, massive stars were manufacturing heavy elements and flinging them back into the cloud by supernova explosions. So as stars were born in succession, each later type had more heavy elements. That enrichment continues today in the disk of the Galaxy.

What of the Galaxy's future? Let's speculate a bit. If we assume that no new gas is added from outside the Milky Way, stellar evolution points to a day when most stars become corpses. Matter that once made up the interstellar medium will be locked up for good. The Galaxy will literally run out of gas; starbirth will halt. Even if density waves still endure, they will have little gas to move around. When the disk of the Galaxy stops evolving, the Galaxy will be essentially defunct. Globular clusters will still swing on their leisurely orbits around the core. The supermassive black hole (if there!) will survive forever. Overall, the Galaxy will be quiet and dull.

■ What does a lower metal abundance indicate about a star's location in the galactic disk?

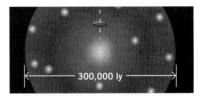

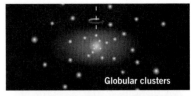

FIGURE 18.24 Schematic sequence of the collapse and condensation of a large cloud of gas and dust to make globular clusters and the Galaxy's disk. Because of its original spin, the matter eventually makes a disk with a central bulge. The stars in the globular clusters form before the disk has developed. The entire process takes less than one billion years. Note how the globular clusters now give the approximate size of the original gas cloud.

Key Concepts

1. The main parts of the Galaxy are the encircling halo, the flat disk, and the central nuclear bulge, which contains the nucleus.

2. Stars of different types inhabit these regions: the disk contains young, metal-rich (few percent) Population I stars; the nuclear bulge has metal-poor (few tenths of a percent) Population I stars; and the halo, metal-poor Population II stars.

3. The disk contains at least two and possibly four spiral arms, which contain concentrations of young stars, gas, and dust. The sun lies on the inner edge of one arm, and we can see pieces of other arms toward and away from the galactic center.

4. We trace out spiral arms by the use of objects found in them: optically, O and B stars, cepheids, and H II regions; with radio, H I clouds, and molecular

clouds. In all cases, the essential problem is to determine the distance to the object observed.

5. The rotation curve shows how fast objects at different distances from the galactic center orbit around it. The failure of the curve to follow Kepler's laws at large distances from the center indicates that a large fraction of the Galaxy's mass lies beyond the sun; this mass has not yet been seen directly.

6. The sun orbits the Galaxy at a distance of some 30,000 ly from the sun at a speed of 220 km/s, with some uncertainty in both numbers.

7. Cepheid variable stars show a period–luminosity relationship: the more luminous the cepheid, the longer the period of its light variation. This relationship is a powerful tool for inferring the distances to cepheid variables, along with the inverse-square law for light.

8. Radio astronomers use the Doppler shift in the molecular clouds and clouds of atomic hydrogen to infer the spiral-arm structure of the Galaxy; to do so, they assume that the clouds move along near-circular orbits (they don't) and that the rotation curve has been well observed (it hasn't).

9. Any model for the evolution of the Galaxy must explain the persistence of spiral arms. The density-wave model does this by having two spiral density waves disrupt the gas of the Galaxy's disk to promote the formation of spiral arms. As the density waves plow through the disk, different material condenses into the spiral arms as the old material dissipates; so the persistence of the original arms is really an illusion.

10. The nucleus of the Galaxy emits intense radio and infrared radiation; it contains supergiant M stars, young massive stars, dust, and gas rotating at high speeds. To account for the motion of the gas requires a concentration of millions of solar masses of material in the inner few light years—perhaps a supermassive black hole.

11. The Galaxy's halo contains globular clusters, some gas, and the invisible objects that make up the mass that shows up in the rotation curve. This invisible material is called dark matter.

12. The Galaxy formed some 15 billion years ago from the gravitational contraction of a large, slowly spinning cloud of gas and dust; the halo formed first, then the disk. We can estimate the relative sequence of formation from the heavy-element content of stars: the more metals they contain, the younger they are.

Key Terms

dark matter

density-wave model

disk (of a galaxy)

galactic rotation curve

halo (of a galaxy)

nuclear bulge

nucleus (of a galaxy)

period–luminosity relationship

spiral arm

spiral tracers

Study Exercises

1. What limits an optical astronomer's investigation of the Galaxy's structure? (Objectives 18-1, 18-2, and 18-3)

2. Why are Population I cepheids good optical spiral-arm tracers? (Objective 18-2)

3. Radio astronomers need the rotation curve of the Galaxy to use 21-cm line observations to establish its spiral structure. Why? (Objectives 18-4 and 18-6)

4. Argue that a spiral arm cannot be a material arm. (Objective 18-7)

5. What are the kinds of celestial object found in spiral arms? (Objectives 18-7 and 18-8)

6. What characteristics of spiral arms does the density-wave model account for? In what respects is the model at present inadequate? (Objective 18-8)

7. What observational evidence do we have that a large fraction of the Galaxy's mass is not in the core, nor, in fact, within the radius of the sun's orbit? (Objective 18-4)

8. Relate the orbits of globulars and their chemical

composition to the birth of the Galaxy. (Objective 18-10)

9. Some of the radio emission from the nucleus of the Galaxy is nonthermal. What does that imply about the physical conditions there? (Objective 18-12)

10. What observational evidence and physical argument can be used to infer that a supermassive black hole may reside in the Galaxy's core? (Objective 18-12)

11. What is the sun's distance from the center of the Galaxy? How is the value of the distance deter-mined? How uncertain is its current value? (Objective 18-5)

12. As the disk of the Galaxy evolves, what do you expect to happen to the percentage of metals found in the interstellar medium? (Objective 18-9)

13. Which region of the galaxy is defined by the orbits of globular clusters? (Objective 18-11)

14. In the future, what do we expect to happen to the molecular clouds that we now see making up so much of the interstellar medium? (Objective 18-13)

Problems & Activities

1. Assume that Kepler's laws apply to the rotation of the Galaxy. What would the orbital period be of a star 150,000 ly from the center? *Hint:* Use the right units!

2. What is the orbital period of the sun around the galactic center?

3. At 3 ly from the Galaxy's center, material orbits at speeds of about 100 km/s. What is the gravitational mass interior to this point?

4. A length of 1 ly at the galactic center corresponds to what angle when seen from the sun?

5. A few globular clusters reach maximum distances of 350,000 ly from the center of the Galaxy. What are their orbital periods?

See for Yourself!

Let's use a selected set of data to find a period–luminosity relation for cepheid variables stars in the Galaxy. Table A.1 gives the information; for each star, we have its period of light variation in days and its average luminosity in units of solar luminosities. Note that these cepheids are much more luminous than the sun.

On a piece of graph paper, lay out a horizontal axis that is the period, ranging from zero to 50 days; try 5-day intervals as the major ticks along the *x*-axis. For the vertical axis, use a range from zero to 35,000 solar luminosities; try major intervals of 5000. With the axes set up, plot a point representing each star.

When all the points have been plotted, you should see a clear trend. How do you describe it? Try drawing by eye a straight line that touches as many points as possible, with outliers falling above and some below the line. Such a line represents roughly the cepheid period–luminosity relation for this set of stars. Note that not every star's point falls right on this line; this scatter in the data gives you some sense of where the actual range within the relation falls. ■

TABLE A.1 CEPHEID VARIABLES

STAR	PERIOD (days)	LUMINOSITY (Sun = 1)
SU Cas	2.00	960
EV Sct	3.10	1,100
CF Cas	4.90	1,800
UY Per	5.40	2,500
CV Mon	5.40	2,300
VY Per	5.50	2,800
V367 Sct	6.30	3,500
U Sgr	6.70	3,900
DL Cas	8.00	3,700
S Nor	9.80	4,500
TW Nor	10.8	2,900
VX Per	10.9	5,900
SZ Cas	13.6	8,500
VY Car	18.9	11,200
T Mon	27.0	18,600
RS Pup	41.4	22,400
SV Vul	45.0	30,200

19-1 Describe the general physical characteristics of spiral, elliptical, and irregular galaxies, including their differences in size, shape, mass, color, stellar types, and amount of interstellar gas and dust.

19-2 Outline the specific methods used to find the bulk physical properties of galaxies.

19-3 Describe how to use the criteria *brightness means nearness* and *smallness means farness* to estimate the relative distances to galaxies.

19-4 Indicate what observations clinched the idea that the spiral nebulas are actually other galaxies.

19-5 Describe briefly the general technique how to find distances to galaxies.

19-6 Outline a contemporary method of finding distances to distant galaxies, starting with the astronomical unit and ending with Hubble's constant.

19-7 Evaluate the weaknesses in the procedure you outlined in Objectives 19-3 and 19-6 so that you can estimate the possible errors in distances to galaxies.

19-8 Show how getting distances and radial velocities for galaxies results in a value for Hubble's constant, and use this value to estimate distances to galaxies.

19-9 State the range of uncertainty in the value of Hubble's constant, and give the implications of this uncertainty.

19-10 Define the term *cluster of galaxies* and describe the layout of a cluster in space.

19-11 Evaluate the evidence for intergalactic medium between and/or within clusters of galaxies.

19-12 Define *supercluster* and describe the general layout of superclusters in space and the voids between them.

19-13 Discuss the evidence, direct or indirect, for dark matter in clusters and superclusters of galaxies.

Central Question
What is the structure and content of galaxies, and how are they distributed throughout the universe?

The Universe of Galaxies

Look, friend, at this universe
with its spiral clusters of stars
flying out all over space
like bedsprings suddenly bursting free. . . .

EDWARD FIELD, "PROLOGUE"

No celestial object is quite as grand as a galaxy. In a large telescope, a bright spiral galaxy is a stunning sight, a star-bright nucleus and a misty swirl of spiral arms. Billions of stars caught in a whirlpool spanning hundreds of thousands of light years.

The galaxies form the basic elements of our modern cosmological vista. Their sheer numbers are beyond our comprehension and beyond the view of today's telescopes. The diversity of structure in these galaxies is also astounding; even more surprising is the fundamental unity found in spite of their wide variety. The fact that galaxies can be sorted into broad divisions hints at a common evolutionary process.

The galaxies form the skeleton of the universe. The crucial problem is the measurement of their distances, for many conclusions of modern cosmology hinge on them. This chapter notes that the notorious difficulties of surveying our own Galaxy are amplified when we try to appraise the vastness of the universe. In spite of present problems, our vision of the universe, underpinned by the theory of general relativity, has a coherence that allows us to draw conclusions about the large-scale structure of spacetime.

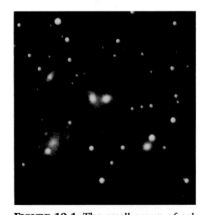

FIGURE 19.1 The small group of galaxies called Stephan's Quintet. Note the various shapes. The stars visible as points of light are foreground stars in the Milky Way. Colors shown here are true ones; reddish-yellow represents old stars, bluish regions of new stars and starbirth. (Courtesy of James D. Wray; McDonald Observatory.)

19.1 The Shapley–Curtis Debate

For almost two centuries, astronomers hotly debated whether the *spiral nebulas* they saw with their large telescopes were simply clouds of gas within the Milky Way or were other galaxies like ours but far beyond it (Fig. 19.1). The controversy came to a head in April 1920, when Harlow Shapley (Section 18.2) and Heber D. Curtis (1872–1942) debated the point publicly.

Some of the Arguments

Shapley believed that the spiral nebulas were distant parts of the Milky Way. Curtis, who claimed that they were galaxies in their own right, relied on the principle of the **uniformity of nature**—that is, similar astronomical objects are assumed to have similar properties until proven otherwise by observations. (In other words, innocent until proven guilty!)

Curtis argued that the wide range of angular diameters of spirals—approximately 4° (for M31, the nearest) to 10 arcmin and less (for the smallest)—required a large range of distances, and so the spirals could not be part of our Galaxy. (Remember that an angular diameter gives you a diameter-to-distance ratio.) Starting from the principle of the uniformity of nature, Curtis assumed that all spirals have roughly the same physical diameter. The range in observed sizes (more or less 10 to 1) implied that the spirals must be enormous distances from the Galaxy. For if they were the same in diameter, the range in apparent size would mean that the ones ten times smaller must be ten times farther off, or about ten times the radius of the Galaxy (Fig. 19.2). So the more distant spirals could not be members of the Milky Way Galaxy, as Shapley argued.

In addition, Curtis noted the so-called **zone of avoidance,** a region near the plane of the Galaxy where very few spirals are visible (Fig. 19.3). Curtis argued that interstellar dust in the galactic plane cuts out the light from the spirals; we see fewer in the zone of avoidance

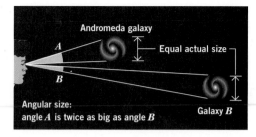

because there we have to look through the disk of the Galaxy. If the spirals were actually associated with the Galaxy, they would be found concentrated in the plane—rather than avoiding it—along with the stars, galactic clusters, and H II regions.

As evidence for this point of view, Curtis cited photographs of spirals showing dark lanes cutting through their planes (Fig. 19.4). If the Milky Way Galaxy and other spirals were similar in structure, Curtis reasoned, then our Galaxy must also have dusty material collected in the plane.

Finally, Curtis pointed out that the spectra of spirals are not bright-line spectra like those of emission nebulas (such as the Orion Nebula); rather, they resemble those from a conglomeration of stars. The spectra show faint dark lines against a bright background—the same spectrum the Galaxy would show if viewed from a great distance.

The Resolution

In 1924 Edwin Hubble (Section 7.4) settled the dispute conclusively by the discovery of cepheid variables (Section

FIGURE 19.2 Distances to galaxies estimated by their angular sizes. If galaxies have roughly the same physical size, the more distant will have smaller angular sizes. Here, Galaxy *B* is about twice as far away as the Andromeda Galaxy, and so half the angular size.

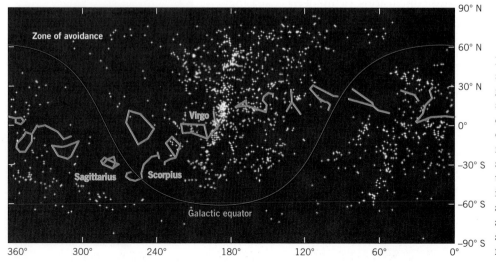

FIGURE 19.3 The "zone of avoidance": this plot shows the brightest galaxies in the sky, the zodiacal constellation outlines, and the *galactic equator* (the plane of the Galaxy) completely around the sky and from north to south. Note how few galaxies are visible above and below the galactic equator. The dust of the Milky Way is settled into the plane of the Galaxy and this blocks the view of distant galaxies around the galactic equator. (Diagram generated by Voyager software for the Macintosh.)

18.3) in the Andromeda galaxy (M31). Although variables had been suspected as early as 1922, Hubble confirmed their existence in the outer arms of M31. He derived a distance of 490,000 ly for M31, far beyond the farthest globular clusters that marked the outer limits of the Galaxy. [Hubble's estimate was too small; current work on M31 establishes a distance of 2.2 million light years (= 2.2 Mly)].

Caution. The principle of the uniformity of nature has great power, but it must be used with caution. Sometimes it works; sometimes it leads you astray. For example, novas in M31 *are* pretty much the same as those in our own Galaxy, because M31 is a spiral galaxy very similar to the Milky Way. But the variable stars in globular clusters are *not*, it turns out, the same as the cepheids in the disk of our Galaxy, for they belong to different populations of stars. You can assume that things are uniform everywhere, and this can lead to new knowledge and insights; but you must always keep looking for new ways to check the assumption, and be prepared to abandon it if inconsistencies develop.

Note. As indicated earlier, in naming some galaxies we use the prefix M; for others we put NGC before the number. The prefixes refer to two catalogs of galaxies. The Messier (M) catalog, compiled by Charles Messier in the eighteenth century, contains star clusters and nebulas in our Galaxy as well as the brighter galaxies visible from midnorthern latitudes. The New General Catalogue (NGC) was compiled in the nineteenth century with larger telescopes. Two supplements to the NGC, called the first and second Index Catalogues (IC), were published in 1895 and 1908.

■ **In what way did Hubble's discovery of cepheid variables help to settle the score between Shapley and Curtis?**

19.2 Normal Galaxies: A Galaxian Zoo

Edwin Hubble pioneered in the field of extragalactic astronomy. He recognized that galaxies have different shapes. To

FIGURE 19.4 Dust in the plane of a spiral galaxy (Messier 104 in the constellation Virgo). The dark band is due to dust cutting out starlight. Note the nuclear bulge at the center and the huge halo of old, yellowish stars. Our Milky Way has a similar layer of dust, creating the zone of avoidance. (Courtesy of James D. Wray; McDonald Observatory.)

catalog the differences in form, Hubble in 1926 proposed his first scheme for the classification of galaxies. Although his initial design is now considered to be too simple, modern classifications still use the fundamental categories of *elliptical, spiral,* and *irregular* galaxies. Most galaxies fall into these categories— *normal galaxies,* whose emission is mostly from starlight. (Chapter 20 describes extraordinary galaxies.) Let's take a look at these basic galaxy types.

Ellipticals

An **elliptical galaxy** (Fig. 19.5) exhibits no spiral structure but does show an elliptical shape. Very little gas or dust appears in elliptical galaxies, and OB stars are also absent. The ellipticals generally have a reddish overall color.

Hubble subdivided the ellipticals in classes from E0 to E7, according to how elliptical they appear. Imagine looking at a circular plate face-on; such is the appearance of an E0 galaxy. Now slowly tilt the plate so that it looks more elliptical and less circular. This flattening of shape presents the same views as the sequence from E0 to E7 galaxies. Be warned that Hubble based the classifications on the appearance of the galaxy, not on its true

FIGURE 19.5 A giant elliptical galaxy, Messier 84 in Virgo. Note the symmetry of the shape and the lack of distinct structure; it is an elliptical of type E1 and contains only old, yellowish stars. (Courtesy of James D. Wray; McDonald Observatory.)

shape. For example, an E7 is really a flat elliptical viewed edge-on, but an E0 may be either a truly spherical galaxy or a flattened galaxy seen face-on.

Ellipticals come in a range of sizes, from supergiants to dwarfs. The largest ellipticals, found in clusters of galaxies (Section 19.6), have diameters of a few *million* light years! In contrast, the smallest dwarf ellipticals span merely thousands of light years in diameter.

Spirals

A **spiral galaxy** (if not seen edge on) displays an obvious spiral structure, usually with two, but sometimes more, spiral arms (Fig. 19.6). One type of spiral has a prominent bar through the nucleus, the spiral arms winding out from the end of the bar (Fig. 19.7). Hubble termed the spirals without a bar *normal* (denoted S) and the others *barred* (denoted SB); our Galaxy is now known to be a barred spiral. He arranged the spiral forms in sequence according to the sizes of their nuclear region, the tightness of the spiral arms, and the degree to which the arms were resolved into patches of stars.

The normal and barred spirals are subdivided further into categories a, b, and c. These types are judged by how tightly the spiral arms wind around (a, the tightest; c, the most open) and the relative size of the nucleus compared to the disk (a, the largest; c, the smallest). For example, the Hubble Sa is a normal spiral with a large nucleus and tightly coiled arms. A few galaxies appear to have the disk of a spiral but no arms. Hubble dubbed those S0. These are now sometimes called **lenticular galaxies** because of their shape, like a convex lens.

Irregulars

Finally, as a catch-all category, Hubble designated as **irregular galaxies** (denoted Irr), those that were devoid of spiral structure or symmetry but were resolvable into distinct patches of stars (Fig. 19.8). These strange beasts are dominated by OB stars and H II regions. Usually there are no conspicuous dust

FIGURE 19.6 A large spiral galaxy with a small nucleus (NGC 253 in the southern constellation Sculptor). It is Hubble type Sc, and one of the dustiest galaxies known. The galaxy appears elongated because we are viewing it only a few degrees above the plane of its disk. Two spiral arms are visible extending from the nucleus; bluish patches of star-forming regions are visible within them. (Courtesy of David Malin, AATB.)

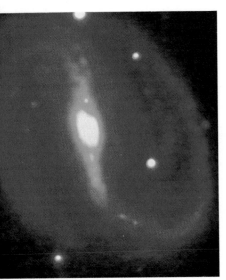

(a)

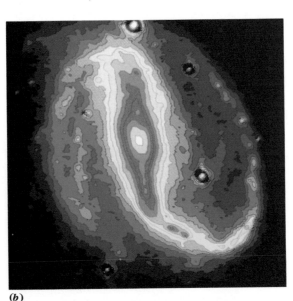

(b)

FIGURE 19.7 A barred spiral galaxy (NGC 7479 in Pegasus). (*a*) False-color image, using red to bring out the spiral structure and bar. The blue shows the disk of the stars. The bar crosses the nucleus and links the two spiral arms; both contain old, yellow stars. Dust lanes run along the leading edge of the bar, and star-forming regions line the dust lanes, as expected from the density-wave model of spiral structure. (*b*) Additional computer processing with more colors highlights the bar, which crosses the nucleus and links the two spiral arms together. (Images by R. Schild; courtesy of Smithsonian Astrophysical Observatory.)

clouds. Radio observations of neutral hydrogen gas tend to show a rotating disk of material. In this respect, they resemble spiral galaxies.

Modern classifications have expanded Hubble's format. I will stick with Hubble's three basic categories: spirals, ellipticals, and irregulars. But this Hubble scheme does not include *all* types of galaxies. Some galaxies stand out as peculiar in shape (Fig. 19.9) and do not fit into the three general Hubble categories. Many of these peculiar galaxies turn out to have evidence of unusual activity. Some appear to be pairs of galaxies close together, interacting gravitationally by tidal forces (Fig. 19.10).

FIGURE 19.8 A dwarf irregular galaxy (NGC 6822 in Sagittarius). Only the most luminous, bluish stars are visible; these were born fairly recently. At the end of a bar lie a few clouds of glowing gas. This galaxy is a member of the Local Group and one of the closest to us, at a distance of 1.8 million light years. (Courtesy of David Malin, AATB.)

Luminosity Classes

Astronomers have recognized that the same Hubble type of galaxy, say Sb, comes in a range of luminosities. So in analogy to stellar luminosity classes (Section 14.5), galaxies also have luminosity classes of I, II, III, IV, and V, with the I the most luminous and the V the least. An Sc I galaxy, for instance, is a very luminous spiral with a small nucleus and spread-out arms. It turns out that luminosity class I galaxies are larger than class II, and so on. So class I galaxies can be thought of as supergiant galaxies.

Of all galaxies that have been *observed* by ground-based telescopes, about 77 percent are spirals, 20 percent ellipticals, and 3 percent irregulars. This sample is dominated by the luminous spirals, however, which are visible at very great distances and so appear to be more common than they really are. Because of this selection of more luminous galaxies, the observed percentages are biased. The *actual* relative numbers in a given volume of space are quite different. A complete survey of a region out to 30 million ly showed that only 34 percent of the galaxies in this volume are spirals, 12 percent are ellipticals, and 54 percent are irregulars. Hence, the majority of galaxies in the nearby universe are irregulars of fairly low luminosity.

But, remember that a telescope is a time machine, and the relative numbers of galaxies may well change with time. A recent observation by the *HST* makes

FIGURE 19.9 The Southern Ring Galaxy, a galaxy with a peculiar shape that does not fit the simple Hubble scheme, is located in the constellation Volans in the southern hemisphere. This galaxy lies about 270 Mly distant; its diameter is about 120,000 ly, about the same as the Galaxy. This one galaxy represents a small class of galaxies called *ring galaxies*. (Courtesy of NOAO/CTIO.)

this point for a group of galaxies that lies at a distance of 4 *billion* light years (Fig. 19.11). The resolution of the telescope clearly shows that long ago, most of the galaxies were *spirals!* That's not the case in the recent past or today. Somehow, the spirals slipped from dominance as the cosmos evolved, but we don't know what caused most of them to disappear.

To group galaxies by shape marks an initial step toward delving into the far depths of the universe. But to probe the

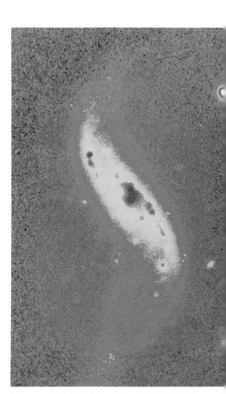

FIGURE 19.10 The galaxy NGC 3187 in the constellation Leo. The arms of this spiral galaxy are twisted by the tidal interaction with a nearby galaxy. (Courtesy of California Association for Research in Astronomy.)

FIGURE 19.11 *HST* image of a group of galaxies at a distance of 4 billion ly. If you examine this image carefully, you will notice that most of the galaxies have spiral shapes, several times as many as are seen today. (Courtesy of Alan Dressler, Carnegie Institution; Augustus Oemler, Yale University; James Gunn, Princeton University; Harvey Butcher, Netherlands Foundation for Research in Astronomy; and NASA.)

physical properties of galaxies—their masses, sizes, and luminosities—ultimately depends on a knowledge of their distances from us, distances that are as difficult to survey as they are vast to imagine.

Spiral Structure Revisited

Although spirals make up a minority of galaxies in the universe today, we pay special attention to them because the Milky Way is a spiral. Recall (Section 18.5) that the density-wave model is a leading contender for an explanation of the evolution of spiral structure, the trigger for star formation from giant molecular clouds (Fig. 19.12). Do observations of spiral galaxies support the density-wave concept? Generally, yes, for galaxies that show a distinct spiral structure (Fig. 19.13).

Some spiral galaxies have structures that the density-wave model cannot explain—those that have a generally patchy appearance. Here supernovas may drive the structure of the interstellar medium, as the shock waves from supernovas plow into the gas, causing it to collapse into clouds. Computer models of such a process produce a realistic spiral structure (Fig. 19.14).

■ **What general type of galaxy is the Milky Way?**

19.3 Surveying the Universe of Galaxies

Only some 70 years have passed since we learned for certain that galaxies are faraway islands of stars. Yet, you may be surprised to discover that we know the distances to galaxies only roughly. For the nearest galaxies, we have measurements that are good to about 10 percent. For the most distant visible galaxies, we are lucky if we know their distances within 50 percent of their actual value. It's a hard but essential astronomical business to survey the distances in the huge universe of galaxies.

Judging Distances

Although the distances to galaxies have continually been revised, the essential techniques remain the same. The initial indicators are the criteria that *brightness means nearness* and *smallness means farness*. Galaxies with the smallest angular size tend to be the most distant. And faint galaxies also tend to be far away. By applying these simple criteria, you can make rough estimates of the relative distances to galaxies.

For example, if you look at two galaxies, one apparently half as large as the other, and if both are in reality approximately the same size, the apparently smaller galaxy is twice as far away as the apparently larger one. A similar argument applies for relative brightness: if one galaxy is a hundred times fainter than another, it is be roughly ten times farther away (recall the inverse-square law for light, Section 14.1).

Refining this rough first approach requires the use of known physical properties of stars and galaxies inferred from theoretical models and careful observations. Each step in surveying the universe applies to certain objects and over a certain range of distances; astronomers work step by step to establish a distance scale.

Now to discuss some steps of the scale. First we need to establish the scale of distances in our Galaxy, starting with the solar system. Parallax measurements are used: heliocentric and spectroscopic

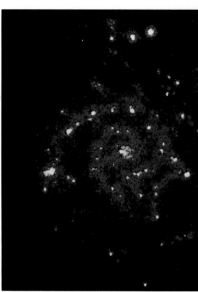

FIGURE 19.12 Ultraviolet image of the spiral galaxy Messier 74 in the constellation Pisces. This ultraviolet view accentuates the young, active regions of star formation, which trace out the main spiral arms. Messier 74 lies 55 Mly away; it is tilted almost face-on to our view. NASA's Ultraviolet Imaging Telescope took this image in 1990. (Courtesy of NASA/Goddard Space Flight Center.)

Galaxies

At a Glance

Types	Spiral, irregular, and elliptical (giant to dwarf)
Relative numbers	Irregulars most common; then spiral and ellipticals
Found in	Clusters, numbering up to a few thousand galaxies
Visible content	Stars, gas, and dust

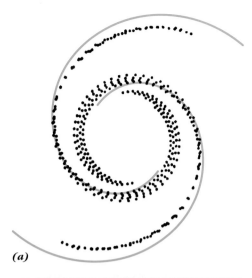

(a)

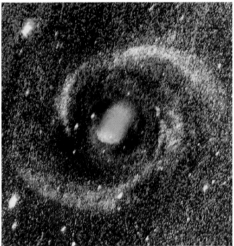

(b)

FIGURE 19.13 Density-wave model of a spiral galaxy. (*a*) Two-arm spiral model of M81 shows the relation of the newest, bluish stars (solid points) and the crest of the density wave (red line). (Based on a diagram by F. Bash, University of Texas at Austin.) (*b*) Computer-processed image of M81 shows broken spiral arms, which offer evidence of density waves acting to trigger star-forming regions. (Courtesy of B. Elmegreen, D. Elmegreen, and P. E. Seiden, IBM Thomas J. Watson Research Center.)

supernovas) is the same for similar objects in other galaxies. Without this necessary assumption, we couldn't get anywhere. It's not a blind, unsupported assumption, however. Observations made so far are consistent with it. For example, cepheids in other galaxies have the same spectra and light curves of the same shape as cepheids of the same type in our Galaxy and follow the same period–luminosity relationship (Section 18.3).

Then, to find distances to galaxies, we must use identifiable objects (within galaxies) whose *luminosities* we know. We compare their fluxes with their luminosities to infer their distances. Unfortunately, even the largest telescopes have a limit, and some objects are too faint to be picked up. So we want to choose the *most* luminous objects in galaxies for which we actually can determine the luminosities.

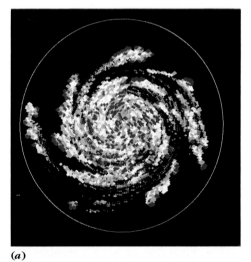

(a)

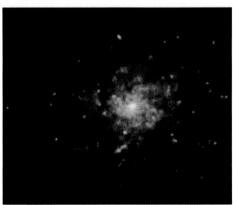

(b)

parallaxes. Along with these are distances using the cepheid period–luminosity relationship. So far, so good, for we can then infer the size of the Galaxy.

Distance Indicators

To bridge the distances to other galaxies, we assume the uniformity of nature: that the essential character of objects in our Galaxy (such as cepheid variable stars or

FIGURE 19.14 Spiral structure shaped by star formation and supernova explosions. (*a*) Theoretical model of self-propagating star formation. Young clusters of stars are colored violet and blue; old clusters yellow and red. The smaller the symbol, the older the cluster. Note that young clusters fall in patches on both sides of the spiral arms. (Courtesy of P. E. Seiden; IBM Thomas J. Watson Research Center.) (*b*) A spiral galaxy (NGC 7793) devoid of the grand design of clear, overall spiral pattern. Note the change from the older stars in the inner disk to the younger stars in the outer one. (Courtesy of James D. Wray; McDonald Observatory.)

As an analogy, imagine that you know that all street lights have the same luminosity, say 500 watts. Then as you look out at night at a city, you can judge the distances to different locations by measuring the flux of the relevant street lights with a light meter. The inverse-square law for light (Section 14.1) allows you to find the distance from the measured flux and the assumed luminosity.

Starting with close galaxies, we apply the period–luminosity relationship to cepheids in other galaxies. That way we find the luminosities of cepheids observed in other galaxies. The cepheids, however, are useful over a very limited range, for they are not especially luminous. Of all visible galaxies, only about thirty are of the right kind and close enough for us to detect cepheids within them. We can use cepheids as standards out to approximately 15 Mly.

To go beyond this limit requires the establishment of other standards whose visibility is greater than that of the cepheids. For instance, supernovas. To use supernovas, recall that they come in two types, distinguished by the shape of their light curves (Section 17.4). Best estimates give, for Type I supernovas, a consistent luminosity at maximum of about 1.3×10^{36} W. So by measuring the flux of a Type I supernova in another galaxy at maximum, we can infer its distance with the inverse-square law for light.

The designation of other distance standards follows the same strategy: select fairly common bright objects, find their luminosities in our own or nearby galaxies, check other galaxies by methods thought to be reliable, and then utilize the standard to the limits of its accuracy.

At distances greater than 100 Mly, we can't see individual objects in galaxies with present ground-based telescopes. What next? We use the luminosities of the galaxies themselves! Galaxies tend to lie in clusters (Section 19.6). To ensure that we study galaxies with the same luminosity, we select one of the brightest galaxies in a cluster rather than picking one at random. The brightest clusters of galaxies have about the same luminosity, *if* you stick to the same *kind* of galaxy (such as large spirals).

Of course, for this method to work, we need a calibration for the luminosities. How to get it? By using cepheid variables to find the distances to nearby spiral galaxies. Then we have a sample over a range of luminosity classes that we hope represents all galaxies. Even with its crippled optics, the *HST* has begun to expand the useful range of cepheids. In a faint spiral galaxy (called IC 4182 in the constellation Canes Venactici), an *HST* team has discovered almost thirty cepheids (Fig. 19.15). With improved optics, many more observations will follow out to distances about ten times greater than currently sampled.

Supergiant spiral galaxies serve as useful, far-reaching standards, in particular, those with small nuclei and spread-out spiral arms—that is, Sc I galaxies. The luminosity of such galaxies is about 25 billion times the sun's luminosity. Visible at great distances, these galaxies are relatively easy to identify because of their distinctive shape. With contemporary telescopes, they can be used as standards to distances of roughly 1.5 billion ly. Beyond that, we rely on the expansion of the universe as described by Hubble's law.

■ What physical law is used to determine the distance of stars based on their flux?

19.4 Hubble's Law and Distances

The universe is expanding, and that expansion is described by Hubble's law (Section 7.4; this is a good place to review that material). Hubble's law states that the farther a galaxy is from us, the greater its radial velocity of recession. The number that relates the recessional speed and the distance is Hubble's constant, *H*. Finding the value of *H* has absorbed the energy of astronomers since the construction of large telescopes. It is the most important single piece of information in cosmology.

Redshifts and Distances

It is relatively easy to measure the redshift of a galaxy, compared with the task

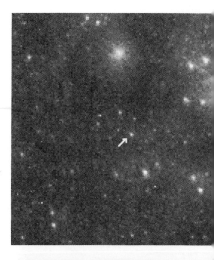

FIGURE 19.15 A cepheid variable observed by *HST* in the galaxy IC 4182, which is 16 Mly away. The star (arrow) is brighter in the frame on the bottom than the one on the top. (Courtesy of Abhijit Saha, STScI and NASA.)

Distance Indicators

At a Glance

OBJECT	CURRENT LIMIT (Mly)
Cepheids	23
Supergiants	32
Novas	65
H II regions	80
Supernovas (Type I)	650
Brightest galaxy in a cluster	1500

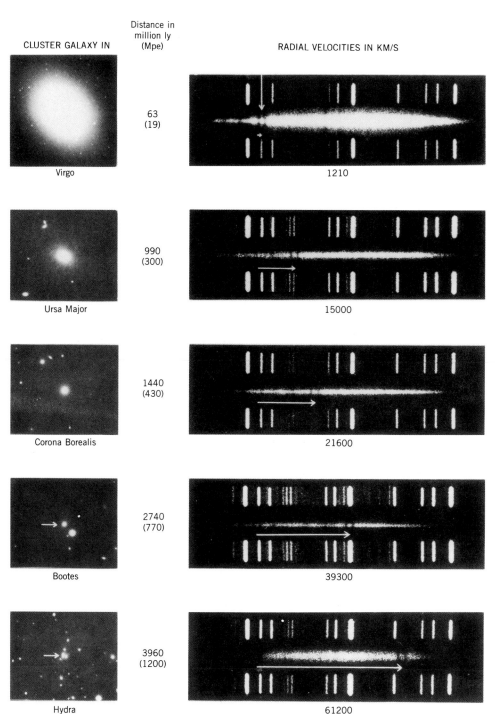

CLUSTER GALAXY IN	Distance in million ly (Mpe)	RADIAL VELOCITIES IN KM/S
Virgo	63 (19)	1210
Ursa Major	990 (300)	15000
Corona Borealis	1440 (430)	21600
Bootes	2740 (770)	39300
Hydra	3960 (1200)	61200

FIGURE 19.16 Measured redshifts and distances for selected galaxies in order of increasing redshift and distance (from top to bottom) from a galaxy in the Virgo cluster to one in the distant Hydra cluster. The redshifts are visible in the spectra on the right, where a white arrow indicates the size of the shift in the H and K lines of calcium; these are the two darkest lines in the spectra. Below each spectrum is the radial velocity in kilometers per second. Above and below each spectrum of the galaxies are emission-line spectra that serve as wavelength markers for a source at rest. The galaxies (at left) are in clusters of galaxies named for the constellation in which they appear. Note, in general, that the farther-away galaxies appear smaller than the closer ones. (Courtesy of Palomar Observatory, California Institute of Technology.)

of finding its distance. A comparison spectrum made at the same time as the galaxy's spectrum affords a direct measurement of the shift in some prominent spectral lines (such as the H and K absorption lines of calcium; Fig. 19.16). The redshift indicates the radial velocity of a galaxy. For example, look at the galaxies pictured and at their respective spectra with redshift indicated (Fig. 19.16). Notice that as the redshifts get larger, the galaxies appear smaller and fainter. Now plot the distances for these galaxies versus their redshifts. You can draw a straight line that represents the trend of these points and so find a value for Hubble's constant, which is the slope of the line (Fig. 19.17). Once we have a value of Hubble's constant, we can use Hubble's law to find the distances to galaxies beyond the limits described in Section 19.3.

In equation form, the Hubble law is

$$v = Hd$$

where v is the radial velocity, in kilometers per second, H the Hubble constant,

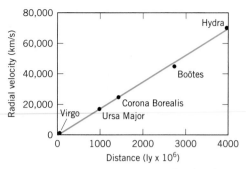

FIGURE 19.17 Hubble plot using the data from Figure 19.16. The straight line is the "best fit" through the points. Its slope is 15 km/s/Mly, which is the value of the Hubble constant derived from these data.

and d the distance, in millions of light years.

Now, we reverse this procedure to find distances from redshifts. Suppose you were given the redshift of a galaxy. You find the distance to which this value corresponds by looking it up on the redshift axis of your plot. If the redshift is larger than measured earlier, you can assume that the line you've drawn for the plotted points may be extended farther out and still be valid. Suppose a galaxy's measured redshift is 40,000 km/s. What is its distance? If H is 20 km/s/Mly, then the plot gives 2 billion ly.

In equation form,

$$d = \frac{v}{H}$$

Looks simple, yes? But a note of caution. This indirect method of distance measurement rests on one crucial fact: that the galaxies with known distances give an accurate value for Hubble's constant! This seems OK but not very secure. Estimates of Hubble's constant range from 15 to 30 km/s/Mly, with some current work favoring the lower figure. The distances derived by this method can vary by a factor of 2, depending on which value of Hubble's constant is used!

Steps to the Hubble Constant

Finding distances to galaxies is one of the toughest jobs for astronomers. Yet it's one of the most important. Let me outline the steps in one contemporary procedure followed by Allan Sandage and G. Tammann, which yielded a value of Hubble's constant of 15 km/s/Mly with an estimated error of 10 percent. There are eight basic steps.

1. Measure the AU using radar reflection from Venus; then the distances to nearby stars using heliocentric parallax with the AU as the baseline.

2. Determine the distance to the Hyades, a galactic cluster, using the motions of its stars, checked for consistency by using heliocentric parallax to compare the brightnesses of Hyades stars with those of nearby stars of known distances.

3. Find the distances to cepheids in our Galaxy. Search out open clusters that contain cepheids, compare the brightness of the stars in these clusters with that of the stars in the Hyades cluster, and find the distances to the clusters with cepheids (there are only a few). We have now calibrated the essential cepheid period–luminosity relationship.

4. Use cepheids to determine the distances to nearby galaxies by the period–luminosity relationship.

5. Measure the angular sizes of H II regions in these nearby galaxies. We find that the actual sizes of H II regions in spiral and irregular galaxies depend on the luminosity of the galaxies. Use nearby galaxies to calibrate that H II region size–luminosity relationship.

6. Extend this calibration to *supergiant,* very luminous Sc I spiral galaxies, for which the nearest, M101, has its distance determined by different methods as a check. We now have the sizes of H II regions for Sc I galaxies. However, we must be careful because the calibration is based on only *one* galaxy, M101.

7. Use the size–luminosity relationship for H II regions to find the distances of galaxies to the limit of this method. (We can use this procedure on about fifty galaxies.) Now we know the luminosities of these galaxies and their relationships to luminosity classes.

8. Look at supergiant spiral (Sc I) galaxies—the objects we can see distinctly at the greatest distances. Use their luminosities (step 7) to find their dis-

tances. Measure their redshifts. Divide their redshifts by their distances. We then have Hubble's constant.

Another Way to *H*

Not all astronomers agree with Sandage and Tammann that *H* is about 15 km/s/Mly. Marc Aaronson, Jeremy Mould, and John Huchra have developed a new technique, independent of the Sandage–Tammann steps, to find a value for *H* of 27 km/s/Mly (with an estimated uncertainty of 5 percent).

The technique rests on a relation between the luminosities of spiral galaxies and the widths of their 21-cm H I emission: the larger the line width, the greater a galaxy's luminosity. The 21-cm emission comes from the neutral gas in a spiral galaxy's disk. If we measure the 21-cm line from a spiral galaxy viewed edge-on, the emission from the gas moving away from us will be redshifted and that moving toward us will be blue-shifted, so the 21-cm line will be broader than would be expected if the galaxy had no rotational motion. In fact, the line width for edge-on galaxies measures the *maximum* rotational velocity in the disk. Most galaxies have maximum rotational velocities of 100 to 400 km/s (depending on their type). So the relation between luminosity and 21-cm line width is really one between luminosity and rotational velocity. It is usually called the **Tully–Fisher relation** after its discoverers, R. Brent Tully and J. Richard Fisher.

If properly calibrated, this relation provides a galaxy's luminosity from a measurement of its 21-cm line width. Infrared observations of nearby galaxies (such as M31 and M33), whose distances are known from cepheids, give the infrared luminosities. These infrared luminosities relate well to the width of the 21-cm lines from local galaxies. Aaronson and colleagues then measured the infrared fluxes and 21-cm line widths for some nearby clusters of galaxies. Again they found a good connection between the fluxes and line widths. Applying the calibration from local galaxies, they got the distances (Fig. 19.18). The measured redshifts then give a Hubble's constant of 27 km/s/Mly.

Note that a value for *H* of 27

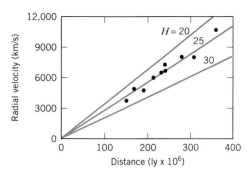

FIGURE 19.18 Hubble plot using the Tully–Fisher relation and infrared magnitudes for nearby calibrating galaxies to estimate the distances to eleven clusters of galaxies, which are indicated here (dots). The outer lines represent the acceptable range in the error of the value of the Hubble constant from 20 to 30 km/s/Mly. (Adapted from a diagram by M. Aaronson.)

km/s/Mly implies a universe no older than about 10 billion years (Focus 7.1). That result poses a problem, for the oldest stars (in globular clusters) are thought to be some 15 billion years old.

Who's right? Both the Sandage–Tammann and Aaronson–Huchra–Mould methods probably contain undetected systematic errors. A very careful study of the calibration procedures by Gérard de Vaucouleurs results in a Hubble constant close to 30 km/s/Mly. An estimate using supernovas gives a value between 20 to 23 km/s/Mly. So *H* lies in the range 15 to 30 km/s/Mly. The actual value may well be near 20 km/s/Mly (which this book generally uses), which wouldn't cause serious problems with the ages of stars.

Warning. Different workers use different values of the Hubble constant. When I relate their results, I will quote the value that was used in each particular investigation. Don't let the differences throw you! We simply do *not* know the value of the Hubble constant very well!

■ Why must we employ caution when using Hubble's constant to find distances to galaxies?

19.5 *General Characteristics of Galaxies*

Let's step back a bit from our cosmological vista and study the characteristics of galaxies (Table 19.1), now that we have methods to find out their distances.

Size

Once you have learned the distance to a galaxy, you can find out its actual size from a measurement of its angular size.

TABLE 19.1 GENERAL PROPERTIES OF GALAXIES

PROPERTY[a]	SPIRALS	IRREGULARS	DWARF ELLIPTICALS	GIANT ELLIPTICALS
Diameter (ly)	90×10^3	20×10^3	30×10^3	150×10^3
Mass (sun = 1)	10^{11}	10^6	10^5 to 10^7?	10^{13}
Luminosity (sun = 1)	10^{10}	10^9	10^8	10^{11}
Color	Bluish (disk), reddish (halo and nucleus)	Bluish	Reddish	Reddish
Neutral gas (fraction of mass)	5%	15%	Less than 1%	Less than 1%
Types of star	Young (disk), old (halo and nucleus)	Young	Old	Old

[a]Mass and luminosity are given in solar masses and solar luminosities, respectively.

The hitch here is that the definition of the edge of a galaxy is more or less arbitrary—and hard to measure well! Different definitions of the edge (which is not simply where the visible stars end) result in different sizes. Also, the galaxy may be tilted to our line of sight.

Despite these problems, we can still make general statements about the sizes of galaxies. Dwarf ellipticals and small irregulars tend to be the smallest galaxies—only 300 to 3000 ly in diameter for some. Giant ellipticals can range up to 200,000 ly in size. To put this into perspective, imagine your height (about 2 m) to be the size of a dwarf galaxy. Then an irregular galaxy would be about twice your size, a spiral ten times your size, and a giant elliptical some twenty times your size!

The very largest galaxies, the supergiant ellipticals, can have radii up to 3 million ly. That's greater than the distance from our Galaxy to the Andromeda Galaxy (M31)! These **supergiant elliptical galaxies** tend to define the gravitational centers of clusters of galaxies (Section 19.6).

Mass

To find a galaxy's total mass is no simple task. The light from a galaxy comes mostly from its stars, but much of its material may not emit visible light—or may not emit at all (black holes, for example). This *dark matter* may inhabit the halos of most (all?) spiral galaxies (as we found for the Milky Way Galaxy—Section 18.7).

The most widely used methods of finding a galaxy's mass are **rotation curves** and **binary galaxies.** The rotation curve method (which can be used for spiral galaxies only) works as follows. A galaxy's material orbits the nucleus in a specific way, so at every distance the orbital velocity has a certain value. This rotation curve comes from the orbital motions, described by Newton's laws, that arise from the distribution of mass within the galaxy. (Recall the rotation curve for the Galaxy, Section 18.2.) So if we observe a galaxy's rotation curve and make up a model for that galaxy's mass distribution, we can work out the galaxy's total mass, including dark masses that we can't see directly. We observe the rotation curve by directing a spectroscope's slit across the galaxy and measuring the Doppler shift at a number of points from the center out to the edge (Fig. 19.19). The same technique can be applied to the 21-cm emission from spirals.

Spiral galaxies hit as high as several trillion (10^{12}) solar masses, although their rotation curves become flat without dropping off out to the extent measured so far (Fig. 19.20). Observations such as

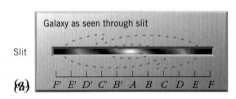

(a)

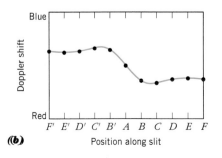

(b)

FIGURE 19.19 Observing the rotation curve of a galaxy. (*a*) A spectroscope is attached to a telescope, and its slit is placed over the image of a spiral galaxy; the positions along the slit are marked. (*F*-*F′*). (*b*) Each position has a different Doppler shift, depending on the radial velocity from the rotation.

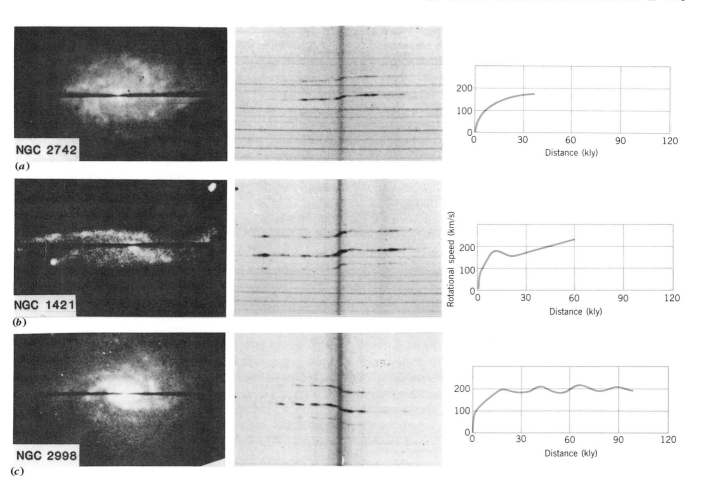

FIGURE 19.20 Actual rotation curves of spiral galaxies NGC 2742 (*a*), NGC 1421 (*b*), and NGC 2998 (*c*): left-most column, the galaxies, with the position of the respective spectroscope slits indicated; center column, actual spectra along each of the slits; right-most column, the measured rotation curves. Note in the spectra the sharp shift of the slope of the lines across the center of the galaxy. (Courtesy of Vera Rubin, Carnegie Institution of Washington.)

these of many spiral galaxies have been pioneered by Vera Rubin and her colleagues.

For binary galaxies, we again make use of the versatile Doppler shift. Imagine two galaxies orbiting about the center of mass of the binary. Assume that the orbits are stable. Then, just as with visual binary stars (Section 14.6), we could apply Newton's form of Kepler's third law to find the masses, if we knew the distance, the angular size of the orbit, the period, and the position of the center of mass.

However, galaxies revolve too slowly for us to see their actual orbits, periods, and relative centers of mass. So we cannot find the individual masses of binary galaxies. All we can measure are the present radial velocities and the separation; we don't know what part of the orbit the galaxies are on, or what the in-

clination is, so we don't know what the true orbital velocities are. But if we examine a large sample of galaxies and assume that their orbits are nearly circular and randomly oriented to our line of sight, we can estimate from these data the *average* masses of the galaxies sampled. An investigation of a few hundred binary systems, mostly spirals, using the Doppler shift of the 21-cm line to get their velocities finds an average mass of 10^{12} solar masses for these spirals (for H = 20 km/s/Mly).

Caution. These two methods—rotation curves and binary galaxies—do not give the same results. Masses from rotation curves are generally smaller than those from the binary galaxy method. This difference has reinforced the idea that spi-

Mass–Luminosity Ratios

At a Glance

OBJECT	RATIO OF MASS TO LUMINOSITY
Sun	1.0
Nucleus: spiral galaxy	1 to 3
Disk: spiral galaxy	10 to 20
Old red main-sequence stars	20
White dwarfs	10^2
Galactic (open) star clusters	1.0
Globular star clusters	1.0

ral galaxies have massive halos surrounding their bright, starry disks. Observations show that the rotation curves for spirals remain flat out to some 300,000 ly from their centers (Fig. 19.20). These flat curves imply directly that these galaxies have massive, invisible halos of dark matter. Otherwise the rotation curves should decline, as expected from Kepler's third law. Our own Galaxy has such a halo (Section 18.7).

Luminosities

If we know the galaxies' distances and fluxes, we can work out their luminosities using the inverse-square law for light. One trouble here is that it's not easy to measure a galaxy's total flux. Why not? Because a galaxy thins out gradually at its edge, making it hard to be sure that you're catching all the light from the galaxy. In addition, corrections must be applied for light absorption: first, for that due to dust in our Galaxy, and second, for absorption due to dust in the galaxy itself (especially for spirals). Finally, since we view most galaxies tilted to our line of sight, we must correct for the fact that we measure only a fraction of their total light output because of extinction by dust.

The luminosities of galaxies range from 10^5 solar luminosities for the smallest dwarf ellipticals to 10^{12} solar luminosities for the largest supergiant ellipticals. The latter types are very rare, however. The Milky Way Galaxy, if we could see it all from space, has a luminosity somewhere near 2.5×10^{10} solar luminosities.

Mass–Luminosity Ratios

Divide the total mass of a galaxy by its luminosity and you have its **mass–luminosity ratio** (abbreviated *M/L*), an indication of the average energy output per unit solar mass from the galaxy. (It is usually expressed in units of solar masses and solar luminosities, so the *M/L* for the sun is 1.) Modern determinations using binary galaxy masses give 35 for the average mass-to-light ratio for spiral galaxies (rotation curves give *M/L* of about 5 for ordinary spirals), and about 70 for giant ellipticals and lenticulars.

For comparison, the *M/L* ratio for stars in the sun's neighborhood is about 1 or a bit larger; that's because these bodies are indeed solar-type stars. A galaxy made of such stars would have the same value. If it contained only B stars, the *M/L* would be about 0.01; if all M stars, about 20. Note that dark matter adds to the mass but not to the luminosity, so it makes the ratio larger. Spiral galaxies show a trend of rising *M/L* as the distance from the center of the galaxy increases. In the nucleus, the value is about 2. At the edge of the visible disk, the value climbs to about 20.

Ellipticals have a larger *M/L* because they contain a greater percentage of low-mass stars with low luminosities—main-sequence stars of class M. This extra abundance of M stars (and the absence of luminous bluish stars) would mean that ellipticals should be redder overall than spirals (see the following paragraphs). Other objects that may contribute to the mass but not to luminosity are neutron stars, black holes (including perhaps a giant one in the nucleus), brown dwarfs, and dark interstellar matter.

Colors

As for stars, we can measure the colors of galaxies by comparing the flux at two different wavelengths. The color of a galaxy depends on the predominant stellar type in its mixture of stars; it is related directly to the kind of stars in the galaxy (Fig. 19.21). For example, a galaxy with many OB stars is bluer than a galaxy with few such stars.

In fact, a direct connection exists between a galaxy's type and its color. Ellipticals tend to be much redder than spirals, and spirals redder than irregular galaxies. Within the spiral group, the galaxies appear redder as their nuclear bulges grow larger and their spiral arms less extensive. The progression of color from the bluer irregulars to the redder ellipticals reflects a trend in the composition of the galaxy's stellar population. The reddish color means that ellipticals and the nuclei of spirals contain an old Population I. In general terms, then, an old Population I predominates in ellipticals, whereas a much younger Population I stands out in the irregulars. The

FIGURE 19.21 The spiral galaxy Messier 83 (NGC 5236; Hubble type Sc). This galaxy appears almost face-on to us, and it lies only 27 Mly away. Note the older, yellow stars in the compact nucleus and the graceful spiral arms with bluish regions of star formation. These areas have been the sites of many supernova explosions in this galaxy. (Courtesy of David Malin, AATB.)

mixture in the spirals is determined by the size of the nuclear bulge (old Population I) compared with that of the spiral arms (young Population I). Population II, which exists mainly in the globular clusters and galactic halo, probably is a minor contributor in all *large* galaxies.

Here is a way to distinguish among galaxy types. Recall that our Galaxy has a halo, a nuclear bulge of reddish stars, and spiral arms of bluish stars. Imagine our Galaxy without the halo and the nuclear bulge. What remains (the arms) is like the stars, gas, and dust found in irregular galaxies. Now imagine our Galaxy stripped of its spiral arms. The remains (nuclear bulge and halo) are typical of the composition of elliptical galaxies. Irregulars have both old and young stars, but ellipticals contain only old stars. Why? Because in irregulars, enough gas and dust remains so that star formation continues; in ellipticals, it halted many years ago.

As you might suspect, a galaxy's color also is related to its overall content of gas and dust. The reddest galaxies, ellipticals, contain almost no gas and dust. The bluest galaxies, irregulars, contain the greatest percentage of gas and dust relative to their total mass.

Spin

The overall shape of a galaxy depends on its spin (actually, its angular momentum—Focus 9.1). The more spin a galaxy has, the flatter its shape. Elliptical galaxies have a small angular momentum. Spirals have a higher angular momentum in the disk than in the halo. In contrast, irregulars have no shape to speak of; their angular momentum is essentially zero. Since these galaxies of different types have different amounts of angular momenta, they cannot evolve by themselves from one type to another if angular momentum is conserved. (However, interacting galaxies can exchange angular momentum and so change their shapes.)

Note that the sequence from ellipticals to irregulars marks a sequence from galaxies in which starbirth ceased long ago (ellipticals: old Population I and some Population II, no gas and dust) to those in which starbirth is carried on very actively (irregulars: young Population I, large percentage of gas and dust). This trend gives a key clue to understanding the evolution of galaxies.

To sum up. Galaxies contain stars, gas, and dust. The mixture of gas, dust, and stars, as well as the stellar types, are related closely to a galaxy's type and structure. Although the masses of galaxies are difficult to measure for distant systems, we know that in general, giant ellipticals are the most massive and dwarf ellipticals the least massive.

■ **The determination of mass by rotation curves works only for what type of galaxy?**

19.6 Clusters of Galaxies

Whenever you get the chance, take a close look at wide-angle photos of the sky (Fig. 19.22). On the photos of regions in which the stars thin out—away from the Milky Way—you can see the tiny forms of galaxies. If you look at enough photos, you will notice that if you find one galaxy, you're likely to see others nearby. You have learned that galaxies tend to come in clusters. In fact, it may be true that *all* galaxies belong to clusters—though many of these clusters may be a simple marriage of two galaxies. Our universe is one of **clusters of galaxies**! But astronomers need to measure the redshifts of individual galaxies to bring depth to the universe—a time-consuming task.

The Local Group

The cluster of galaxies to which the Milky Way Galaxy belongs is called the **Local Group** of galaxies. This small cluster takes up a volume of space nearly 3 Mly across in its long dimension (Fig. 19.23). Our Galaxy is located near one end of the Local Group, and M31 is near the other.

As the most massive objects in the cluster, the Milky Way Galaxy and M31 dominate its motions and secure the other members gravitationally. In fact, the Galaxy and M31 orbit each other. The other members of the Local Group come along for the ride. The Local Group

FIGURE 19.22 A cluster of galaxies in the southern constellation of Pavo, which lies about 300 Mly from the Milky Way. Arrow indicates the supergiant elliptical galaxy near the center of the cluster. Most of the other visible galaxies are spirals. (Courtesy of the Royal Observatory, Edinburgh.)

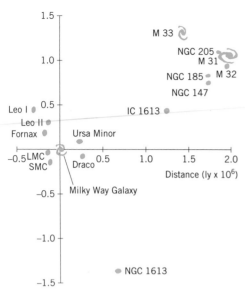

FIGURE 19.23 The most prominent galaxies in the Local Group, as viewed looking down on the plane of the Milky Way Galaxy: most of these are dwarf elliptical or irregular galaxies.

contains about thirty galaxies; they are mostly ellipticals. Some are quite faint; these are the dwarf ellipticals, which contrast so dramatically with the giant ellipticals found in other clusters. The obscuring matter in the Milky Way probably clouds our sight of other members, especially faint dwarf ellipticals. (Radio observations are beginning to find them.) Let's look briefly at some of the more important members of the Local Group.

The Large and Small **Magellanic Clouds** (Fig. 19.24) lie closest to the Gal-axy. The Large Cloud lies at a distance of 170,000 ly; the Small Cloud at 200,000 ly. Both are, in fact, connected to our Galaxy by a bridge of hydrogen gas. The two clouds are physically connected by a large but thin envelope of neutral hydrogen and a thread of stars. Both are distorted by tidal interactions with our Galaxy.

The Large Magellanic Cloud (abbreviated LMC) contains stars totaling about 20 billion solar masses. Spectra of a large number of stars in the LMC show that most stars are similar to those in the solar neighborhood, but more OB stars are there. Radio studies demonstrate that the LMC has large amounts of neutral hydrogen gas, a total of approximately 3 billion solar masses. The LMC is a medium-sized galaxy, with a diameter about half that of the Milky Way and a tenth of its total mass. Its shape is irregular, with a hint of barred spiral structure (perhaps as many as six or seven ill-defined arms) but without an obvious nucleus. (Supernova 1987A exploded in the LMC; Section 17.4.)

The Small Magellanic Cloud (SMC) displays a stellar population similar to that of the LMC. The SMC has a total mass of about one-fourth that of the LMC and is only a little smaller—about one-third the diameter of the Galaxy. Like the LMC, the SMC has an irregular shape with a bar but no nucleus at its center.

M31 is a large spiral galaxy easily visible to the unaided eye on a dark,

FIGURE 19.24 The Large and Small Magellanic Clouds. These are the two closest galaxies to the Milky Way Galaxy, and they interact with it gravitationally. (*a*) The LMC has a hint of a bar; note the many bluish regions of new stars. (*b*) The SMC is an irregular galaxy with no clearly defined structure and relatively fewer young stars than the LMC. (Both courtesy of NOAO/CTIO.)

(*a*)

(*b*)

(a)

(b)

FIGURE 19.25 Messier 31. (*a*) Close-up view of the nucleus of M31, showing its old stars and distinctive dust lanes. (Courtesy of U.S. Naval Observatory.) (*b*) *IRAS* observations of M31; the false colors are blue for emission at 12 μm, green for 60 μm, and red for 100 μm. Star formation is occurring in the bright yellow ring of clouds. The nucleus also shows strong infrared emission from the dust around old stars. (Courtesy of NASA/JPL.)

bees around a hive, in a distribution like that around our galactic system. The Andromeda Galaxy also has companions— a total of seven dwarf ellipticals that orbit it. One of these is called M32, a dwarf elliptical with a diameter of only 1000 ly containing a mere 400 million stars. The *HST* imaged the nucleus of M32 (Fig. 19.26) and found a surprise: a dense concentration of stars in the center that hints at the presence of a supermassive black hole.

M33 is the only other large spiral in the Local Group (Fig. 19.27). It is the closest Sc II galaxy to us, only 2.7 million ly away. In a photo, you can easily trace out wide, open spiral arms. Note that the arms are resolved into stars; most of these are bluish supergiants. Because M33 is so close, we have been able to in-

FIGURE 19.26 *HST* image of the core of the dwarf elliptical galaxy M32. Note the steady increase in brightness in light from the stars toward the center. The density of stars there is 100 million times that in the solar neighborhood. The entire region shown is about 175 ly across. (Courtesy of Tod Lauer, NASA.)

moonless night. With binoculars you find an elliptical, hazy patch of light; the spiral arms are too faint to be seen without a big telescope. The light from M31 arriving now departed from its source 2.2 million years ago!

We can examine directly the nuclear region of M31 (Fig. 19.25*a*), which in optical photographs shows some gas and dust. *IRAS* observations (Fig. 19.25*b*) revealed that the star formation in the nucleus takes place in a distinct ring of clouds—perhaps similar to the ring in the nuclear region of the Milky Way.

M31 tilts 15° to the line of sight. Because of the tilt, dark lanes of obscuring material are plainly visible along with the spiral arms marked by OB stars. A halo of globular clusters surrounds M31 like

FIGURE 19.27 The spiral galaxy Messier 33 in Triangulum is Hubble type Sc; that is, it has a small nucleus and spread-out spiral arms. (Courtesy of Palomar Observatory, California Institute of Technology.)

vestigate in detail its dark dust lanes, H II regions, open star clusters, novas, and cepheid variables. M33 has an overall diameter of about 60,000 ly—only about half the size of the Milky Way Galaxy.

Other Clusters of Galaxies

Other clusters range from compact ones to rather loose arrays of galaxies. The Fornax cluster, one relatively close to us, displays a wide variety of types of galaxies, even though the total number is only sixteen. The huge **Coma cluster** spreads over at least 20 million ly of space and contains thousands of galaxies. Even observations of just the brightest galaxies show how common clustering is for galaxies (Fig. 19.3). From these observations, we find that a typical cluster contains about a hundred galaxies and is separated some tens of millions of light years from its neighboring clusters.

The **Virgo cluster** stands out as one of the most stupendous in the sky (Fig. 19.28; also look back at Fig. 19.3). Of the 205 brightest galaxies in the Virgo Cluster, the four brightest are giant ellipticals, but in all, ellipticals make up only 19 percent compared with the spirals' 68 percent. The Virgo cluster covers about 7° in the sky (fourteen times the diameter of the moon!), which implies that its physical diameter is some 10 Mly at a distance of roughly 50 Mly. The Virgo cluster is so massive and so close that it influences the Local Group gravitationally; we are moving toward the Virgo cluster.

Clusters and the Luminosity of Galaxies

From the description of the Local Group, you might suspect that very luminous galaxies are few in number compared to low-luminosity ones. Only three local galaxies are very luminous (the Milky Way Galaxy, M31, and M33), whereas most galaxies in the Local Group are dwarf, low-luminosity galaxies. Galaxies in clusters have a range of luminosities: there are many faint galaxies and few bright ones in a cluster.

Much of a cluster's mass resides in very faint galaxies. It's difficult to estimate the masses of these clusters because not all the material in them can be seen at visual wavelengths; so adding up all the galaxies gives a lower limit to the cluster's mass. On the other hand, if the cluster is assumed to be bound by gravity, the motions of the galaxies within it establish an upper limit on its mass. (The actual value lies between these two limits.) Masses range from 10^9 to 10^{15} solar masses.

You can use the M/L ratio to consider the amount of dark matter in clusters. Take the Coma cluster as an example; two supergiant galaxies lie near the center. Their huge masses collect matter around them. By studying the velocities of the galaxies in the inner 4 Mly, we estimate a mass of almost 10^{15} solar masses in this region. Yet, when we count up the luminosities of visible galaxies, we find the M/L ratio is 300! (The typical range for clusters is 300 to 500.) Given the types of

Clusters of Galaxies

At a Glance

Number of galaxies	Few to a few thousand
Total mass	10^{14} to 10^{15} solar masses
Extent	Few million light years
Distance to next cluster	Few tens of millions of light years

FIGURE 19.28 The small part of the central region of the Virgo cluster of galaxies. The giant elliptical galaxy M86 lies just to the upper right of center. To its right is another giant elliptical, M84. (Courtesy of NOAO/KPNO.)

galaxy here, we expect a value of 10. So visible stars make up only a few percent of the mass in the Coma cluster. The rest is the so-far mysterious dark matter.

Interacting Galaxies

A remarkable fact about clusters of galaxies is that the spacing of galaxies is pretty close, compared to the sizes of the galaxies themselves. Consider, for instance, planets and stars in terms of sizes and spacing. In the solar system, the planets are spaced out about 100,000 times their diameters. In the Galaxy, stars are spaced out about a million times their diameters. But in a cluster of galaxies, the spacing amounts to only about a hundred times a typical galaxy's diameter. Astronomically speaking, galaxies in a cluster are very crowded together. So you might wonder if they would ever pass close by to one another. Yes, they do!

Consider this additional fact: the most massive galaxies (giant ellipticals) are at least 10 million times more massive than the least massive ones (the dwarf ellipticals). Then it's not hard to imagine tidal forces (Section 9.1) of the largest galaxies disrupting the smallest ones strongly enough to destroy their structure and then pull the pieces in. Astronomers call this devouring of a smaller galaxy by a larger one **galactic cannibalism.**

A striking name, but do we have any evidence to support it? Recent observations show that the supergiant elliptical galaxies do have peculiar properties; they are not simply very massive ellipticals. These properties include (1) extensive halos, up to 3 Mly in diameter, (2) multiple nuclei near their centers, and (3) locations at the center of clusters. These observations, when combined with the motions of supergiant elliptical galaxies within clusters, suggest that these special galaxies formed from galactic cannibalism. How? By close encounters at the centers of clusters that tidally strip material from other galaxies; then the stripped-away material is picked up to promote the growth of a supergiant elliptical galaxy.

Galaxies do not have to actually merge to show the effects of tidal forces.

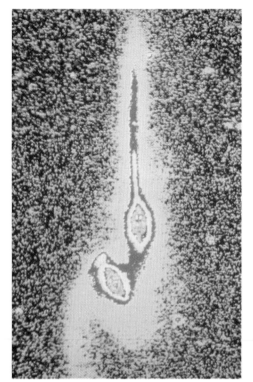

FIGURE 19.29 The galaxies NGC 4676a and 4676b, a tidally-interacting pair. The false colors in this computer-enhanced photograph bring out the details of the gravitational interaction, such as a bridge of material (red) between the two galaxies. Note the tadpolelike tails extending from each galaxy. (Courtesy of NOAO/ KPNO.)

A modestly close encounter would also have observable consequences. As demonstrated by the earth's tidal bulges, matter would be pulled out in tongues from both sides of each galaxy; and because galaxies rotate, their material after an encounter would flow off in arc-shaped streams. So we expect that bridges of material might join two tidally interacting galaxies with tails pointing away from each in opposite directions. Some galaxies with peculiar shapes that do not fall into the standard Hubble categories show indications of tidal interactions (Fig. 19.29). Note a bridge of material between the galaxies and the tadpolelike tails extending from them. Computer simulations show that these forms result naturally from a tidal interaction during a close encounter.

Recent computer simulations indicate the possibility that collisions can transform spiral galaxies into elliptical ones. The interaction of stars, gas and dust, and a halo of dark matter results in the merger of the two galaxies into a single one (Fig. 19.30). The collision drives vast amounts of gas into the center of the

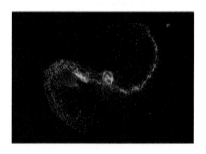

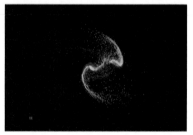

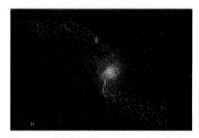

FIGURE 19.30 Computer simulation of a close encounter between two galaxies. Spiral galaxies containing gas (blue), stars (red), and invisible dark matter collide in this computer-generated sequence from top to bottom. The last frame shows a core of dense gas (white) and two long arms—a configuration seen in one kind of elliptical galaxy. (Courtesy of Lars Hernquist and Joshua Barnes; calculations done at the Pittsburgh Supercomputer Center.)

resulting elliptical gas, which might spur the formation of a supermassive black hole there.

Tidal interactions may well trigger the short, large-scale episodes of star formation seen in many irregular galaxies—sometimes called **starburst galaxies.** In particular, the gas and dust throughout a galaxy respond strongly to the tidal forces of a companion, even without an actual merger of the two, such that the gas becomes unstable and starts to clump. A notable example of such a galaxy is the irregular galaxy M82 (Fig. 19.31a), which is thick with dusty clouds of gas in its nuclear region. Infrared observations demonstrate that these are active sites of star formation. Radio data from the VLA (Fig. 19.31b) show radio hot spots—giant H II regions—that coincide with the star-forming regions found in the infrared. The companion to M82 is the spiral galaxy M81, which is close enough to have tidal effects.

IRAS discovered that a few very luminous galaxies emit 95 percent of their energy in the infrared region of the spectrum. *HST* has imaged one of these, called Arp 220, a super example of a starburst galaxy. Arp 220 may have been formed by the merger of two spiral galaxies. The *HST* image reveals gigantic clusters of young stars being produced at a furious rate from the gas and dust of the former galaxies (Fig. 19.32). Other *HST* images show new star clusters, about the size of globular clusters, in the core of a peculiar galaxy that probably is the headon collision of two spirals. Eventually, a giant elliptical galaxy may form. Such mergers of spirals may well explain why spirals are less numerous now than in the past.

■ What defines a cluster of galaxies?

─────

19.7 Superclusters and Voids

Are there clusters of clusters of galaxies? That is, do **superclusters** exist? Yes! Much work on superclustering has been done by Gérard de Vaucouleurs. He finds evidence for a **Local Supercluster** that has a diameter of some 100 Mly. It contains the Local Group as well as the

Virgo cluster among others, for a total mass of some 10^{15} solar masses. The Local Supercluster appears to be somewhat flattened, which may mean that it is rotating, although it is also possible that it simply was formed this way. The center of mass lies in or near the Virgo cluster; our Galaxy and M31 lie on the outskirts.

Brent Tully and Richard Fisher have mapped, in three-dimensions, the Local Supercluster (Fig. 19.33). The map shows a rich, convoluted structure that breaks into two main clouds with streamers (thin, cigar-shaped clouds of galaxies) emerging above and below the central plane. Most of the supercluster is empty space: 98 percent of the visible galaxies are restricted to 5 percent of the volume. These clouds of galaxies outline a disklike structure six times wider than it is thick—a true cosmic pancake of clusters of galaxies!

Astronomers have probed superclusters beyond the local one. Clyde Tombaugh, the discoverer of Pluto (Section 11.7), made the first map of a supercluster. But when he showed it to Edwin Hubble, Hubble refused to believe that galaxies were grouped in superclusters! Before recent observations drove home the reality of superclusters, most astronomers imagined that the universe contained more or less spherical clusters of galaxies embedded in a more or less uniform background of noncluster galaxies. That image has changed drastically in recent years.

For example, the Hercules supercluster (Fig. 19.34) covers more than 600 billion cubic light years of space. (It lies at a distance of 720 Mly, if H is 15 km/s/Mly.) The supercluster itself occupies a broad band, spreading out a distance of 400 to 600 Mly. In front of the supercluster lies a **void** some 330 Mly deep that separates the supercluster from foreground galaxies.

The superclusters that have been well mapped to date exhibit common features. First, they confirm the existence of superclusters as organized structures composed of multiple clusters of galaxies. Second, and this was a surprise, they contain large areas devoid of visible galaxies. These voids must be an integral

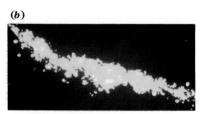

(a)

(b)

FIGURE 19.31 Messier 82 (NGC 3034) in Ursa Major, an example of a starburst galaxy. (*a*) Note, in this optical photograph, the many dust lanes. Infrared observations indicate that star formation is occurring. (Courtesy of Palomar Observatory, California Institute of Technology.) (*b*) VLA map showing many hot spots of radio emission that mark regions of intense star formation. (Courtesy of NRAO/AUI.)

FIGURE 19.32 *HST* true-color image of the central part of the starburst galaxy Arp 220: young, gigantic star clusters of newly formed stars are visible. *HST* images have shown that Arp 220 has two nuclei, one from each of the merging spiral galaxies. (Courtesy of E. Shaya, D. Dowling, University of Maryland and NASA.)

part of the process that forms superclusters—a process about which we have vague ideas right now. Third, streams of galaxies appear to connect the main concentrations in superclusters. Although we have examined only a few percent of the sky in depth, we are beginning to catch a glimpse of the architecture of the cosmos.

Margaret Geller and John Huchra of the Harvard–Smithsonian Center for Astrophysics have mapped a large band of galaxies that they have dubbed the "Great Wall." This structure covers a span of some 500 Mly, has a thickness of no more than 15 Mly, and contains a density of galaxies that is about five times larger than the cosmic average. The total extent of this structure is not yet known, for it has not been completely mapped.

Even the apparent voids may are not so empty. Several dwarf galaxies have been found in a void in the direction of the constellation of Boötes. This gigantic void, covering some 30 million cubic light years of space, had not been known to contain any galaxies before this discovery. The fact that a few, albeit small, galaxies reside within a void has important consequences for any models for the origin of the universe of superclusters (Section 21.7).

Finally, keep in mind that we have so far seen just small slices of the cosmos

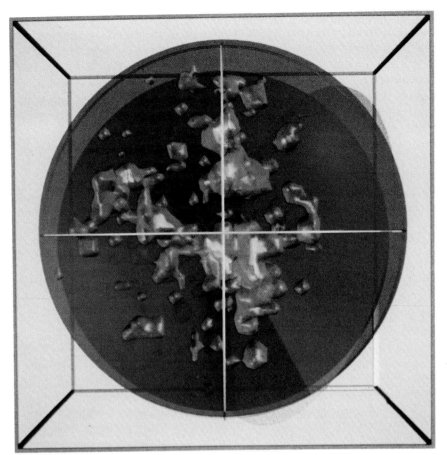

FIGURE 19.33 Distribution of rich clusters of galaxies in space in the Local Supercluster. The view is edge-on to the Local Supercluster; the sphere has a radius of 1100 Mly and contains 382 rich clusters of galaxies. The orange contour is the highest density of clusters per cubic megaparsec; the green contour, the lowest. (Courtesy of Brent Tully, Institute for Astronomy, University of Hawaii.)

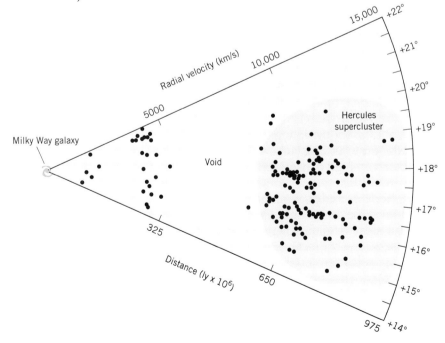

FIGURE 19.34 A slice of the Hercules supercluster. Shown here are the radial velocities (from the red shifts) and so distance (for $H = 15$ km/s/Mly) and positions in the sky for the galaxies in the supercluster, which contains a number of clusters. The Milky Way lies at the vertex of this wedge diagram. Note the clumping of the galaxies and the void of space in front of the supercluster. (Adapted from a diagram by S. A. Gregory and L. A. Thompson.)

in three dimensions. For instance, the Galaxy and the Local Group appear to be moving at speeds of a few hundred kilometers per second toward the Hydra–Centaurus region of the southern sky. This observation has resulted in the proposal of the "Great Attractor"—a conglomeration of galaxies covering about a third of the southern sky. A survey of some 17,000 galaxies here reveals a grouping of galaxies at a redshift of 4000 km/s. This configuration is called the Hydra–Centaurus supercluster; its mass may account for the motion of the Local Supercluster in that direction.

The Cosmic Tapestry

What does the universe of galaxies look like on a grand scale? If we put together a map of the nearby superclusters, we begin to get an idea. The map (Fig. 19.35) shows that galaxies cluster in knots and filaments in a hierarchical fashion. These are chains of galaxies and clusters of galaxies (looking like a chain-link fence). Nearly all galaxies are within these clusters, with huge holes between them devoid of luminous matter.

▪ What is the shape of a supercluster?

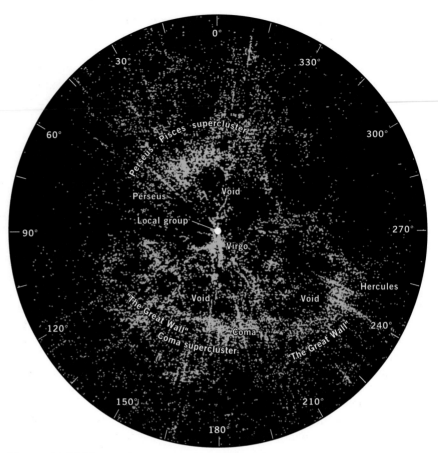

FIGURE 19.35 The nearby universe of more than 14,000 galaxies shown by the layout of nearby superclusters. The Galaxy lies in the center of this schematic representation, which spans 500 Mly in space. Each dot represents a galaxy; at the cores of clusters, it is not possible to show the galaxies individually. The view spans 360° around the sky and has a depth of 36°. The Local Supercluster is dominated by the Virgo cluster. Note the immense voids between us and the great Coma supercluster. (Adapted from a diagram by J. Huchra, M. Geller, and R. Marzke; Harvard–Smithsonian Center for Astrophysics.)

19.8 Intergalactic Medium and Dark Matter

Is intergalactic space empty? Or does an intergalactic medium exist, in analogy to the interstellar medium? If an intergalactic medium is present, it may contain both gas and dust. The gas (probably hydrogen) may be in neutral form, or it may be ionized. We can look for the **intergalactic medium** in two locations: *between* the clusters of galaxies and *within* clusters of galaxies.

To get some idea of how much material might be in an intergalactic medium, imagine the following. Take the matter from all the galaxies we can see and spread it out over the entire volume of space we can observe. This spread-out material would have a density of about 4×10^{-28} kg/m^3. (That's about 2 hydrogen atoms every 10 cubic meters.) For the intergalactic medium to be signif-

icant, it would need to have about this density—equivalent to that of visible matter in the form of galaxies. What is the evidence for such a density of dust, neutral hydrogen, or ionized hydrogen?

Ionized hydrogen (H II) is the most likely candidate for the intergalactic medium. Because intergalactic material does not have a high density, ionized hydrogen would take a very long time to find an electron and recombine. If the gas is hot (in the 10^7 K range), you can search for x-ray or ultraviolet emission. X-ray observations of local superclusters show sources that are contained within superclusters. The sources consist of spots centered on clusters; this implies that a hot gas exists in superclusters, that it is highly clumped, and that little gas

---

Superclusters of Galaxies

At a Glance

Number of clusters	A few tens to a few hundred
Total mass	10^{16} to 10^{18} solar masses
Extent	Few hundred million light years
Shape	Filamentary, pancake
Contain	Voids (millions of cubic light years); roughly spherical

exists between clusters. The gas has a temperature of some 10^8 K.

Another good candidate for matter *within* clusters is (again!) ionized hydrogen gas. X-ray observations back up this idea. Many clusters of galaxies are known to date to emit x-rays (Fig. 19.36); they tend to lie in the central regions of the clusters. The x-ray luminosities of clusters range from 10^{36} to 10^{38} W. The diameters of the x-ray-emitting regions range from 160,000 ly to 5 Mly.

A reasonably confirmed model for this x-ray emission is that it comes from hot, ionized gas. Such a gas tends to settle around the center of mass of a cluster. It has typical temperatures of 10^7 to 10^8 K and densities of about one ion per cubic meter; such properties nicely explain the x-ray observations. So we have reasonable evidence of intergalactic gas in clusters—about equal to the amount of mass in the galaxies themselves. But its density does not appear to be sufficient to bind the cluster gravitationally. Consider again the Coma cluster. Its x-ray emission indicates that no more than 20 percent of the total mass in the inner 4 Mly is in the form of a hot plasma. Recall that the visible galaxies make up just a few percent; the rest is dark matter. So the x-ray-emitting material does *not* solve the dark matter problem for us. Nor does it appear to be sufficient to close the universe. But we are getting close, because the *M/L* for clusters is about 300, and we need about 700 for a closed universe.

We now know that matter resides generally in clusters in a dark form that is undetected so far. That matter may be in the form of elusive particles such as neutrinos (especially if neutrinos have even a very small mass). If such dark matter exists in a proportion of about ten to twenty times that of visible matter, then the dark matter controls the structures of clusters and superclusters, with the luminous matter as the visible tip of the large-scale clustering of matter.

■ **What is the most likely candidate for the hot intergalactic medium?**

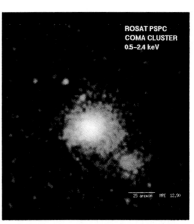

FIGURE 19.36 *ROSAT* map of the x-ray emission from the Coma cluster of galaxies. The x-ray emission comes from hot, ionized gas within the cluster; note how the emission is the strongest around the central region. (Courtesy of *ROSAT*, Max Planck Institute for Extraterrestrial Physics.)

Key Concepts

1. Distances to galaxies are essential to finding out their properties; they can be estimated in a relative sense from brightness and angular sizes.

2. Galaxies come in three main types, based on shape: ellipticals (dwarf, giant, and supergiant), spiral (normal and barred), and irregulars.

3. Measuring distances to galaxies is difficult and relies on indirect schemes. The basic trick is to find very bright objects whose luminosities can be reasonably estimated, identify these in the target galaxies, compare their flux to their luminosity, and use the inverse-square law for light to infer their distance.

4. Measured redshifts and distances of galaxies result in Hubble's law and Hubble's constant, H. The greatest uncertainty in H (its value lies between 15 and 30 km/s/Mly) arises from the uncertainties in distances.

5. Galaxies differ in terms of size, mass, luminosity, mass–luminosity ratio, spin, and colors—all of which reflect their content of stars, gas, and dust. In general, galaxies have mass–luminosity ratios that are greater than one, an indication of dark matter.

6. Galaxies come in clusters, containing from two to thousands, held together (at least for a while) by the gravity between the galaxies. The mass–luminosity ratios are around 100, an indication of a large fraction of dark matter.

7. The Local Group contains around thirty galaxies, mostly of low mass and luminosity, dominated by the Milky Way and Andromeda galaxies. The Local Group spreads over a volume some 3 Mly in diameter.

8. Clusters are grouped in superclusters; for example, the Local Group is one small piece of the Local Supercluster (which is centered on the Virgo cluster). Superclusters seem to be surrounded by voids and to come in long chains; they are the largest entities in the cosmos. At least a few superclusters are known to have a mass–luminosity ratio of a few hundred.

9. A very thin, very hot gas exists between clusters and superclusters; within clusters, hot, thin, ionized gas exists and is observed by the x-rays it emits. This plasma does *not* account for the dark matter in clusters, whose nature is as yet unknown.

Key Terms

binary galaxies

clusters of galaxies

Coma cluster

elliptical galaxy

galactic cannibalism

intergalactic medium

irregular galaxy

lenticular galaxy

Local Group

Local Supercluster

Magellanic clouds

mass – luminosity ratio

rotation curve

spiral galaxy (normal and barred)

starburst galaxies

superclusters

supergiant elliptical galaxies

Tully – Fisher relation

uniformity of nature

voids

zone of avoidance

Study Exercises

1. Which galaxies generally appear redder, ellipticals or spirals? (Objective 19-1)

2. Which galaxies contain more gas and dust relative to their total mass, spirals or irregulars? (Objective 19-1)

3. At the same distance from us, would irregular galaxies appear larger than spiral ones? (Objectives 19-1, 19-2, and 19-3)

4. How do astronomers know that other galaxies are made of stars? (Objective 19-4)

5. In recent years the value of Hubble's constant has been revised by some investigators from 30 to 15 km/s/Mly. How does this change affect the distances to galaxies inferred from redshift and Hubble's constant? (Objectives 19-5, 19-6, and 19-8)

6. Describe how cepheid variable stars can be used to estimate distances to nearby galaxies. State the assumptions and limits of this method. (Objectives 19-6 and 19-7)

7. Why must intergalactic gas be both ionized and hot? (Objective 19-11)

8. Describe the layout of a supercluster of galaxies. (Objective 19-12)

9. How do astronomers see the depth of superclusters in space? What uncertainty exists in their technique? (Objectives 19-6, 19-9, and 19-12)

10. What is the most common type of galaxy found in the regions of space that we can probe? (Objectives 19-1 and 19-10)

11. What is the basic information that implies that a large fraction of the matter in clusters of galaxies is in the form of dark matter? (Objective 19-13)

12. The observed rotation curves of most spiral galaxies become flat at large distances from their centers. What can we infer about the distribution of mass in these galaxies? (Objective 19-2)

13. How do we know that the space between galaxies and clusters of galaxies contains a very small amount of dust? (Objective 19-11)

14. Imagine you observe two spiral galaxies through a telescope. The smaller one appears to have an angular diameter $\frac{1}{4}$ that of the larger one. How far away is the smaller galaxy relative to the larger one? Which galaxy would appear fainter? (Objective 19-3)

Problems & Activities

1. Consider a galaxy whose redshift is 0.10. What is the radial velocity of the galaxy? What is its distance if H is 20 km/s/Mly?

2. Imagine a dwarf galaxy at the edge of the Local Group. What is its orbital period?

3. Consider a hot, opaque cloud of hydrogen gas at a temperature of 10^6 K. At what wavelength would its spectrum peak?

4. If our Galaxy and M31 make up an orbiting pair, what is their orbital period?

5. The typical speed of a cluster of galaxies within a supercluster is about 300 km/s. How long would it

take a cluster to move across a supercluster? How does this compare to the age of the universe?

6. The Coma cluster has a radial velocity from its redshift of about 6650 km/s. What is its distance?

7. The angular diameter of M31 is about 4.5°. What is its diameter-to-distance ratio? What is its diameter?

8. A new spiral galaxy is discovered. Its angular diameter is found to be 10 arcmin. What do you estimate its distance to be?

See for Yourself!

You can make up your own Hubble plot and find a value of Hubble's constant from a sample of bright galaxies. Because they are bright, they are relatively close (in contrast to most of the galaxies in Fig. 19.17). These galaxies are given in Table A.2.

On a graph paper, make the horizontal axis the distance from the closest to the farthest galaxies, in millions of light years. You don't have to go out farther than 100 Mly. These distances are given for each galaxy in the second column in the table. On the vertical axis, you want to use the radial velocity in kilometers per second; that information is given in the third column of the table. Plot the points for the galaxies and draw a straight line through them. DO NOT "CONNECT THE DOTS!" Try to draw a straight line so that about as many galaxies fall above and below the line as well as on the line. Now measure the slope of the line, which is the rise (*y*-axis) over the run (*x*-axis). Use the complete length of this "best-fit" line, not just a part of it. What value do you get? That is

TABLE A.2 SELECTED BRIGHT GALAXIES

GALAXY	DISTANCE (Mly)	RADIAL VELOCITY (km/s)
Fornax A	98	1713
Messier 66	39	593
Messier 106	33	520
NGC 4449	16	250
Messier 87	72	1136
Messier 104	55	873
Messier 64	23	350
Messier 63	36	550
NGC 6744	42	663

your value of Hubble's constant from the information in the figure.

Compare your results to that given in Figure 19.18. Do they look the same? How does your value of the Hubble constant compare to the one for the figure? ■

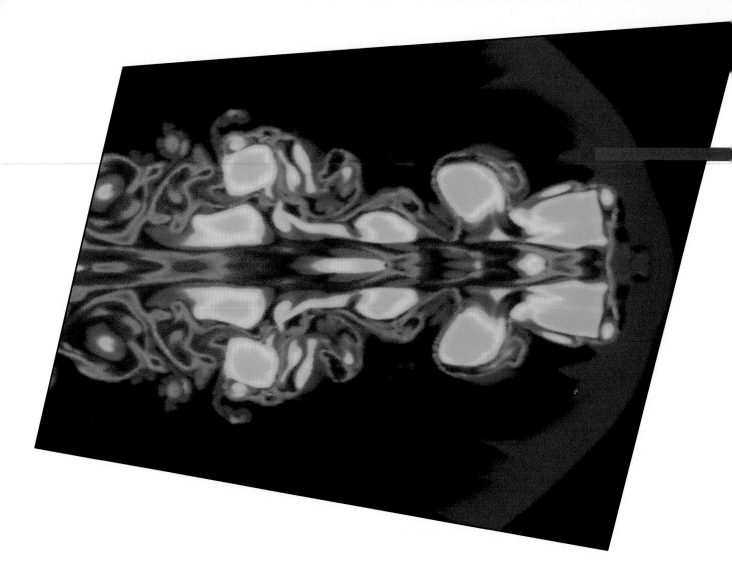

Central Question
What observational evidence do we have for violent activity in objects beyond the Milky Way Galaxy?

Cosmic Violence

If the radiance of a thousand suns
were to burst into the sky,
that would be
the splendor of the Mighty One....

THE BHAGAVAD-GITA

The universe appears calm, caught up in the well-controlled generation of energy in stars and the strict Newtonian dance of matter. Until the middle of the twentieth century, we saw it as a gentle cosmos. Rare outbursts such as novas only occasionally shattered the stillness. The advent of radio astronomy ripped the veil from a violent universe. Radio astronomers found the nucleus of our Galaxy to be a strong radio emitter. They also detected intense radio sources beyond the Galaxy, objects at enormous distances that required a tremendous outpouring of energy. By the 1970s, observational evidence indicated beyond a doubt that violent events commonly occur in the nuclei of galaxies of many types. These are called active galaxies for their energetic properties.

Observations of jets emanating from the nuclei of active galaxies have reinforced this view. These jets appear as cosmic umbilical cords, channeling the power from the nucleus out to immense reservoirs of charged particles. Supermassive black holes in the nuclei of such galaxies may power these energetic, beamed outflows. Quasars, starlike-looking objects, remain a puzzling element in the range of cosmic violence. They have the largest known redshifts of any extragalactic objects. If the redshifts of quasars result from the general expansion of the universe, then the quasars are the most energetic bodies in the universe. At the fringes of the cosmos, quasars represent some of the first-formed objects in the visible universe—and the most powerful. How quasars produce their enormous energies remains a mystery, though observations now imply that quasars are the nuclei of young, hyperactive galaxies.

20.1 Violent Activity in Galaxies

Galaxies appear to be the most serene bodies in the universe. However, as we have uncovered more regions of the spectrum, the nuclei of many galaxies have

459

acquired the aspect of a compact arena of violent events. These have been lumped into the category of **active galaxies,** in contrast to normal galaxies (Chapter 19). The nature of the physical conditions and processes at the heart of an active galaxy remains more unknown than understood.

Evidence of Violence in Our Galaxy

Our home Galaxy's nucleus shows hints of cosmic violence. The nucleus lies at the center of the strong radio source Sagittarius A. One part of it, Sgr A West, appears to mark the actual core of the Galaxy. The central core, Sgr A*, has a size of smaller than 13 AU. It is surrounded by clumps of ionized gas that move at speeds of about 100 km/s. This region exhibits a jetlike structure above and below the galactic plane from a center of activity (Fig. 20.1). (You will see that such a structure typifies some galaxies with notable activity.)

The spectrum of the nucleus is in part nonthermal. That is, some of the emission comes from the synchrotron process. (The thermal emission is from stars.) So high-energy electrons, moving at speeds close to that of light, spiraling in huge magnetic fields, subsist in the nucleus.

Galaxies with nuclei more active than ours display an energy output at a much higher level of violence. In other words, the nucleus of the Galaxy is a scaled-down example of the emission from the nuclei of more active galaxies. The key clues are (1) emission over a wide range of wavelengths (usually nonthermal in character), (2) radio output concentrated in a small space, (3) most of the energy coming out in the infrared, (4) variability in the emission, and (5) strong radio polarization with jetlike structures, all arising from the nucleus (Fig. 20.2).

Synchrotron Emission Revisited

The process of synchrotron emission (Focus 17.1) underlies the basic physics of this chapter so firmly that you *must* have the basic concept in mind. Here it

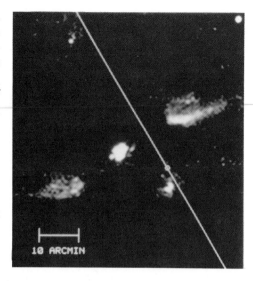

FIGURE 20.1 Jetlike structures in the Milky Way Galaxy. This radio map, at 10.7 GHz, shows the polarized emission in the Sgr A region. The diagonal line marks the plane of the Galaxy; the dot on it lies just above Sgr A. The line at bottom left shows the angular scale. Sources *B* and *C* appear to be lobes of emission extending from source *A*. (Courtesy of F. Yusef-Zadeh; observations with the 100-meter Effelsberg radio telescope by J. H. Seiradakis, A. N. Lasenby, F. Yusef-Zadeh, R. Wieleblinski, and U. Klein.)

is: whenever *charged* particles accelerate, they give off electromagnetic radiation. That's basically how any radio transmitter works; the transmitter forces electrons to accelerate back and forth at the frequency of transmission. These accelerated charges emit at that frequency of oscillation.

Now recall (Section 8.3) that magnetic fields bend the paths of charged particles—electrons are affected more readily than protons because electrons have less mass. So as electrons move through magnetic fields, their paths are bent and the electrons accelerated. (Remember that curved motion is accelerated motion according to Newton's laws.) The electrons emit electromagnetic radiation as they spiral along the magnetic field lines. The magnetic fields do not have to be especially strong (the earth's field strength or less) to do so.

In general, the faster the electrons travel, the more energetic (shorter in wavelength, higher in frequency) the radiation they emit. As the electrons continue to emit, they lose energy and slow down. So to keep a synchrotron source powered up, there must be a supply of electrons moving close to the speed of light.

Synchrotron emission turns out to power most of the strange objects presented in this chapter. We know that this is the case from the shape of the spectrum and polarization of their energy output.

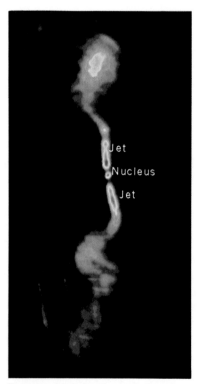

FIGURE 20.2 Radio emission from the active galaxy 3C 449. This VLA false-color map (blue the weakest emission, red the strongest) shows that the emission extends in opposing jets from the nucleus until each bends and ends in a lobe, where the material from the jet piles up. The nucleus of the galaxy contains the engine that drives these high-speed, bipolar outflows. (Observations by J. Burns and R. Perley with the VLA; courtesy of NRAO/AUI.)

Active Galaxies

Radio astronomers have found that many galaxies, the so-called active galaxies defined above, have nonthermal spectra. But how active is "active?" I'll use the term here in contrast to "normal," which applies to our Galaxy. Basically, an active galaxy's spectrum does *not* look like that of a collection of stars. It is mostly nonthermal and has infrared, radio, ultraviolet, and x-ray outputs greater than those in the optical. (Our own Galaxy does emit at all these wavelengths, but not nearly as strongly as an active galaxy.) Figure 20.3 compares the spectra of some active galaxies with those of a normal galaxy. Active galaxies make up only a small percentage of all known galaxies, but they are the most powerful.

If you look carefully at the spectra of galaxies (Fig. 20.3), you'll note that a normal galaxy has a blackbody spectrum that peaks in the optical range of the electromagnetic spectrum—just what you'd expect from a group of stars. In contrast, active galaxies have a strong peak of emission in the far infrared from heated dust. The synchrotron part of the emission crops up most noticeably in the radio. So though the emphasis here is on the synchrotron emission (which can occur at any wavelength), you should not think that it is the *only* source of radiation from active galaxies. Their spectra may well combine synchrotron as well as the continuous emission from stars and dust (which is reradiated in the infrared if the dust grains are heated to temperatures of a few hundred kelvins by the absorption of other forms of electromagnetic radiation). In addition to these forms of continuous emission, we may also find emission lines from hot gases.

Activity originates in the nucleus of active galaxies. So astronomers call these **active galactic nuclei,** or **AGNs** for short. The central issue is, What powers AGNs?

■ What sort of radio structure, exhibited by our Galaxy, typifies an active galaxy?

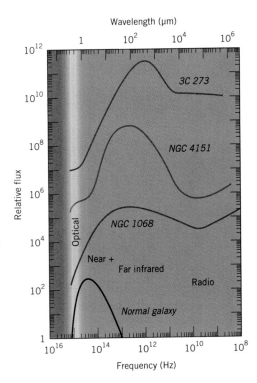

FIGURE 20.3 Comparison of the continuous spectra in the radio, infrared, and optical regions for a quasar (3C 273), two active galaxies (NGC 4151 and NGC 1068), and a typical normal (spiral) galaxy. Whereas the quasar and the active galaxies have a broad peak in their emission at far-infrared wavelengths (thermal emission from dust), the normal galaxy shows a Planck curve peak in the visible from the combined emission of all its stars. The spectrum in the radio region is nonthermal. (Adapted from a diagram by R. Weymann.)

20.2 Radio Galaxies

The largest class of active galaxies consists of the **radio galaxies,** which have strong radio emission. Two principal types of radio galaxy have cropped up: **compact** and **extended.** In an extended galaxy, the radio emission is larger than an optical image of the galaxy, whereas "compact" indicates that the radio emission is the same size or smaller. Compact radio galaxies often display very small radio sources, typically unresolved or no more than a few light years in size, and always coincident with the optical nucleus. Extended radio sources, in contrast, spread far beyond the optically visible galaxy—sometimes in two giant lobes of emission, up to millions of light years in extent, symmetrically balanced on opposite sides of the nucleus. A galaxy may show a compact source at its nucleus and also have the extended lobes; it then usually has **radio jets** connecting the nucleus to the extended radio emission.

We will be dealing with large redshifts in the spectra of the objects discussed in this chapter. Keep in mind that astronomers define the amount of the

Active Galaxies

At a Glance

High luminosity, greater than a billion solar luminosities

Nonthermal emission (usually)

Rapid variability and/or small size of nucleus (a few light years at most)

Very bright nucleus with jetlike projections

Broad emission lines (sometimes); narrow emission lines (less often)

redshift by the *change* in wavelength of a spectral line divided by the line's wavelength *at rest* (Focus 10.1).

A Zoo of Active Galaxies

In a universe of stunning objects, active galaxies take top honors. Let's look at a few examples to see what properties they have in common so that we can begin to construct a model of their unusual activity.

Only 65 Mly away, *Messier 87* is a fine example of a strong radio source and an active galaxy. A giant elliptical galaxy, M87 dominates the Virgo cluster of galaxies. Poking out from the core, a remarkable, optically visible jet fires out into space over a length of some 6000 ly. The jet has a luminosity of roughly 10^7 solar luminosities; its emission is polarized. The jet contains blobs of material (Fig. 20.4), each no more than a few tens of light years in size. Over two decades, the blobs have changed somewhat in flux and polarization.

M87 also emits x-rays with fifty times more energy than its optical emission—about 10^9 solar luminosities in x-rays from the whole galaxy. The jet itself also emits x-rays, in the form of a line of knots. Radio maps confirm that M87's jet radio emission coincides with the optical and x-ray emissions. The radio knots line up with the optical ones. So the jet overall emits over a wide range of frequencies, from radio to x-rays, and each knot of the jet generates this spectrum of energies. The synchrotron process within each knot produces this wide range of emissions (Fig. 20.5).

Other AGNs in elliptical galaxies also possess nuclear jets. In fact, radio jets are common. Almost all radio galaxies of the lowest luminosities have jets. These jets, such as that in M87, terminate in a lobe of radio emission (Fig. 20.5).

Cygnus A, one of the strongest radio sources in the sky and one of the first to be discovered, provides an excellent example of the typical double structure of a luminous, extended radio galaxy. Its radio output, some 10^{11} solar luminosities, comes from two giant lobes set on opposite sides of the optical galaxy (Fig. 20.6). Each lobe has a diameter of 55,000 ly—about half the size of our

Galaxy! The lobes hang roughly 160,000 ly away from the central galaxy and contain a cloud of energetic electrons and vast magnetic fields that store more than 10^{52} J. A needlelike jet extends from the nucleus to one of the lobes. The radio lobes have a wispy appearance, with delicate swirls in which are embedded hot spots of emission.

The optical galaxy of Cygnus A is a supergiant elliptical galaxy (Fig. 20.7)

FIGURE 20.4 The central core of the elliptical galaxy Messier 87, as imaged by the *Hubble Space Telescope* in the near-infrared. Note the steady increase in the overall brightness toward the central region, which shows that the stars are strongly concentrated in the nucleus. The bright spot at center marks the nucleus, which emits synchrotron radiation. The optical jet emerges from the nuclear region; its emission is also synchrotron in origin. The faint starlike sources scattered around the core are globular clusters orbiting within M87. (Image by T. R. Lauer; courtesy of NASA/ESA.)

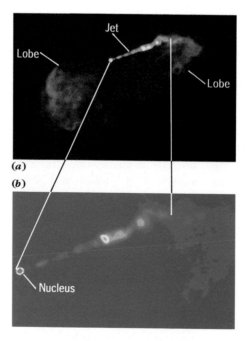

FIGURE 20.5 The active galaxy Messier 87 (also called Virgo A) in the radio region. (*a*) High-resolution, false-color radio map of M87, made at a frequency of 5 GHz. Blue indicates the weakest emission; yellow and red the strongest. The radio jet, which coincides with the optical one, extending to the upper right from the nucleus, ends in the radio lobe at the right. The radio lobe on the opposite side does not have a jet extending toward it. Some radio galaxies show this kind of asymmetry. (Observations by F. Owen; courtesy of NRAO/AUI.) (*b*) False-colors radio map showing the details in the structure of the jet. Note the bright knots along the 6000-ly length, which ends in a weak lobe of radio emission. (Observations by F. Owen; courtesy of NRAO/AUI.)

with a dust lane down its middle. It has an active nuclear region, with a spectrum showing emission lines and a synchrotron continuous emission. But beyond 25,000 ly from the center, the spectrum is just that of a mix of stars. Hence, the emission lines come from the nucleus alone.

Centaurus A (NGC 5128) is a strong radio source also with radio jets. A supergiant elliptical galaxy, Centaurus A is bisected by an irregular dust lane (Fig. 20.8) in which star formation is taking place. At a distance of about 10 Mly, Centaurus A is the closest active galaxy to ours. As viewed with an optical telescope, the galaxy has a diameter of a few tens of thousands of light years. Viewed with a radio telescope, Centaurus A has two huge outer lobes, which span more than 1.5 Mly from tip to tip. Closer to the nucleus, two more radio lobes sit on the edges of the optical galaxy; these are some 15,000 ly in size.

Centaurus A also emits variable x-rays intensely; the small source coincides with the nucleus. The nucleus of Centaurus A also emits infrared and radio strongly. These observations, along with those taken at x-ray energies, show that the nuclear emission is nonthermal.

The nucleus of Centaurus A has a direct connection to one inner radio lobe. X-ray observations show an x-ray jet streaming from the nucleus and consisting of at least seven distinct blobs. A high-resolution radio map (Fig. 20.9) shows radio emission along a jet that extends to one radio lobe. The jet has a bloblike structure that coincides with the x-ray blobs.

High-resolution radio observations reveal in the nucleus itself a very small jet, 4 ly long, that lines up with the larger one. Thin filaments emanate from the knots in the jet (Fig. 20.10); they appear to point in the direction of the presumed flow from the nucleus out to the lobe. The jet has wiggles in its overall form.

What powers the jets of M87 and Centaurus A? Observed at a wide range of wavelengths, the spectra of both jets are nonthermal. This property points to synchrotron process generation of the emission from the jets. The large polarization of the jets' light in M87 (about 25 percent) confirms this idea. The fact that

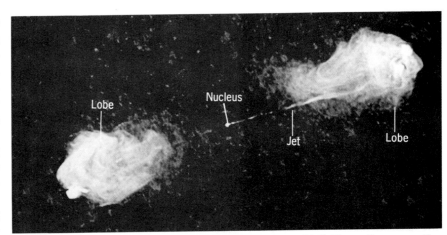

FIGURE 20.6 The radio source Cygnus A. This radio map of Cygnus A was made with the VLA and processed to show the maximum fine detail. Note the thin jets from the nucleus and the filamentary structure in the lobes. (Observations by R. A. Perley, J. W. Dreher, and J. J. Cowan; courtesy of NRAO/AUI.)

FIGURE 20.7 Optical image of the supergiant elliptical galaxy Cygnus A. This CCD image, taken under conditions of excellent seeing, shows the bright core of the galaxy and the galaxy's diffuse halo. (Courtesy of L. A. Thompson, Institute for Astronomy, University of Hawaii.)

FIGURE 20.8 Centaurus A (NGC 5128), a nearby radio galaxy—a supergiant elliptical. The diffuse, circular part of this galaxy contains hundreds of billions of old, yellowish stars. A dust lane across the galaxy's middle obscures the nuclear region; at its ends are small patches of younger, bluish stars and pinkish H II regions. (Courtesy of David Malin, AATB.)

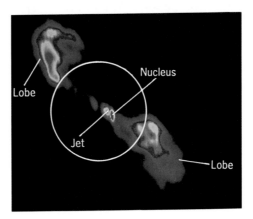

FIGURE 20.9 False-color radio map of the radio emission from Centaurus A: blue indicates the weakest emission; red, the strongest. Note the two inner radio lobes and the radio jet extending from the nucleus to the lobe at the upper left. The circle gives the rough size of the visible galaxy (Fig. 20.8). (VLA observations by J. Burns, E. Schreier, and E. Feigelson; courtesy of NRAO/AUI.)

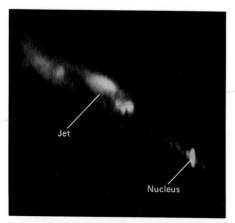

FIGURE 20.10 High-resolution radio map of the Centaurus A jet showing the nucleus and the jet, with a clumpy, twisted structure, emerging from it. This view emphasizes that the jet results from a flow of material from the nucleus. (VLA observations by D. Clarke, J. Burns, and E. D. Feigelson, courtesy of NRAO/AUI.)

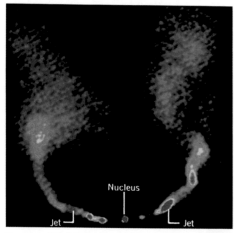

FIGURE 20.11 Radio image, made with the VLA, of the nucleus of the head–tail galaxy NGC 1265. Note how the radio emission curves away and trails from both sides of the nucleus; this is the signature of the narrow-angle-tail class of radio sources. The galaxy is moving at a speed of about 2000 km/s in the Perseus cluster of galaxies. The jets extend some 60,000 ly. (VLA observations by Chris O'Dea and Frazier Owen; courtesy of NRAO/AUI.)

the knots in the jets coincide at different wavelengths implies that the same electrons power all the emission from the knots. The nucleus provides the high-energy electrons, which are expelled either as a fairly constant beam of particles or as a sequence of ionized blobs that are thrown out along a magnetic field. The ionized stream carries magnetic fields

within it, and these help to channel the flows outward. The channel is leaky, which renders the jet's emissions visible but also presents the major problem with this model: since the electrons are losing energy, how do the jets keep up their emission over such a long path?

Structures of Radio Emission

Radio astronomers have now found that when classified by structure, extended radio galaxies fall into three main groups:

1. Doubles (example: Cygnus A): highest luminosities; two lobes aligned through center of galaxy, bright hot spots at ends.
2. Bent doubles (example: Centaurus A): intermediate luminosities; bent through nucleus, tail-like protrusions.
3. Narrow-tailed sources (example: NGC 1265; Fig. 20.11): lowest luminosities; U-shaped, rapidly moving galaxies in a cluster.

The delineation of these radio source structures is important in two ways: by providing insight into the physical processes responsible for the radio emission, and by producing information about the environment in which galaxies are immersed. Low-luminosity tailed or distorted sources make up the majority of extended radio sources. The **head–tail galaxies** are one subgroup. As implied by the name, head–tail radio galaxies have a radio head (around the visible galaxy) and an extended tail. NGC 1265 (Fig. 20.11), in the Perseus cluster of galaxies, is a good example. Not all galaxies with tails have them trailing. Instead, we find a sequence in the amount of bending shown by the tails (Fig. 20.12), which ranges from 180° apart (double source—Fig. 20.12a) to 0° (head–tail source—Fig. 20.12f). The heads usually contain radio jets.

How to explain this structural sequence? Recall (Section 19.8) that clusters of galaxies contain a hot, ionized intracluster medium. Imagine that a galaxy, moving rapidly through this medium, shoots out material (high-speed electrons, for instance) in a jet. As the

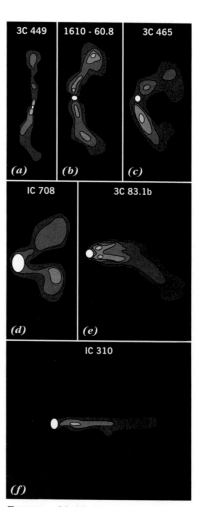

FIGURE 20.12 Sequence for the bending of radio tails around the nuclei of radio galaxies. These images show the extent to which the material from the jets interacts with the surrounding medium and the relative speed of the galaxy through it. The faster the speed, the greater the amount of bending. (Adapted from a diagram by G. Miley.)

galaxy travels along, it leaves behind a radio-visible trail—a fossil record of where it's been. Here's an analogy. Imagine driving a car slowly and blowing smoke out an open window. The air stops the motion of the smoke, and it leaves a trail behind the car. Similarly, material flowing out of a galaxy is decelerated by the intragalactic medium, and the moving galaxy leaves it behind.

How do the AGNs generate these lobes? We don't know all the details yet, but a crucial clue is the radio jets from the nucleus, which are aligned (more or less) with the lobes. These jets suggest that high-speed electrons are channeled by magnetic fields from the nucleus into the medium around the galaxy, where they pile up to form a lobe (Fig. 20.13).

Supercomputer simulations are beginning to reveal the complex interactions of a supersonic jet as it tears into the medium around an active galaxy (Fig. 20.14). Because it is a plasma, the jet carries along a magnetic field that controls the flow of the jet. Still, the jet breaks up in turbulent eddies as it speeds through the medium. Such simulations, repeated under many different conditions, probe the physics and help astronomers to develop a deeper understanding of the processes involved.

To sum up. Many radio galaxies have emission, in the form of lobes, that extends far beyond the visible galaxy and is connected to the nucleus by thin jets. The lobes are energy reservoirs that are a tangle of magnetic fields and high-speed electrons, since the emission is synchrotron radiation. The lobes are powered by the nucleus, with material channeled to them along the jets.

■ **What characteristics do active galaxies exhibit when viewed at radio wavelengths?**

20.3 Seyfert Galaxies and BL Lacertae Objects

I don't want to bog you down with all the strange members of the AGN zoo. But Seyfert galaxies and BL Lacertae objects are two galactic types that provide differ-

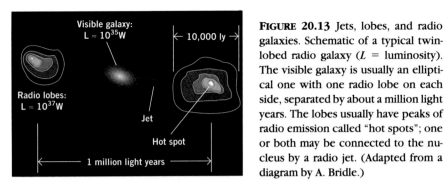

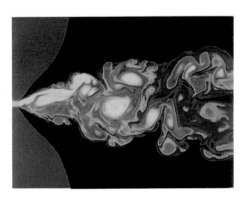

FIGURE 20.13 Jets, lobes, and radio galaxies. Schematic of a typical twin-lobed radio galaxy (L = luminosity). The visible galaxy is usually an elliptical one with one radio lobe on each side, separated by about a million light years. The lobes usually have peaks of radio emission called "hot spots"; one or both may be connected to the nucleus by a radio jet. (Adapted from a diagram by A. Bridle.)

FIGURE 20.14 Supercomputer simulation (in false colors) of the flow of a jet (from left to right) through the medium surrounding a galaxy. Note the breakup into turbulent eddies. (Courtesy of M. Norman, University of Illinois/NCSA.)

ent insights in the AGN phenomenon at lower energy levels and with less drama. They offer nearby clues to build a model of AGNs, and knowledge of them predates by far what we now call active galaxies.

Seyfert Galaxies

In 1943 Carl Seyfert noted that some spiral galaxies showed unusual, broad emission lines. These groupings are now called **Seyfert galaxies.** Later work has added to the collection of Seyfert galaxies; some hundred have been catalogued to date. Seyfert galaxies are close by, so they can be probed in detail. They provide good clues to the physics of AGNs.

What's a Seyfert galaxy? Most galaxies have a bright nucleus, which looks fuzzy when viewed with a telescope; when photographed, a run-of-the-mill galaxy looks obviously fuzzier than a star. In contrast, a Seyfert looks like a bright star surrounded by a faint haze. In short-exposure photographs, the haze isn't seen, and a Seyfert's nucleus passes for a star!

Along with this trademark optical appearance, a Seyfert has a particular spectrum: it shows broad emission lines. Why are they so broad? Consider a gas giving off emission lines. Some of this gas is moving toward us, so its emission is blueshifted. Other parts are moving away, and their emission is redshifted. When we observe a spectrum of the gas, the Doppler-shifted emission is added to the unshifted line at longer and shorter wavelengths. So the emission line appears wider than it would be if the gas were stationary. If we attribute the widths of the lines to Doppler shifts produced by motion in the emitting gas, the gas in the Seyferts has very high random velocities, *thousands* of kilometers per second.

Seyferts are almost always spiral galaxies (Fig. 20.15), only 5 to 10 percent might be ellipticals. (The small angular sizes of some Seyferts make it hard to classify them by form.) Compare this with the fact that most extended radio galaxies are elliptical (Section 20.2). Overall, about 1 percent of all spiral galaxies (ordinary and barred) are Seyferts.

Seyferts show strong Balmer emission lines of hydrogen, which are narrow in some and broad in others (Fig. 20.16). The use of "broad" and "narrow" here is relative. For example, the average Balmer line width is about 3000 km/s. But in some Seyfert galaxies, the Balmer lines have Doppler widths as great as 10,000 km/s. In contrast, other Seyfert galaxies have lines only 1000 km/s wide. To explain the Balmer emission from a Seyfert requires some tens to thousands of solar masses of ionized gas in the nucleus moving at speeds of 1000 to 10,000 km/s.

What ionizes the gas and moves it around? Probably the energy source in the nucleus that generates the synchrotron emission. That source is yet unknown. Also in the nucleus resides a source of high-energy electrons and gas. Observations indicate that the nuclei of Seyferts are small, only a few light years in diameter. Gas moving at 10,000 km/s would flow across such a tiny nucleus in only a century. So the gas must be replaced as it flows out.

To sum up. Seyferts are (all?) spiral galaxies with extraordinary features. The most prominent are as follows.

1. Seyfert galaxies have extremely small and bright nuclei.

2. Their nuclei have spectra that show emission lines not usually seen in the spectra of spiral galaxies. These bright lines come from ionized gas.

3. The emission lines are very wide. Considered as Doppler shifts, the widths of the lines indicate gas motions of 1000 to 10,000 km/s, an indication of violent activity.

4. Many Seyferts have compact, low-luminosity radio sources within their nuclei.

A survey of spiral galaxies shows that Seyferts tend to be in close, binary galaxy systems. Tidal forces (Section 9.1) may then induce the Seyfert activity for a short period of time, making the Seyferts one form of starburst galaxy. Other Seyferts have been found to have *two* nuclei, a possible indication of a collision and merger from too close an interaction between a larger and a smaller galaxy (Fig. 20.17).

BL Lacertae Objects

Our second AGN example consists of objects named after their prototype BL Lacertae; they are called **BL Lacertae objects** (**BL Lac** for short). As a group, the BL Lac objects have the following characteristics.

1. Rapid variability (as fast as one day) at radio, infrared, visual, and other wavelengths.

2. Extremely weak or no emission lines.

3. Nonthermal continuous radiation with most of the energy emitted in the infrared.

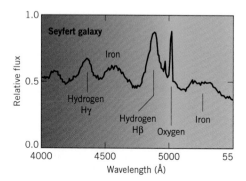

FIGURE 20.15 The Seyfert galaxy NGC 4151, a strong emitter of x-rays. Note how bright the nucleus appears compared to the rest of the galaxy, which has a spiral structure. The yellow nuclear region has a bar structure across it. (Courtesy of James D. Wray, McDonald Observatory.)

FIGURE 20.16 The generic emission-line spectrum of a Seyfert galaxy, showing the strongest lines. Note the especially broad lines of the hydrogen Balmer series and of iron. (Adapted from a figure by D. E. Osterbrock.)

FIGURE 20.17 *HST* image of the double core of the Seyfert galaxy called Markarian 315. The main nucleus is the brighter region on the left; the second piece is separated from the first by about 6000 ly. A merger of two galaxies resulted in this double core. It may also have provided the material to fuel the supermassive black hole that may lie in the center of the main nucleus. (Courtesy of J. MacKenty, STScI, and NASA.)

4. Strong and usually rapidly varying polarization.

5. Generally a starlike appearance; rarely is structure visible.

The BL Lac objects differ the most from other active galaxies in that their emission varies frequently, erratically, and rapidly (Fig. 20.18). For example, BL Lac itself fluctuates in luminosity by twenty times or so. Often BL Lac shows night-to-night variations of 10 to 50 percent in luminosity. That doesn't sound like much, but imagine our Galaxy changing its light output by some 20 percent in a day. That's like 10 billion suns turning on and off simultaneously! A few BL Lac objects have changed their brightness by as much as a hundred times in luminosity.

These rapid variations imply that the emission region of the BL Lac objects is small, no more than a light day or so in diameter. This argument was made in Section 17.6 for pulsars. The same argument applies to active galaxies—the size of the emitting region equals the distance light can travel during the *shortest* variation of its brightness. So a one-day variation means a size of one light day. (This argument only applies if the source emits into all directions in space.)

What puzzles astronomers most

about the BL Lac objects is that their energy variations take place in objects that show almost *no* emission lines in their spectra! The synchrotron emission in the ultraviolet (and even the electrons themselves) should ionize any gas near the nucleus and should produce emission lines. But if the BL Lac objects are powered like other AGNs, where are their emission lines? A very few BL Lac objects have barely visible emission lines, almost drowned out by the strong continuous emission.

Some forty BL Lac objects have been classified to date. They may not all be the same kind of beast. A few are possibly the nuclei of galaxies; some, like BL Lac itself, have a faint surrounding fuzz that might be a galaxy; others look pointlike, without a hint of enveloping material. A few BL Lac objects are found in clusters of galaxies—indirect evidence that they are also galaxies. Indications so far suggest that an elliptical galaxy contains the BL Lac nucleus, although this is not certain.

To compound the uncertainties, we don't have good distance determinations for very many BL Lac objects. Weak absorption features of the nebulosity around BL Lac have a redshift of 0.07, which corresponds to a radial velocity of 21,000 km/s. According to Hubble's law, this velocity corresponds to a distance of 1400 Mly. Many redshifts range from 0.03 to 0.07, though a few are much larger.

To sum up. BL Lac objects are most peculiar because of their rapid variability and usual lack of emission lines. In contrast to Seyferts, they seem to be elliptical galaxies.

■ **What is unique about the spectrum of a Seyfert galaxy?**

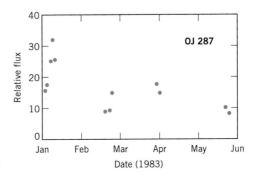

FIGURE 20.18 Variability in the optical fluxes from the BL Lac object called OJ 287 in 1983. Note the strong outburst in January, when the flux increased by about a factor of 3 over the earlier level. (Courtesy of P. Smith.)

20.4 Quasars: Unraveling the Mystery

During the boom of radio astronomy in the late 1950s, radio astronomers—like modern Tycho Brahes—compiled catalogs replete with radio sources that were not identified with any familiar visible objects. Hunting for possible culprits in 1960, Thomas Matthews and Allan Sandage discovered a faint starlike object—hence the name *quasi-stellar object,* or **quasar**—at the position of radio object 3C 48 (object 48 from the Third Cambridge Catalogue of radio sources). This object had a spectrum of broad emission lines that could not be identified, and it emitted more ultraviolet light than an ordinary main-sequence star.

3C 48 remained a unique object until 1963, when the strong radio source 3C 273 was identified with a faint starlike object (Fig. 20.19). The emission lines of 3C 273 were just as puzzling as the emission lines from 3C 48: they coincided with no known atomic lines. Astronomers were baffled.

Quasar Redshifts

Maarten Schmidt finally deciphered the spectral code of 3C 273 by recognizing the prominent emission lines as those of the hydrogen Balmer series, redshifted by 15.8 percent compared to their normal at-rest wavelengths (Fig. 20.20). After Schmidt had decoded 3C 273, Jesse Greenstein applied the same analysis to 3C 48 and found its spectrum to be redshifted by 36.7 percent.

More than 1500 quasars have been identified, and redshifts have been measured for most of these. Many have redshifts that exceed 2.0, and some even 3.0. Note that the redshift is that compared to the wavelength at rest. So a redshift of 2 means that the Lyman line of hydrogen (rest wavelength of 1216 Å) has shifted toward the red by 2 times (1216 Å × 2 = 2432 Å). If interpreted as redshifts from the expansion of the universe, the light from quasars must come from such distances that it would have had to originate billions of years ago. The redshifted quasars are the youngest objects we can see in the universe. Note that "youngest" here refers to objects at a point in their lives far in the past, closer to the Big Bang in time.

Caution. When the measured redshift approaches 1, the simple Doppler formula (Focus 10.1) no longer gives the correct radial velocity. Usually a redshift

FIGURE 20.19 The quasar 3C 273, the first quasar to have its spectrum deciphered. Note the optical jet sticking out of the quasar. North is at top and the angular scale is at the bottom. Note how fuzzy the quasar appears compared to the surrounding stars. (Courtesy of P. Wehinger; observations by S. Wyckoff, P. Wehinger, and T. Gehren with the ESO 3.6-meter telescope.)

FIGURE 20.20 Spectrum of 3C 273, showing its large redshift as indicated by the emission lines (top). A comparison spectrum (bottom) establishes the wavelength reference scale, given for a source at rest with respect to the observer. Compare the hydrogen Balmer lines in the quasar's spectrum to those in the other spectrum. Note how the quasar's lines are shifted to the red end of the spectrum (arrow at top). This redshift amounts to about 16 percent. (Courtesy of M. Schmidt.)

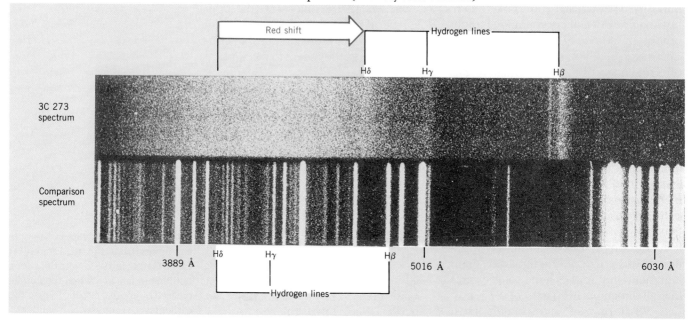

Enrichment Focus 20.1

RELATIVISTIC DOPPLER SHIFT

When the measured redshift is larger than a few tenths, the simple Doppler formula (Focus 10.1) no longer gives the correct radial velocity. For example, if applied to a quasar with a redshift of 4.0 the formula would indicate that this quasar was fleeing at 400 percent the speed of light! A modified formula, based on special relativity, must be used instead.

I won't derive the **relativistic Doppler shift,** but here's the result:

$$z = \frac{\Delta\lambda}{\lambda_o} = \left[\frac{1 + v/c}{1 - v/c}\right]^{1/2} - 1$$

where $\Delta\lambda$ equals $\lambda - \lambda_o$, λ_o is the original wavelength, λ the measured wavelength, v the radial velocity, and c is the speed of light. (The symbol z is often used as an abbreviation for the redshift $\Delta\lambda/\lambda_o$.)

Let's apply the relativistic Doppler shift to a real case. A quasar called PKS 1402 + 044 is a radio source whose spectrum shows the Lyman series from hydrogen Doppler-shifted into the red region of the spectrum. From observations of these emission lines, the object has a redshift (z) of 3.2. By the formula:

$$z = \left[\frac{1 + v/c}{1 - v/c}\right]^{1/2} - 1$$

$$\frac{1 + v/c}{1 - v/c} = (3.2 + 1)^2 = 17.6$$

$$1 + \frac{v}{c} = 17.6\left(1 - \frac{v}{c}\right)$$

$$\frac{18.6\,v}{c} = 16.6$$

$$\frac{v}{c} = 0.89$$

A redshift of 3.2 implies that this quasar is moving away from us at about 89 percent the speed of light!

---

of 0.16, for example, means a radial velocity of 16 percent the speed of light. But then a redshift of 2 would indicate that a quasar was moving away at 200 percent the speed of light! And that is not possible. A modified formula, based on special relativity, must be used instead, and all the radial velocities then come out less than the speed of light (Focus 20.1). We measure redshifts but infer radial velocities.

General Observed Properties

Although most of the early quasars were detected because of their strong radio emission, many uncovered in later optical searches were found to have no detectable radio emission. Quasars are present in about 50 percent of all radio-emitting stellar objects. The observed properties that make them unique and serve as identification tags are (1) starlike appearance with a large redshift, sometimes associated with a radio source (only about 10 percent are known radio sources); (2) broad emission lines in the spectrum with absorption lines sometimes present (usually the redshift of the absorption lines is less than that of the emission lines); (3) often variable luminosity; and (4) in those that are radio sources, aligned, double-lobed structures (like Cygnus A).

The main—and most remarkable—feature of quasars is the size of their redshifts. The expansion of the universe results in the redshift of light from distant galaxies as described by Hubble's law. At first glance, the most natural explanation of the quasars' redshifts is a cosmological one: quasars participate in the expansion of the universe. If so, their enormous redshifts indicate that they are very far from us, and the light we observe was generated when the universe was very young. But if they are as far away as indicated by their redshifts, they must expend vast amounts of energy. For example, 3C 273's redshift of 16 percent, if it is due to the expansion of the universe, implies a distance of 3.1 billion light years, assuming $H = 15$ km/s/Mly. At this distance, 3C 273 emits about 10^{14} solar luminosities, or about forty times as much as the most luminous normal galaxies. (Most of this energy is in the infrared.)

The Light from Quasars

Some quasars emit radio waves intensely, and all emit visible light. What produces this emission? A key clue comes from the spectrum of quasar radiation. Synchrotron radiation requires a continuous supply of clouds of energetic electrons and magnetic fields.

Optical observations of the quasars first discovered as radio sources have added more details to the general physical picture. The continuous optical emission is in part polarized, so synchrotron emission produces some of the optical radiation. Given a common magnetic field, the electrons that produce the quasar's optical radiation must have higher energies than those that emit the radio radiation.

Line Spectra

All quasars have bright lines in their optical spectrum—the emission lines that are used to measure a quasar's redshift. Among the strongest emission lines in a quasar's spectrum are hydrogen's first Lyman series line (Lyman alpha) at 1216 Å and the second Balmer line (Hβ) at 4861 Å (Fig. 20.21).

The emission-line spectra of quasars indicate that a low-density cloud of gas is radiated with photons energetic enough to ionize hydrogen. This radiation probably comes from the synchrotron emission from energetic electrons, which is a plentiful source of ultraviolet photons. So to our central synchrotron source we add clouds or filaments of gas that convert ultraviolet radiation (and some x-rays) from the synchrotron source into visible emission lines.

The emission lines are extremely broad, which implies that the filaments or clouds from which the emission lines originate move rapidly (as was argued for Seyferts in Section 20.3). The broadening of the emission lines arises from motions of the clouds or filaments themselves at radial velocities up to 20,000 km/s. This region, close to the nucleus, makes up the *broad-line region* in a model for quasar. Narrow emission lines are not as obvious as the broad ones. They have widths that correspond to Doppler shifts of some 1000 km/s. So quasars have a *narrow-line region* farther away from the nucleus.

Many (but not all) quasars also have absorption lines in their spectra. Quasars with emission-line redshifts of less than 2.2 typically do not have absorption lines; those with greater redshifts have strong absorption lines. These lines are very narrow compared

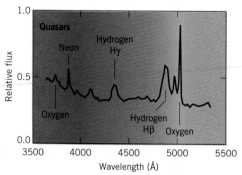

FIGURE 20.21 Composite spectrum of quasars showing the strongest emission lines typical for them. As for Seyfert galaxies (Fig. 20.16), the lines of the hydrogen Balmer series are prominent. (Adapted from a figure by J. S. Miller.)

with the emission lines. So, by Kirchhoff's rules, we expect them to come from cool clouds of transparent gas. In general, the absorption-line redshifts are *less* than the shifts for the emission lines. And, the absorption lines sometimes show more than one redshift.

What produces these absorption lines? Three possibilities are (1) gas clouds in or close to the quasar at temperatures cooler than the emission line regions, (2) absorption by intervening (and otherwise invisible) intergalactic gas clouds, and (3) a halo of gas around an intervening galaxy. The third model seems most likely (see Section 20.5 on the double quasar).

Variability in Luminosity

Light variability was first observed for 3C 48, which showed variations of about 40 percent in 13 months. 3C 273 also has erratic variations over periods of about a year. Most radio-emitting quasars also vary in radio output over periods of years (Fig. 20.22). In contrast, about 20 percent of the quasars exhibit rapid variations in light and radio output with periods on the order of days or weeks. This variability implies that the nuclear power source is very small, as was argued for BL Lac objects (Section 20.3) and any active galaxy with short-period changes in luminosity.

■ Based on the measure of redshifts from quasars, how far away do they tend to be?

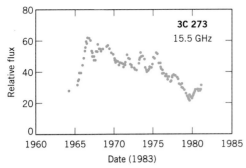

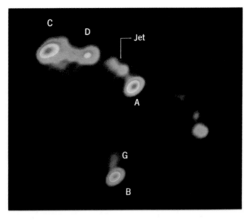

FIGURE 20.22 Variation of the radio emission at a frequency of 15.5 GHz of the quasar 3C 273 from 1965 to 1982. Note the sharp rise in the flux in 1965–1966 and then the gradual decline with some occasional outbursts. (Courtesy of T. Balonek; observations by W. Dent and T. Balonek with the Haystack Observatory radio telescope.)

FIGURE 20.23 VLA map of the radio emission from the double quasar 0957 + 561A and B. The double quasar shows up in the two elliptical regions near the center (marked *A* and *B*). Blobs of emission (marked *C* and *D*) to the left of the upper quasar (*A*) are lobes at the end of a radio jet. The label *G* marks the galaxy, which causes the gravitational-lens effect. (Observations by P. Greenfield, B. Burke, and D. Roberts; courtesy of NRAO/AUI.)

20.5 Double Quasars and Gravitational Lenses

Quasars are rarely close together, yet two quasars, called 0957 + 561A and 0957 + 561B, are only 6 arcsec apart. Even more surprising, the emission-line redshifts of both are essentially the same: 1.41. What was up here?

The virtually identical emission-line redshifts of this double quasar forced astronomers to propose that 0957 + 561A and B were not separate twins but optical images of the *same* quasar. How so? Recall (Section 7.3) that general relativity predicts that masses deflect the paths of light rays—in essence, acting like a lens. If a very small, dense mass (a black hole, for instance) lay along our line of sight to a quasar, its image would be split into two, one above and one below the quasar's actual position. This phenomenon of image making by a mass is called a **gravitational-lens effect.** Such lenses tend to make imperfect, distorted images.

Radio astronomers found a map of the area quite puzzling at first, since the radio sources had complex structures. There were radio blobs (*C* and *D* in Fig. 20.23) to the east of the northern quasar (*A*) that were not visible west of the southern quasar (*B*). It turns out that this emission is that of a jet flowing into two lobes!

Photographs of the quasars on a

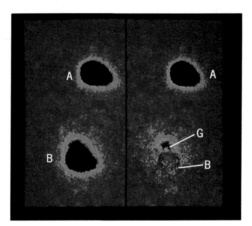

FIGURE 20.24 High-resolution optical photograph of quasars *A* and *B*. The little fuzz sticking out of the top of *B* (lower image) turns out to be the poorly resolved image of the galaxy making the gravitational lens (visible as *G* in the radio map). The image of *B* is superimposed on it by the imperfect imaging. (Courtesy of A. Stockton, Institute for Astronomy, University of Hawaii.)

night of exceptionally good seeing (Fig. 20.24) show quasar *B* with a little bit of fuzz sticking out of it. This fuzz turns out to be the poorly resolved image of a faint elliptical galaxy—the gravitational lens! Since the galaxy is an extended mass, it acts like an imperfect lens and produces a complex pattern of images. (The view is also complicated by the fact that the elliptical galaxy lies in a group of galaxies, and each member helps to bend the light. This cluster is closer to us than the quasar, for the cluster's redshift is only 0.36.) By a quirk of placement, we see only a part of the complete image. Two images are formed by a gravitational lens, an intervening, probably elliptical, galaxy about halfway between us and the quasar (Fig. 20.25). The galaxy needs a mass of some 10^{13} solar masses to bend the light enough to form the images.

This discovery has three important implications.

1. It provides another confirmation of general relativity.

2. It proves that in this case the quasar is

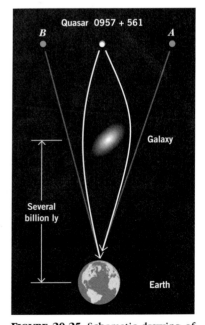

FIGURE 20.25 Schematic drawing of the optical illusion of the double quasar caused by a gravitational-lens effect. The bending of the quasar's light by the massive galaxy (a few billion light years away) makes two images (*A* and *B*) appear on the sky.

more distant than the galaxy, so the quasar's redshift is cosmological.

3. Gas around the galaxy creates the quasar's absorption-line spectrum; this situation may be the case for other quasars as well.

These implications have been strengthened by the discovery of other candidates for **double quasar** status.

We know from the observations of the bending of sunlight in the solar system that masses can, in fact, act like a lens. The light is bent just like light that is refracted when it passes through a lens. Depending on the distribution of the mass doing the lensing, the light from a distant quasar may be split into many images or distorted into rings, incomplete rings visible as arcs, or other distorted shapes.

Now that astronomers have been reminded of the possibility of gravitational imaging, they have found other instances along with the double quasars. An arc is visible in the galaxy cluster Abell 370 (Fig. 20.26), which is dominated by a supergiant elliptical galaxy. The spectrum of the arc is that of a galaxy, so the arc could be the grossly distorted image of a galaxy more distant than the cluster.

Jacqueline Hewitt has discovered a more dramatic example with the VLA, a

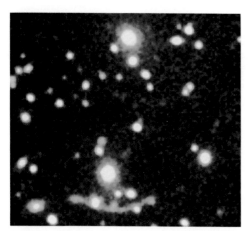

FIGURE 20.26 A giant arc (at bottom) in the galaxy cluster Abell 370. This false-color map was made at the 3.6-meter Canada–France–Hawaii telescope. The more distant galaxy, whose light is bent to make the arc, would lie near the center. (Courtesy of G. Soucail.)

ring of radio emission called MG 1131 + 0456 and dubbed the Einstein ring (Fig. 20.27) because the gravitational imaging is predicted by general relativity. Here we see a fairly symmetrical ring, which implies that the mass doing the lensing lies almost exactly along the line of sight. The masses required to produce such effects are a few times 10^{13} solar masses, just about that for a giant elliptical galaxy.

One of the strangest of the gravitational lens effects is the Einstein cross (Fig. 20.28). The *Hubble Space Telescope* produced an image of a quasar that showed four distinct bright, images and a diffuse central object that is the bright core of the intervening galaxy. The quasar lies about 8 billion ly away. Narrow absorption lines in its spectrum originate in the halo of the galaxy, which is at a distance of only 400 million ly, some twenty times closer. The mass of this galaxy causes the imperfect lensing effect. The *HST* has also picked out the image of a distant galaxy that has been distorted by a closer cluster of galaxies, whose dark matter acts as an imperfect magnifying lens (Fig. 20.29).

◼ **What creates the gravitational-lens effect for 0957 + 561A and 0957 + 561B?**

20.6 Troubles with Quasars

If quasars are actually billions of light years away, then the quasar 3C 273, for example, must emit a total radiative energy in one second equivalent to all the energy produced by the sun in about 3 million years. A typical quasar produces up to 10,000 times as much power as an ordinary spiral galaxy.

Not only do quasars blast out energy at enormous rates, but the energy comes from relatively small regions of space in the centers of quasars—from possibly light hours to light months to no more than light years in diameter. Because quasars' light output varies on time scales of days to years, their energy-emitting regions cannot be larger than light days to light years—smaller than the distance from the sun to the nearest star.

Consider that for a moment. Qua-

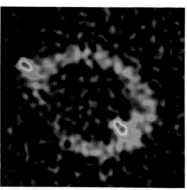

FIGURE 20.27 Another gravitational-lens effect: *Einstein's ring,* formed by a lensing galaxy right along the line of sight. These observations of the radio source MG 1131 + 0456 show the image of a double-lobed radio galaxy distorted by the gravitational lens. (Observations by J. Hewitt and co-workers, Princeton University.)

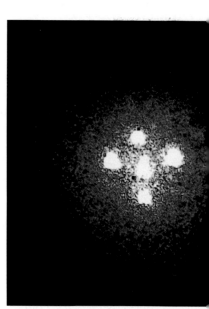

FIGURE 20.28 *Einstein's cross* (gravitational lens G2237 + 0305), an image taken by the Faint Object Camera on the *Hubble Space Telescope.* The four bright outer images are those from a distant quasar. (Courtesy of NASA/ESA.)

sars typically emit about a hundred times the total energy output of our Galaxy from regions no more than light years in diameter!

Intercontinental radio interferometers can resolve very small angular diameters. They show that 3C 273 consists of separate radio components. The smallest parts—if the quasar is at the distance demanded by a cosmological redshift—cannot be more than a few light years across. And the map also shows a radio jet (Fig. 20.30).

High-resolution studies have also revealed the radio structure of other quasars, especially those that are relatively near. These observations show that for the quasars that can be resolved, the radio structures are generally symmetrical doubles, similar to Cygnus A. And some of these show nuclear jets. In a few, the material in the jets appears to travel faster than light. This effect turns out to be another optical illusion caused by jets aligned very close to our line of sight (Focus 20.2).

Energy Sources in Quasars

What energy source lies in the heart of a quasar? The most developed quasar models to date invoke supermassive black holes. One such model relies on a black hole of 100 million solar masses. This model builds on that for binary x-ray sources with black holes (Section 17.8), in which material from the normal star forms an accretion disk around the black hole before the material falls into it. A supermassive black hole in a dense galactic nucleus is fueled by the tidal disruption of passing stars (Fig. 20.31). The stellar material forms an accretion disk and radiates as it spirals into the black hole, thus powering the quasar. The model calculations show that luminosities of 10^{12} solar luminosities, about that of bright quasars, are possible. To feed the black hole requires about 1 solar mass of material a year.

The model must also deal with the nuclear jets now known for some quasars. One way is as follows. The accretion disk can restrict the flow of ionized gas from it, along its spin axis, either by magnetic fields or simply from the formation of a funnel of material around the

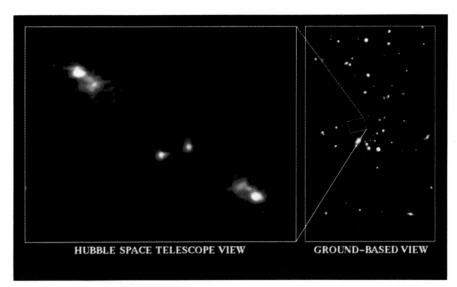

FIGURE 20.29 A gravitational lens in the cluster of galaxies called AC 114. *HST* image (left) shows the distorted, mirror-reflected view of distant galaxies, whose light has been lensed by the visible and dark matter in the foreground cluster of galaxies (right). (Courtesy of Richard Ellis, Durham University and NASA.)

black hole. This initiates a narrow beam that becomes visible as a jet, once it has passed beyond the main body of the quasar.

This model has a number of successful aspects. The supermassive black hole easily generates the level of quasar luminosity in a region of space only a few light years in size (the Schwarzschild radius of a 10^8-solar-mass black hole is only 3×10^8 km, or about 2 AU). And it does the energy conversion (from gravitational to radiative) with high efficiency.

A Generic Quasar and AGN Model

You've probably noticed that quasars and AGNs share some observed (and so

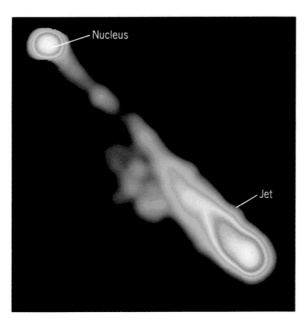

FIGURE 20.30 The radio jet of 3C 273: close-up view of the nucleus and the jet made at a frequency of 408 MHz. This false-color map has white as the strongest radio emission and blue the weakest. (Courtesy of Richard Davis, Jodrell Bank; observations with the MERLIN array.)

some physical) characteristics (Table 20.1). Can a generic model explain them all and link them by common physical processes?

One popular idea interprets active galaxies and quasars as *similar objects viewed from different directions,* so that AGNs and quasars are driven by the same basic physics. This generic model tries to link together the nuclear engine (a supermassive black hole), an accretion disk, both the broad- and narrow-emission-line regions, and the jets—the key observational clues. Let's use these pieces to build a model.

A generic model has as the energy source a **supermassive black hole,** which is surrounded by an accretion disk a few light days in diameter (Fig. 20.32). The supermassive black hole provides the power and the central synchrotron source as it swallows material from the accretion disk. Twin jets emerge from the accretion disk; far out from this central region, they plow into gas to pile up in radio lobes. Surrounding the central engine are two zones of ionized gas. The smaller (few light months in diameter) and closer one has gas moving with high random speeds (some 10,000 km/s); this region produces the broad emission lines. Another ionized region filled with clouds that have slower random motions (only a thousand kilometers a second) extends farther out, up to a few thousand light years. Here the narrow emission lines are generated. Between these two regions lies a thick, dusty ring of molecular gas. It may have a diameter of 10 to 1000 ly.

Note that we can up the power output from this model by simply making the supermassive black hole larger and the rate of mass infall greater. For Seyfert galaxies, a supermassive black hole of 10^6 to 10^7 solar masses would work; for quasars, we'd need some 10^8 to 10^9 solar masses. Other AGNs would fall within these ranges.

How we view this generic model determines what we see. Imagine a view almost edge-on (line *A* in Fig. 20.32). The dusty ring blocks our view of the broad-line region and the central continuum source. So we see narrow emission lines, perhaps the jets, and the radio lobes, and we'd call the object a narrow-line Seyfert or a narrow-line radio galaxy. In contrast, if we observe the model more pole-on (line *B* in Fig. 20.32), we can see the central continuum source, the broad-line region, and some of the narrow-line region, perhaps the jets, and the radio lobes. We'd then call the object a quasar, a broad-line radio galaxy, or a Seyfert

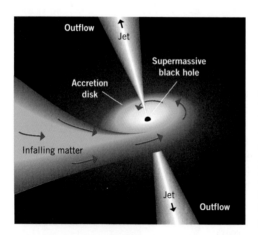

OBJECT	POWER OUTPUT (W)
Cosmic Violence *At a Glance*	
Nucleus of Milky Way	10^{34}
Active galaxy	10^{39}
Quasar	10^{42}

FIGURE 20.31 One model for a supermassive black hole powering a quasar (or active galaxy). An accretion disk, made of infalling material, surrounds the black hole; it is thick close to the hole and thins out away from it. A sharp funnel forms where material from the accretion disk streams into the black hole. The disk is hot, so gas blows off it. The funnel directs the gas up along the rotation axis of the accretion disk, creating a bipolar outflow from both sides of the accretion disk. If the disk wobbles, the jet will also and trace out an "S" shape in space. (Adapted from a diagram by J. Burns and R. M. Price.)

TABLE 20.1 COMPARISON OF ACTIVE GALACTIC NUCLEI AND QUASARS

PROPERTY	RADIO GALAXIES	SEYFERT GALAXIES	BL LAC OBJECTS	QUASARS
Redshifts from spectra	0.01–0.3	0.003–0.06	0.05–0.6	0.2–4+
Continuous spectrum	Nonthermal	Nonthermal	Nonthermal	Nonthermal
Emission lines	Broad and narrow (rare)	Broad and narrow	None or weak	Broad and narrow
Absorption lines	From stars in galaxy	None	None (?)	From intervening gas clouds
Shape (optical)	Elliptical	Spiral	Unclear	Starlike with fuzz
Shape (radio)	Jets and lobes	Weak emission from nucleus	Weak emission from nucleus	Jets and lobes

with mostly broad and some narrow lines. In either case, the dust in the ring can absorb ultraviolet and visual photos, heat up, and radiate in the infrared.

Now imagine that we observed the model very close to or right along the axis of the jets. Then we would observe greater than light speeds and a very concentrated point of emission—a superluminal quasar or BL Lac object. In this orientation, we'd be looking through one radio lobe with the other on the far side of the galaxy. Then we would have a hard time detecting both lobes. At best, we might see a faint halo of radio emission (from the lobe in front) with a strong nuclear source embedded within. We might even see x-rays from the accretion disk.

These various orientations may then explain why sometimes we see only one nuclear jet or one much stronger than the other. The relativistic beaming effect implies that one bipolar jet close to our line of sight will appear to be very strong, the other much weaker (or not at all). So our generic, unified model naturally explains many aspects of AGNs and quasars. AGNs and quasars may be essentially the same kinds of object viewed from different angles.

If we accept a model for the energy output of active galaxies and quasars that imagines a black hole in the nucleus eating up huge amounts of material, the nucleus should show an intense, pointlike source of light (due to a concentration of stars around the black hole); in addition, stars orbiting the center should have high velocities, and emission lines with high Doppler shifts might be visible from the infalling matter. Observations of such effects have been reported for M87: a sharp peak of emission from M87's nucleus (such a peak is not found in the nucleus of normal elliptical galaxies), as well as a dramatic increase in the velocities of stars in the nucleus compared to those outside the nucleus.

A model consistent with these observations is one of a billion-solar-mass black hole hiding in the inner 300 ly of the nucleus. This supermassive black hole would cluster the stars in the nucleus closely around it, and they would orbit at high velocities. In addition to

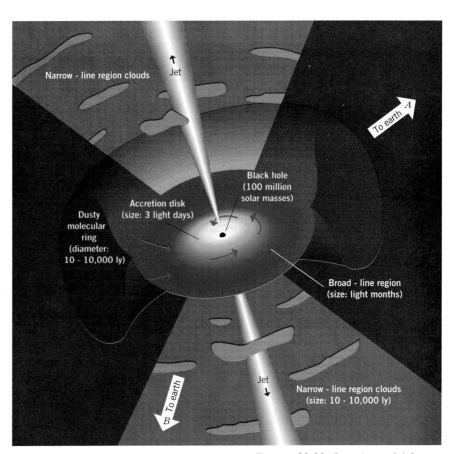

M87, the Andromeda Galaxy needs some 10 million solar masses to explain the light distribution and motions visible in its inner 300 ly. And so it also appears for M32, a companion galaxy to M31.

The *HST* has provided a new image of the core of the spiral galaxy M51 that is highly suggestive of a supermassive black hole in that galaxy (Fig. 20.33). The image shows a dark "X" across the nucleus from absorption by dust in two rings; the horizontal one surrounds the supermassive black hole: an edge-on view of a dusty torus some 100 ly across. In most respects, M51 is a pretty normal spiral galaxy. So it may well be that *every* galaxy has a supermassive black hole in its core. The AGNs simply have the most massive examples.

The Host Galaxy for Quasars

Given that a quasar may be the hyperactive nucleus of a very faraway galaxy, what about the rest of the galaxy? At cosmological distances, the disk of a quasar's parent galaxy, called the *host galaxy,* would be too small and faint to be

FIGURE 20.32 Generic model for an AGN or quasar. The central engine may be a supermassive black hole. It is surrounded by an accretion disk made from infalling material. Close to the disk are blobs of fast-moving hot gas, which make the broad emission lines in the spectrum. Much farther out is a dusty torus of molecular gas. Beyond that lie blobs of slow-moving hot gas; these make the narrow emission lines in the spectrum. If our line of sight is along *A,* the dusty torus blocks out the broad-emission-line region; only the narrow lines are visible. If our line of sight is along *B,* we see both narrow and broad emission lines. The orientation of the AGN in space determines our view.

Enrichment Focus 20.2

FASTER THAN LIGHT?

Intercontinental radio astronomy picks out the finest details of distant radio sources—even quasars. Such observations have detected changes in five quasars that, if taken at face value, lead to the conclusion that parts of these objects are moving at speeds greater than that of light! Impossible! So says special relativity.

Motions that seem faster than light are called *superluminal.* The familiar quasar 3C 273 features a jet that sticks out from its core. Observations from 1977 to 1980 (Fig. F.15) show that a knot in the jet has moved away from the nucleus. If 3C 273 is at the distance implied by its redshift, the observed separation results from a motion across our line of sight of *ten times the speed of light!* What's happening? Like the double quasar, the superluminal effect may simply be an optical illusion.

The sources that exhibit superluminal motions have a common feature: a radio jet from the nucleus. These jets are thought to contain high-energy electrons moving close to the speed of light. If quasars have such jets, and if the jets are almost (but not quite) pointing at us, a blob of material moving out along the jet will *appear* to travel faster than light speed.

The superluminal effect arises from an almost head-on orientation of the jet and the finite speed of light. Here's how. Suppose a jet points at us at an angle of 8° with respect to our line of sight (Fig. F.16). Imagine that a blob of electrons ejected from the nucleus (point N) gets to point A in 101 years and the light emitted from the blob at N takes 100 years to get to point B. For an 8° angle, the separation of A and B is 14 ly. But the light from B is one year ahead of that emitted by the blob when it finally reaches

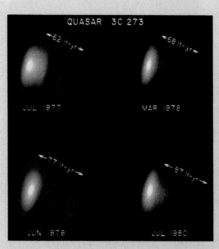

FIGURE F.15 Very-high-resolution radio maps made at a frequency of 10.65 GHz of the quasar 3C 273, which show the radio jet extending from the nucleus. Note how the emission moves out along the jet from 1977 to 1980. In addition, a blob of the jet has moved away from the nucleus from a distance of 62 ly in 1977 to 87 ly in 1980. At face value, the blob has moved faster than the speed of light. (Based on observations by T. J. Pearson, S. C. Urwin, M. H. Cohen, R. P. Linfield, A. C. S. Readhead, G. A. Seielstad, R. S. Simon, and R. C. Walker at the Owens Valley Radio Observatory and NRAO/AUI.)

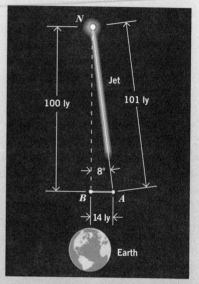

FIGURE F.16 Geometry for superluminal motion in a radio jet.

A. It's taken 100 years for the light from N to reach B, 101 years to reach A.

Many years later, the light that was at B reaches us. Only one year later, that emitted at A arrives. To us, it appears that the source has moved from B to A—a distance of 14 ly—in only one year. It seems to have a speed of fourteen times that of light across our line of sight. In fact, no such physical motion has occurred, only an optical illusion. And the smaller the angle of the jet from our line of sight, the greater the superluminal motion will appear. So almost any faster-than-light speed is possible because some jets will be tilted very close to our line of sight.

easily seen. For example, if the Andromeda Galaxy were placed at a distance of 4 billion ly—equivalent to a redshift of 0.2 (small for a quasar!)—its angular diameter would be a mere 4 arcsec. Remember that the earth's atmosphere limits its seeing to 1 arcsec or somewhat better at best. So a distant galaxy would make an image hard to distinguish from that of a faint star.

We might, however, look for a quasar in a cluster of galaxies—a good sign that we're not looking at stars. A nice example is the quasar 3C 206, which is surrounded by some twenty faint galaxies (Fig. 20.34). The quasar's redshift is 0.206; that of a nearby pair of galaxies, 0.203 (and the same is presumed for the cluster). The essentially same redshifts not only show that the quasar resides in

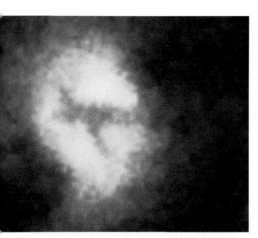

FIGURE 20.33 *HST* image of the nucleus of the spiral galaxy M51. Note the two dust lanes: one horizontal and one at an angle to the other, forming a rough "X." The horizontal lane has been interpreted as the side view of the dusty torus that surrounds the supermassive black hole in the nucleus. The nature of the other dusty region is not known. So "X" may not mark the spot! (Courtesy of Holland Ford, Johns Hopkins University and STScI, and NASA.)

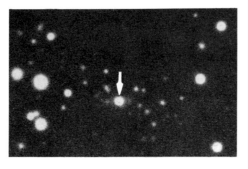

FIGURE 20.34 The quasar 3C 206 (arrow) surrounded by galaxies, which probably all lie in the same cluster of galaxies. Because the quasar and the galaxies have the same redshift, we infer that the quasar lies in the same cluster. (Courtesy of H. Spinrad; observations by S. Wyckoff, P. Wehinger, H. Spinrad, and A. Boksenberg.)

the cluster but also prove that the quasar's redshift is cosmological.

Other observations are beginning to reveal that quasars are surrounded by very faint envelopes that are the hard-to-detect disk of the galaxy. For example, 3C 273, the closest of the high-luminosity quasars, has a fuzzy appearance in computer-processed images and in photographs taken under conditions of excellent seeing. The spectrum of the faint fuzz shows emission lines having the same redshift as the quasar, which implies that the two are the same object rather than two objects that happen to be lined up along the line of sight. In fact, almost all close quasars appear to have an extended structure associated with them—the underlying galaxy! However, we cannot tell whether the galaxies are spiral or elliptical. In general, quasars appear to reside in host galaxies of average luminosity.

Recent observations suggest that the host may be distorted, perhaps as a consequence of tidal interactions with a nearby galaxy. In a sample of nearby quasars (redshifts less than 0.62), more than 30 percent appear to be interacting

with another nearby galaxy. Most of these are located in a small cluster of galaxies, such as 3C 206. Hence, the quasar/AGN phenomenon may be activated by interactions between galaxies in clusters.

An *HST* image of the Seyfert galaxy NGC 1275 brings this point home (Fig. 20.35). NGC 1275 is the largest and brightest galaxy (a giant elliptical) in the Perseus cluster; it is a strong radio and x-ray source. The *HST* view revealed two significant features. One, certain star clusters appear to be globular clusters but have bluer stars, which means they are younger than normal globulars. A reasonable estimate is about a billion years old. Second, the background structure of gas and dust implies that this galaxy formed from a merger of two others. If so, the merger may have provided the material for the "young" globular clusters—and so it took place about a billion years ago.

■ **Why is a supermassive black hole a good possibility for a quasar's energy source?**

FIGURE 20.35 *HST* image (right) of the core of the Seyfert galaxy NGC 1275, a giant elliptical galaxy in the Perseus cluster of galaxies. The ground-based image (left) is a true-color image of the nuclear region of the galaxy. The white dot just above the center is the nucleus. The fainter bluish dots are young globular clusters. NGC 1275 is a strong radio source and a giant elliptical galaxy. It may well have been formed by the merger of two galaxies about a billion years ago, at which time these globular clusters were formed. (Image by J. Holtzman, courtesy of NASA/ESA; ground-based photo from NOAO/KPNO.)

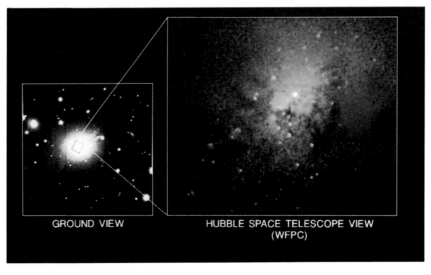

GROUND VIEW   HUBBLE SPACE TELESCOPE VIEW
                            (WFPC)

Key Concepts

1. The nucleus of the Galaxy contains a small, energetic source with a partly nonthermal spectrum; it emits radio, infrared, x-rays, and gamma rays.

2. Continuous spectra can be divided into two types: thermal (blackbody) and nonthermal. A common form of nonthermal emission is synchrotron radiation, produced by high-speed electrons spiraling in magnetic fields. Thermal emission can come from stars or from any hot, opaque material.

3. Active galaxies have much of their power provided by their nuclei, which have nonthermal spectra, at least in part. Some thermal emission continuous emission comes from stars, heated dust, and accretion disks.

4. Radio-emitting galaxies are one type of active galaxy; they fall into two classes: compact (emission from the nucleus) and extended (emission usually in two lobes widely separated from the nucleus).

5. High-resolution radio observations show that in many cases, a jet of emission extends from the galaxy's nucleus to at least one of the lobes of an extended radio galaxy. The synchrotron process may be the source of the emission; it shows that high-speed electrons are somehow channeled along the jet from AGNs out to the lobes.

6. Seyfert galaxies, another type of active galaxy, are spirals, with compact luminous nuclei and very broad emission lines in their spectra.

7. BL Lacertae objects (also active galaxies) have nonthermal spectra; they vary rapidly in luminosity and usually have no emission lines in the spectra of their nuclei.

8. Quasars have a starlike appearance, very large redshifts, and broad emission lines in their spectra (along with narrower dark lines); they sometimes have detectable radio emission with nuclear jets.

9. If the redshifts in the emission lines from quasars arise from the expansion of the cosmos, quasars are very far away—billions of light years. If this is true, their luminosities are huge, up to thousands of times what we find for normal galaxies; yet the emitting region cannot be larger than a few tens of light years. We do not know yet how quasars produce so much energy in so small a space.

10. Quasars and AGNs share some of the same observational characteristics, so they might be the objects of the same types (galaxies) at different evolutionary stages. In particular, a quasar may be a young galaxy with a hyperactive nucleus.

11. We have not yet identified the energy machine in the cores of quasars and AGNs; it may be a supermassive black hole (surrounded by an accretion disk) eating up material in the nucleus.

Key Terms

active galactic nuclei (AGNs)

active galaxies

BL Lacertae (BL Lac) objects

compact galaxies

double quasar

extended galaxies

gravitational-lens effect

head-tail galaxies

quasar

radio galaxies

radio jets

relativistic Doppler shift

Seyfert galaxies

supermassive black hole

Study Exercises

1. What observational evidence do we have that the synchrotron process produces some radiation from AGNs and quasars? (Objectives 20-1 and 20-2)

2. Contrast the evidence for violence in the Milky Way Galaxy with that for an active galaxy. (Objectives 20-2 and 20-7)

3. What evidence do we have that quasars are far away? Hint: How large are their redshifts? (Objectives 20-3 and 20-5)

4. What *observational* evidence indicates that in many cases, the light of quasars must pass through clouds of thin, cool gases? (Objective 20-4)

5. What is thought to power a quasar? (Objectives 20-4 and 20-9)

6. What connection does extended radio emission often have with the nucleus of an active galaxy? (Objectives 20-2 and 20-10)

7. Give one way in which the observed properties of radio galaxies and quasars differ. One way in which they are similar. (Objective 20-7)

8. What physical conditions are required to produce synchrotron emission in AGNs and quasars? (Objective 20-1)

9. Suppose you read an announcement in the paper tomorrow that astronomers have measured Hubble's constant to be twice the value used in this book. What would happen to the distances to quasars? (Objective 20-8)

10. How does the duration of variation in the light from an active galaxy give some indication of the size of the emitting region? (Objective 20-7)

11. Emission lines from quasars and some Seyfert galaxies are very broad. What physical process can account for this observation? (Objective 20-7)

12. In the generic model for an AGN or quasar, what is thought to be the source of power? (Objective 20-9)

13. Suppose we observe the generic model for an AGN side-on so that we are looking right into the dusty ring surrounding the accretion disk and black hole. What would you expect to see? (Objective 20-9)

14. Suppose we observe the generic model for an AGN face-on so that we are looking right onto the accretion disk and black hole. What would you expect to see? (Objective 20-9)

15. Many active galaxies show a broad peak in flux in their spectra at far-infrared wavelengths. What is a likely source of this emission? (Objective 20-1)

16. The spectra of AGN and quasars contain emission lines. What does this observation tell us about the physical conditions of at least part of the nuclear region? (Objectives 20-1 and 20-3)

Problems & Activities

1. The quasar 3C 273 has a redshift of 0.16. What is its radial velocity? Its distance? Hint: Assume the redshift is cosmological.

2. A typical quasar has a luminosity that is about 10,000 times that of the Milky Way Galaxy. If this energy is produced completely by fusion reactions, how much matter is a quasar converting to energy each second?

3. What is the Schwarzschild radius of a 10^8-solar-mass black hole?

4. The quasar named OQ 172 has a redshift of 3.53. What is its radial velocity? What is its distance?

5. What would be the angular diameter of the disk of the Milky Way Galaxy if placed at a distance of 5 billion light years?

See for Yourself!

A few of the active galaxies in the northern sky are bright enough to be seen with binoculars. And all of them vary in brightness, usually by about a factor of 10.

   If you've already have found Algol (Chapter 14), you can easily spot NGC 1275, as long as you are away from a light-polluted region. This Seyfert galaxy is located just about 2° east of Algol in Perseus (Fig. A.6). That's about one-third of the field of 7 × 50 binoculars. You can compare the brightness of the galaxy to nearby stars. If you do this weekly over a few months in the winter, you should be able to notice any significant change in the galaxy's brightness. ■

FIGURE A.6 The location of NGC 1275, a Seyfert galaxy in the constellation of Perseus. The angular distance between Algol and NGC 1275 is a little over 2°. (Map generated by Voyager software for the Macintosh.)

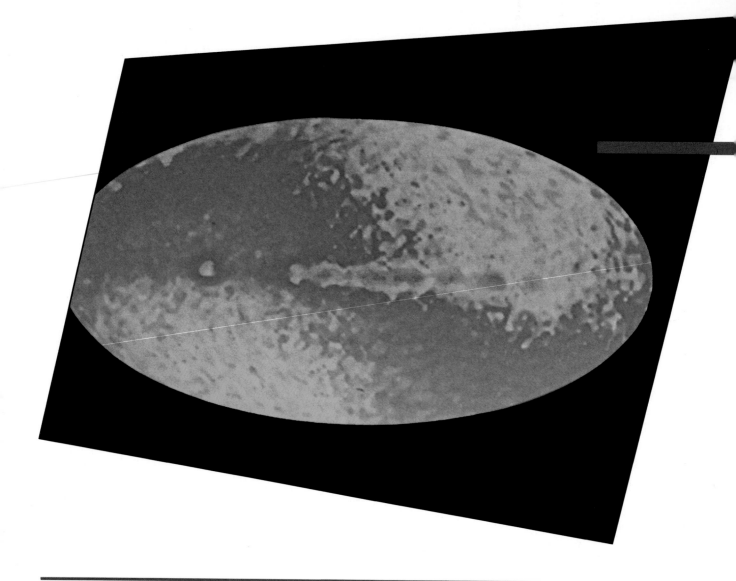

Central Question
How have the physical properties of the universe changed since its origin in the Big Bang?

Cosmic History

I could be bounded in a nutshell and count myself king of infinite space.

WILLIAM SHAKESPEARE, *Hamlet*

Some 20 billion years ago the cosmic bomb exploded. Perhaps you have seen an H-bomb blast on film or tape. Split seconds after detonation, an awesome fireball rips violently through the atmosphere (Fig. 21.1). Our universe was born out of a similar fireball—a cosmic fireball, the Big Bang—in whose violence all that we now see was created.

That, in a nutshell, is a picture of creation accepted by many astronomers today. **Cosmology,** the subject of this chapter, is the study of the nature and evolution of the universe. You can't talk about the evolution of the universe by simply describing what happens to each part; you must consider the universe as a unique whole. That's one of the problems of cosmology: we have only one cosmos to look at! In contrast, we can tell a lot about stars simply because so many stars, at different stages of their lives, are around for us to study.

Cosmologists have been fed a meager diet of observational facts about the universe. Despite (or perhaps because of) this lack, they have been able to dream up many models of the universe. Some of the models have been quite bizarre, but only one has a substantial following now: the Big Bang model. I— and most other astronomers—believe that the present evidence indicates that the universe began in a Big-Bang.

Despite its general acceptance, the Big Bang model does have shortcomings. A new theoretical scenario, called the inflationary universe, has been proposed to deal with some troublesome issues. This inflationary model complements and enhances the standard Big Bang model. Whether future observations will finally prove the inflationary model correct, a main conceptual basis will linger: that the universe is directly linked to elementary particles during its early history. The Big Bang signals the intimate connection of the very small with the very large. That connection at a time deep in the past has controlled the nature of the cosmos. It may mark the time that dark matter formed. And certainly the elementary particles created in the first few minutes shape the universe now.

FIGURE 21.1 A terrestrial analog of the cosmic fireball. The temperature, density, and pressure in this nuclear fireball, fractions of a second after the detonation of a hydrogen bomb, correspond to those in the early moments of the universe. (Courtesy of Eric M. Jones, LANL.)

21.1 Cosmological Assumptions and Observations

Modern astronomy has arrived at models of the cosmos in which the universe *evolves;* it is dynamic, not static. How can we study in detail the evolution of the universe? We need a few fundamental starting assumptions, difficult-to-prove assertions about the nature of the cosmos.

Assumptions

ASSUMPTION 1. *The universality of physical laws.* This assumption covers both local (the earth and solar system) and distant regions and means that we apply the physical laws we uncover here to all localities at all times and to the universe as a whole. Key observations support this assumption. For example, the spectra of distant galaxies contain the same atomic spectral lines as those produced by elements found on the earth. So other galaxies are made of the same elements as here, put together in the same way. Newton's law of gravitation correctly describes the motion of double stars and galaxies.

ASSUMPTION 2. *The cosmos is homogeneous (Fig. 21.2).* This means that matter and radiation are spread out uniformly, with no large gaps or bunches. You know that this assumption is not strictly true, for clumps of matter, such as galaxies and stars, do exist. But the cosmologist assumes that the size of the clumping is much smaller than the size of the universe. It's like looking at the earth from space: bumps, such as mountains, are too small to be seen, so the globe looks smooth.

ASSUMPTION 3. *The universe is isotropic (Fig. 21.3).* This idea relates to a quality of space itself rather than to the matter in it. Here's one way to think of it. Space has the same properties in all directions. So, in accordance with the **cosmological principle,** no direction or place in space can be distinguished from any other by any experiment or observation.

The universe has no center in space, because there is no way to tell if you are there. No direction in space provides special rewards when taken; for example, your mass does not increase as you travel in one direction or decrease when you go another way. Another example: the expansion of the universe is the same in all directions of space. Any observer would see the operation of the same Hubble's law.

These assumptions can be summed up in one sentence: *the universe is uniform.* All irregularities are ironed out. As the cosmologist Edward R. Harrison has quipped, the result is like that of the vanishing Cheshire cat in *Alice in Wonderland:* everything is wiped out except the grin. So cosmological models that rest on these assumptions ignore the structure and substance of planets, stars, and galaxies. Galaxies are considered, but only as tiny particles marking points in space, like a gas of atoms filling the universe. This gas is the cosmic grin. Such a mental simplification has real dangers, however: what lies behind the grin may be crucial. The universe may not obey the laws we lay down or the models we develop for it. Observations must in the end validate our assumptions.

Observations

What is the universe? If we consider all that can be seen with various telescopes, we are considering the **observable universe.** Yet this cannot be *all* of the universe; there are objects too faint and too far away to be seen, regions of the spectrum to which we and our instruments are so far blind, and objects detectable only by their gravitational effects. So there is more to the observable universe: a **physical universe** that includes directly observable matter as well as objects we detect by effects described by the laws of physics (such as dark matter).

The reality of the physical universe rests on the assumption that local physical laws apply to the rest of the universe at all other places and in all other times. That's one of our basic assumptions; without it, we could not conceive of a physical universe at all.

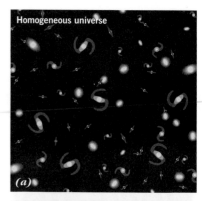

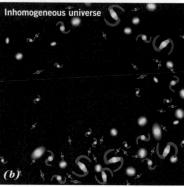

FIGURE 21.2 Distribution of matter in the universe. (*a*) A homogeneous universe. Matter and radiation are spread out uniformly, if you look over large enough distances. (*b*) An inhomogeneous universe. Even at large distances, matter and radiation are not uniform.

A Brief Review of Cosmology

Chapter 7 outlined the rise of relativity and its impact on cosmological ideas. It also presented some fundamental cosmological observations. These observations did not fall into a grand scheme until they were explained by Einstein's general theory of relativity. (This is a good place to review Chapter 7.)

What do we know about the universe? First, the universe *evolves*. Both the whole cosmos and its contents change with time.

Second, matter in the universe is *grouped*. Elementary particles (whatever they are) make up protons, neutrons, and electrons, which make up atoms. Atoms make up gases (molecular and atomic) and dust particles, which form stars, planets, and us. Stars come in clusters of stars, which are found in galaxies. And galaxies are grouped in clusters of galaxies, which in turn congregate in superclusters of galaxies.

Third, the universe is *expanding* (Section 7.4). The observed rate of expansion now lies in the range 15 to 30 km/s/Mly (Section 19.4). From Einstein's theory of general relativity, we know that the rate of expansion—Hubble's constant—relates to the overall density of matter (and energy) in the universe. This average density, in turn, determines the overall geometry of spacetime (Section 7.4). Three possibilities exist for this geometry (Section 7.3):

1. Flat, an open universe with the geometry of Euclid, infinite in space and time.
2. Spherical and closed, the cosmos being finite but unbounded in space and finite in time.
3. Hyperbolic, again an open universe, infinite in space and time, but curved.

Observations have not yet revealed which geometry is the right one. (Einstein had an aesthetic preference for a closed universe.) If enough dark matter exists, the universe could well be closed.

■ The notion of a physical universe rests on what assumption?

21.2 The Big Bang Model

Ingenious theoreticians have devised a bewildering array of cosmological models. Most now fall into the area of the **Big Bang model,** which is the standard model based on Einstein's general theory of relativity.

Almost every astronomers today would agree that the Big Bang model is the standard in use. But be warned that this model is *not* final; it has problems that need to be ironed out. The inflationary universe model (Section 21.8), a revision of the standard Big Bang model, represents one attempt to address these difficulties.

The universe is now expanding. Imagine it running backward. What happens? The galaxies and all matter within and without them eventually come tightly together. The extreme compression heats both matter and radiation to a very high temperature—high enough to break down all structure that had been created, including all the elements fused in stars. The atoms break down into protons and electrons; the matter is completely ionized. In addition, the density of ionized matter is so great that photons can travel only short distances before they are absorbed. As a result, the entire universe is opaque to its own radiation.

Now, when matter is opaque to all radiation, it acts like a blackbody (Focus 13.1). The radiation from a blackbody exhibits a characteristic spectral shape (Fig. 21.4). In an early dense, hot state, the whole universe acts like a blackbody radiator. If you could have been there, you would have seen a bright fog all around—it would have been like sitting in the sun's interior (but with the light at even shorter wavelengths, because of the higher temperature).

Imagine the universe expanding from this infernal state. It is so hot that it expands in a violent rush; hence the name *Big Bang*. As the universe expands, its overall density and temperature decrease. (This is true for the expansion of any gas. Let a gas out of a container. The gas suddenly cools as it expands.) Eventually the temperature drops so low that protons can capture

FIGURE 21.3 Isotropy and the universe. (*a*) An isotropic universe. Space has the same properties in all directions. For example, the distribution of velocities of galaxies in the expanding universe is smooth, uniform, and the same in all directions. A Hubble law is observed. (*b*) An anisotropic universe. The expansion of the universe is different at different places in space. A Hubble law is *not* observed.

electrons to form neutral hydrogen (and helium). This neutral gas is basically transparent to most kinds of light.

This event—the formation of neutral hydrogen from an ionized gas—marks a crucial stage in the evolution of the universe. No longer ionized, the universe becomes transparent to its own radiation; the light is freed of its close interaction with matter. The radiation and matter are no longer connected. The radiation freely speeds throughout space, and the expansion dilutes it. It is also redshifted, just like the light of distant galaxies, for indeed, having been emitted long ago, it comes from far away, and Hubble's law predicts a large redshift. The redshift lowers the temperature of the radiation.

If the universe did in fact begin in a hot Big Bang, debris (both matter and radiation) from that cosmic explosion should now lie all around us. The matter is pretty obvious: you can see planets, stars, and galaxies, and the unformed matter of interstellar gas and dust among all of these. (Not so obvious is the dark matter.) But what about the radiation produced in the Big Bang? It's been redshifted to a fairly low temperature. So the radiation's wavelength will be long compared with that of light. And if the universe expanded uniformly, the radiation's spectrum should show the telltale blackbody shape, because uniform expansion preserves the blackbody properties of the radiation. Have we seen such cosmic radiation? Yes! Its discovery verifies the hot Big Bang model.

■ **When matter is opaque to its radiation, how does the radiation behave?**

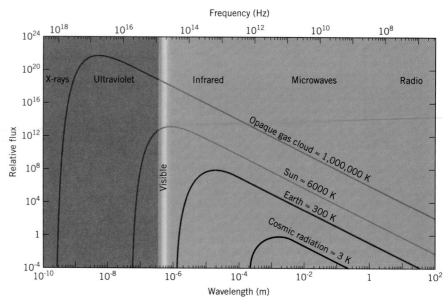

FIGURE 21.4 Comparison of the spectra of blackbodies at different temperatures, from 3 K to 10^6 K. Note that cosmic radiation at about 3 K peaks in intensity at infrared–microwave. Hotter blackbodies, such as the earth and sun, peak at shorter wavelengths. Also note that the overall shape of each spectrum is the same because these are all blackbody emitters whose continuous spectra are Planck curves.

fireball occurred, cosmic blackbody radiation should survive today. Peebles calculated that this fossil radiation should have a blackbody temperature of roughly 10 K. (The American physicist George Gamov and his colleagues Ralph Alpher and Robert Herman had actually developed such ideas in the 1940s.)

Discovery

Just as Roll and Wilkinson were building apparatus to detect the remnants of radiation from the Big Bang, Arno Penzias and Robert Wilson, scientists with Bell Laboratories in New Jersey, detected an annoying excess radiation by means of a

21.3 The Cosmic Background Radiation

In 1964 Robert H. Dicke, P. James E. Peebles, Peter G. Roll, and David T. Wilkinson at Princeton University were pursuing the possible existence of radiation left over from a Big Bang. The Princeton group attacked this problem: What would happen if the universe went through a hot stage, so that high temperatures decomposed any heavy nuclei into elementary particles? If a primeval

The Big Bang Model

At a Glance

Origin	In Big Bang explosion some 15 billion years ago
Evidence	Expansion (redshifts), 3 K background radiation
Nucleosynthesis	During first few minutes
Galaxy formation	After first million years
Future	Expansion forever (the Big Bore), or reverse and collapse (the Big Crunch)

special low-noise radio antenna (Fig. 21.5). They intended to do a sensitive study of the radio emission from the Milky Way. The excess noise they found would affect their results, so they set about to try to eliminate it.

They tuned their radio receiver to 7.35 cm (4080 MHz), where the radio noise from our Galaxy is very small. Still, they picked up the static. Penzias and Wilson further discovered that the noise did not change in intensity with the direction in the sky, the time of day, or the season. Perplexed, they examined the antenna again and found a pair of pigeons roosting inside. These pigeons, oblivious to radio astronomy, had coated a part of the antenna with their droppings. Perhaps the coating generated the excess noise? The birds were moved, their droppings cleaned out. But still the excess noise persisted, with an intensity at 7.35 cm, equivalent to that of a blackbody at 3.5 K.

Penzias eventually called the Princeton group. When he made contact, the Princeton group quickly concluded that the excess noise came from the cosmos—radiation left over from the Big Bang.

This intuitive, risky leap needed verification. After all, the measurement by Penzias and Wilson was at a single wavelength. But to establish the radiation as truly from a hot, dense stage in the universe's past, more observations at different wavelengths were needed to confirm the blackbody shape of its spectrum and its uniform intensity in the sky.

Confirmation

Soon Roll and Wilkinson added another point at another wavelength (3.2 cm) corresponding to a temperature of 2.8 K, close to that observed by the Bell Labs pair, and other experimental groups later contributed additional early evidence to the blackbody nature of the radiation (Fig. 21.6). Later ground-based efforts have pinned down the blackbody temperature of the background radiation quite well: 2.73 K for wavelengths from 10^{-1} to 10^2 cm (frequencies between a few to 100 GHz). For convenience, I'll take the radiation's blackbody temperature as simply 2.7 K.

FIGURE 21.5 Arno Penzias (*right*) and Robert Wilson (*left*) standing in front of the horn antenna with which they discovered the cosmic background radiation. (Courtesy of AT&T Bell Laboratories.)

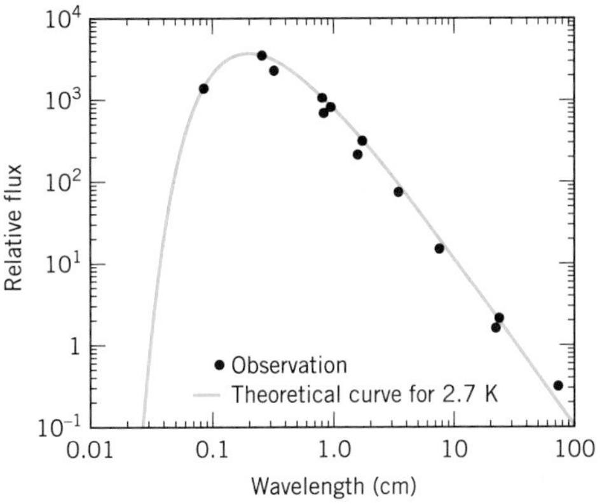

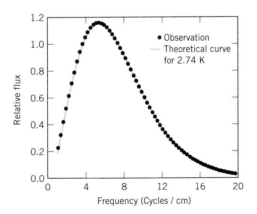

FIGURE 21.6 Some early observations of the cosmic microwave radiation. The points show measured fluxes at radio and infrared wavelengths; the solid curve corresponds to a theoretical blackbody of 2.7 K. (Adapted from a diagram by P. J. E. Peebles.)

FIGURE 21.7 *COBE* measurements of the cosmic background radiation, based on an initial 9 minutes of observing. The solid points are the data; the line is the spectrum from a blackbody at a temperature of 2.74 K. (Adapted from a NASA diagram.)

First results from the *Cosmic Background Explorer (COBE)* satellite confirm the pristine nature of the background radiation. They show that the spectrum is *extremely* smooth, with less than one percent deviation from that of a blackbody (Fig. 21.7). In addition, the spatial extent is very smooth, with no sign of any early variations that might explain the clustering of matter today (Fig. 21.8). Later processing of the data brought a surprise (Section 21.7) and a temperature of 2.726 K with an uncertainty of 0.01K.

To provide further confirmation of its cosmic origin, the radiation should be isotropic; that is, it should have the same measured intensity from all directions in the sky. For if the radiation were from a hot primeval state, the isotropy of the universe at that time would set the isotropy of the radiation. The *COBE* observations back up the notion that the radiation is very isotropic, with a very slight deviation from 2.7 K from our motion relative to the radiation.

Such results imply both that the radiation originated from the early universe and that, in fact, the newborn universe was very nearly isotropic. The radiation is cosmic, existing everywhere in space, so that it arrives at the earth uniformly from every direction. If so, it fills all space at all times. The isotropy observations clinch the interpretation of the radiation as cosmic.

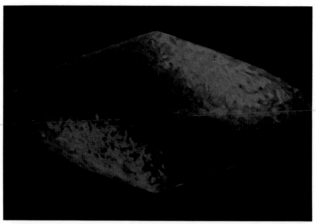

FIGURE 21.8 All-sky map of the cosmic microwave background radiation from the *Cosmic Background Explorer.* At a frequency of 53 GHz, the radiation is very smooth; the variations in the pink and blue regions are very, very small. The colors show the earth's motion relative to the radiation, which results in a Doppler shift and so a change in temperature. The pink false color indicates a "blueshift" (higher temperature); the false color blue a "redshift" (lower temperatures). The dark blue regions are unobserved parts of the sky blocked by the earth (center) and sun (edges). (Courtesy of NASA.)

Properties

The background radiation I have just described is usually given the long-winded name of **cosmic blackbody microwave radiation:** "cosmic" because it comes from all directions in space, "blackbody" because of its spectral shape, and "microwave" because its spectrum peaks at centimeter to millimeter wavelengths. The discovery of this radiation makes it easier to construct models of the universe because it affords new information about the cosmos.

First, the radiation enables astronomers to glimpse the raw, young universe. We can conclude that the initial universe was indeed quite homogeneous and isotropic.

Second, the present temperature (about 2.7 K) and the isotropy of the radiation set severe limits on the thermal history of the universe, that is, the change of the temperature of matter and radiation with time.

Third, the radiation's presence establishes an important marker for galaxy formation. Until the radiation and matter stopped their interaction, matter could not form large clumps. Only after the ionized gas had recombined could matter form clumps that eventually became stars and galaxies (Section 21.6).

Fourth, the radiation provides a reference for measuring the motion of the Galaxy and the Local Group. In the direction of any such motion, the cosmic radiation is blueshifted and so appears to be hotter in that location in the sky. In the opposite direction of the sky, the radiation is redshifted (and so appears to be cooler). The observations indicate that the Local Group and Local Supercluster have a combined motion of about 600 km/s in the direction of a nearby supercluster in the constellations of Hydra and Centaurus. (See the spring star chart in Appendix G for the location of these constellations.)

To sum up. The discovery of the cosmic blackbody microwave radiation has a significance for cosmology as great as that of the discovery of the expansion of the universe. Its measured uniformity backs up our assumptions that the universe is isotropic and homogeneous. And its present measured temperature, combined with Einstein's equation of general relativity, allows us to work out the evolution of the universe as its temperature changed.

■ **What is the key characteristic of the spectrum of background radiation from Big Bang?**

21.4 The Primeval Fireball

Since the discovery of the cosmic microwave background radiation, most astronomers have accepted a hot Big Bang model for the beginning of the universe. With the addition of experimental and theoretical knowledge on how matter behaves under hot, dense conditions, astronomers have been able to develop step-by-step details of what can happen in a Big Bang.

Caution. Don't picture the Big Bang as happening in the "center" of the universe and expanding to fill it. The Big Bang involved the *entire* universe; every place in it was *at* creation, which marked the beginning of time and space.

The Hot Start

Although the present temperature of the cosmic radiation is low, the amount of energy it contributes to the universe is large: each cubic meter contains 4×10^{-14} J, equivalent ($E = mc^2$) to about 4×10^{-31} kg. For comparison, if you took the material contained in the visible galaxies and spread it uniformly around the universe, each cubic meter would contain 4×10^{-28} kg. So for each kilogram of matter, there is approximately 10^{14} J of cosmic radiation. If the energy in the radiation could be used to heat up the matter, the temperature would be greater than 10^{12} K—one clue that the early universe must have been very hot.

This hot beginning is often called the **primeval fireball.** But we don't really know what happened at this time. No one now understands how matter behaves at temperatures greater than 10^{12} K. From Einstein's equations of general relativity we find that a temperature of 10^{12} K corresponds to a time of about 10^{-24} s after creation (time "zero"). By "creation" I mean the actual beginning of the present expansion.

The primeval fireball produced such a rapid expansion that the temperature and density dropped rapidly. The contents of the universe (Table 21.1)—elementary particles and their respective antiparticles—changed and are changing with temperature and time. I will sketch a scenario of the young universe in which temperature plays a crucial, controlling role. Each period of time can be matched with a temperature.

Roughly, the universe's thermal history can be divided into four eras: a **heavy-particle era,** when massive particles and antiparticles dominated; a **light-particle era,** when particles with less mass were made; a **radiation era,** when most particles had vanished and radiation was the main form of energy; and a **matter era,** in which we now live, when the energy of matter dominates that of radiation. (In a cubic meter of space, matter now averages about a thousand times the density of photons,

TABLE 21.1 PARTICLES IN COSMIC NUCLEOSYNTHESIS

PARTICLE AND ANTIPARTICLE	SYMBOL	CHARGE	COMMENTS
Neutrino, antineutrino	ν, $\bar{\nu}$	0, 0	Massless (?) particles that travel at light speed; stable (?)
Proton, antiproton	p, p̄	+1, −1	Nucleus of hydrogen; stable
Electron, positron	e^-, e^+	−1, +1	Particles surrounding the nucleus of an atom; stable
Neutron, antineutron	n, ñ	0, 0	Neutron decays to a proton and an electron in about 10^3 s
Photon	γ	0	Packet of radiation, electromagnetic energy (gamma rays)
Deuteron	^{2}H	+1	Nucleus of deuterium, or "heavy hydrogen"; contains 1 proton, 1 neutron; stable
Helium-3	^{3}He	+2	Nucleus of an unusual type of helium; contains 2 protons, 1 neutron; stable
Helium-4	^{4}He	+2	Nucleus of ordinary helium; contains 2 protons, 2 neutrons; stable
Lithium-7	^{7}Li	+3	Nucleus of most abundant type of lithium; contains 3 protons, 4 neutrons; stable
Beryllium-7	^{7}Be	+4	Nucleus of most abundant type of beryllium; contains 4 protons, 3 neutrons; unstable

considering their energy in the form of mass.) I will consider a proton and the nuclei of elements (such as helium) to be "heavy" particles and electrons to be "light" particles (Table 21.1). As the universe expands, its temperature and density fall *during* an era. The values I give will be representative ones.

Caution. Whenever I say that the universe was "smaller" or "larger," I mean that the *distance* between a pair of objects was smaller or larger. The universe itself may be finite or infinite. That overall geometry does not affect what I'll say about the Big Bang.

Creation of Matter from Photons

Before the story unfolds, you need a little preparation about one key part: the creation of matter and antimatter from photons. (**Antimatter** has the same properties of regular matter except that it has the opposite electrical charge.)

At some time in the primeval fireball, the energy of photons was so high that their collisions produced particles. This process occurs when the energy in the colliding photons equals or exceeds the mass of the particles produced. Sounds bizarre? The result comes directly from Einstein's relation between matter and energy ($E = mc^2$), which does not restrict the *direction* of the transformation: matter can become energy, or energy can become matter.

The creation of matter from light happens in a special way that involves both matter and antimatter. When matter and antimatter collide, they are annihilated and become converted to photons (Fig. 21.9). In reverse, two photons (if they have enough energy) create a matter–antimatter pair when they hit each other. Note that this process always results in *pairs* of particles.

How much energy must the photons have? At least the energy equivalent of the masses of the pair they produce. So to make protons and antiprotons takes more energy than is needed to make electrons and positrons, because protons have more mass than electrons (about 1800 times more). Now we can talk about the energy of photons in the primeval fireball in terms of their temper-

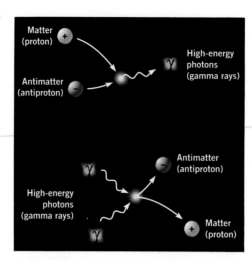

FIGURE 21.9 Matter and antimatter annihilation to make photons; particle and antiparticle production from photons. In this case, the particles are protons; the antiparticles, antiprotons; and the photons, gamma rays. Each gamma ray must have an energy at least equal to that of the mass of a proton for the conversion to work.

ature. To have enough energy to make electrons requires a temperature of at least 1.2×10^{10} K; for protons, at least 2.2×10^{13} K.

Particle production from photons plays a key role in the very young universe. Let's see what the model tells about those times.

Temperature Greater than 10^{12} K, Time Less than 10^{-6} s

Photons create pairs of particles and antiparticles. The photons have enough energy to produce even the most massive elementary particles, such as protons. Annihilation also takes place, and the balance between annihilation and creation fixes the density of particles and antiparticles for the next stage. This balance between the production and destruction of massive particles marks the heavy-particle era. It does not last long, because the cosmos expands rapidly and the temperature declines quickly.

So at the earliest times that we can calculate, the universe is a smooth soup of high-energy light and massive, elementary particles (perhaps of unknown kinds). The nuclei of atoms have yet to be made.

Temperature from 10^{13} to 6×10^9 K, Time from 10^{-6} to 6 s

Annihilation of heavy particles with their antiparticles continues (Fig. 21.10*a*). The remaining photons, however, lack the energy to create new heavy particles. Only light particles—electrons—can be made, because less energy is required for their production from photons. The universe enters the light-particle era. (Keep referring to Table 21.1 as we run through this sequence.)

Protons and electrons interact to generate neutrons (Fig. 21.10b). When the temperature falls to 6×10^9 K, photons can no longer make electron–positron pairs. This temperature marks the end of the light-particle era and the beginning of the radiation era. Neutrons are no longer produced by the interaction of protons with electrons or of antiprotons with positrons (Fig. 21.10c). The neutrons now decay into protons and electrons but no new ones are made. The ones that survive are crucial to the next step.

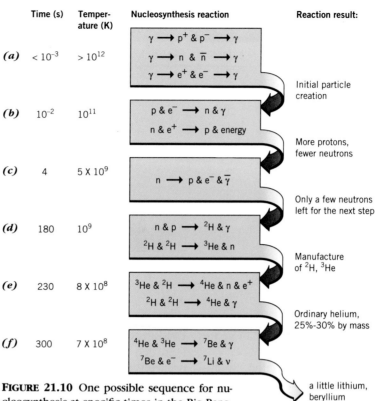

Time (s)	Temperature (K)	Nucleosynthesis reaction	Reaction result:
(a) $< 10^{-3}$	$> 10^{12}$	$\gamma \rightarrow p^+ \& p^- \rightarrow \gamma$ $\gamma \rightarrow n \& \bar{n} \rightarrow \gamma$ $\gamma \rightarrow e^+ \& e^- \rightarrow \gamma$	Initial particle creation
(b) 10^{-2}	10^{11}	$p \& e^- \rightarrow n \& \gamma$ $n \& e^+ \rightarrow p \& energy$	More protons, fewer neutrons
(c) 4	5×10^9	$n \rightarrow p \& e^- \& \bar{\gamma}$	Only a few neutrons left for the next step
(d) 180	10^9	$n \& p \rightarrow {}^2H \& \gamma$ ${}^2H \& {}^2H \rightarrow {}^3He \& n$	Manufacture of ${}^2H$, ${}^3He$
(e) 230	8×10^8	${}^3He \& {}^2H \rightarrow {}^4He \& n \& e^+$ ${}^2H \& {}^2H \rightarrow {}^4He \& \gamma$	Ordinary helium, 25%-30% by mass
(f) 300	7×10^8	${}^4He \& {}^3He \rightarrow {}^7Be \& \gamma$ ${}^7Be \& e^- \rightarrow {}^7Li \& \nu$	a little lithium, beryllium

FIGURE 21.10 One possible sequence for nucleosynthesis at specific times in the Big Bang. Refer to Table 21.1 for the particles. (*a*) Particles and antiparticles are made from photons. (*b*) Neutrons are made and destroyed. (*c*) Neutrons decay; only 16 percent remain from this stage. (*d*) Neutrons and protons make deuterium and tritium. (*e*) Tritium and deuterium make regular helium. (*f*) Helium and tritium form beryllium and lithium. Note that lithium and beryllium are the heaviest elements that can be made in the first few minutes.

Temperature 10^9 K, Time from 6 to 300 s

Now come the key nuclear reactions! The remaining neutrons and protons react to form nuclei of simple elements. The most important reaction involves the combination of a neutron and a proton to form deuterium, ${}^2H$ (Fig. 21.10d). All neutrons, except those that have decayed, end up in deuterium. When the deuterium has been produced, further reactions create ${}^4He$, normal helium (Fig. 21.10e) and also a little tritium (${}^3H$) and ${}^3He$. The net result is about 25 percent helium by mass compared to all forms of matter. A little bit of beryllium and lithium is created by the combination of deuterium and tritium (Fig. 21.10f). Some deuterium is left over. Extremely small amounts of the heavier elements are made.

Helium, once formed, is tough to destroy. So the present helium abundance in the universe rests on a cosmological base. The helium abundance serves as a test of the Big Bang model. It predicts that no celestial object can have a helium abundance of less than 25 percent. (Because stars form helium, the abundance now is greater than this.)

Stop here for a moment and consider what you've just read: a blow-by-blow account of the *first few minutes* of the hot Big Bang model. Doesn't it seem a little unreal to you? It's astounding that

the model allows us to talk of such times with some confidence.

Temperature 3000 K, Time 1 Million Years

So far the temperature has been so high that all atoms in the universe have been ionized. At about 1 million years after creation, the radiation's temperature has plunged to 3000 K, too low to keep matter ionized. The nuclei begin to capture electrons to form neutral atoms—a process called **recombination.** The recombination process happens in a few thousand years; the neutral matter becomes transparent to the radiation. Suddenly light breaks through the now-transpar-

TABLE 21.2 SEQUENCE OF EVENTS IN THE BIG BANG

EVENT	TIME [a]	DENSITY (kg/m^3)	TEMPERATURE (K)	COMMENTS
Creation	0	?	?	Not the province of present science; relativity fails.
Heavy-particle era	10^{-43} s	10^{97}	10^{33}	Photons make massive particles (such as protons) and antiparticles.
Light-particle era	10^{-4} s	10^{17}	10^{12}	Photons have only enough energy to make light particles and antiparticles, such as electrons and positrons; protons and electrons combine to make neutrons.
Radiation era	10 s	10^7	10^{10}	Few particles left in a sea of radiation; these partake in nucleosynthesis of deuterium, helium, lithium, and beryllium.
Matter era	10^6 y	10^{-18}	3000	Ionized hydrogen recombines; cosmos becomes transparent.
Now	10^{10} y	10^{-31} (radiation)	3 (radiation)	Astronomers puzzle about creation.

[a]Times are the approximate midrange of the eras.

ent matter. The matter and radiation are no longer locked together.

Freed from this interaction, the radiation merrily expands with the universe. As the universe expands, the radiation cools down to become the 2.7 K cosmic radiation today. The matter, however, follows a different course because of little bumps in the generally smooth distribution of matter. Clouds of matter condense out of the primeval fireball. The radiation era ends, and the matter era begins. Material clumping could not happen until the radiation and matter had decoupled, as explained in Section 21.6. (Why not? Because the radiation exerted a pressure on the matter that kept if from gathering into clumps.) This event flags the time when galaxy formation could begin (Section 21.7).

To sum up. In a few minutes after creation, the universe expanded and reached temperatures and densities suitable for the formation of deuterium and helium (Table 21.2). The Big Bang model predicts a helium abundance of 25 percent (by mass) and very little formation of heavier elements. Much later, the nuclei captured electrons to form neutral atoms (almost all hydrogen and helium). The universe became transparent. The radiation became the cosmic background radiation and ceased to control the universe's evolution.

Evidence for the Big Bang

How can we judge whether the contemporary Big Bang cosmological picture is correct? What observational evidence supports it?

First and foremost, observations indicate that the background radiation pervades the universe. Within the limits of observational errors, measurements confirm a blackbody spectrum with a temperature of 2.7 K. Measured in many directions of space, the radiation comes to the earth with a pretty uniform intensity. The background radiation has the attributes expected of a cosmic, hot origin.

Second, the model predicts that primeval helium was formed in the first 5 minutes of the universe's history and that the helium abundance should be about 25 percent by mass. The Big Bang model sets this helium abundance as the basement level; any observation of a substantially lower amount would call the model into question. Because helium is formed in stars later, the present helium abundance must be larger than 25 percent—that is, the helium made in the Big Bang plus that formed in stars.

Unfortunately, we can assess the present cosmic helium abundance only indirectly. As Chapter 13 pointed out, the solar photosphere is not hot enough to excite helium lines for direct viewing with a spectroscope, so it's not possible to measure the helium abundance in the solar photosphere. The corona appears to have an abundance of 28 percent. Theoretical models of stellar evolution place the initial helium abundance in the sun's interior at 25 percent.

O and B stars from Population I exhibit helium absorption lines that imply a helium abundance of approximately 35 percent. H II regions surrounding hot stars have helium emission lines at optical and radio wavelengths that give abundances of 28 percent. Planetary nebulas also have strong helium emission lines that imply a helium abundance of greater than 30 percent.

The best objects to search for primeval helium are the oldest stars now surviving, the Population II stars. Unfortunately, most Population II stars are much too cool to accommodate the excitation of helium lines in their spectra. However, indirect evidence from star models indicates that these stars have helium abundances of about 30 percent.

Though the helium values have a considerable range, the helium abundance for a variety of celestial objects falls close to that predicted by the Big Bang model. The match of observations with theory is good. The accumulation of evidence for the hot Big Bang is strong, because it gives a simple, unified picture of the early universe.

■ What are the four eras of the universe's thermal history?

21.5 The End of Time?

So much for the universe's past. What about its future? Recall (Section 7.4) that Einstein's general theory of relativity allows the cosmos to have one of two general geometries: open or closed. If open, the universe will expand forever, and time will never end. If closed, however, the universe must eventually collapse, running backward through the history outlined in the preceding section.

Which fate will be ours? Section 7.5 mentioned one observational test. The current measured value of Hubble's constant ($H = 20$ km/s/Mly) and Einstein's theory give a critical density for a closed universe—it's about 5×10^{-27} kg/m³. If

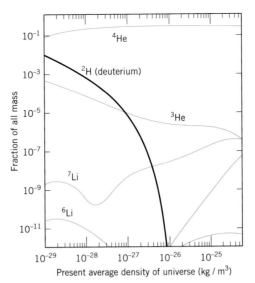

the actual density is less, the universe is open; if more, it's closed. The evidence to date is inconclusive, although it does weakly support an open geometry.

The standard hot Big Bang model provides another test, which is perhaps the strongest. The test rests on the observed present abundance of deuterium.

How can the abundance of deuterium reveal whether the universe is open or closed? To do the Big Bang model calculations, you need to put in the *present* value of Hubble's constant, the *present* temperature of the cosmic radiation, and the *present* average density of the universe. The first two items are reasonably well known, but the third is not. The amount of helium that comes out of the Big Bang calculations does not depend very much on the value used for the present density of the universe. But the amount of deuterium that theoretical calculations predict does depend very sensitively on the value used for the present density (Fig. 21.11).

So we can use the theory to turn the argument around. If we can measure the present cosmic abundance of deuterium, we can check this number against that predicted by the Big Bang model for various densities. The model gives the present average density of the universe, and we can compare that number with the critical density.

Ultraviolet observations of the interstellar medium show lines produced by the molecules containing deuterium.

FIGURE 21.11 Theoretical calculations of the abundance of low-mass elements formed in the Big Bang as related to the average density of the universe now. Note how sharply the abundance of deuterium varies with different present average densities of the universe, while the abundance of ordinary helium changes little. The vertical axis gives the fraction of the element, by mass, compared to all other elements. (Based on calculations by R. V. Wagoner.)

The amount of deuterium (relative to hydrogen) in the interstellar medium is about 2.0×10^{-5} by mass. (Remember that most of the mass in the interstellar medium is hydrogen.)

Now to estimate how much Big Bang deuterium has been burned up in stars. A reasonable guess is about half, so the original deuterium abundance was about twice that observed now, or 4×10^{-5} by mass relative to hydrogen. Look at the Big Bang calculation (Fig. 21.11). A deuterium abundance of 4×10^{-5} implies a present cosmic density of 4×10^{-28} kg/m^3, considerably less than that of the critical density (5×10^{-27} kg/m^3).

Conclusion from this test: the universe is open. It will expand forever. Time will *not* end. But the discovery of enough dark matter could change that conclusion.

■ What type of geometry for the universe is suggested by the evidence to date?

21.6 From Big Bang to Galaxies

The isotropy of the cosmic background radiation implies the matter and radiation in the universe had a very uniform distribution. But that's not the situation now. Even the most spread-out of these systems of matter—clusters of galaxies—have average densities about a hundred times greater than the average density of the universe. So here's the crucial question: How did an originally very smooth universe become clumpy?

A model of galaxy formation must face a critical hurdle; it must operate fast. To date, astronomers have seen objects (the quasars) as far as some *12 billion* ly away. So the matter from the Big Bang must have formed into large clumps well before this time.

The discovery of the cosmic radiation by Penzias and Wilson forced astrophysicists to consider what happens to disturbances in a hot, dense universe filled with matter and radiation. They found that the radiation played a powerful role in inhibiting the growth of disturbances. A dense patch of gas in the early universe will have a high internal pressure because the radiation adds to the pressure force that pushes the patch apart. Only very large disturbances would contract. So disturbances that contained roughly the mass of a galaxy could not contract to become young galaxies.

Just after the universe became transparent, it's a new show. The radiation and the gas no longer interact, so radiation pressure no longer resists gravity. Small disturbances, amounting to only 10^5 solar masses, can condense out of the gas, along with disturbances of greater mass. This result gives some hope, for the large gravitationally bound masses we see now range from 10^5 (globular clusters and dwarf galaxies) to 10^{15} solar masses (clusters of galaxies). Disturbances of this size range can condense just after decoupling. So roughly 1 million years after the Big Bang, marks the point at which the galaxy formation could take place in the young universe.

There's one real weakness in these ideas: even though disturbances can be unstable, they grow slowly—so slowly that the galaxies we see could hardly have formed by now, unless the disturbances were already fairly large in the beginning. But the smoothness of the cosmic background radiation implies that any lumps were small.

One way to see what happened with early galaxy formation is to look very far out in space, and so back in time, searching for the youngest galaxies possible. Such a search has been carried out at optical wavelengths by Anthony Tyson of AT&T Bell Laboratories and Patrick Seitzer of NOAO using the 4-meter telescope at CTIO. They have made very long (hours!) CCD exposures in a direction perpendicular to the plane of the Galaxy. Here stellar images in the Milky Way are at a minimum; what appear are distant galaxies—loads of them! The CCD frames covered an area of only 3 arcmin by 5 arcmin (some 0.02 times the area of the full moon), yet each contained more than a thousand galaxy images (Fig. 21.12). Their redshifts are estimated to fall in the range from 0.3 to 3, or about 7 to 11×10^9 ly away—in the same range as quasars. But these objects are extended, not pointlike. In some frames, they fill up 30 percent of the sky. Tyson believes that these were very young galaxies undergoing their first burst of star formation, shortly after the Big Bang. (Not all astronomers hold this stance, citing the lack of measured redshifts.)

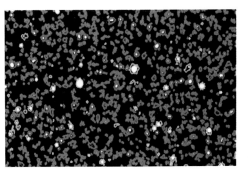

FIGURE 21.12 Long-exposure CCD image of a small section of the southern sky (3 arcmin by 5 arcmin). The image has been processed and color-coded so that most galaxies appear red, those with stellar cores blue, and stars green. Note how many of the galaxy images overlap. (Courtesy of J. A. Tyson, AT&T Bell Laboratories.)

To sum up. Galaxies formed early in the history of the universe.

■ **After decoupling, what happened to radiation pressure?**

21.7 Elementary Particles and the Cosmos

Although accepted as the standard cosmological model, the hot Big Bang model has a number of weaknesses, such as the great dominance of matter over antimatter in the universe today, and the formation of galaxies. Both these problems and others are being tackled now in imaginative theoretical ways that unite the universe of the small—elementary particles—with the universe itself. The connection occurs early in the Big Bang, before the first second had elapsed. To explain this connection, I'll digress a bit into the world of elementary particles and their interactions.

The Forces of Nature

Let's focus on how particles are related—their interactions. We generally think of these relations in terms of forces between particles. You are familiar with two of these: gravitation and electromagnetism. These forces have one property in common: they work over large distances. According to Newton's law of gravitation, the most distant galaxies exert a force on you (and you on them). The same is true of objects that have a net electrical charge. These forces differ in their relative strengths. Electromagnetic forces are *much* stronger than gravity—as you know if you've lifted a nail with a magnet. The gravitational attraction of the entire earth on the nail is weaker than the force of the magnet.

Two other forces operate in nature in the subatomic domain. One, called the **strong force,** holds the nuclei of atoms together. Recall that an atom's nucleus has protons tightly packed together. The electric force of each proton repels the others strongly, especially when so close—only 10^{-15} m apart. The strong force overwhelms this electric repulsion and keeps the nucleus together. The other, called the **weak force,** crops up in radioactive decay. Without it, fission would not take place. Like the strong force, the weak force operates over very short distances, 10^{-17} m and less.

These four forces are all that are known. The strong force is the strongest of them (Table 21.3), electromagnetism second, the weak force next, and gravity takes the bottom as the weakest. Now, although these forces appear to operate very differently, might they have an underlying unity? That quest for a unified theory has tempted physicists for most of this century. And recently they have had some success in struggling to find a **grand unification theory,** fondly known as *GUT.* The development of GUTs (there are more than one) has, curiously enough, modified Big Bang cosmology and helped to shore up some weaknesses.

TABLE 21.3 PROPERTIES OF THE FUNDAMENTAL FORCES

FORCE	STRENGTH (relative to strong)	RANGE (m)
Strong	1	10^{-15}
Electromagnetic	1/137	Infinite
Weak	10^{-5}	10^{-17}
Gravity	6×10^{-39}	Infinite

How is that connection made? First, in the 1970s, theoreticians unified the weak force and the electromagnetic force, introducing the term *electroweak force.* Buoyed by their success, they next took aim at unifying the electroweak and strong forces. As part of that work, they developed the concept of a new elementary particle, called a **quark,** that comprises other so-called elementary particles. For example, a proton consists of three quarks. Since the proton is now a composite particle, it can decay (as the neutron, also a composite particle, decays) with a predicted lifetime of some 10^{30} years. (Experiments are now under way to check this prediction.) GUTs also predict that the unification of strong and electroweak forces won't be apparent until energies of greater than some 10^5 J—the equivalent of converting about 10^{15} protons completely into energy. This unification may occur when the distances between particles are less than 10^{-31} m. Such energies cannot be made in large particle accelerators on earth, but they do occur in the Big Bang model at a time of 10^{-35} s when the temperature was 10^{26} K and the distance between particles was 10^{-40} m. So the conditions then may have been just right for a unification of forces. The Big Bang model serves as a way to test GUTs—the whole universe as a high-energy particle machine!

Note. One result of GUTs has been to simplify our view of elementary particles. We now believe that matter is composed of two classes of elementary particles: *quarks* and *leptons.* Only six particles make up the **lepton** group; they include electrons and neutrinos. (These are low-mass, "light" elementary particles.) Quarks make up all the other particles, such as protons and neutrons and some hundred others. (These are high-mass, "heavy" elementary particles.) Six quarks are known to date. Each has three subtypes, for a total of eighteen. Each of these has an antiparticle, so the total of quarks of all types is thought to be thirty-six. The four basic forces act on all these particles. The aim of GUTs is to reduce all particles to one kind, interacting through one force—truly a grand unification!

GUTs and the Cosmos

We now look at the temperature history of the Big Bang model in a new way: we attempt to determine the temperatures at which specific aspects of particles and their forces freeze into existence. Here I use the word "freeze" in the sense that when water freezes into ice (at 0 °C or 273 K), its state changes abruptly. Water in the form of ice behaves differently from water as a liquid or gas.

I have already discussed two freezings in the early cosmos (Section 21.4): the formation of simple nuclei in the nucleosynthesis era (first few minutes) and of atoms during the recombination time (1 million years). GUTs predict a very special kind of freezing at 10^{-35} s, when the strong nuclear force froze out from its unification with the electroweak force. Before this time, from 10^{-43} s to 10^{-35} s, gravity had frozen out from the other three forces. Before that time, all four forces are presumed to be unified in a single force.

The final freezing takes place at about one second, when the electromagnetic and weak forces split. Now quarks can combine to form particles such as protons, which can then react in the nucleosynthesis era. Before this freezing, the universe contained a hot gas of quarks and electrons. From this time on, the standard Big Bang model applies (Section 21.4).

A special event occurs at the separation of the electroweak force from the strong force at 10^{-35} s. Just as water releases energy in the form of heat when it freezes, the freezing of the strong force released energy, but in enormous amounts. This event pumped energy into the expansion of the universe so that it grew in size by many powers of 10 in just 10^{-32} s. This marks the era of inflation, which is the hallmark of the inflationary universe (Section 21.8).

GUTs and Galaxy Formation

Another aspect of freezing is related to the galaxy-formation problem. Consider the freezing of a pond: the ice sheet does not form all at once but in patches. That is, the freezing process has defects. In cosmological models, these defects have

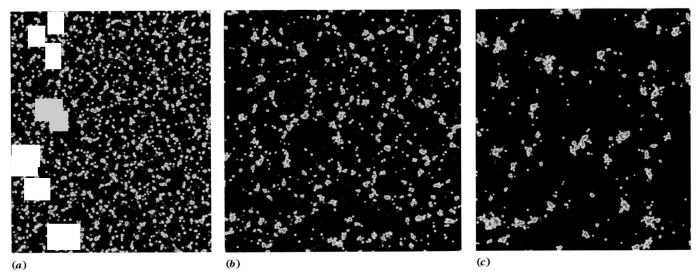

(a)   (b)   (c)

FIGURE 21.13 Supercomputer simulation of a sequence of possible development of large-scale clumping of the universe. This model assumes random fluctuations of density acted on only by gravity and simulates their growth from (a) to (c). The false colors represent density: blue the highest, red the lowest. (Courtesy of Adrian Melott; calculations done at NCSA, University of Illinois.)

mass and survive for a period so long that at the decoupling time, when matter can clump, the defects serve as the cores—the lumps—on which gravitational instability occurs.

Another way to help make galaxies calls for neutrinos. Some GUTs predict that neutrinos should have a very small mass (and a few experiments, yet unconfirmed, support this claim)—perhaps 0.001 percent that of an electron. According to these theories, neutrinos froze out at a time of 1 second, and those relic neutrinos should be with us today. If massless, they would have little effect. If they have even a slight mass, they change the universe. (And neutrinos with mass might solve the solar neutrino problem, outlined in Section 13.6.)

First, neutrinos may contain enough mass in total to close the cosmos. Second, they can aid galaxy formation. Neutrinos with mass can start clumping by gravity well before other particles can. After recombination, atoms would gather around these neutrino clumps and so speed up the gravitational instability process. Third, massive neutrinos may congregate in clusters of galaxies and bind them gravitationally. Fourth, neutrinos can also gather in the halos of galaxies to make up the so-far invisible matter there.

Supercomputer models of clustering show what happens to large disturbances in a mostly smooth distribution of material. These early ripples in the universe may have fostered the formation of clusters and superclusters. The models depict the clusters in elongated structures similar to those observed in the Local Supercluster. Later stages in the simulation show bridges of material still linking the clumps of matter. The final result is a spongelike structure of voids and matter (Fig. 21.13). Most of the matter is assumed to be in the form of cold, dark matter (still invisible today). Small masses form first, and then collide and coalesce into galaxies.

Another supercomputer model that has attacked the same question results in a spongelike structure of the voids and matter (Fig. 21.14). Richard Gott and Changbom Park at Princeton University studied the evolution of 4 million particles to trace the clustering of matter over regions as large as 2 billion ly and for a period as long as 13 billion years. Most of the mass is assumed to be in the form of cold, dark matter; also assumed are small, random ripples started in the first second of the Big Bang. Their results closely imitate the patterns found in the largest redshift maps. Hence, we are sensing that the large-scale complexity

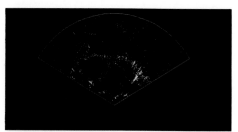

FIGURE 21.14 Results of a supercomputer simulation of the growth of small ripples in the early universe after some 13 billion years of the influence of gravity. Note the stringlike clustering of the matter with large voids in between. The matter here is mostly in the form of cold, dark matter. (Courtesy of C. Park and J. R. Gott, Princeton University.)

grew from simple beginnings: random fluctuations in the early universe, amplified by the action of gravity.

An additional *COBE* result provides a solution to many of these sticky issues. After processing some 70 million observations, the *COBE* team carefully created a map of the sky (Fig. 21.15) that shows some of the cosmic ripples from the Big Bang—wrinkles in the cosmic tapestry that were the likely seeds for the formation of large-scale structure. The map shows minute fluctuations in the cosmic background radiation, variations that amount to only a few parts per million at a time when the cosmos was less than one million years old. In regions of higher temperature, the density of matter at the time was a little higher than average; in the regions of lower temperature, the density was a bit lower. These higher-density variations were large enough to create enough gravity to attract more matter into denser clumps, which eventually contracted into the first-born galaxies. The lower-density regions resulted in the voids. Because the young universe was very smooth but not perfectly so, we have a lumpy universe today.

What kind of matter created these clumps? The *COBE* results do not tell us that key piece of the large-scale puzzle. We will have to rely on models yet to be developed to interpret these startling observations, which reveal the most ancient and largest structures in the universe.

So particle physics can shore up some problems with the simple primeval

fireball. The key point: the connection of the very small to the Big Bang is essential to our understanding of the universe's history. And the information from GUTs enhances the Big Bang model, not only in details but also in the overall picture. The model becomes more aesthetically pleasing (Section 2.1).

■ How do GUTs aid in developing models for the formation of galaxies?

21.8 The Inflationary Universe

The drive to integrate particle physics with cosmology had a new success with a variation of the Big Bang model called the **inflationary universe model.** It copes with serious flaws in the Big Bang picture and so improves it. I'll focus on just two, called the **flatness problem** and the **horizon problem.**

The Flatness Problem
Recall that the universe, in terms of general relativity, can have one of two basic geometries: open or closed. We can evaluate which one applies to the cosmos by examining the ratio of the measured density (of matter and energy) to the crit-

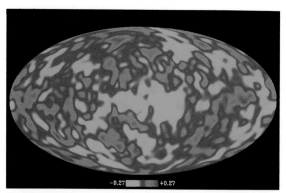

FIGURE 21.15 Another *COBE* all-sky map, made at microwave wavelengths. The Milky Way, if visible, would extend horizontally across the middle of the image. This false-color map is coded so that blue indicates 0.01 percent colder than the average cosmic microwave temperature; red indicates 0.01 percent warmer than the average (see scale at bottom). This radiation is from the universe when it was about 300,000 years old. The red patches represent regions of higher than average density; blue, regions of lower than average density. (Courtesy of NASA/Goddard Space Flight Center.)

ical density predicted from Einstein's general relativity and the value of Hubble's constant. A cosmos whose actual density is exactly the critical density is flat; the ratio is 1. Now if at the Big Bang the ratio were 1, it would remain 1 forever. If, however, the value differed from 1 *ever so slightly,* the ratio would be *much* different from 1 now. Surprisingly, the ratio is believed to have a value between 0.1 and 2, very close to 1. So it must have started very close to 1; otherwise, as the universe evolved, the ratio would have acquired a value much different from 1. The standard Big Bang model has a special starting condition (the geometry of the universe very, very close to flat) but does not give an explanation for it. That is the essence of the flatness problem.

Enter the inflationary cosmos, which deals with times from 10^{-43} to 10^{-32} s, when massive elementary particles dominate the universe. During that interval, the universe undergoes a tremendous spurt of growth, becoming perhaps as much as 10^{50} times larger. Remember, that means that the distance between two particles increased by 10^{50}. To put that increase in perspective, if we inflated the distance from the proton to the electron in a hydrogen atom (about 10^{-10} m) by the same amount as estimated for the early universe, the distance would be 10^{40} times greater—or 10^{24} ly!

This inflationary period neatly and naturally solves the flatness problem (after which the model melds into the standard hot Big Bang model). Imagine that before inflation, spacetime contained strongly curved regions. Inflation would cause these to achieve flatness automatically. As an analogy, consider the curved surface of a partially inflated balloon. The surface is clearly curved, when compared to the overall size of the balloon. Now rapidly blow up the balloon, keeping a close eye on the curvature of the surface. It becomes distinctly flatter (less curved). Similarly, when the universe inflated, curved regions would have become flat. So the ratio of the actual to critical density naturally reached a value very close to 1, without the need of special assumptions. Note that the inflationary model assumes a just-closed universe!

The Horizon Problem

I have already emphasized the uniformity of the cosmic background radiation (Section 21.3), which tells us that the universe was extremely isotropic (Section 21.1) at the time of decoupling. The background radiation deviates from complete uniformity by only a few parts in million! How did this happen? The standard Big Bang model just assumes that it started that way and stayed that way. And with this assumption, the uniformity of the cosmic radiation presents a problem.

Consider a gas in a box. If you add energy to one side of the container, the temperature goes up. But to induce this effect, the particles in the gas must carry information about the addition of energy by moving around at a greater average speed and knocking into each other harder (Section 13.2). A finite time must elapse before these collisions can carry throughout the box the information that energy has been added to it. Now imagine the box expanding much faster than the particles in it, on the average, were moving around. Then only a small region of the box would find out that energy had been added, and this part would have a temperature different from the rest.

Information cannot be communicated faster than the speed of light. Yet, the very early universe expanded so fast that regions of it were rapidly and widely separated. Now in a given time, a light signal can travel some maximum distance, called the **horizon distance.** For example, after one second had elapsed, light could have gone only one second of light travel time, for a horizon distance of about 300,000 km. Yet as a result of the rapid expansion, regions of the universe were separated by almost a hundred times this distance. How could these regions have evolved to the same temperature when they could not communicate with each other? That is the horizon problem.

The inflationary universe model solves the horizon problem by altering the setting of the rapid expansion of the universe: namely, the universe is said to have evolved from a region much smaller (by 10^{50} times or more) than in the standard Big Bang model. According

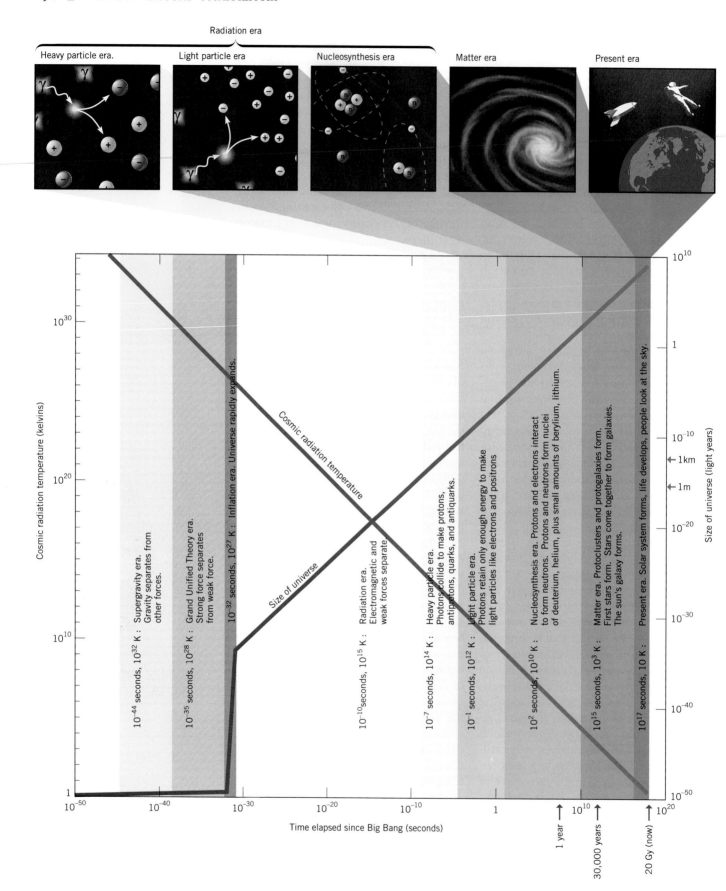

Radiation era

Heavy particle era.   Light particle era   Nucleosynthesis era   Matter era   Present era

FIGURE 21.16 (*Left page*) Schematic, visual history of the universe in the standard Big Bang model with inflation. The horizontal axis is time; the vertical one represents the relative size of the universe (*right*) and the temperature of the background radiation (*left*). Just after the Big Bang, gravity on the smallest scales (supergravity) may describe the universe until a time of 10^{-43} s, when gravity freezes and GUTs describe the physical processes of particle interactions. At 10^{-35} s, inflation begins, prompted by the freezing of the strong force. Electroweak interactions control a soup of quarks and electrons until the quarks form protons, which participate in nucleosynthesis. After matter and radiation have decoupled, the first galaxies form in superclusters. In particular, the cosmic background radiation comes to us from the time of recombination and decoupling. The horizontal axis at top gives the cosmic eras.

to the new model, before the inflationary era began, the universe was much smaller than its horizon distance. Thus all of it could reach the same temperature. Then inflation made it much larger, preserving the uniform temperature. That way we can account for the great uniformity of the cosmic background radiation in the past and today.

To sum up. The inflationary universe model (Fig. 21.16) deals with the earliest times in the universe's history, before a time of 1 second down to a time of 10^{-43} s. Earlier than that, we have no physical model to describe the cosmos.

A new model, which combines quantum ideas with gravity, is being worked on but is not complete. It is called **supergravity** and perhaps can describe the physical conditions then. At 10^{-43} s, according to the supergravity model, gravity froze out from the other forces. At 10^{-35} s, the strong force froze out from the electroweak force and promoted an era of inflation. From 10^{-32} s onward, the model unfolds as the standard Big Bang.

The inflationary universe model grapples with other flaws in the Big Bang model and does so modestly well, in my opinion. It is the best model we have to date that unites the physics of the very large and the very small.

■ **What does the inflationary Big Bang model require for the geometry of the universe?**

Key Concepts

1. Cosmological models assume the universality of physical laws, a homogeneous universe, and an isotropic universe.
2. Cosmological models must explain the evolution of the universe, the grouping of matter, and the expansion of the universe.
3. The main modern cosmological model is the Big Bang model (based on Einstein's theory of general relativity).
4. According to the Big Bang model, the universe began in a hot, dense state at a finite time in the past. The model predicts that radiation from the past pervades now the cosmos at low temperature; this radiation has been observed to be isotropic and to have a temperature of about 2.7 K.
5. The properties of the cosmic blackbody radiation, when combined with Einstein's general relativity and our present knowledge of matter, permit a detailed description of the Big Bang. In it, matter is made from photons and then interacts to form light elements in the first few minutes of the universe's history.
6. The hot Big Bang model predicts a cosmos having no less than 25 percent helium by mass; observations tend to support this prediction.
7. How the universe ends depends on whether it is open or closed. Present observations support an open universe, a conclusion that can be inferred from the cosmological abundance of deuterium.
8. Galaxies could not form in the young universe until the temperature had dropped below a few thousand kelvins (about a million years after the expansion started). Galaxies did form quickly and early on, but they could not simply grow gravitationally from

small disturbances; rather, they needed the help of turbulence from the Big Bang. The *COBE* observations have revealed such ripples.

9. Grand unified theories (GUTs) of elementary particles and forces provide new insights into the Big Bang model and help to solve some of its major problems, such as the matter/antimatter imbalance.

10. According to the inflationary universe model, at the time of the strong force freezing out, the universe went through a stage of rapid expansion; this suggested sequence helps to solve the flatness and horizon problems of the Big Bang model.

Key Terms

antimatter

Big Bang model

cosmic blackbody microwave radiation

cosmological principle

cosmology

flatness problem

grand unification theory (GUT)

heavy-particle era

homogeneous

horizon distance

horizon problem

inflationary universe model

isotropic

lepton

light-particle era

matter era

observable universe

physical universe

primeval fireball quark

radiation era

recombination

strong force

supergravity

universality of physical laws

weak force

Study Exercises

1. State in a few short sentences the assumptions of the Big Bang model. (Objectives 21-1 and 21-3)

2. Make a short list of the fundamental cosmological observations. (Objective 21-2)

3. Interpret the observations in Exercise 2 in the standard Big Bang model. (Objective 21-3)

4. Give *one* observational argument for asserting that the microwave background radiation is cosmic in origin. (Objectives 21-4 and 21-5)

5. How does the discovery of the cosmic microwave background radiation confirm the Big Bang model? (Objective 21-6)

6. List the elements that can be made in a hot Big Bang, and give one reason why no elements heavier than lithium and beryllium are manufactured. (Objective 21-9)

7. What observational evidence do we have to back up the standard Big Bang model? (Objectives 21-3, 21-5, 21-6, 21-7, and 21-9)

8. How can light produce particles? In the Big Bang model, how does it happen that heavy particles are not formed out of light after a certain time? (Objective 21-8)

9. In what sense does the present abundance of anti-

matter in the universe challenge the validity of the Big Bang model? (Objective 21-10)

10. In what sense does the Big Bang model have a flatness problem? How does the inflationary universe model cope with this problem? (Objectives 21-10 and 21-11)

11. The cosmic background radiation has a Planck curve for its continuous spectrum. What does this observation tell us about the conditions in the early universe? (Objectives 21-4 and 21-5)

12. The latest analysis of the *COBE* observations show *small* fluctuations in the temperature of the cosmic background radiation. How do these variations assist in the formation of the first galaxies? (Objective 21-10)

13. In order for the universe to be closed, there must exist a large fraction of dark matter relative to the luminous matter. How is the existence (or not!) of dark matter a problem for Big-Bang cosmology? (Objective 21-12)

14. We know the temperature of the cosmic background radiation now. What is needed in order to determine its temperature at specific times in the past or future? (Objective 21-6)

Problems & Activities

1. What is the peak wavelength of emission of the 2.7 K cosmic background radiation?

2. Gamma rays have frequencies on the order of 10^{22} Hz. In making matter from such photons, what minimum mass of elementary particles could be created by combining the gamma rays?

3. What is the energy equivalent of the mass of an electron? Of a proton?

4. When the universe was last opaque, it had a temperature of about 3000 K. At what wavelength did its emission peak then? How does that compare to the peak now?

See for Yourself!

GO FOR A WALK! Take one outside. As you do, look very carefully at the objects you can see. Touch some of them. Pick some of them up and look at them closely. A rock? Mostly silicon and oxygen. A leaf? Mostly water. The air? Mostly nitrogen and oxygen. Look at your built environment as well as your natural one. Make reasonable guesses at their compositions. You might want to take a notebook with you.

Return to your living space and examine it and its contents carefully, too. Again, make a guess at the com-positions. How is your place heated? Natural gas? Wood? Do you have electricity? How and where is it generated? Nuclear? Water-powered generators? Coal or oil powered?

Now take your list (written or mental) and examine Figure 21.16. Try to place where and when on this time line the elements that made the objects you observed came to be. And perhaps when the natural objects themselves were made (more on that in Chapter 22). ■

Central Question

What are the characteristics and possible origin of life as we know it, and what consequences follow from these for the possibility of life elsewhere in our Galaxy?

Bios and Cosmos

A zygote is a gamete's way of producing more gametes. This may be the purpose of the universe.

ROBERT HEINLEIN, *Time Enough for Love*

In the span of cosmic evolution, the origin of life on the earth involved a *natural* sequence of chemical and biological evolution. The basic material was there. It needed only to be put together in a special way. The special arrangement of molecules that makes a living organism on the earth results naturally from the chemical properties of matter. The origin of life may be both natural and universal.

Are we alone? How life developed here gives us some grasp of whether life exists elsewhere. The question is this: Can we estimate, on some reasonable basis, the chances that life exists elsewhere in our Galaxy or in other galaxies? Yes, from basic physical, chemical, and astronomical knowledge. Admittedly our estimate smacks of speculation, but it rests on a scientific basis.

This chapter will lead you into the realms of molecular biology and biological evolution. These excursions are needed to investigate the essential features of life on Spaceship Earth. They'll show you how the probable origin of life followed this sequence: physical evolution, chemical evolution, and biological evolution. Each stage provides clues to whether life arose elsewhere in the universe.

22.1 The Nature of Life on Earth

What is life? Rather than try to define it, let's accept a useful rule of thumb: *living things are things that reproduce, mutate, and reproduce the mutations.*

Organisms

What does this mean? First, that living things have an *organization* that they pass on when they reproduce (Fig. 22.1). So all living things are **organisms.** Second, reproduction may result in an offspring with a genetic difference—one that exhibits a **mutation**—from its parents. (The genes in a cell's nucleus carry the information code for how an organism is to be put together; if the genes change, the organism changes.) Mutations can provide the possibility of change, which leads to biological *evolution,* the development of organisms of greater complexity, better adapted to their environment. Mutation and evolution are distinguishing features of life.

Proteins and Nucleic Acids

How does an organism pass its organization on? Biochemical discoveries since the 1960s have answered this question. Two basic types of molecule operate in all terrestrial organisms (Fig. 22.2); their interaction results in what we call life. The two are **proteins,** which make up the organism, and **nucleic acids,** which provide the information for the structure of the organism and the means to pass on this information in reproduction.

Both these long molecules have crucial subdivisions. Proteins are built out of **amino acids** (Fig. 22.3). Nucleic acids also consist of smaller subunits called bases. These building blocks consist of simple combinations of the most common chemical elements in the universe: hydrogen, carbon, nitrogen, oxygen, and a few others (Table 22.1).

Energized by sunlight, your cells carry on chemical work using such proteins as **enzymes,** which monitor and facilitate important chemical reactions in a cell. But how does a cell control the functioning of the enzymes? That's the job of the cell nucleus, using the nucleic acid called **deoxyribonucleic acid,** or DNA for short. DNA is an enormous molecule; in your body, DNA strands contain *billions* of atoms. (Yet your whole body contains only one teaspoonful of DNA.) DNA consists of a highly ordered, spaghetti-like arrangement of four chemical substances called bases, plus sugars and phosphates (compounds of phosphorus and oxygen; Fig. 22.4).

DNA serves as the chemical blueprint that informs the protein in cells how to function. DNA is the route by which a cell chemically hands down its blueprint to its offspring. The offspring inherit this information. DNA contains

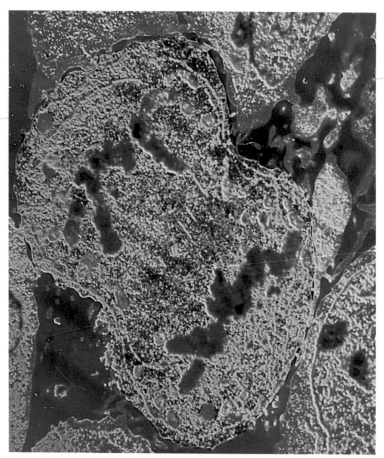

FIGURE 22.1 A human cell dividing. The main body of the cell appears yellow in this false-color image. The genetic material shows as a dark pink. (Courtesy of CNRL, Science Photo Library/Photo Researchers.)

the instructions on how to make the proteins.

All terrestrial organisms have a common chemical composition. And this makeup is closely connected to the physical evolution of the universe, in that those common elements are made in stars or were created in the Big Bang. This common chemistry suggests the central scientific idea for the origin of life: it results naturally in the evolution of the universe. In the view of modern evolutionary biology, the central idea about life's origin is this: *life arose from nonlife.*

■ What are two distinguishing features of life on earth?

22.2 *The Genesis of Life on the Earth*

Modern biology sees life's origin in nonliving material. Life arises as a natural

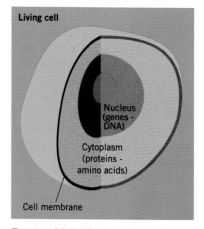

FIGURE 22.2 The key parts of a typical cell. The nucleus contains the genetic material (DNA), the protoplasm is the site of metabolic activity plus the location of most proteins, and the cell membrane allows the passage of selected materials, such as food.

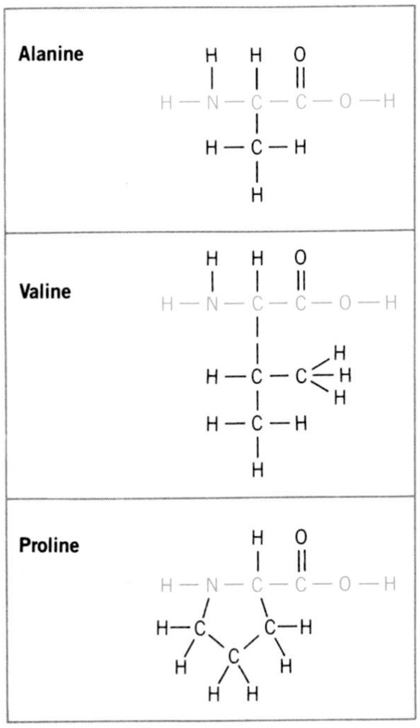

FIGURE 22.3 The chemical structures of alanine, valine, and proline, three common amino acids found in life on the earth. Note that hydrogen, oxygen, carbon, and nitrogen — four of the most common elements in the cosmos — form the chemical basis of these compounds.

patible with its environment. Although Darwin lacked modern knowledge of genetics, he patiently unearthed the general feedback mechanism: **natural selection.**

Put too simply, modern biologists view natural selection as a consequence of successful reproduction. Individuals that are well adapted to a local environment not only survive but, more important, they usually produce offspring. This reproductive success, in turn, spreads their genetic material: the children of such well-adapted pairs tend to survive and reproduce successfully, as well. Eventually their genes dominate those available to a species, and the survival and expansion of new genetic material results in evolution.

Where does this new genetic material come from? Mutations! Natural selection directs the random jumble of mutation into biological evolution. This progression assumes a primeval life form to start biological evolution. At some point biological evolution began, but chemical evolution preceded it.

consequence of the slow processes of chemical and biological evolution. What are these evolutionary processes?

Clues from Biology

In *The Origin of Species,* published in 1859, Charles Darwin (1809–1882) set forth his observations that living species are adapted for survival in their respective environments. No single mutation can generate the evolution of a species. Some feedback mechanism selects mutations that make a species more com-

Clues from Geology

As geologists pare back rock layers, they reveal images of the earth's history. These rocks sometimes trap fossils that show the life of the past. Radioactive dating techniques (Focus 8.1) can fix the age of the rocks and the fossils in them. Careful microscopic inspection of ancient rock samples reveals the remains of bacteria and algae from 1.0 to 3.5 billion years ago. These natural records provide important clues to life's evolution on the earth (Table 22.2).

The oldest set of rocks containing

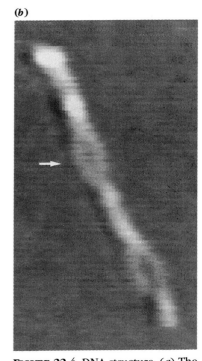

(a)

(b)

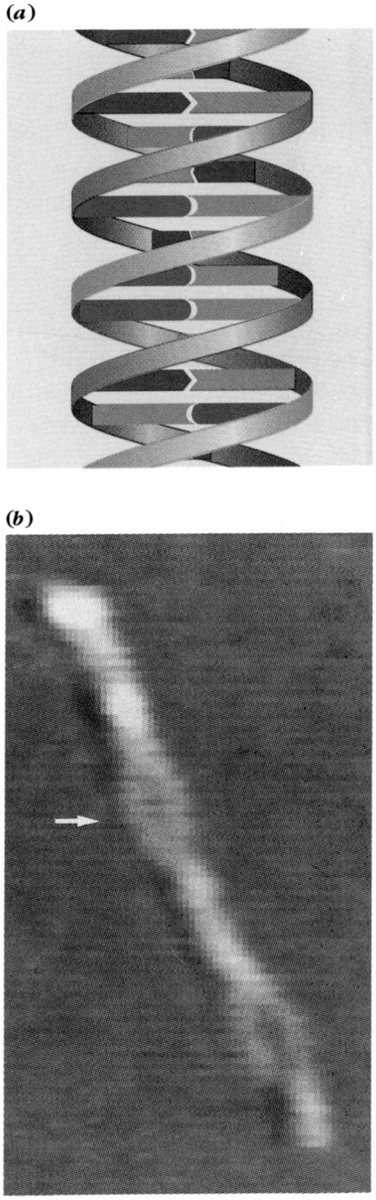

FIGURE 22.4 DNA structure. (*a*) The double-stranded-helix form of DNA. The linkages indicated schematically by horizontal lines are not random; rather, the ordering provides the chemical basis of the DNA information coding. (*b*) Highly magnified image of a DNA molecule. The arrow shows DNA strands twisted upon themselves. (Courtesy of Carlos Bustamante, University of Oregon.)

TABLE 22.1 RELATIVE ABUNDANCES OF ELEMENTS (FRACTION OF ATOMS)

EARTH'S CRUST	SOLAR SYSTEM (Sun)	HUMAN BEINGS	INTERSTELLAR MEDIUM
Oxygen	Hydrogen	Hydrogen	Hydrogen
Silicon	Helium	Oxygen	Helium
Hydrogen	Lithium	Carbon	Carbon
Aluminum	Beryllium	Nitrogen	Nitrogen
Sodium	Boron	Calcium	Oxygen
Iron	Carbon	Potassium	Neon

TABLE 22.2 SIMPLIFIED GEOLOGIC TIME SCALE

ERA	AGE (millions of years)	LIFE
Cenezoic	Now	*Homo sapiens* dominates earth
	2–4	*Homo sapiens* appears
	58	Mammals appear
Mesozoic	63	End of dinosaurs
	135	Flying reptiles
	181	First birds
Paleozoic	239	Dinosaurs appear
	280	First reptiles and insects
	400	First amphibians
	410	First land plant fossils; first insect fossils
	460	First fish fossils
Cambrian	500–600	First plant fossils
Precambrian	3800	Oldest fossils; oldest rocks
	4500–4600	Formation of earth

(a)

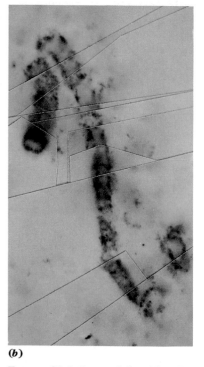

(b)

FIGURE 22.5 Some of the oldest fossils known. (*a*) Layered sedimentary rock, once tidal mud flats. This structure, about 3.5 billion years old, might have contained colonies of photosynthetic microorganisms. (Courtesy of D. I. Groves, University of Western Australia.) (*b*) Microfossils resembling modern bacteria, from the Warrawoona formation near North Pole, Australia. (Courtesy of Biological Photo Research.)

possible microfossils lies in Australia. The rocks are 3.5 billion years old. They contain remains that resemble microorganisms and layered structures that could have been built by colonies of bacteria (Fig. 22.5*a*), which flourished in tidal mud flats at this time. The layers resulted from the accretion of fine grains of sediment from the colonies. Another group of rocks with microfossils some 3.5 billion years old also lies in Australia. These rocks contain evidence of two distinct life forms: rod-shaped structures resembling modern bacteria (Fig. 22.5*b*) and round cells similar to modern blue-green algae. Modern blue-green algae are photosynthetic. So the Australian microfossils suggest the start of photosynthesis at about 1 billion years after the formation of the earth.

The fossils provide vital clues in the investigation of the biological evolution of life. They imply that chemical evolution (that involving just molecules) must have been completed on the primeval earth no more than about 1 billion years after the earth formed. What were the conditions on the young earth that permitted chemical evolution to take place?

Clues from Astronomy

The sun reflects the average chemical composition of material in our Galaxy. The earth ended up with a composition quite different from that of the sun (Table 22.1). How did this happen? Recall that the nebular model for the origin of the solar system (Section 12.5) pictures the protoearth as forming from the accretion of planetesimals. The chemical composition of the planetesimal material depends on the temperature in the solar nebula, as given by the condensation sequence. At 1 AU, it was so hot that only rocky and metallic materials were able to condense.

What about the earth's early atmosphere? If the earth's surface were hot at the time—perhaps even molten—the heat would gasify volatile substances within the earth, which would rise to the surface. That's what happens in volcanoes today (Fig. 22.6). This process is called *outgassing*. Volcanoes and volcanic flows outgas water, carbon dioxide, hydrogen sulfide, methane, and ammonia—but no oxygen. We would expect that pretty much the same materials outgassed from the primitive earth.

How fast this outgassing took place then depended on the crustal temperature. If the earth was originally molten, outgassing of its second atmosphere happened quickly, in 100 million years or less. Although a lot of gas was produced in this way, it still did not make the present atmosphere, for this second atmosphere lacked oxygen. In other words, the second atmosphere probably contained much carbon dioxide and water vapor.

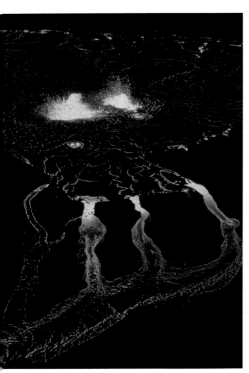

FIGURE 22.6 Volcanic eruptions on Hawaii. Such eruptions spew out gases such as water vapor and carbon dioxide from the earth's interior. (Courtesy of C. W. Stoughton.)

The Spark of Life

The earth started out with an atmosphere of simple molecules. To synthesize complex molecules from simpler ones requires free energy. Photosynthesis in plants now captures solar energy and stores it in the form of chemical bonds. On the young earth, sunlight was available, but plants were not. Where was the energy needed for synthesis of complex molecules obtained? Possible sufficient sources include ultraviolet radiation from the sun and from lightning.

Solar ultraviolet light at wavelengths less than 2200 Å plays a key role because it is absorbed by complex molecules and can cause the formation of still more complex ones. (The ozone layer now filters out most of the ultraviolet radiation. Because of the lack of free oxygen in our second atmosphere, no ozone layer was formed.)

On the earth now, lightning (Fig. 22.7) accounts for almost as much energy as short-wavelength ultraviolet. When the earth cooled enough for rain to fall, lightning storms possibly raged widely over the earth's surface. The en-ergy from lightning may have been more important then than now.

These energy sources came in spurts and bursts rather than at constant rates. They had the capacity to destroy as well as to help synthesize molecules. The balance between creation and destruction determined the number of molecules that could exist, and the kinds.

Synthesis of Simple Organic Molecules

The critical parts of chemical synthesis are energy and an atmosphere that lacks free oxygen, which destroys organic compounds. Laboratory experiments validate this key point. When gaseous mixtures of water, carbon dioxide, methane, and ammonia have energy added to them (in forms such as expected on the young earth), the products include amino acids, some of them commonly found in terrestrial proteins (Fig. 22.8). These experiments, whether with gaseous mixtures or solutions, naturally produce *most* of the amino acids common in protein.

I have emphasized amino acids because they form the building blocks of proteins. Other, somewhat more tentative experiments have tried for the synthesis of hydrocarbons and sugars. But

Life on the Earth

At a Glance

Consists of	Organisms made of cells
Cells	Contain proteins and nucleic acids
DNA	In genes carries structural information
Composition	Mostly hydrogen; carbon plays key role

FIGURE 22.7 Lightning flashes in a summer storm over Tucson, Arizona. The energy released in a flash could have helped to make simple organic molecules during the early times on the earth. (Courtesy of K. Wood, Science Source/Photo Researchers.)

what about the basics of DNA? Experiments to simulate the primitive synthesis of the building blocks of DNA have been less extensive than those for amino acids. They do show that, with phosphates added, amino acids, sugars, and nucleic acid bases result.

The simple organic molecules needed for life form naturally under plausible primitive terrestrial conditions in simulation experiments. The simple molecules can be cooked. What about the actual proteins and nucleic acids?

Synthesis of Complex Molecules

Here we stand on much shakier ground than with the simple molecules (and remember, we aren't yet anywhere near the complexity of living cells). Proteins and nucleic acids are not only huge molecules; they also have a special and precise architecture. How did the first of these giant molecules get together? A number of attempts have been made to make proteins and nucleic acids under prebiologic conditions. I won't bog you down with the chemical details; these experiments seem to be only somewhat successful so far.

To sum up. The precise complexity of proteins and nucleic acids makes their synthesis difficult. We do not yet understand the specific pathways to their original production.

Chemical evolution naturally—and perhaps inevitably—leads to the making of complex organic compounds that are the building blocks of proteins and nucleic acids. Both are needed to join together in a cell. How to make that first cell?

Quite bluntly, *we don't know*. Fossils cannot give information about this crucial time. And chemists have not synthesized anything as complex as a cell. What happened before the introduction of cellular life remains for now a matter for speculation. Biological evolution continues today. The "somehow" that ignited biological evolution is not clearly known. But we do know that it *did* happen. With this scheme in mind, let's turn to the solar system and the Galaxy to investigate the possibility of cosmic neighbors.

■ According to fossil records, when may the first cell have appeared on earth?

22.3 The Solar System as an Abode of Life

So far we know of only one planet in the solar system that harbors carbon-based life: the earth. The environments of the other planets pretty much exclude life—with the slim exception of Mars. The main factor here is water in liquid form.

Mars: The Best Chance

The fate of Martian life hinges on surface water. The Viking missions found the surface pressure of the Martian atmosphere to be about 0.007 earth atmosphere—much less than that on the highest mountains on the earth's surface (in fact, you would have to go about 40 km up into the earth's atmosphere to find pressure so low). At this low pressure, liquid water cannot exist on the surface. As evidenced by the polar caps on Mars, both water and carbon dioxide form solid ice. Even in these regions, the abundant liquid water needed for life probably does not exist. Mars is a very, very dry planet.

The Martian arroyos attest to the flow of liquid water in the Martian past. Some astronomers imagine that about 3.5 billion years ago, Mars had a denser atmosphere capable of holding water vapor sufficient to generate rainfall. A denser atmosphere may have arisen from extensive volcanic activity, which could spew out large volumes of carbon dioxide and water vapor.

Recurring deluges or meltings may explain the origin of the laminated terrain found in the Martian polar regions within 10° of the poles (Fig. 22.9). There, stacks of thin plates of crustal material stand about 10 km tall and up to 200 km across. These regions appear to be the youngest, most evolved parts of the Martian surface. Because they exist only in the polar regions, where carbon dioxide and water ice form annually, the plates may be related to the influx and outgo of these materials.

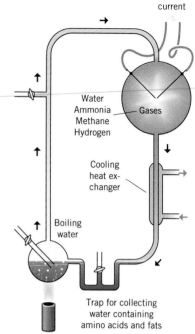

FIGURE 22.8 Schematic diagram of an experiment to produce complex molecules under conditions similar to those on the young earth. Electrical discharges were fired in a gas of water, ammonia, methane, and hydrogen. Output collected at the bottom included amino acids and fats.

The direct test for Martian life came from the Viking landers' biology experiments. What were these results? They proved negative: from the soil that was sampled, life does *not* exist on Mars *now*.

Why such a negative view? Each lander contained an instrument designed to detect and measure organic molecules, the complex chains, secured by carbon, that characterize life. *No large organic molecules were found* at either landing site. The instruments had the sensitivity to detect organic compounds in a concentration of just a few parts in a billion—far below the concentration found in desert soils on the earth.

In light of the lack of complex molecules in the soil, all the apparently positive results from the landers' three biology experiments can be explained by inorganic *chemical* reactions rather than *biological* ones. Our extraterrestrial search for terrestrial life has failed so far.

Amino Acids in Meteorites

Meteorites provide some evidence to support the theories of natural synthesis of organic compounds. Of the three main classes of meteorites (Section 12.3), carbonaceous chondrites, which contain a relatively high percentage of carbon (2 percent), make up a minority. People have regularly speculated that some of the carbon contained in these meteorites might be organic.

On September 28, 1969, at about 11:00 A.M., a meteorite fell in Murchison, Australia. This meteorite, a carbonaceous chondrite, was rushed to the Ames Research Laboratory of NASA and analyzed by a team of scientists headed by Cyril Ponnamperuma. The NASA group discovered five amino acids common to living protein. The quantities were small, only a few micrograms of amino acids in each gram of the meteorite.

Was this terrestrial contamination? Probably not. Organic molecules exist in two distinct forms: right-handed ones and left-handed ones, depending on the direction of the twist of the linkage of the atoms. Almost all terrestrial organic molecules are left-handed, so earth-based contamination is expected to be left-handed. The Murchison meteorite contained just about equal quantities of right- and left-handed molecules, the left-handed forms predominating a little. This evidence strongly points away from terrestrial contamination and toward an extraterrestrial, nonbiological origin of the Murchison organic molecules.

Why nonbiological? When organic molecules are synthesized in a chemistry lab (rather than by an organism), they show an equal number of right-handed and left-handed forms.

In 1969 a Japanese scientific team discovered meteorites in the Antarctic. Since then, more than a thousand samples have been collected (Fig. 22.10). The Antarctic provides a clean, cold environment that is relatively unlikely to contaminate the meteorites with terrestrial materials. One of these meteorites, a carbonaceous chondrite found near Allen Hills, contains amino acids free of terrestrial contamination. But it has only about 10 percent of the total amino acid content of the Murchison meteorite.

The glut of complex molecules in space discovered by radio astronomers (Section 15.1) lends further credibility to extraterrestrial, nonbiological formation of organic substances. Molecules such as formaldehyde, hydrogen cyanide, cyanoacetylene, formic acid, methyl alcohol, and methylacetylene can play a crucial role in organic chemistry. Formaldehyde and hydrogen cyanide, for instance, can be chemically com-

FIGURE 22.9 Laminated terrain near the south polar cap of Mars. These formations of surface soil probably contain water ice. Photo shows region about 500 km across. (Courtesy of NASA.)

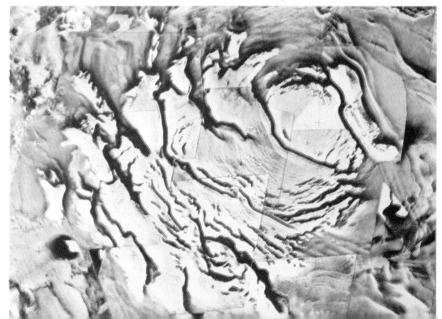

bined to make amino acids. And remember that the solar system formed from an interstellar cloud of gas and dust that probably contained such molecules.

The important conclusion is this: the chemical evolution from simple compounds to complex organic substances occurs so naturally that it takes place even in the hostile environment of space, without biological aid.

Mass Extinctions on Earth?

Meteorites may have an explosive impact on the evolution of life on earth. Roughly 65 million years ago, a sudden trauma swept through life here. In a short time—less than a million years, perhaps as swiftly as a thousand years—mass extinctions hit certain plants and animals. The fossil record shows an abrupt loss of ocean plankton, swimming mollusks and dinosaurs, and land animals with masses greater than 25 kg—most especially, the large walking dinosaurs. This end was good for us, for smaller mammals flourished afterward. What happened then to promote this **mass extinction** of life forms? (We have some weak evidence that the other mass extinctions took place a few times over the geologic time scale.)

Many ideas have been proposed. One seems to be gaining the weight of reasonable evidence: that of the impact of an asteroid-sized body that caused environmental stress, resulting in selective, worldwide extinctions. An object some 10 km in diameter (about the size of a small city) with a mass of some 10^{14} kg could easily penetrate the earth's atmosphere and strike the surface at 11 km/s. Its impact would release some 10^{23} J, equivalent to 10^{14} tons of TNT. (Similar impacts shaped the large basins on the moon.) Temperatures at the impact point would hit 20,000 K. The vaporized object and ground shoot hot gas into the air. A blast wave rockets out at 35,000 km/h and levels everything for a few hundred kilometers around. Some of the material ejected by the impact would plume into space, condense, and shower back onto the earth.

How might such an impact influence the earth's environment? The blast could deposit small dust particles (less

FIGURE 22.10 Meteorite found near Allen Hills, Antarctica. This specimen may well have been ejected from the surface of Mars by a violent impact. (Courtesy of NASA.)

than 1 μm in size) high in the earth's atmosphere, where they would remain for months. It may have caused a darkness like night for well over a month. Dust particles, circulated globally by winds would prevent a significant fraction of the sun's light from reaching the earth's surface, sharply reducing photosynthesis (especially by plants in the oceans) and the general temperature. Animals especially sensitive to temperature changes would not adapt and so disappear.

Astronomical evidence suggests that an object this large collides with the earth once every 100 million years or so, and the asteroid crosses the orbit of the earth. The craters formed by about a hundred such objects are known; they tend to have sizes of a few tens of kilometers. A few astronomers are hunting down more, searching systematically. In January 1991, one asteroid raced by within 170,000 km of the earth—only half the distance between the earth and moon! Collisions among these near-earth asteroids may well supply most of the meteorites that hit our planet.

This idea sounds plausible from an astronomical view. Is there any solid evidence behind it? The main clue comes from the composition of a clay layer deposited about this time. Below it (before it in time), we find the usual range of fossils from the age of the dinosaurs. Above it (afterward), certain fossils no longer appear. The layer itself has a composition that is enriched (relative to the earth's crust) in noble metals such as iridium and gold. The overabundance of iridium in this layer appears to be a worldwide phenomenon. One source of this enrichment could be material from a large asteroid. Its impact would have mixed asteroidal material with terrestrial and could have resulted in the abundances found in the clay layers. So asteroidal-sized bodies colliding with the earth may have had dramatic repercussions on the evolution of life.

One way to confirm this notion is to track down the fossils of such impacts—craters on the earth. But much of the earth's surface is covered with water. And erosion wipes out impact craters fairly fast. Still, as noted above, workers have uncovered scores of old craters, a few with diameters of about a hundred kilometers. That is just about the size expected from the impact of a 10-km asteroid. The trail of clues has revealed the culprit crater, one located in the Yucatán, Mexico. Its name is Chicxulub crater; its has a diameter of about 180 km. Clay pellets found in northeast Mexico and in Haiti (almost 2000 km away) have been chemically linked to materials blasted out from the impact that made Chicxulub crater. And the pellets and the crater both date to the same age—65 million years old. So here may lie the gravestone of the dinosaurs.

The trigger for such impacts may also be astronomical, especially if they are periodic. One speculative idea views the inner solar system as undergoing periodic showers of cometary pieces from the Oort Cloud. The fragments might be prompted to fall inward by the gravitational disturbance from a small body (a mass no smaller than 0.01 solar mass)—dubbed *Nemesis*—supposedly orbiting the sun with a period of 26 million years (Fig. 22.11). The semimajor axis of its orbit would then be (according to Kepler's laws) some 89,000 AU. At perihelion, when Nemesis passes through the Oort Cloud, its gravitational effects activate a shower of comets, a few of which could strike the earth.

■ What was the main result of the Viking lander's analysis of the Martian soil?

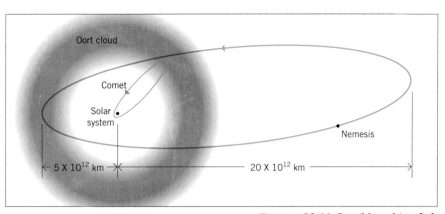

FIGURE 22.11 Possible orbit of the hypothetical Nemesis, a companion to the sun. As Nemesis passes through the Oort comet cloud, it scatters cometary nuclei into the solar system, where they can impact the planets.

22.4 The Milky Way as an Abode of Life

Where might life exist elsewhere in the Galaxy? If the nebular model of planetary formation is correct (Section 12.5), the huge number of stars in the Galaxy implies planets elsewhere. Even if the probability of the genesis of life is slim, the number of possible habitats may be so large that some extraterrestrial creature has viewed the dawn of its day. The elements of life are the most abundant in the cosmos, so there is no lack of proper ingredients. All that is required is the proper construction.

Cosmic Prospecting

How to prospect for life in the Galaxy? Civilizations of living creatures must evolve; that's part of cosmic evolution. So the number of intelligent civilizations in the Galaxy changes with time. At any one time, the number of civilizations depends on the rate at which these civilizations are born and how long they last.

Here's an analogy. Suppose you are locked in a dark room filled with candles; a friend gropes about and lights one candle every 15 minutes—four per hour. Suppose each candle burns for one hour. How many candles are lit at any given time? During the first hour, the number increases from one to two to three to four. But just as the fifth one is lit, the first one goes out. As the sixth is lit, the second goes out. One candle goes out as each new one is lit, leaving four candles burning at any one time. If you think about it, you see that the number of observed candles is equal to the rate of candle lighting, R_c, times the lifetime of one candle, L_c, or

$$N = R_c L_c$$

So, if you know the average lifetime of a single candle and the rate at which the candles begin their life, you can anticipate the number lit at any time. The same reasoning applies to the number of civilizations in the Galaxy at any one time: if R_{ic} is the rate of formation of intelligent civilizations and L_{ic} is their lifetime, then

$$N_{ic} = R_{ic} L_{ic}$$

This relation may be broken down into more specific factors, loosely independent of one another:

$$N_{ic} = R_* P_p P_e N_e P_l P_i L_{ic}$$

This equation was first put together, in somewhat different form, by radio astronomer Frank Drake, so it's called the **Drake equation.**

The meaning of each of these pieces is related directly to an important facet of cosmic evolution. R_* is the rate of star formation averaged over the age of the Galaxy; P_p is the probability that once a star has formed, it will possess planets. The next factor, P_e, is the probability that the star will shine long enough for life to form, and N_e is the number of planets in the region around the star with a suitable range of temperatures. P_l is the probability that a planet in a star's ecosphere will develop life, and P_i is the probability that biological evolution will ultimately lead from life to intelligent life. The final term, L_{ic}, is the lifetime of this intelligent civilization. Note that these factors group into three categories: R_*, P_p, P_e, and N_e relate to astronomy and physical evolution; P_l and P_i relate to biology and chemical evolution; and L_{ic} derives from what I would call speculative sociology.

Astronomical Factors

The Galaxy contains a few times 10^{11} stars. These stars have formed over at least 10 billion years. So the average birthrate of stars from these figures is about 10 per year. I adopt 10 for R_*.

What is the chance that one of these stars will develop a planetary system? Nebular models (Section 12.5) imply that many planets exist in the Galaxy. A contracting gas and dust cloud must form either a star with a planetary system or a multiple-star system, perhaps also with planets. More than 50 percent of the stars in the Galaxy are in binary or other multiple-star systems. If I take a planetary system versus a multiple-star system as an either–or proposition, the P_p equals 0.5.

A star's **ecosphere**—the zone in which planets must lie to have a temperature range suitable for life as we know it—depends primarily on the luminosity of the star. The more luminous the star, the farther out the habitable zone starts (Fig. 22.12). The width of the ecosphere is also greater for luminous stars and thinner for less luminous ones. The ecosphere must persist long enough to allow the genesis and evolution of life. Of course, the ecosphere's location and width changes as a star evolves.

Luminous O and B stars live out their normal lives in about 100 million years, a time much shorter than the one billion years that elapsed while life evolved on the earth. By our standards, these energy spendthrifts are improper parents. So we consider only stars whose life spans are at least equal to the sun's—spectral class G or cooler, a choice that fortunately includes 98 percent of all the normal stars in the Galaxy. Unfortunately, for stars cooler than spectral class K, the ecosphere is too small. If we throw out these cool stars, only about 10 percent of the total remains, so P_e equals 0.10.

How about the number of planets in the ecosphere, N_e? Here we have the only example of our solar system: only the earth lies in that zone now. If the planetary formation processes in the nebular model are universal, we'd expect other planetary systems to more or less resemble the solar system. Is this belief reasonable? Computer models of nebular-style formation result in a regular spacing for the planets, with a few planets orbiting at the right distance from a solar-mass star. So N_e may range from 1 to 4 or so; I'll use 1.

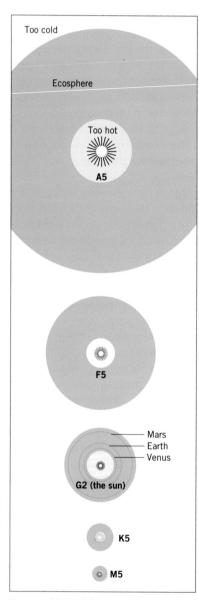

FIGURE 22.12 The sizes of stellar ecospheres for main-sequence stars of different spectral types. Note that the cooler the star, the thinner the ecosphere and the closer it lies to the star. For the sun, three planets lie in this zone (Venus, Earth, and Mars), with the earth near the center.

Biological Factors

Here's another either–or proposition. Either the existence of terrestrial life is unique and the probability of life elsewhere is zero, or the earth is typical, the normal result of cosmic evolution, and the probability of any planet's developing life under favorable astronomical conditions is 1.

Appealing to the uniformity of physical laws, let's assume that we are typical. Lab experiments have shown the natural start of chemical evolution. Because it seems that the nature of the universe makes the start of chemical evolution inevitable, I choose P_l equal to 1.

Is intelligence inevitable? The ability to learn aids in survival, and it appears at even simple levels. The adaptive powers of a thinking organism appear so great that I think that if it is at all possible genetically, intelligence is very likely to be the ultimate result of natural selection. So I choose P_i equal to 1. This choice assumes that a comfortable environment persists for intelligence to develop. However, its value may actually be *very* small.

Speculative Sociological Factors

How long can an advanced, technological civilization survive? By our own example to date, the lifetime of an intelligent civilization may be only a few thousand years. But if it is possible that every advanced civilization successfully manages its problems, it should survive as long as the parent star. The lifetimes of civilizations encircling a G-type star may be about 10^{10} years. But their lifetimes may also be much, *much* shorter.

The Numbers Game

As I have progressed through the astronomical, biological, and sociological factors needed for a rough estimate, the footing has become shakier. I have also ignored some important factors in the analysis, such as the possible stable planetary orbits in a binary star system. I did not intend to give precise results, but rough estimates, because exact answers are not yet possible. The point is to get a feel for reasonable exclusions in the

enormous range of values each element might take.

Not evaluating L_{ic}, I come up with

$$N_{ic} = R_*P_pP_eN_eP_lP_iL_{ic}$$
$$\approx 10 \times 0.5 \times 0.1 \times 1 \times 1 \times 1 \times L_{ic}$$
$$\approx 0.5\,L_{ic}$$

The result depends critically on L_{ic}—how long our candle remains lit. If we assume that we are at the brink of destruction, then L_{ic} is approximately 1000 and N_{ic} about 500. Intelligent civilizations are few and far between. If we survive as long as the sun shines, then L_{ic} is roughly 10^{10} and N_{ic} some 50×10^9. In this case, many a G or K star in the Galaxy has fostered an intelligent civilization!

How seriously can you take these results? Not very—basically they are speculation and should be viewed very skeptically. Our own example, life on the earth, may be more special than we have been willing to admit (especially if mass extinctions were caused by astronomical events).

■ Why is it so difficult to assign a firm number to the variable L_{ic}?

22.5 Neighboring Solar Systems?

What evidence do we have of other planetary systems? Because a planet shines by reflected light from its parent star, because it is small, and because it lies very close to its local sun, as seen from the earth, a planet's gleam would be lost in the stellar glare. So we cannot *directly* observe other planets outside the solar system with earth-based telescopes.

Center-of-Mass Motions

Instead of searching for the light from very large planets, we can hunt for the motion around the *center of mass* of the planet–star system (Section 4.5). As a result of this seesaw effect, the visible star wobbles from side to side about the center of mass, if a massive planet orbits it (Fig. 22.13). From the observed stellar wobble and an estimate of the stellar mass, we can estimate the mass of the invisible planetary companion by the same method used to measure binary star masses. Or we can examine the spectrum of a star to look for the slight Doppler shift caused by the orbital motion about the center of mass. From these observations, and an estimate of the stellar mass, we can estimate the mass of the invisible planetary companion by the same methods used to measure binary star masses (Section 14.6).

The observations required to detect planets around nearby stars are *extremely* difficult to make. The wiggles sought for are only about 0.001 arcsec, or about one one-hundredth the size of a star's best image observed from the ground. Such minuscule changes are dramatically affected by changes in the telescopes themselves, whether produced by self-aging or conscious effort (such as cleaning). One analysis of the errors in such observations concluded that *no* good evidence supports the existence of Jovian-mass planets.

Doppler Shift Detections

For now the Doppler technique seems to be a more promising approach to detection of planetary companions to stars. Two new techniques developed in recent years give high-precision radial velocity measurements of stars, hence the ability to detect small variations in velocity due to low-mass companions. In the approach taken by David Latham and colleagues at the Harvard–Smithsonian Center for Astrophysics in Cambridge, Massachusetts, a computer finds the Doppler shift that gives the best fit of a standard spectrum to the observed spectrum of a star, in effect measuring the average shift of thousands of lines. The Latham group measured radial velocities for more than 1500 stars during an 8-year period, yielding many spectroscopic binary orbits. One solar-type star, some 100 ly distant, seems to have a velocity variation of some 500 m/s with a period of 84 days (Fig. 22.14). The analysis suggests a companion with a mass ten times the mass of Jupiter or a little less. It may be a brown dwarf.

The other technique, developed by Bruce Campbell of the University of Vic-

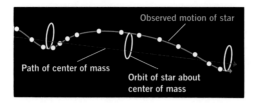

FIGURE 22.13 The path of a star in space with orbital motion around the center of mass of the star–planet system.

toria, Canada, measures Doppler shifts to a precision of about 10 m/s. (Ordinary spectrographs are capable of a precision of about 1 km/s, a hundred times larger.) Campbell and his colleagues have monitored eighteen stars over several years. However, only one, Gamma Cephei, has been followed for an entire period. Its companion may have a mass of about 1.6 Jupiter masses. As with all spectroscopic binaries that do not eclipse, the inclination of the orbits of these suspected companions is uncertain, and their masses could be significantly higher.

The foregoing discussion refers to Jovian-sized planets. We have little hope of detecting terrestrial-sized planets from the earth. Not only are the gravitational effects smaller, because of the smaller mass, but the effects also would have to be disentangled from the effects of larger bodies in the same planetary system.

Other Evidence

If brown dwarfs (Section 17.1) are discovered by infrared observations, then the suggestion that many stars in the Galaxy have very-low-mass companions increases in plausibility. In this context, a very low mass is one about a hundred times the mass of Jupiter. Recall that it takes a mass of about 80 times that of Ju-

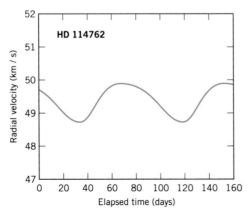

FIGURE 22.14 Cycle of the radial velocity from the Doppler shift observed for the star named HD 114762. Note from the horizontal axis that one complete period takes 84 days. This cyclical pattern of the Doppler shift is typical of that from a binary system. (Adapted from a diagram by D. W. Latham, T. Mezah, R. P. Stefanik, M. Mayor, and G. Burki.)

piter to make the smallest possible main-sequence star. In fact, the lowest measured mass of any star—the companion to one called Ross 614—is 0.085 solar mass.

A curious fact emerges from such investigations: the sun is the only star known *for certain* to have a planetary system that does *not* have a companion star of significant mass.

■ What is the essential problem with using center-of-mass motions to detect planetary systems?

22.6 Where Are They?

Whether we should search for other solar systems depends on how many technologically advanced civilizations exist in the Galaxy *now*. (For this discussion, I take "technologically advanced" to mean populated by creatures who can manipulate their environment at least to the extent that we can, so they have electricity, radios, telescopes, and so on.) If the number is large, then on the average such civilizations must be closer together than if the number is small.

As I pointed out, the key element in estimating this number, N_{ic}, is the *lifetime*, L_{ic}, of technological civilizations. For example, if the lifetime is about a century (which is about how long we've had a technologically advanced human culture on earth), then the average distance between galactic civilizations is roughly 10,000 ly. That makes communication practically impossible. Why? If we tried to signal by radio, for example, by sending out a message just at the moment our technology permitted, our civilization would have died while our words were still in transit. Communication is possible only if the number of civilizations is large and their lifetimes are long.

Note that N_{ic} is *not* a fixed number. It changes with time as the Galaxy evolves. For instance, for the first billion years of the Galaxy's existence, N_{ic} was probably zero, because life had not yet had time to evolve. So our estimates for now need not apply to the past or future.

How Far to Our Galactic Neighbors?

If we are it, we don't have any neighbors within 10^5 ly—the size of the Galaxy. That's for $N_{ic} = 1$. If N_{ic} is very large, say 10^9 to 10^{10}, then our neighbors are only a few tens of light years away. If N_{ic} is 10^6—a compromise guess—our nearest neighbors live within a few hundred light years of us—just within reach. Note that each of these choices implies a value for L_{ic}. If N_{ic} is very small, L_{ic} is at most a few hundred years. We are then probably on the verge of extinction. If N_{ic} is very large, L_{ic} is 10^9 to 10^{10}. Civilizations then last as long as their solar-type suns.

To sum up. We have no evidence for other advanced civilizations in the Galaxy.

Are We Alone?

You could take a pessimistic stance and argue that there are no other advanced civilizations in our Galaxy. Michael Hart is one astronomer who holds to the "we are alone" view. His opinion is based partly on computer calculations of the evolution of the earth's atmosphere. He finds that a most delicate balance must be maintained to keep temperatures in a moderate range. Hart notes that if the earth had an orbit of 0.95 AU, a run-away greenhouse effect would have turned us into a Venus. On the other hand, if the earth had orbited at 1.01 AU from the sun, glaciation would have iced up the earth 1.7 billion years ago. Our planet never would have gotten warm enough to foster the evolution of life. The key point is this: early conditions on the earth may have been so special, balanced between freezing and steaming, that the chances of a similar balance elsewhere in the Galaxy are very small.

Frank Tipler, who also has argued against the optimistic view (expressed by people such as Carl Sagan and Frank Drake), makes two points. One, that biological evolution from one-celled creatures to beings that think is so improbable that we *are* the only intelligent species to exist. (I find this point ethnocentric.) Two, that low-speed interstellar space travel is easy and cheap, especially for robot probes. A civilization just somewhat more advanced than us could probe the Galaxy. Yet, we have no evidence so far that such visits have taken place in the solar system. So such civilizations do not exist. (This argument is stronger, but it can be countered by observing that absence of evidence is not evidence of absence.)

Still, the evidence (rather than the speculation) so far indicates that we may very well be alone.

The Search for Extraterrestrial Intelligence

Many people feel that the astronomical, chemical, and biological data add up to a strong case for the plausibility of other intelligent civilizations. Some therefore feel it is reasonable to search for evidence of such civilizations, to undertake **SETI,** the *S*earch for *E*xtra*t*errestrial *I*ntelligence. How might this be done?

I'll take the optimistic assumption for N_{ic}, so that our neighbors are fairly close—tens to hundreds of light years. How can we reach or find civilizations tens to hundreds of light years away? The best way is radio communication. Radio astronomers have already constructed telescopes specially designed to detect very weak radio signals.

But radio covers a wide band of the electromagnetic spectrum. What range of frequencies is best? The choice of frequencies hinges on what part of the radio spectrum has the least background of natural noise, because we will try to detect weak signals. (By noise is meant, for example, the incessant jumble you hear when you tune your AM radio to a spot between stations.)

Astronomy and physics naturally define a low-noise band. At the low-frequency end (0.1 to 1 GHz), noise from the Galaxy (mostly synchrotron emission from high-speed electrons) dominates. At the high-frequency end (100 to 1000 GHz), noise in radio receivers, which comes from the quantum nature of matter and so cannot be eliminated, picks up. Between these two noise hills lies a valley of relative quiet from 1 to 100 GHz—part of the microwave region of the radio spectrum. The earth's atmo-

sphere fills in a bit more of this microwave noise valley at frequencies greater than 10 GHz. Any receiver on a planet like the earth would have the same low-noise window available—1 to 10 GHz.

How do we begin SETI with the instruments at hand? First, we spend our limited time listening. Next, we need to decide on the search mode: do we make a *targeted search* toward specific stars or scan the *whole sky,* hunting for sources. Remember that a radio telescope sees only a small fraction of the sky at a time—roughly a circle a few arcminutes in size (called the telescope's beam). It takes roughly 10^8 circles an arcminute in size to cover the entire sky. Changing a telescope's sky position once a second, we could cover the sky completely in only a few years, but that would give little time to listen at each position. In a whole-sky search you spend a lot less time on any area—therefore you can detect only strong sources. To start SETI, a mix of target and whole-sky search modes seems like the best strategy for success.

How can we effectively search the microwave band? Remember, we do not know the exact frequencies of our hypothetical transmitters. When we search the radio band, we need to do so in small intervals of wavelength or frequency. The range in frequency around a tuned-in frequency that is detected by a radio receiver is called its *bandwidth*. The problem here is that if we choose very narrow bandwidths, we have many individual channels to tune on. For example, a 1-GHz bandwidth means ten channels from 1 to 10 GHz. A 1-MHz bandwidth means 10^4 channels, a 1-kHz bandwidth 10^7 channels.

With a narrow bandwidth, you must spend time receiving at each of many channels. You have a wide range to search, and you cover only a small fraction of it at each channel you tune to. Most radio telescopes can now operate at many channels at a time; a few can search a few thousand channels simultaneously. What is needed for SETI is a receiver that can operate a few million channels simultaneously. Recent advances in electronics make such receivers possible.

If SETI can be done to a modest degree with present equipment, have any searches been done? Yes. Have any been successful? No. Other searches have been carried out sporadically by astronomers in the United States and the former Soviet Union since 1960. A few hundred stars have been observed. *No extraterrestrial signals have been detected to date.* The lack of results is understandable. Less than 0.01 percent of the sky and the terrestrial microwave window has been searched.

The revived SETI work in the United States, started in October 1992 and supported by NASA, consists of the two complementary parts of the strategy. One is an all-sky survey using the 34-meter antenna at the Goldstone complex of the Jet Propulsion Laboratory's Deep Space network. This search will span the frequency range of 1 to 10 GHz, with receivers that work at some million channels at a time. The antenna points to each position in the sky for few seconds and can detect emissions out to about 25 ly; it will take about 6 years to complete the project. The other is a targeted search using the 305-meter Arecibo antenna. Specific nearby stars will be observed for longer periods of time so that fainter signals can be detected. The problems are no longer ones of technology or tactics. They boil down to whether NASA can maintain financing, through political support for these projects over a time that seems an eternity in the vision of politicians.

■ **What may be the chief reason for our inability to detect other intelligent life in the Galaxy?**

22.7 *The Future of Humankind*

Before the sun dies, if we are still alive, we probably will have left our home planet—perhaps for others, perhaps not for planets at all. Predicting the future is always a dubious enterprise, but I think that we will have to leave the earth—long before the sun dies. Why? Because Spaceship Earth is a finite resource. If we do not achieve a stable population for the human race—for moral, ethical, or political reasons—we will have no choice but to leave.

Growth

The problem here is one of growth, constant growth that looks small at any given time but adds up quickly. Here's an example to show how constant growth rapidly gets out of hand. Steady increases lead to rapid increases in numbers, even for small fractional growth rates. In 1975 the growth rate of the human population was 1.9 percent a year. So the doubling time is roughly 37 years. That doesn't sound like much, but *if* the rate were to remain constant, the mass of people would equal the mass of the earth in only 1600 years!

Constant increases result in the rapid use of finite resources. Consider bacteria that double every minute. Suppose we place a bacterium in a bottle and note that the bottle is full up at 12 noon. When was the bottle half-full? One doubling time (one minute) earlier: 11:59. When was it one-quarter full? Two doubling times earlier: 11:58. Suppose that at 11:58 some far-sighted bacteria leaders got together and intensively searched for more living space. At 11:59, they find an empty bottle on the shelf, which doubles their total available space. When will that new bottle be filled? 12:01! When consumption grows steadily, even enormous increases in resources are consumed in short times.

The earth is a small planet. Given present rates of consumption, even if zero population growth were achieved tomorrow, our resources would be exhausted in only a few human lifetimes. We will need to leave our home planet: for space, for energy, and for natural resources.

Space Colonization

When we leave the earth, where will we go? The traditional science fiction view had us journeying to and colonizing other planets in our solar system. We now realize that with the possible exception of Mars, the other worlds in the solar system are not habitable planets for us. Even if they were, don't forget the lesson of the bacteria in the bottle. Human population growth now doubles roughly every 37 years. Suppose we fill the earth to the limit. How long would it take us to fill another planet, say Mars? Right—*one doubling time*, 37 years or so, if our present growth rate continues. In only 1500 years, we would have enough people to populate 10^{11} planets, one for every star in the Galaxy!

But there's no good reason to restrict ourselves to living on planets at all. This fact has encouraged Gerard K. O'Neill to revive and develop an older idea (some aspects were foreseen by the Russian physicist Konstantin Tsiolkowsky almost 100 years ago): human habitation in space, often known as space colonies. O'Neill has aimed at making the dream of space colonization a reality with available technology.

I won't detail his plans here, but let me sketch the broad outlines. Stripped of luxuries, people need energy, air (oxygen), water, land, and (probably) gravity to live a comfortable life. With space colonies in orbit around the earth, somewhere between the earth and the moon, all these are available: energy from the sun, oxygen and raw materials from the moon, and water from the earth. (Water might also be collected from the asteroids or from the moons of Jupiter and Saturn.) What about gravity? It can be simulated by rotating the space colony.

The first space colonies and solar satellite stations would be built with resources from the earth. Expansion to many colonies requires cutting Mother Earth's umbilical cord. For raw resources for development, we can turn to the moon and mine it. The moon's surface, as we know from the Apollo missions, contains abundant aluminum, titanium, oxygen, and silicon. (It lacks water, so we'd have to bring hydrogen from the earth to combine with lunar oxygen to make water.) A solar power station can support mining activities.

So a lunar base marks a key step in space colonization. NASA has recently investigated concepts and operational details for such a base—a permanently inhabited facility for scientific observations, with some capability of self-support. The base would have a primarily scientific function to explore the moon and do unique forms of astronomy (such as using an earth–moon radio interferometer). It could also serve to focus efforts for planetary exploration and space colonization.

What might a space colony look like (Fig. 22.15)? The simplest design is a cylinder. One some 6 km in diameter and 32 km long, spinning once every 2 minutes to simulate gravity, could support upward of 200,000 persons on its inner surface. The inside would alternate strips of land and windows. The windows would have shutters to simulate the seasons and day and night by controlling the influx of sunlight. The colony craft would be constructed of aluminum and titanium from the moon.

How long would this high frontier accommodate our population growth? At present growth rates, the local space could handle 400 to 500 years of population doubling. That may appear like a long time, but it's only about seven human lifetimes. And continued population growth would eventually fill up the local space. For example, imagine that we fill all the space between the sun and the earth with people jammed together. How long until this region is saturated? About 3000 years, at present growth rates.

The moral. The human race will overwhelm the solar system a few thousand years from now. Unless we change our ways.

FIGURE 22.15 Artist's view of a twenty-first-century space colony as envisioned by G. K. O'Neill. Mirrors and windows control the sunlight to simulate night and day. The rotation of the spacecraft would simulate gravity. (Courtesy of NASA.)

Key Concepts

1. Life as we know it consists of organisms that reproduce and may mutate (undergo changes in their genetic structure) and reproduce those mutations. These organisms contain proteins (the material of their construction) and nucleic acids (the blueprint of their construction); their chemical composition (except for helium) reflects the chemistry of the cosmos.

2. Natural selection shapes mutations and so drives biological evolution. Mutations appear as changes in the structure of DNA.

3. Fossils show that simple organisms existed on the earth at least 3 billion years ago and that significant evolution took billions of years more. Life arose from nonlife; physical and chemical evolution preceded biological evolution.

4. Lab experiments adding energy (sparks, ultraviolet light) to simple compounds in gaseous form (carbon dioxide, water, methane, ammonia, and so on) result in the natural synthesis of amino acids, most of which are common in life.

5. Meteorites rich in carbon and water contain amino acids made by chemical (that is, nonbiological) processes in the early solar system.

6. No life has been found on Mars. It is doubtful that life exists anywhere else in the solar system.

7. We can roughly estimate the number of technologically advanced civilizations in the Galaxy now by the Drake equation; that number could range from 1 to 10 billion; the greatest uncertainty arises from our lack of knowledge of the lifetimes of such civilizations.

8. Some nearby stars may have dark companions; these may be planetary systems or very-low-mass stars or brown dwarfs.

9. The number of technologically advanced civilizations now sets the average distance between them; if that number is 1 million, the average distance is a few hundred light years. Any search for such civilizations, say by listening at microwave bands, relies on an optimistic assumption of the lifetimes.

10. Unrestrained population growth and wastage of natural resources will force the human race to leave Spaceship Earth and journey to the stars.

Key Terms

amino acids

deoxyribonucleic acid (DNA)

Drake equation

ecosphere

enzymes

mass extinctions

mutation

natural selection

nucleic acids

organism

proteins

SETI

Study Exercises

1. Life on earth centers on carbon. Where did the carbon come from, and how did it get here? (Objectives 22-1, 22-2, and 22-3)

2. How do supernovas play a crucial role in the origin of life? (Objectives 22-2 and 22-3)

3. What are the chances for life elsewhere in the solar system now? (Objective 22-7)

4. O stars have the largest ecospheres around them, yet they are not good suns for fostering life. Why not? (Objective 22-8)

5. Criticize the book's estimate of the number of intelligent civilizations in the Galaxy. (Objective 22-9)

6. In what sense is it astronomically correct to say that we are children of the stars? (Objectives 22-3 and 22-6)

7. Eventually, when the interstellar gas is depleted, star formation in the Galaxy will stop. What effect will that have on an estimate of the number of civilizations in the Galaxy? (Objective 22-9)

8. How have the results from the Viking landers reinforced the pessimistic view that we may be alone in the Galaxy? (Objective 22-7)

9. Why is the microwave region of the spectrum the best one to search for signals from extraterrestrial civilizations? (Objective 22-10)

Problems & Activities

1. Find the location of the center of mass of the Jupiter–sun system. Would the center-of-mass motion of the sun be visible from the nearest star, using present technology?

2. If Nemesis has an orbital period of 26 million years, what is the size of the semimajor axis of its orbit?

3. Using the Drake equation, come up with your own number of advanced civilizations in the Galaxy. Be sure to make your assumptions clear!

4. What is the orbital period of HD 114762 (Fig. 22.14) and its invisible companion? Using Kepler's laws, what is the semimajor axis of the orbit? The star HD 114762 is a G V star; make an estimate of its mass. What might be the mass of the companion, if the radial velocity amplitude is 500 m/s? To simplify this estimate, assume that we view the orbit edge-on (even though we have no evidence of eclipses).

See for Yourself!

Make your own ecospheres around stars! Get a large sheet of paper and place a "star" at the center. Lay out distances in AU, out to 10 AU at the edge of the paper.

Proper temperature is perhaps the most important consideration for the existence of the simplest life forms. Living cells have been found in hot springs, almost at the boiling point of water. Though individual cells are destroyed when the water in them freezes, organisms that provide their own heat can survive at much lower temperatures. Let's take the temperature range from 200 to 373 K as the range suitable for life.

The zone around a star where the range of temperatures on a planet is suitable for life is called the star's *ecosphere*. How thick is the ecosphere around a star? That depends on its luminosity. We can estimate its thickness roughly as follows. A rapidly rotating airless planet, completely absorbing all light that hits it, at distance d from a star with luminosity L has a temperature

$$T \text{(kelvins)} = 279 \left(\frac{L}{d^2}\right)^{1/4}$$

To use this equation, we turn it around to have

$$d \text{(AU)} = (279)^2 \frac{L^{1/2}}{T^2}$$

At the inner edge of the ecosphere, $T = 373$ K; at the outer edge, 200 K. For the solar system ($L = 1$),

$$d_{inner} = \frac{(279)^2(1)^{1/2}}{(373)^2} = 0.6 \text{ AU}$$

$$d_{outer} = \frac{(279)^2(1)^{1/2}}{(200)^2} = 1.9 \text{ AU}$$

and the thickness of the ecosphere is 1.3 AU. On your graph paper, lay out circles at 0.6 and 1.9 AU. Color in the region between; that's the sun's ecosphere.

For a star more luminous than the sun, the ecosphere is farther out and thicker. For Sirius A for instance, which has a luminosity twenty-five times that of the sun:

$$d_{inner} = \frac{(279)^2(25)^{1/2}}{(373)^2} = 2.8 \text{ AU}$$

$$d_{outer} = \frac{(279)^2(25)^{1/2}}{(200)^2} = 9.7 \text{ AU}$$

and its thickness is 6.9 AU. On the same graph paper, draw circles at 2.8 and 9.7 AU. Color this region in also, so you can visually compare this ecosphere to that for the sun. ■

PART FOUR *Epilogue*

Assume that life as we know it is typical in the sense that it has arisen naturally in the course of cosmic evolution. Then we can hope to find the trail of that evolution, which falls into three interconnected stages: physical, chemical, and biological evolution.

We can hunt down the traces of biological evolution on the earth. Fossil evidence implies that with time, life has grown more complex on the earth and that the origin of life took about a billion years to get to the first cell.

Astronomical ideas underpin our understanding of physical evolution. For life as we know it, we need a planet (the earth), a star (the sun), and the proper elements (hydrogen, carbon, nitrogen, oxygen, and some others). Where did these come from? The earth, from the dust of the interstellar medium; the sun, from the gases—both dust and gas are the material lost by earlier stars, mostly in their violent ends. These explosions and normal fusion reactions in stars manufactured the chemical elements, except for hydrogen and helium, which were made in the first few minutes of the Big Bang.

This sequence in general seems appropriate for all the observable universe. So if life is typical, it must be common. If so, our Galaxy and the entire universe teem with life. This life may not be restricted to planets. Every civilization at some stage may face the decision: grow or die. And growth may require leaving the home planet for the high frontier of space exploration. We may well meet our neighbors in our vagabond wanderings.

Astronomy teaches that we are creatures of the universe, children of the stars. We are products of cosmic evolution. We are also part of its process. Perhaps we serve to make the universe aware of itself. When you and I look into space, we see the source of ourselves. And to those wide-open spaces we add hope, fear, imagination, and love.

I have tried in this last part to show you the cosmic connections that touch us all. Reading this book, you have touched me. Years from now, I hope that you'll remember that when you touch your sister, brother, parent, lover, friend—you touch the stars. ■

The Unifying View

ASTRONOMICAL CONCEPTS

All we see directly are angular relationships among objects on an apparently two-dimensional sky.
> We add depth by finding distances; for galaxies, this means measuring the redshifts in their spectra.

Measuring distances is crucial to all aspects of astronomy:
> Angular sizes are size-to-distance ratios; smaller objects of the same type tend to be farther away, so generally, the smaller the angular diameter of a galaxy, the more distant it is.

> Galaxies are arrayed in clusters that are arranged in a supercluster.

Light conveys to us information about astronomical objects; we are always viewing the past and so the history of the cosmos.

The cosmos is expanding, as seen in the redshift-distance relation for galaxy (Hubble's law):
> The cosmos began at a finite time in the past.

> Its rate of expansion now (Hubble's constant) is not well known.

> Unless much dark matter exists, the cosmos will expand forever.

INFORMATION CONCEPTS

We perceived the world through our senses, sometimes aided by instruments:
> Telescopes, on ground and in space, can use special detectors to probe the universe at great distances.

Data has no meaning until it becomes information
> *COBE* observations of the temperature of the cosmic background radiation have very small uncertainties.

522

Model building is the essential process by which science discovers knowledge and reveals information:

The current model of the Galaxy views it as a disk with spiral structure.

AGNs and quasars share a generic model powered by a supermassive black hole; the angle at which they are viewed strong effects what is seen.

The Big-Bang model, modified by inflation, is the current cosmological model accepted by most astronomers.

For the origin of life, the basic model views simple life forms as naturally arising early-on in the earth's history.

ENERGY CONCEPTS Energy comes in many forms such as potential, kinetic, thermal, radiative:

Gravitational potential energy powers quasars and AGNs.

Radiative energy carried by photons, the greater with higher frequency; low-energy photons tell us about the early universe, high-energy photons about active galaxies.

Mass is energy ($E = mc^2$), and energy is mass; in the early universe, high-energy light made matter.

Radiative energy decreases in flux as the inverse-square of the distance, so that more galaxies appear to be fainter than closer ones.

Transformations occur among the various forms of energy, and energy can be transferred from one location to another:

Atoms and molecules absorb and emit light, usually at special energies, so we can identify elements in distant objects, such as quasars.

Mass and energy warp spacetime and may close the universe.

MOTION AND FORCE CONCEPTS Newton's laws of motion plus his law of gravitation explain and predict the motion of the masses in galaxies in a unified way.

Knowledge of these laws permit us to measure the masses of galaxies.

Kepler's laws are an alternative expression of forces and gravitation and sometimes are easier to apply.

The concept of escape speed can be applied to the cosmos to understand what it means to be open or closed.

Einstein unified space and time in the cosmos:

The general theory of relativity predicted an expanding cosmos.

This prediction is confirmed by the Hubble law from measuring the redshifts and distances to galaxies.

From the general theory, we can understand the past and predict the future of the universe.

QUANTUM CONCEPTS On a very small scale, matter is made of discrete particles:

At very early times in the cosmos, we do not know what form of elementary particles were in existence.

We believe that the forces that determine interaction of elementary particles were unified at some early time in the universe.

Physical phenomena are governed by a few basic interactions of matter and energy, often at the quantum level:

The cosmic background radiation comes from a time in the universe's history when matter and light were interacting strongly.

General relativity and quantum physics have yet to be unified.

523

Improve Your Night Vision

SPRING

Fred Schaaf

The warmer weather of spring doesn't beckon us out to study the stars alone but also a number of other special sky phenomena. For instance, by day the spotty showers of spring, which occur in much of the United States and Canada, are ideal for producing rainbows. Look for a second, higher, wider bow with reversed order of colors, and brighter sky below the main or primary bow.

By almost night—in other words, twilight—there are other wonders to see. At midnorthern latitudes, the ecliptic (midline of the zodiac and apparent yearly path of the sun in the sky) is steepest in the west at dusk in the spring—and thus planets and a crescent moon are seen highest above the sun's afterglow. The best evening appearance of

the elusive planet Mercury almost always occurs with the particular greatest elongation of it from the sun, which takes place within a month or so of the spring equinox.

If you wish to try to see the most breathtakingly slender moon possible, check the day after new moon low in the west about 20 to 40 minutes after sunset. The next few nights you may also get your best looks at glow on the dark part of the moon—*earthshine,* the glow of our own planet shining on the moon's night side. Keep records of when earthshine is brightest and if you live in the contiguous United States or southern Canada check TV weather coverage like that on the Weather Channel to see if these times occur when there is much cloudiness over the Pacific

524

(clouds reflect more light out to space than water, so a cloudy daytime side of the earth shines more brightly upon the moon).

Twilight, or right around sunset (or sunrise) actually, is the best time to glimpse sharp detail on the moon with the naked eye. Test this for yourself and see if you can figure out why. The best phases at which to see a lot of detail on the moon with binoculars and telescopes are first quarter and last quarter because at these times the terminator lies along the very rough central meridian of the moon. The *terminator,* the line separating light and dark on the moon, is the place at which the greatest amount of detail is visible because features are not washed out by sunlight from high above; rather, they are etched sharply with shadows cast by the low sun as seen from that part of the moon. If you want to test all this out with your binoculars or telescope, spring is a good time to look at a high first quarter moon after nightfall.

When night falls, the Big Dipper is the classic illustrator of how, unlike more southerly constellations, stars near the north celestial pole (the point in the heavens right over earth's own north pole) never have their vast circlings of the celestial pole cut off into arcs by the east and west horizons. In early spring at midevening, the Big Dipper is upside-down and at its highest (a position in which the line drawn through the pointer stars on the outside of its bowl should be extended *down* to show the location of the pole and of Polaris, the North Star). Later in the night in spring, or at the same time of night later in the year, check to see where the swing of the Big Dipper has placed this familiar sight. You'll see why it has been called a compass and clock (or hand of a clock) in the sky.

For people with fairly good eyesight under fairly good sky conditions, the star at the bend in the handle of the Big Dipper is seen to be a *double star*—that is, a point of light which, upon closer examination, proves to consist of two or more components. The naked-eye companion star of the bright Mizar is called Alcor—but a telescope shows that Mizar itself can be further split into components.

In most cases, the members of a double-star system really are going through space together, perhaps circling around a common center of gravity (occasionally, though, we have an "optical double" in which what we are seeing is just one star much farther away than another, which happens to lie along about the same line of sight). Spring offers many beautiful and colorful doubles (often the colors are accentuated by contrast with each other). Most are too close together for naked eye or binoculars to split, but there are exceptions. Also, later in the night, check out the double stars in Lyra and Cygnus, including those mentioned in the Summer essay.

The most colorful spring star to the naked eye is the brightest of spring, orange Arcturus. An arc extended from that of the Big Dipper's handle leads you to Arcturus (you can then "drive a spike" to Spica in mostly dim, huge Virgo). Most of the stars we see in the sky are traveling around the center of the Galaxy pretty much in the equatorial plane of the Galaxy, as we are. The *apex,* or point toward which the sun seems to be headed, is located not far from the star Vega, already coming up in the northeast on spring evenings. But after you have looked in this direction toward which we are heading, gaze up at Arcturus and consider that this is a gypsy star, which is dropping through the equatorial plane of the galaxy on an orbit highly inclined to that plane. This means that Arcturus will plummet to the south and disappear from naked-eye view in the near astronomical future—only about half a million years from now.

In these essays for the seasons, we have mentioned repeatedly the problem of city light pollution, which is due to poor shielding and poor lighting practices. Astronomers define *light pollution* as excessive or misdirected outdoor lighting. It wastes energy and money, glares dangerously in the eyes of motorists and homeowners, seriously disrupts nocturnal flora and fauna—and is threatening to steal most of the direct view of the universe from most people on earth.

Fortunately, there is at last action against light pollution and major laws are being passed to combat it. The best way to find out more about the issue is to write to the International Dark-Sky Association (IDA), 3545 North Stewart, Tucson, AZ 85716. IDA each year asks observers to count the Pleiades from different locations at various distances from (and within) cities, collecting valuable data while stimulating people to go out and see the problem of light pollution for themselves. You can participate in autumn, winter, or early spring. In doing so, you may be helping to save our view of this lovely Seven Sisters cluster—and much else of the heavens—for us, our children, and generations to come. Write to the IDA and ask for information on the Star-Watching program for the Pleiades.

Units

Powers of Ten

Astronomers deal with quantities that range from the microcosmic to the macrocosmic. To avoid having to read and write numbers such as 149,597,890,000 meters (the average earth–sun distance), we use powers-of-ten notation. That simply means that we write a number as a fraction between one and ten times the appropriate power of ten rather than with a string of leading or trailing zeros. A positive exponent tells how many times to multiply by 10. For example:

$$10^1 = 10$$
$$10^2 = 100$$
$$10^3 = 1000$$

and so on. A negative exponent, in contrast, tells how many times to *divide* by 10. For instance:

$$10^{-1} = 0.1$$
$$10^{-2} = 0.01$$
$$10^{-3} = 0.001$$

in analogy to the positive powers. The only trick here is to remember that 10 to the zero power is 1:

$$10^0 = 1$$

In powers-of-ten notation the earth–sun distance is 1.4959789×10^{11} meters.

Significant Figures

Significant figures are the meaningful digits in a number. In experimental or observational results, the accuracy of the procedure or technique limits the number of significant figures. No more significant figures than that number can be stated as the true value of the result (which should always have a range of error attached to it). In arithmetical calculations, we are sometimes interested in an approximate result to some specified degree of accuracy. Approximate ("back of the envelope") calculations using just one or two significant figures prove useful in science, especially in astronomy, because they are sufficient to illustrate a point, or because the quantities are simply not well known.

Here are some examples:

0.0045676	five significant figures
4.5676×10^{-1}	five significant figures
4.5×10^{-4}	two significant numbers
2.17	three significant figures

What about the number 2,170,000? It is not clear whether there are three or seven significant figures. But when we use powers-of-ten notation, the ambiguity disappears: 2.17×10^6 has three significant figures; 2.170000×10^6 has seven significant figures.

The general rule in doing calculations is that the result cannot have more significant figures than the *smallest* number of significant figures in the quantities that are used in the computation. It may have even fewer—for example, if the calculation involves subtraction of nearly equal numbers.

Rounding

To round off numbers, look at the last digit. If it is greater than 5, round up; round down if it is less than 5. If it is 5, then round up if the next-to-the-last digit is even, and down if it is odd. Most of the numbers in the main text of this book have been rounded off to two significant figures.

The English and Metric Systems

You are probably familiar with the fundamental units of length, mass, and time in the English system: the yard, the pound, and the second. Other common units in this system are often strange multiples of these fundamental units. For example: the ton is 2000 pounds, the mile is 1760 yards, and the inch is 1/36 of a yard. Most of the world uses the much more rational metric system, which applies powers-of-ten relationships to units, making them easy to multiply and divide.

The fundamental metric system units are:

Length:
1 meter (m)
Mass:
1 kilogram (kg)
Time:
1 second (s)

This is often called the meter-kilogram-second, or *mks* system. An older system, often used by astronomers, is the centimeter-gram-second, or *cgs* system, with the following fundamental units:

Length:
1 centimeter (cm)
Mass:
1 gram (g)
Time:
1 second (s)

TABLE A.1 SELECTED SI BASIC UNITS

Length: meter (m)	Work and energy: joule (J)
Time: second (s)	Power: watt (W)
Mass: kilogram (kg)	Frequency: hertz (Hz)
Current: ampere (A)	Charge: coulomb (C)
Temperature: kelvin (K)	Magnetic field: tesla (T)
Force: newton (N)	Pressure: pascal (Pa)

TABLE A.2 CONVERSION TO SI UNITS

Length

1 ft = 12 in. = 30.48 cm

1 yd = 3 ft = 91.44 cm

1 light year = 9.461×10^{15} m

1 Å = 0.1 nm

1 μm = 10^{-6} m

Area

$1 m^2 = 10^4 cm^2$

$1 km^2 = 0.3861 mi^2$

$1 in.^2 = 6.4516 cm^2$

$1 ft^2 = 9.29 \times 10^{-2} m^2$

$1 m^2 = 10.75 ft^2$

Volume

$1 m^3 = 10^6 cm^3$

1 liter (L) = $1000 cm^3 = 10^{-3} m^3$

1 gal = 3.786 L = $231 in.^3$

Time

1 h = 60 min = 3.6 ks

1 day = 24 h = 1440 min = 86.4 ks

1 year = 365.24 days = 31.56 Ms

Speed

1 km/h = 0.2778 m/s = 0.6125 mi/h

1 mi/h = 0.4470 m/s = 1.609 km/h

Density

$1 g/cm^3 = 1000 kg/m^3 = 1 kg/L$

Force

1 N = 0.2248 lb = 10^5 dynes

1 lb = 4.4482 N

Pressure

$1 Pa = 1 N/m^2$

1 atm = 101.325 kPa = $14.7 lb/in.^2$ = 760 mm Hg

Energy

1 kW • h = 36 MJ

1 eV = 1.602×10^{-19} J

1 erg = 10^{-7} J

Power

1 hp = 550 ft • lb/s = 745.7 W

1 W = 1.341×10^{-3} hp

Magnetic field

1 G = 10^{-4} T

1 T = 10^4 G

The current international standard for physical units is called the *Système International* (in French) or *SI*. This system is based on the mks system with many modifications. Many astronomers do not yet use SI units, though they should, because it is now the international standard. I have tried to stick with SI units in this book but have occasionally used units of convenience, such as atmospheres for pressure. Table A.1 gives most of the fundamental SI units, and Table A.2 some conversions.

Some multiples of the metric system and their associated prefixes are:

10^{-15}	femto-
10^{-12}	pico-
10^{-9}	nano-
10^{-6}	micro-
10^{-3}	milli-
10^{-2}	centi-
10^{-1}	deci-
10	deka-
10^2	hecto-
10^3	kilo-
10^6	mega-
10^9	giga-
10^{12}	tera-
10^{15}	peta-

Some relationships between the metric and English system are:

Length:

1 kilometer (km) = 1000 m = 0.6214 mile

1 meter (m) = 1.094 yd = 39.37 inches

1 centimeter (cm) = 0.01 m = 0.3937 inch

1 millimeter (mm) = 0.001 m = 0.03937 inch

1 mile = 1.6093 km

1 inch = 2.5400 cm

Mass:

1 metric ton = 10^6 g = 1000 kg = 2.2046×10^3 pounds

1 kilogram (kg) = 10^3 g = 2.2046 pounds

1 gram (g) = 0.0353 oz = 0.0022046 pound

1 milligram (mg) = 0.0001 g = 2.2046×10^{-6} pound

1 pound = 453.6 g

1 ounce = 28.3495 g

Temperature Scales

The freezing and boiling points of water (at sea level) set the fundamental points on common temperature scales. In the United States, the Fahrenheit (F) system dominates. Water freezes at 32°F and boils at 212°F. Most of the world uses the Celsius (C; formerly centigrade) system. On it, water freezes at 0°C and boils at 100°C.

Astronomers mostly employ the Kelvin (K) system, which uses the same size degrees as the Celsius system but starts from absolute zero, the temperature at which random molecular motion reaches a minimum (0 K = −273.15°C). At absolute zero, no more thermal energy can be extracted from a body. Water freezes at about 273 K and boils at about 373 K. Note that the degree mark is not used with temperatures on the Kelvin scale and that temperatures are often referred to as "kelvins," not "degrees kelvin."

To convert between systems, recognize that 0 K = −273°C = −460°F and that Celsius and Kelvin degrees are larger than Fahrenheit degrees by 180/100 = 9/5. The relationships between systems are:

$$K = C + 273$$
$$C = \frac{5}{9}(°F - 32)$$

Kelvins are never negative, because absolute zero is the lowest point on any temperature scale.

Astronomical Distances

Although astronomers do use the metric system, they encounter distances so large that other measures are often used as well. In the solar system, the natural scale is the average distance from the earth to the sun, the astronomical unit (AU), which is about 1.5×10^8 km.

Beyond the solar system, the AU is too small compared to stellar distances. Here astronomers use the parsec and the light year. One parsec (pc) equals 206,265 AU or about 3.1×10^{13} km. A light year (ly) is the distance that light travels in one year, about 9.5×10^{12} km. Note that 1 pc is about 3.3 ly.

For even larger distances, astronomers most often talk in large multiples of parsecs: a kiloparsec (kpc) is 10^3 pc, and a megaparsec (Mpc) is 10^6 pc. Equivalent units in light years are abbreviated kly and Mly.

Other Physical Units

The SI unit of force is the *newton* (N). It is the force needed to accelerate an object having a mass of one kilogram by one meter per second squared.

The SI unit of energy is the *joule* (J). It takes about one joule to lift an apple off the floor to over your head. Table A.3 compares the energy of some familiar events and some astronomical ones.

Power is the rate of energy transfer, so it has units of joules per second. Astronomers often use the term *luminosity* in place of power. A convenient unit for luminosity or power is the *watt* (W), defined as one joule per second. Table A.3 includes some power comparisons.

When talking about stars, astronomers often measure quantities relative to the sun—for example, in units of a solar mass, a solar luminosity, or a solar radius. The mass of Sirius is 2.3 solar masses, its radius is 1.8 solar radii, and its luminosity is 23 solar luminosities.

Astronomers use both the *gauss* (G) and *tesla* (T) as the unit of magnetic field strength (1 T = 10^4 G); the tesla is the SI unit. The earth's magnetic field is about 4×10^{-3} T, or 0.4 G.

The SI unit of pressure is the *pascal* (Pa), which is one newton per meter squared. You can get a feel for a pascal next time you check the air in your car's tires: a typical pressure of 28 pounds per square inch (psi) equals 196 kPa (kilopascals). Astronomers also use the *atmosphere* (atm) as a pressure unit; one atmosphere is about 100 kPa.

TABLE A.3 COMPARATIVE ENERGY AND POWER OUTPUTS

ENERGY OR POWER SOURCE	APPROXIMATE TOTAL ENERGY (J) OR POWER (W)
Earth's rotational kinetic energy	10^{29} J
Sunlight	10^{26} W
Sunlight (1 y)	10^{34} J
Earth's daily input of solar energy	10^{22} J
H-bomb	10^{17} J
100% conversion of 1 gram of matter	10^{14} J
World's current energy consumption	10^{13} W
1 barrel of oil	10^{10} J
1 power plant	10^{9} W
1 medium pizza	10^{7} J
1 flashlight battery (D cell)	10^{4} J
1 small person (M.Z.)	100 W
1 mosquito pushup	10^{-7} J
Fission of one atom of uranium-235	10^{-11} J

Adapted in part from a table in the *American Journal of Physics*, vol. 60, June 1992, p. 596.

Planetary Data

TABLE B.1 PLANETARY ROTATION RATES AND INCLINATIONS

PLANET	ROTATION PERIOD (equatorial)	INCLINATION OF EQUATOR TO ORBITAL PLANE
Mercury	58.65 days	0°
Venus	243.01 days (retrograde)	117° 18′
Earth	23 h 56 min 4.1 s	23° 27′
Mars	24 h 37 min 22.6 s	25° 12′
Jupiter	9 h 50.5 min	3° 07′
Saturn	10 h 14 min	26° 44′
Uranus	17 h 14 min (retrograde)	97° 52′
Neptune	16 h 3 min	29° 34′
Pluto	6.39 days (retrograde)	98°

TABLE B.2 DISTANCES AND PERIODS OF THE PLANETS

PLANET	SEMIMAJOR AXIS (AU)	SEMIMAJOR AXIS ($\times 10^6$ km)	SIDEREAL PERIOD (tropical years)	SIDEREAL PERIOD (days)	SYNODIC PERIOD (days)	ORBITAL ECCENTRICITY
Mercury	0.387	57.9	0.241	87.97	115.9	0.206
Venus	0.723	108.2	0.615	224.7	583.9	0.007
Earth	1.000	149.6	1.000	365.26		0.017
Mars	1.524	227.9	1.881	686.98	779.9	0.094
Jupiter	5.203	778.4	11.86	4332	398.9	0.048
Saturn	9.522	1424	29.46	10,761	378.1	0.054
Uranus	19.20	2872	84.01	30,685	369.7	0.048
Neptune	30.07	4499	164.8	60,195	367.5	0.007
Pluto	39.72	5943	248.6	90,471	366.7	0.253

NOTE: A tropical year is the year of seasons. See the Glossary for the definitions of sidereal and synodic periods.

TABLE B.3 PHYSICAL DATA OF THE PLANETS

PLANET	AVERAGE RADIUS (km)	RADIUS (earth radii)	ALBEDO	TEMPERATURE (K)	ESCAPE SPEED (km/s)
Mercury	2,439	0.38	0.06	100–700	4.3
Venus	6,052	0.95	0.76	700	10.4
Earth	6,378	1.00	0.40	250–300	11.2
Moon	1,738	0.27	0.07	120–390	2.4
Mars	3,397	0.53	0.16	210–300	5.0
Jupiter	71,492	11.19	0.51	110–150	59.6
Saturn	60,268	9.45	0.50	95	35.6
Uranus	25,559	4.01	0.66	58	21.3
Neptune	25,269	3.96	0.62	56	23.8
Pluto	1,140	0.18	0.50	40	1.2

TABLE B.4 MASSES AND DENSITIES OF THE PLANETS

PLANET	MASS (earth masses)	MASS (kg)	DENSITY (kg/m³)	SURFACE GRAVITY (earth = 1)
Mercury	0.0562	3.30×10^{23}	5430	0.38
Venus	0.815	4.87×10^{24}	5240	0.91
Earth	1.000	5.974×10^{24}	5520	1.00
Moon	0.012	7.35×10^{22}	3340	0.16
Mars	0.1074	6.42×10^{23}	3940	0.39
Jupiter	317.9	1.899×10^{27}	1330	2.54
Saturn	95.1	5.68×10^{26}	690	1.07
Uranus	14.56	8.66×10^{25}	1270	0.90
Neptune	17.24	1.03×10^{26}	1640	1.14
Pluto	0.0018	1.1×10^{22}	2100 (?)	0.06

TABLE B.5 ATMOSPHERIC GASES OF THE PLANETS AND SOME MOONS

PLANET	GASES (in order of relative abundance)
Mercury	Sodium, potassium, helium, hydrogen
Venus	Carbon dioxide, carbon monoxide, hydrogen chloride, hydrogen fluoride, water, argon, nitrogen, oxygen, hydrogen sulfide, sulfur dioxide, helium
Moon	Helium, argon
Earth	Nitrogen, oxygen, water, argon, carbon dioxide, neon, helium, methane, krypton, nitrous oxide, ozone, xenon, hydrogen, radon
Mars	Carbon dioxide, carbon monoxide, water, oxygen, ozone, argon, nitrogen
Jupiter	Hydrogen, helium, methane, ammonia, water, carbon monoxide, acetylene, ethane, phosphine, germane
Saturn	Hydrogen, helium, methane, ammonia, acetylene, ethane, phosphine, propane
Titan	Nitrogen, methane, ethane, acetylene, ethylene, hydrogen cyanide
Uranus	Hydrogen, helium, methane
Neptune	Hydrogen, helium, methane, ethane
Pluto	Methane

TABLE B.6 SATELLITES OF THE TERRESTRIAL PLANETS

PLANET	SATELLITE	DISTANCE ($\times 10^3$ km)	ORBITAL PERIOD (days)	RADIUS (km)	MASS (planet = 1)	BULK DENSITY (kg/m³)
Earth	Moon	384	27.32	1738	0.012	3340
Mars	Phobos	9.38	0.3189	$14 \times 11 \times 9$	1.5×10^{-8}	2200
	Deimos	23.46	1.262	$8 \times 6 \times 6$	3.1×10^{-9}	1700

TABLE B.7 SATELLITES OF JUPITER

NAME	DISTANCE ($\times 10^3$ km)	DISTANCE (Jupiter radii)	ORBITAL PERIOD (days)	RADIUS (km)	MASS (planet = 1)	BULK DENSITY (kg/m³)
Metis	128	1.79	0.294	20	0.5×10^{-10}	
Andrastea	129	1.80	0.298	$12 \times 10 \times 8$	0.11×10^{-10}	
Amalthea	181	2.55	0.498	$135 \times 83 \times 75$	38×10^{-10}	3000
Thebe	222	3.11	0.674	55×45	4×10^{-10}	
Io	422	5.95	1.77	1815	4.7×10^{-5}	3550
Europa	671	9.47	3.55	1569	2.5×10^{-5}	3040
Ganymede	1,070	15.10	7.15	2631	7.8×10^{-5}	1930
Callisto	1,883	26.60	16.7	2400	5.7×10^{-5}	1830
Leda	11,094	156	239	8	0.03×10^{-10}	
Himalia	11,480	161	251	93	50×10^{-10}	~1000
Lysithea	11,720	164	259	18	0.4×10^{-10}	
Elara	11,737	165	260	38	4×10^{-10}	
Ananke	21,200	291	631R	15	0.2×10^{-10}	
Carme	22,600	314	692R	20	0.5×10^{-10}	
Pasiphae	23,500	327	735R	25	1×10^{-10}	
Sinope	23,700	333	758R	18	0.4×10^{-10}	

NOTE: Distances are from the center of Jupiter. An "R" indicates retrograde rotation. If the radius has more than one dimension, the object is not spherical.

TABLE B.8 SATELLITES OF SATURN

SATELLITE	DISTANCE ($\times 10^3$ km)	DISTANCE (Saturn radii)	ORBITAL PERIOD (days)	RADIUS (km)	MASS (planet = 1)	BULK DENSITY (kg/m^3)
Pan	133.58	2.22	0.575	10		
Atlas	137.67	2.28	0.602	20×10		
Prometheus	139.35	2.31	0.613	$70 \times 50 \times 40$		
Pandora	141.70	2.35	0.628	$55 \times 45 \times 35$		
Epimetheus	151.42	2.51	0.694	$70 \times 60 \times 50$		
Janus	151.42	2.51	0.694	$110 \times 100 \times 80$		
Mimas	185.54	3.08	0.942	196	8.0×10^{-8}	1170
Enceladus	238.02	3.95	1.370	250	1.3×10^{-7}	1240
Tethys	294.66	4.88	1.887	530	1.3×10^{-6}	1260
Telesto	294.66	4.88	1.887	$17 \times 14 \times 13$		
Calypso	294.66	4.88	1.887	$17 \times 11 \times 11$		
Dione	377.40	6.26	2.736	560	1.85×10^{-6}	1440
Helene	377.40	6.26	2.736	$18 \times 16 \times 15$		
Rhea	527.04	8.74	4.517	765	4.4×10^{-6}	1330
Titan	1,221.8	20.25	15.945	2575	2.36×10^{-4}	1880
Hyperion	1,481.1	24.55	21.276	$205 \times 130 \times 110$	3×10^{-8}	
Iapetus	3,561.3	59.02	79.330	730	3.3×10^{-6}	1210
Phoebe	12,952	214.7	550.4R	110	7×10^{-10}	

NOTE: Distances are from the center of Saturn. An "R" indicates retrograde rotation. If the radius has more than one dimension, the object is not spherical.

TABLE B.9 SATELLITES OF URANUS, NEPTUNE, AND PLUTO

PLANET	SATELLITES	DISTANCE ($\times 10^3$ km)	ORBITAL PERIOD (days)	RADIUS (km)	MASS (planet = 1)	BULK DENSITY (kg/m^3)
Uranus	Cordelia	49.75	0.3350	13		
	Ophelia	53.77	0.3764	16		
	Bianca	59.16	0.4346	22		
	Cressida	61.77	0.4636	33		
	Desdemona	62.65	0.4737	29		
	Juliet	64.63	0.4931	42		
	Portia	66.10	0.5132	55		
	Rosalind	69.93	0.5585	27		
	Belinda	75.25	0.6235	34		
	Puck	86.00	0.7618	77		
	Miranda	129.41	1.413	240	0.2×10^{-5}	1260
	Ariel	191.0	2.520	579	1.8×10^{-5}	1650
	Umbriel	266.3	4.144	586	1.2×10^{-5}	1440
	Titania	435.9	8.706	790	6.8×10^{-5}	1590
	Oberon	583.5	13.463	762	6.9×10^{-5}	1500
Neptune	Naiad	48.23	0.2944	29		
	Thalassa	50.07	0.3115	40		
	Despoina	52.53	0.3347	79		
	Galatea	61.95	0.4287	74		
	Larissa	73.55	0.5547	96		
	Proteus	117.64	1.1223	208		
	Triton	354.8	5.877R	1350	2.1×10^{-4}	
	Nereid	5513	360	170	2.0×10^{-7}	
Pluto	Charon	19.1	6.387	600	0.2	1000?

Physical Constants, Astronomical Data

Physical Constants

Gravitational constant
$$G = 6.673 \times 10^{-11} \text{ N} \cdot \text{m}^2/\text{kg}^2$$

Speed of light in a vacuum
$$c = 2.9979 \times 10^8 \text{ m/s}$$

Planck's constant
$$h = 6.62608 \times 10^{-34} \text{ J} \cdot \text{s}$$

Wien's constant
$$\sigma_{\text{W}} = 0.002898 \text{ m} \cdot \text{K}$$

Boltzmann's constant
$$k = 1.3806 \times 10^{-23} \text{ J/K}$$

Stefan–Boltzmann constant
$$\sigma = 5.6697 \times 10^{-8} \text{ W/m}^2 \cdot \text{K}^4$$

Electron mass
$$m_{\text{e}} = 9.10939 \times 10^{-31} \text{ kg}$$

Proton mass
$$m_{\text{p}} = 1.6726 \times 10^{-27} \text{ kg} = 1836.1 \ m_{\text{e}}$$

Neutron mass
$$m_{\text{n}} = 1.6749 \times 10^{-27} \text{ kg}$$

Mass of hydrogen atom
$$m_{\text{H}} = 1.6735 \times 10^{-27} \text{ kg}$$

Astronomical Data

Astronomical unit
$$\text{AU} = 1.4959789 \times 10^{11} \text{ m}$$

Parsec
$$\text{pc} = 206{,}264.806 \text{ AU}$$
$$= 3.2616 \text{ ly}$$
$$= 3.0856 \times 10^{16} \text{ m}$$

Light year
$$\text{ly} = 9.46053 \times 10^{15} \text{ m}$$
$$= 6.324 \times 10^4 \text{ AU}$$

Sidereal year
$$\text{y} = 3.155815 \times 10^7 \text{ s}$$

Mass of sun
$$M_{\text{sun}} = 1.989 \times 10^{30} \text{ kg}$$

Luminosity of sun
$$L_{\text{sun}} = 3.90 \times 10^{26} \text{ W}$$

Solar constant
$$S = 1370 \text{ W/m}^2$$

Radius of sun
$$R_{\text{sun}} = 6.96 \times 10^5 \text{ km}$$

Mass of earth
$$M_{\text{earth}} = 5.9742 \times 10^{24} \text{ kg}$$

Equatorial radius of earth
$$R_{\text{earth}} = 6.37814 \times 10^3 \text{ km}$$

Mass of moon
$$M_{\text{moon}} = 7.348 \times 10^{22} \text{ kg}$$

Radius of moon
$$R_{\text{moon}} = 1.738 \times 10^3 \text{ km}$$

Hubble constant
$$H \approx 20 \text{ km/s/Mly}$$

Nearby Stars

NAME	PARALLAX (arcsec)	DISTANCE [pc (ly)]	SPECTRAL TYPE	PROPER MOTION (arcsec/y)	APPARENT VISUAL MAGNITUDE	ABSOLUTE VISUAL MAGNITUDE
Sun			G2 V		−26.7	4.85
α Cen C (Proxima)	0.772	1.3 (4.2)	M5.5e V	3.85	11.0	15.5
α Cen A	0.750	1.3 (4.3)	G2 V	3.68	−0.01	4.37
α Cen B	0.750	1.3 (4.3)	K1 V	3.68	+1.33	5.71
Barnard's star	0.546	1.8 (6.0)	M3.8 V	10.3	+9.5	13.2
Wolf 359	0.419	2.4 (7.7)	M5.8e V	4.70	+13.5	16.7
BD + 36° 2147 (Lalande 21185)	0.397	2.5 (8.2)	M2.1e V	4.78	+7.5	10.5
Luyten 726-8A	0.387	2.7 (8.8)	M5.6e V	3.36	+12.5	15.5
Luyten 726-8B (UV Ceti)	0.387	2.7 (8.8)	M5.6e V	3.36	+13.0	16.0
Sirius A	0.377	2.6 (8.6)	A1 V	1.33	−1.46	1.42
Sirius B	0.377	2.6 (8.6)	wd	1.33	+8.3	11.2
Ross 154	0.345	2.9 (9.4)	M3.6e V	0.72	+10.5	13.1
Ross 248	0.314	3.2 (10.4)	M4.9e V	1.60	+12.3	14.8
ε Eri	0.303	3.3 (10.8)	K2e V	0.98	+3.73	6.14
Ross 128	0.298	3.3 (10.9)	M4.1 V	1.38	+11.1	13.5
61 Cyg A	0.294	3.4 (11.1)	K3.5e V	5.22	+5.22	7.56
61 Cyg B	0.294	3.4 (11.1)	K4.7e V	5.22	+6.03	8.37
ε Ind	0.291	3.4 (11.2)	K3e V	4.70	+4.68	7.00
Luyten 789-6A	0.290	3.4 (11.2)	M5e V	3.26	12.2	15.0
Luyten 789-6B		3.4 (11.2)	?	3.26	?	15.6
Procyon A	0.285	3.5 (11.4)	F5 IV	1.25	+0.37	2.64
Procyon B	0.285	3.5 (11.4)	wd	1.25	+10.7	13.0
BD + 59° 1915A (S 2398 A)	0.282	3.6 (11.6)	M3 V	2.29	+8.9	11.2
BD + 59° 1915B (S 2398 B)	0.282	3.6 (11.6)	M3.5 V	2.29	+9.7	11.9
CD −36° 15693	0.279	3.6 (11.7)	M1.3e V	6.90	+7.35	9.58
G 51-15	0.278	3.6 (11.7)	M6.6 V	1.27	+14.8	17.0
τ Ceti	0.273	3.6 (11.8)	G8 V	1.92	+3.5	5.72
BD + 5° 1668	0.266	3.8 (12.3)	M3.7 V	3.77	+9.82	11.9
Luyten 725-32 (YZ Ceti)	0.261	3.8 (12.5)	M4.5e V	1.32	+12.0	14.1
CD −39° 14192 (Lacaille 8760)	0.260	3.8 (12.5)	K5.5e V	3.46	+6.66	8.74
Kapteyn's star	0.256	3.9 (12.7)	M0 V	8.72	+8.84	10.9
Kruger 60 A	0.253	4.0 (12.9)	M4	0.86	+9.85	11.9
Kruger 60 B	0.253	4.0 (12.9)	M5e	0.86	+11.3	13.3

NOTE: An *e* after the spectral type indicates emission lines in the spectrum; *wd* indicates a white dwarf. The apparent and absolute visual magnitude is that over the visual range of the spectrum. These terms, and proper motion, are defined in the Glossary.

The Brightest Stars

STAR	NAME	APPARENT VISUAL MAGNITUDE	SPECTRAL TYPE	ABSOLUTE VISUAL MAGNITUDE	DISTANCE [pc (ly)]	PROPER MOTION (arcsec/y)
α CMa A	Sirius	−1.46	A1 V	+1.4	2.7 (8.7)	1.32
α Car	Canopus	−0.72	A9 II	−2.5	23 (74)	0.034
α Boo	Arcturus	−0.04	K1.5 III	0.2	10 (34)	2.28
α Cen A	Rigil Kentaurus	−0.01	G2 V	+4.4	1.3 (4.3)	3.68
α Lyr	Vega	+0.03	A0 V	+0.6	7.7 (25)	0.35
α Aur AB	Capella	+0.08	G6 III G2 III	−0.4	13 (41)	0.43
β Ori A	Rigel	+0.12	B8 Ia	−8.1	430 (1400)	0.004
α CMi A	Procyon	+0.38	F5 IV−V	+2.7	3.4 (11)	1.25
α Eri	Achernar	+0.46	B3 V	−1.3	21 (69)	0.108
α Ori	Betelgeuse	+0.5	M2 Iab	−7.2	430 (1400)	0.028
β Cen AB	Hadar	+0.63	B1 III	−4.4	98 (320)	0.030
α Aql	Altair	+0.77	A7 V	+2.3	5.1 (16)	0.662
α Tau A	Aldebaran	+0.86	K5 III	−0.3	18 (60)	0.200
α Sco A	Antares	+0.92	M1.5 Ia	−5.2	160 (520)	0.024
α Vir	Spica	+1.0	B1 V	−3.2	67 (220)	0.054
β Gem	Pollux	+1.14	K0 III	+0.7	10.7 (35)	0.629
α PsA	Fomalhaut	+1.16	A3 V	+2.0	6.9 (22.6)	0.373
α Cyg	Deneb	+1.25	A2 Ia	−7.2	460 (1500)	0.005
β Cru	Becrux	+1.28	B0.5 III	−4.7	140 (460)	0.042
α Leo A	Regulus	+1.35	B7 V	−0.3	21 (69)	0.248

Periodic Table of the Elements

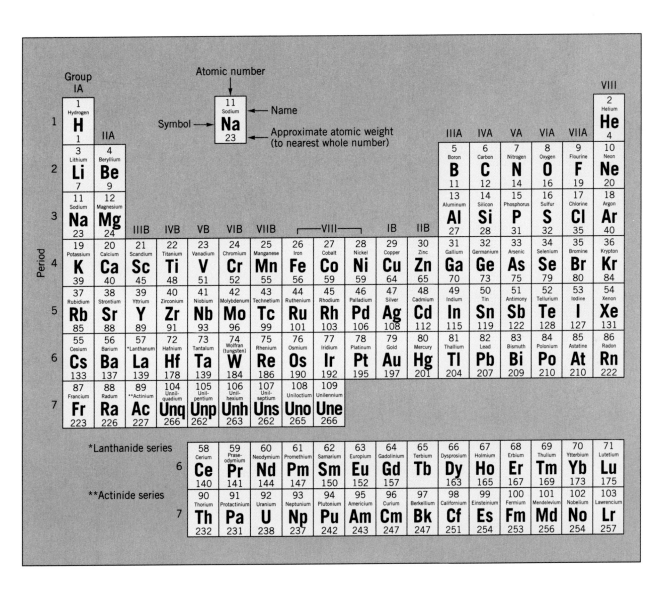

Seasonal Star Charts

NORTH CIRCUMPOLAR SKY

Vega

Draco

"Big Dipper"

Ursa Major

Ursa Minor

Deneb

Polaris

Lynx

Cepheus

Camelopardalis

Lacerta

Auriga   Capella

Cassiopeia

Perseus

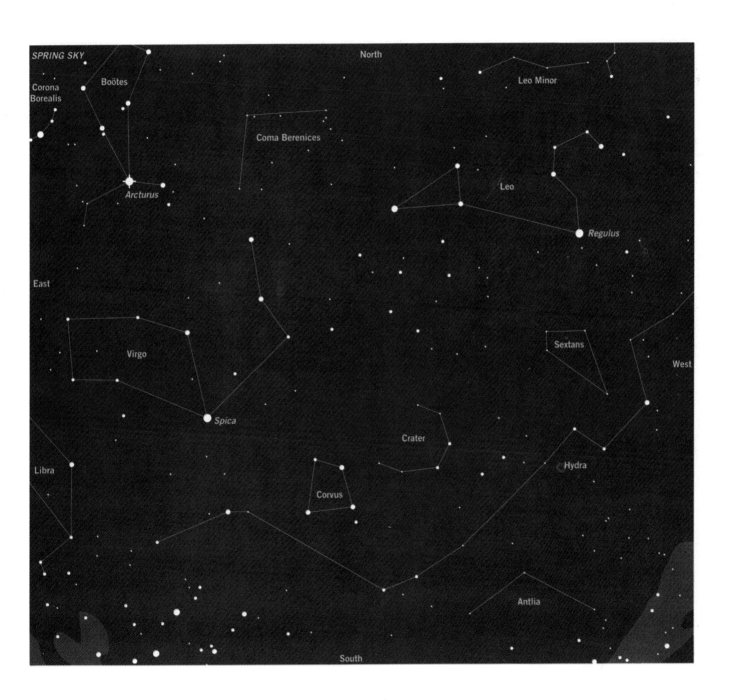

SPRING SKY

North

Corona
Borealis

Boötes

Leo Minor

Coma Berenices

Leo

Arcturus

Regulus

East

Sextans

West

Virgo

Spica

Crater

Hydra

Libra

Corvus

Antlia

South

AUTUMN SKY

North

Triangulum

Aries

Pegasus

Pisces

Equuleus

East

West

Mira

Aquarius

Cetus

Eridanus

Capricornus

Fomalhaut

Piscis Austrinus

Sculptor

Fornax

Microscopium

Grus

South

Alnair

Seasonal Star Charts (continued)

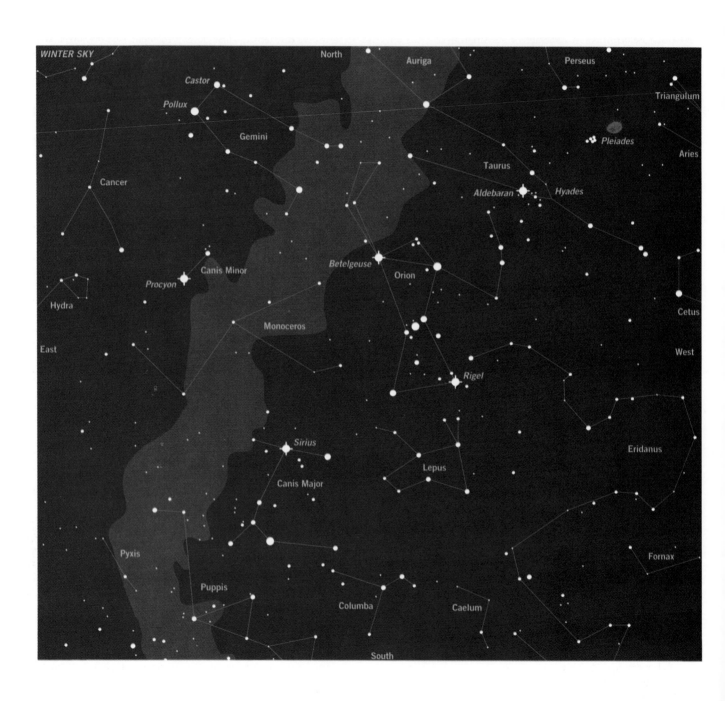

Expanded Glossary

absolute magnitude A measure of the brightness a star would have if it were to be placed at a standard distance of 10 parsecs (32.6 light years) from the sun.

absorption (dark) lines Discrete colors missing in a continuous spectrum because of the absorption of those colors by atoms.

absorption-line spectrum Dark lines superimposed on a continuous spectrum.

acceleration The rate of change of velocity with time.

accretion The colliding and sticking together of small particles to make larger masses.

accretion disk A disk made by infalling material around a mass; the conservation of angular momentum results in the disk shape.

active galactic nuclei (AGNs) The nuclei of galaxies that have a nonthermal continuous spectrum over a wide range of wavelengths and signs of unusual, energetic activity such as radio jets; include Seyferts, BL Lac objects, and quasars.

active galaxies Galaxies characterized by a nonthermal spectrum and a large energy output compared to a normal galaxy.

active regions Areas on the sun (and other stars), in which magnetic fields are concentrated; these generate sunspots and flares.

aerobe An organism that is dependent on free oxygen for its metabolism.

albedo A measure of an object's reflecting power; the ratio of reflected light to incoming light for a solid surface, where complete reflection gives an albedo of 1.0 or 100 percent.

almanac A set of tables giving positions of the sun, moon, and planets at various times, as well as other astronomical information; an ephemeris.

Alpha Centauri The closest star to the sun; a triple-star system; the component Alpha Centauri A has almost the same luminosity, mass, and surface temperature as the sun.

alpha particle A helium nucleus emitted in the radioactive decay of heavy elements.

altitude In the horizon system, the angle along a vertical circle from the horizon to a point on the celestial sphere.

Amalthea A small asteroidal moon of Jupiter; elongated and irregular in shape with a reddish, cratered surface.

amino acids The building blocks of proteins, consisting mostly of carbon, nitrogen, oxygen, and hydrogen atoms.

Andromeda Galaxy (M31) The closest spiral galaxy to the Milky Way Galaxy at a distance of 680 kpc (2.2 Mly); it has a diameter of about 50 kpc (160 kly).

anaerobe An organism that does not depend on free oxygen for its metabolism.

angstrom (Å) A unit of length, equal to 10^{-12} m, often used to measure the wavelength of visible light.

angular diameter The apparent diameter of an object in angular measure; the angular separation of two points on opposite sides of the object.

angular distance The apparent angular spacing between two objects in the sky.

angular momentum The tendency for bodies, because of their inertia, to keep spinning or orbiting.

angular separation The observed angular distance between two celestial objects, measured in degrees, minutes, and seconds of angular measure.

angular speed The rate of change of angular position of a celestial object viewed in the sky.

anorthosite A basaltic mineral composed of calcium and sodium with aluminum silicate; the predominant mineral of the lunar highlands.

antapex The direction in the sky from which the sun appears to be moving relative to local stars; located in the constellation Columba.

antimatter Elementary particles having their electric charge or other property reversed compared to ordinary matter.

apex The direction in the sky toward which the sun appears to be moving relative to local stars; located in the constellation Hercules.

aphelion The point on the orbit of a body orbiting the sun, that is farthest from the sun.

Aphrodite Terra A large highland region on Venus.

apogee The point in its orbit at which an earth satellite is farthest from the earth.

apparent magnitude The brightness of a star (or any other celestial object) as seen from the earth; an astronomical measure of the object's flux.

arroyo A dry channel carved in the ground by sporadic water flows.

asterism A conspicuous pattern of stars with a popular name that does not make up a constellation; the Big Dipper is an example.

asteroid Minor planet; one of several thousand very small members of the solar system that revolve around the sun, generally between the orbits of Mars and Jupiter.

asteroid belt The region lying between the orbits of Mars and Jupiter, containing the majority of asteroids.

astronomical unit (AU) The average distance between the earth and the sun; 149.6×10^6 km or 8.3 light minutes.

astrophysical jets Collimated beams of material (usually ions and electrons) expelled from astrophysical objects, such as the nuclei of active galaxies.

astrophysics Physical concepts applied to astronomical objects.

asymptotic giant branch (AGB) The path on the H–R diagram that a star traverses during the phase of evolution when it has both a hydrogen-burning and a helium-burning shell, after its main-sequence phase.

atmosphere A gaseous envelope surrounding a planet, or the visible layers of a star; also a unit of pressure (abbreviated *atm*) equal to the pressure of air at sea level on the earth's surface.

atmospheric absorption or extinction The decrease in light caused by passage through the atmosphere.

atmospheric escape The process by which particles at the exosphere with greater than escape speed and unhindered by collisions leave a planet.

atmospheric reddening Preferential scattering of blue light over red by air particles, which results in an object, such as the setting sun, appearing to be redder than it actually is.

atom The smallest particle of an element that exhibits the chemical properties of the element.

AU Abbreviation for astronomical unit(s).

aurora Light emission from atmospheric atoms and molecules excited by collisions with energetic, charged particles from the magnetosphere.

autumnal equinox The fall equinox; see equinox.

axis One of two or more reference lines in a coordinate system; also, the straight line, through the poles, about which a body rotates.

azimuth Angular position along the horizon, measured clockwise from north.

Balmer jump The large change in opacity at 3650 Å due to bound–free transitions from the second energy level of hydrogen atoms.

Balmer series The set of transitions of electrons in a hydrogen atom between the second energy level and higher levels; also the set of absorption or emission lines corresponding to these transitions that lies in the visible part of the spectrum, the first of which is the hydrogen-alpha (Hα) line.

bandwidth The range of frequencies detected simultaneously by a radio receiver (or other electromagnetic sensor).

barred spirals A subclass of spiral galaxies that have a bar across the nuclear region.

baryon Name for group of elementary particles, usually of large mass, such as protons and neutrons.

basalt An igneous rock, composed of olivine and feldspar, that makes up much of the earth's lower crust.

basins Large, shallow lowland areas in the crusts of terrestrial planets created by asteroidal impact or plate tectonics.

Becklin–Neugebauer object An infrared source associated with the Orion Nebula; probably a very young, massive star.

belts Regions of downflow, hence low pressure, in the atmosphere of a Jovian planet.

beta decay A process of radioactive decay in which a neutron disintegrates into a proton, an electron, and a neutrino.

beta particle An electron or positron emitted by a nucleus in radioactive decay.

Beta Regio A highland region on Venus containing at least two large shield volcanoes.

Betelgeuse A red supergiant star having a luminosity roughly 8000 times that of the sun.

Big Bang model A picture of the evolution of the universe that postulates its origin, in an event called the Big Bang, from a hot, dense state that rapidly expanded to cooler, less dense states.

binary accretion model A model for the origin of the moon in which the moon and the earth formed by accretion of material from the same cloud of gas and dust.

binary galaxies Two galaxies bound by gravity and orbiting a common center of mass.

binary stars Two stars bound together by gravity that revolve around a common center of mass.

binary x-ray source A binary system containing an x-ray emitter, which is usually a collapsed object surrounded by a hot accretion disk giving off x-rays.

biological evolution The natural development over time from simple to complex organisms, generated by mutations that change the gene structure and directed by natural selection of individuals well-adapted to the environment.

bipolar outflows High-speed outflows of gas in opposite directions from a young stellar object (YSO); the result of a magnetized accretion disk around the YSO, collimating its stellar wind.

blackbody A (hypothetical) perfect radiator of light that absorbs and reemits all radiation incident upon it; its light output depends only on its temperature.

blackbody spectrum The continuous spectrum emitted by a blackbody; the flux at each wavelength is given by the formula known as Planck's law.

black dwarf The cold remains of a white dwarf after all its thermal energy has been exhausted.

black hole A mass that has collapsed to such a small radius that the escape speed from its surface is greater than the speed of light; thus light is trapped by the intense gravitational field.

BL Lacertae (BL Lac) objects A type of active galaxy whose nonthermal emission from the nucleus varies rapidly (one day or so) and is highly polarized.

blueshift A decrease in the wavelength of the radiation emitted by an approaching celestial body as a consequence of the Doppler effect; a shift toward the short-wavelength (blue) end of the spectrum.

Bohr model of the atom A simple picture of atomic structure in which electrons have well-defined orbits (energy levels) about the nucleus of the atom.

bolometer A detector of infrared radiation, usually a small chip of semiconductor material cooled to a few kelvins; absorption of infrared radiation causes a change in its resistance, which can be measured in an electronic circuit.

bolometric magnitude The magnitude of an object measured over all wavelengths of the electromagnetic spectrum.

bolometric luminosity The total energy output per second at all wavelengths emitted by an astronomical object.

Boltzmann's constant The number that relates pressure and temperature, or kinetic energy and temperature in a gas; the gas constant per molecule (see Appendix C).

bound–bound transition A transition of an electron between two bound energy states of an atom or ion.

bound–free transition A transition of an electron between a bound and an unbound (free) state.

bow shock wave The shock wave created by the interaction of the solar wind with a planetary magnetosphere.

breccias Rock and mineral fragments cemented together; a common part of the lunar surface.

bright-line spectrum See emission-line spectrum.

bright nebula See diffuse nebula.

brightness An ambiguous term, usually meaning the energy per unit area received from or emitted by an object—its flux—but sometimes used to refer to an object's luminosity.

brown dwarf A very-low-mass object (roughly 0.05 solar mass) of low temperature and luminosity; its core never becomes hot enough to ignite thermonuclear reactions.

C-type asteroids Dark asteroids with low albedos (around 0.04), which probably contain a large percentage of carbon materials.

Callisto Second largest of the Galilean moons of Jupiter and the farthest out; has a density about twice that of water; a heavily cratered, icy surface with multiringed basins.

Caloris basin A large multiringed basin on Mercury, surrounded by a mountain range, located near a "hot" longitude of 180°.

canali Italian term for "channels" term used by Giovanni Schiaparelli to describe dark linear features seen on the surface of Mars.

capture model A model of the moon's origin proposed about 1955 that pictures the moon as having been captured by the earth's gravity, after which it spiraled in toward the earth, reversed orbital direction, and spiraled outward.

carbon–nitrogen–oxygen (CNO) cycle A series of thermonuclear reactions taking place in a star's core in which carbon, nitrogen, and oxygen aid the fusion of hydrogen into helium; in the sun it is a secondary process of energy production, but in high-mass, main-sequence stars it is the major process.

carbonaceous chondrites A class of meteorites that contain chondrules embedded in a matrix material with a large percentage of carbon (about 4 percent).

Carina arm A local segment of the spiral arms of the Milky Way Galaxy, located inward from the sun toward the constellation Carina.

Cassegrain reflector The design for reflecting telescopes in which the secondary mirror directs the beam to a focus through a hole in the center of the primary mirror.

Cassini's division A gap about 2000 km wide in Saturn's rings, discovered in 1675 by Giovanni Cassini; now known to contain many small ringlets.

catastrophic models Models for the origin of the solar system in which an improbable event involving a large mass (usually collision with another star) led to the expulsion and then to the collection of gaseous materials that became the planets.

causally connected A description of events in spacetime that can be affected by a given event; all the events within the light cone of a given event are causally connected to it.

cD galaxies See supergiant elliptical galaxies.

celestial coordinate system A system for specifying positions on the celestial sphere, similar to the system of latitude and longitude used to specify positions on the earth.

celestial equator An imaginary projection of the earth's equator onto the celestial sphere; declination is zero along the celestial equator.

celestial meridian See meridian.

celestial pole An imaginary projection of the earth's pole (north or south) onto the celestial sphere; a point about which the apparent daily rotation of the stars takes place.

celestial sphere An imaginary sphere of very large radius, centered on the earth, on which the celestial bodies appear fastened and against which their motions are charted.

cell Most common form of a living organism on the earth.

Centaurus arm A spiral arm of the Milky Way Galaxy located about 12,000 ly away in the direction of the constellation Centaurus.

Centaurus X-3 A binary x-ray source in the constellation Centaurus; contains a low-mass neutron star (the x-ray source) orbiting a blue giant star.

center of mass The balance point of a set of interacting or connected bodies.

central force A force directed along a line connecting the centers of two objects.

centripetal acceleration The acceleration of a body toward the center of a circular path.

centripetal force A force required to divert a body from a straight path into a curved one, directed toward the center of the curve.

cepheid variables (cepheids) Stars that vary in brightness as a result of a regular variation in size and temperature; a class of variable stars for which the star Delta Cephei is the prototype.

Ceres The first observed asteroid, discovered by Father Giuseppe Piazzi in 1801.

CETI Acronym for communication with extraterrestrial intelligence.

Chandrasekhar limit The maximum mass for a white dwarf star (about 1.4 solar masses); this amount leads to the highest density and smallest radius for a star made of a degenerate electron gas; if more than this mass is present, the star cannot be supported by the electron gas pressure and so collapses gravitationally.

charge-coupled device (CCD) A small chip of semiconductor material that emits electrons when it absorbs light; the electrons are trapped in small regions called pixels; the pattern of charges is read out in a way to preserve the image striking the chip.

Charon Moon of Pluto, with a diameter of about 1200 km.

chemical condensation sequence The sequence of chemical reactions and condensation of solids that occur in a low-density gas as it cools at specific densities and temperatures.

chemical evolution The natural development from simple to complex molecules, such as proteins and nucleic acids, which are the building blocks of life.

chondrite A stony meteorite characterized by the presence of small, round silicate granules (chondrules).

chondrules Round silicate granules lacking volatile elements; found in chondritic meteorites, or chondrites, they are believed to be primitive solar system materials.

chromosphere The part of the sun's atmosphere just above the photosphere; hotter and less dense than the photosphere; it creates the flash spectrum seen during eclipses.

circular velocity The speed at which an object must travel to maintain uniform circular motion around a gravitating body.

circumpolar stars For an observer north of the equator, the stars that are continually above the northern horizon and never set; for a southern observer, the stars that never set below the southern horizon.

closed geometry See spherical geometry.

cluster of galaxies A grouping of galaxies containing a few to thousands of galaxies.

CNO cycle See carbon–nitrogen–oxygen cycle.

collisional de-excitation Loss of energy by an electron of an atom in a collision so that the electron drops to a lower energy level.

collisional excitation Forcing an electron of an atom to a higher energy level by a collision.

color excess The difference between the actual color (of a star) and its observed color; usually redder because of interstellar dust along the line of sight.

color index The difference in the magnitudes of an object measured at two different wavelengths; a measure of the color, hence the temperature, of a star.

color temperature Temperature inferred from color usually by fitting a Planck curve to the continuous spectrum of a star at two wavelengths).

coma The bright, visible head of a comet.

Coma cluster A cluster of galaxies located in the sky in the direction of the constellation Coma Berenices; it contains thousands of galaxies spread out over millions of light years.

comets Bodies of small mass that revolve around the sun, usually in highly elliptical orbits; in the dirty snowball model, comets consist of small, solid particles (probably of rocky material) embedded in frozen gases.

compact (radio) galaxies Active galaxies whose nucleus contains a small, strong radio source.

compound A substance composed of the atoms of two or more elements bound together by chemical forces.

condensation The growth of small particles by the sticking together of atoms and molecules.

condensation sequence See chemical condensation sequence.

conduction Transfer of thermal energy by particles colliding into one another.

conjunction The time at which two celestial objects appear closest together in the sky.

conservation of angular momentum The principle stating that with no applied torques, the total angular momentum of an isolated system is constant.

conservation of energy A fundamental principle in physics that states that the total energy of an isolated system remains constant regardless of internal changes that may occur.

conservation of magnetic flux The physical principle stating that under certain circumstances, the number of magnetic field lines passing through an area remains constant even as the area changes in size.

conservation of momentum The physical principle stating that with no outside net forces, the total momentum of a system is constant.

constellation An apparent arrangement of stars on the celestial sphere, usually named after an ancient god, hero, animal, or mythological being; now an agreed-upon region of the sky containing a group of stars.

continental drift The theory that the present continents were at one time a unified landmass that broke up, such that the fragments drifted apart.

continuous spectrum A spectrum showing emission at all wavelengths, unbroken by either absorption lines or emission lines.

contour map A diagram showing how the intensity of some kind of radiation varies over a region of the sky; lines on such a map connect points of equal flux; closely spaced lines mean that the flux changes rapidly over a small distance, widely spaced lines mean it changes more slowly.

convection The transfer of energy by the moving currents of a fluid.

coplanar Lying in the same plane.

core (of the earth) The central region of the earth; it has a high density, is probably liquid, and is believed to be composed of iron and iron alloys.

core (of the Galaxy) The inner few light years of the nucleus, which contains a small, nonthermal radio source and fast-moving clouds of ionized gas.

core (of the sun) The inner 25 percent of the sun's radius, where the temperature is great enough for thermonuclear reactions to take place.

core–mantle grains Interstellar dust particles with cores of dense materials (such as silicates) surrounded by a mantle of icy materials (such as water).

corona The outermost region of the sun's atmosphere, consisting of thin, ionized gases at a temperature of about 1×10^6 K; large magnetic loops give the corona its structure.

coronal holes Regions in the sun's corona that lack a concentration of high-temperature plasma; here magnetic field lines extend out into interplanetary space and mark the source of the solar wind.

coronal interstellar gas High-temperature interstellar plasma made visible by its x-ray emission.

cosmic blackbody microwave radiation Radiation with a blackbody spectrum at a temperature of about 2.7 K; the remains of the primeval fireball in which the universe was created; this radiation permeates the universe, offering an observational confirmation of the Big Bang model.

cosmic rays Charged atomic particles moving in space with very high energies (the particles travel close to the speed of light); most originate beyond the solar system, but some low-energy cosmic rays are produced in solar flares.

cosmological principle The statement that the universe, averaged over a large enough volume, appears the same from any location.

cosmology The study of the nature and evolution of the physical universe.

cosmos The universe considered as an orderly and harmonious system.

Crab Nebula A supernova remnant, located in the constellation Taurus, produced by the supernova explosion visible from earth in A.D. 1054; a pulsar in the nebula marks the corpse of the exploded star.

crater A circular depression of any size, usually caused by the impact of a solid body or by a surface eruption.

cratered terrain Landscape with an abundance of craters, which implies that it is old and unevolved.

crescent The phase during which a moon or planet has less than half of its visible surface illuminated.

critical density In cosmology, the density that marks the transition from an open to a closed universe; the critical density that provides enough gravity to just bring the expansion to a stop after infinite time.

crust The thin, outermost surface layer of a planet; on the earth, it is composed of basaltic and granitic rocks.

curvature of spacetime See spacetime curvature.

cyclone Spiral flows in a planet's atmosphere produced by the planet's rotation.

Cygnus arm A segment of one of the spiral arms in the outer part of our Galaxy, about 14 kpc (45 kly) from the center.

Cygnus X-1 A binary x-ray source in the constellation Cygnus; it contains a probable black hole orbiting a blue supergiant star.

dark cloud An interstellar cloud of gas and dust that contains enough dust to blot out the light of stars behind it (as seen from the earth).

dark-line spectrum See absorption-line spectrum.

dark matter Matter in an unknown form detectable by its gravitation effects on luminous matter.

declination In the equatorial coordinate system, the angle measured north or south along an hour circle from the celestial equator to a point on the celestial sphere.

decoupling The time in the universe's history when the density and opacity became so low that matter and light stopped interacting.

deferent An ancient geometric device used to account for the apparent eastward motion of the planets; a large circle, usually centered on the earth, that carries around a planet's epicycle.

degenerate electron gas An ionized gas in which nuclei and electrons are packed together as much as possible, filling all possible low-energy states, so that the perfect ideal gas law relating pressure, temperature, and density no longer applies.

degenerate gas pressure A force exerted by degenerate matter that depends mostly on how dense the matter is and very little on its temperature.

degenerate neutron gas Matter made up of neutrons packed together as tightly as possible.

Deimos The smaller of the two moons of Mars; it has a dark, cratered surface.

density The amount of mass per volume in an object or region of space.

density-wave model A model for the generation of spiral structure in galaxies in which spiral density waves (similar to sound waves) are pictured as plowing through the interstellar matter and sparking star formation.

deoxyribonucleic acid (DNA) The basic genetic material of life as we know it; the DNA molecule is very large, consisting of subunits called nucleotides.

detector Any device, sensitive to electromagnetic radiation, placed at the focus of a telescope.

differential rotation The tendency of a fluid, spherical body to rotate faster at the equator than at the poles.

differentiated Layered interior of a planet or a moon, generally with the less dense materials atop more dense ones.

diffraction The spreading out of light waves as they pass the edge of an opaque body.

diffuse (bright) nebula A cloud of hot ionized gas, mostly hydrogen, with an emission-line spectrum.

dipole field A magnetic field configuration like that of a bar magnet with opposed north and south poles.

dirty snowball comet model A model for comets that pictures the nucleus as a compact solid body of frozen materials, mixed with pieces of rocky matter, that turn into gases as a comet nears the sun, creating the head and tail.

disk (of a galaxy) The flattened wheel of stars, gas, and dust outside the nucleus of a spiral galaxy.

dispersion The effect that causes pulses of radiation emitted simultaneously at different frequencies to arrive at different times after traversing the interstellar medium.

D lines A pair of dark lines in the yellow region of the spectrum, produced by sodium.

DNA See deoxyribonucleic acid.

Doppler shift A change in the wavelength of waves from a source reaching an observer when the source and the observer are moving with respect to each other along the line of sight; the wavelength increases (redshift) or decreases (blueshift) according to whether the motion is away from or toward the observer.

double quasar An optical illusion, produced on the sky by a gravitational-lens effect, of two adjacent images of a single quasar.

Drake equation A relationship of the important factors for estimating the number of civilizations in the Galaxy now, devised by radio astronomer Frank Drake.

dust tail The part of a comet's tail containing dust particles, pushed out by radiation pressure from sunlight.

dwarf A star of relatively low light output and relatively small size; a main-sequence star of luminosity class V.

dynamo model A model for the generation of a planet's (or star's) magnetic field by the circulation of conducting fluids in its core (or corrective zone).

Earth Our home planet; model for the other terrestrial planets: the moon, Mercury, Venus, and Mars.

earthshine A faint illumination of the dark (night) regions of the moon facing the earth; an effect especially noticeable when the moon is a thin crescent; arises from sunlight reflected from the earth onto the moon.

eccentric An ancient geometric device used to account for nonuniform planetary motion; a point offset from the center of circular motion.

eccentricity The ratio of the distance of a focus from the center of an ellipse to its semimajor axis.

eclipse The phenomenon of one body passing in front of another, cutting off its light.

eclipsing binary system Two stars that revolve around a common center of mass, the orbits lying edge-on to the line of sight, so that each star periodically passes in front of the other to cause a decrease in the amount of light.

ecliptic From the earth, the apparent yearly path on the celestial sphere of the sun with respect to the stars; also, the plane of the earth's orbit.

ecosphere The region around a star in which an orbiting planet's surface temperature is suitable for life as we know it.

Eddington luminosity For a given mass of an object, the luminosity that would just provide sufficient radiation pressure on surrounding material to prevent gravity from pulling it in.

Eddington mass For a given luminosity of an object, the mass that would produce enough gravity to just overcome radiation pressure on surrounding material and allow it to fall inward.

effective temperature The temperature a body would have if it were a blackbody of the same size radiating the same luminosity.

Einstein X-Ray Observatory An earth-orbiting x-ray telescope, with a 58-cm objective, launched in 1978; the telescope ceased functioning in 1981, but observational data are contained in archives that can be examined now.

Einstein ring An optical illusion created by highly symmetrical gravitational imaging of a radio source.

electromagnetic radiation A self-propagating electric and magnetic wave, such as light, radio, ultraviolet, or infrared radiation; all types travel at the same speed and differ in wavelength or frequency, which is a function of the energy.

electromagnetic spectrum The range of all wavelengths of electromagnetic radiation.

electron A lightweight, negatively charged subatomic particle.

electron volt (eV) A convenient unit of energy for atomic physics; the energy gain of an electron when it is accelerated by one volt; $1 \text{ eV} = 1.60 \times 10^{-19}$ J.

electroweak force The combination of the weak and electromagnetic forces at temperatures greater than 10^{15} K.

element A substance that is made of atoms having the same chemical properties and cannot be decomposed chemically into simpler substances.

ellipse A plane curve drawn so that the sum of the distances from a point on the curve to two fixed points is constant.

elliptical galaxy A gravitationally bound system of stars that has rotational symmetry but no spiral structure and contains mainly old stars, with little gas or dust.

elongation The angular separation of an object from the sun as seen in the sky.

emission (bright) lines Light of specific wavelengths or colors emitted by atoms; sharp energy peaks in a spectrum caused by downward electron transitions from a discrete quantum state to another discrete state.

emission-line spectrum A spectrum containing only emission lines.

emission nebula See diffuse nebula; a hot cloud of mostly hydrogen gas whose visible spectrum is dominated by emission lines.

empirical Derived from experiment or observation.

energy The ability to do work.

energy level One of the possible quantum states of an atom, with a specific value of energy.

enzyme A protein that brings about or accelerates reactions at body temperatures without itself undergoing destruction.

ephemeris (pl., ephemerides) A table that gives the positions of celestial objects at various times.

epicycle A small circle whose center lies on a larger one (the deferent), used by ancient astronomers, such as Ptolemy, to account for the westward retrograde motion and other irregular motions of the planets.

epoch The time for which the stellar positions (which change as a result of precession) given in an almanac or ephemeris are calculated.

equant An ancient geometrical device invented by Ptolemy to account for variations in planetary motion; essentially an eccentric in which the center of the circle is *not* the center of uniform motion.

equation of state A relationship that describes the conditions in a gas, such as an equation relating the pressure, temperature, and density of a gas.

equatorial bulge The excess diameter, about 43 km, of the earth through its equator compared with the diameter through its poles; any planet will have a large equatorial bulge if it has a fluid interior and rotates rapidly.

equatorial system A celestial coordinate system based on the celestial equator and the celestial poles, in which two angles, right ascension and declination, are used to specify the position of a point on the celestial sphere.

equilibrium A state of a physical system in which there is no overall change.

equilibrium temperature The temperature that has been achieved when a body emits the same energy per second that it absorbs.

equinox Time of year of equal length of day and night; the two times of the year when the sun crosses the celestial equator; spring (vernal) equinox occurs about March 21, and fall (autumnal) equinox about September 22.

escape speed The speed a body must achieve to break away from the gravity of another body and never return to it.

escape temperature The temperature that particles in an atmosphere need to enter space with the escape velocity.

Euclidean (flat) geometry The geometry in which only one parallel line can be drawn through a point near another line; the sum of the angles in a triangle drawn on a flat surface is always 180°.

Europa The smallest of the Galilean moons of Jupiter; has a density about three times that of water; its icy, cracked surface is devoid of impact craters.

event A point in four-dimensional spacetime.

evolution Changes in the physical properties of a body with time.

evolutionary track On a temperature–luminosity (H–R) diagram, the path made by the points that describe how the temperature and luminosity of a star change with time.

excitation The process of raising an atom to a higher energy level.

exosphere The topmost region of a planet's atmosphere, from which particles in the atmosphere can escape into space.

expanding (3-kpc) arm A segment of spiral-arm structure encircling the center of our Galaxy at a distance of about 3 kpc (10 kly); it appears to be moving toward us and away from the Galaxy's center.

expansion of the universe The expansion of spacetime, as indicated by markers such as the distances between galaxies; believed to be a result of the Big Bang.

exponential decay A process, such as radioactive decay, for which the rate of change is directly proportional to the quantity present, so that the quantity remaining is given by an exponential function of time.

extended radio galaxies Active galaxies that show extended radio emission, usually in the form of two lobes on either side of the nucleus.

extinction The dimming of light when it passes through some medium, such as the earth's atmosphere or interstellar material.

extragalactic Outside the Milky Way Galaxy.

eyepiece A magnifying lens used to view the image produced by the main light-gathering lens of a telescope.

f-ratio The ratio of the focal length of a lens or mirror to its diameter.

first dredge-up Convection acting to bring to a star's surface material processed by hydrogen burning during the star's first episode as a red giant.

first quarter The phase of the moon a quarter of the way around its orbit (eastward) from new moon, when it looks half-illuminated when viewed from the earth.

fission See nuclear fission.

fission model The earliest of the major models for the origin of the moon, suggesting that a young, rapidly spinning, molten earth lost a piece that spiraled out into orbit and cooled down to form the moon.

flash spectrum The spectrum that appears immediately before the totality of a solar eclipse as the normal absorption spectrum is replaced briefly by the chromosphere's own emission spectrum.

flat geometry See Euclidean geometry.

flatness problem In the Big Bang model, the fact that the geometry of the universe is flat (or very close to it) and remains flat as the universe evolves.

flocculent galaxies Spiral galaxies that show a puffy structure in their disks rather than well-defined spiral arms.

fluorescence The process by which a high-energy photon is absorbed by an atom and reemitted as two or more photons of lower energy.

flux The amount of energy flowing through a given area in a given time.

focal length The distance from a lens (or mirror) to the point at which it brings light to a focus for a distant object.

focus (pl., foci) The point at which light is gathered in a telescope.

forbidden line An emission line from an atom produced by a transition with a low probability of occurrence.

force A push or pull; a measure of the strength of a physical interaction.

forced motion Any motion under the action of a net force.

frame of reference A set of axes with respect to which the position or motion of something can be described or physical laws can be formulated.

Fraunhofer lines The name given to strongest absorption lines in the spectrum of a star, especially the sun.

free–bound transition A transition of an electron between a free energy state and one bound to an atom; results in the atom adding an electron and emitting a photon; the reverse process is a bound–free transition.

free-fall Gravitational collapse under the condition of no resisting internal pressure caused by collisions among the particles.

free-fall collapse time The time for a cloud to collapse gravitationally under free-fall conditions; uniquely a function of the initial density of the cloud.

free–free absorption or emission Photons absorbed or emitted when electrons and ions interact without the capture of the electron by the ion.

free–free transition A transition of an electron between two different free states; if energy is lost by the electron, the process is free–free emission; if gained, it is free–free absorption.

frequency The number of waves that pass a particular point in some time interval (usually a second); usually given in units of hertz. One hertz (Hz) is one cycle per second.

full moon The lunar phase during which the moon is opposite the sun in the sky and looks fully illuminated as seen from the earth.

fusion See nuclear fusion.

G-band Absorption band in stellar spectra caused by metals.

galactic cannibalism A model for galaxy interaction in which more massive galaxies strip material, by tidal forces, from less massive galaxies.

galactic center (core) The innermost part of the Galaxy's nuclear bulge; contains a cluster of stars, ionized gas clouds, and a rotating ring of material.

galactic (open) cluster A small group (about ten to a few hundred) of gravitationally bound stars of Population I, found in or near the plane of the Galaxy.

galactic equator The great circle along the line of the Milky Way, marking the central plane of the Galaxy.

galactic latitude The angular distance north or south of the galactic equator.

galactic longitude The angular distance along the galactic equator from a zero point in the direction of the galactic center.

galactic rotation curve A description of how fast an object some distance from the center of a galaxy revolves around it; the orbital speed as a function of distance.

galaxy A huge assembly of stars (between a million and hundreds of billions), plus gas and dust, that is held together by gravity; the Milky Way Galaxy, our own galaxy, containing the sun, is designated the Galaxy.

Galilean moons The four largest satellites of Jupiter (Io, Europa, Ganymede, Callisto), discovered by Galileo with his telescope.

gamma ray A very-high-energy photon with a wavelength (10^{-12} m) shorter than that of x-rays.

gamma-ray bursters Astronomical sources of short bursts of gamma rays; thought to originate from highly magnetic neutron stars.

Ganymede The largest of the Galilean moons of Jupiter, with a density about 1.9 times that of water; has both heavily cratered and faulted terrain.

gas tail The part of a comet's tail that consist of ions and molecules; it is shaped by interaction with the solar wind.

gauss (G) A physical unit measuring magnetic field strength.

general theory of relativity The idea developed by Albert Einstein that mass and energy determine the geometry of spacetime and that any curvature of spacetime shows itself by what we commonly call gravitational forces; Einstein's theory of gravity.

geocentric Centered on the earth.

geomagnetic axis The axis that connects the earth's magnetic poles; it is inclined about 12° from the geographic spin axis and does not pass through the earth's center.

giant stars of luminosity class III; typically have radii between 10 and 100 times the sun's.

giant impact model A scenario for the moon's origin in which a Mars-sized object strikes the young earth with a glancing blow; materials from the colliding object and the earth form a disk around the earth out of which the moon accretes.

giant molecular clouds Large interstellar clouds, with sizes up to tens of light years and containing 100,000 solar masses of material; found in the spiral arms of the Galaxy, giant molecular clouds are the sites of massive star formation.

gibbous The phase of the moon occurring between first quarter and full moon, when the moon is more than half-illuminated as seen from the earth.

globular cluster A gravitationally bound group of about 10^5 to 10^6 Population II stars (of roughly solar mass), symmetrically shaped, found in the halo of the Galaxy and orbiting the galactic center.

gnomon An ancient instrument for measuring time; most simply, a stick stuck vertically into the ground whose shadow is used to indicate the sun's position with respect to the horizon.

grand design spirals Spiral galaxies that have a well-defined spiral-arm structure.

grand unification theories (GUTs) Physical theories that attempt to unite the elementary particles and the four forces in nature as the actions of one particle and one force.

granite The type of rock making up the continental regions of the earth's crust.

granules Hot regions that appear briefly (3 to 10 min) as a rough texture on the solar photosphere; they mark the tops of areas of convective upflow from the convection zone.

gravitation In Newtonian terms, a force between masses that is characterized by their acceleration toward each other; the size of the force depends directly on the product of the masses and inversely on the square of the distance between them; in Einstein's terms, the curvature of spacetime.

gravitational bending of light The effect of gravity on the usually straight path of a photon.

gravitational collapse The unhindered contraction of any mass from its own gravity; internal pressure can slow down a collapse to *gravitational contraction*.

gravitational field The property of space having the potential for producing gravitational force on objects within it; characterized by the acceleration of free masses.

gravitational focusing The directing of the motions of small masses by a larger one so that their paths cross, which enhances their accretion onto the larger mass.

gravitational instability The tendency for a disturbed region in a gas to undergo gravitational collapse.

gravitational-lens effect The bending effect of a large mass on light rays, which causes them to form an imperfect image of the source of light.

gravitational mass The mass of an object as determined by the gravitational force it exerts on another object.

gravitational potential energy Potential energy related to a body's position in a gravitational field.

gravitational redshift The change to longer wavelengths that marks the loss of energy by a photon that moves from a stronger to a weaker gravitational field.

great circle Shortest distance between two points on a sphere.

Great Red Spot A large, long-lived, high-pressure storm in Jupiter's atmosphere.

greenhouse effect The production of an increased equilibrium temperature at the surface of a planet that occurs when the opacity of its atmosphere in the infrared results in the trapping of outgoing heat radiation; increasing the opacity will likely increase the equilibrium temperature.

ground state the lowest energy level of an atom.

GUTs See grand unification theories.

H I region A region of neutral hydrogen in interstellar space; emits photons at a wavelength of 21 cm.

H II region A zone of hot, ionized hydrogen in interstellar space; it usually forms a bright nebula around a hot, young star or cluster of hot stars.

habitable zone See ecosphere.

Hadley cells Large regions of convective flow in the earth's or any planet's atmosphere.

half-life The time required for half the radioactive atoms in a sample to disintegrate by fission.

Halley's comet The periodic comet (orbital period about 76 years) whose orbit was first worked out by Edmond Halley from Newton's laws; has a small (about 10 km diameter), dark, irregular nucleus and emits jets of gas and dust.

halo (of a galaxy) The spherical region around a galaxy, not including the disk or the nucleus, containing globular clusters, some gas, a few stray stars, and an invisible form of matter.

Hα line The first or alpha (α) line of the Balmer series, the set of transitions in a hydrogen atom between the second energy level and levels with higher energy; it lies in the red part of the visible spectrum.

head–tail (radio) galaxy An active galaxy whose radio lobes have been swept back to form a tail because of interaction with the surrounding medium.

heat The total random energy of motion of particles in an object; thermal energy.

heavy-particle era In the hot Big Bang model, the time up to 0.001 second when gamma rays collide to make high-mass particles, such as protons.

heliocentric Centered on the sun.

heliocentric stellar parallax An apparent shift in the positions of nearby stars (relative to more distant ones) due to the changing position of the earth in its orbit around the sun; the size of the shift can be used to measure the distances to close stars; see trigonometric parallax.

helioseismology The study of the interior of the sun from an analysis of vibrations that appear at its surface.

helium burning Fusion of helium into carbon by the triple-alpha process; requires a temperature of at least 10^8 K.

helium flash The rapid burst of energy generation with which a star initiates helium burning by the triple-alpha process in the degenerate core of a low-mass red giant star.

hertz (Hz) A physical unit of frequency equal to 1 cycle per second.

Hertzsprung–Russell (H–R) diagram A graphic representation of the classification of stars according to their spectral class (or color or surface temperature) and luminosity (or absolute magnitude); the physical properties of a star are correlated with its position on the diagram, so a star's evolution can be described by its change of position on the diagram with time (see evolutionary track).

highlands Regions of higher than average elevation on a planet or satellite; usually refers to the older, cratered region of the moon's surface.

high-velocity clouds Clouds of gas associated with the Galaxy that move at speeds of hundreds of kilometers per second.

high-velocity stars Stars in the Galaxy having velocities greater than 60 km/s relative to the sun; these stars have orbits with high eccentricities, often at large angles with respect to the galactic plane.

homogeneous Having a consistent and even distribution of matter, the same in all parts.

Homo sapiens Humankind.

horizon The intersection with the sky of a plane tangent to the earth at the location of the observer.

horizon distance In cosmology, the maximum distance that light can travel in some epoch of the universe.

horizon problem In the Big Bang model, the problem that arises from the rapid expansion of the early universe, during which different regions could not communicate.

horizon system The celestial coordinate system based on the horizon plane and the zenith, in which altitude and azimuth are the two angles specifying a point on the celestial sphere.

horizontal branch The portion of the Hertzsprung–Russell diagram reached by Population II stars of low mass after the red giant stage and typically found in a globular cluster; it ranges from yellowish to reddish stars all having about the same luminosity (about 100 times that of the sun).

hour circle A great circle on the celestial sphere passing through the celestial poles.

H–R diagram A Hertzsprung–Russell diagram.

Hubble constant The proportionality constant relating radial velocity and distance in the Hubble law; the value, now around 75 km/s/Mpc (20 km/s/Mly), changes with time as the universe expands.

Hubble's law A description of the expansion of the universe, such that the more distant a galaxy lies from us, the faster it is moving away; the relation $v = Hd$, between the expansion velocity (v) and distance (d) of a galaxy, where H is the Hubble constant.

Hubble Space Telescope (HST) A space telescope launched in 1990 by the U.S. space shuttle into a low earth orbit; with a 2.4-meter mirror, *HST* can observe faint objects with high resolution, especially in the ultraviolet.

hydrogen burning Any fusion reaction that converts hydrogen (protons) to heavier elements, such as helium.

hydrogen corona A shell of hydrogen (emitting Lyman-alpha) gas around the earth, coming from the atmospheric escape of hydrogen produced by the dissociation of water.

hydrostatic equilibrium An equilibrium characterized by the absence of mass motions, when pressure balances gravity.

hyperbola A curve produced by the intersection of a plane with a cone; the shape of the orbit of a body with more than escape velocity.

hyperbolic geometry An alternative to Euclidean geometry, constructed by N. I. Lobachevski on the premise that more than one parallel line can be drawn through a point near a straight line; the sum of the angles of a triangle drawn on a hyperbolic surface is always less than 180 degrees.

ideal gas law The pressure of a gas is directly proportional to its temperature and density.

igneous rock A rock formed by the cooling of molten lava.

image Light rays gathered at the focus of a lens or mirror in the same relative alignment as the real object.

image processing The computer manipulation of digitized images to enhance aspects of them.

impact (model of) cratering The idea that craters are formed by the impact of solid objects onto a surface.

inertia The resistance of an object to a force acting on it because of its mass.

inertial mass Mass determined by subjecting an object to a known force (not gravity) and measuring the acceleration that results.

inferior conjunction For a planet orbiting interior to another, the alignment with the sun when the interior planet lies on same side of the sun as the outer planet.

inflationary universe model A modification of the Big Bang model in which the universe undergoes an early, brief interval of rapid expansion.

Infrared Astronomical Satellite (IRAS) An infrared space telescope (joint project of the United States, the United Kingdom, and the Netherlands) with a 57-cm mirror; *IRAS* mapped almost all the entire sky at far-infrared wavelengths; its data are available in a computer archive.

infrared cirrus Patches of interstellar dust, emitting infrared radiation, which look like cirrus clouds on the images of the sky produced by *IRAS*.

infrared telescope A telescope, optimized for use in the infrared part of the spectrum, fitted with an infrared detector.

injection velocity The speed and direction at which a satellite is placed into an orbit; a specific speed at right angles to the radius of the orbit gives a circular velocity, whereas somewhat larger or smaller speeds result in elliptical orbits.

instability strip The locus on the H–R diagram of cepheid variable stars; these stars are burning helium in their cores.

intergalactic medium The gas and dust found between the galaxies.

interferometer See radio interferometer.

International Ultraviolet Explorer (IUE) A space telescope with a 45-cm mirror designed to record spectra in the ultraviolet region; its data are available in computer archives.

interstellar dust Small, solid particles in the interstellar medium.

interstellar extinction curve The amount of extinction from interstellar dust as a function of wavelength.

interstellar gas Atoms, molecules, and ions in the interstellar medium.

interstellar medium All the gas and dust found between stars.

inverse beta decay The process in which electrons and protons are forced together to form neutrons and neutrinos; the reverse process of neutron decay.

inverse-square law for light The decrease of the flux of light with the inverse square of the distance from the source.

Io Third largest and most dense Galilean moon of Jupiter, with a density about 3.5 times that of water; a volcanically active body with a sulfurous surface; internally heated by tidal stresses.

ion An atom that has become electrically charged by the gain or loss of one or more electrons.

ionization The process by which an atom loses or gains electrons; collisions and absorption of energetic photons are the two most common processes in astrophysics.

ionization energy The minimum energy required to ionize an atom.

ionized gas A gas that has been ionized so that it contains free electrons and positively charged ions; a plasma.

ionosphere A layer of the earth's atmosphere ranging from about 100 to 700 km above the surface in which oxygen and nitrogen are ionized by sunlight, producing free electrons.

ion tail The part of a comet's tail that contains ionized gas.

iron meteorites One of the three main types of meteorite, typically made of about 90 percent iron and 9 percent nickel, with traces of other elements.

irregular cluster (of galaxies) A cluster of galaxies lacking spherical symmetry and having no obvious single center.

irregular galaxy A galaxy without spiral structure or rotational symmetry, containing mostly Population I stars and abundant gas and dust.

Ishtar Terra A large upland plateau in the northern hemisphere of Venus, bounded on the east by Maxwell Montes, a shield volcano.

isotopes Atoms with the same number of protons but different numbers of neutrons.

isotropic Having no preferred direction in space.

Jeans length The minimum size a disturbance in a gas must have to result in gravitational contraction; it depends on the pressure, temperature, and density of the medium.

Jeans mass The mass contained in the Jeans length; the minimum mass a disturbance in a gas must have to grow larger.

jet streams Latitudinal, coherent flows at high speed in the upper atmosphere of the earth or another planet.

joule (J) A physical unit of work and energy in the SI system.

Jovian planets Planets with physical characteristics similar to Jupiter: large mass and radius, low density, mostly liquid interior.

kelvin The unit of temperature in the SI system.

Kepler's laws of planetary motion The three laws of planetary motion, propounded by Johannes Kepler, which describe the properties of elliptical orbits with an inverse-square force law.

Keplerian motion Orbital motion that follows Kepler's laws.

kiloparsec (kpc) One thousand parsecs.

kinematic distance In our Galaxy, the distance to an object inferred from its Doppler shift, direction, and the Galaxy's rotation curve.

kinetic energy The ability to do work because of motion.

Kirchhoff's rules Empirical descriptions of the physical conditions under which the main types of spectrum originate.

Kleinmann–Low Nebula A diffuse dust cloud with a temperature of about 70 K near the Orion Nebula, emitting strongly at far-infrared wavelengths.

KREEP A lunar material composed of potassium (K), rare-earth elements (REE), and phosphorus (P).

Large Magellanic Cloud A small galaxy, irregular in shape, about 50 kpc (160 kly) from the Milky Way; it contains abundant gas, dust, and young stars.

last quarter The phase during which the moon is three-quarters of the way around its orbit from new moon, when it looks half-illuminated when viewed from the earth.

latitude Angular distance north or south of the equator.

LAWKI Acronym for life as we know it.

laws of motion The physical descriptions of the nature of forced and natural motion; see Newton's laws of motion.

least-energy orbit The orbit connecting two points in the solar system (or any other gravitating system) that requires the least change of energy to move from one point to the other; the least-energy path connecting two circular orbits is the ellipse just tangent to both of them.

lens A curved piece of glass designed to bring light rays to a focus.

lenticular galaxies Galaxies of Hubble type S0, with a disk like a spiral galaxy, but lacking spiral arms and without gas or dust.

lepton An elementary particle that has low mass, such as an electron.

light curve A graph of a star's changing brightness with time.

light-gathering power The ability of a telescope to collect light as measured by the area of its objective.

light-particle era In the hot Big Bang model, the interval from 0.0001 to 4 seconds when gamma rays can collide to make low-mass particles, such as electrons.

light pollution An increase in the background level of the night sky brightness caused by the scattering of light from artificial sources; as light pollution increases, fewer faint astronomical sources are visible.

light rays Imaginary lines in the direction of propagation of a light wave.

light year (ly) The distance light travels in a year, about 3.09×10^{13} km.

lighthouse model For a pulsar, a rapidly rotating neutron star with a strong magnetic field; the rotation provides the pulse period, and the magnetic field generates the electromagnetic radiation.

limb darkening The apparent darkening of the sun along its edge caused by an optical depth effect.

line profile The variation of a spectral line's flux as a function of wavelength; greatest at the center of the line.

local celestial meridian An imaginary line through the north and south points on the horizon and the zenith overhead.

Local Group (of galaxies) A gravitationally bound group of about thirty galaxies to which our Milky Way Galaxy belongs.

local standard of rest A frame of reference that participates in the average motion of the sun and nearby stars around the center of the Milky Way Galaxy.

Local Supercluster The supercluster of galaxies in which the Local Group is located; spread over 100 million light years, it contains the Virgo cluster.

longitude Angular distance, east or west, along the equator; on the earth, the reference longitude is that of Greenwich, England; a similar reference is used on other planets.

longitudinal wave A sound wave that moves in a push–pull motion through solids, liquids, and gases with a velocity that depends on the density of the medium.

long-period comets Comets with orbital periods greater than 200 years.

lowlands Regions of below-average elevation on a planet or satellite; usually refers to the younger, impact basins of the moon.

low-velocity stars Stars with close to circular orbits in the plane of the Galaxy; they travel at less than 60 km/s with respect to the sun.

luminosity The total rate at which radiative energy is given off by a celestial body, over all wavelengths; the sun's luminosity is about 4×10^{26} W.

luminosity class The categorization of stars that have the same surface temperatures but different sizes, resulting in different luminosities; based on the widths of dark lines in a star's spectrum, giant stars having narrower lines than dwarf stars of the same surface temperature.

luminosity function (of galaxies) The number of galaxies as a function of their luminosity in a cluster of galaxies.

luminosity function (of stars) The number of stars in a limited region of the Galaxy as a function of their luminosity.

lunar eclipse The cutoff of sunlight from the moon, when the moon lies on the line between the earth and sun so that it passes through the earth's shadow; a lunar eclipse can occur only at full moon, and when the full moon lies very close to the ecliptic.

lunar occultation The passage of the moon in front of a star or planet.

lunar soil The fine particles created by the bombardment of the lunar surface by meteorites that, with larger rock fragments, compose the lunar soil.

Lyman series Transitions in a hydrogen atom to and from the lowest energy level; they involve large energy changes, corresponding to wavelengths in the ultraviolet part of the spectrum; also, the set of absorption or emission lines corresponding to these transitions.

M-type asteroids Asteroids having albedos of about 10 percent and resembling metals in their reflective properties.

Mach number The ratio of an object's speed in a medium to the speed of sound in that medium; traveling faster than Mach 1.0 creates a shock wave in the medium.

Magellanic Clouds Two neighboring galaxies, the Large Magellanic Cloud (LMC) and the Small Magellanic Cloud (SMC), visible in the Southern Hemisphere to the unaided eye; gravitationally bound companions to our Galaxy.

magnetic field The property of space having the potential of exerting magnetic forces on bodies within it.

magnetic field lines A graphic representation of a magnetic field showing its direction and, by the degree of packing of the lines, its intensity.

magnetic flux The number of magnetic field lines passing through an area.

magnetic reconnection The sudden connection of magnetic field lines of opposite polarity; a process that releases energy stored in the magnetic field.

magnetometer A device to measure the strength of a magnetic field.

magnetosphere The region around a planet in which particles from the solar wind are trapped by the planet's magnetic field.

magnifying power The ability of a telescope to increase the apparent angular size of a celestial object.

magnitude An astronomical measurement of an object's brightness; larger magnitudes represent fainter objects.

main sequence The principal series of stars in the Hertzsprung–Russell diagram; such stars are converting hydrogen to helium in their cores by the proton–proton process or by the carbon–nitrogen–oxygen cycle; this is the longest stage of a star's active life.

main-sequence lifetime The duration of time that a star spends fusing hydrogen to helium in its core; the longest phase of its evolution.

major axis The larger of the two axes of an ellipse.

mantle The major portion of the earth's interior below the crust, made of a plastic rock probably composed of olivine.

mare (pl., maria; Latin for "sea") A lowland area on the moon that appears darker and smoother than the highland regions, probably formed by lava that solidified into basaltic rock about 3 to 3.5 billion years ago.

mare basalts Basaltic rocks found on the surface of the lunar maria; tend to be the youngest (most recently formed) rocks on the lunar surface, with ages around 3.2 billion years.

mascons High-density concentrations of mass beneath the lunar maria; they have been detected by their effect on the orbits of moon-orbiting satellites.

mass A measure of an object's resistance to change in its motion (inertial mass); a measure of the strength of gravitational force an object can produce (gravitational mass).

mass extinction The sudden disappearance from the fossil record of a large number of species; thought to result from catastrophic changes in the earth's environment.

mass loss The rate at which a star loses mass, usually by a stellar wind, per year.

mass–luminosity ratio For galaxies, the ratio of the total mass to the luminosity; a rough measure of the kind of stars in the galaxy; a large ratio implies a large fraction of dark matter.

mass–luminosity relation An empirical relation, for main-sequence stars, between a star's mass and its luminosity, roughly proportional to the fourth power of the mass; for stars of other types the numerical value of the power is different.

matter era In the Big Bang model, the time interval from about one million years after the Big Bang to now, in which matter dominates the universe.

maximum elongation The greatest angular distance (east or west) of an object from the sun.

mean lifetime The time it takes for the number of atoms of a radioactive substance to decrease by one factor of e.

mechanics A branch of physics that deals with forces and their effects on bodies.

megaparsec (Mpc) 1 million parsecs, or about 3.26 million light years.

megaton An explosive force equal to that of 1 million tons of TNT (about 4×10^{15} J).

meridian (celestial) An imaginary line drawn through the north and south points on the horizon and through the zenith.

mesosphere Region of the earth's atmosphere between 50 and 100 km, where the temperature falls rapidly.

Messier 17 (M17) A nearby H II region that is the site of recent massive star formation at one end of a giant molecular cloud.

metal-poor stars Stars with metal abundances much less than the sun's.

metal-rich stars Stars with metal abundances like that of the sun (about 1 to 2 percent of the total mass).

metallic hydrogen A state, reached at high pressures, in which hydrogen is able to conduct electricity.

meteor The bright streak of light that occurs when a solid particle (a meteoroid) from space enters the earth's atmosphere and is heated by friction with atmospheric particles; sometimes called a falling star.

meteor shower A rapid influx of meteors that appear to come out of a small region of the sky, called the radiant.

meteor stream A uniform distribution of meteoroids along an orbit around the sun.

meteor swarm Meteoroids grouped in a localized region of an orbit around the sun; the source of meteor showers.

meteor trail The visible path of a meteor through the atmosphere, created by ionization of the air and vaporization of the meteoroid.

meteorite A solid body from space that survives a passage through the earth's atmosphere and hits the ground.

meteorite fall A meteorite seen in the sky and recovered on the ground.

meteorite find A recovered meteorite that was not seen to fall.

meteoroid A very small solid body moving through space in orbit around the sun.

micrometeorites Very small meteorites (about 0.1 to 1 μm in diameter) that cool and solidify before they hit the ground.

micrometer (μm) A millionth of a meter (10^{-6}); common unit of measurement of the wavelength of light.

midoceanic ridge The almost continuous submarine mountain chain that extends some 64,000 km through the earth's ocean basins.

Milky Way The band of light that encircles the sky, caused by the blending of light from the many stars lying near the plane of the Galaxy; also sometimes designates the Galaxy to which the sun belongs.

millisecond pulsar Generic name given to any pulsar with a pulse period of a few milliseconds.

minimum resolvable angle The smallest angle a telescope can clearly show.

minute of arc (arcmin) One-sixtieth of a degree $1°/60$.

molecular clouds Large, dense, massive clouds in the plane of a spiral galaxy; they contain dust and a large fraction of gas in molecular form.

molecular maser Microwave amplification by stimulated emission of radiation from a molecule.

molecule A combination of two or more atoms bound together electrically; the smallest part of a compound that has the properties of that substance.

momentum The product of an object's mass and velocity.

moving-cluster method A method for finding a distance to a cluster of stars by determining their radial velocities and the convergent point of their proper motions.

mutation A basic change in gene structure.

nadir The point on the celestial sphere directly below the observer, opposite the zenith.

nanometer (nm) A billionth of a meter (10^{-9} m); common unit of measurement of the wavelength of light.

narrow-tailed radio galaxies Radio galaxies that show a U-shaped tail behind the nucleus; they are fast-moving galaxies in a cluster of galaxies.

natal astrology The belief that treats the supposed influence of the stars and planets on human affairs by the use of their positions and relationships at the time of an individual's birth.

natural motion Motion without forces.

natural selection The process by which individuals with genes that produce characteristics that are best adapted to their environment have greater genetic representation in future generations.

nebula (Latin for "cloud") A cloud of interstellar gas and dust.

nebular model A model for the origin of the solar system in which an interstellar cloud of gas and dust contracted gravitationally to form a flattened disk out of which the planets formed by accretion.

neutrino An elementary particle with no (or very little) mass and no electric charge that travels at the speed of light and carries energy away during certain types of nuclear reaction.

neutron A subatomic particle with about the mass of a proton and no electric charge; one of the main constituents of an atomic nucleus; the union of a proton and an electron.

neutron star A star of extremely high density and small size (10 km) that is composed mainly of very tightly packed neutrons; cannot have a mass greater than about 3 solar masses.

new moon The phase during which the moon is in the same direction as the sun in the sky, appearing almost completely unilluminated as seen from the earth.

newton (N) The SI unit of force.

Newton's laws of motion The three laws describing motion from Newton's viewpoint; they are the inertial, force, and reaction laws.

Newtonian reflector A reflecting telescope designed so that a small mirror at a 45° angle in the center of the tube brings the focus outside the tube.

nonthermal radiation Emitted energy that is not characterized by a blackbody spectrum; in astronomy, usually designates synchrotron radiation.

noon Midday; the time halfway between sunrise and sunset when the sun reaches its highest point in the sky with respect to the horizon.

Norma arm A segment of a spiral arm of our Galaxy, about 4000 pc (13,000 ly) from the sun toward the center of the Galaxy in the direction of the constellation Norma.

north magnetic pole One of the two points on a star or planet from which magnetic lines of force emanate and to which the north pole of a compass points.

nova (Latin for "new") A star that has a sudden outburst of energy, temporarily increasing its brightness by hundreds to thousands of times; now believed to be the outburst of a degenerate star in a binary system; also used in the past to refer to some stellar outbursts that modern astronomers now call supernovas.

nuclear bulge The central region of a spiral galaxy, containing mostly old Population I stars.

nuclear fission A process that releases energy from matter: a heavy nucleus hit by a high-energy particle splits into two or more lighter nuclei whose combined mass is less than the original, the missing mass being converted into energy.

nuclear fusion A process that releases energy from matter by the joining of nuclei of lighter elements to make heavier ones; the combined mass is less than that of the constituents, the difference appearing as energy.

nucleic acid A huge spiral-shaped molecule, commonly found in the nucleus of cells, that is the chemical foundation of genetic material.

nucleosynthesis The chain of thermonuclear fusion processes by which hydrogen is converted to helium, helium to carbon, and so on, through all the elements of the periodic table.

nucleus (of an atom) The massive central part of an atom, containing neutrons and protons, about which the electrons orbit.

nucleus (of a comet) Small, bright, starlike point in the head of a comet; a solid, compact (diameter a few tens of kilometers) mass of frozen gases with some rocky material embedded in it as dust.

nucleus (of a galaxy) The central portion of a galaxy, composed of old Population I stars, some gas and dust, and, for many galaxies, a concentrated source of nonthermal radiation.

number density The number of particles per unit volume.

OB association Loose groupings of O and B stars in small subgroups; they are not bound by gravity and so dissipate in a few tens of millions of years.

OB subgroup A small collection of about ten O and B stars, a few tens of light years across, within an OB association.

objective The main light-gathering lens or mirror of a telescope.

observable universe The parts of the universe that can be detected by the light they emit.

occultation The eclipse of a star or planet by the moon or another planet.

Olbers' paradox The statement that if there were an infinite number of stars distributed uniformly in an infinite space, the night sky would be as bright as the surface of a star, in obvious contrast to what is observed.

Olympus Mons A large shield volcano on the surface of Mars in the Tharsis ridge region; probably the largest volcano in our solar system.

Oort Cloud A cloud of comet nuclei in orbit around the solar system, formed at the time the solar system formed; the reservoir for new comets.

Oort's constant In Oort's equations describing differential galactic rotation, the number relating the sun's distance from the center of the Galaxy and the change in angular speed as a function of radius; current value is 15 km/s/kpc.

opacity The property of a substance that hinders (by absorption or scattering) light passing through it; opposite of transparency.

open cluster Same as galactic cluster.

open geometry See hyperbolic geometry.

opposition The time at which a celestial body lies opposite the sun in the sky as seen from the earth; the time at which it has an elongation of 180°.

optics The manipulation of light by reflection or refraction.

orbital angular momentum The angular momentum of a revolving body; the product of a body's mass, orbital velocity, and the distance from the system's center of mass.

orbital inclination The angle between the orbital plane of a body and some reference plane; in the case of a planet in the solar system, the reference plane is that of the earth's orbit; in the case of a satellite, the reference is usually the equatorial plane of the planet; for a double star, it is the plane perpendicular to the line of sight.

organic Relating to the branch of chemistry concerned with the carbon compounds of living creatures.

organism An ordered living creature.

Orion Nebula The hot cloud of ionized gas that is a nearby region of recent star formation, located in the sword of the constellation of Orion; also called Messier 42 (M42).

outgassing Release of gases from nongaseous materials; extrusion of gases from the body of a planet after its formation.

ozone layer (ozonosphere) A layer of the earth's atmosphere about 40 to 60 km above the surface, characterized by a high content of ozone, (O_3).

parabola A geometric figure that describes the shape of an escape speed orbit.

parallax The change in an object's apparent position when viewed from two different locations; specifically, half the angular shift of a star's apparent position as seen from opposite ends of the earth's orbit.

parent bodies Small solid bodies, a few hundreds or thousands of kilometers in size, believed to be the source of nickel–iron meteorites; these objects formed early in the history of the solar system and broke up through collisions.

parsec (pc) The distance an object would have to be from the earth to attain a heliocentric parallax of 1 second of arc; equal to 3.26 light years; a kiloparsec (kpc) is 10^3 parsecs, and a megaparsec (Mpc) is 10^6 parsecs.

pascal (Pa) Unit of pressure in the SI system.

Pauli exclusion principle The statement that no two electrons can be in the same quantum state at the same time.

perfect cosmological principle The statement that the universe appears the same to an observer at all locations and at all times.

perigee The point in its orbit at which an earth satellite is closest to the earth.

perihelion The point at which a body orbiting the sun is nearest to it.

period The time interval for some regular event to take place; for example, the time required for one complete revolution of a body around another.

period–luminosity relationship For cepheid variables, a relation between the average luminosity and the time period over which the luminosity varies; the greater the luminosity, the longer the pulsational period.

periodic comets Comets that have relatively small elliptical orbits around the sun, with periods of less than 200 years.

periodic (regular) variables Stars whose light varies with time in a regular fashion.

Perseus arm A segment of a spiral arm that lies about 3 kpc (10 kly) from the sun in the direction of the constellation Perseus.

perturbations Small changes in the motions of a mass because of the gravitational effects of another mass.

phases of the moon The monthly cycle of the changes in the moon's appearance as seen from the earth; at new, the moon is in line with the sun and so not visible; at full it is in opposition to the sun and we see a completely illuminated surface.

Phobos The larger of the two moons of Mars.

photodissociation The breakup of a molecule by the absorption of light with enough energy to break the molecular bonds.

photometer A light-sensitive detector placed at the focus of a telescope; used to make accurate measurements of small photon fluxes.

photometry Measurement of the intensity of light.

photon A discrete amount of light energy; the energy of a photon is related to the frequency f of the light by the relation $E = hf$, where h is Planck's constant.

photon excitation Raising an electron of an atom to a higher energy level by the absorption of a photon.

photosphere The visible surface of the sun; the region of the solar atmosphere from which visible light escapes into space.

physical universe The parts of the universe that can be seen directly plus those that can be inferred from the laws of physics.

pitch angle The angle between a spiral-arm's direction and the direction of circular motion about the Galaxy.

pixel The smallest picture element in a two-dimensional detector.

Planck curve The continuous spectrum of a blackbody radiator.

Planck's constant The number that relates the energy and frequency of light; it has a value of 6.63×10^{-34} J · s.

planet From the Greek word for "wanderer;" any of the nine (so far known) large bodies that revolve around the sun; traditionally, any heavenly object that moved with respect to the stars (in this sense, the sun and the moon were also considered planets).

planetary nebula A thick shell of gas ejected from and moving out from an extremely hot star; thought to be the outer layers of a red giant star thrown out into space, the core of which eventually becomes a white dwarf.

planetesimals Asteroid-sized bodies that, in the formation of the solar system, combined to form the protoplanets.

plasma A gas consisting of equal numbers of ionized atoms and electrons.

plate tectonics A model for the evolution of the earth's surface that pictures the interaction of crustal plates driven by convection currents in the mantle.

Pluto The last major planet discovered in the solar system; actually a double-planet system with its moon, Charon, a small body of low mass with a thin atmosphere, an icy surface, and a density about twice that of water.

polar caps Icy regions at the north and south poles of a planet.

Polaris The present north pole star; the outermost star in the handle of the Little Dipper.

polarization A lining up of the planes of vibration of light waves.

polarized light Light waves whose planes of oscillation are the same.

polodial magnetic field Magnetic field configuration with two poles and field lines running along meridians.

Population I stars Stars found in the disk of a spiral galaxy, especially in the spiral arms, including the most luminous, hot, and young stars, having a heavy-element abundance similar to that of the sun (about 2 percent of the total); an old Population I is found in the nucleus of spiral galaxies and in elliptical galaxies.

Population II stars Stars found in globular clusters and the halo of a galaxy; somewhat older than any Population I stars and containing a smaller abundance of heavy elements.

positron An antimatter electron; essentially an electron with a positive charge.

potential energy The ability to do work because of position; it is storable and can be converted into other forms of energy.

Poynting–Robertson effect The tendency for small particles in the solar system to spiral into the sun as a result of a drag produced by solar radiation pressure acting in a slightly nonradial direction; the force due to radiation appears to have a component in the direction opposite the particle's motion, just as vertical raindrops appear to come from the direction toward which a person runs.

PP chain See proton–proton chain.

precession of Mercury's orbit The turning, with respect to the stars, of the major axis of Mercury's orbit at a rate of 43 arcsec per century.

precession of the equinoxes The slow westward motion of the equinox points on the sky relative to the stars of the zodiac, due to the wobbling of the earth's spin axis.

pre-main-sequence star The evolutionary phase of a star just before it reaches the main sequence and starts hydrogen core burning.

pressure Force per unit area.

pressure gradient The rate of change of pressure along a direction.

primary The more luminous of the two stars in a binary system.

prime focus The direct focus of an objective without diversion by a lens or mirror.

primeval fireball The hot, dense beginning of the universe in the Big Bang model, when most of the energy was in the form of high-energy light.

principle of equivalence The fundamental idea in Einstein's general theory of relativity; the statement than one cannot distinguish between gravitational accelerations and accelerations of other kinds, or, equivalently, a statement about the equality of inertial mass and gravitational mass; a consequence is that gravitational forces can be made to vanish in a small region of spacetime by choosing an appropriate accelerated frame of reference.

prominences Cool clouds of hydrogen gas above the sun's photosphere in the corona; they are shaped by the local magnetic fields of active regions.

proper motion The angular displacement of a star on the sky from its motion through space.

protein A long chain of amino acids linked by hydrogen bonds.

protogalaxies Clouds with so much mass that they are destined to contract gravitationally into galaxies.

proton A massive, positively charged elementary particle; one of the main constituents of the nucleus of an atom; a baryon.

proton–proton (PP) chain A series of thermonuclear reactions that occur in the interiors of stars, by which four hydrogen nuclei are fused into helium; this process is believed to be the primary mode of energy production in the sun.

protoplanet A large mass formed by the accretion of planetesimals; the final stage of formation of the planets from the solar nebula.

protoplasm The fluid within a cell.

protostar A collapsing mass of gas and dust out of which a star will be born (when thermonuclear reactions turn on); its energy comes from gravitational contraction.

pulsar A radio source that emits signals in very short, regular bursts; thought to be a highly magnetic, rotating neutron star.

quantum (pl., quanta) A discrete packet of energy.

quantum number In quantum theory, one of the four special numbers that determine the energy structure and quantum state of atoms.

quantum state The quantum description of the arrangement of electrons in an atom; allowed quantum states are filled starting with those of lowest energy first.

quark An elementary particle that makes up others, such as protons.

quasar or quasi-stellar object An intense, pointlike source of light and radio waves that is characterized by large redshifts of the emission lines in its visible spectrum.

radar mapping The surveying of the geographic features of a planet's surface by the reflection of radio waves from the surface.

radial velocity The component of relative velocity that lies along the line of sight.

radian (rad) A unit of angular measurement; 1 rad = 57.3°; 2π rad = 360°.

radiant The point in the sky from which a meteor shower appears to come.

radiation Usually refers to electromagnetic waves, such as light, radio, infrared, x-rays, ultraviolet; also sometimes used to refer to atomic particles of high energy, such as electrons (beta radiation) and helium nuclei (alpha radiation).

radiation belts In a planet's magnetosphere, regions with a high density of trapped solar wind particles.

radiation era In the Big Bang model, the time in the universe's history in which the energy in the universe is dominated by radiation.

radiative energy The capacity to do work that is carried by electromagnetic waves.

radio galaxies Galaxies that emit large amounts of radio energy by the synchrotron process, generally characterized by two giant lobes of emission situated on opposite ends of a line drawn through the nucleus and connected to it by jets; they are divided into two types, compact and extended.

radioactive dating A process that determines the age of an object by the rate of decay of radioactive elements within the object; for rocks, gives the date of the most recent solidification.

radioactive decay The process by which an element is converted by fission into lighter elements.

radio interferometer A radio telescope that achieves high angular resolution by combining signals from at least two widely separated antennas.

radio jets Astrophysical jets that are visible to radio telescopes; generally, the jets extending from the nuclei of active galaxies.

radio recombination-line emission Sharp energy peaks at radio wavelengths caused by low-energy transitions in atoms from one very high energy level to another nearby level following recombination of an electron with an ion.

radio telescope A telescope designed to collect and detect radio emissions from celestial objects.

rapid process (r-process) Formation of very heavy elements by the rapid addition of neutrons to a nucleus followed by beta decay.

ray On the moon or other satellite, a bright streak formed by material ejected from an impact crater.

recombination The joining of an electron to an ion; the reverse of ionization.

recombination line Emission line from an electron following the process of recombination.

red giant A large, cool star with a high luminosity and a low surface temperature (about 2500 K), which is largely convective and has fusion reactions going on in shells; lies on the red giant branch or asymptotic giant branch in an H–R diagram.

reddening The preferential scattering or absorption of blue light by small particles, allowing more red light to pass directly through.

redshift An increase in the wavelength of the radiation received from a receding celestial body as a consequence of the Doppler effect; a shift toward the long-wavelength (red) end of the spectrum.

red variables A class of cool stars variable in light output.

reference frame A set of coordinates by which position and motion may be specified.

reflecting telescope A telescope that has a uniformly curved mirror as its primary light gatherer.

reflection The return of a light wave at the interface between two media.

reflection nebula A bright cloud of gas and dust that is visible because of the reflection of starlight by the dust.

refracting telescope A telescope that uses glass lenses to gather light.

refraction Bending of the direction of a light wave at the interface between two media, such as air and glass.

regolith The pulverized surface soil of the moon (or any airless body) caused by meteorite impacts.

regular cluster (of galaxies) A cluster of galaxies with definite symmetry and a well-defined core.

relativistic Doppler shift Wavelength shift from the radial velocity of a source as calculated in special relativity, because of which very large redshifts do not imply that the source moves faster than light.

relativistic jet A beam of particles moving at speeds close to that of light; such jets contained charged particles channeled by magnetic fields.

resolution The ability of a telescope to separate close stars or to pick out fine details of celestial objects.

retrograde motion The apparent anomalous westward motion of a planet with respect to the stars, which occurs near the time of opposition (for an outer planet) or inferior conjunction (for an inner planet).

retrograde rotation Rotation from east to west.

revolution The motion of a body in orbit around another body or a common center of mass.

rift valley A depression in the surface of a planet created by the separation of crustal masses.

ring system The complete set of rings around a planet; these consist of smaller ringlets.

ringlets The subsections of a planetary ring system, containing many small particles orbiting together and obeying Kepler's laws.

Roche limit The minimum distance from the center of a planet that a satellite can orbit and not undergo tidal disruption; roughly 2.5 times the radius of the planet.

Roche lobe In a binary star system, the region in the space around the pair in which the stars' gravitational fields on a small mass are equal.

rotation The turning of a body, such as a planet, on its axis.

rotation curve The relation between rotational speed of objects in a galaxy and their distance from its center.

RR Lyrae stars A class of giant, pulsating variable stars with periods of less than one day; they are Population II objects and commonly found in globular clusters.

runaway accretion The process by which a planetesimal that starts out with an escape speed greater than its neighbors grows very rapidly.

Rydberg constant A number relating to the spacing of the energy levels in a hydrogen atom.

S-type asteroids Asteroids whose albedos indicate a surface made of silicates.

Sagittarius A (Sgr A) Radio sources at the center of the Galaxy; Sgr A West is a thermal radio source (H II region), Sgr A East a nonthermal source, and Sgr A* a pointlike source that may mark the Galaxy's core.

Sagittarius arm A portion of spiral-arm structure of the Galaxy that lies about 2 kpc (6 kly) from the center of the Galaxy in the direction of the constellation Sagittarius.

scarp A long, vertical wall running across a flat plain.

scattering (of light) The change in the paths of photons without absorption or change in wavelength.

Schwarzschild radius The critical size that a mass must reach to be dense enough to trap light by its gravity, that is, to become a black hole.

scientific model A mental image of how the natural world works, based on physical, mathematical, and aesthetic ideas.

seafloor spreading The lateral motion of the ocean basins away from oceanic ridges.

second dredge-up The process by which convection brings the products of helium burning to the surface of a massive star during the second time it becomes a red giant.

second of arc (arcsec) 1/3600 of a degree (0.000278°), or 1/60 of a minute of arc (0.0167 arcmin).

secondary The less luminous of the two stars in a binary system.

secular parallax A method of determining the average distance of a group of stars by examining the components of their proper motions produced by the straight-line motion of the sun through space.

seeing The unsteadiness of the earth's atmosphere that blurs telescopic images.

seismic waves Sound waves traveling through and across the earth that are produced by earthquakes.

seismometer An instrument used to detect earthquakes and moonquakes.

semimajor axis Half of the major axis of an ellipse; distance from the center of an ellipse to its farthest point.

sequential star formation A model for the formation of massive stars from a giant molecular cloud in which the hot, ionized gas around a small cluster of OB stars creates a shock wave that initiates the collapse of another part of the molecular cloud to give birth to another small cluster of OB stars until the cloud's material is used up.

SETI Acronym for the search for extraterrestrial intelligence.

sexagesimal system A counting system based on the number 60, such as 60 minutes in an hour, or 60 minutes of arc in one degree.

Seyfert galaxy A galaxy having a bright nucleus showing broad emission lines in its spectrum; Seyferts probably are spiral galaxies; one type of AGN.

shepherd satellites Small moons that confine a planet's ring in a narrow band; one is located on the inside edge of the ring, one on the outside.

shield volcano A large volcano with gentle slopes formed by the slow outflow of magma.

shock wave A discontinuity in a medium created when an object travels through it at a speed greater than the local sound speed.

short-period comets Comets with orbital periods less than 200 years, probably captured from longer period orbits by an encounter with a major planet.

sidereal month The period of the moon's revolution around the earth with respect to a fixed direction in space or a fixed star; about 27.3 days.

sidereal period The time interval needed by a celestial body to complete one revolution around another with respect to the background stars.

signs of the zodiac The twelve angular 30° segments into which the ecliptic is divided; each corresponds to a zodiacal constellation.

silicate A compound of silicon and oxygen with other elements, very common in rocks at the earth's surface.

singularity A theoretical point of zero volume and infinite density to which any mass that becomes a black hole must collapse, according to the general theory of relativity.

Sirius B The white dwarf companion to Sirius A; Sirius A and B make up a binary star system.

slow process (s-process) Formation of very heavy elements by the slow addition of neutrons to the nucleus followed by beta decay.

Small Magellanic Cloud (SMC) The smaller of the two companion galaxies to the Milky Way; it is an irregular galaxy containing about 2 billion solar masses.

solar core Region of the sun's interior in which temperatures and densities are high enough for fusion reactions to take place.

solar cosmic rays Low-energy cosmic rays generated in solar flares.

solar day The interval of time from noon to noon.

solar dynamo model See dynamo model.

solar eclipse An eclipse of the sun by the moon, caused by the passage of the new moon in front of the sun.

solar flare Sudden burst of electromagnetic energy and particles from a magnetic loop in an active region; probably triggered by magnetic reconnection.

solar mass The amount of mass in the sun, about 2×10^{30} kg.

solar nebula The thin disk of gas and dust around the young sun, out of which the planets formed.

solar wind A stream of charged particles, mostly protons and electrons, that escapes into the sun's outer atmosphere (from coronal holes) at high speeds and streams out into the solar system.

solstice The time at which the day or the night is the longest; in the Northern Hemisphere, the summer solstice (around June 21) is the time of the longest day and the winter solstice (around December 21) the time of the shortest day; the dates are reversed in the Southern Hemisphere.

south magnetic pole A point on a star or planet from which the magnetic lines of force emanate and to which the south pole of a compass points.

space A three-dimensional region in which objects move and events occur and have relative direction and position.

space astronomy Astronomy done by instruments in orbit above the earth's atmosphere.

spacetime Space and time unified; a continuous system of one time coordinate and three space coordinates by which events can be located and described.

spacetime curvature The bending of a region of spacetime because of the presence of mass and energy.

spacetime diagram A diagram with one axis representing the three dimensions of space and the other time; it shows the relation of events and worldlines.

space velocity The total velocity of an object through space, combining the components of radial and transverse velocities.

special theory of relativity Einstein's theory describing the relations between measurements of physical phenomena as viewed by observers who are in relative motion at constant velocities.

spectral line A particular wavelength of light corresponding to an energy transition in an atom.

spectral sequence A classification scheme for stars based on the strength of various lines in their spectra; the sequence runs O-B-A-F-G-K-M, from hottest to coolest.

spectral type (or class) The designation of the type of a star based on the relative strengths of various spectral lines.

spectrograph or spectrometer An instrument used on a telescope to obtain and record a spectrum of an astronomical object; a spectrograph records on a photographic plate, a spectrometer uses another type of spectral detector, such as a CCD.

spectroscope An instrument for examining spectra; also a spectrometer or spectrograph if the spectrum is recorded and measured.

spectroscopic binary Two stars revolving around a common center of mass that can be identified by periodic variations in the Doppler shift of the lines of their spectra.

spectroscopic distances A technique for measuring distance by comparing the brightnesses of stars with their actual luminosities, as determined by their spectra.

spectroscopy The analysis of light by separating it into wavelengths (colors).

spectrum (pl., spectra) The array of colors or wavelengths obtained when light is dispersed, as by a prism; the amount of energy given off by an object at every different wavelength.

speed The rate of change of position with time.

spherical (closed) geometry An alternative to Euclidean geometry, constructed by G. F. B. Riemann on the premise that no parallel lines can be drawn through a point near a straight line; the sum of the angles of a triangle drawn on a spherical surface is always greater than 180°.

spicules Spears of hot gas that reach up from the sun's photosphere into the chromosphere.

spin angular momentum The angular momentum of a rotating body; the product of a body's mass distribution, rotational speed, and radius.

spiral arm Part of a spiral pattern in a galaxy: a structure composed of gas, dust, and young stars, which winds out from near the galaxy's center.

spiral galaxy A galaxy with spiral arms; the presumed shape of our Milky Way Galaxy.

spiral nebula An older term for a spiral galaxy as it appeared visually through a telescope.

spiral tracers Objects that are commonly found in spiral arms and so are used to trace spiral structure; for example, Population I cepheids, H II regions, molecular clouds, and OB stars.

spontaneous emission The emission of a photon by an excited atom in which an electron falls to a lower energy level.

spontaneous generation The supposed natural origin of living things from lifeless matter.

sporadic meteor A meteor that occurs at random and so is not associated with a shower.

stadium (pl., stadia) An ancient Greek unit of length, probably about 0.2 km.

standard candle An astronomical object of known luminosity used to estimate distances to galaxies.

starburst galaxies Active galaxies that show evidence of large-scale star formation of massive stars.

star counting A technique to measure the extent of the Galaxy, first used by William Herschel, in which it is assumed that the directions in space in which more stars are found (in a specific area) mark regions of greater extent of the Galaxy.

star model See stellar interior model.

starspots Dark, magnetic active regions in the photospheres of stars in analogy to sunspots.

statistical parallax A parallax, and so a distance, determined from the average proper motions of selected groups of stars.

steady-state model A theory of the universe based on the perfect cosmological principle, in which the universe looks basically the same to all observers at all times; largely discredited by current observations of the cosmic microwave background radiation.

Stefan–Boltzmann law For a blackbody radiator, the relation between temperature and power emitted per unit area of surface; the flux goes as the fourth power of the temperature.

stellar corona The hot (1×10^6 K), thin outer atmosphere of a star in analogy to the sun's corona; visible by the emission of x-rays.

stellar flares Flares associated with active regions on stars, analogous to solar flares.

stellar interior model A table of values of the physical characteristics (such as temperature, density, and pressure) as a function of position within a star for a specified mass, chemical composition, and age, calculated from theoretical ideas of the basic physics of stars.

stellar lifetime The total duration of a star's lifetime from birth to death; the sun's lifetime is expected to be about 10 billion years.

stellar parallax See heliocentric stellar parallax.

stellar spectral sequence See spectral sequence.

stellar surface temperature The temperature of the photosphere of a star.

stellar wind The outflow, both steady and sporadic, of gas at hundreds of kilometers per second from the corona of a star.

stimulated emission Radiation produced by the effect of a photon stimulating an atom in an excited state to emit another photon of the same wavelength.

stochastic star formation A model for the generation of spiral arms by the random formation of stars in a molecular cloud, triggered by a supernova explosion, which then are drawn out into a spiral pattern by differential galactic rotation.

stones Meteorites made of light silicate materials.

stony-iron meteorite A type of meteorite that is a blend of nickel–iron and silicate materials.

straight line The shortest distance between two points in any geometry.

stratosphere A layer in the earth's atmosphere in which temperature changes with altitude are small and clouds are rare.

strong force One of the four forces of nature; the strong force acts over short distances to keep the nuclei of atoms together.

summer solstice See solstice.

sunspot A temporary cool region in the sun's photosphere, associated with an active region, with a magnetic field of some 0.1 tesla (a few thousand gauss).

sunspot cycle The 11-year number cycle and a 22-year magnetic polarity cycle of the formation of sunspots.

superclusters Clusters of clusters of galaxies; superclusters appear as long, filamentary chains with voids in between.

supergiant A massive star of large size and high luminosity; red supergiants are thought to be the progenitors of Type II supernovas.

supergiant elliptical (cD) galaxies The largest and most massive elliptical galaxies, sometimes with more than one nucleus; found at the core of a rich cluster of galaxies; they may have been formed by mergers.

supergravity A model that combines quantum ideas with gravity to describe the unification of all forces of nature during the first 10^{-43} second of the universe's history.

superior conjunction A planetary configuration in which an inner planet lies in the same direction as the sun but on the opposite side of the sun as viewed from the outer planet.

superluminal motion Motion apparently faster than the speed of light.

supermassive black hole A black hole with a mass of a million solar masses or more; probably powers active galaxies and quasars.

supernova A stupendous explosion of a star, which increases its brightness hundreds of millions of times in a few days; Type II results from the core implosion of a massive star at the end of its life; Type I may originate from a carbon-rich white dwarf in a binary system.

supernova remnant An expanding gas cloud from the outer layers of a star, blown off in a supernova explosion; detectable at radio wavelengths; moves through the interstellar medium at high speeds.

superwind A very strong stellar wind.

surface temperatures photospheric temperature of a star.

synchronous rotation A solid body that rotates and revolves at the same rate; for a satellite, this means that the body keeps the same face to its parent planet.

synchrotron radiation Radiation from an accelerating charged particle (usually an electron) in a magnetic field; the wavelength of the emitted radiation depends on the strength of the magnetic field and the energy of the charged particles.

synodic month The time interval between similar configurations of the moon and the sun (for example, between full moon and the next full moon); about 29.5 days.

synodic period The interval between successive similar lineups of a celestial body with the sun (for example, between oppositions).

temperature A measure of the average random speeds of the microscopic particles in a substance.

temperature gradient The change in temperature over a unit change in distance.

terminator The fairly sharp boundary between the portions of a planet illuminated and unilluminated by sunlight; easily visible on the moon.

terrestrial planets Planets similar in composition and size to the earth: Mercury, Venus, Mars, and the moon.

tesla (T) In the SI system, a unit of measure of magnetic flux.

Tharsis ridge A highland region on Mars containing a cluster of volcanoes, including Olympus Mons.

theoretical resolution A telescope's resolving power based on its optics alone; ground-based optical telescopes have a resolution limited by the seeing of the atmosphere.

thermal energy The internal energy of an object from the random motions of the particles that make it up.

thermal equilibrium Steady-state situation characterized by an absence of large-scale temperature changes.

thermal pulses Bursts of energy generation from the triple-alpha process in the shell of a red giant star.

thermal radiation Electromagnetic radiation from a body that is hot; often characterized by a blackbody spectrum.

thermal speed The average speed of the random motion of particles in a gas.

thermosphere A layer in the earth's atmosphere, above the mesosphere, heated by x-rays and ultraviolet radiation from the sun.

threshold temperature The temperature at which photons have enough energy to create a given type of particle–antiparticle pair.

tidal force The difference in gravitational force between two points in a body caused by a second body, which may result in the deformation of the second body.

tidal friction Friction caused by the tidal motion of the water in the ocean basins.

time A measure of the flow of events.

Titan Saturn's largest satellite, the first satellite detected to have an atmosphere.

Titius–Bode law A nonphysical formula that gives the approximate distances of the planets from the sun in AU.

ton (metric) 1000 kg.

torodial magnetic field A magnetic field configuration in which the field lines run parallel to the equator.

torque A twisting force applied through a level arm.

total energy The sum of all forms of energy attributed to an isolated body or a system of bodies; usually just the sum of the kinetic and potential energies.

transition (in an atom) A change in the electron arrangements in an atom, which involves a change in energy.

transition region In the sun's atmosphere, the region of rapid temperature rise lying between the chromosphere and the corona.

transverse fault A crack in a solid surface where the ground has moved sideways.

transverse velocity The component of an object's velocity that is perpendicular to the line of sight.

transverse wave A wave in which the oscillatory motion is perpendicular to the direction of propagation; such waves cannot travel through liquids.

Trapezium cluster A small cluster of young massive stars located in the Orion Nebula; the hot stars ionize the gas around them.

trigonometric parallax A method of determining distances by measuring the angular position of an object as seen from the ends of a baseline having a known length; see heliocentric parallax.

triple-alpha reaction A thermonuclear process in which three helium atoms (alpha particles) are fused into one carbon nucleus.

Triton The largest moon of Neptune; it has a thin atmosphere.

tropopause The boundary in the earth's atmosphere between the troposphere and the stratosphere.

troposphere The lowest level of the earth's atmosphere, reaching 10 km from the surface; the area in which most of the weather takes place.

Tully–Fisher relation The relation between the luminosity of a galaxy and the width of its 21-cm emission line; the greater the luminosity, the wider the line.

turbulence Irregular and sometimes violent convective motion.

turbulent viscosity The property of a gas (or any fluid) by which turbulent flow in one part affects the flow of a nearby part; an important effect in the transfer of angular momentum outward in the solar nebula.

turnoff point The point on the H–R diagram of a cluster at which the main sequence appears to terminate at the high-luminosity end.

T-Tauri stars Newly formed stars of about 1 solar mass; usually associated with dark clouds; some show evidence of flares and starspots, others have circumstellar disks; one type of young stellar object.

21-cm line The emission line, at a wavelength of 21.11 cm, from neutral hydrogen gas; it is produced by atoms in which the directions of spin of the proton and electron change from parallel to opposed.

two-sphere universe The basic premise of the celestial coordinate systems that the universe is composed of two concentric spheres, the earth and the celestial sphere.

Type I, Type II supernovas Classification of supernovas by their light curves and spectral characteristics: Type I show a sharp maximum and slow decline with no hydrogen lines; Type II have a broader peak and a very sharp decline after 100 days with strong hydrogen lines in the spectrum.

ultraviolet telescope A telescope optimized for use in the ultraviolet region; must be used above the earth's atmosphere.

uniformity of nature The assumption that astronomical objects of the same type are the same throughout the universe.

universal law of gravitation Newton's law of gravitation; see gravitation.

universality of physical laws The assumption, borne out by some evidence, that the physical laws understood locally apply throughout the universe and perhaps to the universe as a whole.

universe The totality of all space and time; all that is, has been, and will be.

upland plateaus Large highland masses on the surface of Venus.

Valles Marineris Extensive canyonlands region near the equator of Mars.

Van Allen radiation belts Belts of charged particles (from the sun) concentrated and trapped in the earth's lower magnetosphere.

variable star Any star whose luminosity changes over a short period of time.

vector A quantity that expresses magnitude and direction; for example, forces and accelerations are vector quantities.

velocity The rate and direction in which distance is covered over some interval of time.

velocity dispersion The range of velocities around an average velocity for a group of objects, such as a cluster of stars or galaxies.

vernal equinox The spring equinox; see equinox.

vertical circle On the celestial sphere, any great circle through the zenith.

Very Large Array (VLA) A radio interferometer located in New Mexico; it consists of 27 antennas arranged in a Y-shaped pattern.

Very-Large-Baseline Array (VLBA) A radio interferometer with antennas spread across the United States; the processing and control center is in New Mexico.

Very-Long-Baseline Interferometry (VLBI) Radio interferometry carried out by telescopes at widely separated locations (across continents and oceans!); the signals are brought together and processed by computer; the VLBA is a special form of VLBI.

Virgo cluster The nearest large cluster of galaxies containing thousands of galaxies; it appears to lie in the direction of the constellation of Virgo and makes up a large part of the Local Supercluster.

virial theorem The statement that the gravitational potential energy is twice the negative of the kinetic energy of a system of particles in equilibrium.

visual binary Two stars that revolve around a common center of mass, both of which can be seen through a telescope, allowing their orbits to be plotted.

visual flux The flux from a celestial object measured across the visual part of the electromagnetic spectrum.

visual luminosity The luminosity from a celestial object measured across the visual part of the electromagnetic spectrum.

vis viva equation A relation between orbital speed and distance from the focus for a body moving in an elliptical orbit; a form of the law of conservation of energy for a body moving in a gravitational field.

voids Regions between superclusters that contain no large concentrations of luminous matter; they have somewhat a spherical shape and extend over millions of light years.

volatiles Materials, such as helium or methane, that vaporize at low temperatures.

volcanic model The formation of craters as cones left over from lava eruptions.

volcanism All processes by which any material is expelled from a body's interior to its crust.

W Virginis stars Pulsating variable stars; Population II cepheids.

watt (W) An SI unit of power; one joule expended per second.

wavelength The distance between two successive peaks or troughs of a wave.

weak force A short-range force that operates in radioactive decay.

weight The total force on some mass produced by gravity.

weightlessness The condition of apparent zero weight, produced when a body is allowed to fall freely in a gravitational field; in general relativity, weightlessness signifies motion on a straight line in spacetime, that is, natural motion.

white dwarf A small, dense star that has exhausted its nuclear fuel and shines from residual heat; such stars have an upper mass limit of 1.4 solar masses, and their interior is a degenerate electron gas.

Widmanstätten figures Large crystal patterns that appear on the surfaces of iron meteorites when they are polished and etched; they are formed by slow cooling of the material.

Wien's law The relation between the wavelength of maximum emission in a blackbody's spectrum and its temperature; the higher the temperature, the shorter the wavelength at which the peak occurs.

winter solstice See solstice.

worldline A series of events in spacetime.

x-rays High-energy electromagnetic radiation with a wavelength of about 10^{-10} m.

x-ray burster An x-ray source that emits brief, powerful bursts of x-rays; probably occurs from accretion onto a neutron star in a binary system.

young Population I stars A class of stars found in the disk of a spiral galaxy, especially in the spiral arms; they have a metal abundance similar to the sun's and are the youngest stars in the galaxy.

young stellar object (YSO) Generic name for any star prior to its main-sequence phase.

ZAMS Acronym for zero-age main sequence.

Zeeman effect The splitting of spectral lines because of strong magnetic fields.

Zeilik An American form of a Ukrainian family name.

zenith The point on the celestial sphere that is located directly above the observer at 90° angular distance from the horizon.

zero-age main sequence (ZAMS) The position on an H−R diagram reached by a protostar once it has come to derive most of its energy from thermonuclear reactions rather than from gravitational contraction.

zodiac The traditional twelve constellations through which the sun travels in its yearly motion, as seen from the earth; a thirteenth constellation, Ophiuchus, is actually part of the zodiac.

zodiacal light Sunlight reflected from dust in the plane of the ecliptic.

zone A region of high pressure in the atmosphere of a Jovian planet.

zone of avoidance A region near the plane of the Galaxy in which very few other galaxies are visible because of obscuration by dust.

Index

Photo Credits for Chapter Opening Images

Chapter 1 Annular eclipse over San Diego, CA in 1992; *Courtesy of Phil Lohmann*

Chapter 2 16th century Arab astronomer; *Courtesy of the Grainger Collection*

Chapter 3 Johannes Kepler and Käiser Rudolf II; *Courtesy of the Grainger Collection*

Chapter 4 Sir John Flamsteed at the Old Royal Observatory; *Courtesy of the Grainger Collection*

Chapter 5 Spectral image; *Courtesy of David Parker/Science Photo Library*

Chapter 6 Deployment of Hubble Space Telescope; *Courtesy of NASA/JPL*

Chapter 7 NGC 5754 and 5755; *Courtesy of NOAO*

Chapter 8 Aurorae over the North Pole as seen from space; *Courtesy of NASA*

Chapter 9 Earth and moon seen from space by Magellan; *Courtesy of NASA/JPL*

Chapter 10 Valles Marineris; *Courtesy of Paul Geissler, University of Arizona*

Chapter 11 Jupiter family portrait; *Courtesy of NASA*

Chapter 12 Comet Swift-Tuttle; *Courtesy of USNO*

Chapter 13 Sun's analemma; *Courtesy of Frank Zullo*

Chapter 14 Colors of stars in Orion; *Courtesy of AAO; photo by D. Malin*

Chapter 15 CG-4; *Courtesy of AAO; photo by D. Malin*

Chapter 16 NGC 7293; *Courtesy of AAO; photo by D. Malin*

Chapter 17 Eskimo Nebula; *Courtesy of CARA*

Chapter 18 Milky Way Galaxy; *Courtesy of IRAS*

Chapter 19 NGC 1365; *Courtesy of AAO; photo by D. Malin*

Chapter 20 Numerical simulation of an extragalactic radio jet; *Courtesy of M. L. Norman, J. L. Burns, D. A. Clarke/NCSA*

Chapter 21 Full sky microwave map-DMR; *Courtesy of COBE/NASA/Goddard Space Flight Center*

Chapter 22 "Distant Rendezvous" (Computer-generated art); *Courtesy of Boris Starosta*

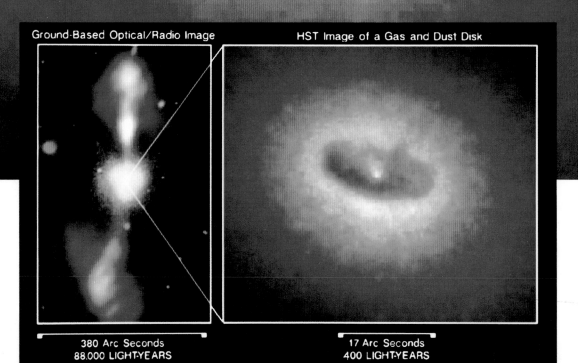

A supermassive black hole?
The HST presents strong visual evidence.
Courtesy of Walter Jaffe/Leiden Observatory,
Holland Ford/JHU/STScI, and NASA

Ground-Based Optical/Radio Image

HST Image of a Gas and Dust Disk

380 Arc Seconds
88,000 LIGHT-YEARS

17 Arc Seconds
400 LIGHT-YEARS